THE DYNAMIC EARTH

THE DYNAMIC EARTH

An Introduction to Physical Geology

Fourth Edition

BRIAN J. SKINNER
Yale University

STEPHEN C. PORTER
University of Washington

JOHN WILEY & SONS, INC.

New York   Chichester   Weinheim   Brisbane   Singapore   Toronto

Acquisitions Editor	Cliff Mills
Developmental Editor	Marian Provenzano
Marketing Manager	Susan Lyons
Production Editor	Sandra Russell
Cover and Text Design	Harold Nolan
Illustration Editor	Edward Starr & Anna Melhorn
Illustrations	Precision Graphics, Inc.
Photo Editor	Jill Tatara
Photo Research and Editing	Alexandra Truitt & Jerry Marshall
Cover Photo	©Denis Finnin/American Museum of Natural History

This book was typeset in Times Ten Roman by Color Associates, Inc. and printed and bound by Von Hoffmann Press, Inc. The cover was printed by Von Hoffmann Press, Inc.

The paper in this book was manufactured by a mill whose forest management programs include sustained yield harvesting of its timberlands. Sustained yield harvesting principles ensure that the number of trees cut each year does not exceed the amount of new growth.

This book is printed on acid-free paper. ⊖

Library of Congress Cataloging in Publication Data:
Skinner, Brian J.
The dynamic earth: an introduction to physical geology / Brian J. Skinner, Stephen C. Porter — 4th edition
Includes bibliographical references and index.

ISBN 0-471-16118-7 (paper/CD-DROM : alk. paper)
1. Physical geology. I. Porter, Stephen C. II. Title.

QE28.2 .S55 1999
550—dc21   99-055475

Printed in the United States of America

10 9 8 7 6 5 4 3 2

Preface

We live on an extraordinary planet, the only one known to support life, that has abundant oxygen in its atmosphere, that rains water from its clouds. What we humans are and why our environment has the form it has are the result of innumerable interactions among the Earth's solid rocks and soil beneath our feet, the water and ice on land and in the oceans, the atmosphere that surrounds us, and living matter.

The science that studies the Earth and all its interactions is geology, and those involved in the studies are geologists. It is a unique and fascinating science because its laboratory is the environment in which we live. Geologists find it very difficult to carry out controlled experiments in their "geological laboratory"—the scales of space and time needed for such experiments are simply too large. Even if the space and time problems could be handled, there is always the chance that the experiments would cause the environment to change in some unfortunate way. Geologists must study the Earth as it exists. From their assembled observations they draw conclusions about the processes that are shaping the Earth today and events that have shaped it over the past 4.6 billion years. Increasingly, geologists are called upon to use their understanding of the Earth to suggest what changes might be expected in the future and how our collective human activities may be causing some of the changes.

A SCIENCE IN FERMENT

Revolutionary advances in the breadth and depth of our knowledge of the Earth and the other planets of the solar system have occurred during the past 50 years. Never before have so many dramatic discoveries been made in such a short time, and the revolutionary discoveries continue unabated. For example, samples of ice drilled from glaciers in Greenland, Antarctica, China, and elsewhere have been found to carry samples of air trapped since the ice formed from falling snow thousands to tens of thousands of years ago. To everyone's surprise, this research has revealed that large and rapid changes in the composition of the atmosphere have occurred; humans played no role in them. So far there is no adequate explanation for these changes.

Another example of a recent, intriguing discovery concerns a vast submarine lava field in the western Pacific Ocean. Ocean-going geologists mapped and then drilled into the lava field. The size of the field and the discovery that the lavas were erupted with great

rapidity 120 million years ago, just when the Earth's climate became very warm, have led geologists to suggest that gigantic volcanic "burps" released enough carbon dioxide into the atmosphere every now and then to completely change the climate. The most remarkable part of the hypothesis is that the volcanic "burps" seem to be controlled by the Earth's molten core, implying that the core must play a role in determining the climate.

Geology, then, is a science in ferment, a science laced with challenging excitement. New discoveries, new insights, and new theories heighten the excitement almost every day. It is not a science that has grown distant from everyday life—it is the science of the world around us, a science in which everyone can participate.

KNOWING THE EARTH

The science of geology is like all other sciences in that it is based on observations. We have tried to write this book so that readers can sense both the fun of making their own observations and the challenge of drawing conclusions; the reader does not need a geological background to read the text and apply the principles it introduces.

We believe that everyone, when given a chance to do so, quickly becomes intrigued to learn how the Earth works. We hope that the reader of this book will become actively involved in the science of geology. The *Dynamic Earth* was written to provide that chance. It is a book for everyone.

Organization

The book has 21 chapters that follow a more or less traditional sequence. The first two chapters introduce the Earth, its overall properties, and plate tectonics. The next six chapters discuss minerals and their properties, igneous rocks and magmatic processes, weathering and soils, sedimentation, metamorphism, and geological time. Chapters 9 through 14 deal with surficial processes—mass wasting, streams and drainage, groundwater, glaciation, deserts, and the oceans. Then follow three chapters dealing with internal processes, rock deformation, earthquakes, and global tectonics. Chapter 18 is concerned with the changing face of the land as a result of interactions between internal and external processes. Chapter 19 ties together the many environmental themes that were introduced in earlier chapters, and the book closes with a chapter on resources and one on planets and planetary processes.

Each chapter in the book is written to stand alone. Teachers who prefer to teach a subject in an order that differs from the text can readily do so.

Themes

We stress four major themes throughout the book. The first is plate tectonics—the slow, lateral motions of fragments of the Earth's outermost 100 km (called plates) at rates of up to 20 cm per year. The shapes and locations of continents and oceans, the locations of mountains and volcanoes, and the violence of earthquakes are all determined by plate tectonics. We have used plate tectonics as a framework within which to integrate most geological processes. Plate tectonics is the link between the Earth's internal and external activities, and as a result the topic appears in most chapters.

The second theme concerns the influence that the human race is having on the Earth's external processes. These influences, which are often given such names as environmental change, are emphasized throughout the book because it is essential that we all understand what is happening to the human race and to the planet on which we live. We humans are now so numerous (about 6.0 billion and increasing by approximately three people a second) that our daily activities are having measurable effects on such things as rainfall, climate, and rates of erosion. To understand how the Earth works today, therefore, we must appreciate the part played by humans in geological processes.

This second theme is so important to us that we devote Chapter 19, "Our Changing Planet," to a topic that has received too little attention in physical geology textbooks, namely, how human activity is changing the Earth's climate. We emphasize again that the Earth is a complex, interactive, and dynamic system in which a change in one part is likely to change other parts, often in unanticipated ways. As we burn fuels to run society, we pollute the atmosphere and alter its chemical composition. In doing so, we unwittingly have contributed to changes in the Earth that may have serious effects on all of us in the years ahead. Geology provides a record of billions of years of natural environmental change on our planet. It therefore has an extremely important role to play in our attempt to understand how both natural and human-induced changes may affect our planet and its inhabitants. In Chapter 19 we introduce the ozone hole, the greenhouse effect, and other global environmental changes in which each of us plays a role.

The third and fourth themes also concern the human race. The third is the human use of natural resources. Each of us uses, on average, about 10 tons of mineral resources each year. (In North America the figure is closer to 20 tons per person.) Finding where to dig those resources is one of the greatest challenges faced by geologists. Understanding the consequences of using mineral resources at the rate we do is another great challenge for geologists. We have, therefore, integrated the topic of natural resources throughout the book, while Chapter 20, a short chapter devoted to mineral and energy resources, covers those economic and technical issues of resources that do not fit readily into other chapters. The fourth theme emphasizes the understandings gained from studying the Earth as a closed system and points to the many new tools and approaches used to measure changes in the Earth system.

The Fourth Edition

In preparing the fourth edition, we have retained the overall topic structure and themes that served us well in the past three editions. We have, however, added some topics in response to suggestions from users of the text, we have changed the order of chapters a little, and we have recast some features to emphasize important environmental themes. In Chapter 1, for example, we introduce the concept of the Earth as a closed system that is comprised of many interacting open systems; then in each subsequent chapter there is an essay on some aspect of change to the environment. The major environmental issues are addressed in greater detail in Chapter 19.

We have retained the expanded Chapter 14 on "The Oceans and Their Margins" from the third edition. The chapter includes a discussion of deep water and the role that the oceans play in determining the Earth's climate. We believe that one cannot truly understand the functioning of the Earth if one does not understand the key role played by the oceans. Finally, of course, we have updated the text throughout, adding discussion of recent events and the most recent data wherever available, for example, on such topics as the ozone hole, global warming, soil orders, and the dating of magnetic reversals.

Chapter 18, "The Changing Face of the Land," is new to the book. The major components of the Earth system meet and interact at the land surface, and through those interactions the Earth's amazing variety of landscapes has developed. This chapter discusses the competing geological forces that determine the shape of the Earth's surface.

The sequence of chapters has been changed in response to requests from users. The topics of "Weathering and Soils" (Chapter 5) now precedes "Sediments and Sedimentation" (Chapter 6). The two topics that users tell us are less used than others, "Resources of Minerals and Energy," and "Beyond Planet Earth," are now Chapters 20 and 21, respectively.

Several features of the first three editions of *The Dynamic Earth* that drew favorable comments from users and reviewers have been retained and expanded:

Chapter Essays

Each chapter opens with a short essay as before, but many of these essays are new. Most of the essays deal with our environment, or the effect the environment has on us. Chapter 11, for example, opens with an essay on water problems in Tucson, Arizona, and Chapter 16 opens with an essay on new research results that identified covered faults beneath Los Angeles.

Each chapter also contains one or more essays dealing with issues concerned with "Using Resources," "Understanding Our Environment," and "Measuring Our Earth."

New Media

Our goal has been to bring the powers of interactive media to bear on the exploration of physical geology. At the end of appropriate chapters, references are made to two new media components: GEOSCIENCES IN ACTION CD-ROM, where students find themselves as "virtual interns" exploring certain problems as a geologist would, as well as viewing animations and videos showing key geologic processes; and GEOSCIENCES TODAY: AN INTERACTIVE CASEBOOK ON THE WWW, where students can take virtual tours exploring cases in geoscience.

The Artwork

The art program has been revised and simplified, and photographic research has been extensive. About 30 percent of the illustrations in the book are new.

Student Review Material

Pedagogical material such as chapter summaries, questions for review, and short lists of important words and phrases have been retained, with the number of technical terms kept to a minimum. A glossary of important terms is found at the end of the book; it includes terms identified as key terms in the chapters, as well as many other terms for student reference. Finally, the appendices at the end of the book contain useful information for students on units and their conversions, the properties of chemical elements and their isotopes, and the properties of common minerals.

SUPPLEMENTS

A full range of supplements to accompany The *Dynamic Earth, 4e* is available to assist both the instructor and the student.

Geosciences in Action. This CD-ROM accompanies each text and allows students to become "virtual interns" in geology, whether exploring the source of a certain pollutant or determining the volcanic hazard at an island resort. These "virtual internships" were authored by David DiBiase, Thomas Bell, and Hobart King, and developed by the Deasy Geographic Labs at Pennsylvania State University. In addition, the CD-ROM contains animations and videos showing key geologic processes.

Geosciences Today. This casebook and interactive web site provides students with eight cases from around the world in which to see and explore the interaction of people and their environment. This casebook was authored and developed by Robert Ford, Westminster University and James Hipple, University of Missouri.

Making Earth: An Interactive Guide to the Planet. A Study Guide to accompany The Dynamic Earth, 4e. This manual integrates the use of independent multiple choice questioning throughout the course. It encourages students to actively participate in the study of physical geology. The topic order reflects that of *The Dynamic Earth, 4e* and provides answers to the questions that appear at the end of each chapter in *The Dynamic Earth, 4e.* The guide is authored and developed by Michael and Susan Kimberly of North Carolina State University. An interactive CD accompanies the manual, which provides self-quizzes and evaluations for all topics.

Geoscience Laboratory, 2/e. This manual contains 19 labs covering the major topics in physical geology. It is authored by Tom Freeman of the University of Missouri–Columbia.

Instructor's Resource Manager CD-ROM. This CD-ROM, free to adopters of the text, contains all of the line illustrations and many of the photos from the text for lecture projection, as well as several animations and videos showing key geologic processes.

Instructor's Manual and Test Bank. This manual includes a test bank, chapter overviews and outlines, and lecture suggestions. The Test Bank is also available in a computerized format.

Transparencies. The transparencies include 150 full-color textbook illustrations, resized and edited for maximum effectiveness in large lecture halls.

PowerPoint Slides on CD-ROM. This full-color slide presentation highlights key figures from the text as well as many additional lecture outlines, concepts, and diagrams. The lecture notes can be modified so instructors may tailor the slide presentation to their specific course.

Take Note! This supplement for students contains all of the line illustrations from the text in a black-and-white format for students to use to take notes.

Web Site. A web site has been developed in support of *The Dynamic Earth, 4/e.* The Site is available at *http://www.wiley.com/college/skinner* and offers a range of information, which includes the following special resources for instructional enrichment:

American Museum of Natural History website linking this text to the Museum's permanent exhibition, the Gottesman Hall of Planet Earth, Rose Center for Earth and Space. The website features animations, images and other features from the Museum's extensive collection and website. This site will be updated and evolve to include new content. All links are keyed to the text of *The Dynamic Earth, 4/e.*

Interactive case abstracts, virtual tours, and additional resources for *Geosciences Today.*

ACKNOWLEDGMENTS

The third edition of *The Dynamic Earth* came to fruition under the guidance of Chris Rogers, then the Earth Sciences editor at John Wiley & Sons, and the publisher, Kaye Pace. The fourth edition has been guided by Earth Sciences editor Clifford Mills. Our tasks were made infinitely more enjoyable and considerably easier through the skilled attention and careful guidance of developmental editor Marian Provenzano.

The professional skills and competence of the staff at John Wiley & Sons and of the freelance experts they found to work with us are outstanding: Sandra Russell oversaw production; Marjorie Graham, Alexandra Truitt, and Jill Tatara sought exactly the photos we needed; Edward Starr and Anna Melhorn managed the illustration program; Harry Nolan designed the text; Jennifer Yee and Jennifer Cerceillo obtained the text's supplements; Sysan Lyons developed the marketing program; and Caroline Ryan and Yolanda Pagan provided editorial assistance. They and everyone else at Wiley were always cordial and always helpful, no matter how badly our travel schedules upset book schedules. Above all, we benefited greatly from the careful and detailed reviewing of the text by Dr. J. Marion Wampler of Georgia Institute of Technology. He not only caught mistakes of fact and reasoning, but he also pointed out many places where our words were ambiguous or confusing.

We are also indebted to the many people who provided elegant colored photographs that appear in the book. Most of the photographers are geologists, and their discerning eyes can be sensed through the beautiful photos they took. The names are listed in the Photo Credits in the back of the book, but being so placed is no reflection on their importance.

We especially thank the thoughtful and dedicated teachers who commented on the previous texts or reviewed the present text. These fine people not only helped us keep a reasonable balance to the book, they also helped us keep the volume as up-to-date as possible without downplaying the great geological discoveries of the past. In particular, we thank the reviewers who have helped us prepare the fourth edition:

Gary C. Allen
University of New Orleans

Thomas B. Anderson
Sonoma State University

Charles W. Barnes
Northern Arizona University

J. Allan Cain
University of Rhode Island

George Clark
Kansas State University

Timothy W. Duex
University of Southwestern Louisiana

William R. Dupré
University of Houston

Julian W. Green
University of South Carolina at Spartanburg

Dr. Katherine Giles
New Mexico State University

Richard A. Heimlich
Kent State University

Dr. Michael M. Kimberley
North Carolina State University

Tim K. Lowenstein
State University of New York at Binghamton

Bart S. Martin
Ohio Wesleyan University

Stephen A. Nelson
Tulane University

Loren A. Raymond
Appalachian State University

Mary Jo Richardson
Texas A&M University

Dr. Paul A. Schroeder
University of Georgia

Donald Schwert
North Dakota State University

Dr. Thomas Sharp
Arizona State University

Dr. John F. Shroder
University of Nebraska at Omaha

Brad Singer
University of Wisconsin-Madison

Hubert C. Skinner
Tulane University of Louisiana

Robert K. Smith
University of Texas at San Antonio

William A. Smith
Grand Valley State University

Piyush Srivastav
University of Nebraska

Stephen D. Stahl
Central Michigan University

Dr. John Stimac
Lane Community College

Dr. Sam Swanson
University of Georgia

Harve S. Waff
University of Oregon

Earlier editions were reviewed by:

Gary C. Allen
University of New Orleans

J. C. Allen
Bucknell University

N. L. Archbold
Western Illinois University

Philip Brown
University of Wisconsin

Collete D. Burke
Wichita State University

Alan Cain
University of Rhode Island

Robert A. Christman
Western Washington University

George Clark
Kansas State University

Nicholas K. Coch
Queens College

Kristine Crossen
University of Alaska

John Diemer
*University of North Carolina
at Charlotte*

Grenville Draper
Florida International University

M. Ira Dubins
*State University of New York
at Oneonta*

John Ernissee
Clarion University of Pennsylvania

Stewart Farrar
Eastern Kentucky University

Mike Follo
*University of North Carolina
at Chapel Hill*

C. B. Gregor
Wright State University

Ann G. Harris
Youngstown State University

Robert L. Hopper
University of Wisconsin

Robert Horodyski
Tulane University

Albert Hsui
*University of Illinois at Urbana-
Champaign*

Allan Kolker
University of Nebraska-Lincoln

Peter L. Kresan
University of Arizona

Albert M. Kudo
University of New Mexico

Judith Kusnick
California State University

Peter B. Leavens
University of Delaware

Nancy Lindsley-Griffin
University of Nebraska

William W. Locke
Montana State University

David N. Lumsden
Memphis State University

Gerald Matisoff
Case Western University

Robert McConnell
Mary Washington College

Bruce Nocita
University of South Florida

Anne Pasch
University of Alaska-Anchorage

Gary Peters
*California State University, Long
Beach*

John Renton
West Virginia University

Jack Rice
University of Oregon

Mary Jo Richardson
Texas A&M University

Robert W. Ridky
University of Maryland at College Park

Donald Ringe
Central Washington University

Len Saroka
St. Cloud State University

Fred Schwab
Washington & Lee University

Christian Teyssier
University of Minnesota

Graham Thompson
University of Montana

Charles P. Thornton
Pennsylvania State University

Page Twiss
Kansas State University

James B. Van Alstine
University of Minnesota

Neil Wells
Kent State University

Monte Wilson
Boise State University

Margaret Woyski
California State University-Fullerton

Anne Wyman
University of Nevada-Las Vegas

A CLOSING THOUGHT

After digesting a beginning geology textbook, the reader may well come away with a feeling that we geologists have all the answers, that the major principles are known, and that the significant challenges have been resolved. We hope this will not happen with this book. We have tried to show that we do not have all the answers. In fact, it is because we have so many important and challenging questions remaining before us that geology is such a dynamic and exciting science in which to work. Those of us who have seen the remarkable advances of the past 50 years have not doubted that the next 50 will produce even more startling discoveries about our dynamic Earth.

Brian J. Skinner   Stephen C. Porter

13 Wind Action and Deserts, 341

Opening Essay:
The Dust Bowl: Could It Happen Again?, 341

14 The Oceans and Their Margins, 369

Opening Essay:
Our Changing Coastlines, 369

15 Deformation of Rock, 403

Opening Essay:
Geology and World War I, 403

Boxed Essays

About the Authors

The authors of this book have been privileged to work as geologists all of their professional lives. They have attempted in this book to share the excitement and wonder gained from the study of the intricacies of our planet Earth. In writing this book they have drawn extensively on their own experience, as well as the experience of numerous fellow geologists whose collective geological knowledge spans the field of physical geology. Brian Skinner's research has focused on the physical properties of minerals and on the genesis of base metal deposits. He has worked in Australia, Africa, and North America, and with students in Asia and Europe. Stephen Porter's professional career has been largely concerned with studies of glaciation in many of the world's major mountain systems and with the history of the climatic changes that their deposits record. He also has studied the evolution of midocean and continental volcanoes, the products of their prehistoric eruptions, and how volcanic eruptions may have influenced the Earth's climate. With foreign colleagues he has studied the hazards of large rockfalls in the Alps and the thick, extensive deposits of wind-blown dust in central China that provide one of the longest continuous records we have of climatic change during the last several million years.

Between them, the authors have carried out geologic field work on all of the continents. This global perspective is reflected in the book by examples and illustrations from around the world, for it is important to emphasize that geology is a global science, a science that recognizes no political boundaries. Only by studying the Earth in its entirety can we hope to understand how our amazing planet works.

Brian J. Skinner and Stephen C. Porter, with the San Juan Islands in the background.

THE DYNAMIC EARTH

1

Measuring
Our Earth

Understanding
Our Environment

GeoMedia

Jagged spires of rock tower above El Nido on the northern tip of Palawan Island, Philippines. Coral reefs can be seen in the shallow water of the bay. The photograph emphasizes the interactions between the parts of the Earth system.

A First Look at Planet Earth

Earth Sense and the Earth System

We live in an extraordinary age; it is a time when scientific discoveries let us see and measure things as never before. Astronomers study the sky through a telescope that orbits the Earth far above the atmosphere, and they have discovered that planets circle around other suns far out in space. From space we now keep an eye on Planet Earth and witness the birth of a hurricane, the erosion of soil, the destruction of a forest, and dust spreading from a volcano. For the first time in history we can see what is happening everywhere on the surface of the Earth.

We cannot actually "see" what is happening inside the Earth, but information such as the speed of earthquake waves, volcanic eruptions, and how heat escapes tells a great deal, in an indirect way, about the unseen world beneath our feet.

More than a century ago scientists realized that the waters of the world, the atmosphere, plants and animals, and solid rocks all interact with each other. For example, by chemical actions of the atmosphere and rainfall, rock is broken down to soil; plants then extract moisture and nutrients and, by a process called photosynthesis, convert water and carbon dioxide to carbohydrate and in the process release oxygen to the atmosphere. In short, the Earth is an assemblage of parts, all of which interact, and it is this marvelous system of interactions—the *Earth system*—that creates the environment in which life, including human life, lives and prospers. Now, with satellites in the sky, deep-diving submarines in the oceans, and myriad other devices, it is possible to continually monitor many of the interactions between parts of the Earth system, and thereby has arisen a new branch of science—Earth system science.

Some interactions between parts of the Earth system are still incompletely understood, and because some of those interactions influence the human life support system, a lot of scientific research is devoted to bettering our understanding of just how the Earth works. Human activities are strongly influenced by climate, for example, and the climate in turn is influenced by the composition and properties of the atmosphere. We know that human activities are changing the composition of the atmosphere, but how those changes will affect the climate are poorly understood.

We rely on the availability of usable water, and now, in at least forty nations, there are crises concerning clean water supplies. The concrete, copper, steel, and phosphate fertilizers that we mine from the minerals and rocks near the Earth's surface amount to an annual consumption rate of about 10 tons of rock for every person on the Earth. In aggregate, we humans use more of the Earth's resources than all of the sediment carried to the sea each year by all the rivers of the world.

Finally, soil, the thin surface layer of Earth that supports much of the photosynthetic biosphere, combines with water and sunlight to provide our food. We now know that we must grow an increasingly large food supply to feed a rapidly growing world population, yet the Worldwatch Institute tells us that we are losing soil at a rate of nearly 1 percent per year. If this statistic is true, it indicates that we could be headed for a food disaster. (In some places, such as the Sudan and Somalia, the crisis is already here.)

No one wants to live in a state of crisis and potential disaster. The antidote to crisis is *Earth Sense*—appreciating how the Earth system works, understanding that the system has limits, and accepting a shared responsibility to see that the limits are not transgressed.

GEOLOGY AND GEOLOGISTS

The Earth is always changing. Small, slow changes are continuous, while massive but rapid changes, like those produced by hurricanes, are sporadic. Fast or slow, large or small, continuous or sporadic, the key word is *change*. The Earth is never still.

The discipline that studies these changes, past and present, is **geology**. The word comes from two Greek roots: *geo-*, meaning of the Earth, and *-logia*, meaning study or science. Scientists who study the Earth are called **geologists**.

Geologists work in every corner of the world, from ice-covered peaks, to erupting volcanoes, and to the depths of the ocean. They seek to understand all the processes that operate on the Earth and to document the Earth's long, complex history. They pay special attention to bodies such as rivers and lakes that are being altered by human activities and to processes such as volcanic eruptions that can threaten human life. For a discussion about an inland sea that is being changed by human actions, see the essay on *The Disappearing Sea*. In their work geologists study directly any place they can reach. To investigate places they cannot reach, they drill deep holes in solid rock. Beyond the reach of drill holes, geologists must rely on indirect observations. As a doctor listens for noise inside our body with a stethoscope, so a geologist employs sensitive measuring devices to "listen" to the rumbles of distant earthquakes and sense the pulse of activities deep inside the Earth. From their observations, geologists try to understand the history of Planet Earth and the origin of its complex landscapes. They try, too, to predict where new oil fields lie, whether a well will strike water, and where rich ore deposits are hidden.

Geology is traditionally divided into two broad topic areas with related but differing aims. **Physical geology** is concerned with understanding (1) the *processes* that operate at or beneath the surface of the Earth, and (2) the *materials* on which those processes operate. What causes volcanoes to erupt or how earthquakes, landslides, and floods happen are examples of processes. Examples of materials are soils, sands, rocks, air, and seawater. **Historical geology** has as its goal the chronology of the *events*, both physical and biological, that have occurred in the past. Historical geology seeks to resolve questions such as when the oceans formed, when dinosaurs first appeared, when the Rocky Mountains rose, and when and where the first trees appeared.

Physical geology, the subject of this book, serves as a starting point for studying the Earth. Most people seek to know about the Earth because we humans are inherently interested in the things around us. But there is also a very practical reason for studying the Earth. If we are to understand the environment in which we live and be able to make predictions about changes that might lie ahead, we must understand how the Earth works. In order to obtain that understanding, and especially to understand how we humans may be affecting the Earth, we need to examine both materials and processes.

Understanding Our Environment

The Disappearing Sea

The Aral Sea in Uzbekistan, in central Asia, is shrinking so rapidly that once-prosperous fishing villages are now 50 km from the shore. Human activities caused the change. The story of that change is a lesson for us all: when we alter the balance of nature, unforeseen side effects almost always crop up.

Thirty years ago, the Aral Sea was the fourth largest lake in the world after the Caspian Sea, Lake Superior, and Lake Victoria. The sea covered 68,000 km^2, had an average depth of 16 m, and yielded 45,000 tons of fish a year. Today the sea is only the sixth largest lake. It now covers 40,000 km^2, has an average depth of 9 m, is so salty the fishing industry is dead, and is disappearing so fast it will be a waterless desert by 2010 (Fig. B1.1).

The Aral Sea is fed by two large rivers, the Amu Dar'ya and the Syr Dar'ya, which carry meltwater across the desert from the snowy mountains of northern Afghanistan. Water leaves the sea by evaporation, so the size of the sea is a balance between evaporation and river inflow. The sea is shrinking because inflow has declined.

A small part of the problem is climatic; there were a number of years in the 1970s when snowfalls were light. The largest part of the problem, however, is the irrigation that has been practiced in the river valleys for millennia. In modern times, and especially during the years when Uzbekistan was part of the former Soviet Union, the extent of irrigation increased dramatically. By 1960, so much irrigation water was taken from the two rivers that inflow to the Aral Sea had declined to a trickle. The sea has been shrinking steadily ever since.

The people who planned the irrigation systems expected the Aral Sea to shrink. What they did not anticipate were the side effects. The sea, it is now realized, exerts a major influence on the local climate. Because it is shrinking, local rainfall is declining, the average temperature is rising, and wind velocities are increasing. Most of the newly exposed sea bottom is covered with salt. The wind blows the salt around and has created withering salt storms. Potable water supplies have declined, and various diseases, especially intestinal diseases, are afflicting the local population at alarming levels.

The situation could, of course, be reversed by simply reducing the amount of irrigation. The problem is that the irrigated area is now one of central Asia's most prosperous, so the ultimate solution will probably have to be somewhere between returning the sea to its original size and keeping all the irrigated land.

Figure   B1.1   The Shrinking Sea   Fishing boats stranded in the desert by shrinkage of the Aral Sea.

The word *rock* is an important one, and because it is used frequently in this book, we must define it carefully. **Rock** is any naturally formed, nonliving, firm, and coherent aggregate mass of solid matter that constitutes part of a planet. Note that the definition specifies a coherent aggregate, which means that all the rock particles are locked together to make a solid mass. A pile of loose sand grains is not rock because the grains are not locked together—not coherent. A tree is not a rock, even though it is solid, because it is living. But coal, which is a compressed and coherent aggregate of twigs, leaves, and other bits of dead plant matter, is a rock.

THE SCIENTIFIC METHOD

All the processes geologists study obey the fundamental laws of nature discovered by physicists, chemists, and mathematicians. In a sense, then, one might call geology a derivative science. But geology is also a special and very practical science because it is the science of the planet on which we live, the science of our own environment. Geologists investigate our environment using a general research strategy known as the **scientific method**. The core of the scientific method of investigation is evidence that can be seen and tested. Based on that tested evidence, geologists draw conclusions and so advance our understanding of the way the Earth works.

Although it is not always a clear-cut process, the scientific method can be viewed as consisting of the following steps:

1. *Observation and measurement:* Scientists acquire evidence that can be measured and observed. For example, geologists measure the thickness and extent of rock layers (Fig. 1.1).

2. *Formation of a hypothesis:* Scientists try to explain their observations by developing a **hypothesis**—an unproved explanation for the way things happen. For example, a geologist might hypothesize that the horizontal layering in the rocks in Figure 1.1 is due to sediment being deposited in water.

3. *Testing of a hypothesis and formation of a theory:* Hypotheses are used to make predictions about new observations. A comparison of the predictions with the new observations is a *test* of the hypothesis. When a hypothesis has been examined and found to withstand numerous tests, scientists become more certain about it and it becomes a **theory**, which is a generalization about nature. A prediction arising from the hypothesis that the horizontal layering in Figure 1.1 is a consequence of sediment being deposited in water would be that fossilized remains of aquatic plants or animals might be present in the rocks. Theories cannot be considered the final word, they are always open to more testing.

4. *Formation of a Law or Principle:* Eventually, a theory or a group of theories may be formulated into a *Law* or a *Principle*. Laws and Principles are statements that some aspects of nature are always observed to happen in the same way and no deviations have ever been seen. The **Law of Original Horizonality** states that water-laid sediments are deposited in layers that are horizontal or nearly horizontal and parallel or nearly parallel to the Earth's surface.

5. *Continual reexamination:* The assumption that underlies all of science is that everything in the material world is governed by scientific laws. Because hypotheses and theories are open to questions when new evidence is found, hypotheses and theories are continually reexamined. In fact, the key to the scientific method is *testability*. Any theory that cannot be tested and, at least possibly, be disproved if the tests fail, is not scientific.

Catastrophism and Uniformitarianism

Among the many important questions that have faced geologists is one that concerns the importance of small, slow changes like erosion caused by a single rainstorm, as opposed to massive, catastrophic changes, like earthquakes and floods, that are infrequent and sporadic but cause rapid, dramatic changes to the landscape. During the seventeenth and eighteenth centuries, before geology became the scientific discipline it is today, people believed that all the Earth's features—mountains, valleys, and oceans—had been produced by a few great catastrophes. The catastrophes were thought to be so huge that they could not be explained by ordinary processes; the supernatural had to be called upon. This concept came to be called **catastrophism**. Not only were the catastrophes thought to be gigantic and sudden, but they were also thought to have occurred relatively recently and to fit a chronology of catastrophic events recorded in the Bible.

During the late eighteenth century, the hypothesis of catastrophism was reexamined, compared with geological evidence, and found wanting. The person who assembled much of the evidence and proposed a counter hypothesis was James Hutton (1726–1797). Hutton, a Scot, was a physician and gentleman farmer. He was intrigued by what he saw in the environment around him, especially in Edinburgh where he studied and lived. He wrote about his observations and used logical arguments to draw conclusions based on evidence. In short, he investigated his environment using the scientific method. Hutton is widely regarded today as the

Figure 1.1 Geologist A geologist studying strata of sedimentary rocks in Deadman Wash, Wupatki National Monument, Arizona. The rocks are interbedded sandstones, shales, and mudstones of the Moenkopi Formation.

father of modern scientific geology. In 1795 he published a two-volume work titled *Theory of the Earth, with Proofs and Illustrations* in which he introduced his counter hypothesis to catastrophism.

Hutton observed the slow but steady effects of erosion: the transport of rock particles by running water and their ultimate deposition in the sea. He reasoned that mountains must slowly but surely be eroded away, that new rocks must form from the debris of erosion, and that the new rocks in turn must be slowly thrust up to form new mountains. Hutton didn't know the source of the energy that caused mountains to be thrust up, but everything, he argued, moved slowly along in a repetitive, continuous cycle.

His ideas evolved into what we now call the **Principle of Uniformitarianism**, which states that the same external and internal processes we recognize in action today have been operating throughout the Earth's history. The Principle of Uniformitarianism provides a first and very significant step in understanding the Earth's history. We can examine any rock, however old, and compare its characteristics with those of similar rocks forming today in a particular environment. We can then infer that the old rock very likely formed in the same sort of environment. For example, in many deserts today we can see gigantic sand dunes formed from sand grains transported by the wind. Because of the way they form, the dunes have a distinctive internal structure (Fig. 1.2A). Using the Principle of Uniformitarianism,

A

B

Figure 1.2 Sand Dunes Then and Now The intricate pattern characteristic of wind-deposited sand. A. Layers seen in a hole dug in a modern sand dune near Yuma, Arizona. B. A similar pattern can be seen in sandstone rocks millions of years old, in Zion National Park, Utah. Using the Principle of Uniformitarianism we can infer that these ancient rocks were once sand dunes.

we can safely infer that any rock composed of cemented grains of sand and having the same distinctive internal structure as modern dunes (Fig. 1.2B) is the remains of an ancient dune.

Hutton was especially impressed by evidence he saw at Siccar Point in Scotland (Fig. 1.3). There, he could see ancient sandstone layers, originally horizontal, now standing vertical and capped by layers of younger sandstone. The boundary between the layers, he realized, was an ancient surface of erosion. The now-vertical layers are composed of debris eroded from an ancient landmass. Transported by streams, deposited on the seafloor, and there formed into new rocks, they were uplifted, tilted, and eroded in turn. Eventually, when erosion had formed a new, flat surface on the old, tilted, sandstone layers, a pile of younger erosional debris was deposited on the new erosion surface. Eventually, the younger debris became rock, and uplift occurred again. The cycle of uplift, erosion, transport, deposition, solidification into rock, and renewed uplift that could be deduced from this visible evidence impressed Hutton immensely. There is, he wrote, "no vestige of a beginning, no prospect of an end" to the Earth's geological cycles.

Geologists who followed Hutton have been able to explain the Earth's features in a logical manner by using the Principle of Uniformitarianism. But in so doing

they have made an outstanding discovery—the Earth is incredibly old. It is clear that most processes of erosion are exceedingly slow. An enormously long time is needed to erode a mountain range down, for instance, or for huge quantities of sand and mud to be transported by streams, deposited in the ocean, then cemented into new rocks, and the new rocks deformed and uplifted to form a new mountain. Slow though it is, the cycle of erosion, formation of new rock, uplift, and more erosion has repeated many times during the Earth's long history.

Uniformitarianism is a powerful principle, but should we abandon catastrophism as a totally incorrect hypothesis? Recent discoveries of thin but very unusual rock layers at many places around the world suggest that random, catastrophic events did cause massive changes in the geological record. These were not the catastrophes perceived by seventeenth-century biblical scholars, but they do seem to have caused sudden massive changes in the Earth's external appearance. If the discoveries are proved correct, we will have to conclude that catastrophism has played a role in the Earth's history.

The kind of catastrophe suggested by the unusual rock layers that have led geologists to think again about the validity of catastrophism is the sudden impact of a huge meteorite somewhere on the Earth (Fig. 1.4). Many meteorites have hit the Earth during human his-

Figure 1.3 Evidence of Ancient Erosion Siccar Point, Berwickshire, Scotland. The vertical layers of sedimentary rocks on the right, originally horizontal, were lifted up into their vertical position. Erosion developed a new land surface that became the surface on which the now gently sloping layers of younger sediments were laid. The gently sloping layers, which are named the Old Red Sandstone, are 370 million years old. At this locality, in 1788, James Hutton first demonstrated that the cycle of deposition, uplift, and erosion is repeated again and again.

Figure 1.4 Scar from an Ancient Impact Meteor Crater, near Flagstaff, Arizona. The crater was created by the impact of a meteorite about 50,000 years ago. It is 1.2 km in diameter and 200 m deep. Note the raised rim and the blanket of broken rock debris thrown out of the crater. Many impacts larger than the Meteor Crater event are believed to have occurred during the Earth's long history.

tory, but fortunately there have not been any really big impacts in recent times. The peculiar rock layers mentioned above are rich in the uncommon metal iridium, which is much more abundant in meteorites than in the Earth's common rocks. The unusual iridium-rich rocks have been discovered in Italy, Denmark, and other places around the world (Fig. 1.5). They suggest that a catastrophic impact did occur about 66 million years ago and that many forms of life, including the dinosaurs, became extinct as a result. The hypothesis is that a gigantic meteorite impact threw so much debris into the atmosphere that most animals and many plants could not survive. When the debris settled, it formed a thin, iridium-rich layer wherever sediments were being deposited around the world. Even more dramatic extinctions than the 66-million-year-old one have occurred at other times in the past. The geologic record indicates that one event, about 245 million years ago, sent almost 90 percent of all living plants and creatures to extinction. We still have not discovered the cause of that extinction, but surely it was a catastrophic disaster. Such infrequent massive events fall somewhere between uniformitarianism and catastrophism. When we view the Earth's history as a series of such repeated but sporadic events, there is absolutely no evidence to suggest that similar events will not occur again. Nor is there any evidence to

Figure 1.5 Subtle Record of a Global Disaster This thin, dark layer of rock (marked by the coin) is rich in the rare chemical element iridium and looks out of place in the thick sequence of pale-colored limestones above and below. The iridium-rich layer, here seen in the Contessa Valley, Italy, has been identified at many places around the world and is believed to have formed as a result of a world-circling dust cloud formed by a great meteorite impact about 66 million years ago. It is thought that the impact and its aftermath was the event that sent the dinosaurs to extinction.

suggest when another might occur.

A fascinating but frightening suggestion has been made that a disaster of a different kind may already be happening. The suggestion is that our collective human activities may be changing the Earth so rapidly, and so massively, that plants and animals may be becoming extinct at a rate that is similar in magnitude to some of the major extinctions in the geological record. At present the suggestion is only a hypothesis. It remains to be tested and thereby proved or disproved. Nevertheless, the very fact that serious scientists are concerned that the hypothesis might be true emphasizes an important fact: Geology and the welfare of the human race are indissolubly linked.

THE SOLAR SYSTEM

Once, during a spaceflight, Russian cosmonaut Vitali Sevastyanov reported to ground control, "Half a world to the left, half to the right; I can see it all. The Earth is so small." Cosmonauts and astronauts report that the strongest impression they bring back from their space trips is the aloneness, the smallness, the seeming vulnerability of the little blue planet suspended in the vast emptiness of space. The Earth is not actually alone in space because it is part of the **solar system**, which is all the matter that is gravitationally retained by the Sun.

In addition to the Sun, the solar system consists of nine planets, 61 known moons, a vast number of asteroids, millions of comets, and innumerable small fragments of rock and dust called meteoroids. All of the objects in the solar system move through space in orbits controlled by gravitational attraction. The planets, asteroids, and meteoroids all circle the Sun, while the moons circle the planets.

The distances between the planets are so immense it is difficult to comprehend them. To put the solar system into perspective, think of the Sun as a basketball. The nearest planet, Mercury, would be a grain of sand about 12 m away. The Earth would be a granule about 2.5 mm in diameter and 30 m away, and Saturn would look like a grape 2.5 cm in diameter and nearly 300 m away.

The Terrestrial Planets

The planets can be separated into two groups based on their densities and closeness to the Sun (Fig. 1.6). The innermost planets, Mercury, Venus, Earth, and Mars, are small, rocky, and dense. Each has a density of 3 g/cm^3 or more. They are similar in composition and are called the *terrestrial planets* because the other three are similar to the Earth (*terra* is the Latin word for Earth).

The Jovian Planets

The planets farther from the Sun than Mars are much larger than the terrestrial planets (with the exception of Pluto), yet much less dense. The masses of Jupiter and Saturn, for example, are 317 and 95 times the mass of the Earth, but their densities are only 1.3 and 0.7 g/cm^3, respectively. These *jovian planets*—Jupiter, Saturn, Uranus, Neptune, and Pluto—take their name from *Jove*, an alternative designation for the Roman god Jupiter. They all probably have solid centers that resem-

Figure 1.6 Family Portrait of the Solar System The solar system's nine planets, shown in proper relative size against the Sun. (Note, however, that the distances between the planets are much greater than shown, and the planets never line up so neatly.) The terrestrial planets are the four small, rocky ones closest to the Sun. The jovian planets are the large gas-rich bodies distant from the Sun. Images from space vehicles reveal ring systems around all four jovian planets, though only Saturn's rings are bright enough and large enough to depict on a diagram of this scale. Pluto is the "odd planet out"; it is much smaller and lacks the gaseous envelope characteristic of the jovian planets.

ble terrestrial planets, but, with the exception of Pluto, most of their planetary mass is contained in thick shells that consist mostly of the very light elements hydrogen and helium. The predominance of hydrogen and helium keeps the densities of the jovian planets low. Other, less volatile substances condense to form the clouds we can see in the atmosphere of these planets (Fig. 1.7A and B).

The Origin of the Solar System

How did the solar system form? We may never know the exact answer to this question, but we can discern the outlines of the process from evidence obtained by astronomers, from our knowledge of the solar system today, and from the laws of physics and chemistry.

The birth throes of our Sun and its planets were similar to those of many other suns. Birth began with space that was not entirely empty. Space was not empty because earlier stars had exploded in what astronomers call supernovas. The explosions scattered atoms of various elements everywhere through a huge volume of space. Most of the atoms were hydrogen and helium, but small percentages of all the other chemical elements were present too. Even though thinly spread, the atoms formed a tenuous, turbulent, swirling cloud of cosmic gas. Over a very long period the gas thickened as a result of a slow gathering of all the thinly spread atoms. The gathering force of the gas was gravity, and as the atoms slowly moved closer together, the gas became hotter and denser. Near the center of the gathering cloud of gas, the temperature became so great that hydrogen atoms eventually became so tightly pressed and so hot that they began to fuse to form helium. When, in the gas cloud that formed the solar system, fusion of hydrogen commenced, the Sun was born. The time is estimated to have been about 4.6 billion years ago.

At some stage the outer portions of the cosmic gas cloud became cool enough and dense enough to allow solid objects to condense, in the same way that ice condenses from water vapor to form snow (Fig. 1.8). The solid condensates eventually became the planets, moons, and the other solid objects of the solar system.

Planets and moons nearest the Sun, where the temperatures were highest, are built largely of substances that contain only compounds that can condense at high temperatures (Fig. 1.9). Those substances, known as refractory substances, consist of chemical elements such as iron, silicon, magnesium, and aluminum, mostly as strongly bound chemical compounds with oxygen.

Planets and moons distant from the Sun, where temperatures were lower, contain some refractory substances, but also large quantities of volatile substances that do not condense at high temperatures; for example,

A

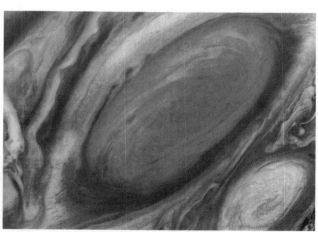

B

Figure 1.7 Turbulent Atmosphere of Jupiter Jupiter, largest of the planets, is a gas-shrouded giant that conceals its presumably solid interior from view.   A. An image of Jupiter taken February 5, 1979, from a distance of 28 million km by spacecraft *Voyager 2*. The banded pattern is due to turbulence in Jupiter's atmosphere. A particularly violent storm is visible in the lower left-hand corner.   B. A gigantic hurricane-like storm has raged for centuries in the atmosphere of Jupiter. First reported by Galileo, the Great Red Spot, as the storm is called, has a diameter twice that of Earth. There is no explanation yet to explain either the huge size or the long life of the storm. This extraordinary image was recorded by the *Voyager 2* on July 6, 1979, from a distance of 2,635,000 km.

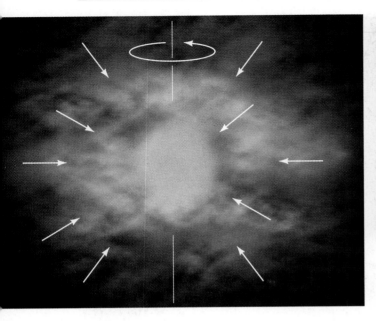

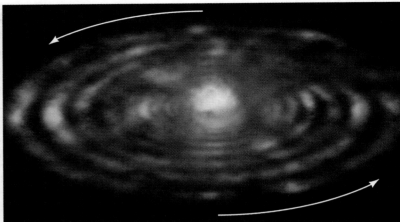

Figure 1.8 Birth of the Solar System The gathering of atoms in space created a rotating cloud of dense gas. The center of the gas cloud eventually became the Sun; the planets formed by condensation of the outer portions of the gas cloud.

the elements hydrogen and sulfur form compounds with oxygen that are solid only at low temperatures (Fig. 1.10A). The farther away from the Sun the condensation occurred, the lower was the temperature and the greater the fraction of volatile substances. One striking demonstration of this fact is the large amount of frozen water in the moons of the jovian planets (Fig. 1.10B).

Condensation of solids from the gas cloud is only the first part of the birth story for the terrestrial planets. Condensation formed innumerable small rocky fragments, but the fragments had still to somehow be joined together to form a terrestrial planet. This happened by impacts between fragments drawn together by gravitational attraction. The largest masses slowly swept up more and more of the condensed rocky fragments, grew larger, and became the terrestrial planets. Meteorites, such as those in Figure 1.11, still fall on the Earth, proving that even now some ancient, condensed rocky fragments still exist in space. Some meteorites resemble rocks formed on the Earth, and these meteorites are believed to have been ejected from a planet or a moon as a result of a large meteorite impact, such as the one that formed Meteor Crater (see Fig. 1.4). Meteorites and the scars of ancient impacts provide evidence of the way the terrestrial planets grew to their present sizes.

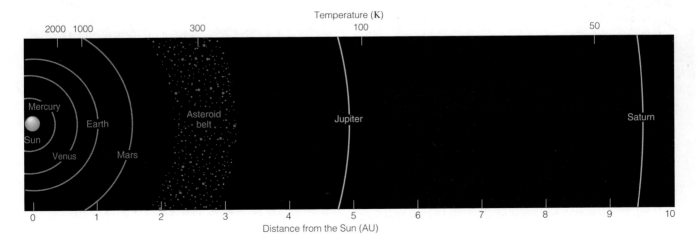

Figure 1.9 Temperature Gradient Temperature gradient in the planetary nebula. Close to the Sun temperatures reached 2000K and the materials that condensed were largely oxides, silicates, and metallic iron and nickel. Farther away, in the region of Jupiter and Saturn, temperatures were low enough for ices of water, ammonia, and methane to condense. (See Appendix A for an explanation of the Kelvin temperature scale. Distances are measured in astronomical units (AU); one AU is the distance from the Sun to the Earth.)

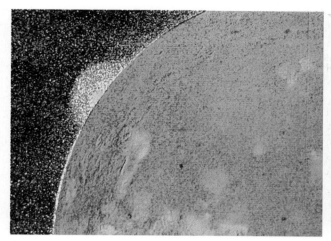

A

B

Figure 1.10 Volatile-rich Moons Two moons of Jupiter that are rich in volatile substances.   A. Io is red because it is rich in sulfur. The image shows a volcanic eruption on Io. The volcanic plume is mostly gas, but small solid particles are also distributed by the gas. The plume rises to a height of 100 km above the surface of Io and is believed to be largely sulfur dioxide (SO_2). Several sites of active volcanism have been discovered on Io.   B. Europa, smallest of the four large moons of Jupiter. Europa has a low density, indicating it contains a substantial amount of ice. The surface is mantled by ice to a depth of 100 km. The fractures indicate that some internal process must be disturbing and renewing the surface of Europa. The dark material (here appearing in red) in the fractures apparently rises up from below. The cause of the fracturing is not known. The image was taken by *Voyager 2* in July 1979.

Figure 1.11 Messenger from Space Meteorites are messengers from space that carry some of the history of the earliest days of the solar system. This boy is examining a large meteorite in the American Museum of Natural History, New York.

The growth process—a gathering of more and more bits of solid matter from surrounding space—is called *planetary accretion.*

Planetary Accretion: Ongoing!

Has planetary accretion ceased? Clearly, the answer is no, because a large (but not gigantic) impact formed Meteor Crater only 50,000 years ago, and in 1912 a very large explosion occurred in a remote area of Siberia called Tunguska—it was the largest explosion due to impact in historic times, and so it is fortunate that it occurred in a relatively uninhabited corner of the world. What such impacts prove is that the Earth and the other planets, even today, can be hit by bits of ancient space debris. Fortunately, large-impact events are rare, but large impacts in the early history of the Earth were much more frequent affairs.

The velocities of small meteorites entering the Earth's atmosphere have been measured between 4 and 40 km/s. If a large meteorite had such a velocity, the amount of energy released on impact would be enormous. It has been calculated, for example, that a meteorite 30 m in diameter and traveling at a speed of 15 km/s would, on impact, release as much energy as the explosion of 4 million tons of TNT. The resulting impact crater would be the size of Meteor Crater in Arizona—1200 m across and 200 m deep. Cratering is a very rapid geological process; the Meteor Crater event is estimated to have lasted about 1 minute. We discuss impact cratering in greater detail in Chapter 21.

THE INTERNAL STRUCTURE OF THE EARTH

Scientists believe that as accretion continued and the terrestrial planets grew larger, their temperatures must have risen. The reason is straightforward: Energy can be changed from one form to another (from electricity to heat, for example), but it cannot be destroyed. A moving object has energy of motion (called *kinetic energy*), and when a meteorite impacts a planet, the kinetic energy is transformed to heat. As planet Earth grew larger and larger, the continual impacts would necessarily have raised its temperature.

In addition, heat must continually have been added from another source. Among the many chemical elements in the Earth are several that are naturally radioactive; they have atoms that spontaneously transform to another element. Examples are uranium, thorium, and potassium. But every time a radioactive transformation occurs, a tiny amount of heat is also produced. Therefore, radioactivity continued to heat the Earth even as the frequency of impacts declined. Eventually, the Earth began to melt. Lighter melted materials, rich in silicon, aluminum, sodium, and potassium, rose toward the surface. Rocks at the Earth's surface are still rich in these elements. Denser melted materials, such as molten iron, sank to the center of the planet. The melting released volatile substances, and these escaped as gases through volcanoes. It was the escaped gases, mainly water vapor, but also smaller amounts of gaseous substances containing carbon, nitrogen and sulfur—that gave rise to the Earth's atmosphere. From the same source came the water we now find in the Earth's oceans. Partial melting changed the Earth from an originally homogeneous planet to a compositionally layered one.

Layers of Differing Composition

Planet Earth has three main parts below its solid surface; they are distinguished from one another by differences in composition (Fig. 1.12). At the center is the densest of the three parts, the **core**. The core is a spherical mass, composed largely of metallic iron, with lesser amounts of nickel and other elements.

The thick shell of dense, rocky matter that surrounds the core is called the **mantle**. The mantle is less dense than the core but denser than the outermost layer. Above the mantle lies the thinnest and outermost layer, the **crust**, which consists of rocky matter that is less dense than the rocks of the mantle below.

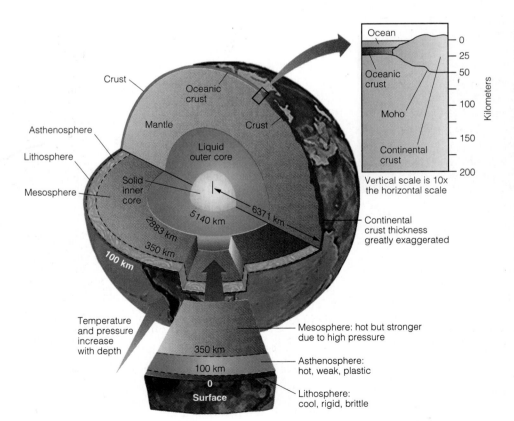

Figure 1.12 Inside the Earth A sliced view of the Earth reveals layers of different composition and zones of differing rock strength. The compositional layers, starting from the inside, are the core, the mantle, and the crust. Note that the crust is thicker beneath the continents than under the oceans. Note, too, that boundaries between zones of differing strength—lithosphere (outermost), asthenosphere, mesosphere—do not coincide with compositional boundaries.

The crust is far from uniform and differs in thickness from place to place by a factor of about ten. The crust beneath the oceans, the **oceanic crust**, is as thin as 5 km in places and has an average thickness of about 8 km, whereas the **continental crust** averages about 45 km and ranges in thickness from 25 to 70 km.

We cannot see either the core or the mantle, and so it is valid to ask how we know anything about their composition. The answer is that indirect measurements are used. One way to get information about composition is to measure the times required for earthquake waves to travel through the Earth by different paths. From these observations, the speeds of the waves in all parts of the interior can be determined. Wave speed depends on the density and elastic properties of the material the waves travel through (see Chapter 16), which in turn depend on the composition of the material. At some depths abrupt velocity changes indicate sharp changes in density and elastic properties. From the sharp changes, we can infer that the Earth's interior does not have a uniform composition but must instead consist of layers with different characteristics. From what we know of these characteristics we can estimate what the compositions of the different layers must be.

Slight compositional variations probably exist within the mantle, but we know little about them. We can see and sample the crust, however, and the sampling shows that even though it is quite varied in composition, the crust's overall composition and density are very different from those of the mantle, and the boundary between them is distinct.

The composition of the core presents the most difficulty. The temperatures and pressures in the core are so great that materials there probably have unusual properties. Some of the best evidence concerning core composition comes from iron meteorites. Such meteorites are believed to be fragments from the core of a small terrestrial planet that was shattered by a gigantic impact early in the history of the solar system. Scientists hypothesize that this now-shattered planet had compositional layers similar to those of the Earth and the other terrestrial planets.

Layers of Differing Physical Properties

In addition to compositional layering, other changes occur within the Earth. Most important, there are changes of physical properties such as rock strength and solid versus liquid. Changes in physical properties are largely controlled by temperature and pressure rather than rock composition. The places where physical properties change do not coincide exactly with the compositional boundaries between the crust, mantle, and core shown in Figure 1.12.

The Inner Core and the Outer Core

Within the core an inner region exists where pressures are so great that iron is solid despite its high temperature. The solid center of the Earth is the **inner core**. Surrounding the inner core is a zone where temperature and pressure are such that the iron is molten and exists as a liquid. This is the **outer core**. The difference between the inner core and the outer core is not primarily one of composition (the compositions are believed to consist mostly of elemental iron). Instead, the difference lies in the physical states of the two: one is a solid, the other a liquid.

The Mesosphere

The strength of a solid is controlled by both temperature and pressure. When a solid is heated, it loses strength; when it is compressed, it gains strength. Differences in temperature and pressure divide the mantle and crust into three strength regions. In the lower part of the mantle, the rock is so highly compressed that it has considerable strength even though the temperature is very high. Thus, a solid region of high temperature but also relatively high strength exists within the mantle from the core-mantle boundary (at 2883 km depth) to a depth of about 350 km and is called the mesosphere ("intermediate, or middle, sphere") (Fig. 1.12).

The Asthenosphere

Within the upper mantle, from 350 to between 100 and 200 km below the Earth's surface, is a region called the **asthenosphere** ("weak sphere"), where the balance between the effects of temperature and pressure is such that rocks have little strength. Instead of being strong, like rock in the mesosphere, rock in the asthenosphere is weak and easily deformed, like butter or warm tar. As far as geologists can tell, the composition of the mesosphere and the asthenosphere is the same. The difference between them is one of physical properties; in this case the property that changes is strength.

The Lithosphere

Above the asthenosphere is the outermost strength zone, a region where rocks are cooler, stronger, and more rigid than those in the plastic asthenosphere. This hard outer region, which includes the uppermost mantle and all of the crust, is called the **lithosphere** ("rock sphere"). It is important to remember that despite the fact that the crust and mantle differ in composition, it is rock strength, not rock composition, that differentiates the lithosphere from the asthenosphere.

The differences in strength between rock in the lithosphere and rock in the asthenosphere is a function of temperature and pressure. At a temperature of 1300°C and the pressure reached at a depth of 100 km, rocks of all kinds lose strength and become readily deformable. This is the base of the lithosphere beneath the oceans, or, as it is more colloquially termed, the *oceanic lithosphere*. The base of the *continental lithosphere*, by contrast, is about 200 km deep where the temperature is about 1350°C. The reason for the difference between the two kinds of lithosphere is the way temperature increases with depth.

GEOTHERMAL GRADIENTS AND CONVECTION

Among the fundamental laws of nature are the three Laws of Thermodynamics, which deal with the movement of heat. A very important consequence of the laws is that heat always flows from a hot place to a cold one. Thus, heat must flow outward from the hot interior of the Earth toward the cool surface. Careful measurements made in mines and drill holes around the world show that the rate of temperature increase with depth (called the **geothermal gradient**) varies from place to place, ranging from 5°C/km to 75°C/km. Gradients lessen with depth; below 200 km the gradient is thought to be only 0.5°C/km. By extrapolation, we calculate that the temperature at the center of the Earth must be at least 5000°C, which is approximately the same as the temperature of the surface of the sun.

The process by which heat moves through solid rock, or any other solid body, without deforming the solid, is called **conduction**. Heat conduction is a familiar process.

It is the way heat moves along the handle of a hot saucepan. Conduction does not cause the movement of hot material from one place to another. But because volcanoes involve the physical movement of hot material, we have to conclude that heat can move in the Earth by another process in addition to conduction. **Convection**, unlike conduction, does cause movement. Convection is the process by which hot, less dense materials rise upward and are replaced by cold, downward- and sideways-flowing materials to create a **convection current** (Fig. 1.13). Wind is an example of convection. Because volcanoes erupt hot lava, at least some heat energy must therefore move inside the solid Earth by convection. The conclusion that convection must occur in a seemingly rigid solid body like the Earth may seem odd, but it is a very important one. Tests show that rock doesn't have to melt before it can flow. Rock, if sufficiently hot, can flow like sticky liquid, although the rate of flow is exceedingly slow. The higher the temperature, the weaker a rock is and the more readily it will flow. Slow convection currents are possible deep inside the Earth because the interior is very hot. Convection currents bring hot rocks upward from the Earth's interior. The currents of hot rock flow slowly up, spread sideways, and eventually sink downward as they become cooler and more dense.

Because rock in the lithosphere is too strong and too rigid for convection to happen, heat moves through the lithosphere primarily by conduction. Conduction

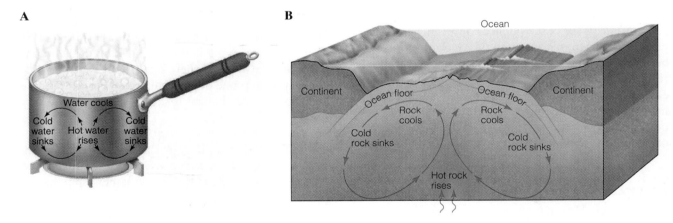

Figure 1.13 Convection in the Mantle Convection shapes the Earth's surface. A. Convection in a sauce-pan full of water. Water that is heated expands and rises. As it rises, it starts to cool, flows sideways, and sinks, eventually to be reheated and pass again through the convection cell. B. Convection as it is thought to occur in the Earth. Though much slower than convection in a saucepan, the principle is the same but there is a difference in detail. In the Earth it is thought that heat loss

causes rocks to become cool enough to sink and that cooling from the top rather than heating from below, starts convective motions. Once sinking starts, hot rock rises slowly from deep inside the Earth to balance the downward flow. Eventually the upward flowing rock flows sideways, cools, and sinks. The rising hot rock and sideways flow are believed to be the factors that control the positions of ocean basins and continents.

is a slow process, however, so thermal gradients in the lithosphere are steep—that is, temperatures change rapidly with depth. The temperature at the base of the lithosphere—that is, at the lithosphere-asthenosphere boundary—varies from about 1300°C beneath the oceans to 1350°C beneath the continents.

The average temperature at the top of the lithosphere is close to 0°C. Since the oceanic lithosphere is about 100 km thick, the average geothermal gradient in the oceanic lithosphere is 1300°C/100 km, or 13°C/km. By contrast, the continental lithosphere is about 200 km thick, so the average geothermal gradient in the continental lithosphere is about 1350°C/200 km, or 6.7°C/km (Fig. 1.14).

Within the asthenosphere and mesosphere temperatures are everywhere high enough, and rocks everywhere weak enough, so that convection can occur. As a result, in the deeper portions of the Earth, convection is a far more important process for the movement of heat than is conduction.

Consider what happens when a mass of hot rock, deep inside the Earth, starts to rise convectively. Because pressure increases with depth, a rising mass will experience progressively lower pressures. As the pressure decreases, the rising mass will expand somewhat and, like air expanding as it escapes from a balloon, the temperature will drop. Even though the temperature drops, no heat energy is lost from the rising mass; the process is called *adiabatic* expansion, which means without loss or gain of heat. The thermal gradient owing to adiabatic expansion is approximately 0.5°C/km. Below a depth of 200 km, therefore, the oceanic and continental geothermal gradients merge and remain roughly constant at 0.5°C/km to the core-mantle boundary. The estimation that the temperature of the Earth's core at the coremantle boundary is 5000°C is based on an assumption of an adiabatic gradient throughout the asthenosphere and the mesosphere.

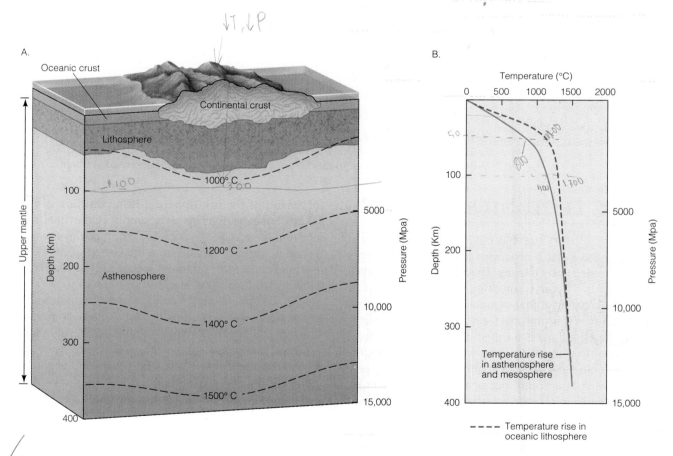

Figure 1.14 Deeper and Hotter A. Temperature increases with depth in the Earth, as shown here. The dashed lines are isotherms, lines of equal temperature. Note that the temperature increases more slowly with depth under the continents than under the oceans, where it gets quite hot at a shallow depth. B. This shows the same information as in (A) but in graph form. The Earth's surface is at the top, so depth (and corresponding pressure) increases downward. Temperature increases from left to right. The solid curve shows how temperature increases with depth under the oceans. The dotted curve shows how temperature increases more slowly with depth under the continents.

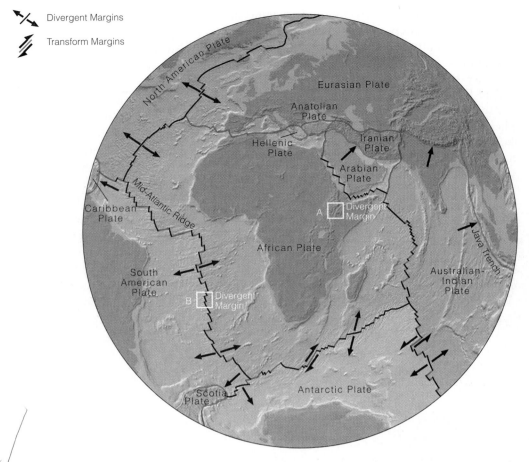

Figure   1.15   Plates of Lithosphere   Six large plates and a number of smaller ones comprise the Earth's surface. They are moving very slowly in the directions shown by the arrows. The labels A, B, C, D, and E correspond to the different kinds of plate margins discussed in Chapter 2.

PLATE TECTONICS

A remarkable story that has emerged from geological studies of the Earth's processes is that the continents themselves are slowly moving. They are drifting sideways at rates up to 12 cm/yr, sometimes bumping into each other and creating a new mountain range by collision, and sometimes splitting apart so that a new ocean basin forms. The Himalaya is a range of geologically young mountains that began to form when the Indian sub-continent collided with Asia about 45 million years ago. The Red Sea is a young ocean that started forming about 30 million years ago when a split developed between the Arabian Peninsula and Africa as the two landmasses began to move apart.

But it is not just the continents that move, it is the lithosphere. The continents and large blocks of seafloor rock are moving along like passengers on large rafts; the rafts are huge plates of lithosphere that float on the asthenosphere. As a result, all the major features on the Earth's surface, whether submerged beneath the sea or exposed on land, arise as either a direct or indi- rect result of the lithosphere drifting on the astheno- sphere. Such motions involve complicated events, both seen and unseen, all of which are embraced by the term **tectonics**, which is derived from the Greek word, *tekton*, meaning carpenter. Tectonics is the study of the move- ment and deformation of the lithosphere.

The lithosphere is not a continuous layer. Instead, like ice on a lake during a thaw, the lithosphere consists of a number of separate plates. Unlike the ice on a lake, however, the plates of the lithosphere are huge and distinctly curved (Fig. 1.15). The Earth's internal con- vection is always moving the plates of the lithosphere and changing the Earth's surface. Mountains like the Alps or the Appalachians that seem changeless to us are only transient wrinkles when viewed in geological time. Mountain ranges grow where plates of lithosphere con- verge and heave masses of twisted and deformed rock upward; then the ranges are slowly worn away, leaving only the eroded roots of an old mountain range to record the ancient collision. The Earth's long dynamic history is recorded in rocky scars such as those shown in Fig 1.16.

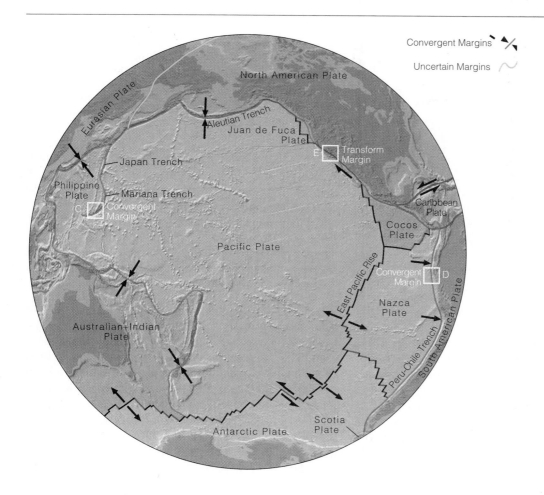

Convergent Margins

Uncertain Margins

North American Plate

Eurasian Plate

Aleutian Trench

Juan de Fuca Plate

Japan Trench

E Transform Margin

Philippine Plate

Mariana Trench

C Convergent Margin

Caribbean Plate

Cocos Plate

Pacific Plate

Convergent Margin D

Nazca Plate

East Pacific Rise

Australian+Indian Plate

South American Plate

Peru-Chile Trench

Scotia Plate

Antarctic Plate

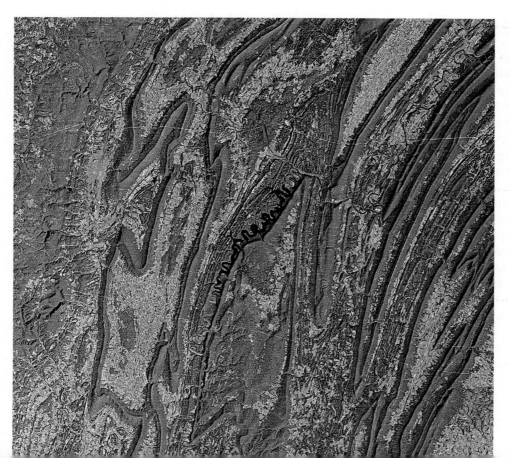

Figure 1.16 Scars of a Collision Between Continents The scar of an ancient collision. Layers of rock, once horizontal, were twisted and contorted as a result of a collision between two plates. These eroded roots of a once much grander mountain range are today's central Appalachians in Pennsylvania.

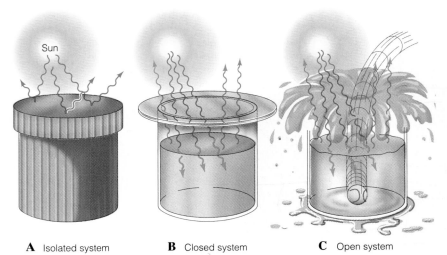

A Isolated system   **B** Closed system   **C** Open system

Figure 1.17 Kinds of Systems An isolated system allows neither energy nor matter to cross its boundaries. Isolated systems don't really exist in nature. A closed system allows energy but not matter to cross it boundaries. Closed systems are rare; the Earth is a natural example of a closed system. An open system allows both energy and matter to cross its boundaries. Open systems are common in nature; most of the Earth's subsystems are open systems.

The special branch of tectonics that deals with the processes by which plates of lithosphere are moved laterally over the asthenosphere is called plate tectonics. Plate tectonics was proposed as a theory only in the 1960s. Some details are still inadequately understood and still being investigated. But the discoveries and new understanding that have already come from studies made to test the plate tectonics theory are so profound that the theory has sparked a modern geological revolution. Necessarily, a lot of attention is paid to plate tectonics in this book, and in the next chapter (Chapter 2), there is a detailed discussion of the way plates form, move and interact.

THE SYSTEM CONCEPT

The system concept is a way to break down any large, complex problem into smaller, more easily studied pieces. A **system** can be defined as any portion of the universe that can be isolated from the rest of the universe for the purpose of observing and measuring changes. By saying that *a system is any portion of the universe*, we mean that the system can be whatever the observer defines it to be. That is why a system is only a concept; you choose its limits for the convenience of your study. It can be large or small, simple or complex. You could choose to observe the contents of a beaker in a laboratory experiment. Or you might study a flock of nesting birds, a lake, a small sample of rock, an ocean, a volcano, a mountain range, a continent, or even an entire planet. A leaf is a system, but it is also part of a larger system (a tree), which in turn is part of an even larger system (a forest).

The fact that a system has been *isolated from the rest of the universe* means that it must have a boundary that sets it apart from its surroundings. The nature of the

boundary is one of the most important defining features of a system, leading to the three basic kinds of systems—isolated, closed, and open—as shown in Figure 1.17. The simplest kind of system to understand is an **isolated system**; in this case the boundary is such that it prevents the system from exchanging either matter or energy with its surroundings. The concept of an isolated system is easy to understand because, although it is possible to have boundaries that prevent the passage of matter, in the real world it is impossible for any boundary to be so perfectly insulating that energy can neither enter nor escape.

The nearest thing to an isolated system in the real world is a **closed system**; such a system has a boundary that permits the exchange of energy, but not matter, with its surroundings. An example of a nearly closed system is the space shuttle, which allows the material inside to be heated and cooled but is designed to minimize the loss of any material. The third kind of system, an **open system**, is one that can exchange both energy and matter across its boundary. An island on which the rain is falling is a simple example of an open system: Some of the water runs off via streams or seeps downward to become groundwater, while some is absorbed by plants or evaporates back into the atmosphere (Fig. 1.18).

To a close approximation, the Earth system is a closed system. Energy enters and leaves, but except for a few meteorites arriving from space and a tiny amount of gas leaking away from the atmosphere, the Earth has a nearly constant mass. The Earth is comprised of four vast reservoirs of matter, and each is an open system because both matter and energy flow back and forth between them (Fig. 1.19). The four reservoirs are:

1.   The **atmosphere**, which is the mixture of gases—predominantly nitrogen, oxygen, argon, carbon dioxide, and water vapor—that surrounds the Earth.

2.   The **hydrosphere**, which is the totality of the Earth's

A

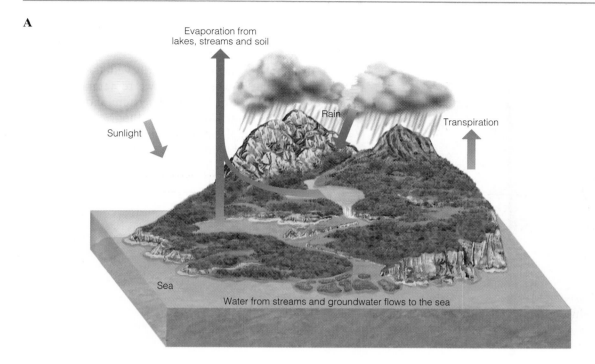

B

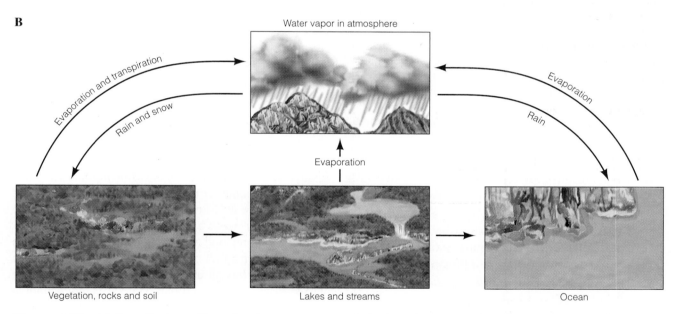

Figure 1.18 An Open System Example of an open system. A. Energy (sunlight) and water (rainfall) reach an island from external sources. The energy leaves the island as long-wavelength radiation; the water either evaporates or drains into the sea. B. Depiction of the open system above by a box model to illustrate how materials flow from one subsystem to another.

water, including oceans, lakes, streams, underground water, and all the snow and ice, but exclusive of the water vapor in the atmosphere.

3. The **biosphere**, which is all of the Earth's organisms as well as any organic matter not yet decomposed.

4. The **geosphere**, which is the solid Earth, and is composed principally of rock and **regolith** (the irregular blanket of loose, uncemented rock particles that covers the solid Earth).

Each of the four systems can be further subdivided into smaller, more manageable study units. For example, we can divide the hydrosphere into the

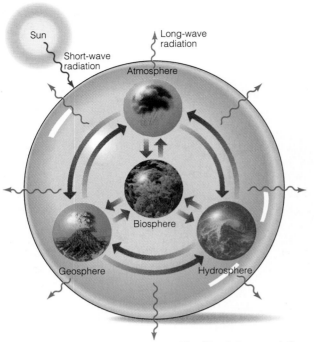

Figure 1.19 A Closed System The Earth is essentially a closed system. Energy reaches the Earth from an external source and eventually returns to space as long wavelength radiation. Smaller systems within the Earth, such as the atmosphere, biosphere, hydrosphere, and geosphere are all open systems.

ocean, glacier ice, streams, and groundwater. All of the smaller systems within the Earth System are open systems.

This book is concerned primarily with the geosphere. It is important, however, that we remember that the geosphere is part of the Earth system and that many important processes involve the other reservoirs of the system. For example, **erosion**, which is the breakdown of rock by the combined effects of rain, wind, snow, ice, plants, and animals, involves the atmosphere, hydrosphere, and biosphere.

ABOUT THIS BOOK

The scope of this text is all of physical geology, but four main themes are emphasized:

1. The connective link between internal convection and the Earth's external features through plate tectonics.

2. The insights gained from studying the Earth as a system of interdependent parts, and the new tools used to measure the Earth's processes.

3. The influence of the human race on the Earth's external processes and environments.

4. The need for humans to be wise in their use of the Earth's limited store of natural resources.

The first theme, plate tectonics, emphasizes the most important scientific theory to arise from geological investigations in the twentieth century. The second concerns new advances to our understanding of interactions between the many parts of the Earth system, and new tools that allow us to make measurements of

the interactions. An example of new insights arising from the use of new tools is discussed in the boxed essay *"The Warmest Year of the Millennium."* The third and fourth themes concern the human race. The third focuses on the well-being of the environment, and the fourth on our needs and the supplies of resources.

Each chapter opens with an essay, and most of the essays concern either our environment or the effect the environment has on us. Incorporated into the text are boxed essays on special topics related to Understanding Our Environment, Using Resources, and Measuring Our Earth.

Each chapter ends with three features designed to help you assimilate the material that has been covered: a brief summary, a list of key terms, and a series of questions based on material in the chapter. The following are examples of these features for this first chapter.

Measuring Our Earth

The Warmest Year of the Millennium

Recent measurements confirm what many people suspected. 1998 was the warmest year of the millennium. A study by Michael E. Mann of the University of Massachusetts and colleagues combined information recorded in "natural thermometers"—tree rings and layers of glacial ice-with measurements by modern instruments. The result is a reconstruction of temperatures in the Northern Hemisphere going back to the beginning of the year 1000.

A previous analysis of temperatures had shown that the decade of the 1990s was the warmest decade of the past six centuries, but scientists questioned whether there might have been warmer decades in earlier centuries. It was known, for example, that in the first three centuries after the year 1000, a time known as the medieval warm period, Europe enjoyed an especially warm climate. But the new study shows that while the North Atlantic region was comparatively warm, the rest of the Northern Hemisphere was not.

According to Mann, the current warm climate is unprecedented. Although there are uncertainties in the estimates for the earlier part of the record, Mann says he is 99 percent certain that 1998 was the warmest year of the millennium and that the 1990s is the warmest decade of the past thousand years.

Tree-ring researchers are not entirely convinced by the evidence and by Mann's conclusions. They point out that there is too much uncertainty in the data to warrant such firm conclusions. As is often the case, more research is needed.

Source: "1998: Warmest Year of Past Millennium," *Science News* (March 20, 1999), p. 191.

SUMMARY

1. Geology is the study of changes, past and present, that happen to the Earth. Scientists who study the Earth are called geologists.

2. Science is a system of learning and understanding that advances by application of the scientific method: observation, formation of a hypothesis, testing of the hypothesis, formation of a theory, more testing, and, in some instances, formation of a law.

3. The Principle of Uniformitarianism states that the internal and external processes operating today have been operating throughout Earth's history.

4. Random massive disasters, such as gigantic meteorite impacts, appear to have played an important role in the Earth's history. These events cause catastrophic change in the Earth's appearance, and to many forms of life, but are not attributed to supernatural forces the way the events of the outdated concept of catastrophism were.

5. There are nine planets in orbit around the Sun. The four innermost, Mercury, Venus, Earth, and Mars, are small, dense, rocky bodies. Four of the outer plants, Jupiter, Saturn, Uranus, and Neptune, are larger, less dense objects with thick atmospheres of hydrogen and helium. Pluto, the planet farthest from the Sun, is small, not gassy, and does not fit the mold of either terrestrial or jovian planet.

5. The planets formed by a two-step process. First, small rocky fragments condensed from a disc-shaped envelope of gas that rotated around the Sun. Then the rocky fragments started accreting into ever larger masses. The largest, today's planets, had all formed by 4.55 billion years ago.

7. The heat released by decay of naturally radioactive chemical elements was sufficient, early in the Earth's history, to cause a fraction of the Earth to melt. Heavy materials sank to the center, and lighter ones rose, giving the Earth a compositionally three-layered structure: core, mantle, and crust.

8. The crust consists of two parts: oceanic crust with an average thickness of 8 km and continental crust with an average thickness of about 45 km.

9. The Earth is also layered with respect to its physical properties, in particular strength. The lithosphere, the outer zone of the solid Earth, consists of rock that is strong and relatively rigid. Beneath the oceanic crust, the lithosphere is about 100 km thick; beneath the continental crust, it is 200 km thick. Below the lithosphere, down to a depth of 350 km, is the asthenosphere, a region where high temperatures make rock weak and easily deformed. Beneath the asthenosphere is the mesosphere, where rocks become gradually stronger. Within the core there are also two regions differing in physical properties but with the same composition: the inner core is solid, the outer core molten.

10. The lithosphere consists of six large and many small plates that slide slowly over the asthenosphere at rates up to 12 cm/yr, as a result of a process called plate tectonics.

11. The Earth's internal heat reaches the surface by conduction and convection. Convection is the dominant way heat moves below the lithosphere. Within the lithosphere heat moves mainly by conduction.

12. The Earth can be considered a system of four vast, interdependent reservoirs: the geosphere, the atmosphere, the hydrosphere, and the biosphere.

13. Material and energy move back and forth from one of the Earth's reservoirs to another in continuing cycles.

THE LANGUAGE OF GEOLOGY

asthenosphere (p. 15)
atmosphere (p. 20)

biosphere (p. 21)

catastrophism (p. 6)
closed system (p. 20)
conduction (p. 16)
continental crust (p. 15)
convection (p. 16)
convection current (p. 16)
core (p. 14)
crust (p. 14)

erosion (p. 22)

geologists (p. 4)
geology (p. 4)
geosphere (p. 21)
geothermal gradient (p. 16)

historical geology (p. 4)
hydrosphere (p. 20)
hypothesis (p. 6)

inner core (p. 15)
isolated system (p. 20)

Law of Original Horizontality (p. 6)
lithosphere (p. 15)

mantle (p. 14)

oceanic crust (p. 15)
open system (p. 20)
outer core (p. 15)

physical geology (p. 4)
plate tectonics (p. 20)
principle of uniformitarianism (p. 7)

regolith (p. 21)
rock (p. 5)

scientific method (p. 6)
solar system (p. 10)
system (p. 20)

tectonics (p. 18)
theory (p. 6)

QUESTIONS FOR REVIEW

1. Suggest three human activities that affect the Earth's external processes in a detectable way.

2. Identify three human activities in the area where you live that are causing big changes to the environment.

3. What is the scientific method? Illustrate your answer with an example of the scientific method in practice.

4. How does the Principle of Uniformitarianism help us understand the history of the Earth?

5. Briefly describe the steps by which scientists believe planets form from a huge cloud of gaseous material.

6. Describe the Earth's compositional layers. Discuss how the layers developed from an originally homogeneous Earth.

7. The Earth is layered with respect to its physical properties. Describe the major physical property layers and discuss how they arise.

8. What are the relationships between the crust, the mantle, and the lithosphere?

9. Why and how does the thickness of the lithosphere vary?

10. Describe the way heat escapes from the Earth's interior. How and why does the geothermal gradient through the oceanic lithosphere differ from the geothermal gradient in the asthenosphere and mesosphere?

11. What are the differences between a closed and an open system?

12. What consequences arise from the fact that the Earth is a closed system? Would the Earth still be a closed system if we started a colony on Mars and started trading with the colony?

For an interactive case abstract, virtual tours, activities, and additional learning resources, go to
GEOSCIENCES TODAY: **www.wiley.com/college/skinner**

Saline Lakes and Global Climate Change

The Aral Sea, on the border of Kazakhstan and Uzbekistan, provides a classic example of human-caused ecological disaster. Once the fourth largest lake in the world, the Aral Sea provided a moderating influence on the climate of the region, served as a major transportation corridor, and supported a thriving fishing industry that employed more than 70,000 people. Today the sea is only the sixth largest lake and is disappearing so fast it will be a waterless desert by 2010. A part of the problem is climate, but the largest part of the problem has been caused by human activities.

OBJECTIVE: The goal of this case is to develop an appreciation for the saline lake—an important ecosystem. It will discuss saline lakes in general, look at some of the world's most well-known saline lakes, and explore the effects of human intervention on the Aral Sea.

The Human Dimension: How can the saga of the Aral Sea serve as an example of the effects of human intervention on sensitive natural ecosystems?

Questions to Explore:

1. What is a saline lake, and what attributes do saline lakes share?

2. What are some of the world's most well-known saline lakes, how were they formed, and what can they tell us about the Earth's geologic and climatic history?

3. Why are saline lakes important?

2

**Measuring
Our Earth**

**Understanding
Our Environment**

**Using
Resources**

GeoMedia

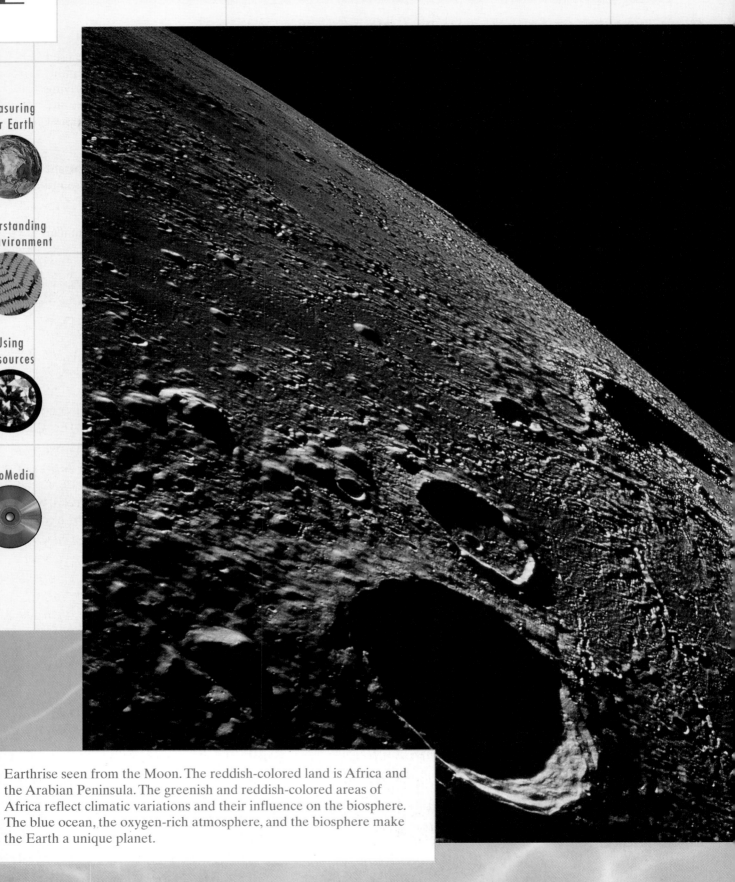

Earthrise seen from the Moon. The reddish-colored land is Africa and
the Arabian Peninsula. The greenish and reddish-colored areas of
Africa reflect climatic variations and their influence on the biosphere.
The blue ocean, the oxygen-rich atmosphere, and the biosphere make
the Earth a unique planet.

The Dynamic Earth

The Special Planet

Visits to the Moon by astronauts and images of distant planets sent back by unmanned spaceships have astonished us. But surely the most remarkable spaceship image brought back by the astronauts is of Earth itself. For the first time all of us can see our planet in one sweeping view; see its clouds, oceans, polar ice caps, and continents at the same time. Planet Earth, we can see, is just a small planet in orbit around an ordinary, medium-sized star. But the Earth is a special planet, and that too can be seen. First, the Earth has an overall blue and white complexion as is shown in the chapter-opening photograph. The blueness is due to sunlight being scattered by atmospheric gases, predominantly nitrogen, oxygen, argon, and water vapor. No other planet or moon in the solar system has such an atmosphere. In the Earth's atmosphere are also white clouds of condensed water vapor. The clouds form because water continually evaporates from the ocean, from water on land, and from leaves of green plants.

Another special thing about the Earth that can be seen from space is

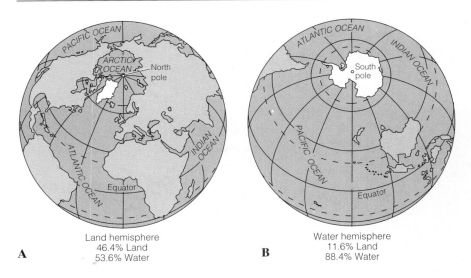

A   Land hemisphere
46.4% Land
53.6% Water

B   Water hemisphere
11.6% Land
88.4% Water

Figure 2.1 Land and Sea The unequal distribution of land and sea is emphasized by these two views of the Earth.   A. View of the world from a point directly above Spain; 65 percent of the world's landmass is in the northern hemisphere.   B. View of the world from a point directly above New Zealand. Only 35 percent of the world's landmass is in the southern hemisphere. As a result, nearly 90 percent of the hemisphere is ocean.

evidence of life. Viewed from space, the biosphere is most dramatically revealed by blankets of green plants on some of the landmasses. Some parts of the landmass are brown rather than green. These brown, weather-beaten places are also special; they are evidence of weathering, which is the chemical alteration and physical breakdown of rock as a result of exposure to the atmosphere, hydrosphere, and biosphere. In short, what the view of the Earth from space makes abundantly clear is that the Earth is not just a special planet; it is unique, and it is a closed system.

The fact that the Earth system is closed has a profound significance for those who live here: although materials like water, potassium, and carbon cycle through various complex pathways through the geosphere, biosphere, atmosphere, and hydrosphere, the *amount* of matter is fixed. This means that, unless we are prepared to mount major resource-hunting expeditions in outer space, the material resources available to us are fixed in quantity and we must learn to live with what we have.

A further consequence of living in a closed system made up of interacting open systems is that a change in one part of the system necessarily causes changes somewhere else in the system. For many, many years it was

believed that human activities only amounted to a drop in the bucket compared to the grand scale of the Earth's global activities. The results of ongoing research have demolished this old belief and are now bringing revelation after revelation. Not only do our collective human activities change things both locally and globally, but also the changes are happening in a much shorter time than we ever dreamed possible.

THE SHAPE AND FACE OF THE EARTH

Seen from space, the Earth seems to be a nearly perfect sphere, but careful measurements tell a different story. The Earth bulges around the equator and is slightly flattened at the poles. The bulge is not large—the radius at the equator is 6378.2 km, and at the poles it is 6356.8 km, a difference of 21.4 km. The shape is a consequence of the Earth's spinning on its axis. It has been calculated that a body the size of the Earth made entirely of water, and rotating at the same speed as the Earth, would have almost exactly the same bulging shape as the

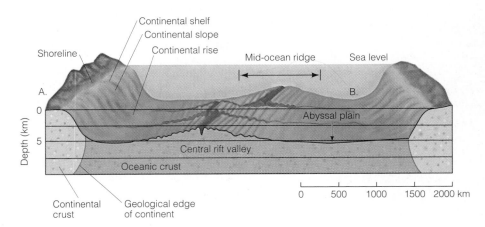

Figure 2.2 Topography of the Seafloor Section across the northern part of the Atlantic Ocean showing the major topographic features.

Earth. This interesting calculation points to an important conclusion: The Earth, overall, has little strength, and when a deforming force is applied for a long time, such as the outward pull due to rotation, the seemingly solid Earth can be readily deformed.

Distribution of Land and Sea

The primary division of the world's surface is into land and sea. The seas cover 71 percent of the Earth's surface, and the land occupies the remaining 29 percent. The land is not evenly distributed. A convenient way to view the unequal distribution of land and sea is to compare the view of the Earth from a spot directly above Spain with the view from a spot directly above New Zealand (Fig. 2.1). About 65 percent of the land is in the northern hemisphere (see Fig. 2.1A), and only 35 percent of the land is in the southern hemisphere (see Fig. 2.1B).

Modern shorelines don't coincide exactly with the boundaries between the continental crust and the oceanic crust. In fact, the present shoreline is of little significance except to communities of marine animals and plants that live close to the shore. The sea level fluctuates with time as more or less water is locked up in glaciers and ice caps. When water is withdrawn and stored in ice caps, sea level falls and the shoreline retreats. When ice caps melt and sea level rises, shorelines advance. Throughout the twentieth century shorelines have been advancing because sea level is slowly rising.

At present there is more water in the ocean than is needed to fill the ocean basins, and as a result some of the ocean spills out of the ocean basin onto the continent (Fig. 2.2). The geological edge of the ocean basin is not the shoreline: rather, it is the place where the oceanic crust joins the continental crust. The boundaries between continental and oceanic crust are therefore hidden from view because they are covered by water, and today's shorelines are actually on the continents. As a result, each continent is surrounded by a flooded margin of variable width known as the **continental shelf**. The geological edge is at the bottom of the **continental slope**, a pronounced slope beyond the seaward margin of the continental shelf. If, instead of today's shoreline, we take the bottom of the continental slope to be the boundary of the continents, only 60 percent of the Earth's surface is occupied by ocean basins, while 40 percent is occupied by continents. Thus, 25 percent of the continental crust is covered by seawater (Fig. 2.3).

The **continental rise** lies at the base of the continental slope. It is a region of gently changing slope where the floor of the ocean basin meets the margin of the continent. The rise is actually part of the floor of the ocean basin, but it is a distinctive part because it is underlain by oceanic crust and covered by a thick pile of erosional debris shed from the adjacent continent.

Beyond the continental slope and continental rise lies the strange, rarely seen world of the deep ocean floor. Teams of oceanographers and seagoing geologists, using new devices, have sounded and sampled the ocean bot-

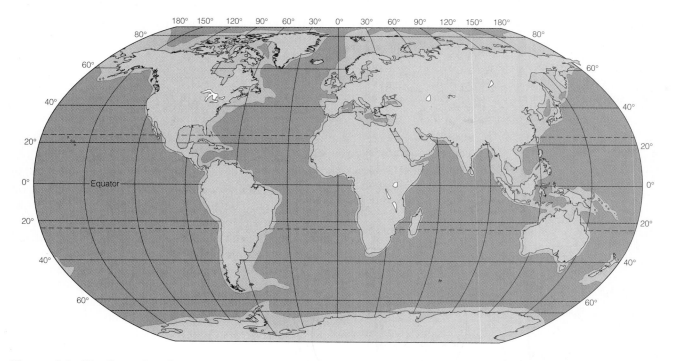

Figure   2.3   The Ocean Overflows   The continental shelves and slopes (shown in light blue) form about 25 percent of the mass of the continental crust. The shelves and slopes are water covered because there is more water in the ocean than just sufficient to fill the ocean basins.

tom during submarine dives of limited duration. As a result of this work, today we know a great deal about the seafloor, though not yet as much as we know about the land surface.

The large, flat areas known as the *abyssal plains* are a major topographic feature of the seafloor and lie adjacent to the continental rise (Fig. 2.2). They generally are found at depths of 3 to 6 km below sea level and range in width from about 200 to 2000 km. Plains are most common in the Atlantic and Indian oceans, which have large, mud-laden rivers entering them. Abyssal plains form as a result of sediment that has settled on the continental shelf becoming unstable and slipping down the slope, over the continental rise, and out to the deep ocean where it buries the original seafloor topography beneath a blanket of fine debris.

If it were possible to remove all the water from the ocean and then view the dry Earth from a spaceship, we could contrast the ocean basins and the continents. We would then see that the continents stand, on average, about 4.5 km above the floor of the ocean basins (Fig. 2.4). The continents stand higher than the ocean basins because the thick continental crust is relatively light (density 2.7 g/cm^3), while thin oceanic crust is relatively heavy (density close to 3.0 g/cm^3). Because the lithosphere is floating on the asthenosphere, those portions of the lithosphere capped by the thick, light continental crust stand high, while those capped by the thin, heavy oceanic crust sit lower. The mechanism by which portions of the lithosphere rise or subside until the mass is buoyantly supported is known as the **Principle of Isostasy**.

Isostasy is a very important property of the lithosphere because it explains why the lithosphere can move up and down as well as sideways. For example, as erosion wears away a high spot on the continental crust, such as a mountain, the crust will rise, just as an

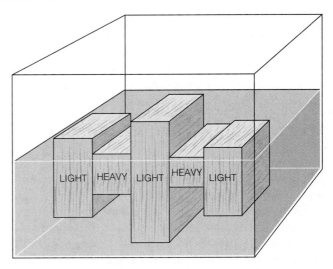

Figure 2.5   Why Topography is Bimodal   Diagram showing how isostasy works. Heavy, thin wooden blocks float low in the water; light, thick blocks float high. The heavy blocks are equivalent to the oceanic crust (density 3.0 g/cm^3); the light blocks are equivalent to continental crust (density 2.7 g/cm^3).

iceberg floating in the ocean will rise when the top of the berg melts. Because of isostasy, all parts of the lithosphere are in a floating balance, like so many blocks of wood, each having a different density. Low-density blocks float high and have deep roots, whereas high-density blocks float low and have shallow roots (Fig. 2.5). The combination of plate tectonics with its sideways motions of plates of lithosphere, and isostasy with its vertical motions, can explain most of the major features of the Earth's surface.

PLATE MOTIONS

As we learned in Chapter 1 and saw in Figure 1.15, the lithosphere currently consists of six large plates and numerous smaller ones, all moving at speeds ranging from 1 to 12 cm a year. As a plate moves, everything on it moves too. If the plate is capped partly by oceanic crust and partly by continental crust, then both the ocean floor and the continent move with the same speed and in the same direction.

The hypothesis that the ocean floor might be moving was first proposed in the early 1960s and was one of the key steps that led to the theory of plate tectonics in 1967. But the suggestion that continents move goes back to the early years of the twentieth century. The idea of continental movement was most forcefully proposed by a German scientist, Alfred Wegener. The concept came to be called **continental drift**. When first proposed, the idea did not receive widespread support because at the

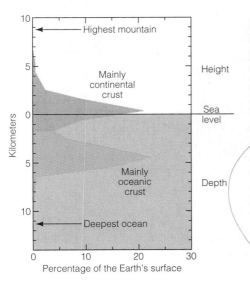

Figure   2.4 Bimodal Topography Distribution of the areas of the Earth's solid surface above and below sea level, expressed as a percentage. Note that areas underlain by continental crust are considerably higher than areas underlain by oceanic crust.

time no adequate explanation could be offered as to how it could happen. Plate tectonics provided the answer.

The original suggestion for continental drift was that continents must somehow slide across the floor of the ocean. Scientists soon realized, however, that friction would prevent such motions. Rocks on the ocean floor are too rigid and strong for continents to slide over them. Eventually, following the discoveries that the oceanic crust on the floor of the ocean also moves and that the asthenosphere is weak and easily deformed, geologists realized that the entire lithosphere must be in motion, not just the continents, and that plates of hard lithosphere must be sliding across the top of the soft, plastic asthenosphere.

The first clear evidence that seafloor and continent on the same plate of lithosphere move at the same velocity, and in the same direction, came from studies of the magnetic properties of rocks. This evidence is discussed in Chapter 17. Recently, however, a series of remarkable measurements have provided an even more convincing body of evidence. The new evidence of plate motion comes from satellites. To a close approximation, plates of lithosphere behave as rigid bodies. This means that plates do not stretch and shrink the way rubber sheets do. The distance between, say, New York City and Chicago, both on the North American Plate, remains fixed, even though the plate may flex and warp up and down. Of course, the distances between places on adjacent plates—Los Angeles on the Pacific Plate and San Francisco on the North American Plate, for instance—do change because of plate motions. Figure 2.6 shows the inferred relative motions of plates today, based on past velocities calculated from magnetic measurements. The motions recorded by magnetic measurements can be inferred to be today's motions only if actual measurements show the plates really are in motion. The space age has made it possible to get that proof. Using laser beams bounced off satellites, we can measure the distance between two points on the Earth with an accuracy of about 1 cm. By making distance measurements several times a year, therefore, we can measure present-day plate velocities directly. As seen in Figure 2.6, velocities based on these satellite measurements agree very closely with the velocities calculated from magnetic measurements. The agreement implies that the plates move steadily rather than by starts and stops.

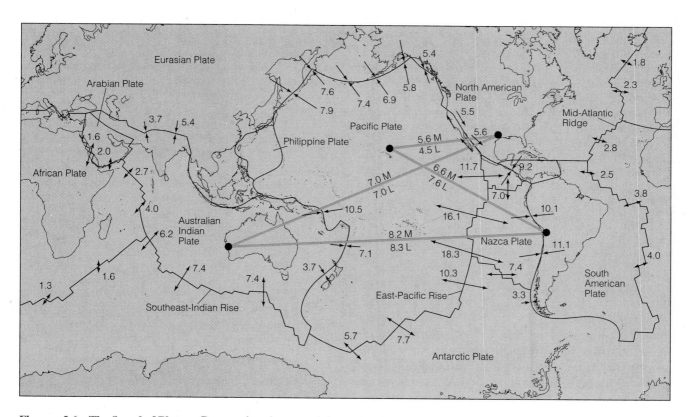

Figure 2.6 The Speed of Plates Present-day plate speeds in centimeters per year, determined in two ways. Numbers along the midocean ridges are average speeds indicated by magnetic measurements. A speed of 16.1 cm, as shown for the East Pacific Rise, means that the distance between a point on the Nazca Plate and a point on the Pacific Plate increases, on average, by 16.1 cm each year in the direction of the arrows. The long red lines connect stations used to determine plate speeds by means of laser ranging (L) techniques. The measured speeds between stations are very close to average speeds estimated from magnetic measurements (M).

Understanding Our Environment

Is Any Place in the USA Safe?

Each year, more than 700,000 Americans die of heart disease; more than 40,000 are killed in motor-vehicle accidents, and another 20,000 die as a result of homicides and police actions. Fewer than 350 lives are lost as a result of floods, lightning, tornadoes, hurricanes, earthquakes, avalanches, tsunamis, volcanic eruptions, and hailstones. Yet if the popularity of movies like *Twister* is any indication, Americans are both fascinated and deeply fearful of these kinds of natural disasters.

Despite these fears, growing numbers of Americans are choosing to live in areas that are prone to earthquakes, floods and tornadoes and especially in hurricane-prone coastal areas. The population of hurricane-prone Florida, for example, has more than doubled since 1970. Vacation homes on North Carolina's Outer Banks, also at high risk from hurricanes, are in demand. Southern California's benign climate continues to attract millions, despite the well-known fact that the area lies above the unstable San Andreas Fault, with its frequent earthquakes.

Those who live in other parts of America are not necessarily safer than those who live in Florida and Southern California. Residents of Texas, Oklahoma, and Kansas, live in an area known for a high occurrence of twisters, hence the name "Tornado Alley." People living on the floodplains of major midwestern rivers have seen their homes washed away by enormous floods. Even the central part of the nation is not entirely safe. Beneath southwestern Missouri lies the New Madrid Fault, a potential source of earthquakes in a region where structures are not built to withstand them.

Ironically, it is the very things that make life possible—the atmosphere, the Earth's crust, the oceans and rivers—that create the natural catastrophes we fear. In fact, we are surrounded by danger. And in recent years the number and intensity of natural disasters seem to have increased. In the past

decade the total cost of damage caused by natural disasters in the United States has doubled, from $25 to $50 billion a year.

It is popularly believed that this increase is due to phenomena such as El Niño. Scientists point out, however, that by exposing ourselves to increased risks, human activity is at least partly to blame. As more people build homes and businesses in vulnerable areas such as coastlines or along flood-prone rivers, the costs of natural disasters increase proportionately.

Most of the time we believe we can prevent the worst effects of these disasters—for example, by building levees along the Mississippi River. However, such measures may provide a false sense of security; levees were inadequate to withstand the devastating Mississippi River floods of 1993. Improved prediction holds out greater promise for saving lives, if not property. This is especially true in the case of hurricanes, since historically most deaths from natural disasters have resulted from flooding caused by hurricane-driven storm surges. Fortunately, some progress is being made in increasing the accuracy of predictions; the National Hurricane Center in Miami reports an annual increase of 1 percent in forecast accuracy.

Disaster professionals also point to the need for what they term *mitigation*: applying what has been learned about disasters in order to mitigate their effects the next time they occur. In places where earthquakes are likely, for example, buildings can be bolted down and bridge supports fitted with steel jackets. Where floods are frequent, zoning regulations can be changed to prevent new building on floodplains.

In the end, however, whether any place is safe will depend on human nature. As time passes and memories of disasters fade, people become less interested in preparing for them. They become less likely to heed warnings or to invest in mitigation efforts. But as the costs of disasters to both victims and taxpayers rise, the need for improved mitigation becomes increasingly evident.

Source: Michael Parfit: "Living with Natural Hazards," *National Geographic*, July 1998, pp. 8-38.

PLATE MARGINS

Plates move as individual units, and interactions between plates occur along their edges. Plate interactions are most distinctively expressed by earthquakes and volcanism because a majority of the Earth's volcanoes and earthquakes occur along plate margins. One might conclude that places close to plate margins must be dangerous spots to live. Some plate margins *are* dangerous places, but we can ask the question addressed in the boxed essay *Is Any Place in the United States Safe?* It has been through studies of these phenomena, particularly earthquakes, that geologists have been able to decipher the boundaries and shapes of today's plates.

Plates have three kinds of margins (Fig. 2.7):

1. **Divergent margins**, which are also called **spreading centers** because such margins are breaks in the lithosphere where new lithosphere continually forms as two plates move apart.

2. **Convergent margins**, where two plates move toward each other. Along convergent margins, one plate must either sink beneath the other, in which case we refer to the margin as a **subduction zone**, or the two plates must collide, in which case we refer to the margin as a **collision zone**.

3. **Transform fault margins**, which are fractures in the lithosphere where two plates slide past each other, grinding and abrading their edges as they do so. Earthquakes are frequent along most transform fault margins.

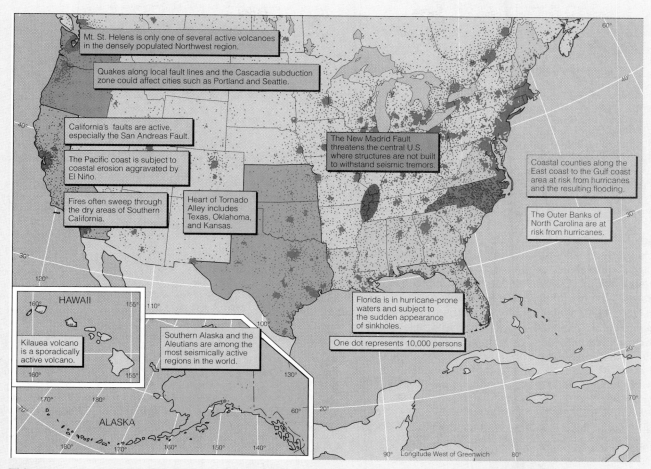

Mt. St. Helens is only one of several active volcanoes in the densely populated Northwest region.

Quakes along local fault lines and the Cascadia subduction zone could affect cities such as Portland and Seattle.

California's faults are active, especially the San Andreas Fault.

The Pacific coast is subject to coastal erosion aggravated by El Niño.

Fires often sweep through the dry areas of Southern California.

Heart of Tornado Alley includes Texas, Oklahoma, and Kansas.

The New Madrid Fault threatens the central U.S. where structures are not built to withstand seismic tremors.

Coastal counties along the East coast to the Gulf coast area at risk from hurricanes and the resulting flooding.

The Outer Banks of North Carolina are at risk from hurricanes.

HAWAII

Kilauea volcano is a sporadically active volcano.

Southern Alaska and the Aleutians are among the most seismically active regions in the world.

Florida is in hurricane-prone waters and subject to the sudden appearance of sinkholes.

One dot represents 10,000 persons.

ALASKA

Longitude West of Greenwich

This map of the United States and southern Canada depicts population and the potential for major hazards. One dot represents 10,000 people.

Spreading Centers

When we examine how a plate moves, a good analogy is a conveyor belt. In a conveyor, the belt rises from below, moves along a certain length, and then turns down and passes from sight. Although broad and irregular rather than long and narrow, a plate of lithosphere acts like the top of a slowly moving conveyor belt.

Each plate moves away from a spreading center as if it were a continuous belt rising up at the spreading center from the mantle below. The analogy is only partly correct because the plate is not rising as a solid ribbon. New material is being added to the plate by formation of new crust at the spreading center, as well as by cooling of the

originally hot rock of the new crust and the uppermost mantle beneath it as the rock moves away from the spreading center. Another disparity in the analogy is that the two plates are moving apart in opposite directions. A more accurate analogy would be two conveyor belts moving in opposite directions.

When a divergent margin occurs in oceanic crust, it coincides with a midocean ridge. We can't see into the mantle beneath the midocean ridges, but it is possible to infer what must be happening. Convection currents bring up hot rock from deep in the mantle, and as a result the lithosphere-asthenosphere boundary reaches very close to the seafloor and local portions of the asthenosphere become hot enough to start melting.

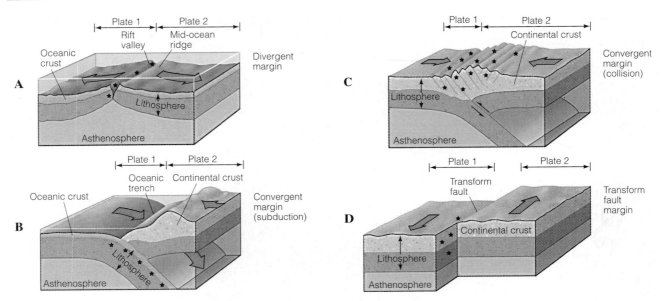

Figure 2.7 Plate Margins The various kinds of plate margins, through schematic diagrams. Locations of earthquake centers shown by stars. A. Divergent margin, also called a spreading center, for which the topographic expression is a midocean ridge. B. Convergent subduction margin for which the topographic expression is a seafloor trench. C. Convergent collision margin for which the topographic expression is a mountain range. D. Transform fault margin, which does not produce a consistent topographic expression but is often marked by a long, thin valley due to preferential erosion along the fault.

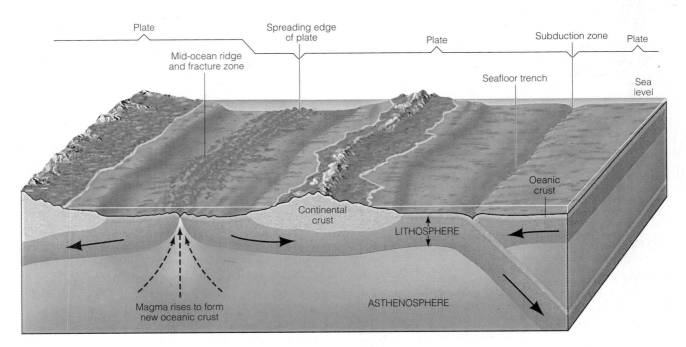

Figure 2.8 Plates Section through the Earth's outer layers showing how magma (dashed arrows) moves from the asthenosphere upward into spreading centers in the ocean floor and cools there to form new lithosphere capped by oceanic crust. To accommodate the new material, the lithosphere carrying the oceanic crust (solid arrows) moves away from the spreading center and eventually sinks slowly down into the asthenosphere.

Lithospheric Plates

Figure 2.9 Spreading Center Splitting a Continent The African Rift Valley, which extends from the Red Sea in the north to Malawi in the south, is a gigantic rent in the Earth's surface marking the place where a spreading center is splitting Africa into two pieces. This photo is of a portion of the Rift Valley in central Kenya. To the east (right-hand side) is a plateau bounded by a jagged fracture. The two hills (rear and left-hand edge) are volcanoes.

Molten rock is **magma**. The magma that forms in the asthenosphere beneath the midocean ridge rises upward to the top of the mantle, where it cools and hardens to form new oceanic crust (Fig. 2.8).

When a spreading center splits continental crust, an interesting sequence of events occurs. First, a great rift is formed; the African Rift Valley that runs from Ethiopia through Kenya, Tanzania, and Malawi is a modern example. As the two fragments of continental crust move apart, volcanism commences, as shown in Figure 2.9. Continued movement allows the rift to widen and deepen, and eventually the sea enters to form a long, narrow body of water; the Red Sea is a modern example. Eventually, the fragments of continental crust move far apart, new oceanic crust separates them, and a new ocean, like the Atlantic, has been formed.

Two hundred and fifty million years ago, there was no Atlantic Ocean. Instead, the continents that now border it were joined together into a single huge continent (Fig. 2.10). The place that is now New York was then as far from the sea as central Mongolia is today. About 200 million years ago, new spreading centers split the huge continent. We do not yet fully understand why this occurred, but presumably it involved new convection currents in the asthenosphere and mesosphere. The new spreading centers split the lithosphere and in the process broke the ancient continent into the pieces we see today. These fragments, today's continents, then drifted slowly into their present positions. At first the Atlantic Ocean was a narrow body of water that separated North America from Europe and North Africa. As movement continued, the ocean widened and length-ened, splitting South America from Africa and then growing to its present size. The Atlantic is still growing wider by about 5 cm each year.

Evidence is abundant to mark where the torn margins were formerly fitted together. If the pieces are reassembled, the continental slopes on each side of the ocean fit like the matched pieces of a jigsaw puzzle (Fig. 2.10). The line of match follows the spreading center, the present Mid-Atlantic Ridge.

Subduction Zones

Near a spreading center, the lithosphere is thin and its boundary with the asthenosphere comes close to the surface (Fig. 2.11). The lithosphere is thin near a spreading center because rising magma heats the surrounding rock and only a thin layer near the top is cool enough to have the strength that is characteristic of the lithosphere.

As rock of the crust and uppermost mantle moves away from the spreading center, it cools and becomes denser. Because of the cooling, the lithosphere gradually becomes thicker and the boundary between the lithosphere and the asthenosphere moves deeper. Finally, about 1000 km from the spreading center, the lithosphere reaches a constant thickness and is so cool that it is more dense than the hot, weak asthenosphere below it and starts to sink downward. Old lithosphere with its capping of oceanic crust sinks into the asthenosphere and eventually into the mesosphere. The process by which lithosphere sinks into the asthenosphere is called **subduction**, and the margins along which plates are subducted are called subduction zones, as noted above.

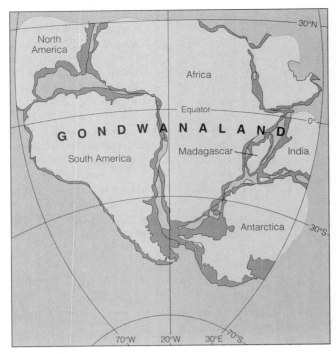

200 million years ago

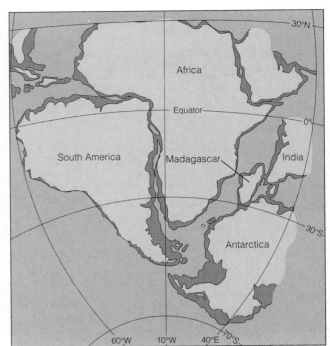

120 million years ago

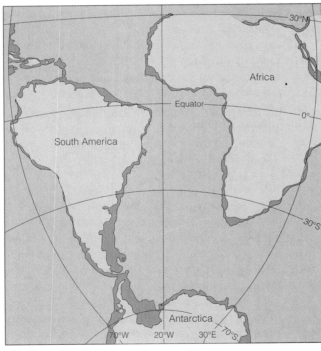

56 million years ago

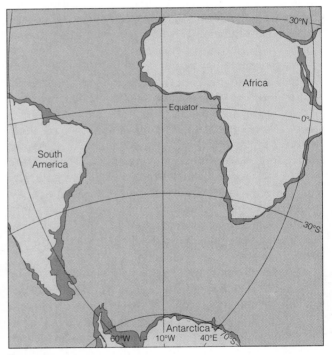

Present

Figure 2.10 Continent Breaks Apart Breakup of a supercontinent. The present southern hemisphere continents were joined together 200 million years ago as the southern half of a supercontinent. The whole supercontinent was called Pangaea, and the southern half shown here was named Gondwanaland. (The northern half, not shown, was Laurasia.) Magnetic data obtained from the oceanic crust were used to plot the opening of the southern part of the Atlantic Ocean as South America and Africa drifted apart. When the continents are fitted back together along a line 2000 m below sea level, as shown in the upper left-hand corner, the match is very close. Notice how the continents have moved relative to the equator and the way Antarctica slowly moved southward.

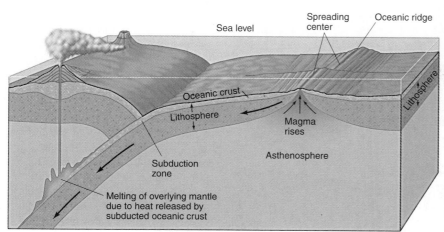

Figure 2.11 Anatomy of a Plate Schematic diagram showing some major features of a plate. Near the spreading center, where the temperature is high because of rising magma, the lithosphere is thin. Away from the spreading center the lithosphere cools, becomes thicker, and so the lithosphere-asthenosphere boundary is deeper. As the lithosphere sinks into the asthenosphere at a subduction zone, it starts to be reheated. At a depth of about 100 km, water and other volatiles released from the oceanic crust cause the mantle rock overlying the sinking slab of lithosphere to start to melt. The magma that is formed rises and creates a belt of volcanoes parallel to the subduction zone.

As the moving strip of lithosphere sinks slowly through the asthenosphere, it passes beyond the region where geologists can study it directly. Consequently, what happens next is partly conjecture. On one point, however, we can be quite certain: The lithospheric plate does not turn under, as a conveyor belt does, and reappear at the spreading edge; rather, it is heated and may be slowly mixed with the material of the mantle. As the crust sinks, it eventually becomes hot enough to release water. The water acts as a flux, and the hot mantle rock immediately above the sinking slab starts to melt. Some of the magma formed by melting of the overlying mantle reaches the surface to form volcanoes. As a result, subduction zones are marked by an arc of volcanoes parallel to, but about 150 km from, the edge of the plate (Fig. 2.12).

Collision Zones

Continental crust is not recycled into the mantle; it takes a shorter trip that ends more suddenly.

Continental crust is lighter, that is, less dense, than even the hottest regions of the mantle. As a result, continental crust is too buoyant to be dragged downward into the mantle. So, in continent-sized pieces, such crust floats on plates of lithosphere from place to place on the Earth's surface. Movement eventually stops when two fragments of continental crust collide, but the actual collision process may last for tens of millions of years. Such collisions can happen only after there has been subduction of oceanic crust beneath one of the colliding fragments. Because the plate being subducted also carries a fragment of continental crust, a collision will inevitably occur when the two pieces of continental crust meet along the subduction zone (Fig. 2.13). Such collision zones form spectacular mountain ranges. The Alps, the Himalaya, and the Appalachians are the results of past continental collisions. Because continental crust cannot sink down into the mantle, much of the evidence concerning ancient plates and their motions is recorded in the bumps and scars of past continental collisions.

Transform Faults

Besides spreading centers and the subduction and collision zones of convergent margins, there is a third kind of plate margin—transform fault margins—along which

Figure 2.12 Volcanoes Above a Subducting Plate This chain of volcanoes in Oregon and Washington sits above the subducting Juan de Fuca Plate where it sinks below the western edge of the North American Plate. Six snow-capped volcanoes are visible in this aerial photograph. In the foreground is Bachelor Butte, behind are the Three Sisters, and in the distance are Mounts Jefferson and Hood.

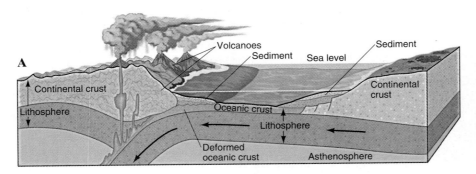

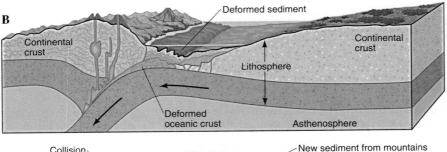

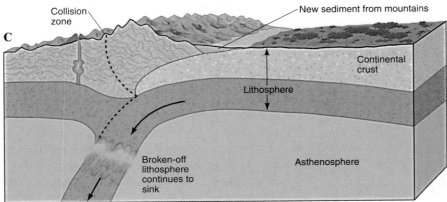

Figure 2.13 Collisions Form Mountains Mountains formed by plate-tectonic collision between two masses of continental crust. A. Subducting oceanic lithosphere compresses and deforms sediments at the edge of continent (left). Sediments at the edge of approaching continent (right) are undeformed. B. Collision. Sediment at the edge of approaching continent (right) starts to become deformed and welded onto already deformed continental crust (left). C. After collision. The leading edge of the subducting plate breaks off and continues to sink. The two continental masses are welded together with a mountain range standing where once there was ocean.

plates simply slip past each other. These margins of slipping are great vertical fractures (discussed in more detail in Chapter 17) that cut right down through the lithosphere. One transform fault much in the public eye because of the threat of earthquakes along it is the San Andreas Fault in California (Fig. 2.14). This fault, which runs approximately north-south, separates the North American Plate on the east side, on which San Francisco sits, from the Pacific Plate west of the fault, on which Los Angeles sits. The Pacific Plate is moving northward and the North American Plate southward. As the two plates grind and scrape past each other, Los Angeles is slowly moving north and San Francisco is moving south. At times the plate edges grab and lock, and as the plates move, the rock on both sides of a locked section flexes; that is, it bends slightly. When the locked section breaks free, the bent rock abruptly straightens as the rock slips along the fault and an earthquake occurs. To see whether your home might be affected by transform faults (or other hazards), see Understanding Our Environment.

Topography of the Ocean Floor

Because of plate tectonics, the ocean floor has a very distinctive topography. Two particularly prominent features are (1) **oceanic ridges** (also referred to as **midocean ridges** or **oceanic rises**), which are rocky ridges on the ocean floor, tens of thousands of kilometers long, many hundreds of kilometers wide, and standing at heights of 0.6 km or more above the seafloor; and (2) **trenches**, which are long, narrow, deep basins in the seafloor (Fig. 2.15).

The oceanic ridge system is a continuous chain of mountains some 84,000 km in length that twists and branches in a complex pattern through the ocean basins. This great mountain chain would be one of the most impressive features we would see if we could view a dry Earth from out in space.

The midocean ridges are, of course, the spreading centers that separate the two plates. A narrow valley, or rift, runs down the center of all oceanic ridges. The rifts are characterized by intense volcanic activity—in fact, the midocean ridge is the most volcanically active feature on

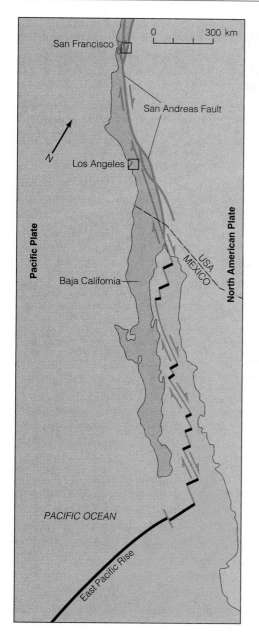

Figure 2.14 Transform Fault The San Andreas Fault is a transform fault that separates the Pacific Plate from the North American Plate. Directions of motion are shown by the arrows. Los Angeles, on the Pacific Plate, is moving northward, while San Francisco is moving in the opposite direction, bringing the two cities ever closer together.

the Earth. At several places around the world, the oceanic ridge with its central rift reaches sea level and forms volcanic islands. The largest of these is Iceland, which lies on the center of the Mid-Atlantic Ridge (Fig. 2.16).

Trenches are the places on the seafloor where the lithosphere sinks into the mantle. The deepest places in the ocean are in trenches, but when trenches are close to land they tend to be filled with sediment.

INTERACTIONS BETWEEN THE INTERNAL AND EXTERNAL SPHERES

We walk on the regolith, breathe the atmosphere, feel the rain; the evidence is clear that the Earth's external layers are places of intense and continual activity. Water and air penetrate the regolith and far into the crust. Chemical and physical disintegration of rock goes on because the atmosphere, biosphere, and hydrosphere combine to alter and break down the crust.

Figure 2.15 Three Plates Simplified diagram of a section across a portion of the Pacific Ocean margin of southern South America at about the latitude of Puerto Monte, Chile. Note that the side of the trench adjacent to the continent (to the right) is steeper than the oceanic side.

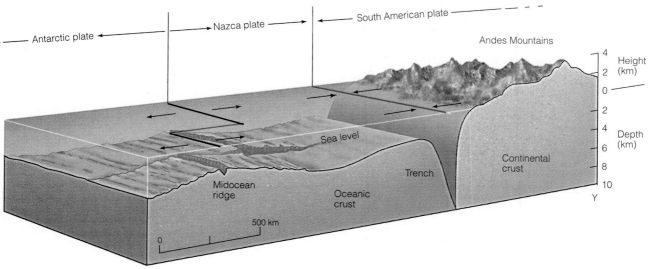

Figure 2.16 Midocean Ridges on Land Long, deep fractures that split Iceland mark the center of a midocean ridge. Iceland is on the Mid-Atlantic Ridge and is one of the few places in the world where the midocean ridge can be seen above sea level.

Cyclic Movements

It is useful to envision the interactions between the Earth's internal and external spheres as processes that facilitate the movement of materials and energy among the Earth's reservoirs. The movement of materials is cyclic, meaning it is continuous; this feature provides a convenient framework that helps us study how materials and energy are stored and how they are cycled among the principal reservoirs, the geosphere, hydrosphere, atmosphere, and biosphere, by the Earth system discussed in Chapter 1 (Fig. 1.19). There are two key aspects to cycles: (1) the reservoirs in which materials reside; and (2) the flows, or fluxes, of materials from reservoir to reservoir. For example, when raindrops form, they dissolve gases from the atmosphere (a reservoir) and carry them down to the Earth's surface (part of the geosphere and therefore another reservoir). The transport of water and dissolved gases from the atmosphere to the Earth's surface is a flow. The water and its dissolved gases react with minerals at the Earth's surface and convert some of the constituents of the minerals into new minerals, while other constituents dissolve and remain in solution. Those more soluble products of weathering are transported by streams (a flow) to become concentrated in the ocean (a reservoir). This is the origin of many of the salts in seawater.

How long do materials reside in reservoirs? The storage time can differ greatly. For example, an individual molecule of water may spend 200,000 years in a great ice sheet like the one that covers Antarctica, 40,000 years in the ocean, 1000 years in a deep underground reservoir, 10 years in a lake, 10 days in the atmosphere, and 10 hours in an animal's body. Water is transferred between these and other reservoirs via innumerable pathways and at greatly differing rates (Fig 2.17). A cycle can therefore include reservoirs of greatly differing size and involve processes that operate on many different time scales.

The three most important cycles that involve the solid Earth, and therefore the most important cycles for the study of physical geology, are

1. The **hydrologic cycle**, which is the day-to-day and long-term cyclic changes and movements of water in the Earth's hydrosphere, which are made evident by such elements as rain, snow, and running streams.

2. The **rock cycle**, which describes all of the various processes by which rock is formed, modified, decomposed, and reformed by the internal and external processes of the Earth.

3. The **tectonic cycle**, which deals with the movements and interactions of lithospheric plates, and the internal processes of the Earth's deep interior that drive plate motions.

The rock cycle, tectonic cycle, and hydrologic cycle are closely linked through physical, chemical, and biological processes. The three cycles involve most of the topics discussed in this book. There are, of course, many other cycles. For example, a useful group of cycles is the **biogeochemical cycles**, which trace the movement of chemical elements that are essential to life, including carbon, oxygen, nitrogen, sulfur, hydrogen, and phosphorus (see the boxed essay on *Cycling of Elements Important for Life*).

The Hydrologic Cycle

Hydrologic Cycle

The hydrologic cycle (Fig. 2.17) is powered by heat from the Sun, which evaporates water from the ocean and land surface. The water vapor thus produced enters the atmosphere and moves with the flowing air. Some of the water vapor condenses and is precipitated as rain or snow back into the ocean, or onto the land. Rain falling on land may drain off into streams, percolate into the ground, or be evaporated back into the air, where it is

A

Figure 2.17 Cycling Water Between Reservoirs Water in the environment. A. Example of the cycling of water. The Sun heats the ocean (a reservoir) and causes water to evaporate into the atmosphere (a reservoir). Water vapor in the atmosphere condenses to form clouds, rain, and snow. Compacted snow forms ice in a glacier (a reservoir). The glacier flows slowly down to the sea where huge bergs break off and slowly melt, rejoining the ocean. In this photo, icebergs are calving from the Hubbard Glacier in Alaska. B. The hydrologic cycle for the conterminous United States showing the amounts of water exchanged between reservoirs, in millions of cubic meters a day.

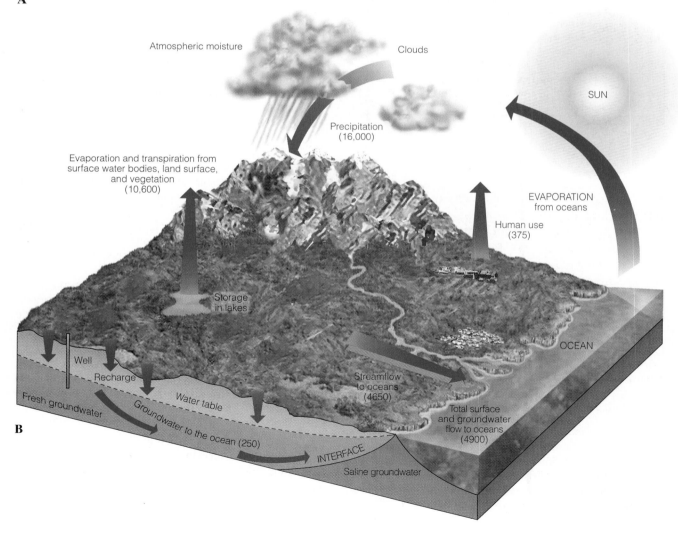

B

Using Resources

Cycling of Elements Important for Life

Plants extract carbon dioxide from the atmosphere as well as water from the soil and, through the process of photosynthesis, combine the two to make carbohydrates and oxygen; this is the source of oxygen in the atmosphere (Fig. B2.1). Photosynthesis is just one among many steps by which carbon, an element essential for life, cycles between the biosphere, atmosphere, hydrosphere, and geosphere. The natural cycle describing the movements and interactions of a chemical element essential to life is called a biogeochemical cycle. The most important elements for life are carbon, oxygen, sulfur, hydrogen, nitrogen, and phosphorus. The biogeochemical cycle of carbon is discussed in detail in Chapter 19. Here we describe the cycle of nitrogen as an example of a biogeochemical cycle.

The Biogeochemical Cycle of Nitrogen

Amino acids are essential constituents of all living organisms. They are given the name *amino* because they contain amine groups (NH_2), and the key element in amines is nitrogen. As a consequence, nitrogen is essential for all forms of life.

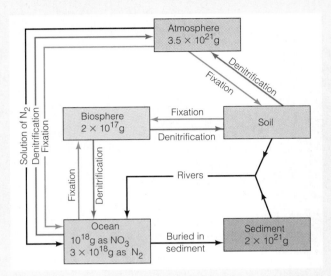

Figure B2.2 The Nitrogen Cycle A box diagram showing the biogeochemical cycle of nitrogen. The boxes are reservoirs, and the arrows represent the flows of material between the reservoirs. Numbers are the estimated amount of nitrogen in a given reservoir, in grams.

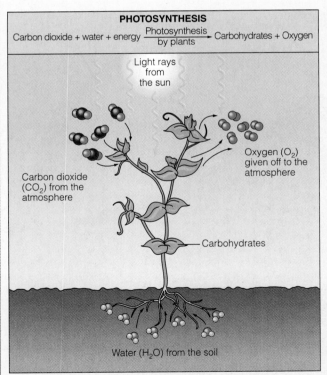

Figure B2.1 Source of Oxygen Through photosynthesis, plants combine carbon dioxide (CO_2) and water (H_2O) to make carbohydrate, a basic food and energy resource. The process releases a waste gas, oxygen, to the atmosphere. Humans and animals require oxygen to survive.

Nitrogen in nature exists in three forms: in its elemental form (N_2), in a reduced form such as ammonia (NH_3), and in oxidized forms such as nitrate (NO_3). It is only in the reduced form that nitrogen is incorporated into the newly formed organic matter of living organisms.

Most of the Earth's nitrogen is N_2 in the atmosphere. The key to the nitrogen cycle is understanding how reduction (also called *fixation*) and oxidation (also called *denitrification*) take place and the way nitrogen moves between five major reservoirs—the atmosphere, biosphere, ocean, soil, and sediment. As you read the following discussion, please study Figure B2.2 carefully; the figure is a cartoon showing the reservoirs, the estimated number of grams in each reservoir, and the paths by which nitrogen moves between the reservoirs.

The nitrogen cycle is dominated by the atmosphere and by the fact that N_2 cannot be directly used by organisms. Nitrogen is removed from the atmosphere and made accessible to organisms in three ways:

1. By solution of N_2 in the ocean.

2. By combination of N_2 and O_2 (oxidation) to form NO_3 by lightning discharges, in which form it is rained out of the atmosphere and into the soil and the sea. Plants can reduce NO_3 to NH_3, thus making nitrogen assimilable to the biosphere.

3. By reduction of N_2 to NH_3 through the actions of nitrogen-fixing bacteria in the soil or the sea. The reduced nitrogen is quickly assimilated by the biosphere.

The nitrogen cycle is interesting because of its complexity but interesting too because parts of the cycle must

have evolved over geologic time. The ancient atmosphere did not contain oxygen, so path 2 above must have evolved after the atmosphere became oxygenated. The atmosphere became oxygenated long after primitive life appeared on the Earth. Because organisms cannot use N_2 directly, either some reduced nitrogen must have been available when life arose or the earliest organisms had the ability to reduce N_2. Anaerobic nitrogen-fixing bacteria are certainly very ancient, and the fixation chemistry that evolved with them will not work in the presence of oxygen. Such bacteria must, therefore, have evolved before the atmosphere contained oxygen. Today these bacteria live only in oxygen-free environments. A few nitrogen-fixing bacteria have developed an oxygen tolerance, even though they still use the old, anaerobic fixation chemistry. They perform this trick by making sure that the sites in their cells where fixation occurs are carefully guarded from oxygen.

As the oxygen content of the atmosphere slowly increased through geological time, the amount of nitrate rained into the soil must have increased too. This provided new opportunities for organisms that learned to reduce NO_3 to NH_3. Many of the higher plants have this ability to use nitrate. Those that cannot reduce nitrate act as hosts to symbiotic nitrogen-fixing bacteria to which they supply energy in exchange for fixed nitrogen (Fig. B2.3).

Upon death of an organism, the reduced nitrogen of tissues will either be reused by other organisms and remain in the biosphere, or it will be oxidized back to N_2 and returned to the atmosphere. The main route by which nitrogen returns to the atmosphere, however, is the reduction of nitrate to form N_2. This route is kept open by bacteria that use the oxygen in nitrate in order to oxidize carbon compounds during metabolism.

Are human activities upsetting the biogeochemical cycle of nitrogen? The cycle is so complicated that it is difficult to know all the effects we are causing. What is known is that the amount of nitrogen fixed as ammonia and used in fertilizers equals the amount fixed naturally by bacteria. There is no concern about depleting the N_2 reservoir in the atmosphere because the reservoir is so large. The concerns center on the effects of ammonium salts carried into lakes, rivers, and the seas, and the sudden blooming of aquatic plants.

Figure B2.3 Nitrogen fixers Root nodules on a white clover plant produced by colonies of nitrogen-fixing bacteria.

further recycled. Part of the water in the ground is taken up by plants, which return water to the atmosphere through their leaves by a process called transpiration. Snow may remain on the ground for one or more seasons until it melts and the meltwater flows away. Snow that nourishes glaciers remains locked up much longer, through many tens or even thousands of years, but eventually it too melts or evaporates and returns to the oceans. The many reservoirs and pathways of the hydrologic cycle are shown in Figure 2.18.

Water is a crucial resource; without it, life on the Earth would not be possible. The processes that control the movement and distribution of water are also important in everyday life. When rainfall is high, floods may occur. During extended periods with below-average rainfall, there are droughts. Water also performs services, such as diluting or carrying away wastes. This service can be troublesome, however; for example, rain falling on a landfill may infiltrate and dissolve noxious chemicals, carrying them away to contaminate a stream or groundwater

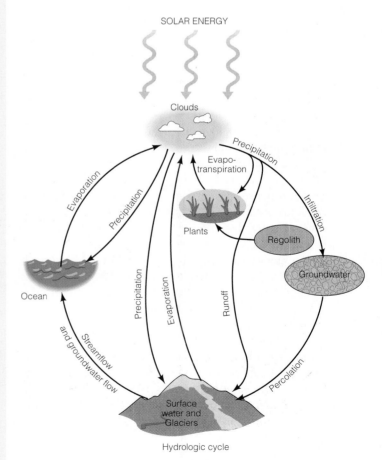

Figure 2.18 The Water Cycle The hydrologic cycle (or water cycle) describes the pathways by which water moves between the principal reservoirs, the ocean, ice, groundwater, and the atmosphere.

supply. Moving water also carries sediment, which sometimes creates problems: we lose valuable topsoil, or a stream channel that is needed for transportation becomes clogged with sediment. But sometimes the movement of sediment by water is beneficial. For example, before the Nile River was subjected to heavy engineering for flood-control and irrigation purposes, its annual flooding deposited mineral-rich sediment. This created one of the most fertile agricultural regions on the Earth.

Bodies of water—especially the ocean—control the weather and influence the distribution of climatic zones. The varied landscapes we see around us are another important consequence of the hydrologic cycle. The erosional and depositional work of streams, waves, and glaciers, combined with tectonic movement, volcanism, and deformation of crustal rocks, have produced a diversity of landscapes that makes the Earth's surface unlike that of any another planet in the solar system. In all these ways and more, the hydrologic cycle influences our everyday lives.

The Rock Cycle

To discuss the rock cycle, we must first introduce the major kinds of rocks. There are three large families, each defined by the process that forms the rocks.

The Three Rock Families

The first of the three major rock families is the **igneous rock** family (named from the Latin *igneus*, meaning fire). Igneous rocks are formed by the cooling and solidification of magma.

Some products of weathering are soluble in water and are carried away in solution by streams, but most weathering products are loose particles in the regolith. Particles of the regolith that are transported by water, wind, or ice will sooner or later settle to form deposits of **sediment** ("settling"). Sediment eventually becomes **sedimentary rock**, which is any rock formed by consolidation of sediment, including sediment formed by chemical precipitation of materials that had been carried in solution to the sea by solution in stream water. Sedimentary rocks constitute the second rock family.

The final major rock family is metamorphic rock (from the Greek *meta*, meaning change, and *morphe*, meaning form: hence, change of form). **Metamorphic rocks** are those rocks whose original form has been changed as a result of high temperature, high pressure, or both. Metamorphism, the process that forms metamorphic rocks from sedimentary or igneous rocks, is analogous to the process that occurs when a potter fires a clay pot in an oven. The tiny mineral grains in the clay undergo a series of chemical reactions as a result of the increased temperature. New compounds form, and the formerly soft clay molded by the potter becomes hard and rigid.

The Cycle

When mountain ranges rise as a result of plate tectonics, or when lava and volcanic ash erupt from volcanoes, the newly exposed or newly formed rocks are quickly attacked by water, wind, and ice. These constantly modify the Earth's surface, cutting away material here, depositing material there, and in the process creating the landscapes we see around us. Through weathering and erosion, the atmosphere, hydrosphere, and biosphere continually react with and change the surface of the solid Earth and are thereby closely linked to the rock cycle. The rock cycle describes all the processes by which rock is formed, modified, transported, decomposed, and reformed as a result of the Earth's internal and external processes.

Cycles are continuous and have no beginnings or ends. The rock cycle, which principally involves the continental crust, is like a merry-go-round, endlessly turning, powered by the Earth's internal heat (Figure 2.19) energy and by incoming energy from the Sun. In order to discuss the rock cycle, we have to jump in somewhere, so let's start at the top of Figure 2.19 with the uplift of continental crust and the exposure of crustal rocks. Exposed crustal rocks are vulnerable to weathering and are transformed into regolith by processes involving the atmosphere, hydrosphere, and biosphere. The combined processes of erosions then transport the regolith and eventually deposit it as sediment. After deposition, the sediment is buried and compacted, eventually becoming sedimentary rock. The processes of plate motion and crustal uplift lead to rock deformation. Deeper burial turns sedimentary rock into metamorphic rock, and even deeper burial may cause some of the metamorphic rock to melt, forming magma from which new igneous rock will form. At any stage in the cycle, tectonic processes might elevate the crust and expose the rocks to weathering and erosion. As a consequence of the rock cycle, material of the continental crust is being constantly recycled.

Because erosion of land is generally very slow and the mass of the continental crust is large, the average time rock spends in the continental crust is very long. Time estimates vary, and they are difficult to make, but the average age of all rock in the continental crust seems to be about 650 million years.

The Tectonic Cycle

The tectonic cycle and the rock cycle are closely connected. However, because the rock cycle is mainly a phenomenon of the continental crust and is dominated

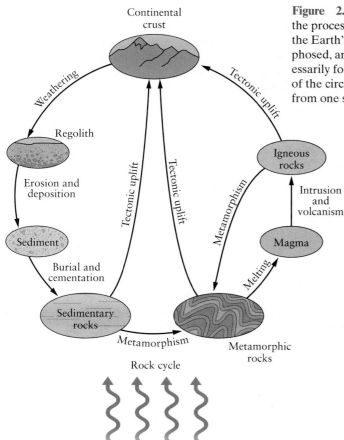

Rock cycle

GEOTHERMAL ENERGY

Figure 2.19 The Rock Cycle The rock cycle traces the processes whereby materials within and on top of the Earth's crust are weathered, deposited, metamorphosed, and even melted. Note that rocks don't necessarily follow the path that leads around the outside of the circle; they can follow any of the short circuits from one stage to another.

by interactions with the atmosphere, hydrosphere, and biosphere, it is driven in large part by solar energy. The tectonic cycle, by contrast, mainly involves oceanic crust and is dominated by processes deep inside the Earth that are driven by the Earth's geothermal energy.

The tectonic cycle is depicted in Figure 2.20. When magma rises from deep in the mantle, it forms new oceanic crust at spreading centers. Old oceanic crust returns to the mantle at subduction zones. The lifetime of oceanic crust is shorter than the lifetime of the continental crust. The most ancient oceanic crust of the ocean basins is only about 180 million years old, and the average of all oceanic crust is only 60 million years.

An important interaction between the oceanic crust and the continental crust involves volcanism. When sinking lithosphere carries old oceanic crust back down into the mantle, volatile constituents of the oceanic crust cause mantle rock overlying the sinking lithosphere to melt. The magma so formed rises to form vol-

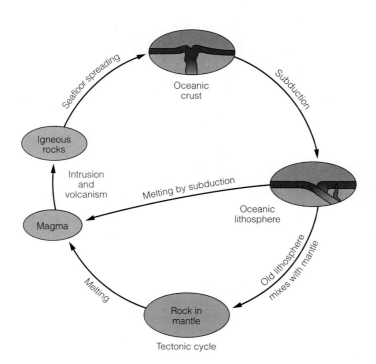

Tectonic cycle

Figure 2.20 The Tectonic Cycle Processes driven by the Earth's geothermal energy power the tectonic cycle. Rock in the mantle melts and the magma rises to make new oceanic crust. Eventually, the oceanic crust is subducted, which causes partial melting of the mantle rocks in contact with the subducting lithosphere capped by oceanic crust. This magma, in turn, rises to create volcanoes.

canoes, as we learned earlier in this chapter. Thus, there is a continual addition of new mantle-derived material to the continental crust.

Another important interaction involving oceanic crust concerns the oceans and the composition of seawater. The magma that rises to form new oceanic crust at spreading centers forms hot igneous rocks that react with seawater. In the reaction, some constituents in the hot rock, such as calcium, are dissolved in the seawater and constituents already in the seawater, such as magnesium, are deposited in the igneous rock. Because the magma that forms oceanic crust comes from the mantle, the reactions between hot crust and seawater are one way in which the mantle plays a role in determining the composition of seawater, and a very important way that the materials and processes of the tectonic cycle interact with those of the hydrologic cycle.

Uniformitarianism and Rates of Geologic Processes

The great contributions James Hutton made to geology were mentioned in Chapter 1. It was Hutton who recognized that the same external and internal processes we recognize in action today have been operating throughout the Earth's long history, and it was Hutton's associates and successors who codified his observations into the Principle of Uniformitarianism.

During the nineteenth century, geologists tried to estimate how long the processes of the rock cycle had been operating by using the thickness of a pile of sedimentary rock as a way to estimate how long it took for the sediment from which it formed to accumulate. They assumed that the Principle of Uniformitarianism applied to rates of geologic processes as well as to the processes themselves, and hence that rates of deposition have always been constant and equal to today's rates. Thus, they thought, it would be possible to estimate the time needed to produce all the sedimentary rock still present on the Earth. The figure so obtained, they argued, would be a minimum age for the Earth. The results, we now know, were greatly in error. One reason for the error was the assumption that geological rates were constant.

The more we learn of the Earth's history and the more accurately we determine the timing of past events through radiometric dating (Chapter 8), the clearer it becomes that the rates of geologic processes have not always been the same. The evidence is strongly against constancy; some rates were once more rapid, others much slower.

One reason the rates of processes involved in the rock cycle have changed through time is that the Earth is very slowly cooling down as its internal heat is lost. The Earth's internal temperature is maintained, in part, by natural radioactivity. Early in the Earth's history, more radioactive atoms were present than there are today, and so more heat must have been produced than is produced today. Internal processes, which are all driven by the Earth's internal heat, must have been more rapid than they are today. It is possible that a billion years ago oceanic crust was created at a faster rate than it is now, and that continental crust was uplifted and eroded at a faster rate. Either or both actions would cause the rock cycle to speed up.

At the same time, the rates of external processes have also varied. Long-term changes in the rates have occurred because of slow increases in the heat output of the sun and also because of the gradual slowing in the rate of rotation of the Earth on its axis—scientists estimate that 600 million years ago there were 400 days in the year, and 2 billion years ago there were 450 days a year, for instance. Short-term effects on external process have also arisen because of changes in the orientation of the Earth's axis of rotation and changes in the Earth's orbit. (These changes are discussed more fully in Chapter 12.) It is clear, therefore, that even though the cycles have been continuous, the processes involved in the cycles have not maintained a constant rate through time.

The conclusion that rates of geological processes have differed in the past from today's rates is an important one. It means that the relative importance of different geological processes has probably differed in the past. For example, just because glaciation is an important process today, we cannot assume it has been equally important through geological time. But we can assume that when glaciation did affect the Earth in geologically remote times, the processes and effects were the same as the processes and effects of glaciation we observe in Antarctica today.

Connections Between the Cycles

The three geologic cycles discussed in this chapter, the hydrologic cycle, rock cycle, and tectonic cycle, have been continuous since there was continental crust and oceans of seawater on the Earth—at least 4 billion years. The rates of the processes involved in the cycles may have changed, but the processes themselves have been continuous. The interactions we observe between the many cyclic processes must therefore have been going on at least 4 billion years. Today, as discussed in the boxed essay *New Science, New Tools,* new instruments and new ways of handling data let us monitor contemporary changes in the geologic cycles.

In Figure 2.21 you will see how the processes of the three main geologic cycles interact. This figure is a convenient diagram to bring Chapter 2 to a close because most of the subsequent chapters in the book discuss one or more of the flows of material and energy between reservoirs.

Measuring Our Earth

New Science, New Tools

New tools open the way to new science. One need only think of modern astronomy for an example; astronomy could not have developed as it has without the invention of the telescope. A new science arises because new tools make possible new kinds of observation and measurements, and these in turn lead to new ways of thinking about some phenomena. A familiar example is the germ theory of disease, which owes its existence to observations made using a new tool, the microscope.

A rapidly advancing contemporary science that is highly dependent on new tools is Earth system science. Earth system science involves observation and measurements of the Earth at all scales from the largest to the smallest. The huge amounts of data that are gathered come from many different locations and require special data handling techniques. Important new tools that facilitate Earth system science include satellite remote sensing, small deep-sea submarines, and geographic information systems.

More than any other way of gathering evidence, satellite observations continually remind us that each part of the Earth interacts with and is dependent on all other parts. Earth system science was born from the realization of that interdependence. Satellite remote sensing makes possible observations at large scales, and in many cases, measurements of factors that could not otherwise have been measured. For example, the ozone hole over Antarctica—the

decrease in the concentration of ozone high in the atmosphere—is measured by remote sensing, as are changes in deserts, forests, and farmlands around the world. Such measurements can be used in many areas of specialization beside Earth system science. Archaeology, for example, has benefited from satellite observations that reveal the traces of ancient trade routes across the Arabian desert.

New tools for exploring previously inaccessible areas of the Earth have also added greatly to our knowledge of the Earth system. Small deep-sea submarines like ALVIN allow scientists to travel to the depths of the ocean. There they have discovered previously new species and ecosystems thriving near deep-sea vents that emit heat, gases, and mineral-rich water.

Just as important as new methods of measurement and exploration are new ways to store and analyze data about the Earth system. Computer-based software programs known as geographic information systems, or GIS, allow a large number of data points to be stored along with their locations. These can be used to produce maps and to compare different sets of information gathered at different times. For example, a satellite remote sensing image of a forest can be converted to represent stages in the forest's growth. Two such images made at different times can be overlaid and compared, and the changes that have taken place can be represented in a new image.

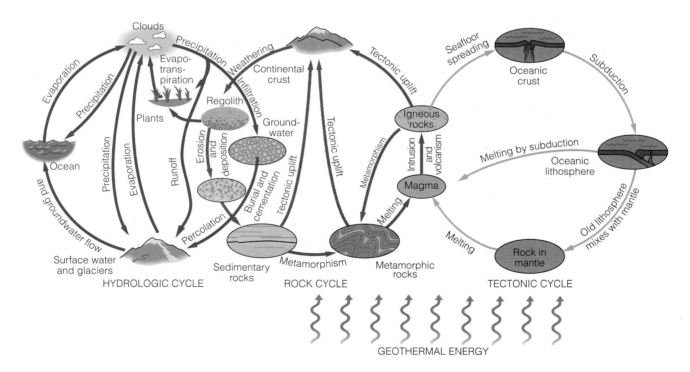

Figure 2.21 Overview of the Cycles The hydrologic, rock, and tectonic cycles are interconnected, and together they comprise a major part of the Earth system.

SUMMARY

1. The Earth is not a perfect sphere. As a consequence of rotation, the Earth bulges at the equator and is slightly flattened at the poles.

2. At present, 71 percent of the Earth's surface is covered by seas and only 29 percent is land. Most of the land (65 percent) is in the northern hemisphere.

3. Shorelines do not coincide with the join between continental crust and oceanic crust. The join is at the base of the continental slope and is everywhere covered by water.

4. The continents stand, on average, 4.5 km above the floor of the ocean basins. They do so because the lithosphere floats on the asthenosphere and, by the Principle of Isostasy, light, thick, continental crust floats high, while heavy, thin oceanic crust floats low.

5. Plates of lithosphere have three kinds of margins: divergent (called spreading centers), convergent (called subduction zones or collision zones), and transform faults.

6. When a spreading center is in oceanic crust, it is marked by a midocean ridge. When a spreading center splits, a continent is marked by a great rift in the continental crust.

7. When oceanic lithosphere capped by oceanic crust is far enough away from a hot spreading center, it cools sufficiently to become more dense than the asthenosphere; therefore the old, cold lithosphere sinks back into the mantle, creating a subduction zone.

8. At a depth of about 100 km or more, subducted oceanic crust becomes hot enough to release water. The water acts as a flux, and the hot mantle overlying the sinking slab of lithosphere starts to melt, creating magma. The magma rise to form a chain of volcanoes.

9. On the seafloor, spreading centers are marked by midocean ridges, and subduction zones are marked by trenches. Trenches are the deepest parts of the ocean.

10. Movement of material and energy between the various spheres of the Earth is continual, and because it is repeated endlessly, the movements are cyclic.

11. The three most important cycles that describe activities of the solid earth are the hydrologic cycle, the rock cycle, and the tectonic cycle.

12. The hydrologic cycle, driven by heat energy from the Sun, is the cyclic movement of water between ocean, air, and land. The movement occurs through evaporation, wind transport, precipitation of snow and rain, streamflow, and percolation.

13. The rock cycle arises from the interactions of the Earth's internal and external processes. Rock is eroded, creating sediment, which is deposited in layers that become sedimentary rock. Burial may lead to changes in temperature and pressure, forming metamorphic rock. Eventually, temperatures and pressures may become so high that rock melts and forms new magma. The magma rises, forms new igneous rock, and the cycle is repeated.

14. The tectonic cycle deals with the movements and interactions of lithospheric plates and the internal processes of the Earth's deep interior.

15. The rock cycle mainly involves the continental crust; the tectonic cycle involves the oceanic crust.

THE LANGUAGE OF GEOLOGY

biogeochemical cycles (p. 40)

collision zone (p. 32)
continental drift (p. 30)
continental rise (p. 29)
continental shelf (p. 29)
continental slope (p. 29)
convergent margin (p. 32)

divergent margin (p. 32)

hydrologic cycle (p. 40)

igneous rock (p. 44)
Isostasy, Principle of (p. 30)

magma (p. 35)
metamorphic rock (p. 44)
midocean ridge (p. 38)

oceanic ridge (p. 38)
oceanic rise (p. 38)

rock cycle (p. 40)

sediment (p. 44)
sedimentary rock (p. 44)
spreading centers (p. 32)
subduction (p. 35)
subduction zone (p. 32)

tectonic cycle (p. 40)
transform fault margin (p. 32)
trenches (oceanic) (p. 38)

QUESTIONS FOR REVIEW

1. The Earth is a closed system. What consequences does that fact have for human beings?

2. What reason can you give for the fact that shorelines all lie on the continental crust and that joins between oceanic crust and continental crust are everywhere covered by water?

3. Sketch a section through a margin of a continent, showing the shoreline continental shelf, continental slope, and continental rise. Also mark the approximate position of the join between continental and oceanic crust.

4. How are abyssal plains thought to have formed?

5. What is the Principle of Isostasy?

6. Why are ocean basins low spots on the Earth's surface and continents high places?

7. Briefly describe the three kinds of plate margins.

8. Describe what happens when two plates topped by oceanic crust converge. Compare your description with what happens when the converging is between two plates capped by continental crust.

9. Identify the major topographic features of the ocean floor, and state how they are related to tectonic plates.

10. Briefly describe the hydrologic cycle. Where does the energy that drives the cycle come from?

11. What is the rock cycle? Describe two ways that processes of the hydrologic cycle interact with processes of the rock cycle.

12. What is the tectonic cycle? Describe a way in which processes of the tectonic cycle influence the composition of the ocean.

13. Have the processes of all the cycles been constant through geologic time? Explain your answer.

3

Measuring
Our Earth

Understanding
Our Environment

Using
Resources

GeoMedia

Waste pile of blue asbestos (crocidolite), Wittenoom Gorge, Hamersley Range, Western Australia. People who worked in the mining, processing, and use of blue asbestos have a high incidence of a form of cancer called mesothelioma.

Atoms, Elements, and Minerals

Asbestos: How Risky Is It?

With every breath of air we inhale tiny particles of suspended mineral matter. The particles are too small to be seen without a microscope, but they are everywhere present—we call them dust. Over a human lifespan each of us inhales billions of mineral particles.

Most of the inhaled particles are expelled when we breathe out, but a few always lodge in the bronchial tubes or lung. The human body has several ways of preventing such foreign bodies from causing harm. The lung handles the problem by enveloping the foreign body in a sort of scar tissue, but can only handle so much—if the load of particles becomes too great, the lung's capacity starts to decline.

A contentious scientific debate surrounds the long-term effects of fibrous mineral particles—asbestos—in the lung. Based on the experience of asbestos workers, some scientists argue that asbestos is a major hazard and that natural fibers pose severe health effects. Others insist that the issue has been blown out of proportion and that it is far from clear that workplace evidence applies to the general living environment. The differences of opinion have arisen for several reasons.

The very term *asbestos* is the source of some of the uncertainty. The fact that mineral fibers can be woven into a fireproof cloth was known to the ancient Greeks, and they called such fibers *asbestos*, which means *unaffected by fire*. Many different minerals can, under suitable circumstances, grow as long, flexible fibers. Asbestos, therefore, is not a specific mineral. Rather, it is a term that describes any mineral found as a strong, flexible fiber. If one kind of asbestos is a health hazard, does it mean that all fibrous minerals are equally dangerous?

Six different minerals are mined and used as asbestos. The majority of commercial asbestos (95%) is a white mineral called chrysotile, which is a member of a group of minerals called serpentines. The five remaining asbestos minerals of commerce all belong to a different mineral group called the amphiboles. Some of the most compelling evidence linking cancer and asbestos involves a blue-colored amphibole called crocidolite, mined mainly in South Africa and Australia. Compositionally, the amphiboles and serpentines are quite different. Can evidence derived from blue asbestos be used to evaluate the effects of white asbestos? Therein lies the source of some of the differences of opinion. If it is the shape and size of the particles that matters, then mineral differences may be unimportant. But if it is composition that matters, then we may be denying ourselves use of a valuable material by saying that all asbestos, regardless of composition, is dangerous.

Just how much asbestos is dangerous? Asbestos workers exposed to high levels of any kind of asbestos dust can indeed develop scarred lungs and can eventually develop severe lung disorders. Those of us who are not exposed in the workplace may well inhale a few fibers of chrysotile because it is such a common mineral, but at some level the body is probably able to handle the fibers so that they don't pose a threat. What that level may be is a matter of research and therefore another source of opinion differences.

The issue of asbestos and health has long since moved beyond science and into the political arena. As a result, and regardless of what kind of asbestos is involved, a large new industry has emerged to remove asbestos from buildings. A separate legal specialty has also emerged—legal costs associated with asbestos issues have grown to gargantuan proportions. It will probably be a long time before we hear the end of the asbestos story; perhaps we may never hear the end of the dangerous mineral story because suspicions are now being voiced about hazards posed by other common minerals. To appreciate the debates of the future, it is essential to understand enough about minerals to comprehend underlying issues.

MINERALS AND THEIR CHEMISTRY

The word **mineral** has a specific connotation in geology; it is any naturally formed, solid, chemical substance that has a specific composition and a characteristic crystal structure. Quartz is a mineral. It is naturally formed, it is a solid, it contains atoms of silicon and oxygen in the ratio of 1:2 so that it has a specific composition (SiO_2), and the atoms are packed together in a regular geometric array called its crystal structure. Granite is mostly silicon and oxygen too, but it is not a mineral. Granite is a mixture of several different minerals that differ in amount from sample to sample, so different granites differ somewhat in composition. Granite does not have a specific composition. It is a kind of rock.

Rocks are aggregates of minerals together with organic debris, glass, and other natural materials; they are nature's books and in them is recorded the story of the way the Earth works. Rocks tell such stories as the way continents move, how mountains form and slowly erode away, and why volcanoes are located where they are. The words used in nature's books are minerals, and in order to be able to read the words we must investigate the branch of geology that deals with the properties and distribution of minerals. The easiest way to introduce the subject of minerals is to examine the two most important characteristics of minerals:

1. **Composition,** which is the chemical elements present and their proportions; and

2. **Crystal structure,** which is the way in which the atoms of the chemical elements are packed together in a mineral.

Because most minerals contain several chemical elements, it is helpful to commence our discussion by briefly reviewing the way in which chemical elements combine to form compounds.

Elements and Atoms

Chemical Elements

If you were a chemist and you were asked to analyze a mineral or rock, you would report your findings as the kinds and amounts of the chemical elements present. *Chemical elements* are the most fundamental substances into which matter can be separated by chemical means. For example, table salt is not an element because it can be separated into sodium and chlorine. But neither sodium nor chlorine can be broken down further chemically, so each is an element.

Each element is identified by a symbol, such as H for hydrogen and Si for silicon. Some symbols, such as that for hydrogen, come from the element's name in English. Other symbols come from other languages. For example, iron is Fe from the Latin *ferrum*, copper is Cu from the Latin word *cuprum* which in turn comes from the Greek *kiprios*, and sodium is Na from the Latin word *natrium*. The naturally occurring elements and their symbols are listed in Appendix B.

A piece of a pure element—even a tiny piece no bigger than a pin's head—consists of a vast number of particles called *atoms*. An **atom** is the smallest individual particle that retains the distinctive properties of a given chemical element. Atoms are so tiny that they can only be seen by using the most powerful microscopes; even then the image is imperfect because individual atoms are only about 10^{-10} m in diameter.

Atoms

Atoms are built up from *protons* (which have positive electrical charges), *neutrons* (which, as their name suggests, are electrically neutral), and *electrons* (which have negative electrical charges that balance exactly the positive charges of protons). Protons and neutrons are dense but very tiny particles, and they join together to form the core, or *nucleus*, of an atom. Electrons are even tinier particles than protons or neutrons; they move, like a distant and diffuse cloud, in orbits around the nucleus (Fig. 3.1).

Protons give a nucleus a positive charge, and the number of protons in the nucleus of an atom is called the **atomic number**. The number of protons in the nucleus is what gives the atom its special chemical characteristics and what makes it a specific chemical element. Elements are catalogued by atomic number, beginning with hydrogen, which has an atomic number of 1 because hydrogen atoms contain one proton. Hydrogen atoms are followed by helium atoms, which have two protons, and so on.

Uranium, atomic number 92 (with 92 protons in the nucleus), is the naturally occurring element with the highest atomic number. Scientists have made synthetic elements with atomic numbers higher than 92; the highest, reported early in 1999, has an atomic number of 114.

What roles do neutrons play in the nucleus? Neutrons act like a glue that holds the nucleus together. The sum of the number of neutrons and protons in an atom is called the **mass number**. All atoms of a given element have the same atomic number. But atoms of an element can have different mass numbers because they can have different numbers of neutrons in their nuclei. Atoms with the same atomic number but different mass numbers are called **isotopes**. For example, there are three natural isotopes of carbon: carbon-12, carbon-13, and carbon-14. Each of the isotopes of carbon has six protons per atom and thus the same atomic number—six. But the three isotopes contain different numbers of neutrons: six, seven, and eight per atom, respectively. Most of the common chemical elements are mixtures of two or more isotopes. Mass numbers of isotopes are written as superscripts; thus, the three isotopes of carbon are denoted ^{12}C, ^{13}C and ^{14}C, respectively.

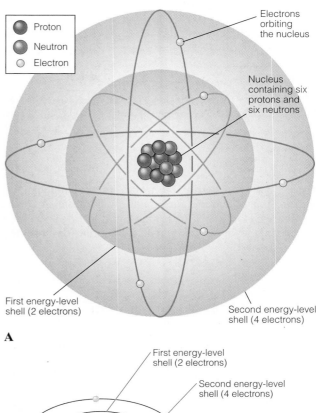

Figure 3.1 Structure of an Atom Diagram of an atom of carbon-12 (^{12}C). A. The nucleus contains six protons and six neutrons. Electrons circle the nucleus in complex paths called orbitals, so the diagram is only schematic. Two electrons are in orbitals close to the nucleus, called energy-level shell 1. Four electrons are in more distant orbitals in energy-level shell 2. B. Two-dimensional representation of a ^{12}C atom.

Energy-Level Shells

Electrons move around the nucleus of an atom in complex three-dimensional patterns called *orbitals*. Figure 3.1 is a schematic diagram of an atom of carbon-12; it is schematic because the paths of the electrons are too complex to show. Note that two electrons are in orbitals close to the nucleus, and four electrons are in more distant orbitals. The two groupings of orbitals are called **energy-level shells**. The maximum number of electrons that can have orbitals in a given energy-level shell is fixed. Shell 1, closest to the nucleus of an atom, can accommodate only 2 electrons; shell 2, however, can accommodate up to 8 electrons; shell 3, 18; and shell 4, 32.

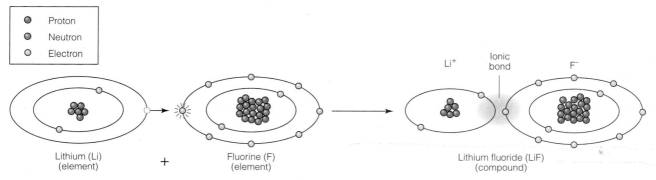

Figure   3.2   Transferring Electrons   To form lithium fluoride, an atom of lithium combines with an atom of fluorine. The lithium atom transfers its lone electron in shell 2 in order to fill the fluorine atom's shell 2, creating an Li^{1+} cation and a F^{1-} anion in the process. The electrostatic force that draws the lithium and fluorine ions together is an ionic bond.

Ions

When an energy-level shell is filled with electrons, it is very stable, like an evenly loaded boat. To fill their energy-level shells and so reach a stable configuration, atoms share or transfer electrons among themselves. An atom is electrically neutral because it has the same number of protons and orbiting electrons. But when transfer of an electron occurs, the balance of electrical forces is upset. An atom that loses an electron has lost a negative electrical charge and is left with a net positive charge. An atom that gains an electron has a net negative charge. An atom that has excess positive or negative charges caused by electron transfer is called an **ion**. When the charge is positive (meaning that the atom gives up electrons), the ion is called a **cation**; when negative (meaning an atom adds electrons), it is called an **anion**.

The convenient way to indicate ionic charges is to record them as superscripts. For example, Li^{1+} is a cation (lithium) that has given up an electron, while F^{1-} is an anion (fluorine) that has accepted an electron. Be careful you don't confuse mass numbers and ionic charges. Both are recorded as superscripts, but mass numbers are on the left while ionic charges are on the right. For example, sodium, which has a single electron in its second energy-level shell, forms a cation with a charge of 1+ by giving up the electron. Sodium has a mass number of 23, so a sodium cation is written $^{23}Na^{1+}$. In practice, scientists only record the mass number when it is necessary to do so.

Compounds

Chemical compounds form when atoms of one or more elements combine with atoms of another element in a specific ratio. For example, lithium and fluorine combine to form lithium fluoride (written LiF), a compound used in making ceramics and enamels. Writing the formula LiF indicates that for every Li atom there is one F atom. Similarly, the compound H_2O forms when hydrogen combines with oxygen in the ratio of two atoms of hydrogen to one of oxygen.

The formula of a compound is written by putting the element that tends to form cations first and the element that tends to form anions second. The relative number of atoms is indicated by subscripts, and for convenience the changes of the ions are usually omitted. Thus, for water, we write H_2O rather than $H_2^{1+}O^{2-}$.

An example of the way electron transfer leads to formation of a compound is shown in Figure 3.2 for lithium and fluorine. A lithium atom has energy-level shell 1 filled by two electrons but has only one electron in shell 2, even though shell 2 can accommodate eight electrons. The lone outer electron in shell 2 can easily be transferred to an element such as fluorine, which already has seven electrons in shell 2 and needs only one more to be completely filled. In this fashion, if the lithium and fluorine are in close proximity, both a lithium cation and a fluorine anion finish with filled shells, and the resulting positive charge on the lithium and the negative charge on the fluorine draw the two ions together.

Properties of compounds are quite different from the properties of their constituent elements. The elements sodium (Na) and chlorine (Cl) are highly toxic, for example, but the compound sodium chloride (NaCl, table salt) is essential for human health. The manner in which electrons are transferred or shared between atoms leads to several distinctive ways by which atoms join together.

The smallest unit that has the distinctive chemical properties of a compound is called a **molecule**. Do not confuse a molecule and an atom; the definitions are similar, but a molecular compound always consists of two or more kinds of atoms held together. The force that holds the atoms together in a compound is called **bonding**. There are several different kinds of bonding, and because bonding determines the physical and chemical properties of a compound, it is helpful to briefly review the subject.

Bonds

As we mentioned previously, an energy-level shell that is filled with its quota of electrons is very stable. To fill shells and reach a stable configuration, atoms transfer or share electrons, and it is the transferring and sharing of electrons that forms the strongest bondings between atoms. There are four important kinds of bondings:

1. *Ionic Bonding:* Electron transfers between atoms produce cations and anions that can exist as free entities, but even so an electrostatic attraction draws these negatively and positively charged ions together, forming an **ionic bonding**. The chemical bond between lithium and fluorine shown in Figure 3.2 is an ionic bond.

Compounds with ionic bonds tend to have moderate strength and moderate hardness. Table salt (NaCl) has ionic bonds. When you eat salt and it dissolves in your mouth, the NaCl separates into Na^{1+} and Cl^{1-} ions. The elements Na and Cl are toxic, but their ions in solution are not; it is the ions that create the familiar salty flavor.

2. *Covalent Bonding:* Some atoms share electrons rather than transferring them. Electron sharing doesn't form ions, but it does create a strong bonding known as **covalent bonding**. One important substance in which covalent bonding is present is diamond, a form of carbon. The second energy-level shell of carbon has four electrons but requires eight for maximum stability. To reach stability, each carbon atom shares two electrons with four other carbon atoms, as shown in Figure 3.3.

Elements and compounds with covalent bonding tend to be strong and hard. Covalent bonds also give diamond and other substances special optical properties. The sparkle that makes diamonds attractive gems is due to covalent bonding.

One very important substance in which covalent bonding is present is water (H_2O). The second energy-level shell of oxygen has six electrons but requires eight for maximum stability. A hydrogen atom has only one electron in the first energy-level shell but requires two for maximum stability. Thus, both kinds of atoms satisfy their electron needs by sharing as shown in Figure 3.4.

3. *Metallic Bonding:* Metals have a special kind of bonding that is found in a small group of minerals. In **metallic bonding**, the atoms are closely packed; electrons in higher energy-level shells are shared between several atoms, and because they are loosely held they can readily drift from one atom to another. The drifting electrons give metals their distinctive properties—for example, metals are opaque, malleable, and good conductors of heat and electricity. Gold, silver, iron, copper, mercury, and platinum are examples of naturally formed metals with metallic bonding.

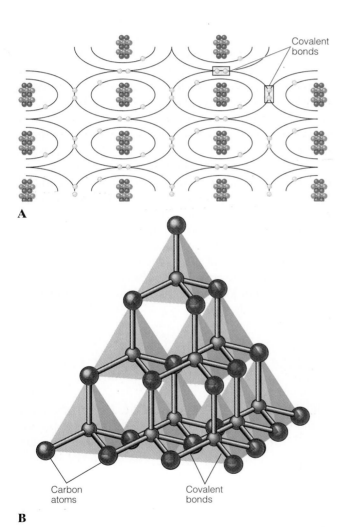

A

B

Figure 3.3 Covalent Bonding in Diamond A. Schematic diagram showing how each carbon atom in diamond shares its four electrons in energy-level shell 2 with four other carbon atoms so that all atoms have eight electrons in shell 2. Each shared electron pair creates a covalent bond. B. The three-dimensional geometric arrangement of carbon atoms in diamond. Note that each atom is surrounded by four others. The actual covalent bonding in diamond is three dimensional, not two dimensional, as shown for simplification in A.

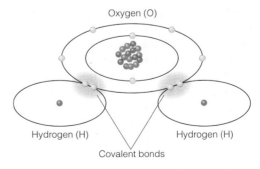

Figure 3.4 Covalent Bonding in Water Two atoms of hydrogen form covalent bonds with an oxygen atom through sharing of electrons. The oxygen atom thereby has its most stable configuration with eight electrons in shell 2, and each of the hydrogen atoms fills shell 1 with two electrons, making the compound H_2O.

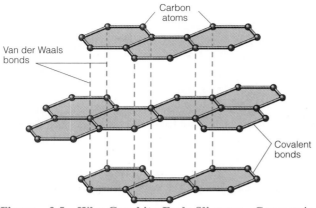

Figure 3.5 Why Graphite Feels Slippery Geometric arrangement of atoms in graphite. Bonding within the sheets of atoms is covalent; bonding between sheets is van der Waals. The weak van der Waals bonds are easily broken, so when graphite is rubbed between your fingers, the covalently bonded sheets of carbon readily slide past one another.

4. *Van der Waals Bonding:* The last common kind of bonding arises because of a weak secondary attraction between certain molecules formed by transferring electrons.

Van der Waals bonding is weak—much weaker than ionic, covalent, or metallic bondings—but it plays an important role in the structure of certain minerals, of which graphite is a good example. Graphite contains only atoms of carbon, and it has a sheetlike structure in which each carbon atom has three nearest neighbors at the corners of an equilateral triangle (Fig. 3.5). Carbon atoms are bonded covalently within the sheets, and as a result the sheets are very strong, but because they are sheets they are flexible. The graphite used in a tennis racket or a golf club makes use of the strong covalent bonds in the sheets. Adjacent sheets of graphite are held together by weak van der Waals bonding. The van der Waals bonding is so weak it is easily broken. Graphite feels slippery when you rub it between your fingers because the rubbing breaks the bonding and the sheets slide easily past each other. Graphite is used as a lubricant for very high temperature purposes, in which case it is the property of van der Waals bonding that is being exploited. Talcum powder is another example of a substance with van der Waals bonding. Talc, the mineral in talcum powder, has a sheet structure analogous to a graphite sheet, and just like graphite, the sheets are held together by van der Waals bonding. Talcum powder feels smooth and slippery because the van der Waals bonding is so easily broken.

Complex Ions

Sometimes two or more kinds of ions form such strong covalent bonds that the combined atoms act as if they were a single entity. Such a strongly bonded unit is called a **complex ion**. Complex ions act in the same way as single ions, forming compounds by ionic bonding with other atoms. For example, carbon and oxygen combine to form the complex carbonate anion $(CO_3)^{2-}$. Other important examples of complex ions are the sulfate $(SO_4)^{2-}$, nitrate $(NO_3)^{1-}$, and silicate $(SiO_4)^{4-}$ anions.

Periodic Table of the Chemical Elements

More than two centuries ago chemists started to group the chemical elements on the basis of similarities of chemical properties, such as their ability to combine with other chemical elements. Although the early chemists did not know it, the groupings reflected the specific numbers of electrons in the different energy-level shells. The person most directly responsible for working out the way chemical elements can be grouped into a sequence based on properties was the Russian scientist Dmitri Mendeleev (1834–1907). A modern version of the periodic table of chemical elements is shown in Figure 3.6. The periodic table organizes the elements into rows and columns. Within rows, elements increase in atomic number from left to right. Elements within a given column have the same number of electrons in their outermost energy-level shell. All elements in column one readily give up the lone outer-shell electron and so form cations. Elements in columns on the right-hand side of the periodic table tend to gain electrons and therefore form anions. The final column, on the right-hand side of the table, contains the six elements that have full energy-level shells. The six elements with full shells are called the *noble gases* because they have no tendency to gain or lose electrons and thus no tendency to form compounds.

ATOMIC ORDER IN SOLIDS

All minerals are solids. Thus, whether a given chemical element or compound can be called a mineral is controlled by its state, by which we mean whether it occurs as a solid, liquid, or gas (see *The Three States of Matter*). Ice in a glacier is a mineral, but water in the ocean and water vapor in the atmosphere are not minerals.

Crystal Structure

Whereas the molecules and atoms in gases and liquids are randomly jumbled, the atoms in most solids are organized in regular, geometric patterns, like eggs in a carton, as shown in Figure 3.7A. The geometric pattern that atoms assume in a solid is called the **crystal structure**, and solids that have a crystal structure are said to

1 H Hydrogen																	2 He Helium
3 Li Lithium	4 Be Beryllium											5 B Boron	6 C Carbon	7 N Nitrogen	8 O Oxygen	9 F Fluorine	10 Ne Neon
11 Na Sodium	12 Mg Magnesium											13 Al Aluminum	14 Si Silicon	15 P Phosphorus	16 S Sulfur	17 Cl Chlorine	18 Ar Argon
19 K Potassium	20 Ca Calcium	21 Sc Scandium	22 Ti Titanium	23 V Vanadium	24 Cr Chromium	25 Mn Manganese	26 Fe Iron	27 Co Cobalt	28 Ni Nickel	29 Cu Copper	30 Zn Zinc	31 Ga Gallium	32 Ge Germanium	33 As Arsenic	34 Se Selenium	35 Br Bromine	36 Kr Krypton
37 Rb Rubidium	38 Sr Strontium	39 Y Yttrium	40 Zr Zirconium	41 Nb Nobelium	42 Mo Molybdenum	43 Tc Technetium	44 Ru Ruthenium	45 Rh Rhodium	46 Pd Palladium	47 Ag Silver	48 Cd Cadmium	49 In Indium	50 Sn Tin	51 Sb Antimony	52 Te Tellurium	53 I Iodine	54 Xe Xenon
55 Cs Cesium	56 Ba Barnum	57* La Lanthanum	72 Hf Hafnium	73 Ta Tantalum	74 W Tungsten	75 Re Rhenium	76 Os Osmium	77 Ir Iridium	78 Pt Platinum	79 Au Gold	80 Hg Mercury	81 Tl Thallium	82 Pb Lead	83 Bi Bismuth	84 Po Polonium	85 At Astatine	86 Rn Radon
87 Fr Francium	88 Ra Radium	89† Ac Actinium															

* Lanthanides

58 Ce	59 Pr	60 Nd	61 Pm	62 Sm	63 Eu	64 Gd	65 Tb	66 Dy	67 Ho	68 Er	69 Tm	70 Yb	71 Lu

† Actinides

90 Th	91 Pa	92 U	93 Np	94 Pu	95 Am	96 Cm	97 Bk	98 Cf	99 Es	100 Fm	101 Md	102 No	103 Lr

Elements present in continental crust in amounts equal to or greater than 0.1% weight.

Figure 3.6 Periodic Table of Chemical Elements Chemical elements arranged in rows in order of increasing atomic number, from left to right, and in vertical columns so that all the elements in a column have the same number of electrons in their outermost energy-level shell, and therefore similar chemical properties. The group of elements, numbers 58 to 71, called the lanthanides or rare earths, and numbers 90 to 103, called the actinides, are shown separately. Lanthanides and actinides have the same number of electrons in their outermost shells but differ from each other in the electrons in their inner shells. The most abundant and geologically important elements in the Earth are indicated.

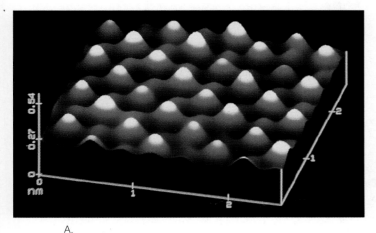

A.

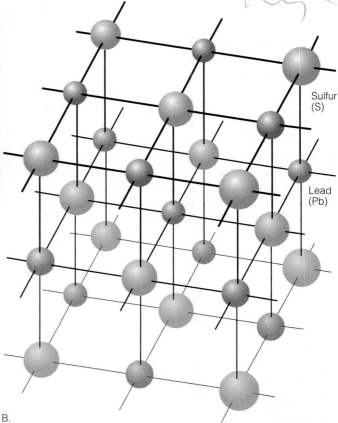

Sulfur (S)

Lead (Pb)

B.

Figure 3.7 Seeing Atoms in a Mineral The arrangement of atoms of lead (Pb) and sulfur (S) in galena (PbS), the most common lead mineral. Bonding between lead and sulfur is ionic; lead forms a cation Pb^{2+}, sulfur an anion, S^{2-}, so to maintain a charge balance there must be equal numbers of Pb atoms and S atoms in the structure. Atoms are so small that a cube of galena 1 cm on an edge contains about 10^{22} atoms each of Pb and S.   A. Atoms at the surface of a galena crystal revealed with a scanning-tunneling microscope. S atoms are the large bumps, Pb atoms the smaller ones.   B. The packing arrangement of atoms in a galena crystal. The atoms are shown pulled apart along the black lines to demonstrate how they fit together.

Measuring Our Earth

The Three States of Matter

Solid, liquid, and gas are the three states in which compounds and chemical elements exist at the Earth's surface. The state of a substance is determined by its temperature and pressure. For example, when ice is heated, a temperature will be reached where the ice will melt to water—a change of state has occurred. If the temperature is raised still further, the water will evaporate and become water vapor. By contrast, if water vapor is cooled or compressed under pressure, it will condense to either ice, as in snow, or water, as in rain. Figure B3.1 shows the regions of temperature and pressure in which the different states of H_2O are stable. Similar diagrams can be drawn for most compounds and elements.

A change of state involves energy. The transition from ice to water, for example, requires the addition of energy in the form of heat. In the reverse case, when water freezes to ice, heat is released. The amount of heat per gram released or absorbed during a change of state is known as *latent heat* (from the Latin, *latens,* meaning hidden, hence hidden heat). The latent heat corresponding to changes of state in H_2O are shown in Figure B3.2.

Changes in the state of H_2O at the Earth's surface play an important role in the distribution of heat energy in the atmosphere, hydrosphere, and lithosphere, and thereby in the Earth's climate.

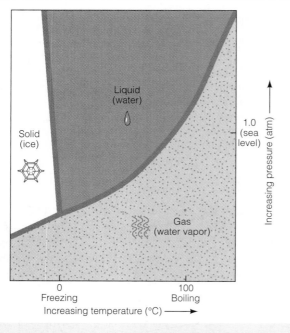

B 3.1 The States of Water The state in which the H_2O occurs is determined by the temperature and pressure at any given time and place. Along the boundary lines between any two states, both states can exist. The Earth is the only body in the solar system in which H_2O exists in all three states at the surface.

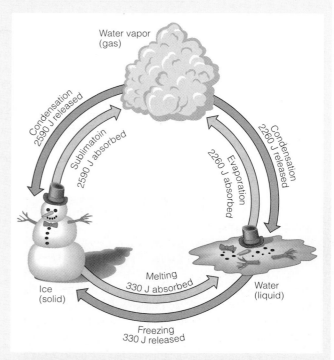

Figure B3.2 Latent Heat The amounts of heat absorbed by, or released from, one gram of H_2O during a change of state.

be **crystalline**. Solids that lack crystal structures are *amorphous* (a term of Greek origin meaning without form). Glass and amber are examples of solids that are amorphous. All minerals are crystalline, and the crystal structure of a mineral is a unique property of that mineral. All specimens of a given mineral have identical crystal structure.

Ionic Substitution

The crystal structure of the mineral galena, PbS, a common lead mineral, is shown in Figure 3.7B. The bonding is ionic—lead is the cation (Pb^{2+}), sulfur the anion (S^{2-}). Note that the ions in Figure 3.7B are arranged in a cubelike grid in which each sulfur is surrounded by six leads and each lead by six sulfurs.

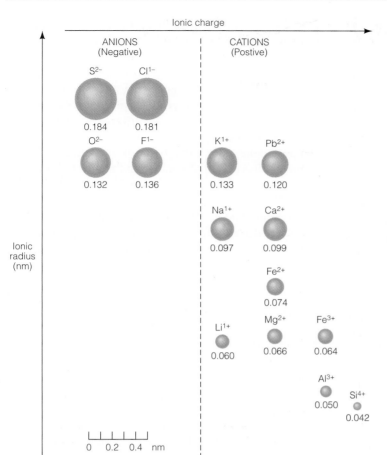

Ionic charge

ANIONS (Negative)

CATIONS (Postive)

Ionic radius (nm)

0 0.2 0.4 nm

Figure 3.8 Different Sized Ions Ionic radii of some geologically important ions range from Si^{4+} at lower right to S^{2-} at upper left. Ions are arranged in vertical groups based on charge, from $^{2-}$ at left to $^{4+}$ at right. Note that the anions tend to have larger ionic radii than cations. Ions in each of the pairs Si^{4+} and Al^{3+}, Mg^{2+} and Fe^{2+}, and Na^{1+} and Ca^{2+} are about the same size and commonly substitute for each other in crystal structures. Radii are expressed in nanometers (nm).

The bonding present in most common minerals is ionic, and for this reason it is helpful to briefly discuss two important properties of ions. The first property is the size of ions. Note that the anions in Figure 3.7B are larger than the cations. The size of ions is commonly stated in terms of the **ionic radius**, which is the distance from the center of the nucleus to the outermost shell of orbital electrons. Anions tend to have large radii because the addition of extra electrons to fill an energy-level shell means that the pull exerted on each orbital electron by the protons in the nucleus is slightly reduced. As a result, the electrons move a bit farther away from the nucleus. By contrast, cations tend to be small because they lose electrons and their remaining electrons are more tightly held. Most of the volume of a crystal structure is taken up by the largest ions, the anions, and as a result the crystal structures of minerals are determined largely by the packing arrangements of anions. The radii of some common ions are shown in Figure 3.8.

The second important property of ions concerns their electrical charges. It is apparent from Figure 3.8 that certain ions have the same electrical charge and are nearly alike in size. For example, Fe^{2+} has a radius of 0.074 nm, while Mg^{2+} has a radius of 0.066 nm. Because of their similarity in size and charge, ions of Fe^{2+} can substitute for ions of Mg^{2+} in magnesium-bearing minerals. The crystal structure of the magnesium mineral is not changed as a result of the substitution. The substitution of one ion for another in a random fashion throughout a crystal structure is **ionic substitution**.

The way ionic substitutions are indicated in chemical formulas can be demonstrated with the mineral olivine, Mg_2SiO_4. When Fe^{2+} substitutes for Mg^{2+} in olivine, the formula is written $(Mg,Fe)_2SiO_4$, which indicates that the Fe^{2+} substitutes for the Mg^{2+}, but not for any other atoms in the structure.

DEFINITION OF A MINERAL

Before discussing specific minerals and their properties, let's review what is meant by the term *mineral*. To be called a mineral, a substance must meet four requirements:

1. It must be *naturally formed*. This excludes the vast numbers of substances produced in the laboratory.

2. It must be a *solid*. This excludes all liquids and gases. This requirement is based on the state of the material, not its composition.

3. It must have a *specific chemical composition*. The requirement that a mineral have a specific chemical composition has several implications. Most importantly, it means that minerals are either chemical elements (gold, copper, and diamonds are examples) or chemical compounds in which atoms are present in specific ratios. Quartz, with the formula SiO_2, is an example of a chemical compound. The ratio of Si to O in quartz is always 1 to 2. Many minerals have more complicated formulas than quartz; for example, the chemical formula for the mineral phlogopite, a variety of mica, is $KMg_3AlSi_3O_{10}(OH)_2$. Other minerals have even more complicated formulas, but in all cases the elements present are combined in specific ratios. Even compounds in which ionic substitution is present obey the rule of specific combining ratios. For example, Fe^{2+} can substitute for Mg^{2+} in phlogopite, so that the formula becomes $K(Mg,Fe)_3AlSi_3O_{10}(OH)_2$, but the ratio of K to (Mg+Fe) remains 1 to 3.

4. It must have a *characteristic crystal structure*. This excludes amorphous materials, such as glass and amber.

The term *mineral group* is used to describe a mineral that displays extensive ionic substitution without changing the cation–anion ratio. Special names are given to the ideal, unsubstituted compositions of a mineral group. For example, olivine is a mineral group with the formula $(Mg,Fe)_2SiO_4$. Forsterite is the name given to Mg_2SiO_4, while Fe_2SiO_4 is called fayalite. It is important to remember the common mineral groups because we use them repeatedly throughout this book and because they are commonly used names in science. It is not so important to remember the names used for ideal compositions.

Mineraloids

Some naturally occurring solid compounds do not fulfill the definition of a mineral because they lack either a definite composition, a characteristic crystal structure, or both. Examples are natural glasses and resins, both of which have wide and variable composition ranges and are amorphous. Another example is opal, which has a more or less constant composition but is amorphous. The term **mineraloid** is used to describe such mineral-like materials.

Polymorphs

Each mineral has a unique crystal structure. Some elements and compounds are known to form two or more

different minerals because the atoms can be packed to form more than one kind of crystal structure. Diamond and graphite provide an example of an element that forms two different minerals. Both consist of pure carbon, but as shown in Figures 3.3 and 3.5, they have quite different structures. $CaCO_3$ provides an example of a compound that forms two different minerals. One is *calcite*, the mineral of which limestone and marble are composed; the other is *aragonite*, most commonly found in the shells of clams, oysters, and snails. Calcite and aragonite have identical compositions but entirely different crystal structures.

Elements and compounds that occur in more than one crystal structure are called **polymorphs** (many forms). Some common mineral polymorphs are listed in Table 3.1.

Properties of Minerals

The properties of minerals are determined by composition and crystal structure. Once we know which properties are characteristic of which minerals, we can use those properties to identify the minerals. To discover a mineral's identity it is not necessary, therefore, to analyze it chemically or to determine its crystal structure. The properties most often used to identify minerals are obvious ones such as color, external shape, and hardness—as well as some less obvious properties, such as luster, cleavage, and specific gravity. Each property is briefly discussed below, and an extensive table of individual mineral properties is given in Appendix C.

Crystal Form and Growth Habit

Crystal Form
Ice fascinated the ancient Greeks. When they saw glistening needles of ice covering the ground on a frosty morning, they were intrigued by the fact that the needles were six-sided and had smooth, planar surfaces. Greek philosophers made many discoveries about the branch of mathematics called geometry, but they could not explain how three-dimensional, geometric solids could apparently grow spontaneously. The Greeks called ice *krystallos;* the Romans latinized the name to *crystallum*. Eventually, the word **crystal** came to be applied to any solid body that grows with planar surfaces. The planar surfaces that bound a crystal are called **crystal faces**, and the geometric arrangement of crystal faces, called the **crystal form**, became the subject of intense study during the seventeenth century.

Seventeenth-century scientists discovered that the crystal form can be used to identify minerals. But certain features were difficult for them to explain; for example,

TABLE 3.1	Examples of Polymorphs, with the Most Common Mineral Listed First
Composition	Mineral Name
C	Graphite
	Diamond
$CaCO_3$	Calcite
	Aragonite
FeS_2	Pyrite
	Marcasite
SiO_2	Quartz
	Cristobalite

Figure 3.10 Irregular Growth—The Usual Case Quartz grains that grew in an environment where other quartz grains prevented development of well-formed crystal faces. The amber-colored grains are iron carbonate ($FeCO_3$). Compare with Figure 3.9, which shows crystals that grew in open spaces, unhindered by adjacent grains.

why did the sizes of crystal faces differ widely from sample to sample? Under some circumstances, a mineral may grow as a long, thin crystal; under others, the same mineral may grow as a short, fat one, as Figure 3.9 shows. Superficially, the two crystals of quartz in Figure 3.9 look very different. It is apparent from the figure that the overall size of crystals and crystal faces is not a unique property of a mineral. The person who solved the mystery was a Danish physician, Nicolaus Steno. In 1669 Steno demonstrated that the unique property of a crystal of a given mineral is not the relative face sizes but rather the angles between the faces. The angle between any designated pair of crystal faces is constant, he wrote, and is the same for all specimens of a mineral, regard-

less of overall shape or size. Steno's discovery that interfacial angles are constant is made clear by the numbering in Figure 3.9. The same faces occur on both of the quartz crystals. The sets of faces are parallel; therefore, the angle between any two equivalent faces must be the same on each crystal.

Steno and other early scientists suspected that the capacity for crystals to form and for interfacial angles to be constant must depend on some kind of internal order, but the ordered particles—atoms—were too small for them to see, so they could only speculate. Proof that crystal form reflects internal order was finally achieved in 1912. In that year the German scientist Max von Laue demonstrated, by use of X rays, that crystals must be made up of atoms packed in fixed geometric arrays, as shown in Figure 3.7.

Crystals can form only when mineral grains can grow freely in an open space. Unfortunately, crystals turn out to be uncommon in nature because most minerals do not form in open, unobstructed spaces. Compare Figures 3.9 and 3.10. Figure 3.9 shows crystals of quartz that grew freely into an open space and well-developed crystal faces were able to form. But in the second case, Figure 3.10, quartz grew as irregularly shaped grains in an environment restricted by the presence of other quartz grains. By making use of von

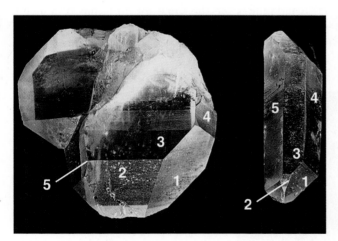

Figure 3.9 Crystal Faces and Angles in Quartz Two quartz crystals showing equivalent crystal faces. Although the sizes of the equivalent faces differ markedly between the two crystals, it is apparent that each numbered face on one crystal is parallel to an equivalent face on the other crystal. It is a fundamental property of crystals that, as a result of the internal crystal structure, the angles between adjacent faces are identical for all crystals of the same mineral.

Laue's discovery, it is easy to show that in both a quartz crystal and an irregularly shaped quartz grain all the atoms present are packed in the same strict geometric arrangement; that is, both the quartz crystals and the irregular quartz grains are crystalline. The word *crystalline*, rather than crystal, is therefore used in the definition of a mineral.

Growth Habit

Every mineral has a characteristic crystal form. Some have such distinctive forms that we can use the property as an identification tool without having to measure angles between faces. For example, the mineral pyrite (FeS_2) is commonly found as intergrown cubes (Fig. 3.11) with markedly striated faces. A few minerals can even develop distinctive growth habits when they grow in restricted environments. These, too, can be used as an aid to identification. For example, Figure 3.12 shows chrysotile asbestos, a variety of the mineral serpentine that characteristically grows as fine, elongate threads.

Cleavage

A mineral's tendency to break in preferred directions along bright, reflective planar surfaces is called **cleavage**.

If you break a mineral with a hammer or drop a specimen on the floor so that it shatters, you will probably see that the broken fragments are bounded by surfaces that are smooth and planar, just like crystal faces. In exceptional cases, such as the halite (NaCl) fragments shown in Figure 3.13A, all of the breakage surfaces are smooth planar surfaces. Don't confuse crystal faces and cleavage surfaces, however, even though the

Figure 3.11 Pyrite; Also Known as Fool's Gold Distinctive crystal form of pyrite, FeS_2. One of the growth habits of pyrite is as cube-shaped crystals with pronounced striations on the cube face. The largest crystals in the photograph are 3 cm on an edge. The specimen is from Bingham Canyon, Utah.

two often look alike. A cleavage surface is a breakage surface, whereas a crystal face is a growth surface.

The planar directions along which cleavage occurs are governed by the crystal structure (Fig. 3.13B). They are planes along which the bonding between atoms is relatively weak. Because the cleavage planes are direct expressions of the crystal structure, the angles between cleavage planes are the same for all grains of a given

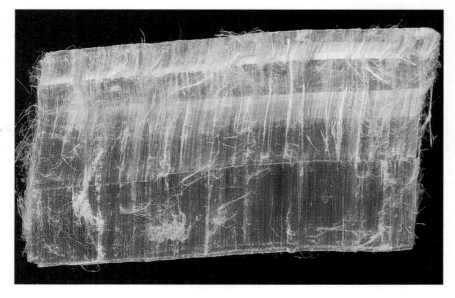

Figure 3.12 Fibers of Chrysotile Asbestos Some minerals have distinctive growth habits even though they do not develop well-formed crystal faces. The mineral chrysotile sometimes grows as fine, cottonlike threads that can be separated and woven into fireproof fabric, in which case it is referred to as asbestos. Chrysotile is one of several minerals that have asbestiform growth habits and are mined and commercially processed for asbestos. In commerce, chrysotile asbestos is commonly called "white asbestos."

A

B

Figure 3.13 Mineral Cleavage Relation between crystal structure and cleavage. A. Halite, NaCl, has three well-defined cleavage directions and breaks into fragments bounded by perpendicular faces. B. The crystal structure in the same orientation as the cleavage fragments shows that the directions of breakage are planes in the crystal between equal numbers of sodium and chlorine atoms.

Figure 3.14 A Book of Mica Perfect cleavage of muscovite (a mica), shown by thin, planar flakes into which this specimen is being split. The cleavage flakes suggest leaves of a book, a resemblance embodied in the name "book of mica."

mineral. Just as interfacial angles of crystals are constant, so are the angles between cleavage planes constant. Cleavage, therefore, is a valuable guide for the identification of minerals. Many common minerals have distinctive cleavage planes. One of the most distinctive is found in mica (Fig. 3.14). Clay minerals also have distinctive cleavage, and it is an easy cleavage direction that makes them feel smooth and slippery when rubbed between the fingers. Other minerals with distinctive cleavages are fluorite (CaF_2), which breaks along four planar directions, and potassium feldspar ($KAlSi_3O_8$), which breaks along two planar directions that are perpendicular, creating fragments that are approximately rectangular (Fig. 3.15).

Figure 3.15 Cleavage in Two Common Minerals Two common minerals with distinctive cleavages. Fluorite (A) breaks along four planar directions. Potassium feldspar (B) breaks along two planar directions that are perpendicular. The fluorite is from England and the feldspar is from Maine.

A

B

A   **B**   **C**

Figure 3.16 Luster Three minerals with different lusters.   A. Quartz has a glassy luster.
B. Sphalerite has a resinous luster—it resembles dried pine-tree resin.   C. Talc has a pearly luster.

Luster

The quality and intensity of light reflected from a mineral produce an effect known as **luster**. Two minerals with almost identical color can have quite different lusters. The most important lusters are described as *metallic*, like that on a polished metal surface; *vitreous*, like that on glass; *resinous*, like that of resin; *pearly*, like that of pearl; and *greasy*, as if the surface were covered by a film of oil (Fig. 3.16).

Color and Streak

The color of a mineral is often a striking property, but unfortunately color is not a very reliable means of identification. Color is determined by several factors, but the main cause is chemical composition. Some elements can create strong color effects even when they are present only in very small amounts by ionic substitution. For example, the mineral corundum (Al_2O_3) is commonly white or grayish, but when small amounts of Cr^{3+} replace Al^{3+} by ionic substitution, corundum is blood red, to which the gem name *ruby* is given. Similarly, when small amounts of Fe^{2+} and Ti^{4+} are present, the corundum is deep blue and another prized gem, *sapphire*, is the result (Fig. 3.17).

One of the reasons that color can be a confusing property is weathering. Water and oxygen will react with a mineral such as pyrite (FeS_2), forming a surface of alteration that differs in color from the unaltered mineral. Weathering reactions such as that described for pyrite are common and, as discussed in the boxed essay on the *Long Term Danger of Abandoned Mines*, can have important consequences for the environment.

Streak
Color in opaque minerals with metallic lusters can be very confusing because the color is partly a property of the size of the individual mineral grains. One way to

Figure 3.17 Same Mineral, Different Colors Two specimens of the same mineral, corundum (Al_2O_3), with distinctly different colors. The red crystal, ruby, is from Tanzania and is about 2.5 cm across. The blue crystal, sapphire, comes from Newton, New Jersey, and is about 3 cm across.

reduce errors of judgment where color is concerned is to prepare a *streak*, which is a thin layer of powdered mineral made by rubbing a specimen on a nonglazed porcelain plate. The powder gives a reliable color effect because all the grains in a powder streak are very small and the grain size effect is reduced. Red streak characterizes hematite (Fe_2O_3), even though the specimen itself looks black and metallic (Fig. 3.18).

Hardness

The term **hardness** refers to a mineral's relative resistance to scratching. It is a distinctive property of minerals. Hardness, like crystal form and cleavage, is

Understanding Our Environment

Long Term Danger of Abandoned Mines

Weathering is a general term for all of the reactions between minerals, air, and water. Most minerals are susceptible to weathering reactions, but the most susceptible family of minerals are the sulfide minerals. The two common iron sulfide minerals, pyrite (FeS_2) and pyrrhotite (FeS), in particular, weather rapidly at the Earth's surface. Both pyrite and pyrrhotite react and combine with oxygen in the air or dissolved in water. The reaction products are relatively insoluble iron hydroxide compounds and sulfuric acid. The acid thus produced will attack more resistant sulfide minerals such as sphalerite (ZnS), galena (PbS), and chalcopyrite ($CuFeS_2$), producing more sulfuric acid and solutions with high concentrations of heavy metals. The oxidation of sulfide minerals can proceed inorganically, but certain bacteria can speed up the process hundreds of times faster.

In nature, fresh rock is exposed relatively slowly, so the rate at which acid is released by oxidation is very slow. But when mining operations expose fresh rocks by blasting and digging, a lot of sulfide minerals can be exposed rapidly. The mining of sulfide ores usually proceeds so rapidly that acid solutions are not a problem. But when mining ceases and workings fill with water, bacteria start to proliferate and oxidation proceeds rapidly. The waters seeping from old mines can be so acidic that no plants or animals can live in them. Released into streams, acid mine waters can cause devastating damage.

Although no mine that contains sulfide minerals is entirely free of acid water, the problem is particularly serious in abandoned coal mines. All coal contains some iron sulfide minerals, and large coal mines can leave tens or even hundreds of miles of openings for water and air to enter. Of course, climate plays a role in acid mine drainage. In areas of plentiful rainfall, such as Pennsylvania, old mines fill quickly with water so that acid mine drainage is a continual problem.

So far no one has found a way to prevent oxidation and the formation of acid mine waters. But forward planning for the design of mine openings, and the control of effluents once mining has ceased, can greatly ameliorate the problem.

An Abandoned Mining Mess
Acid drainage seeping from an abandoned 19th century silver mine, Silverton, Colorado. The acidity is due to the oxidation of pyrite.

Figure 3.18 Streak of a Mineral Color contrast between hematite and a hematite streak. Massive hematite is opaque, has a metallic luster, and appears black. On a porcelain plate, hematite gives a red streak.

governed by crystal structure and by the strength of the bondings between atoms. The stronger the bonding, the harder the mineral.

Relative hardness values can be assigned by determining whether one mineral will scratch another. Talc, the basic ingredient of most body ("talcum") powders, is the softest mineral known, and diamond is the hardest. A scale called the *Mohs' relative hardness scale* is divided into 10 steps, each marked by a distinctive mineral (Table 3.2). These steps do not represent equal intervals of hardness, but the important feature of the hardness scale is that any mineral on the scale will scratch all minerals below it. Minerals on the same step of the scale are just capable of scratching each other. For convenience, we often test relative hardness by using a common object such as a penny or a penknife as the scratching instrument, or glass as the object to be scratched.

Density and Specific Gravity

Another obvious physical property of a mineral is its density, which in practical terms means how heavy it feels. We know that two equal-sized baskets have different weights when one is filled with feathers and the other with rocks. The property that causes this difference is *density*, or the average mass per unit volume. The units of density are grams per cubic centimeter (g/cm^3). Minerals with a high density, such as gold, contain atoms with high mass numbers that are closely packed.

Minerals with a low density, such as ice, have loosely packed atoms.

Gold has a density of 19.3 g/cm^3 and feels very heavy, but many others such as galena (PbS) and magnetite (Fe_3O_4), which have densities of 7.5 g/cm^3 and 5.2 g/cm^3, respectively, also feel heavy by comparison with many common minerals that have densities of 2.5–3.0 g/cm^3.

Density is difficult to measure accurately. We usually measure a property called specific gravity instead. **Specific gravity** is the ratio of the weight of a substance to the weight of an equal volume of pure water. Specific gravity is a ratio of two weights, so it does not have any units. Because the density of pure water is 1 g/cm^3, the specific gravity of a mineral is numerically equal to its density. Specific gravity can be compared by holding different minerals in the hand and comparing their weights. Metallic minerals feel heavy, whereas nearly all others feel light.

Mineral Properties and Bond Types

Mineral properties depend strongly on the kinds of bonding present (Table 3.3). Strength and hardness are determined by the weakest bonding present because, like the weakest link in a chain determining the strength of the chain, so the weakest bonding present in a mineral determines the strength of that mineral. Ionic and covalent bonds are strong, and their presence makes minerals

TABLE 3.2	Mohs' Scale of Relative Hardness*		
	Relative Number in the Scale	Mineral	Hardness of Some Common Objects
	10	Diamond	
	9	Corundum	
	8	Topaz	
	7	Quartz	
	6	Potassium feldspar	
			Pocketknife; glass
Decreasing	5	Apatite	
	4	Fluorite	
			Copper penny
	3	Calcite	
			Fingernail
	2	Gypsum	
	1	Talc	

*Named for Friedrich Mohs, an Austrian mineralogist, who chose the 10 minerals of the scale.

TABLE 3.3	Examples of Mineral Properties That Depend on Bond Type				
Bond Type	Mineral	Strength	Hardness	Electrical Conductance	Solubility in Water and Weak Acids
Ionic	Calcite ($CaCO_3$) Halite (NaCl)	High	Moderate to high	Very low	High
Covalent	Diamond (C) Sphalerite (ZnS)	Very high	High	Very low	Very low
Mixed ionic and covalent	Olivine (Mg_2SiO_4) Muscovite $KAl_2(Si_3Al)O_{10}(OH)_2$	Very high	Moderate to high	Very low	Low
Metallic	Gold (Au) Copper (Cu)	Moderate	Low	High	Very low
van der Waals	Graphite (C) Sulfur (S)	Very low	Very low	Low	Low

hard and strong. By comparison, metallic and van der Waals bonds are much weaker, and minerals containing them tend to be soft and easily deformed.

COMMON MINERALS

Scientists have identified approximately 3600 minerals. Most occur in the crust, but a few have been identified in meteorites and two new ones were discovered in the Moon rocks brought back by the astronauts. The total number of minerals may seem large, but it is tiny by comparison with the astronomically large number of ways chemists can combine naturally occurring elements to form synthetic minerals. The reason for the disparity between nature and chemical experiment becomes apparent when we consider the relative abundances of the chemical elements. As Table 3.4 shows, only 12 elements occur in the continental crust in amounts greater than 0.1 percent by weight. Together these 12—usually referred to as the abundant elements—make up 99.23 percent of the crust mass. The crust, therefore, is constructed mostly, but not entirely, of a limited number of minerals in which one or more of the 12 abundant elements is an essential ingredient.

Rather than forming distinct minerals, many of the scarce elements tend to occur by ionic substitution. For example, the mineral olivine [$(Mg,Fe)_2SiO_4$], contains, in addition to Mg, Fe, Si, and O, trace amounts of Cu, Ni, Co, Mn, and many other elements as ionic substitutes for the Mg or Fe.

TABLE 3.4	Average Composition of the Continental Crust

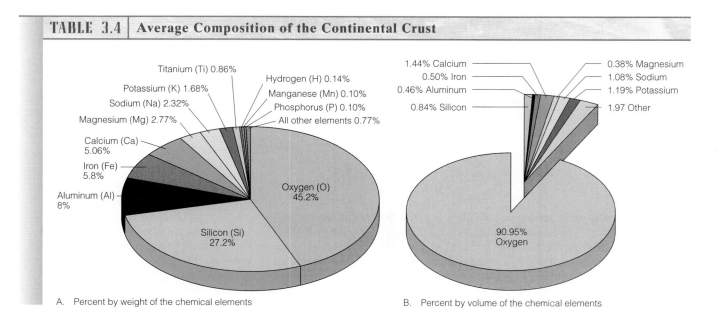

A.   Percent by weight of the chemical elements

B.   Percent by volume of the chemical elements

Minerals containing scarce elements certainly do occur but only in small amounts, and those small amounts only form under special and restricted circumstances. A few scarce elements, such as hafnium and rhenium, are not known to form minerals under any circumstances: they occur only by ionic substitution.

As Table 3.4 shows, two elements, oxygen and silicon, make up more than 70 percent of the weight of the continental crust. Oxygen forms a simple anion, O^{2-}, and compounds that contain the O^{2-} anion are called oxides. Silicon forms a simple cation, Si^{4+}, but oxygen and silicon together form an exceedingly strong complex ion, the **silicate anion** $(SiO_4)^{4-}$. Minerals that contain the silicate anion are complex oxides, and to distinguish them from simple oxides they are called **silicates**, or **silicate minerals**. For example, MgO is an oxide, but Mg_2SiO_4 is a silicate.

Silicate minerals are the most abundant of all naturally occurring, inorganic compounds, and simple oxides are the second most abundant group. Other mineral groups, all important, but less abundant than silicates and oxides, are sulfides, which contain the simple anion S^{2-}, carbonates $(CO_3)^{-2}$, sulfates $(SO_4)^{-2}$, and phosphates $(PO_4)^{-3}$.

The Silicate Minerals

The Silicate Tetrahedron

The four oxygen atoms in a silicate anion are tightly bonded to the single silicon atom. The bonding, which is very strong, is largely covalent. Oxygen has a large ionic radius (Fig. 3.8), while silicon has a small ionic radius. In the silicate anion, the oxygen ions pack into the smallest space possible for four large spheres. As can be seen in Figure 3.19, the four oxygens sit at the corners of a tetrahedron, and the small silicon atom sits in the space between the oxygens at the center of the tetrahedron. Therefore, the shape of the silicate anion is a tetrahedron. The structures and properties of all silicate minerals are determined by the way the $(SiO_4)^{4-}$ silicate tetrahedra pack together in the crystal structure.

Each silicate tetrahedron has four negative charges. Silicon has four electrons in shell 2, and oxygen has six. The silicon atom in a silicate tetrahedron shares one of its electrons with each of the four oxygens. Each of the four oxygens, in turn, shares one of their electrons with the silicon. This leaves the silicon with a stable shell 2 of eight electrons, but each of the four oxygens still requires an additional electron for a stable octet. The four oxygens can each attain a stable shell 2 of eight electrons in two ways:

1.   They can each accept an electron from, or share an electron with, other atoms. An example of this is found in olivine (Mg_2SiO_4), in which Mg atoms transfer their two outer shell electrons to the oxygens, forming ionic bonds in the process.

2.   An oxygen atom can bond with two Si atoms at the same time. The shared oxygen is then covalently bonded to each of two silicons. The second electron shell of the shared oxygen is now filled because it shares an electron with each silicon, and the two tetrahedral-shaped silicate anions, now joined at a common apex,

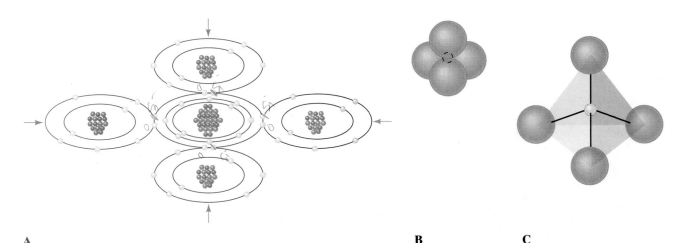

A   **B**   **C**

Figure 3.19   Silicate Tetrahedron   Bonding and structure of the silicate anion.   A. Silicon has four electrons in energy-level shell 2. Each of the four electrons is shared with one oxygen, and each oxygen in turn shares an electron with the silicon. The silicon finishes with eight electrons in shell 2, each oxygen with seven electrons. Bonding between the silicate anion and other ions occurs when the four oxygens share

or accept an additional electron, thereby filling their second energy-level shells.   B. Tetrahedral-shaped silicate anion with oxygens touching each other in natural positions. Silicon (dashed circle) occupies central space.   C. Expanded view showing large oxygens at the four corners, equidistant from a small silicon atom. The lines show bonds between silicon and oxygen; shading outlines the tetrahedron.

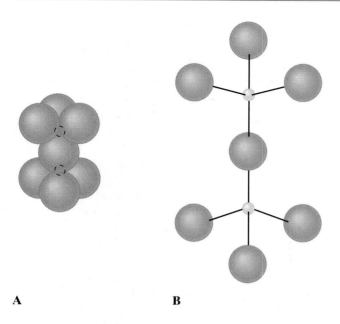

A   **B**

Figure 3.20 Linking Silicate Tetrahedra Two silicate tetrahedra share an oxygen, thereby satisfying some of the unbalanced electrical charges and, in the process, forming a larger
and more complex anion group.   A. The arrangement of
oxygen and silicon atoms in a double tetrahedron, giving the
complex anion $(Si_2O_7)^{6-}$. B. An expanded view of two tetrahedra sharing an oxygen.

form an even larger complex anion with the formula
$(Si_2O_7)^{6-}$, as shown in Figure 3.20. The large $(Si_2O_7)^{6-}$
anion forms compounds by accepting or sharing elec-
trons with other atoms in the same way that the smaller
$(SiO_4)^{4-}$ anion does. Only one common mineral, epidote,
contains an $(Si_2O_7)^{6-}$ anion formed by the simplest form
of oxygen sharing. However, just as more and more
beads can be strung on a necklace, so can **polymerization**, the process of linking silicate tetrahedra by oxygen
sharing, be extended to form huge anions (Fig. 3.21A).

Many common silicate minerals contain very large
anions as a result of polymerization. As shown in Figure 3.21B, if a tetrahedron shares more than one oxygen
with adjacent tetrahedra, the complex anionic structures so formed can have large circular shapes, endless
chains, sheets, and even three-dimensional networks of
tetrahedra. No matter how big an anion becomes by
polymerization, any oxygen that does not have a stable
outer shell of eight electrons can form bonds with
cations by accepting or sharing electrons.

Observe in Figures 3.21A and B that an important
restriction to the polymerization process is always met—
two adjacent tetrahedra never share more than one
oxygen. Stated another way, tetrahedra only join at
their apexes, never along the edges or faces. The common polymerizations, together with the rock-forming

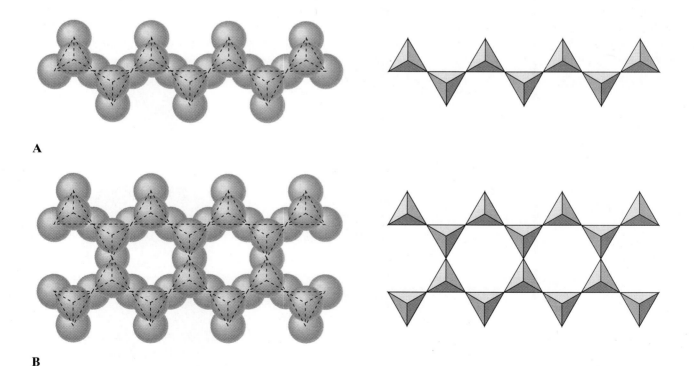

A

B

**Figure Figure 3.21 Large Anions Formed by Polymer-
ization** Formation of complex silicate anions by polymerization.   A. Polymerization of silicate anions to form a
continuous chain in which each silicate anion shares two
oxygens with adjacent anions. A geometric representation of
the chain is on the right. Formula of the complex anion is
$(SiO_3)^{2n-}$.   B. Double chain of a polymerized silicate anions
with formula $(Si_4O_{11})^{6n-}$.

Silicate Structure	Mineral/formula	Cleavage	Example of a specimen
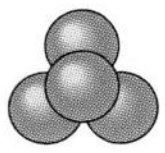 Single tetrahedron	Olivine Mg_2SiO_4	None	
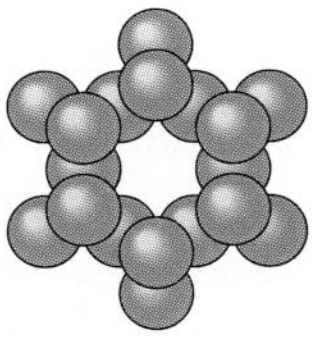 Hexagonal ring	Beryl (Gem form is emerald) $Be_3Al_2Si_6O_{18}$	One plane	
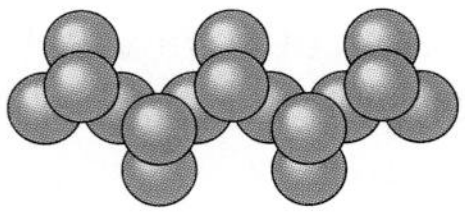 *Pyroxene = single chain* Single chain	Pyroxene group $CaMg(SiO_3)_2$ (variety: diopside)	Two planes at 90°	
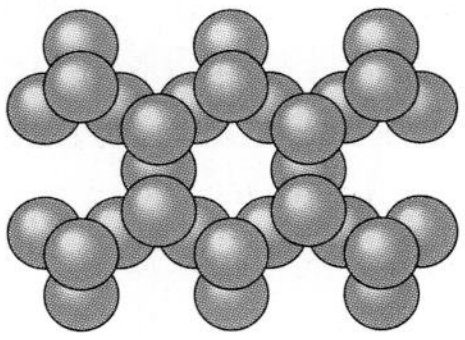 Double chain	Amphibole group $Ca_2Mg_5(Si_4O_{11})_2(OH)_2$ (variety: tremolite)	Two planes at 120°	
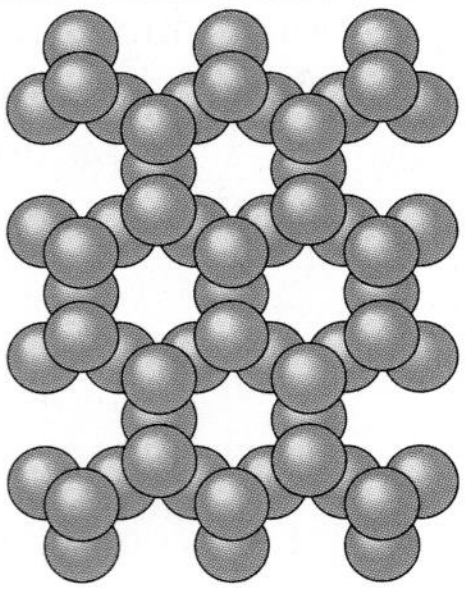 Sheet	Mica $KAl_2Si_3O_{10}(OH)_2$ (variety: muscovite) $K(Mg,Fe)_3AlSi_3O_{10}(OH)_2$ (variety: biotite)	One plane	
Too complex to draw. Three-dimensional network	Feldspar $KAlSi_3O_8$ (variety: orthoclase)	Two directions at 90°	
	Quartz SiO_2	None	

Figure 3.22 Common Silicate Minerals Summary of the ways in which silicate anions can link together to form the common silicate minerals. Typical examples of each type are listed and shown in small photographs. Note that there are many other minerals in each polymerization category—far too many to show on a diagram like this. Silicate linkages other than those shown are also possible but do not occur in common minerals.

minerals containing them, are illustrated in Figure 3.22 and discussed below in order of increasing complexity of polymerization.

The Olivine Group

Two important mineral groups contain isolated silicate tetrahedra. The first is the olivine group, which is a glassy-looking group of minerals that are usually pale green in color. As previously discussed, Fe^{2+} can substitute readily for Mg^{2+} in olivine, giving rise to the group formula $(Mg,Fe)_2SiO_4$. It is the Fe^{2+} that gives olivine its green color. Olivine is one of the most abundant mineral groups in the Earth, being a very common constituent of igneous rocks in the oceanic crust and the upper part of the mantle. In rare cases, olivine occurs in such flawless and beautiful crystals that it is used as the gem called peridot.

The Garnet Group

The second important mineral group with isolated silicate tetrahedra is the garnet group. As with olivine, ionic substitution gives garnet a range of compositions. But with garnets the range is much wider. The garnet group has the complex formula $A_3B_2(SiO_4)_3$, where A can be any of the cations Mg^{2+}, Fe^{2+}, Ca^{2+}, and Mn^{2+}, or any mixture of them, while B can be either Al^{3+}, Fe^{3+}, Cr^{3+} or a mixture of them. Garnet is characteristically found in metamorphic rocks of the continental crust, but it can also be found in certain igneous rocks. One of the most striking features of garnet is its tendency to form crystals, some of which can be cut to make beautiful gems (Fig. 3.23). An important combination of properties of garnet is the hardness and the lack of cleavage; the combination makes garnet a useful abrasive material for grinding and polishing.

The Pyroxene and Amphibole Groups

Pyroxenes and amphiboles are two silicate mineral groups that contain long, chainlike anions. Pyroxenes contain a polymerized chain of tetrahedra, each of which shares two oxygens, so the chain has the general formula $(SiO_3)^{2n-}$, where n is any very large number. Amphiboles are built from double chains of tetrahedra equivalent to two pyroxene chains in which half the tetrahedra share two oxygens and the other half share three oxygens, giving a general anion formula of $(Si_4O_{11})^{6n-}$. Both the pyroxene and amphibole chains are bonded together by cations such as Ca^{2+}, Mg^{2+}, and Fe^{2+} which form bonds with oxygens in adjacent chains.

The general formula for pyroxene is $AB(SiO_3)_2$, where A and B can be any of a number of cations, the most important of which are Mg^{2+}, Fe^{2+}, Ca^{2+}, Mn^{2+}, Na^{1+}, and Al^{3+}. The pyroxenes are most abundantly found in igneous rocks of the oceanic crust and mantle, but they also occur in many igneous and metamorphic rocks of the continental crust. The most common pyroxene is a shiny black variety called *augite*, with the approximate formula $Ca(Mg,Fe,Al)[(Si,Al)O_3]_2$.

Amphiboles have perhaps the most complicated formulas of all the common minerals. The general formula is $A_2B_5(Si_4O_{11})_2(OH)_2$, in which A is most commonly either Ca^{2+}, Mg^{2+}, Fe^{2+}, or Na^{1+} and B is usually Mg^{2+}, Fe^{2+}, Fe^{3+} or Al^{3+}.

The wide range of possible compositions is illustrated by the five varieties of commercial amphibole asbestos mentioned in the essay at the beginning of this chapter. The five are: actinolite, $Ca_2(Fe^{2+},Mg)_5(Si_4O_{11})_2(OH)_2$; anthophyllite, $Mg_2Mg_5(Si_4O_{11})_2(OH)_2$; crocidolite, $Na_2Fe_3^{2+}Fe_2^{3+}(Si_4O_{11})_2(OH)_2$; grunerite (commercially called amosite) $Fe_2^{2+}Fe_5^{2+}(Si_4O_{11})_2(OH)_2$; and tremolite, $Ca_2Mg_5(Si_4O_{11})_2(OH)_2$.

Even the complicated general formula and the possible substitutions referred to above do not completely describe the composition of the most abundant variety of amphibole, *hornblende*, a dark green to black mineral that looks rather like augite but can be distinguished from augite because the angles between cleavage surfaces differ in the two cases, as shown in Figure 3.24.

The Clays, Micas, Chlorites, and Serpentines

Clays, micas, chlorites, and serpentines are related mineral groups because they all contain polymerized sheets of silicate tetrahedra. The sheets are formed by each tetrahedron sharing three of its oxygens with adjacent tetrahedra to give a general anion formula $(Si_4O_{10})^{4n-}$. This leaves a single oxygen in each tetrahedron to which cations such as Al^{3+}, Mg^{2+}, and Fe^{2+} can bond and thereby hold the polymerized sheets together in the

Figure 3.23 Gem Garnets A collection of garnets cut and polished for use as gemstones. The range of colors is an indication of the wide range of compositions possible as a result of ionic substitutions.

A. Pyroxene

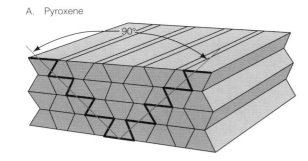

B. Amphibole

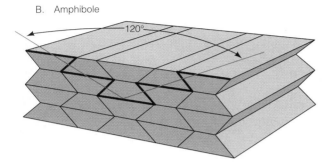

Figure 3.24 Cleavage in the Chain Silicates Cleavage in pyroxene and amphibole. A. Stacking of pyroxene chains, viewed end-on. Bonds that hold the chains together break along the wide black line. The angle between the two cleavage directions (red line) is close to 90°. B. Stacking of amphibole double chains. The angle between cleavage directions is 120°.

crystal structure. The Si–O bonds in the polymerized sheets are much stronger than are the bonds between the other cations and oxygen. As a result, the clays, micas, and chlorites all display a pronounced cleavage parallel to the sheets (Fig. 3.25).

In the case of the clays, the oxygens with unfilled shells bond with Al^{3+} cations, leading to the formula $Al_4Si_4O_{10}(OH)_8$ for the clay mineral *kaolinite*. The clays are among the most common minerals in the regolith.

With the micas, both Al^{3+} and K^{1+} are the bonding cations. There are two common kinds of mica: muscovite and biotite. The variety called *muscovite* has the formula $KAl_2(Si_3Al)O_{10}(OH)_2$. Muscovite is a clear, almost colorless mica that takes its name from muscovy, an old term for Russia, which was famous for producing large cleavage sheets of muscovite that could be used for windows. *Biotite*, which has the formula $K(Mg,Fe)_3(Si_3Al)O_{10}(OH)_2$, is a dark-brown mica whose color is due to the iron. The micas are common minerals in both igneous and metamorphic rocks.

The *chlorite group* has the formula $(Mg,Fe,Al)_6(Si,Al)_4O_{10}(OH)_8$. It is usually green in color, and it derives its name from *chloros*, a Greek word meaning green. Chlorite is a common alteration product from other minerals such as olivine, biotite, hornblende, and augite, which contain iron and magnesium. Igneous rocks of the oceanic crust, for example, commonly contain olivine and augite. When such minerals are in contact with seawater, they alter to chlorite. The (OH) in chlorite is one of the ingredients that promotes the melting of oceanic crust during subduction discussed in Chapter 2.

The *serpentine group* consists of three polymorphs with the formula $Mg_6Si_4O_{10}(OH)_8$: chrysotile, antigorite, and lizardite. The three minerals commonly occur together as fine-grained greenish masses formed by the alteration of olivine or other magnesium silicates. Chrysotile is the white asbestos of commerce (Fig. 3.13). Chrysotile fibers form when the polymerized silicate sheets roll up, like a tightly wound carpet.

The name *serpentine* is an ancient one derived from the Latin word for serpent. No one is quite sure how the name came to be given to the mineral group. Some suggest that a spotted serpentine-containing rock called *verde antique* bears a fanciful resemblance to a spotted serpent, whereas others think the name derives from the Greek writer Dioscorides who, in A.D. 50, recommended serpentine for the prevention of snake bite. There is no evidence that serpentine has any curative powers for snake bites.

Quartz

The only common mineral composed exclusively of silicon and oxygen is quartz, SiO_2. It provides an example of polymerization filling all the second energy-level

Figure 3.25 Why Clay Is Good for Pottery Making Minuscule crystals of the clay mineral kaolinite as seen through a scanning electron microscope. Pronounced cleavage parallel to the polymerized sheets of silicate anions is apparent. Clay absorbs water between the polymerized sheets making the wet clay weak, slippery, and easy to mold into pots.

Figure 3.26 The Colors of Quartz Trace amounts of atoms other than silicon and oxygen give rise to a wide range of colors in quartz.

shells and in the process forming a three-dimensional network of tetrahedra.

Quartz characteristically forms six-sided crystals (Fig. 3.9) and is found in many beautiful colors, several of which are prized as gemstones (Fig. 3.26). The colors come from minute amounts of iron, aluminum, titanium, and other elements present by ionic substitution. Quartz occurs in igneous, metamorphic, and sedimentary rocks and is one of the most widely used gem and ornamental minerals.

Certain specimens of quartz—those formed by precipitation from cool water solutions—are so fine-grained that they almost appear amorphous, and we can demonstrate the presence of an internal crystal structure characteristic of minerals only through the use of high-powered microscopes, X-ray machines, and other research tools. The name given to these fine-grained forms of quartz is *chalcedony*. Varietal names for chalcedony are *agate* if it has color banding (Fig. 3.27), or *flint*

(gray) and *jasper* (red) if the color is uniform.

The Feldspar Group

The name *feldspar* is derived from two Swedish words, *feld* (field) and *spar* (mineral). Early Swedish miners were familiar with feldspar in their mines and found the same mineral group in the abundant rocks they had to clear from their fields before they could plant crops. They were so struck by the abundance of feldspar that they chose a name to indicate that their fields seemed to be growing an endless crop of the minerals. Feldspar is indeed the most common mineral in the Earth's crust. It accounts for about 60 percent of all minerals in the continental crust, and together with quartz it constitutes about 75 percent of the volume of the continental crust. Feldspar is also abundant in rocks of the seafloor.

Feldspar, like quartz, has a structure formed by polymerization of all the oxygen atoms in the silicate tetrahedra. Unlike quartz, however, some of the tetrahedra contain Al^{3+} substituting for Si^{4+}, and so another atom must be added to the structure to donate an electron and balance the ionic charges. The cations that are present in minerals of the feldspar group are K^{1+}, Na^{1+}, or Ca^{2+}.

Feldspars are a complex mineral group that have a wide range of compositions. The names and compositions of common feldspars are *potassium feldspar*, $KAlSi_3O_8$; and *plagioclase*, $(Na,Ca)AlSi_3O_8$.

Potassium feldspar has several polymorphs—but the structural differences between them are subtle. They are sometimes pink or green due to ionic substitution of small amounts of Fe^{3+} for Al^{3+} (Fig. 3.28).

The most important ionic substitution in feldspar is

Figure 3.27 Banded Agate Agate, a color-banded microcrystalline variety of quartz formed by precipitation of SiO_2 in an open space from groundwater. Color banding is due to minute amounts of impurities. The sample is 10 cm across.

Figure 3.28 Green Feldspar The two most common minerals in the continental crust are feldspar and quartz. This specimen from Pike's Peak, Colorado, is a group of green potassium feldspar crystals and dark gray quartz grains.

the substitution of Ca^{2+} for Na^{1+} in plagioclase. This substitution is possible because, as can be seen in Figure 3.8, Ca^{2+} and Na^{1+} are much closer in ionic radius than either is to the radius of K^{1+}. However, Ca^{2+} and Na^{1+} have different charges, and so the substitution involves the coupled substitution of two ions: $(Na^{1+} + Si^{4+})$ for $(Ca^{2+} + Al^{3+})$. The substitution is so effective that plagioclase can range in composition from *albite* ($NaAlSi_3O_8$) to *anorthite* ($CaAl_2Si_2O_8$).

The Carbonate, Phosphate, and Sulfate Minerals

Carbonates

The complex carbonate anion $(CO_3)^{2-}$ forms three common minerals: *calcite, aragonite,* and *dolomite.* We have already seen that calcite and aragonite have the same composition, $CaCO_3$, and are polymorphs. Calcite is much more abundant than aragonite. Dolomite has the formula $CaMg(CO_3)_2$.

Both calcite and dolomite are abundant minerals, and they look similar. They have the same vitreous luster and the same distinctive cleavage (Fig. 3.29), both are relatively soft, and both are common in sedimentary rocks—calcite in limestone, dolomite in dolostone. One simple way to distinguish between them is with dilute hydrochloric acid (HCl). Calcite reacts vigorously, bubbling and effervescing, whereas dolomite reacts very slowly with little or no effervescence.

Figure 3.29 Confusing Carbonates   Calcite ($CaCO_3$) on the left and dolomite $CaMg(CO_3)_2$ on the right have similar crystal structures and, as a result, similar cleavages. Both cleave in three directions that are not perpendicular, yielding rhombohedral-shaped fragments. One quick way to distinguish between calcite and dolomite is to put a drop of diluted hydrochloric acid on the mineral—calcite will fizz and effervesce, while dolomite will slowly dissolve but not effervesce.

Phosphates

Apatite is by far the most important phosphate mineral. It contains the complex anion $(PO_4)^{3-}$ and has the general formula $Ca_5(PO_4)_3(F,OH)$. Apatite is the substance from which our bones and teeth are made. It is also a common mineral in many varieties of igneous and sedimentary rocks and is the main source of the phosphorus used for making phosphate fertilizers.

Sulfates

All sulfate minerals contain the sulfate anion, $(SO_4)^{2-}$. Although many sulfate minerals are known, only two are common, and both are calcium sulfate minerals: *anhydrite* ($CaSO_4$) and *gypsum* ($CaSO_4.2H_2O$). Both form when seawater evaporates—anhydrite when temperatures are high, gypsum at lower temperature. Gypsum is the raw material used for making plaster. Plaster of Paris got its name from a quarry near Paris where a very desirable, pure white form of gypsum was mined centuries ago.

The Ore Minerals

The term *ore mineral* is used for minerals that are sought and processed for their valuable metal contents. Such minerals tend to be elements, sulfides, or oxides. The important ore minerals are listed in Appendix C; for the importance of the ore minerals, see the boxed essay, *Are We Running Out of Minerals?*

Sulfides

The common sulfide minerals all have metallic lusters and high specific gravities. The two most common, *pyrite* (FeS_2) and *pyrrhotite* (FeS), are not actually mined for their iron content, but even so they are commonly referred to as ore minerals. Most of the world's lead is won from *galena* (PbS), most of the zinc from *sphalerite* (ZnS), and most of the copper from *chalcopyrite* ($CuFeS_2$) (Fig. 3.30). Other familiar metals won from sulfide ore minerals are cobalt, mercury, molybdenum, and silver.

Oxides

Because iron is one of the most abundant elements in the crust, the iron oxides *magnetite* (Fe_3O_4) and *hematite* (Fe_2O_3) are the two most common oxide minerals. Magnetite takes its name from the Greek word *Magnetis*, meaning stone of *Magnesia*, an ancient town in Asia Minor. Magnetis had the power to attract iron particles, and so the terms *magnet* and *magnetite* eventually joined our vocabulary. Hematite refers to its red color when powdered, the Greek word for red blood

A　　　　　　　　　　**B**　　　　　　　　　　**C**

Figure 3.30 Ore Minerals Example of three common ore minerals.　A. Sphalerite (ZnS), the principal zinc mineral, from Joplin, Missouri.　B. Galena (PbS), the main lead mineral, from Joplin, Missouri.　C. Chalcopyrite ($LuFeS_2$), an important copper mineral, from Ugo, Japan.

being *haima*. Magnetite and hematite are the main ore minerals of iron.

Other oxide ore minerals are *rutile* (TiO_2), the principal source of titanium; *cassiterite* (SnO_2), the main ore mineral for tin; and *uraninite* (U_3O_8), the main source of uranium. Other metals won from oxide ore minerals are chromium, manganese, niobium, and tantalum.

MINERALS AS INDICATORS OF THE ENVIRONMENT OF THEIR FORMATION

Minerals should not be regarded merely as objects of beauty or sources of economically valuable materials. Contained within their makeup are the keys to the conditions under which they (and the rocks they are in) formed. The study of minerals, therefore, can provide insight into the chemical and physical conditions in regions of the Earth we cannot observe and measure directly.

Our understanding of the growth environments of minerals has come largely through studying minerals in the laboratory. By suitable experiments, for example, scientists have been able to define the temperatures and pressures at which a diamond will form rather than its polymorph, graphite (Fig. 3.31). Because we can infer how temperature and pressure increase with depth in the Earth, we can state with certainty that rocks in which diamonds are found have been subjected to pressures equivalent to those in the mantle at least 150 km below the Earth's surface. Another example concerns weathering. The minerals that form in the regolith during weathering are controlled by the climate—cold and wet versus hot and dry, for example. Past climates can therefore be deciphered from the kinds of minerals preserved in sedimentary rocks. The composition of seawater in past ages can also be determined from the minerals formed when the seawater evaporated and deposited its salts. Rather than elaborating many examples at this point, we move next, in Chapter 4, to an examination of rocks and, in later chapters, to the way both rocks and minerals can be used to understand past environments.

ROCK

As we saw in Chapter 2, rocks can be grouped into three large families based on the way they formed. The three families are *igneous rocks*, formed by solidification

Using Resources

Are We Running Out of Minerals?

Most minerals that are abundant in the Earth's crust have neither commercial value nor any particular use. Ore minerals that are the raw materials of industry tend to be rare and hard to find. From the ore minerals we get the metals to make our machines and the ingredients to formulate our chemicals and fertilizers. Our modern society is totally dependent on an adequate supply of ore minerals. As Figure B3.3 makes clear, each of us uses, directly or indirectly,

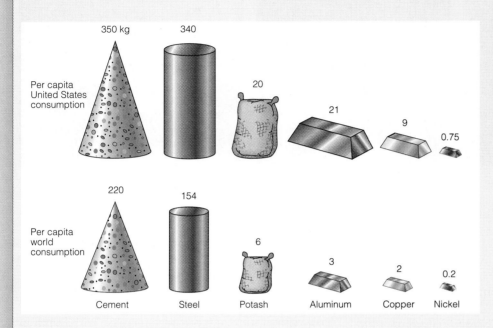

Figure B3.3 Relying on Minerals The average amounts of mined natural resources that are consumed per person (called the per capita consumption) are greater in industrially advanced countries such as the United States or Canada than they are for the world as a whole.

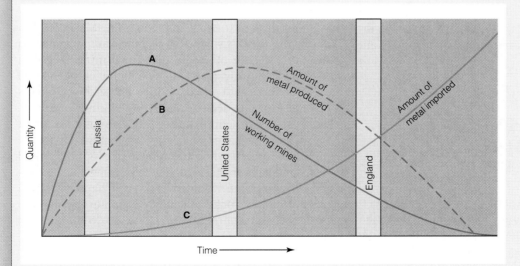

Figure B 3.4 History of Mining Changing metal production illustrated by the histories of three countries. The number of mines rises rapidly (curve A) but declines when the rate of mine exhaustion exceeds the discovery rate. The amount of metal produced (curve B) also rises, then falls when mines become exhausted. Curve C represents the growing imports of metal when internal production fails to meet needs. As time passes, the position of a country moves from left to right. In the late nineteenth century, Britain was about where the United States is today, at which time the United States was at about the present position of Russia.

a very large amount of material derived from ore minerals. Without the needed supplies we could not build planes, cars, televisions, or computers. Industry would falter, and living standards would decline.

Can the ore minerals in the Earth's crust sustain both a growing population and a high standard of living for everyone? This difficult question has many experts worried. The minerals they worry most about are those used as sources of such important metals as lead, zinc, and copper. Metals, the experts point out, begin the chain of resource use. Without metals, we cannot make machines. Without machines, we cannot convert the chemical energy of coal and oil to useful mechanical energy. Without mechanical energy, the tractors that pull plows must grind to a halt; trains and trucks must stop running; and indeed our whole industrial complex must become still and silent.

Experts have no way of telling how long the Earth's supplies of ore mineral will last. Optimistic experts point to the great success our technological society has enjoyed over the past two centuries as ever more remarkable discoveries have been made. If mineral supplies become limited, they suggest, we will find ways to get around the limits by recycling, by substitution, and by the discovery of new technologies. Many geologists have an opposite and more pessimistic opinion. Technologically advanced societies have faced mineral resource limits in the past, they point out, but the solution has always been to import new supplies from elsewhere rather than trying to develop substitutes or effective recycling measures.

England, for instance, was once a great supplier of metals (Fig. B3.4). Today, the minerals are mined out, most of its mines are closed, and English industry runs on raw materials imported from abroad. The United States, too, was once self-sufficient in most minerals and an exporter of many minerals. Slowly the situation has changed, so that now the United States is a net importer and has to rely on supplies from such countries as Australia, Chile, South Africa, and Canada. The only large industrial country that can still supply most of its own mineral needs is Russia, but eventually even the Russian mines will be depleted of their minerals, and so too will the mines of Australia and other countries. Where then will society turn?

The answer to the question just posed is not obvious, but it is one that must be answered in the foreseeable future. It is highly likely that within the lifetimes of the people who read this book, mineral limitations may occur. Which minerals, and therefore which metals, would first be in short supply is still an open question. That limitations will eventually happen, however, is no longer an open question. How society will cope and respond, and when it will have to do so, are just two of the great social and scientific issues still to be solved.

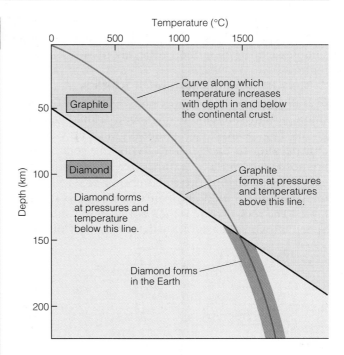

Figure 3.31 Diamonds from the Mantle Line separating regions of temperature and pressure (here plotted as depth) in which the two polymorphs of carbon—graphite and diamond—grow. At a pressure equal to that of a depth of 150 km, the line separating the graphite and diamond regions intersects the curve depicting the geothermal gradient in and below the continental crust. The diamonds found at the Earth's surface have formed at depths of 150 km or greater.

of magma; *sedimentary rocks*, formed by sedimentation of materials transported in solution or suspension; and *metamorphic rocks*, formed by the alteration of preexisting sedimentary or igneous rocks in response to increased pressure and temperature. Although we group rocks by their mode of formation, only rarely can we actually see a rock being formed. We may observe lavas solidifying to form igneous rocks, but we cannot see magma solidifying deep inside the Earth. Nor can we see metamorphism happening. In most cases, we have to use the evidence we can actually see and measure in order to deduce how a given rock is formed.

Texture

At first glance rocks seem confusingly varied. Some are either platy or distinctly layered and have pronounced, flat surfaces covered with mica. Others are coarse and evenly grained and lack layering; yet, they may still contain the same kinds of minerals as those present in the platy, micaceous rock. Studying a large number of rock specimens soon makes it clear that no matter what kind of rock is being examined—sedimentary, metamorphic, or igneous—the differences between samples can be described in terms of two kinds of features.

The first feature is **texture**, by which is meant the overall appearance that a rock has because of the size, shape, and arrangement of its constituent mineral grains. For example, the mineral grains may be flat and parallel to each other, giving the rock a pronounced platy, or flaky, texture—like a pack of playing cards. In addition, the various minerals may be unevenly distributed and concentrated into specific layers. The rock texture is then both layered and platy. Specific textural terms are used for each of the rock

families and will be introduced in the appropriate places in the next three chapters, where the three rock families are discussed.

Mineral Assemblage

The second feature of a rock is the kind of minerals that are present. A few kinds of rock contain only one mineral and natural glasses don't contain any minerals, but most rocks contain two or more minerals. The varieties and abundances of minerals present in rocks, commonly called the **mineral assemblages** of the rocks,

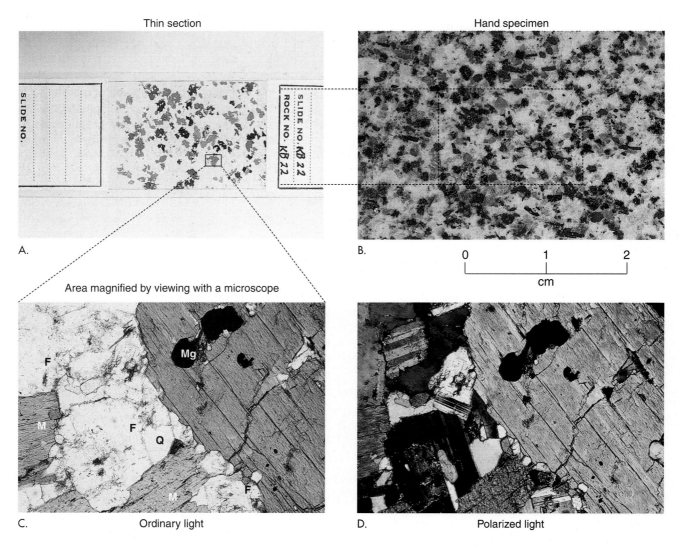

Figure 3.32 Igneous Rock Seen Through a Microscope Polished surfaces and thin slices reveal textures and mineral assemblages to great advantage. The specimen here is an igneous rock containing quartz (Q), feldspar (F), amphibole (A), mica (M), and magnetite (Mg).   A. A thin slice mounted on glass. The slice is 0.03 mm thick, and light can pass through the minerals.   B. A polished surface. The dashed rectangle indicates the area used to make the thin slice shown in part A.   C. An area of the thin slice as viewed under a microscope. The magnification is 25×.   D. The same view as in part C seen through polarizers in order to emphasize the shapes and orientations of individual grains.

are important pieces of information for interpreting how a rock formed.

A systematic description of a rock involves both the mineral assemblage and texture. Two additional terms, *megascopic* and *microscopic*, are very useful. Megascopic refers to those textural features of rocks that we can see with the unaided eye or with the eye assisted by a simple lens that magnifies up to ten times. Microscopic refers to those textural features of rocks that require high magnification in order to be viewed.

Examination of a microscopic texture commonly requires the preparation of a special *thin section* of rock that must be viewed through a microscope. A thin section is prepared by first grinding a smooth, flat surface on a small piece of rock. The flat surface is glued to a glass slide, and then the rock is ground away until the fragment glued to the glass is so thin that light passes through it easily. An example of a thin section is shown in Figure 3.32.

Percentages of Rock Types in the Crust

The internal processes that form magma and that in turn lead to the formation of igneous rock interact with external processes through erosion. When rock erodes, the eroded particles form sediment. The sediment may eventually become cemented, usually by substances carried in water moving through the ground, and thereby converted into new sedimentary rock. In places where such sedimentary rock forms, it can reach depths at which pressure and heat cause new compounds to form, so that the sedimentary rock becomes metamorphic rock. Sometimes metamorphic rock settles so deep that the high temperatures melt it and magma is formed. The new magma can then move upward through the crust, where it can cool and form another body of igneous

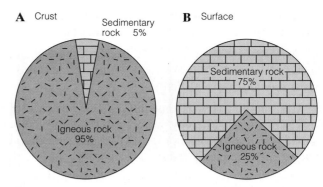

Figure 3.33 Frequency of Rock Types Relative amounts of sedimentary and igneous rock. (Metamorphic rocks are considered to be either sedimentary or igneous, depending on their origin.)   A. The great bulk of the crust consists of igneous rock (95%), but sedimentary rock (5%) forms a thin covering at and near the surface.   B. The extent of sedimentary rock cropping out at the surface is much larger than that of igneous rock, so 75 percent of all rock seen at the surface is sedimentary, and only 25 percent is igneous.

rock. Eventually, the new body of igneous rock can be uncovered and subjected to erosion, the eroded particles start once more on their way to the sea, sediment is laid down, and the cycle repeats. As James Hutton first recognized, the cycle has occurred again and again throughout the Earth's long history.

The crust is 95 percent igneous rock or metamorphic rock derived from igneous. However, as seen in Figure 3.33, most of the rock that we actually see at the Earth's surface is sedimentary. The difference arises because sediments are products of weathering, and as a result they are draped as a thin veneer over the largely igneous crust below. The distribution of rock types is one consequence of the rock cycle.

Because igneous rocks are the most abundant kind of rocks in both the continental and oceanic crusts, they are the first rock family to be discussed, which we do in the next chapter.

SUMMARY

1. The nucleus of an atom contains protons (positive electrical charge) and neutrons (no electrical charge). Moving in orbitals around the nucleus are electrons (negative electrical charge). The positive electrical charge on a proton is equal but opposite to the negative charge on an electron. The number of protons in a nucleus is the atomic number of an element.

2. Orbital electrons have different energy levels, and the number of electrons that can occupy a specific

energy-level shell is fixed: 2 in shell 1, 8 in shell 2, 18 in shell 3, 32 in shell 4.

3. The forces that hold atoms together in minerals are called bondings, and there are four different kinds of bondings. Ionic bondings arise when atoms transfer to, or accept orbital electrons from, other atoms. Covalent bondings involve the sharing of electrons between atoms. Metallic bondings involve the sharing of electrons in higher energy-level shells between several atoms. Van der

Waals bondings are weak, secondary attractions that form as a result of the sharing or transfer of electrons.

4. Minerals are naturally formed, solid chemical elements or compounds that have a definite composition and a specific crystal structure.

5. The compositions of some minerals vary because of ionic substitution, whereby one ion in a crystal structure can be replaced by another ion having a like electrical charge and a like ionic radius.

6. Some compounds have the same composition but different crystal structures. Each structure is a separate mineral. Minerals that have the same compositions but different structures are called polymorphs.

7. The principal properties used to characterize and identify minerals are crystal form, growth habit, cleavage, luster, color and streak, hardness, and specific gravity.

8. Approximately 3600 minerals are known, but of these about 20 make up more than 95 percent of the Earth's crust.

9. Silicates are the most abundant minerals, followed by oxides, carbonates, sulfides, sulfates, and phosphates.

10. The basic building block of silicate minerals is the silicate tetrahedron, a complex anion in which a silicon atom is covalently bonded to four oxygen atoms. The four oxygen ions sit at the apexes of a tetrahedron, with the silicon at its center. Adjacent silicate tetrahedra can bond together to form larger complex anions by sharing one or more oxygens. The process is called polymerization.

11. The feldspars are the most abundant group of minerals in the continental crust, comprising approximately 60 percent of the volume. Quartz is the second most common mineral in the continental crust.

12. Rocks are grouped into three families according to the way they are formed. We rarely observe rocks being formed, however, so rocks are usually described in terms of their textures and mineral assemblages.

13. Igneous rocks, or metamorphic rocks derived from igneous rocks, account for 95 percent of all rocks in the Earth's crust; sedimentary rocks account for 5 percent.

THE LANGUAGE OF GEOLOGY

anion (p. 54)
atom (p. 53)
atomic number (p. 53)

bonding (p. 54)

cation (p. 54)
cleavage (p. 62)
complex ion (p. 56)
composition (p. 52)
covalent bonding (p. 55)
crystal (p. 60)
crystal faces (p. 60)
crystal form (p. 60)
crystalline (p. 58)
crystal structure (p. 56)

energy-level shells (p. 53)

hardness (p. 64)

ion (p. 54)
ionic bonding (p. 55)
ionic radius (p. 59)
ionic substitution (p. 59)
isotope (p. 53)

luster (p. 64)

mass number (p. 53)
metallic bonding (p. 55)
mineral (p. 52)
mineral assemblage (p. 78)

mineraloid (p. 60)
molecule (p. 54)

polymerization (p. 69)
polymorphs (p. 60)

silicate (p. 68)
silicate anion (p. 68)
silicate mineral (p. 68)
specific gravity (p. 66)

texture (of a rock) (p. 78)

van der Waals bonding (p. 56)

MINERAL NAMES TO REMEMBER

agate (p. 73)
albite (p. 74)
amphibole group (p. 71)
anorthite (p. 74)
apatite (p. 74)
aragonite (p. 74)
augite (p. 71)

biotite (p. 72)

calcite (p. 74)
cassiterite (p. 75)
chalcedony (p. 73)
chalcopyrite (p. 74)
chlorite group (p. 72)
clay group (p. 71)
corundum (p. 64)

diamond (p. 55)
dolomite (p. 74)

feldspar group (p. 73)

flint (p. 73)

galena (p. 74)
garnet group (p. 71)
graphite (p. 56)
gypsum (p. 74)

halite (p. 63)
hematite (p. 65)
hornblende (p. 71)

ice (p. 58)

jasper (p. 73)

kaolinite (p. 72)

magnetite (p. 74)
muscovite (p. 72)

olivine group (p. 71)
*opal (p. 60)

plagioclase (p. 73)
potassium feldspar (p. 73)
pyrite (p. 62)
pyroxene group (p. 71)
pyrrhotite (p. 74)

quartz (p. 72)

rutile (p. 75)

serpentine group (p. 72)
sphalerite (p. 74)

talc (p. 64)

uraninite (p. 75)

*A mineraloid

QUESTIONS FOR REVIEW

1. Oxygen has an atomic number of 8, and there are three isotopes of oxygen—mass numbers 16, 17, and 18. For each isotope list the number of protons and neutrons in the nucleus.

2. How many electrons occupy energy-level shell 2 of an oxygen atom?

3. What is the difference between an atom and an ion? between an atom and a molecule?

4. Sketch the structure of a silicon atom and show how the silicate anion is formed by electron sharing with oxygen atoms.

5. What is a mineral? Give three reasons why the study of minerals is important.

6. Describe three ways chemical elements bond together to form compounds. Name two minerals that are examples of each bond type.

7. What properties besides composition and crystal structure can be used to identify minerals?

8. What are polymorphs? What are the mineral names of the two polymorphs of $CaCO_3$? of C?

9. What mineral has a crystal structure in which Pb and S atoms alternate at the corners of a three-dimensional, right-angled geometric grid?

10. What is ionic substitution? Illustrate your answer with two examples.

11. Approximately how many common minerals are there? What is the most abundant one?

12. Name five common minerals that are found in the area in which you live.

13. Describe how silicate anions join together to form silicate minerals.

14. Describe the polymerization of silicate anions in the pyroxenes, the micas, and the feldspars.

15. Name five common minerals that are not silicates, and name the anion each contains.

16. Are any minerals mined in the area in which you live? What are they and what are they mined for?

17. Why are igneous rocks so much more abundant than sedimentary rocks, yet sedimentary rocks cover most of the continental crust and are therefore the most visible?

4

Measuring
Our Earth

Understanding
Our Environment

Using
Resources

GeoMedia

Mount St. Helens erupting, July 22, 1980. The dormant volcano in the
rear is Mount Rainier. Both volcanoes are in Washington.

The Fire Within: Volcanoes and Magmas

Eruptions!

During the summer of 1883, an apparently dormant Indonesian volcano called Krakatau started to emit steam and ash. Krakatau was an island off the western end of Java. On Sunday, August 26, activity increased, and on the next day Krakatau blew up: the island disappeared. As a telegram of the time tersely reported, "Where once Mount Krakatau stood, the sea now plays." Noise from the paroxysmal explosion was heard on an island in the Indian Ocean, 4600 km away. As the volcano blew apart, it created *tsunami* that moved out from the site of the explosion and crashed into the shores of Java and Sumatra, the two closest Indonesian islands. Thirty-six thousand people lost their lives.

The effect of Krakatau was felt around the world. About 20 km^3 of volcanic debris was ejected during the eruption, some blasted as high as 50 km into the stratosphere. Within 13 days the stratospheric dust had encircled the globe, and for months there were strangely colored sunsets—sometimes green or blue and other times scarlet, or flaming orange. One November sunset over

New York City looked so like the glow from a massive fire that fire engines were called out. The suspended dust made the atmosphere look so opaque to the Sun's rays that the temperature around the Earth dropped an estimated 0.5° during 1884. It was five years before the atmosphere cleared and the climate returned to normal.

In March 1980, the sudden, violent eruption of Mount St. Helens, a long quiescent volcano in the state of Washington, reminded us once again of the enormous magnitude of volcanic forces. Mount St. Helens was known to have been active in historic times, but it had not erupted for more than 200 years. Then, early in 1980 people living near Mount St. Helens began reporting frequent small earthquakes. On March 27 steam and volcanic ash puffed from the summit.

Monitoring by geologists working for the U.S. Geological Survey quickly revealed that Mount St. Helens was bursting like a balloon: the north face was watched especially closely because by early May it was bulging outward at a rate of 1.5 m/day. From an observation post several kilometers north of the volcano, geologists stationed at Vancouver, Washington, mounted a round-the-clock watch. On Sunday, May 18, 1980, David A. Johnston was on duty, and at 8:32 A.M. he shouted into his microphone, "Vancouver, Vancouver, this is it!" They were his last words. A devastating eruption was under way. A gigantic mass of volcanic particles and very hot gases blasted out sideways, directly toward David Johnston. No trace of him or the observation post has ever been found. At least 63 other people were killed by the eruption, but the total would have been much higher had not the authorities heeded early warnings by geologists and kept people far away. Mount St. Helens didn't exactly disappear the way Krakatau did, but it was certainly beheaded. Originally a little more than 2900 m high, it is only 2490 m high today.

A more recent example of volcanic activity occurred on the island of Montserrat in the British West Indies. Although Montserrat is a volcanic island, the eruption of its Soufrière Hills volcano in August 1995 took residents by surprise. Soon after the eruption began the southern portion of the island was evacuated because so much volcanic ash was falling. The volcano remained active throughout the next two years, eventually making most of the island uninhabitable.

Geologists studying the Soufrière eruption observed repeated cycles of earthquakes, ground deformation, degassing, and explosive eruptions. Over time they were able to identify the causes of these cycles, and this in turn enabled them to improve the procedures they used to monitor the volcano's activity. In this way they became better able to forecast the timing and nature of eruptions and to issue appropriate warnings.

Scientists recognize several kinds of volcanoes, each characterized by a distinct kind of magma and a distinct eruption style. Magma is the molten material that forms when rock melts. What kind of rock melts, in which part of the Earth's interior the melting happens, and what kind of volcanic edifice marks the place where the magma reaches the surface reveal much about the processes taking place deep inside the Earth.

It is important to study volcanism and the deep-seated processes that give rise to it because volcanism plays an important role in the Earth system. Volcanic eruptions, particularly large ones such as the Krakatau eruption of 1883, can change climates around the world. Through volcanism, events that happen deep inside the Earth influence what happens on its surface. And through volcanism, new fresh rock is brought to the Earth's surface, where it becomes weathered and forms new, rich soil.

MAGMA

All igneous rocks form by the solidification of magma, so we first need to define what is meant by the term. **Magma** is molten rock, together with any suspended mineral grains and dissolved gases, that forms by melting of rock in the crust or mantle. When magma reaches the Earth's surface, it does so through a **volcano**, which is a vent from which magma, solid rock debris, and gases are erupted. The term *volcano* comes from the name of the Roman god of fire, Vulcan, and it immediately conjures up visions of streams of **lava**—magma that reaches the Earth's surface—pouring out over the landscape. Some lava does flow as hot streams, but magma can also be erupted as clouds of tiny red-hot fragments, as was the case when Mount St. Helens in the state of Washington erupted in March 1980. Magma, volcanoes, and eruption processes are much more varied than is commonly realized.

Volcanoes are the only places we can actually see and study magma, so we start this chapter by gaining some insight into volcanoes and the properties of magma, then proceed to a discussion of the kinds of igneous rock formed by the cooling of magma, and finally to a discussion of the way magma is thought to form.

By observing the eruption of lava, we can draw three important conclusions concerning magma.

1. Magma is characterized by a *range of compositions* in which silica (SiO_2) is always predominant.

2. Magma is characterized by high temperatures.

3. Magma has the ability to flow. This is true even though some magma flows so slowly it almost seems to be stationary. Most magma is a mixture of liquid (often referred to as melt) and mineral grains.

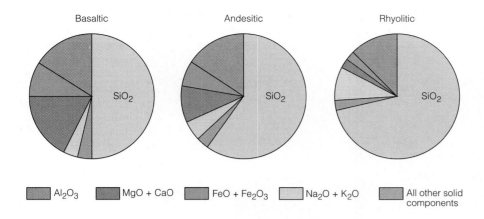

Basaltic **Andesitic** **Rhyolitic**

Legend: Al_2O_3 | $MgO + CaO$ | $FeO + Fe_2O_3$ | $Na_2O + K_2O$ | All other solid components

Figure 4.1 The Three Main Magmas The average compositions (in weight percent) of the three principal kinds of magma. The analyses are of rocks that solidified from the magmas. In addition to the solid materials the magmas contained dissolved gases, most of which escaped during solidification. Basaltic magma has a low content of dissolved gas; andestic and rhyolitic magmas tend to be very gas-rich.

Composition

The composition of magma is controlled by the most abundant elements in the Earth—Si, Al, Fe, Ca, Mg, Na, K, H, and O. Because magmas consist almost entirely of substances in which other elements are chemically combined with oxygen, it is usual to express compositional variations of magmas in terms of oxides, such as SiO_2, Al_2O_3, CaO, and H_2O. The most abundant component is SiO_2.

Three distinct types of magma are more common than all others. The first type contains about 50 percent SiO_2, the second about 60 percent, and the third about 70 percent. The three magmas are *basaltic*, *andesitic*, and *rhyolitic*, and the names of the common igneous rocks derived from them are **basalt**, **andesite**, and **rhyolite** (Fig. 4.1). The three magmas are not formed in equal abundance. Approximately 80 percent of all magma erupted by volcanoes is basaltic, while andesitic and rhyolitic magmas are each about 10 percent of the total. Magma from Hawaiian volcanoes such as Kilauea and Mauna Loa are basaltic, those from Mount St. Helens and Krakatau are usually andesitic, and those that were erupted from volcanoes that once were active at Yellowstone Park were mostly rhyolitic. As we shall see later in this chapter, the distribution of the different kinds of volcanoes is closely related to plate tectonics.

Gases Dissolved in Magma

Small amounts of gas (0.2 to 3% by weight) are dissolved in all magma, and even though present in only small amounts, these gases strongly influence the properties of magma. The principal gas is water vapor which, together with carbon dioxide, accounts for more than 98 percent of all gases emitted from volcanoes. Other substances, which are rarely present in volcanic gas in amounts exceeding 1 percent, include sulfur compounds, hydrogen chloride, nitrogen, and argon.

Temperature

The temperature of magma is difficult to measure, but it can sometimes be done during volcanic eruptions. As we have seen, volcanoes are dangerous places. Because scientists who study them are not eager to be roasted alive, measurements must be made from a distance by using optical devices. Magma temperatures determined in this manner during eruptions of volcanoes, such as Kilauea in Hawaii and Mount Vesuvius in Italy, range from 1000° to 1200°C. Experiments on synthetic magmas in the laboratory suggest that, under some conditions, magma temperatures might even be as high as 1400°C.

Viscosity

Dramatic pictures of lava flowing rapidly down the side of a volcano prove that some magmas are very fluid. Basaltic lava moving down a steep slope on Mauna Loa in Hawaii has been clocked at 16 km/h. Such fluidity is rare, and flow rates are more commonly measured in meters per hour or even meters per day. As seen in Figure 4.2, which shows basaltic lava—a kind of lava that often flows quite rapidly—destroying a house in Hawaii, flow rates are usually slow enough that people are not endangered. Magma containing 70 percent or more SiO_2 and very little dissolved gas flows so slowly that movement can hardly be detected.

The internal property of a substance that offers resistance to flow is called **viscosity**. The more viscous a magma, the less easily it flows. Viscosity of a magma depends on temperature and composition (especially the silica and dissolved-gas contents).

Effect of Temperature on Viscosity

The higher the temperature, the lower the viscosity, and the more readily a magma flows. A very hot magma

Figure 4.2 A House Is No Match for Lava An advancing tongue of basaltic lava setting fire to a house in Kalapana, Hawaii, during an eruption of Kilauea Volcano in June 1989. Flames at the edge of the flow are due to burning lawn grass.

Figure 4.3 Same Lava, Different Flows The way lava flows is controlled by viscosity. Two strikingly different flows are visible. They have the same basaltic composition. The lower flow, on which the geologist is standing, is a pahoehoe flow formed from a low-viscosity lava like that shown in Figure 4.2. The upper flow (the one being sampled by the geologist), which is very viscous and slow moving, is an aa flow erupted from Kilauea Volcano in 1989. The pahoehoe flow was erupted in 1959.

erupted from a volcano may flow readily, but it soon begins to cool, becomes more viscous, and eventually slows to a complete halt. In Figure 4.3, the smooth ropy-surfaced lava, locally called *pahoehoe* in Hawaii, formed from a very hot, very fluid lava. The rubbly, rough-looking lava, by contrast, formed from a cooler lava with a high viscosity. Hawaiians call this rough lava *aa*.

Effect of Silica Content on Viscosity

The $(SiO_4)^{4-}$ anions that occur in silicate minerals (Chapter 2) are also present in magmas. As they do in minerals, these anions polymerize by sharing oxygens. Unlike the anions in silicate minerals, however, those in magma form irregularly shaped groupings of silicate tetrahedra. The size of such a grouping depends on the number of silicate tetrahedra it contains. As the average number of silicate tetrahedra in the polymerized groups becomes larger, the magma becomes more viscous—in other words, more resistant to flow.

The number of tetrahedra in the groups depends on the silica content of the magma. The higher the silica content, the larger the polymerized groups. For this reason rhyolitic magma is always more viscous than basaltic magma, and andesitic magma has a viscosity that is intermediate between the two.

ERUPTION OF MAGMA

Magma, like most other liquids, is less dense than the solid rock from which it forms. Therefore, once formed, the lower density magma will exert an upward push on the enclosing higher density rock and will slowly force its way up. There is, of course, a pressure on a rising mass of magma due to the weight of all the overlying rock. The pressure is proportional to depth. As a magma rises upward, therefore, the pressure must decrease.

Pressure controls the amount of gas a magma can dissolve—more at high pressure, less at low. Gas dissolved in a rising magma acts the same way as gas dissolved in soda water. When a bottle of soda is opened, the pressure inside the bottle drops, gas comes out of solution, and bubbles form. Gas dissolved in an upward-moving magma also comes out of solution and forms bubbles. What happens to the bubbles once they are formed is determined by the viscosity of the liquid.

Nonexplosive Eruptions

It is understandable why people tend to regard any volcanic eruption as a hazardous event and active volcanoes as dangerous places that should be avoided. However, geologists have discovered that some volcanoes are comparatively safe and relatively easy to study. Nonexplosive eruptions, like those we can witness in Hawaii, are relatively safe compared to violent, explosive events like the 1980 eruption of Mount St. Helens in Washington and the 1982 eruption of El Chichón in Mexico, each of which caused substantial destruction and loss of life. The differences between nonexplosive and explosive eruptions depend largely on magma viscosity and dissolved-gas content. Nonexplosive eruptions are favored by low-viscosity magmas and low dissolved-gas contents.

Even nonexplosive eruptions may appear violent during their initial stages. Gas bubbles in a low-viscosity basaltic magma will rise rapidly upward, like the gas bubbles in a glass of soda water. If a basaltic magma rises rapidly, which means the pressure drop will be fast, gas can bubble so rapidly out of solution that spectacular fountaining will occur (Fig. 4.4). Bits of falling lava spatter when they strike the ground and can pile up as a *spatter cone* or *spatter rampart* beside the vent. When fountaining dies down, hot, fluid lava emerging from the vent will flow rapidly downslope (Fig. 4.5). Because heat is lost quickly at the top of a flow, the surface forms a crust beneath which the liquid lava will continue to flow downslope along well-defined channels. These enclosed lava tubes inhibit upward loss of heat and enable low-viscosity lava to move along just below the surface for great distances from the vent. As the lava cools and continues to lose dissolved gases, its viscosity increases and the character of flow changes. The very fluid lava initially forms thin pahoehoe flows, but with increasing viscosity the rate of movement slows and the stickier lava may be transformed into a rough-surfaced aa flow that now moves very slowly. Thus, during a single, nonexplosive, Hawaiian-type eruption, spatter cones, pahoehoe, and aa may be formed from the same batch of magma.

Figure 4.4 Taking a Volcanic Temperature Fountaining starts an eruption of Kafla, a basaltic volcano in Iceland. Use of a telephoto lens foreshortens the field of view. The geologist in a protective suit has been making measurements several hundred meters away from the fountain.

Figure 4.5 Fast-Flowing Lava This stream of low-viscosity, basaltic lava moving smoothly away from an eruptive vent demonstrates how fluid and free-flowing lava can be. The initial temperature of the lava was about 1100°C. The eruption occurred in Hawaii in 1983.

Figure 4.6 Captured Bubbles Vesicular basalt, China Lake, California.

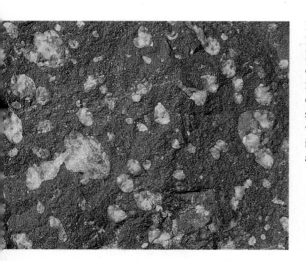

Figure 4.7 Filled Bubbles Holes Amygdaloidal basalt formed when secondary minerals such as calcite fill vesicles. The specimen is 4.5 cm across.

As a basaltic lava cools and the viscosity rises, the gas bubbles find it increasingly difficult to escape. When the lava finally solidifies to rock, the last-formed bubbles become trapped and their forms preserved. These bubble holes are called *vesicles*, and the texture they produce in an igneous rock is said to be *vesicular* (Fig. 4.6). Vesicular basalts are common. In many vesicular basalts the vesicles have been later filled with calcite, quartz, or some other mineral deposited by heated groundwater. Vesicles filled by secondary minerals are *amygdules* (Fig. 4.7).

Explosive Eruptions

In viscous andesitic or rhyolitic magmas, gas bubbles can rise only very slowly. As such a magma moves upward toward the surface, the change in confining pressure causes gas to form and expand within the magma, which can lead to explosive eruptions. When confining pressure drops quickly, the gas in a magma can expand into a froth of innumerable, glass-walled bubbles called *pumice*. Some pumice has such a low density that it will float in water. Beaches on mid-Pacific Islands, for example, are often littered with pumice that has floated in on currents from distant volcanic eruptions. In many instances, instead of forming pumice, small bubbles expanding within a huge mass of sufficiently gas-rich, viscous magma will shatter the magma into tiny fragments called *volcanic ash*. Volcanic ash is the most abundant product of explosive eruptions.

The word *ash* is somewhat misleading because it means, strictly, the solid matter left after something flammable, such as wood, has burned. However, the fine debris thrown out by volcanoes looks so much like true ash that it has become customary to use the word for this material also.

When little or no dissolved gas is present, a magma will be erupted as a lava flow regardless of composition. If dissolved gas is present, however, it must escape somehow, and the higher the viscosity, the greater the likelihood that the escaping gas will cause an explosive eruption.

Pyroclasts and Tephra

A fragment of rock ejected during a volcanic eruption is called a **pyroclast** (named from the Greek words *pyro*, meaning heat or fire, and *klastos*, meaning broken; hence, hot, broken fragments). Rocks formed from pyroclasts are **pyroclastic rocks**. Geologists also commonly refer to a deposit of pyroclasts as **tephra**, a Greek term for ash that was originally used by Aristotle and was revived in recent years by the Icelandic volcanologist Sigurdur Thorarinsson. Tephra is employed as a collective term for all airborne pyroclasts, including fragments of newly solidified magma as well as fragments of older broken rock. It includes individual pyroclasts that fall directly to the ground and those that move over the ground as part of a hot, moving flow. The terms used to describe tephra particles of different sizes—*bombs*, *lapilli*, and *ash*—are listed in Table 4.1 and illustrated in Figure 4.8.

Eruption Columns and Tephra Falls

The largest and most violent explosive eruptions are associated with silica-rich magmas having a high dissolved-gas content. As the rising magma approaches the surface, rapid decompression causes the gases to form bubbles and expand; this, in turn, produces a violent upward thrust of a dense mixture of hot gas and tephra. The hot, turbulent mixture rises rapidly in the

TABLE 4.1	Names for Tephra and Pyroclastic Rock	
Average Particle Diameter (mm)	Tephra (unconsolidated) material	Pyroclastic Rock (consolidated) material
>64	Bombs	Agglomerate
2–64	Lapilli	Lapilli tuff
<2	Ash	Ash tuff

Figure 4.8 Tephra of Various Sizes A. Large spindle-shaped bombs up to 50 cm in length cover the surface of a tephra cone on Haleakala Volcano, Maui. B. Intermediate-sized tephra called lapilli cover the Kau Desert, Hawaii. The coin is about 1 cm in diameter. C. Volcanic ash, the smallest-sized tephra, covers leaves in a garden in Anchorage, Alaska, following an eruption of Mount Spurr.

A

B

cooler air above the vent to form an *eruption column* that may reach as high as 45 km in the atmosphere (Fig. 4.9). The rising, buoyant column is driven by heat energy released from hot, newly formed pyroclasts. At a height where the density of the material in the column equals that of the surrounding atmosphere, the column begins to spread laterally to form an anvil-shaped cloud of the type so familiar in thunderstorm clouds.

As an eruption cloud begins to drift with the upper atmospheric winds, the particles of debris fall out and eventually accumulate at the ground surface as tephra deposits. During exceptionally explosive eruptions, tephra can be spread over distances of 1000 to 2000 km in sufficient amounts to form noticeable accumulations of ash. Fine volcanic dust is carried much farther. Some eruption columns reach such great heights that high-level winds are able to transport the fine debris and associated sulfur-rich gas completely around the world. Such atmospheric pollution, by blocking incoming solar energy, can lower average temperatures at the land surface by as much as 1°C for a period of a year or longer and cause spectacular sunsets as the Sun's rays are refracted by the airborne particles.

Animation
Pyroclastic

Pyroclastic Flows
A hot, highly mobile flow of tephra that rushes down the flank of a volcano during a major eruption is called a **pyroclastic flow**. These are among the most devastating and lethal forms of volcanic eruptions. Analysis of the worldwide geologic record of prehistoric pyroclastic flows shows that they can travel 100 km or more from source vents, and historic observations indicate that they can reach velocities of more than 700 km/h. One of the most destructive, on the Caribbean island of Martinique in 1902, rushed down the flanks of Mount Peleé Volcano and overwhelmed the city of St. Pierre, instantly killing some 29,000 people. Such a pyroclastic flow can be caused by the explosion of a mass of hot magma near the top of the volcano which produces a denser-than-air mixture of hot gases and ash, often with some larger pyroclastic materials as well. Geologists call the resulting poorly sorted deposit an *ignimbrite*. Pyro-

C

clastic flows can also be generated by the partial or continuous collapse of an eruption column. During the 1980 eruption of Mount St. Helens, for example, a number of hot (850°C) pyroclastic flows, likely caused by column collapse, traveled up to 8 km down the north side of the mountain and covered an area of about 15 km².

Lateral Blasts

The 1980 eruption of Mount St. Helens, which was briefly discussed in the opening essay of this chapter, displayed many of the features of a typical large explosive eruption. Nevertheless, the magnitude of the event caught geologists by surprise. The events leading to this eruption are shown diagrammatically in Figure 4.10. As magma moved upward under the volcano, the north flank of the mountain began to bulge upward and outward. Finally, on May 18, 1980, a strong earthquake caused a large mass of rock on the unstable slope to break loose and quickly move downslope as a gigantic landslide. The landslide exposed the mass of hot magma in the core of the volcano. With the lid of rock removed, the gas-rich magma underwent such rapid decompression that a mighty blast resulted which blew a mixture of magmatic gas and ash, along with pulverized rock, sideways as well as upward.

Within the devastated area that extends as much as 30 km from the Mount St. Helens crater and covers some 600 km², trees in the formerly dense forest were blasted to the ground and covered with hot debris. Although Mount St. Helens provides the best documented recent example of a lateral blast, a closely similar 1956 eruption of Bezmianny Volcano in Kamchatka produced a devastating directed blast, a high eruption column, and associated ash flows. For a discussion of the return of life to the devastated area around Mount St. Helens, see the boxed essay titled *After the Blast: Life Returns to Mount St. Helens!*

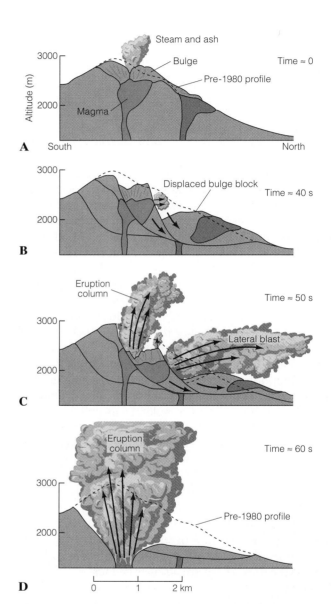

Figure 4.9 Plinian Eruption Column An eruption column of hot gas and fine tephra rising above the summit of Mount St. Helens during the eruption event of May 1980. The cloud, a mixture of hot gas and tiny pyroclasts, rises, expands, and cools. When the column density equals that of the surrounding air, the column stops rising and starts spreading sideways.

Figure 4.10 Eruption of Mount St. Helens Sequence of events leading to the eruption of Mount St. Helens in May 1980. A. Earthquakes and then puffs of steam over a period of several months indicated that magma was rising; a small crater formed, and the north face of the mountain bulged alarmingly. B. On the morning of May 18, an earthquake shook the mountain, the bulge broke loose, and a rock mass slid downward. This reduced the pressure on the magma and initiated the lateral blast that killed David Johnston. C. The violence of the eruption caused a second block to slide downward, exposing more of the magma, and initiated an eruption column. D. The eruption increased in intensity. The eruption column carried volcanic ash as high as 19 km into the atmosphere.

Understanding Our Environment

After the Blast: Life Returns to Mount St. Helens

On May 18, 1980, Mount St. Helens blew up. The volcano had been dormant for over a century, but on that morning a magnitude 5.2 earthquake triggered an explosive eruption. The north face of the mountain collapsed, and the resulting avalanche buried Spirit Lake and the headwaters of the Toutle River. Hot pumice and ash flowed down the flanks of the mountain; mudflows swept along nearby rivers. Winds with speeds up to 1000k/h and temperatures of 600 degrees Celsius blew down trees and scorched the soil. More than 620 square kilometers of forest were leveled, and the ground near the volcano was covered with more than three feet of ash.

It was not the first time Mount St. Helens had devastated its surroundings. Over the course of many millennia the volcano has spewed out millions of tons of lava, pumice, and ash, enough to form the mountain that erupted in 1980. Evidence of those eruptions can be seen in ancient mudflow deposits along riverbanks and layers of ash and pumice along steep roadsides. After each eruption the surrounding ecosystem has slowly recovered. Today it is also recovering, but this time scientists are studying the process in great detail.

Perhaps the most significant conclusion of these studies is that natural disturbances like the Mount St. Helens eruption are essential to certain ecological processes: They initiate forest succession and create new habitats. The process begins with the plants and animals that survived the disturbance. At Mount St. Helens these included below-ground dwellers such as pocket gophers and deer mice, ground dwelling insects and spiders, and animals that were hibernating at the time of the eruption. In addition, there were buried roots and bulbs, seedlings and shrubs covered by snowbanks, and the few trees that remained standing.

The process of recovery began when the snow melted and the plants that had been covered by snowbanks started to spread. Seeds in soil that had been buried in snow sprouted. Wind-blown seeds of plants like firewood took root. As the plant life recovered, wildlife gradually returned to the blast zone.

An unexpected effect of the eruption and its aftermath was the creation of a number of different types of habitats, which attracted a mix of species that had not previously lived in the area. This was especially true of birds, with high-country species joining the species that are common to the Great Basin and those seen in lowland meadows and wetlands. Ten years after the blast, willows, alders and cottonwoods growing along streams created a deciduous woodland, and more new bird species began to appear—in particular, "neotropical" migrating species that have been declining in recent decades.

In 1982 Congress created the Mount St. Helens Volcano Monument and placed it under the management of the Forest Service. With an ample food supply and little human disturbance, some species have thrived. Today a wide variety of plant and animal species can be seen in the 110,000-acre monument. Birds have been especially successful, but there are many kinds of mammals as well. Herds of elk and deer have expanded dramatically. Small mammals such as mice and voles are multiplying, followed by their predators— weasels, hawks and the like. While the blast was lethal to most insects, ant colonies survived and resumed their activities soon after the eruption. The most conspicuous plants are small fir and hemlock trees and weedy plant species such as fireweed and lupine.

Scientists view the recovery of vegetation at Mount St. Helens as the first step in a long-term sequence that over the course of the next 200 to 500 years will eventually transform an open, ash-covered blast zone to an old-growth forest. As different habitats are created and filled with new or returning species, ecologists will refine their knowledge and use their findings to develop better strategies for maintaining natural diversity in managed areas.

A

B

Devastation at Mount St. Helens   Life returns to Mount St. Helens.   A. Trees snapped off and killed by the volcanic blast of May 18, 1980.   B. Revegetation by Pacific silver fir and mountain hemlock trees, 19 years after the blast. Meta Lake Trail, Mount St. Helens National Volcanic Monument, Washington.

Figure 4.11 Shield Volcano Mauna Kea, a 4200-m-high shield volcano on Hawaii, as seen from Mauna Loa. Note the gentle slopes formed by highly fluid basaltic lava. The view is almost directly north. A pahoehoe flow is in the foreground on the northeast flank of Mauna Loa.

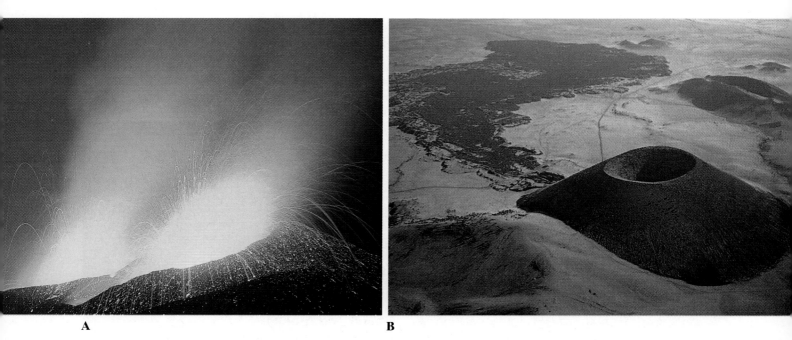

A   B

Figure 4.12 Tephra Cones A. Two small tephra cones forming as a result of an eruption in Kivu, Zaire. Arcs of lights are caused by the eruption of red-hot lapilli and bombs. B. Tephra cone in Arizona built from lapilli-sized tephra. Note the small basaltic lava flow coming from the base of the cone.

VOLCANOES

Shield Volcanoes

The kind of volcano that is easiest to visualize is one built up of successive flows of very fluid lava. Such lavas are capable of flowing great distances down gentle slopes and of forming thin sheets of nearly uniform thickness. Eventually, the pile of lava builds up a **shield volcano**, which is a broad, roughly dome-shaped formation with an average surface slope of only a few degrees (Fig. 4.11).

The slope of a shield volcano is slight near the summit because the magma is hot and very fluid; it will readily run down a very slight slope. The further the lava flows down the flank, the cooler and more viscous it becomes and the steeper a slope must be in order for it to flow. Observe that a steepening of the slope of Mauna Kea can be seen in Figure 4.11. Slopes on young, growing shield volcanoes, such as Kilauea in Hawaii, typically range from less than 5° near the summit to 10° on the flanks.

Shield volcanoes are characteristically formed by the eruption of basaltic lava—the proportions of ash and other tephra are small. Hawaii, Tahiti, Samoa, the Galapagos, and many other oceanic islands are the upper portions of large shield volcanoes.

Tephra Cones

When gases continually bubble out of a magma that has become exposed within a volcanic vent, a relatively gentle shower of pyroclastic debris may continue for many days. As the debris showers down, a **tephra cone** builds around the vent (Fig. 4.12). The slopes of tephra cones are typically about 30° because the loose, coarse pyroclastic fragments of which they are typically composed will not be stable on a steeper slope.

Stratovolcanoes

Large, long-lived volcanoes, particularly those of andesitic composition, emit a combination of lava flows and tephra. **Stratovolcanoes** are defined as volcanoes that emit both tephra and viscous lava and build up steep conical mounds. The volume of tephra may equal or exceed the volume of the lava.

The slopes of stratovolcanoes, which may be thousands of meters high, are steep like those of tephra cones. Near the summit of a stratovolcano, the slope may be as steep as 40°. Toward the base, the slopes of stratovolcanoes flatten to about 6° to 10°. The steep slopes near the summit of a stratovolcano are due in part to the short, viscous lava flows that are erupted and in part to the tephra. The lava flows are a major distinguishing factor between tephra cones and stratovolcanoes. As a stratovolcano develops, lava flows act as a cap to slow down erosion of the loose tephra, and thus the volcano becomes much larger than a typical tephra cone.

The beautiful, steep-sided cones of stratovolcanoes are among Earth's most picturesque sights (Fig. 4.13). The snow-capped peak of Mount Fuji in Japan has inspired poets and writers for centuries. Mount Rainier

Figure 4.13 Classic Stratovolcano Mount Fuji, Japan, a snow-clad giant that towers over the surrounding countryside, displays the classic profile of a stratovolcano.

A

B

Figure 4.14 Volcanic Landscapes Features unique to volcanic terrains. A. Tephra cones on the flanks of Mauna Kea, Hawaii. The cone in the foreground is called Nohonaohae, and the large shield volcano in the background is Kohala. Nohonaohae was built about 20,000 years ago. B. Volcanic gas streams from several vents following a 1965 fissure eruption in Hawaii. Sulfur condensed from the acrid-smelling gases has killed vegetation and covers the ground with a yellow coating of elemental sulfur.

Figure 4.15 Where the Geyser Got Its Name The great Geysir in Iceland, from which all geysers take their name.

and Mount Baker in Washington and Mount Hood in Oregon are majestic examples in North America.

Craters, Calderas, and Other Volcanic Features

Certain features give volcanic terrains a unique character. Fractures may split the flanks of a large shield volcano or stratovolcano so that flank eruptions of lava and/or tephra can occur. Small cones of spattered lava or tephra may then develop above the fractures, peppering the slope of the main volcano like so many small pimples (Fig. 4.14A).

Gases that bubble up from magma far below the surface may emerge either from a central volcanic vent or from small, satellite vents. The emitted gases tend to be mostly water vapor, but a certain amount of evil-smelling sulfurous gas may be present too. Such emerging gases can alter and discolor the rocks with which they come into contact (Fig. 4.14B).

When active volcanism finally ceases, rock in and near an old magma chamber may remain hot for hundreds of thousands of years. Descending groundwater that comes into contact with the hot rock becomes heated and tends to rise again to the surface along rock fractures to form *thermal springs*. Thermal springs at many volcanic sites in, for example, Italy, Iceland, Japan, and New Zealand, have become famous health spas and also sources of energy (see the boxed essay on *Harnessing the Heat Within*).

A thermal spring equipped with a system of plumbing and heating that causes intermittent eruptions of water and steam is a *geyser*. The name comes from the Icelandic word *geysir*, meaning to gush, for Iceland is the home of many geysers (Fig. 4.15). Most of the world's geysers outside Iceland are in New Zealand and in Yellowstone National Park.

Video

Geysers

Craters

Near the summit of most volcanoes is a **crater**, a funnel-shaped depression opening upward from which gases, tephra, and lava are ejected.

Calderas

Many volcanoes, both shield and stratovolcanoes, are marked near their summit by a striking and much larger depression than a crater. This is a **caldera**, a roughly circular, steep-walled basin a kilometer or more in diameter. Calderas originate through collapse following eruption and the partial emptying of a magma chamber. Rapid ejection of magma during a large lava or tephra eruption can leave the magma chamber empty or partly empty. The unsupported roof of the emptying chamber sinks slowly under its own weight, like a snow-laden roof on a shaky barn, dropping downward on a ring of steep vertical fractures. Subsequent volcanic eruptions commonly occur along these fractures, thus creating roughly circular rings of small cones. Crater Lake, Oregon, occupies a circular caldera 8 km in diameter, formed during a great tephra eruption about 6600 years ago (Fig. 4.16). The volcano that erupted has been posthumously called Mount Mazama. The tephra can still be seen at many places in Crater Lake National Park and over a vast area of the northwestern United States and adjacent parts of Canada (Fig. 4.17). During the outpouring of about 75 km^3 of tephra, what remained of the roof collapsed

into the magma chamber (Fig. 4.18). Yellowstone National Park contains several overlapping giant calderas that formed during gigantic tephra eruptions 2.0, 1.2, and 0.6 million years ago.

Resurgent Domes

A volcano does not necessarily cease activity following the formation of a caldera. Often, more magma enters the chamber and in the process causes the uplifting of the collapsed floor of a caldera to form a structural

Figure 4.17 Remains of a Long-ago Eruption The Pinnacles, Crater Lake National Park. Striking erosional forms developed in the thick tephra blanket left by the eruption of Mount Mazama 6600 years ago.

Figure 4.16 Crater Lake Crater Lake in Oregon occupies a caldera 8 km in diameter that crowns the summit of what remains of a once lofty stratovolcano, posthumously called Mount Mazama. Crater Lake is the deepest lake in North America.

Using Resources

Harnessing the Heat Within

Iceland is an unusual place. It lies even farther north than the northernmost part of Hudson Bay, and, as the name suggests, it's a very cold place. But Iceland straddles the Mid-Atlantic Ridge, and so it is also a place of active volcanism. The people who live in Iceland have found many clever ways to harness volcanic heat in order to combat the cold climate. Using water warmed by hot volcanic rocks, they heat their houses, grow tomatoes in hothouses even though the temperature outside may be below freezing, and swim year-round in naturally heated pools. Icelanders also generate most of the electricity they need by using volcanically produced steam. Everywhere, the Earth's temperature increases with depth. In theory, at least, everyone, not just Icelanders, should be able to drill water circulation holes deep enough so that the Earth's internal heat could be tapped. In practice it turns out to be very difficult to do so. In order to be used efficiently, geothermal steam should be 200°C or hotter. In most parts of the world, holes must be 5 to 7 km deep in order to reach rock that is 200°C. For the practical development of *geothermal energy*, as the Earth's store of heat is called, we need to be able to reach rock temperatures of 200°C or higher within 3 kilometers of the surface. So far at least, it is also necessary for there to be a natural reservoir of underground water that is heated by the hot rocks, as shown in Figure B4.1. The places on the Earth where these conditions are met are places of recent volcanic activity. Most of the world's volcanic and magmatic activity is close to plate margins, and it is here, in places such as New Zealand, the Philippines, Japan, Italy, Iceland, and the western United States, that geothermal power is being used.

Unfortunately, the total amount of energy that can be recovered from natural geothermal fields is not very large. Today geothermal power is of local importance only. It is an interesting question whether the situation may change. Many volcanoes, like Mount St. Helens, are too active and too dangerous to be considered as geothermal energy sources. Some places, like Yellowstone National Park, could produce a large amount of energy, but if we drilled and pumped out the steam and hot water reservoirs beneath Yellowstone, the famous hot springs and geysers would soon be dry.

Interesting geothermal experiments are now being conducted in New Mexico. The goal is to create our own geothermal fields. In the Jemez Mountains on the edge of an extinct (but still hot) volcano, scientists drilled two holes deep into the hot rock, as shown in Figure B4.2. Then they shattered the hot rock with explosives to create an artificial reservoir and pumped water through to produce steam. The first tests have been only partly successful. A major difficulty was that water did not flow uniformly through the hot rock but instead flowed mostly through the wider flow paths, and the rocks that lined them soon cooled down. Further tests are also proceeding not only in the Jemez Mountains but also in France, England, and other countries.

Hot dry rocks are much more abundant than geothermal steam fields. If the water flow troubles encountered in the Jemez Mountains experiments can be overcome, geothermal energy may someday play a major role in meeting the energy needs of society.

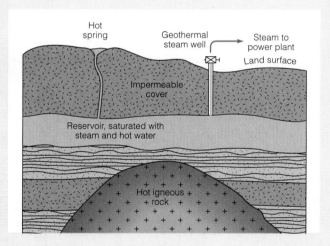

Figure B4.1 Geothermal Reservoir   A typical geothermal steam reservoir. Water in a permeable aquifer, such as a tuff, is heated by magma or hot igneous rock. As steam and hot water are withdrawn through the well, cold water flows into the reservoir through the aquifer.

---

dome. Such a feature is called a resurgent dome. Subsequently, small tephra cones and lava flows build up in the interior of the caldera. Wizard Island, in Crater Lake, is a cone with such an origin.

Lava Domes

When lava is extruded following a major volcanic eruption, it tends to have very little dissolved gas left in it. If the lava is very viscous, it squeezes out to form a *lava dome*. A growing lava dome more than 200 m high now

sits at the center of the crater of Mount St. Helens (Fig. 4.19).

Fissure Eruptions

Some lava reaches the surface via elongate fissures through the crust. Extrusion of lava along an extended fracture is called a *fissure eruption* (Fig. 4.20). Such eruptions, which are often very dramatic, are characteristically associated with basaltic magma, and the lavas that emerge from a fissure eruption on land tend

9336 people died. There is good evidence to prove that larger eruptions have occurred in prehistoric times. The Roza flow, a great sheet of basaltic lava in eastern Washington State, can be traced over 22,000 km^2 and shown to have a volume of 650 km^3.

Pillow Basalts

The most extensive volcanic system on the Earth lies beneath the sea. There the fissures that split the centers of the midocean ridges are channelways for the rising basaltic magma that forms new oceanic crust at the spreading margins of plates of lithosphere.

Seawater cools the basaltic magma so rapidly that a very distinctive lava form is observed. Close to a submarine volcanic fissure, where the lava temperature is highest, thin lava flows with rapidly quenched, glassy surfaces form. The flows build up piles of basalt in which each sheet may only be 20 cm or so thick. Farther away from a vent, where the temperature of the flowing lava has decreased, pillow structure develops. The term **pillow basalt** describes a structure characterized by discontinuous, pillow-shaped masses of basalt, ranging in size from a few centimeters to a meter or more in greatest dimension (Fig. 4.21).

Pillow structure forms when the surface of the basaltic lava is chilled quickly. The brittle, chilled surface cracks, making an opening for the still molten magma inside to ooze out like a strip of toothpaste. In turn, the newly oozed strip chills, its surface cracks, and the process continues. The end result is a pile of lava pillows that resemble a jumbled pile of sandbags. Most of the lavas of the oceanic crust are pillow basalts.

Volcanic Hazards

Volcanic eruptions are not rare events. Every year about 50 volcanoes erupt somewhere on the Earth. Eruptions of basaltic shield volcanoes are usually not dangerous but tephra eruptions from andesitic or rhyolitic stratovolcanoes, such as Mount St. Helens and Krakatau, can be disastrous. Millions of people live on or close to stratovolcanoes, and eruptions present five kinds of hazards.

1. Hot, rapidly moving pyroclastic flows and laterally directed blasts may overwhelm people before they can run away. The tragedies of Mount Pelée in 1902 and Mount St. Helens in 1980 are examples.

2. Tephra and hot poisonous gases may bury or suffocate people. Such a tragedy occurred in A.D. 79 when Mount Vesuvius, a supposedly dormant volcano in southern Italy, burst to life. Hot, poisonous gases killed people in the nearby Roman city of Pompeii and then tephra buried them (Fig. 4.22).

3. Tephra can be dangerous long after an eruption has ceased. Rain or meltwater from snow can loosen

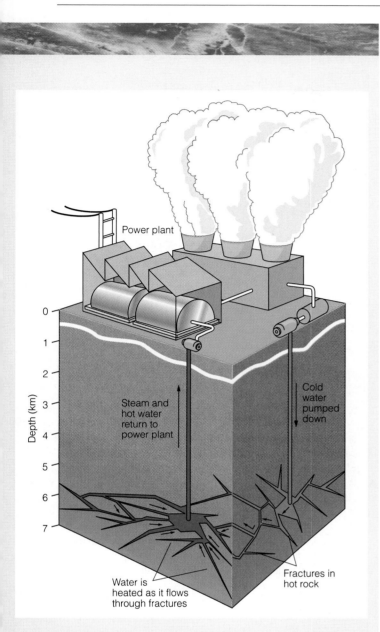

Figure B4.2 Power from Hot Rocks Geothermal energy from hot dry rocks. Cold water is pumped through fractures at the bottom of a 6-km-deep well, becomes heated, then flows back up to geothermal power plant, where heat energy is converted to electricity.

to spread widely and to create flat lava plains called **plateau basalts**. An eruption in Iceland in 1783, known as the Laki eruption, occurred along a fracture 32 km long. Lava flowed 64 km outward from one side of the fracture and nearly 48 km outward from the other side. Altogether it covered an area of 588 km^2. The volume of the lava extruded was 12 km^3, making this the largest lava flow in historic times. It was also one of the most deadly. The flow destroyed homes and food supplies, killed stock, and covered fields. Famine followed and

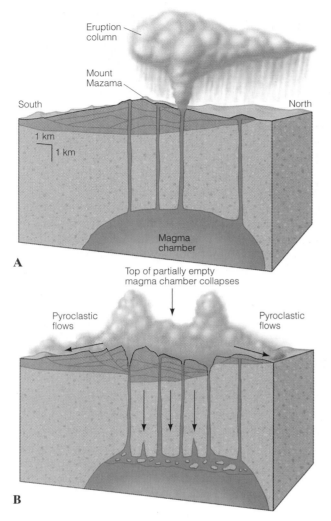

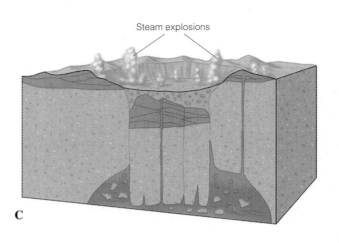

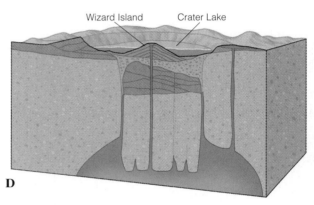

Figure 4.18 The Eruption of 6600 Years Ago Sequence of events that formed Crater Lake following the eruption of Mount Mazama 6600 years ago. A. Eruption column of tephra rises from the flank of Mount Mazama. B. The eruption reaches a climax. Dense clouds of ash fill the air, the hot pyroclastic flows sweep down the mountain side as the central part of the mountain subsides into the emptying magma chamber. It was at this stage that the deposits shown in Figure 4.17 were formed. C. The top of Mount Mazama collapses still further into the magma chamber, forming a caldera 8 km in diameter. D. During a final phase of eruption, Wizard Island formed. The water-filled caldera is Crater Lake, shown in Figure 4.16.

Figure 4.19 Lava Dome A lava dome in the crater of Mount St. Helens, Washington, in May 1982. The plume rising above the dome is steam.

tephra piled on a steep volcanic slope and start a deadly mudflow sweeping down the mountainside. In 1985, following a small and otherwise nondangerous eruption of the Colombian Volcano Nevado del Ruíz, massive mudflows formed when glaciers at the summit melted, moved swiftly down the mountain, and killed 20,000 people.

4. Violent undersea eruptions can cause giant sea waves called *tsunami*. Set off by the eruption of Krakatau, tsunamis killed more than 36,000 coast dwellers on Java and other Indonesian islands.

Figure 4.20 Fissure Eruption Aerial view of a fissure eruption, Mauna Loa, Hawaii, in 1984. Basaltic lava is erupting from a series of parallel fissures. Note tephra cone (upper left) formed during an earlier eruption.

5. A tephra eruption may wreak such havoc on agricultural land and livestock that people die from famine.

Since A.D. 1800 there have been 18 volcanic eruptions in which a thousand or more people died (Table 4.2). It is certain that other violent and dangerous eruptions will occur in the future. To some extent, volcanic hazards can be anticipated. Geologists who study volcanoes gather data before, during, and after eruptions so that they can be better prepared to anticipate future eruptions. Geologists who are so prepared can then advise civil authorities when to implement hazard warnings and when to move endangered populations to areas of lower risk.

Figure 4.21 Lava on the Seafloor Tubular-shaped pillows of basalt photographed in the central rift of the east Pacific Rise.

Figure 4.22 Ancient Volcanic Disaster Evidence of an ancient disaster. Casts of bodies of five citizens of Pompeii, Italy, who were killed during the eruption of Mount Vesuvius in A.D. 79. Death was caused by poisonous gases, then the bodies were buried by lapilli. Over the centuries the bodies decayed away but the body shapes were imprinted in the tephra blanket. When excavators discovered the imprints, they carefully recorded them with plaster casts.

IGNEOUS ROCK

All igneous rocks are formed by the cooling and solidification of magma, but important differences exist between **extrusive igneous rocks**, which are formed by solidification of lava and tephra on the Earth's surface, and **intrusive igneous rocks**, which are formed by solidification of magma within the crust. Both extrusive and intrusive igneous rocks are classified and named on the basis of rock texture and the mineral assemblage.

Texture

The two most obvious textural features of an igneous rock are the size of the mineral grains of which it is comprised and the way the mineral grains are packed together. Size and packing of mineral grains are the two most important features in determining the texture of an igneous rock.

Sizes of Mineral Grains

Intrusive igneous rocks tend to be coarse grained because magma that solidifies in the crust cools slowly and has sufficient time to form large mineral grains. Figure 4.23A is an example of coarse-grained igneous

TABLE 4.2 | Volcanic Disasters Since A.D. 1800 in Which a Thousand or More People Lost Their Lives

Volcano	Country	Year	Fatalities According to Primary Causes			
			Tephra Eruption	Mudflow	Tsunami	Famine
Mayon	Philippines	1814	1,200			
Tambora	Indonesia	1815	12,000			80,000
Galunggung	Indonesia	1822	1,500	4,000		
Mayon	Philippines	1825		1,500		
Awu	Indonesia	1826		3,000		
Cotopaxi	Ecuador	1877		1,000		
Krakatau	Indonesia	1883			36,417	
Soufriere	St. Vincent	1902	1,565			
Mount Peleé	Martinique	1902	29,000			
Santa Maria	Guatemala	1902	6,000			
Taal	Philippines	1911	1,332			
Kelud	Indonesia	1919		5,110		
Merapi	Indonesia	1930	1,300			
Lamington	Papua-New Guinea	1951	2,942			
Agung	Indonesia	1963	1,900			
El Chichón	Mexico	1982	1,700			
Nevado del Ruíz	Colombia	1985		25,000		

A   **B**

Figure 4.23 Phaneritic Igneous Rock Large, clearly visible mineral grains are characteristic of phaneritic igneous rocks. A. Quartz (light gray), potassium feldspar (pink), and plagioclase (white) are clearly visible in this sample of granite from Death Valley National Park, California. The pink potassium feldspar crystals are about 1 cm in length. B. Huge crystals in a pegmatite. In this outcrop of pegmatite of granite composition from the Black Hills, North Dakota, very large crystals of biotite (dark), potassium feldspar (pink), and quartz (white) are visible. Potassium feldspar grains are 10 cm or more in length.

rock. A coarse-grained rock is called a **phanerite**, meaning that its individual mineral grains are readily visible to the unaided eye. Practically, that means that the mineral grains are at least 2 mm in their largest dimension.

An intrusive igneous rock that contains unusually large mineral grains is called a *pegmatite* (Fig. 4.23B). The term is used for rocks in which average grain diameters are 2 cm or larger. Sometimes individual mineral grains in pegmatites can be huge. Single grains of muscovite and feldspar several meters across have been reported. Extrusive igneous rocks, by contrast with intrusive igneous rocks, solidify rapidly and tend to be fine-grained or even

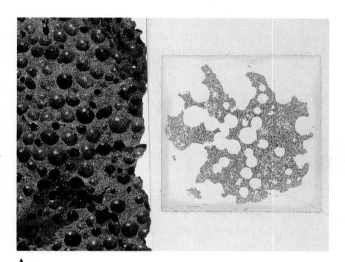

A

B

Figure 4.24 Aphanitic Igneous Rock. Magnifier Needed. In aphanitic igneous rocks, it may not be possible to see individual mineral grains without the help of a microscope. A. Specimen of a vesicular basalt in which mineral grains can barely be resolved. A thin section is adjacent to the specimen and is about 3 cm x 3 cm. B. Photograph of an area of the thin section, 3 mm x 2 mm, as viewed through a microscope. Minerals visible are plagioclase (white), pyroxene (mottled brown), and olivine (dark brown).

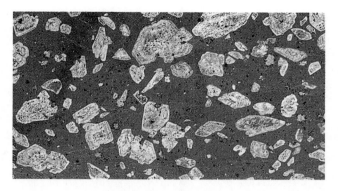

Figure 4.25 Porphyritic Texture Porphyritic texture is characterized by a bimodal distribution in the sizes of the mineral grains. This photo of a porphyry from Oslo, Norway, shows large phenocrysts of feldspar in a fine-grained matrix. The rock is a porphyritic andesite, and the field of view is about 8 cm across.

glassy. Fine-grained igneous rocks are called **aphanites**, meaning that individual mineral grains can only be seen clearly by using some sort of magnification (Fig. 4.24A and B). Practically, that means that the mineral grains are less than 2 mm in the maximum dimension.

One special igneous rock texture involves a distinctive mixture of large and small grains. Rock of such a texture is called a **porphyry**, meaning an igneous rock consisting of coarse mineral grains scattered through a mixture of fine mineral grains, as shown in Figure 4.25. The isolated large grains in porphyry are called *phenocrysts*, and they form in the same way mineral grains in coarse-grained igneous rocks do—by slow cooling of magma in the crust. The fine-grained groundmass that encloses phenocrysts provides evidence that a partly solidified magma moved quickly upward. In the new setting, the magma cooled rapidly, and as a result the later mineral grains, which form the groundmass, are all tiny. Many extrusive igneous rocks are porphyries.

Glassy Rocks

Lava may sometimes cool and solidify so rapidly that its atoms do not have time to organize themselves into minerals. Glass, a mineraloid, forms instead. Extrusive igneous rocks that are largely or wholly glassy are called **obsidian**. Such rocks display a distinctive fracture pattern on a broken surface. The fracture pattern consists of a series of smooth, curved surfaces like a conch shell (Fig. 4.26). Another common variety of glassy igneous rock is pumice which, as previously discussed, is a mass of glassy bubbles of volcanic origin. Volcanic ash is also mostly glassy, and many smaller pyroclasts are wholly or partly glassy because the fragments of magma cooled so rapidly following eruption.

The Way Mineral Grains Pack in Igneous Rock

A little more than two centuries ago, while making field

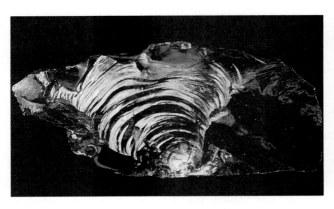

Figure 4.26 Glassy Rock Obsidian from the Jemez Mountains, New Mexico, is almost entirely glassy. The composition is rhyolitic. The curved ridges are typical of fracture patterns observed in glass that is broken by a sharp blow. The specimen is 10 cm across.

Figure 4.27 A Natural Jigsaw Puzzle Photograph of a granite, taken in polarized light through a microscope. Note how the mineral grains interlock to produce a tough structure. This texture is typical of igneous rocks. Minerals visible are muscovite (green), amphibole (yellow), quartz (colorless), and feldspar (gray and pale blue). The specimen is 1 cm across.

observations in Scotland, James Hutton saw that bodies of coarse-grained rock with a distinctive texture intruded and cut across the layering of sedimentary rocks. The texture he observed was an interlocking of randomly oriented mineral grains, like tightly fitting pieces of a jigsaw puzzle. Hutton recognized that the interlocking texture was similar to the texture produced by slow crystallization of molten substances in the laboratory. Hutton concluded that the cross-cutting rock masses must once have been molten and that the distinctive interlocking of numerous grains resulted from solidification of magma.

Almost all phaneritic igneous rocks have the characteristic texture of interlocking mineral grains observed by Hutton (Fig. 4.27).

Mineral Assemblage

Except for those extrusive igneous rocks that are glassy or largely glassy, the assemblage of minerals that forms when magma solidifies is a major factor used in identifying igneous rocks. For a magma of a given composition, the mineral assemblage that forms is the same for intrusive and extrusive igneous rocks. The differences between the two groups are textural. If the texture of an igneous rock has been determined, the name of the kind of rock can then be divided on the basis of its mineral assemblage. To see how this is done, it is convenient to employ the diagram shown in Figure 4.28. All

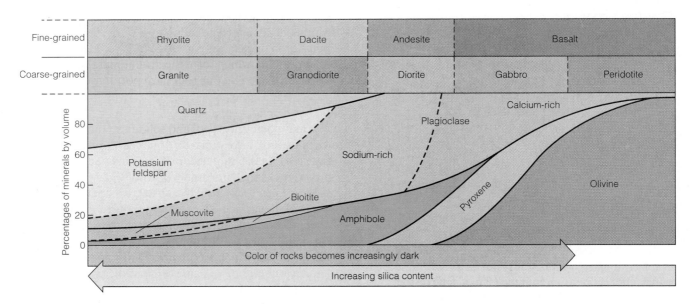

Figure 4.28 Naming Igneous Rocks The proportions of minerals in the common igneous rocks. Boundaries between rock types are not abrupt but gradational, as suggested by the broken lines between rock names. To determine the composition range for any rock type, project the broken lines vertically downward; then estimate the percentages of the minerals by means of the numbers at the edge of the diagram. Quartz- and feldspar-rich rocks are said to be *felsic* and are light in color (see Figure 4.29). Rocks rich in pyroxene, amphibole, and olivine are called *mafic* igneous rocks and are dark in color.

common igneous rocks consist largely of minerals or mineral groups named in Figure 4.28: quartz, feldspar (both potassium feldspar and plagioclase), mica (both muscovite and biotite), amphibole, pyroxene, and olivine. The vertical axis in Figure 4.28 records mineral percentages. When the percentage of each mineral in an igneous rock has been estimated, it will usually be possible to find a vertical line in Figure 4.28 along which those percentages are approximately represented. (Do not expect a perfect match; natural variations in mineral assemblages cannot all be represented on a two-dimensional diagram.) The position of the vertical line will correspond to two rock names. The fine- or coarse-grained texture is selected, whichever is appropriate.

For example, suppose you have a rock specimen that is 30 percent olivine, 30 percent pyroxene, and 40 percent plagioclase. To name this specimen, you estimate the point on Figure 4.28 where the curve separating olivine from pyroxene cuts a horizontal line drawn through 30 percent. Draw a vertical line through the point. You will see that this vertical line cuts the curve between pyroxene and plagioclase at a point close to 60 percent. Olivine (30%) plus pyroxene (30%) equal 60 percent. Plagioclase accounts for the remaining 40 percent. If your rock specimen is coarse grained, it is called gabbro; if fine grained, basalt.

When a rock has a porphyritic texture, we use the name determined by the mineral assemblage as an adjective and the term for the texture of the groundmass for the noun. For example, if the groundmass is fine grained, we will call it a rhyolite porphyry, say, or a dacite porphyry; if the groundmass is coarse grained, we will call it a granite porphyry or a granodiorite porphyry.

Whether a rock is light or dark in color is a useful piece of information. Quartz, feldspar, and muscovite are light; biotite, amphibole, and pyroxene are iron-bearing minerals and are dark. Olivine is also an iron-bearing mineral, but it is not as dark as the others. Note that rocks to the left in Figure 4.28 are light colored and those to the right are increasingly dark. The boundary line between diorite and gabbro is drawn where the dark-colored minerals reach 50 percent of the total and exceed the light-colored feldspars. The boundary is carried through between the fine-grained equivalents of diorite and gabbro—andesite and basalt, respectively—on the basis of color.

Varieties of Intrusive Igneous Rock

Granite and Granodiorite
Feldspar and quartz are the chief minerals in granite and granodiorite (Figs. 4.28 and 4.29). A mica, either muscovite or biotite, is usually present, and many granites contain scattered grains of hornblende.

The name *granite* is applied only to quartz-bearing rocks in which potassium feldspar is at least 35 percent by volume of the total feldspar present. The name *granodiorite* applies to similar rocks in which plagioclase is 65 percent or more of the total feldspar present. Without special equipment the differences in feldspars are not always easily recognized, and in a general study the term *granitic* is extended to this whole group of rocks.

Diorite
The chief mineral in *diorite* is plagioclase. Quartz and mica are usually absent, but either or both amphibole and pyroxene are invariably present (Figs. 4.28 and 4.29). Diorite is a common igneous rock but not as common as granite and granodiorite.

Gabbro and Peridotite
Dark-colored diorite grades into *gabbro* (Figs. 4.28 and 4.29), in which the dark-colored minerals pyroxene and olivine exceed 50 percent of the volume of the rock. A coarse-grained igneous rock in which olivine is the most abundant mineral is called a *peridotite*.

Varieties of Extrusive Igneous Rock

Rhyolite and Dacite
An aphanitic igneous rock with the composition of a granite is a *rhyolite* (Figs. 4.28 and 4.29); if the composition is that of a granodiorite, the aphanitic rock is a *dacite*. Both rhyolites and dacites are quartz-bearing; the difference between them, as with granites and granodiorites, is in the predominance of potassium feldspar in rhyolites and plagioclase in dacite.

Because it is commonly impossible to identify fine-grained feldspars in aphanitic igneous rocks without special microscopes, dacites are very difficult to distinguish from rhyolites. Many geologists, when in doubt, simply call both rocks rhyolites, or even *rhyodacites*.

Fortunately, many rhyolites and dacites have porphyritic textures in which phenocrysts of quartz, and sometimes of feldspar, are set in an aphanitic groundmass. When the phenocrysts are about 5 percent or more of the volume of the rock, the name *rhyolite porphyry* or *dacite porphyry* should be used, as appropriate.

The colors of rhyolites and dacites are always pale—ranging from nearly white to shades of gray, yellow, red, or purple. Many obsidians have the same compositions as rhyolites or dacites. Such obsidians may appear dark, even black, and seem to contradict the rule that igneous rocks with a high silica content are light colored. But rhyolitic obsidian chipped to a thin edge appears transparent with very little color, like slightly smoky glass. The dark appearance results from a small amount of dark mineral matter distributed evenly in the glass.

Silica Content of Magma	Approx. Melting Temperature of Rock	Viscosity of Magma	Color of Mineral Assemblage	Hand Samples of	
				Resulting Plutonic Rocks	Resulting Volcanic Rocks
High (≈70–75%)	Low (≈800°C)	High (not runny)	Felsic (light-colored)	Granite	Rhyolite
Intermediate (≈60%)	Medium (≈1000°C)	Medium	Intermediate	Diorite	Andesite
Low (≈45–50%)	High (≈1200°C)	Low (runny)	Mafic (dark-colored)	Gabbro	Basalt

Figure 4.29 Common Igneous Rocks Comparison of coarse-grained and fine-grained igneous rocks. Note that mafic igneous rocks are darker in color than felsic igneous rocks. Each specimen is about 7 cm in length.

Andesite

An igneous rock similar in appearance to a dacite, but lacking quartz, is an *andesite* (Figs 4.28 and 4.29). Named for the Andes, the major mountain system of western South America, andesite is equivalent in composition to a diorite and is a common rock. Common colors are shades of gray, purple, and even dark green. Many andesites are porphyritic, with phenocrysts of amphibole, pyroxene, or plagioclase, but not quartz. When phenocrysts are abundant, the name andesite porphyry should be used.

Basalt

The dominant rock of the oceanic crust is *basalt*, a fine-grained igneous rock, sometimes porphyritic, that is always dark gray or black (Figs. 4.28 and 4.29). Compositionally equivalent to gabbro, basalt is the most common kind of extrusive igneous rock. Phenocrysts, when present in a basalt porphyry, can be either plagioclase, pyroxene, or olivine.

Varieties of Pyroclastic Rocks

There is a saying among geologists that tephra is igneous when it goes up but sedimentary when it comes down. As a result, pyroclastic rocks are transitional between igneous and sedimentary rocks, and the names of pyroclastic rock types reflect this fact. As we will see in the next chapter, the names of one class of sedimentary rocks are determined by the sizes of the sediment fragments. So, too, with pyroclastic rocks: They are called **agglomerates** when tephra particles are bomb sized, or

tuffs when the particles are either lapilli or ash (Fig. 4.30). As seen in Table 4.1, the appropriate rock names are *lapilli tuff* and *ash tuff*.

Tephra can be converted to pyroclastic rock in two ways. The first, and most common, way is through the addition of a cementing agent such as quartz or calcite introduced by groundwater. Figure 4.30A is an example of a rhyolitic lapilli tuff formed by cementation. The second way tephra is transformed to pyroclastic rock is through welding of hot, glassy, ash particles. When ash is very hot and glassy, the individual particles can fuse together to form a glassy pyroclastic rock. Such a rock is called **welded tuff** (Fig. 4.30B).

PLUTONS

Beneath every volcano there lies a complex of chambers and channelways through which magma reaches the surface. Naturally, we cannot study the magmatic channels of an active volcano, but we can look at ancient channelways that have been laid bare by erosion, as seen in Figure 4.31. What we find is that these

Figure 4.30 Pyroclastic Igneous Rocks Two kinds of tuff. A. Lapilli tuff, formed by cementation of lapilli and ash, from Clark County, Nevada. B. Welded tuff from the Jemez Mountains, New Mexico. The dark patches are glassy pyroclasts that were flattened during welding. Both samples are 4 cm across.

A

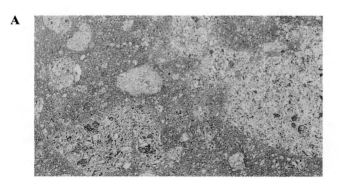

B

A

B

Figure 4.31 Volcanic Neck Shiprock, New Mexico. A. The conical tephra cone that once surrounded this volcanic neck has been removed by erosion. B. Diagram of the way the original volcano may have appeared prior to erosion.

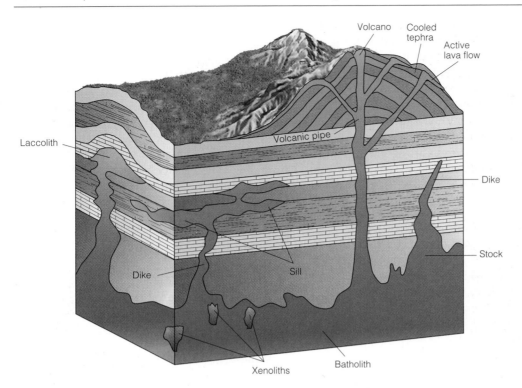

Volcano Cooled tephra

Active lava flow

Volcanic pipe

Laccolith

Dike

Dike

Sill

Stock

Xenoliths

Batholith

Figure 4.32 Plutons
Diagrammatic section of part of the crust showing the various forms assumed by plutons. Many plutons were once connected with volcanoes, and a close relationship exists between intrusive and extrusive igneous rocks.

ancient channelways are filled by intrusive igneous rock because they are the underground sites where magma solidified.

All bodies of intrusive igneous rock, regardless of shape or size, are called **plutons** after Pluto, the Greek god of the underworld. The magma that forms a pluton did not originate where we now find the pluton. Rather, the magma was intruded upward into the surrounding rock from the place where it formed.

Minor Plutons

Plutons are given special names depending on their shapes and sizes. The common plutons are illustrated in Figure 4.32, and each is briefly discussed below.

Dikes

The most obvious and familiar evidence of past igneous activity is a **dike**, a tabular sheetlike body of igneous rock that cuts across the layering of the rock it intrudes (Fig. 4.33A). By sheetlike we mean that the dikes are much more extensive in lateral dimensions than they are thick. Several dikes are visible in the photograph of Shiprock, New Mexico, shown in Figure 4.31. A dike forms when magma squeezes into a fracture. The erupting fissure in Figure 4.20, for example, is probably now filled by a dike of gabbro, formed as a result of the cooling of the basaltic magma that filled the fissure at the end of the eruption.

Sills

Tabular, sheetlike bodies of intrusive igneous rock that are parallel to the layering of the rocks they intrude are called **sills** (Fig. 4.33B). Commonly, dikes and sills occur together as part of a network of plutons, as shown in Figure 4.32. Both dikes and sills can be very large. For example, the Great Dike in Zimbabwe is a mass of gabbro nearly 500 km long and about 8 km wide, with parallel, vertical walls. The Great Dike fills what must once have been a huge fracture in the crust. An example of a large and well-known sill can be seen in the cliffs of the Palisades that line the Hudson River opposite New York City. The Palisades Sill reaches a thickness of about 300 m, and like the Great Dike is gabbro. The sill was intruded between layers of ancient sedimentary rock about 200 million years ago. The sill is visible today because tectonic forces raised that portion of the crust upward and erosion then removed the covering sedimentary rocks.

Laccoliths

A variation of a sill is a *laccolith*, an igneous body intruded parallel to the layering of the rocks it intrudes and above which the layers of the intruded rocks have been bent upward to form a dome.

Volcanic Pipes and Necks

A *volcanic pipe* is the approximately cylindrical conduit that once fed magma upward to a volcanic vent and that became filled with igneous rock or rock fragments after eruption ended. If erosion strips away the surrounding rock, the body of rock that filled a pipe is called a *vol-*

A

B

Figure 4.33 Dike and Sill Sheetlike plutons. A. Dike of gabbro cutting across horizontally layered shales in Hance Rapid, Grand Canyon National Park, Arizona. B. Sill of gabbro (dark brown) intruded parallel to layering of sedimentary rocks above and below, in Big Bend National Park, Texas.

canic neck. Figure 4.31 shows a famous example of a volcanic neck, together with associated dikes, at Shiprock, New Mexico.

Major Plutons

Batholiths

A **batholith** is the largest kind of pluton. It is an intrusive igneous body of irregular shape that cuts across the layering of the rock it intrudes. Most batholiths are composite masses that comprise a number of separate intrusive bodies of slightly differing composition. The differences reflect variations in composition of the magma from which the batholith formed. Some batholiths exceed 1000 km in length and 250 km in width; the largest in North America is the Coast Range Batholith

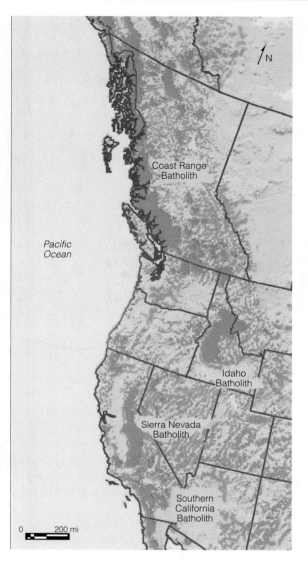

Figure 4.34 Giant Intrusions The Idaho, Sierra Nevada, and Southern California batholiths, largest in the conterminous United States, are dwarfed by the Coast Range Batholith in southern Alaska and British Columbia. Each of these giant batholiths formed from magma generated by the partial melting of continental crust, and each intrudes metamorphosed rocks.

of British Columbia and southern Alaska, which has a length of about 1500 km (Fig. 4.34).

Figure 4.32 does not show what the bottom of a batholith looks like. Where it is possible to see them, the walls of batholiths tend to be nearly vertical. This observation led to a once commonly held perception that batholiths extend downward to great depths—possibly even to the base of the crust. Geophysical measurement and studies of deeply eroded bodies of igneous rock now suggest that this perception is incorrect. Batholiths seem to be at the most only 20 to 30 km thick, which is rather thin compared to their great widths and lateral extents.

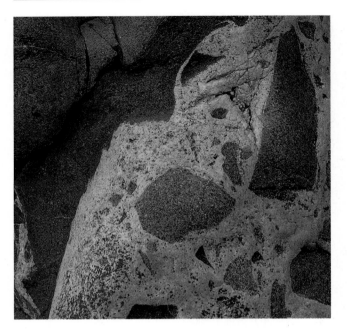

Figure 4.35 Xenoliths Xenoliths dislodged during intrusion of the East Range Granite. The outcrop is along Yakutat Bay, Tongass National Forest, Alaska.

Batholiths do not form by magma squeezing into fractures because there are no fractures large enough to account for batholiths. Despite the huge volumes involved, the magmas that form batholiths do move upward. Even though intruded rocks can be pushed upward by the slowly rising magma, some other process must also operate.

The rising magma apparently dislodges fragments of the overlying rock by a process known as *stoping*. Dislodged blocks are more dense than the rising magma and

will therefore sink. As they sink, those fragments may react with and be partly dissolved by the magma.

Some stoped fragments may dissolve completely, but more commonly such fragments retain their identity even though they partially react with the magma. Some may sink all the way and reach the floor of the magma chamber, but others sink so slowly through the viscous magma that they end up suspended in the solidifying rock of the batholith. Any fragment of rock still enclosed in a magmatic body when it solidifies is known as a *xenolith* (Fig. 4.35) (from the Greek words *xenos*, meaning stranger and *lithos*, meaning stone).

Stocks

Like batholiths, **stocks** are irregularly shaped intrusives no larger than 10 km in maximum dimension. As is apparent from Figure 4.32, a stock may merely be a companion body to a batholith or even the top of a partly eroded batholith.

THE ORIGIN OF MAGMA

We come now to the most difficult but also one of the most interesting questions concerning magmas and volcanoes—how and where do magmas form, and why are there three major kinds of magmas—basaltic, andesitic, and rhyolitic? Many clues to the question of origin can be learned from the distribution of the kinds of volcanoes from which the three different magma types erupt. A summary of present thinking about the distribution of the kinds of volcanoes is presented in Figure 4.36 and in the following discussion.

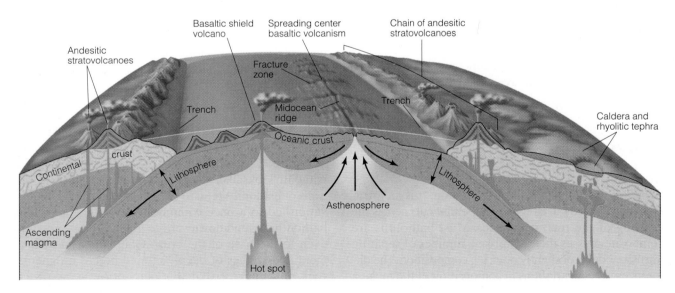

Figure 4.36 Kinds of Volcanoes Diagram illustrating the locations of the major kinds of volcanoes in a plate-tectonic setting.

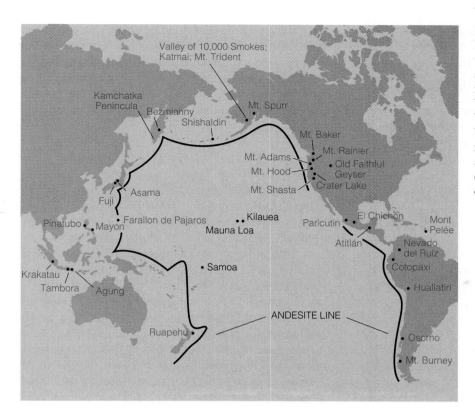

Figure 4.37 Where Andesites Are Found The Andesite Line surrounds the Pacific Ocean basin and separates areas where andesitic magma is not found from areas where it is common. Volcanoes that are inside the Pacific basin, such as Mauna Loa, erupt basaltic magma but not andesitic magma. Those outside the line, such as Mount Shasta and Mount Fuji, may erupt basaltic magma too, but they also erupt andesitic magma.

Distribution of Volcanoes

It has long been known that volcanoes that erupt rhyolitic magma are found abundantly on the continental crust. There are a few places in the ocean where andesitic volcanism is building new crust that is like continental crust, and volcanoes in these areas erupt some rhyolitic magma. These are important observations because they suggest that the processes that form rhyolitic magma must be restricted to the continental crust, including those places in the ocean where new crust of continental character is forming, but they do not occur in oceanic crust. Nor, presumably, do the processes that form rhyolitic magma occur in the mantle, because if they did, the magma would be expected to rise to the surface regardless of the kind of crust above.

Volcanoes that erupt andesitic magma are found on both the oceanic crust and the continental crust. This suggests that andesitic magma forms in the mantle and rises up regardless of the overlying crust. An additional piece of information comes from the geographic distribution of andesitic volcanoes around the Pacific Ocean, shown in Figure 4.37A. A line surrounds the Pacific separating andesitic volcanoes from those that erupt only basaltic lava. The *Andesite Line* is generally parallel to the plate subduction margins shown in Figure 1.15. As we shall see later in this chapter, andesitic magma is probably formed as a result of the heating of old oceanic crust that has been subducted back into the mantle.

Volcanoes that erupt basaltic magma also occur on both the oceanic and the continental crust. The source of basaltic magma, therefore, must also be the mantle. The geographic distribution of basaltic volcanoes does not coincide with a specific kind of crustal feature, however, which suggests that basaltic magma must be formed by the melting of the mantle itself regardless of the kind of overlying crust.

Two observations concerning basaltic volcanoes do suggest something about the origin of basaltic magma, however. First, everywhere along the midocean ridges volcanoes erupt basaltic magma. Midocean ridges coincide with plate spreading margins, so we must consider the possibility that plate motion and the generation of spreading-center magma might somehow be connected. The second observation concerns large basaltic volcanoes that are not located along midocean ridges. An example is found in the volcanoes of Hawaii, which sit on oceanic crust in the middle of the Pacific Plate, far from any plate edges. The Hawaiian volcanoes that are active today, Mauna Loa, Kilauea, and Loihi (a submarine volcano), are just the youngest members of a chain of mostly extinct volcanoes. The volcanic rocks in the Hawaiian chain are progressively older to the northwest.

The Hawaiian volcanic chain is believed to have formed as the Pacific Plate moved slowly northwestward across a midocean hot spot above which frequent and voluminous eruptions built a succession of volcanoes. The hot spot is thought to have been building basaltic volcanoes on the moving plate for at least 70 million years. This must mean that somewhere deep in the mantle there is a long-lived source of hot rock from which the basaltic magma can form. The exact causes of hot-spot magmas are still conjectural, but as we shall see in later chapters, both spreading-center magmas and hot-spot magmas are believed to play important roles in plate tectonics and are caused by convection in the mantle.

The Origin of Basaltic Magma

The first question to be answered when the origin of magma is discussed is whether the rock that melted to form the magma was wet or dry. The presence of water lowers the temperature at which melting begins. Two additional questions must also be answered—what kind of rock melted, and did the rock melt completely or only partially? The kind of rock that melts must obviously control the composition of the magma that forms. Partial versus complete melting is not such an obvious question. A mineral melts at a specific temperature. An assemblage of minerals (a rock) melts over a temperature range of as much as 200° C, and within the melting range a mixture of unmelted mineral grains and magma coexist. The process of forming a magma through the incomplete melting of rock is known as **chemical differentiation by partial melting**. For a more detailed discussion of melting, see *How Rock Melts*.

The dominant minerals found in basalt and gabbro are olivine, pyroxene, and plagioclase. None contains water in their formulas. This fact suggests that basaltic magma is probably either a dry or a water-poor magma. Indeed, all evidence from observations of basaltic lava during eruption suggests that the water content of basaltic magma rarely exceeds 0.2 percent. Since there is not enough water to have much of an effect on the melting of silicate minerals, it may be concluded that basaltic magma originates by some sort of virtually dry partial-melting process, and, as we had earlier concluded, the process must occur in the mantle.

Much debate has centered on the question of the exact chemical composition of the mantle, but it appears that the upper portion contains rocks rich in olivine and garnet called *garnet peridotites*. Laboratory experiments on the dry, partial-melting properties of garnet peridotite show that at pressures and temperatures reached in the asthenosphere (100 km deep), a 10 to 15 percent partial melt is a magma of basaltic composition. It is also important to note that 10 to 15 percent partial melting is the point where a melt is able to separate from the residual solid and move upward. While this leaves unanswered, for the moment, the question of a heat source, and why basaltic magma should develop in some parts of the asthenosphere but not others, we can nevertheless confidently conclude that basaltic magma forms by dry partial melting of rocks in the upper mantle. We will return to the questions of a heat source and why melting occurs in Chapter 17 when convection in the mantle is discussed further.

The Origin of Andesitic Magma

The chemical composition of andesitic magma is close to the average composition of the continental crust. Igneous rocks formed from andesitic magma are commonly found in the continental crust. From these two facts it might be supposed that andesitic magma forms by the complete melting of a portion of the continental crust. Some andesitic magma may indeed be generated in this way, but andesitic magma is also extruded from volcanoes that are far from the continental crust. In those cases, the magma must be developed either from the mantle or from the oceanic crust. Laboratory experiments provide a possible answer.

In the laboratory, wet partial melting of mantle rock under suitably high pressure yields a magma of andesitic composition. An interesting hypothesis suggests how this might happen in nature. When a moving plate of lithosphere plunges back into the asthenosphere, it carries with it a capping of basaltic oceanic crust saturated by seawater. The plate heats up, and eventually the crust starts to release water, which is then thought to promote partial melting in the mantle above the downgoing plate. Wet partial melting that starts at a pressure equivalent to a depth of 80 km produces a melt having the composition of andesitic magma. It is also possible that some of the downgoing oceanic crust may partially melt.

A number of details concerning the wet partial melting of mantle rock and oceanic crust remain to be deciphered, but two pieces of evidence support the idea that much of the andesitic magma forms in this manner. The first concerns active volcanoes and the Andesite Line (Fig. 4.37). The Andesite Line corresponds closely with plate subduction margins. The second piece of evidence comes from the distribution of andesitic volcanoes with respect to the subduction zone. On the seafloor, a subduction zone is marked by the presence of a deep-sea trench. Beyond the trench, the lithosphere of one plate sinks into the asthenosphere, carrying with it its capping of wet oceanic crust. A horizontal distance of about 250 km, equivalent to a depth of about 80 km, marks the edge of a curved belt of volcanoes. The situation is nicely demonstrated by the andesitic stratovolcanoes of Japan, as shown in Figure 4.38.

Measuring Our Earth

How Rock Melts

The making of a magma requires that temperatures be very high, so high, in fact, that the idea of a rock melting is a difficult one for many people to accept. So high, too, that one of the important turning points in the history of geology was the demonstration by a Scot, James Hall, almost 200 years ago, that common rocks can be melted and that such melts have the same properties as magma erupted from volcanoes.

No longer is there a question as to whether or not a rock can melt. Rather, the kinds of questions we ask now are, what kind of rock melted, did all of the rock melt or only part of it, how does temperature increase with depth, and how far below the Earth's surface did melting occur?

As can be seen in Figure 1.14, temperatures beneath the oceanic crust and the continental crust rise to about 1000°C at rather shallow depths (120 km or less) in the upper part of the mantle. Measurements made on lava prove that magma is fluid at 1000°C, so an immediate question is, "Why isn't the Earth's mantle entirely molten?" The answer is that pressure influences melting temperatures. As the pressure rises, the temperature at which a silicate mineral melts also rises. For example, the feldspar albite ($NaAlSi_3O_8$) melts at 1104°C at the Earth's surface, where the pressure is 0.1 Mpa. At a depth of 100 km, however, where the pressure is 35,000 times greater, the melting temperature is 1440°C (Fig. B4.3A). Therefore, whether a particular rock melts and forms a magma at a specified depth in the Earth depends on both the temperature and the pressure at that depth.

The Effect of Water on Melting

The effect of pressure on melting is straightforward if no other substances are present. However, if water is present, under sufficient pressure to contain the water vapor at high temperatures, a complication enters. At any given pressure of water vapor, a mineral will begin to melt at a lower temperature than it would in the absence of water. The effect is the same as that of salt and ice. Salt can melt the ice on an icy road. This happens because a mixture of ice and salt melts to a salt solution at a lower temperature than pure ice melts to water. In the same way, a mineral and water mixture melts to magma at a lower temperature than the melting temperature of the pure mineral. Furthermore, as the pressure rises, the effect of water on the melting temperature increases. This is so because the higher the pressure, the greater the amount of water that will dissolve in the melt. Therefore, the temperature at which the mineral begins to melt in the presence of water decreases as pressure increases (exactly the opposite of what happens with a dry mineral), as Figure B4.3B shows.

A rock that is a composite of several minerals, as nearly all rocks are, does not melt at one specific temperature as a pure mineral would (at a fixed pressure). Instead, as shown in Figure B4.4, a rock melts over a temperature interval. Melting begins at a temperature lower than the temperature at which any of the minerals would melt by itself. The early-formed melt is a mixture of substances from different minerals, but it will contain relatively more of the substance of the mineral that has the lowest melting temperature. Thus, the melt and the mixture of unmelted minerals will almost always differ in composition, and both will differ from the starting composition of the parent rock.

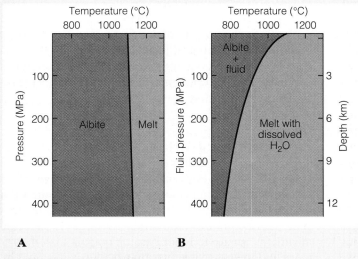

A **B**

Figure B4.3 Effect of Pressure on Melting Influence of pressure on the melting temperature of albite ($NaAlSi_3O_8$). A. Dry-melting curve. Increasing pressure raises the melting temperature. B. Wet-melting curve. Water dissolves in the melt and decreases the melting temperature.

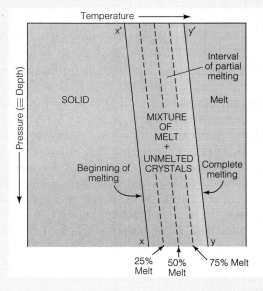

Figure B4.4 Fractional Melting Dry melting of rock containing several kinds of minerals. The pressure effects are similar to those shown in Figure B4.3A. Line x–x' marks the onset of melting, curve y–y' the completion of melting. Between the two lines is a region in which melt and a mixture of unmelted crystals coexist.

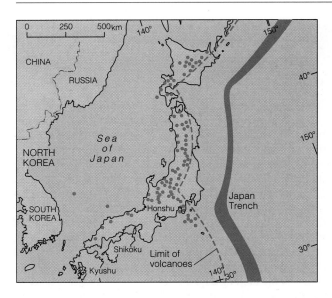

Figure 4.38 Volcanoes in Japan Relations between ocean trenches and arcs of volcanoes erupting andesitic magma. Arc-shaped Japanese islands are parallel to the Japan Trench. Andesitic volcanoes (dots) active during the last million years are also confined behind arc-shaped boundaries.

The Origin of Rhyolitic Magma

Two observations suggest a continental origin for rhyolitic magma:

1. Volcanoes that extrude rhyolitic magma are confined to the continental crust or to regions of andesitic volcanism.

2. Volcanoes that extrude rhyolitic magma give off a great deal of water vapor, and intrusive igneous rocks formed from rhyolitic magma (granite) contain significant quantities of OH-bearing minerals such as mica and amphibole. The OH must have come from water dissolved in the magma.

These two points suggest that the generation of rhyolitic magma probably involves some sort of wet partial melting of rock with the composition of andesite, that is, with the average composition of the continental crust. Laboratory experiments bear out this suggestion. When, in the laboratory, water-bearing rocks having the average composition of the continental crust have partially melted, the composition of the magma that forms is rhyolitic. The source of heat for such melting to occur in the crust is the mantle. The actual situation is thought to be that hot rock, or magma upwelling in the mantle, brings up the heat that causes crustal rock to partially melt.

Once a rhyolitic magma has formed, it starts to rise. However, the magma rises slowly because it has a high SiO_2 content (70%, as we learned earlier) and is therefore very viscous. As it slowly rises, the pressure on the magma decreases. The effectiveness of water in reducing the melting temperature is diminished by reduced

pressure. Unless there is some way to heat it, therefore, a rising magma formed by wet partial melting will tend to solidify and form an intrusive igneous rock underground. But a rising magma traverses cool rock, and there is no source of heat to cause a temperature increase. As a result, the depth-temperature path of a rising body of rhyolitic magma brings it closer and closer to solidification. Therefore, most rhyolitic magma solidifies underground and forms *granitic batholiths* rather than reaching the surface and forming lava or tephra.

SOLIDIFICATION OF MAGMA

Literally hundreds of different kinds of igneous rock can be found. Most are rare, but the fact that they exist suggests an important point: A magma of a given composition can crystallize into many different kinds of igneous rock.

Solidifying magma forms several different minerals, and those minerals start to crystallize from the cooling magma at different temperatures. The process is just the opposite of partial melting, and the temperature interval across which solidification occurs is simply the reverse of the melting interval discussed in the boxed essay and illustrated in Figure B4.4. If at any stage during crystallization the melt becomes separated from the crystals, a magma with a brand-new composition results, while the crystals left behind form an igneous rock with a composition that is quite different from the composition of the magma.

Crystal–melt separations can occur in a number of ways. For example, compression can squeeze melt out of a crystal–melt mixture. Another mechanism involves the sinking of the dense, early crystallized minerals to the bottom of a magma chamber, thereby forming a solid mineral layer covered by melt. However a separation occurs, the compositional changes it causes are called **magmatic differentiation by fractional crystallization**.

Bowen's Reaction Series

It was a Canadian-born scientist, N. L. Bowen, who first recognized the importance of magmatic differentiation by fractional crystallization. Because basaltic magma is far more common than either rhyolitic or andesitic magma, he suggested that basaltic magma may be primary and the other magmas may be derived from it by magmatic differentiation. At least in theory, Bowen argued, a single magma could crystallize into both basalt and rhyolite because of fractional crystallization. Such extreme differentiation rarely happens, we now know, but fractional crystallization is nevertheless an exceedingly important phenomenon in producing a wide range of rock compositions.

Bowen knew that plagioclases crystallized from basaltic magma are usually calcium-rich (anorthitic), whereas those formed from rhyolitic magma are commonly sodium-rich (albitic). Andesitic magma, he observed, tends to crystallize plagioclases of intermediate composition.

Bowen's experiments provided an explanation for these observations. He discovered that even though the composition of the first plagioclase that crystallizes from a basaltic magma is anorthitic, the composition changes toward albitic as crystallization proceeds and the ratio of crystals to melt increases. This means that all the plagioclase crystals, even the ones formed earliest, must be continually changing their compositions as the magma cools. A chemical balance between crystals and melt, if maintained, is referred to as chemical equilibrium.

As explained in Chapter 3, the plagioclases involve a coupled ionic substitution in which $Ca^{2+} + Al^{3+}$ in the crystal are replaced by $Na^{1+} + Si^{4+}$. Bowen recognized that such substitution allows plagioclase to undergo a continuous change in composition in a cooling magma. Bowen called such a continuous reaction between crystals and melt a *continuous reaction series*, by which he meant that even though the composition changed continuously, the crystal structure remained unchanged. The speed at which the change occurs is controlled by the rates at which the four ions can diffuse through the plagioclase structure. To maintain equilibrium, some Ca^{2+} and Al^{3+} must diffuse out of the early formed plagioclase and be replaced by Na^{1+} and Si^{4+} that diffuse into the crystal structure from the melt. Such diffusion is exceedingly slow. Equilibrium is rarely attained because crystals usually grow fast enough that diffusion doesn't maintain equilibrium between the surface and interior of a crystal. As a result, compositionally zoned crystals of anorthite are formed, with anorthite-rich inner zones and albite-rich outer zones. The anorthite-rich inner zones are out of chemical equilibrium with the albite-rich outer zones and with the residual magma.

Bowen also knew that plagioclase grains in many igneous rocks have concentric zones of differing compositions such that the innermost core (and therefore the earliest formed part of the crystal) is anorthitic and successive layers are increasingly albite rich (Fig. 4.39A). He pointed out an important implication for the existence of zoned crystals of plagioclase. When anorthite-rich cores are present, the residual melt is necessarily richer in sodium and silica than it would have been if equilibrium had been maintained. According to Bowen, the anorthite-rich cores are an example of **magmatic differentiation by fractional crystallization**. If, in a partially crystallized magma containing zoned crystals, the melt were somehow squeezed out of the crystal mush, the result would be

A

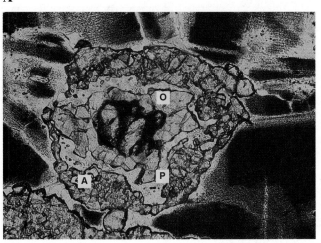

B

Figure 4.39 Proof That Equilibrium Is Not Maintained Textures illustrating Bowen's reaction series. A. Zoned plagioclase crystal in andesite, proof that a continuous reaction occurred but that equilibrium was not maintained. Photographed in polarized light to enhance the zoning. Bands near center are anorthite-rich, progressing to albite-rich at the rim. The crystal is 2 mm long. B. A grain of olivine (O) in a gabbro surrounded by reaction rims of pyroxene (P) and amphibole (A) demonstrates discontinuous reaction. The diameter of the outer rim is 3 mm.

an albite-rich magma and the residue would be an anorthite-rich rock.

Bowen identified several sequences of reactions besides the continuous reaction series of the feldspars. One of the earliest minerals to form in a cooling basaltic magma is olivine. Olivine contains about 40 percent SiO_2 by weight, whereas a basaltic magma contains 50 percent SiO_2. Thus, crystallization of olivine will leave the residual liquid a little richer in silica. Eventually, the solid olivine reacts with silica in the melt to form a more silica-rich mineral, pyroxene (Fig. 4.39B). The pyroxene in turn can react to form amphibole, and then the amphi-

MAGMA COMPOSITION MINERALS FORMING ROCK TYPE

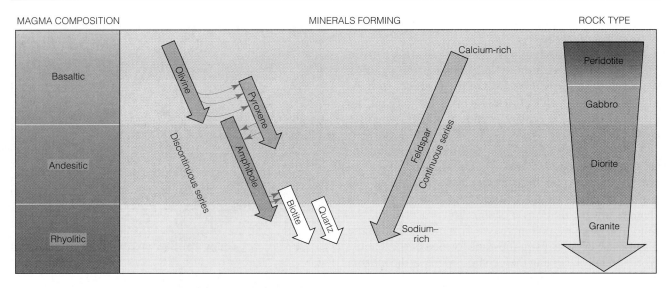

Figure 4.40 Bowen's Reaction Series Bowen demonstrated how the cooling and crystallization of a primary magma of basaltic composition, through reactions between mineral grains and magma, followed by separation of mineral grains from the remaining magma, can change from basaltic to andesitic to rhyolitic. Bowen identified two series of reactions: a continuous series in which plagioclase changes from anorthitic to albitic in composition, and a discontinuous series in which minerals change abruptly from one kind to another, for example, from olivine to pyroxene.

bole can react to form biotite. Such a series of reactions, where early formed minerals form entirely new minerals through reaction with the remaining liquid, is called a *discontinuous reaction series*.

If a core of olivine is shielded from further reactions by a rim of pyroxene, the remaining liquid will be more silica-rich than it would be if equilibrium were maintained and all the olivine were converted to pyroxene. Bowen reasoned that if partial reactions occurred in both continuous and discontinuous reaction series, differentiation by fractional crystallization in a basaltic magma could eventually produce a residual magma with a rhyolitic composition (Fig. 4.40).

The answer to the question that Bowen posed—whether large volumes of rhyolitic magma can form from basaltic magma by fractional crystallization—is negative. The problems with Bowen's idea are (1) if equilibrium is maintained, crystallization of a magma is complete long before a residual magma is siliceous enough to have a rhyolitic composition, (2) calculations show that only a tiny percentage of the volume of a basaltic magma could ever be differentiated to rhyolitic magma, and (3) rhyolitic magma forms in the continental crust. If rhyolitic magma formed by fractional crystallization of basaltic magma, we would surely expect to find a lot of rhyolite in the oceanic crust, for it is there that basaltic magma is most common.

Although rhyolitic magma forms principally through partial melting of the continental crust, the importance of Bowen's reaction series has been demonstrated many times. Careful study of almost any igneous rock reveals evidence that fractional crystallization played a role in its formation.

Magmatic Mineral Deposits

The processes of partial melting and fractional crystallization in magmas sometimes lead to the formation of large and potentially valuable mineral deposits. Because magma is involved in the formation process, such deposits are called *magmatic mineral deposits*.

When a magma undergoes differentiation by fractional crystallization, the residual melt becomes progressively enriched in any chemical element that is not removed by the early crystallizing minerals. Separation and crystallization of the residual melt produces an igneous rock that contains the concentrated elements.

An important example of this kind of concentration process is provided by pegmatites, especially those formed as a result of the crystallization of rhyolitic magma. Some pegmatites form by magmatic differentiation during the formation of granitic stocks and batholiths. Pegmatites may contain significant enrichments of rare elements such as beryllium, tantalum, niobium, uranium, and lithium, which, if sufficiently rich, can be profitably mined.

Another form of magmatic differentiation by fractional crystallization occurs when early formed dense minerals sink and accumulate on the floor of a magma chamber. The process is called *crystal settling*, and in some cases the segregated minerals make desirable

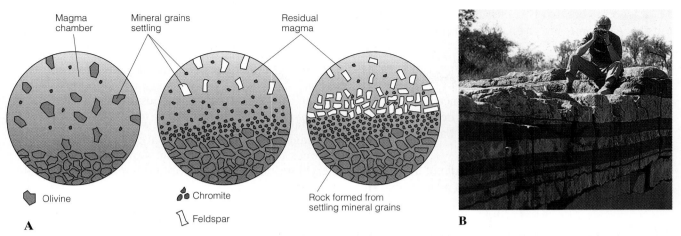

Figure 4.41 Magmatic Differentiation Fractional crystallization. A. Grains of three minerals—olivine, chromite (chromium iron-oxide), and feldspar—settle one after the other to the bottom of a magma chamber, producing three kinds of rocks whose compositions differ considerably from that of the parent magma. B. Layers of plagioclase (light gray) and chromite (black) formed by fractional crystallization in the Bushveld Igneous Complex, South Africa.

ores. Most of the world's chromium ores were formed in this manner by accumulation of the mineral chromite ($FeCr_2O_4$) (Fig. 4.41). The largest known chromite deposits are in South Africa, Zimbabwe, and the former Soviet Union. Similarly, vast deposits of ilmenite ($FeTiO_3$), a source of titanium, were formed by magmatic differentiation. Large deposits occur in the Adirondack Mountains.

A form of concentration similar to crystal settling occurs when, for reasons not clearly understood, certain magmas separate, as oil and water do, into two immiscible liquids. One, a sulfide liquid that is rich in copper and nickel, sinks to the floor of the magma chamber because it is denser. After cooling and crystallization, the resulting igneous rock has a copper or nickel ore at the base. The world's greatest known concentration of nickel ore, at Sudbury, Ontario, is believed to have formed in this fashion. Other great nickel deposits in Canada, Australia, and Zimbabwe formed in the same manner.

SUMMARY

1. Three kinds of magma predominate: basaltic, andesitic, and rhyolitic. Approximately 80 percent of all magma erupted by volcanoes is basaltic.

2. The principal controls on the physical properties of magma are temperature, SiO_2 content, and, to a lesser extent, dissolved-gas content. High temperature and low SiO_2 content result in fluid magma (basaltic). Lower temperature and high SiO_2 contents result in viscous magma (andesitic and rhyolitic magma). Dissolved gas reduces viscosity, but the main controls on viscosity are the SiO_2 content and temperature.

3. Basaltic magma forms by dry partial melting of rock in the mantle. Andesitic magma forms during subduction by wet partial melting of mantle rock as a result of water vapor released from heated oceanic crust. Rhyolitic magma forms by wet partial melting of rock in the continental crust.

4. The sizes and shapes of volcanoes depend on the kind of material erupted, viscosity of the lava, and explosiveness of the eruptions.

5. Viscous magmas erupt a lot of tephra and tend to build steep-sided tephra cones or stratovolcanoes.

6. Low-viscosity magmas low in SiO_2 tend to be erupted as fluid lavas that build gently sloping shield volcanoes or lava plateaus.

7. Magma forms by the partial melting of rock. (Complete melting, if it happens at all, is rare.)

8. Igneous rock forms by the solidification and crystallization of magma.

9. Igneous rock may be intrusive (meaning it formed within the crust) or extrusive (meaning it formed on the surface). The grain sizes of igneous rocks indicate how and where the rocks formed.

10. Igneous rocks rich in quartz and feldspar, such as granite, granodiorite, and rhyolite, are characteristically found in the continental crust. Basalt, which is rich in pyroxene and olivine, is derived from magma formed in the mantle and is common in the oceanic crust.

11. All bodies of intrusive igneous rock are called plutons. Special names are given to plutons based on shape and size.

12. Modern volcanic activity is concentrated along plate margins. Andesitic volcanoes are found at subduction margins, and basaltic volcanoes are concentrated along spreading margins. Rhyolitic volcanoes occur at collision margins and in places where rising basaltic magma causes the continental crust to melt.

13. Volcanism that occurs in the middle of an oceanic plate is apparently due to sources of hot material that are deep in the mantle from which basaltic magma can form. As oceanic plates move over hot spots, long chains of basaltic volcanoes are formed.

14. Processes that separate remaining melt from already formed crystals in a cooling magma lead to the formation of a wide diversity of igneous rocks.

15. Magmatic mineral deposits, which form as a result of magmatic differentiation, are the world's major sources of nickel, chromium, vanadium, platinum, beryllium, and a number of other important industrial metals.

THE LANGUAGE OF GEOLOGY

aphanite (p. 101)

batholith (p. 107)

caldera (p. 95)
chemical differentiation by partial melting (p. 110)
crater (p. 95)

dike (p. 106)

extrusive igneous rocks (p. 99)

intrusive igneous rocks (p. 99)

lava (p. 84)

magma (p. 84)
magmatic differentiation by fractional crystallization (p. 112)

phanerite (p. 101)
pillow basalt (p. 97)
plateau basalt (p. 97)
pluton (p. 106)
porphyry (p. 101)
pyroclast (p. 88)
pyroclastic flow (p. 89)
pyroclastic rocks (p. 88)

shield volcano (p. 93)
sill (p. 106)
stock (p. 108)
stratovolcano (p. 93)

tephra (p. 88)
tephra cone (p. 93)

viscosity (p. 85)
volcano (p. 84)

IMPORTANT ROCK NAMES TO REMEMBER

agglomerate (p. 105)
andesite (p. 105)

basalt (p. 105)

dacite (p. 103)
diorite (p. 103)

gabbro (p. 103)
granite (p. 103)
granodiorite (p. 103)

obsidian (p. 101)

pegmatite (p. 101)

peridotite (p. 103)

rhyolite (p. 103)

tuff (p. 105)

welded tuff (p. 105)

QUESTIONS FOR REVIEW

1. What controls the grain size and texture in igneous rocks? Would you expect solidified lava flows to be basalt or gabbro?

2. What are the distinguishing features of pyroclastic rocks? How might you tell the difference between a rhyolite that flowed as a lava and a rhyolitic tuff?

3. Do porphyries always contain phenocrysts? What minerals might you find as phenocrysts in a rhyolite porphyry? a basalt porphyry?

4. What is the major difference between the mineral assemblage of a diorite and a granodiorite? between granite and granodiorite? between a gabbro and a peridotite?

5. How is pumice formed and of what is it made? Can you suggest why pumice will float on water?

6. How does a lapilli tuff differ from a welded tuff? Could both rocks form as a result of eruptions from the same volcano?

7. Is the major oxide component of magma SiO_2, MgO, or Al_2O_3? Briefly describe the effect of the SiO_2-content on the fluidity of magma. What effect does temperature have on viscosity?

8. What does the term *partial melting* mean, and what role does it play in forming of basaltic magma? Where in the Earth does basaltic magma form?

9. How does the magma-forming process of dry partial

melting differ from wet partial melting? Can you suggest an example of a magma type formed by each kind of melting?

10. What is the origin of andesitic magma such as that erupted by Mount St. Helens? With what kind of volcanoes are andesitic eruptions associated? What is the distribution of andesitic volcanoes with respect to today's tectonic plates?

11. How might it be possible for fractional crystallization to produce more than one kind of igneous rock from a single magma? Comment on the role of fractional crystallization in the formation of mineral deposits.

12. Comment on the importance of Bowen's reaction series for understanding differentiation processes in cooling magmas. How does a continuous reaction series differ from a discontinuous series?

13. Why does a shield volcano like Mauna Loa in Hawaii have a gentle surface slope, while a stratovolcano such as Mount Fuji in Japan has steep sides?

14. How does a lava dome such as the one that now lies in the crater of Mount St. Helens form? How, if at all, does a lava dome differ from a resurgent dome?

15. Describe the distinctive kind of basaltic lava flows formed by submarine fissure eruptions. How do such flows form?

16. How does a dike, like the Great Dike of Zimbabwe, differ from a sill, such as the Palisades Sill?

17. How big can batholiths be, and by what mechanism are granitic magmas believed to move upward in the crust?

18. Briefly describe the plate-tectonic settings of andesitic and basaltic volcanoes. Where does Kilauea Volcano in Hawaii sit with respect to the Pacific Plate? Where does Mount St. Helens sit with respect to a plate boundary?

Virtual Internship: You Be the Geologist

Your CD-ROM contains a virtual internship through which you will become a geologist as you assess volcanic hazards in the imaginary island of Paradiso.

5

**Understanding
Our Environment**

**Using
Resources**

GeoMedia

Base of a tower excavated at the Ash Shisur water hole in Oman. The
tower and the other ruins being excavated are believed to be the
remains of the legendary city of Ubar.

Weathering and Soils

Remote Sensing and the Rediscovery of Ubar

More than 4700 years ago, the city of Ubar was built somewhere on the Arabian peninsula. For nearly three thousand years Ubar was the center of the frankincense trade. Frankincense, used as incense and in perfumery, embalming, and fumigation, is an aromatic gum resin from trees found in the Middle East and Africa. Thousands of years ago it was also a symbol of great holiness and wealth. Ubar was renowned for its architecture, fruit trees, trees yielding frankincense, and opulence. Frankincense was so valuable that it was called "gold" in Sumerian literature. Then suddenly, around two thousand years ago, Ubar disappeared. Its remains were covered by shifting, blowing sands. Its location was forgotten, but its allure was not.

Where is Ubar? Only death kept T.E. Lawrence (Lawrence of Arabia) from seeking it. Another British explorer, Bertrand Thomas, searched the Rub'al-Khali desert for Ubar in 1929–1930. Guided by historical documents and local lore, he actually came near it, possibly even walking over what is now believed to be the site. But, missing the big picture of the area he was searching, he failed to identify the city. In 1981, Los Angeles filmmaker Nicholas Clapp became intrigued with finding Ubar. He learned about NASA's ability to make images of the earth from space using radio waves. Because of its very low dielectric constant, sand allows radio signals to penetrate several meters through it. The radio waves then reflect off of harder surfaces below the ever-changing desert floor. Clapp recruited a team of experts in several fields to join the search, including scientists at NASA's Jet Propulsion Laboratory in Pasadena, California, who had developed the imaging radar system for the space shuttle. In a 1984 shuttle flight, the imaging radar took aim on the southern Arabian peninsula, the same area explored by Thomas (and also described in vague terms as the location of Ubar by geographer Claudius Ptolemy in the second century A.D.).

The radar images were digitized and combined by computer with infrared and optical images taken by the French SPOT and American Landsat satellites. The data from the infrared and visible radiation provided information about the soil and sand, such as the fact that camels have ground the sand into fine-grained pieces different from the rest of the desert. The trampled sand reflects different amounts of infrared and visible radiation than does loosely packed sand. As a result, the combined images of the desert showed many previously unknown caravan trails. While the satellite images did not reveal the ruins of a city, the group of physicists and geologists working on the project at the Jet Propulsion Laboratory observed that many of the old caravan trails converged on the Ash Shisar water hole in Oman, inland from the south-central coast of the Arabian peninsula.

In 1991, using satellite positioning equipment, archaeologists, geologists, and their support team traveled to the area and located some of the caravan trails identified from space. They found themselves near the area Thomas had explored, but unlike him, they knew from the satellite images that they were at the hub of a vast trading center. Persisting in their exploration and digging, the team brought together by Nicholas Clapp began uncovering a vast, elaborate city. While it is still uncertain whether this is Ubar, the evidence in the form of architectural wonders, pottery, and metal fragments mounts. If this is Ubar, why did it vanish? The answer comes from below the city; the explorers found that it was built over a vast limestone cave that eventually collapsed, plunging the city down into the earth. The ubiquitous Arabian sand quickly covered it over, where it has remained for nearly two millennia.

OUR DECAYING BUILDINGS AND MONUMENTS

In order to construct buildings, monuments, tombstones, and other structures, people have long sought stone that would be both attractive and durable. More than two millennia ago, the Greeks and Romans discovered that limestone and marble make ideal building materials, for they are solid, reasonably strong, and relatively easy to carve and produce aesthetically pleasing structures. However, what they failed to appreciate was how vulnerable these carbonate rocks would be to decay.

In many large industrial cities, both new and ancient structures built of marble or limestone are slowly disappearing under the vigorous attack of airborne pollutants produced by the burning of fossil fuels. Especially damaging is sulfur dioxide gas, the main contributor to the acid rain that causes the calcite composing these rocks to dissolve. As a result, many European public statues and historic buildings, some dating to medieval times, have been extensively disfigured. Even in the United States, where most public buildings are less than two centuries old, widespread decay has begun to obliterate the surface of structures built of carbonate rocks.

Early nineteenth-century marble tombstones in New England churchyards are so severely corroded that inscriptions have become illegible. Most of this damage has taken place since the beginning of the Industrial Revolution when steam engines and coal fires came into widespread use. In the twentieth century, increasing combustion of fossil fuels has pumped vast amounts of pollutants into the air, thereby greatly accelerating the rate at which the perfectly natural geologic process of rock decay takes place. In the eastern United States, studies using freshly quarried stone have shown that the surfaces of limestone and marble slabs exposed at relatively nonpolluted sites are dissolving at an average rate of 0.001 to 0.002 cm per year. At highly polluted sites, rates are expected to be far higher. Clearing the air of harmful pollutants, an important challenge for all nations in the coming decades, will lead not only to healthier air to breathe, but to a healthier environment and longer life for buildings and monuments as well.

WEATHERING

Whether it occurs rapidly or slowly, the physical and chemical alteration of rock takes place throughout the zone where materials of the lithosphere, hydrosphere, biosphere, and atmosphere mix. The zone extends downward into the ground to whatever depth air, water, and microscopic forms of life can penetrate. Within it, the rock constitutes a porous framework, full of fractures, cracks, and other openings, some of which are very small but all of which make the rock vulnerable. This open framework is continually being attacked, both chemically and physically, by water solutions and by microscopic life. Given sufficient time, the result is conspicuous alteration of the rock.

When exposed at the Earth's surface, no rock (whether bedrock or structures built of stone) escapes the effects of **weathering**, the chemical alteration and mechanical breakdown of rock and regolith when exposed to air, moisture, and organic matter. Weathering is the group of processes by which rock is converted to regolith. The results of weathering are often seen in landslide scars and large excavations that expose bedrock. In Figure 5.1, loose, unorganized, earthy regolith in which the texture of the bedrock is no longer apparent grades downward into rock that has been altered but still retains its organized appearance, and

Figure 5.1 Soil Profile Long-continued leaching has produced this soil profile in North Carolina. Note that plant roots are largely confined to the uppermost brownish layer and that the profile grades down to layered rocks which, although partly altered, are recognizable as metamorphic rocks.

then into practically unaltered bedrock. It is evident from such exposures that alteration of the rock progresses from the surface downward.

A close look at the bedrock in Figure 5.1 would show that near the bottom of the exposure, the cleavage surfaces of feldspar grains are bright and reflective, while higher up such surfaces are lusterless and stained. Soft, earthy material near the top of the outcrop no longer resembles the feldspar that was once present there and that now has largely decomposed. The changes that have occurred result mainly from **chemical weathering**, the decomposition of rocks and minerals as chemical reactions transform them into new chemical combinations that are stable at or near the Earth's surface.

Sometimes regolith consists of fragments identical to the adjacent bedrock. The mineral grains are fresh or only slightly altered. Piles of loose rock fragments are commonly found at the base of bedrock cliffs from which the debris quite obviously has been derived. When compared with the bedrock, the coarse rock fragments show little or no evidence of chemical weathering, which implies that bedrock can be broken down not only chemically but also physically. Although we consider **physical weathering**, which is the disintegration (physical breakup) of rocks, as being distinct from chemical weathering, the two processes generally work hand in hand, and their effects are inseparably blended.

Physical Weathering

Physical weathering of rock is common in nature and takes place when crystals of ice or salt grow in rock fractures, when rock is heated by fire, and when the growth of plant roots disrupts rock.

Development of Joints
Most rocks are subject to a kind of mechanical fracturing that plays an important role in weathering. Rocks in the upper half of the crust are brittle, and like any other brittle material they break at the weak spots when they are twisted, squeezed, or stretched by tectonic forces. The timing and origin of such tectonic forces are not always obvious, but they leave their footprints in the form of **joints**, which are fractures in a rock along which no appreciable movement has occurred. Joints form as a result of tectonic forces while rocks are buried deeply in the crust. Through uplift and erosion the rocks slowly rise toward the surface, and as they do the weight of the overlying rocks decreases and the joints open slightly, allowing water, air, and microscopic forms of life to enter.

Removal of overlying rocks by erosion can lead to further breakage and formation of additional joints as a result of unloading. Rock adjusts to unloading by expanding upward. As it does so, closely spaced fractures approximately parallel to ground surface can develop,

Figure 5.2 Sheet Joints Sheet jointing in massive granite forms a steplike surface in Yosemite National Park, California.

giving the rock a steplike appearance (Fig. 5.2). Generally, joints formed by unloading disappear below a depth of about 50 to 100 m, as the pressure of the overlying rock becomes too great for joints to form, but joints can form as a result of tectonic forces at depths of thousands of meters.

Rarely do joints occur singly. Most commonly, they occur as a widespread set of parallel joints (Fig. 5.3). Intersecting joints strongly influence the way a rock breaks apart. Once formed, joints act as passageways by which rainwater can enter a rock and promote both physical and chemical weathering.

One class of joints is restricted to tabular bodies of igneous rock—such as dikes, sills, lava flows, and welded tuffs—that cooled rapidly at or close to the land surface. When such a body of igneous rock cools, it contracts and may fracture into pieces, in much the same way that a very hot glass bottle contracts and shatters when plunged into cold water. Unlike shattered glass, however, cooling fractures in igneous rock tend to form regular patterns. For joints that split igneous rocks into long prisms or columns, the term *columnar joints* is used (Fig. 5.4).

Crystal Growth

Groundwater moving slowly through fractured rocks contains ions that may precipitate out of solution to form salts. The force exerted by salt crystals growing either within rock cavities or along grain boundaries can be very large and can result in the rupture or disaggregation of rocks. The effects of such physical weathering can often be seen in desert regions, where salt crystals

Figure 5.3 Joint Sets Two well-developed sets of vertical joints combine with a horizontal bedding plane to cause this red sandstone in Brecon Beacons, Wales, to break into roughly rectangular blocks.

Figure 5.4 Columnar Joints Columnar jointing in igneous rock near San Miguel Regla, Mexico, offers a challenge to a rock climber jambing his way up a crack between two adjacent columns.

Figure 5.5 Weathering in Antarctica Granitic bedrock on the flank of Gondola Ridge in Antarctica has been so strongly weathered that it resembles Swiss cheese. Such cavernous weathering results from granular disintegration as salts crystallize in small cavities and along grain boundaries.

grow as rising groundwater evaporates and its dissolved salts are precipitated.

In the ice-free valleys of coastal Antarctica, bizarre cavernous landforms have been produced by the physical weathering of granite boulders and bedrock outcrops (Fig. 5.5). When salts crystallize out of solutions confined within the pores and fine cracks between mineral grains in the granite of the ice-free valleys, the rock slowly disintegrates. The resulting debris is transported away by strong winds blowing off the nearby ice sheet, leaving an unusual landscape in which the exposed rocks resemble Swiss cheese.

Frost Wedging

Wherever temperatures fluctuate about the freezing point for part of the year, water in the ground periodically freezes and thaws. When water freezes to form ice, its volume increases by about 9 percent. The high stresses resulting from this volume increase lead to disruption of rocks. As freezing occurs in the pore spaces of a rock, water is strongly attracted to the growing ice, thereby increasing the stresses against the rock. The

process leads to a very effective type of physical weathering known as **frost wedging**, the formation of ice in a confined opening within rock, thereby causing the rock to be forced apart. Frost wedging is strong enough to force apart not only tiny particles but huge blocks, some weighing many tons (Fig. 5.6). Frost wedging probably is most effective at temperatures of −5° to −15° C. At higher temperatures ice pressures are too low to be very effective, and at lower temperatures the rate of ice growth drops because the water necessary for crack growth is less mobile.

Frost wedging is responsible for most of the rock debris seen on high mountain slopes. At lower altitudes this process is likely to be most effective wherever the number of yearly freeze–thaw cycles is high.

Effects of Heat

Some geologists have speculated that daily heating of rock in bright sunlight followed by cooling each night should cause the physical breakdown of rocks because the common rock-forming minerals expand by different amounts when heated. Surface temperatures as high as 80° C have been measured on desert rocks, and daily temperature variations of more than 40° have been recorded on rock surfaces. Highest temperatures are achieved by dark-colored rocks, like basalt, and by rocks that do not easily conduct heat inward. Nevertheless, despite a number of careful tests, no one has yet demonstrated that such daily heating and cooling cycles have noticeable physical effects on rocks. These tests, however, have been carried out only over relatively brief time

Figure 5.6 Frost Wedging An extensive field of granite blocks covers the summit of Mount Whitney, the highest peak in California's Sierra Nevada. Frost wedging, the result of melting snow that percolates downward into cracks where it refreezes, has disrupted the granite bedrock, producing this vast litter of angular boulders.

Figure 5.7 Heat Spalling A fast-moving forest fire in the pine forests of Yellowstone National Park, Wyoming, caused mechanical weathering of a boulder of igneous rock. Fresh, bright-colored spalls, which flaked off as heat caused the outermost part of the rock to expand rapidly, litter the fire-blackened ground.

intervals. Possibly, thermal fracturing occurs only after repeated, extreme natural temperature fluctuations over long periods of time.

Fire, on the other hand, can be very effective in disrupting rocks, as anyone knows who has witnessed a rock beside a campfire become overheated and shatter explosively. Because rock is a relatively poor conductor of heat, an intense fire heats only a thin outer shell, which expands and breaks away as a **spall**. The heat of forest fires and brush fires can lead to the spalling of rock flakes from exposed bedrock or boulders (Fig. 5.7). Studies of fire history in forested regions show that large natural fires, most started by lightning, may recur every several hundred years. Therefore, over long intervals of geologic time, fires may contribute significantly to the physical breakdown of surface rocks.

Figure 5.8 Root Wedging A Ponderosa pine tree that began growing in a crack on this bedrock outcrop has caused a large flake of rock to break away, thereby exposing the tree's expanding root system.

Plant Roots

Seeds germinate in cracks in rocks to produce plants that extend their roots farther into the cracks. The roots of trees growing in cracks may wedge apart adjoining blocks of bedrock (Fig. 5.8). In much the same way, roots also disrupt stone pavements, sidewalks, garden walls, and even buildings. Large trees swaying in the wind can cause cracks to widen, and, if blown over, can pry rock apart. Although it would be difficult to measure, the total amount of rock breakage done by plants must be very large. Much of it is obscured by chemical weathering, which takes advantage of the new openings as soon as they are created.

Chemical Weathering

All of the common rock-forming minerals, and indeed the majority of all minerals whether rock-forming or not, are chemically unstable when exposed to the weather at the Earth's surface. Such minerals break down, and their components form new, more stable minerals. A very few minerals, of which diamond and gold are examples, are so nearly impervious to chemical attack by the weather that they can exist at the Earth's surface for hundreds of millions of years. Such minerals are rare exceptions.

The principal agents of chemical weathering are water solutions that behave as weak acids. The effects of chemical weathering are most pronounced in regions where both precipitation and temperature are high, two conditions that enhance the rapidity of chemical reactions.

Chemical Weathering of Rock-Forming Minerals

As rainwater forms in the atmosphere, it dissolves small quantities of carbon dioxide, producing **carbonic acid**. As the slightly acid water moves downward and laterally through the soil, additional carbon dioxide is dissolved from decaying vegetation. The carbonic acid ionizes to form hydrogen ions and bicarbonate ions (Table 5.1, reaction 1). The hydrogen ions are so small they can enter a crystal and replace other ions, thereby changing the composition.

Hydrolysis. The involvement of the H^{1+} ion in decomposing minerals is illustrated by the way potassium feldspar, a common rock-forming mineral, is decomposed by carbonic acid (Table 5.1, reaction 2). Hydrogen ions and water interact with the aluminum-silicate crystal structure of the feldspar surface to slowly release alumina and silica into solution. Once in solution, the alumina and silica combine with water and precipitate out of solution as the clay mineral **kaolinite**. Kaolinite was not present in the original rock and was created by the chemical reaction. Kaolinite, a common member of the group of relatively insoluble minerals that constitute clay, forms a substantial part of the regolith. Any chemical reaction, such as the one describing the decomposition of potassium feldspar, in which water is a reactant, is called a **hydrolysis** reaction. Hydrolysis is one of the chief processes involved in the chemical breakdown of common rocks.

Leaching. Another common process of chemical weathering is **leaching**, the continued removal, by water solutions, of soluble matter from bedrock or regolith. For example, when silica is released from rocks by chemical weathering, some of it remains in the clay-rich regolith and some is removed into solution by water moving through the ground. Many of the potassium ions weathered from the rock also escape in solution. Soluble substances leached from rocks during weathering are present in all surface water and groundwater. Sometimes their concentrations are high enough to give the water an unpleasant taste.

Oxidation. **Oxidation** is a process in which a chemical element loses electrons and the oxidation state of the element increases. Iron is a normal constituent of many common rock-forming minerals, including biotite, augite, and hornblende. When any of these minerals is chemi-

TABLE 5.1	Common Chemical Weathering Reactions

1. Production of carbonic acid by solution of carbon dioxide, and the ionization of carbonic acid:

$$H_2O + CO_2 \rightleftarrows H_2CO_3 \rightleftarrows H^{1+} + HCO_3^{1-}$$

Water   Carbon   Carbonic   Hydrogen   Bicarbonate
dioxide   acid   ion   ion

2. Hydrolysis of potassium feldspar:

$$4KAlSi_3O_8 + 4H^{1+} + 2H_2O \rightarrow 4K^{1+} + Al_4Si_4O_{10}(OH)_8 + 8SiO_2$$

Potassium   Hydrogen   Water   Potassium   Kaolinite   Silica
feldspar   ions   ions

3. Oxidation and hydrolysis of iron (Fe^{2+}) compounds to form ferric hydroxide:

$$4FeO + 6H_2O + O_2 \rightarrow 4Fe(OH)_3$$

Iron   Water   Oxygen   Ferric
oxide   hydroxide

4. Dehydration of ferric hydroxide to form goethite:

$$Fe(OH)_3 \rightarrow FeO.(OH) + H_2O$$

Ferric   Goethite   Water
Hydroxide

5. Dehydration of goethite to form hematite:

$$2FeO \cdot OH \rightarrow Fe_2O_3 + H_2O$$

Goethite   Hematite   Water

6. Dissolution and hydrolysis of carbonate minerals by carbonic acid:

$$CaCO_3 + H_2CO_3 \rightarrow Ca^{2+} + 2(HCO_3)^{1-}$$

Calcium   Carbonic   Calcium   Bicarbonate
carbonate   acid   ion   ions

cally weathered, iron is released and, in the presence of oxygen, rapidly oxidized from Fe^{2+} to Fe^{3+}. Typically, this results in the formation of a reddish-yellow precipitate of ferric hydroxide [$Fe(OH)_3$], through **hydration**, which is the term for the incorporation of water into the structure of a solid chemical compound (Table 5.1, eq. 3). The ferric hydroxide will soon **dehydrate**, meaning it loses some water, in which case it forms goethite ($FeO.OH$), (Table 5.1, eq. 4). Goethite may later dehydrate still further to form **hematite** (Fe_2O_3), a brick-red mineral (Table 5.1, eq. 4). The intensity of the colors of ferric hydroxide, goethite, and hematite in weathered rocks and soils can provide clues to how much time has elapsed since weathering began and to the degree or intensity of weathering.

Dissolution. A few soluble minerals will simply dissolve in water, a process called **dissolution**. Halite ($NaCl$) is an example of a mineral that is removed by dissolution. Dissolution plays a part in virtually all chemical weathering processes and is usually accompanied by hydrolysis and leaching. Consider the reaction of calcite in rainwater. Calcite is only slightly soluble in water, and

when it dissolves, Ca^{2+} and CO_3^{2-} ions are formed in the solution. If carbonic acid is present, the process is speeded up and CO_3^{2-} ions react with H^+ ions in solution to form bicarbonate ions (HCO_3^-). The effects of dissolution and hydrolysis are combined in equation 6 of Table 5.1. The effects of these processes are widely seen in the distinctive landscapes underlain by carbonate rocks (Chapter 11).

Effects of Chemical Weathering on Common Rocks

The minerals and soluble ions that result when an igneous rock weathers chemically depend on the original mineral composition of the rock. Granite has a higher silica content than basalt and a different mineralogical makeup. A typical granite contains quartz, which is relatively inactive chemically, potassium-bearing minerals such as muscovite and potassium feldspar, plagioclase feldspar, and minerals rich in iron and magnesium (called ferromagnesian minerals). When a granite decomposes, it does so by the combined effects of dissolution, hydrolysis, and oxidation. The feldspar, mica, and ferromagnesian minerals weather to clay minerals and soluble Na^{1+}, K^{1+}, and

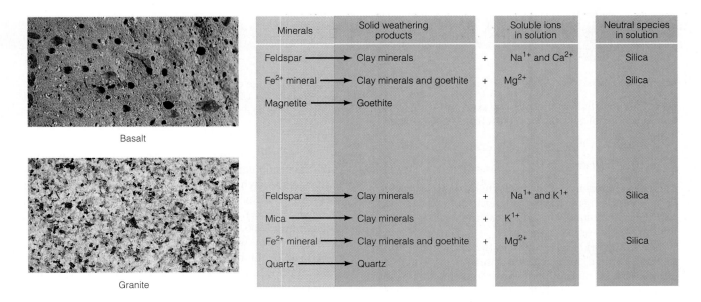

Minerals	Solid weathering products		Soluble ions in solution	Neutral species in solution
Feldspar ⟶	Clay minerals	+	Na^{1+} and Ca^{2+}	Silica
Fe^{2+} mineral ⟶	Clay minerals and goethite	+	Mg^{2+}	Silica
Magnetite ⟶	Goethite			
Feldspar ⟶	Clay minerals	+	Na^{1+} and K^{1+}	Silica
Mica ⟶	Clay minerals	+	K^{1+}	
Fe^{2+} mineral ⟶	Clay minerals and goethite	+	Mg^{2+}	Silica
Quartz ⟶	Quartz			

Basalt

Granite

Figure 5.9 When a Rock Weathers When a basalt weathers chemically, its silicate minerals and magnetite are converted to clay minerals, goethite, soluble cations, and silica. The weathering products of a granite not only include clay minerals, goethite, soluble cations, and silica in solution from the breakdown of feldspar and ferromagnesian minerals, but grains of quartz that are resistant to chemical breakdown.

Mg^{2+} ions (Fig. 5.9). The quartz grains, being relatively inactive chemically, remain essentially unaltered. Because basalt lacks quartz, mica, and potassium feldspar, quartz grains and K^{1+} ions are not among its chemical weathering products. Like granite, the plagioclase feldspar and ferromagnesian minerals in a basalt weather to clay minerals and soluble ions (Na^{1+}, Ca^{2+}, and Mg^{2+}), while the iron from ferromagnesian minerals, together with iron from magnetite, forms goethite.

When limestone, the most common sedimentary rock that contains calcium carbonate, is attacked by dissolution and hydrolysis, it is readily dissolved, leaving behind only the nearly insoluble impurities (chiefly clay and quartz) that are always present in small amounts in the rock. Therefore, as limestone weathers chemically, the residual regolith that develops from it consists mainly of clay and quartz.

Concentration of Stable Minerals

Not only quartz but also a number of other minerals are so resistant to chemical attack that they can persist for long periods of time at the Earth's surface. Minerals such as gold, platinum, and diamond, as previously mentioned, persist during weathering; such persistent minerals can be carried away during erosion of the regolith and become sediment. Because some of these minerals have higher specific gravities than common minerals such as quartz, they concentrate at the beds of streams or along ocean beaches (Chapters 10 and 14). Some may end up sufficiently concentrated to form mineral deposits

of economic value. Valuable mineral deposits may also form as a result of chemical weathering reactions other than concentration of residual minerals (see the boxed essay on *Mineral Deposits Formed by Weathering*).

Weathering Rinds

If you crack open a cobble of weathered basalt, you usually will see a light-colored rind surrounding a darker core of unaltered rock (Fig. 5.10). The rind is composed

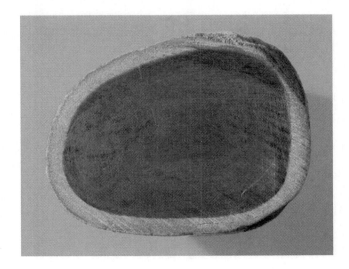

Figure 5.10 Weathering Rind A basaltic stone from the eastern Cascade Range in Washington displays a well-developed weathering rind about 2 cm thick surrounding a black, unweathered core.

Using Resources

Mineral Deposits Formed by Weathering

By removing soluble minerals and concentrating less soluble ones, chemical weathering can form valuable mineral deposits.

Lateritic Concentrations

Laterite is an example of enrichment by weathering (see p. 137). The starting materials for laterite are ordinary rocks, and it is the common chemical elements iron and aluminum that become concentrated.

Goethite (FeO·OH) is among the least soluble of the minerals formed during chemical weathering. Under conditions of high rainfall in a warm, tropical climate, most other minerals are slowly leached out of a soil, leaving an insoluble, iron-rich limonitic crust of laterite at the surface (Fig. B5.1). In a few places, such as West Africa, laterites are rich enough to be mined for iron.

Bauxites

Although iron-rich varieties are by far the most common laterites, the most important as far as mineral exploitation is concerned are aluminous laterites called **bauxites** (Fig. B5.2). Bauxites are formed by the leaching of clay minerals, a process in which silica (SiO_2) is removed in solution and a residue of gibbsite [$Al(OH)_3$] remains. Gibbsitic bauxites are the source of most of the world's aluminum.

Secondary Enrichment

When a preexisting mineral deposit is chemically weathered, the results can sometimes be a spectacular upgrading in the metal content. The process is called **secondary enrichment**. Some of the world's most valuable deposits of iron, manganese, nickel, and copper formed in this way. Iron-rich sedimentary rocks (Chapter 6) may contain as much as 25 percent iron by weight. Such rocks are too lean to be mined at a profit, but if chemical weathering removes silica in solution, a secondarily enriched deposit of goethite and hematite containing as much as 66 percent iron can result (Fig. B5.3).

Manganese-rich sediments that form as a result of submarine hot-spring actions (Chapter 6) can, like iron-rich sedimentary rocks, become enriched by chemical weathering because the oxide mineral of fully oxidized manganese is virtually insoluble in surface waters. Secondary enrichment deposits of manganese, as the residual mineral *pyrolusite* (MnO_2), are mined in many tropical countries.

Nickel is always present in gabbros and peridotites (Chapter 3) in trace amounts. When such rocks are subjected to chemical weathering, the pyroxene, olivine, and

Figure B5.1 Laterite Red, iron-rich laterite near Khao Yai, Thailand. Originally a conglomerate, the rock is now a mixture of hematite and clay as a result of extreme tropical leaching.

Figure B5.2 Bauxite Bauxite mining in Venezuela. The bauxite is a 6-m-thick capping of aluminous laterite developed on granite.

Figure B5.3 Iron Ore Leaching of silica from an iron-rich sedimentary rock leads to the formation of a secondarily enriched mass of limonite and hematite. This outcrop, in the Hamersley Range of Western Australia, developed by secondary enrichment from the kind of iron-rich sediment shown in Figure 6.13. Remnants of the original silica-rich bands can be seen to the right of the hammer.

plagioclase feldspar break down, silica and most other constituents are removed in solution, and in the limonite residue there is a mixture of nickel hydroxide minerals called garnierite. One of the largest and most important nickel deposits ever discovered, on the island of New Caledonia, was formed by secondary enrichment of a nickeliferous peridotite.

Secondary enrichment is striking in many hydrothermal deposits containing sulfide minerals. One example is the great copper deposits found in the arid southwestern United States and the Chilean Andes (Fig. B5.4). The primary min-

erals are *pyrite* (FeS_2) and *chalcopyrite* ($CuFeS_2$). These minerals are oxidized by rainfall and the atmosphere to form a copper-bearing solution of sulfuric acid and a residue of limonite. The acidic solution percolates down and reacts with unoxidized pyrite and chalcopyrite by dissolving iron and precipitating copper to form *chalcocite* (Cu_2S). The chalcocite-rich blanket lies below the weathered residue and above the leaner primary mineral deposit below.

Deposits formed by weathering are all vulnerable to erosion. They are not found in glaciated regions, for example, because overriding glaciers have scraped off the soft surface materials. The vulnerability of such deposits means that most are geologically young. More than 90 percent of all known bauxites, for example, formed during the last 60 million years, and all of the very large deposits formed less than 25 million years ago. A few are still forming today.

Figure B5.4 Secondary Enrichment Secondary enrichment produced a rich ore (bluish-colored lowest level) from a lean primary ore at the Silver Bell deposit, Arizona. The acid solutions produced by oxidation removed copper from the oxidized capping (brown) and deposited it in the secondary enrichment zone.

of the solid products resulting from chemical weathering and is called a **weathering rind**. Weathering begins at the freshly exposed surface of a dark, unaltered cobble and proceeds slowly inward. Commonly, this involves oxidation of iron-rich minerals to produce goethite (Table 5.1, eq. 4), which imparts a light brownish color to the developing rind. Such a rind forms on all but the most chemically stable rock types when chemical weathering attacks them. An exception is carbonate rocks, which dissolve under chemical attack (Table 5.1, eq. 5), thereby generally preventing the formation of an obvious rind.

In some rock types, the altered mineral matter in a developing rind crumbles away as the rock slowly weathers. In others, like fine-grained basalt, the rind tends to remain coherent and becomes thicker as weathering moves inward, attacking the solid, unaltered core. Geologists have found that the thickness of rinds on stones can be a useful measure of the relative ages of different bodies of sediment that contain the same rock types and occur in similar climatic settings.

Exfoliation and Spheroidal Weathering

During weathering, concentric shells of rock may spall off from the outside of an outcrop or a boulder, a process known as **exfoliation** (Fig. 5.11). Sometimes only a single exfoliation shell is present, but 10 or more shells may develop, thereby giving a rock the layered appearance of an onion.

Exfoliation is caused by differential stresses within a rock that result mainly from chemical weathering. For example, when feldspars weather to clay, the volume of weathered rock is greater than the volume of original rock. Stresses thus produced cause thin shells of rock to separate from the main mass of unweathered rock.

Beneath the ground surface, chemical weathering frequently will give rise to a halo of decayed rock around an unaltered rock core. Starting with a cube of solid, fresh rock, as water moves along joints and attacks the rock from all sides, the volume of unaltered rock will slowly decrease in size and become more spherical (Fig. 5.12), a process known as **spheroidal weathering**. The results are often seen in fresh roadcuts where rounded boulders produced by such progressive decomposition often are found in rows running in several directions. The pattern results from intersecting joint sets that control the slow movement of water through the rock.

At this point, two important relationships should be noted. First, the effectiveness of chemical weathering increases as the surface area exposed to weathering increases. Second, surface area increases simply from the subdivision of large blocks into smaller blocks. Subdividing a cube, while adding nothing to its volume, greatly increases the surface area (Fig. 5.13). Repeated subdi-

Figure 5.11 Exfoliation Exfoliating granite boulders (The Devil's Marbles) in central Australia. Thin, sheetlike spalls flake off the rock as it weathers, gradually causing the boulders to increase in sphericity.

Figure 5.12 A Rounded Landscape
Spheroidal weathering of granitic bedrock in the northern Sierra Nevada of California produces boulders of solid granite that are surrounded by a core of disintegrated rock. Although the boulders resemble rounded stream gravel, their form is entirely the result of weathering.

A.

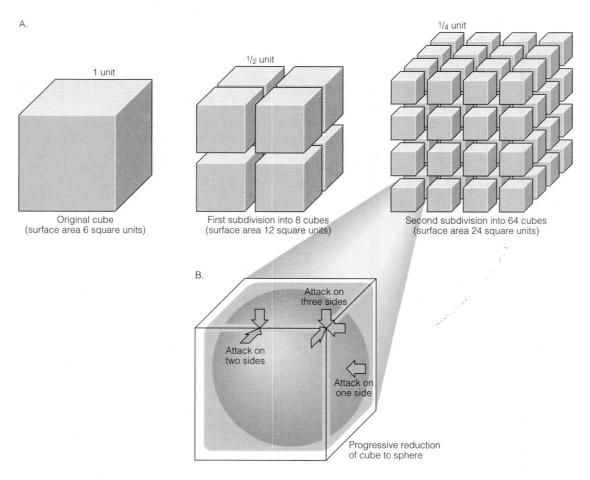

1 unit

Original cube
(surface area 6 square units)

1/2 unit

First subdivision into 8 cubes
(surface area 12 square units)

1/4 unit

Second subdivision into 64 cubes
(surface area 24 square units)

B.

Attack on three sides

Attack on two sides

Attack on one side

Progressive reduction of cube to sphere

Figure 5.13 Increasing Surface Area Subdivision and weathering of rock cubes. A. Each time a cube is subdivided by slicing it through the center of each of its edges, the aggregate surface area doubles. This greatly increases the speed of chemical reaction. B. Solutions moving along joints separating nearly cubic blocks of rock attack corners, edges, and sides at rates that decline in that order, because the numbers of corresponding surfaces under attack are 3, 2, and 1. Corners become rounded, and eventually the blocks are reduced to spheres. Once a spherical form is achieved, the energy of attack becomes uniformly distributed over the whole surface, so that no further change in form occurs.

vision leads to a remarkable result: one cubic centimeter of rock has a surface area of 6 cm^2; when subdivided into particles the size of the smallest clay minerals, the total surface area increases to nearly 40 million cm^2. Chemical weathering therefore leads to a dramatic increase in surface area.

Factors That Influence Weathering

Rock Type and Structure

Because different minerals react differently to weathering processes, the mineralogical assemblage of a rock, and therefore the rock type, clearly must influence decomposition. Quartz is so resistant to chemical breakdown that rocks rich in quartz are also resistant. In the Appalachian Mountains, resistant quartzite strata form ridges that stand prominently above valleys that are underlain by more erodible rocks containing less quartz. Granite also is resistant to weathering, for it consists of minerals (such as quartz, muscovite, and potassium feldspar) that are more resistant to chemical breakdown than are most other silicate minerals (Fig. 5.9). Like quartzite, granite typically forms hilly or mountainous terrain.

The rate at which a rock weathers is also influenced by its texture and structure. Even a rock that consists entirely of quartz may break down rapidly if it contains closely spaced joints or other partings that make it susceptible to frost action.

Contrasts in local topography may be due to *differential weathering*, which means that weathering has occurred at different rates as a result of variations in the composition and structure of rocks (Fig. 5.14). In a sequence of alternating shale and quartz sandstone, for instance, the shale is likely to weather more easily, leaving the sandstone beds standing out in relief. If the beds are horizontal, the result is likely to be a stepped topography, with the sandstone forming abrupt cliffs between more gentle slopes of shale. If the bedding is inclined, the sandstone will stand as ridges separated by linear depressions underlain by shale.

Slope

A mineral grain loosened by weathering on a steep slope may be washed downhill by the next rain. On such a slope the solid products of weathering move quickly away, continually exposing fresh bedrock to renewed attack. As a result, weathered rock seldom extends far beneath the surface. On gentle slopes, however, weathering products are not easily washed away and in places may accumulate to depths of 50 m or more.

Climate

Moisture and heat promote chemical reactions. Not surprisingly, therefore, weathering is more intense and gen-

Figure 5.14 Differential Erosion Differential erosion of Miocene marine strata at Nelson, New Zealand. Weathering has etched away erodible mudstone from between layers of siltstone, leaving siltstone standing as ridges above a wave-eroded platform.

erally extends to greater depths in a warm, moist climate than in a cold, dry one (Fig. 5.15). In moist tropical lands, like Central America or Southeast Asia, obvious effects of chemical weathering can be seen at depths of 100 m or more. By contrast, in cold, dry regions like northern Greenland and Antarctica, chemical weathering proceeds very slowly. Instead, the effects of physical weathering are generally obvious, for bedrock surfaces typically are littered with rubble dislodged by frost action.

Dramatic contrasts in weathering can be seen in the case of carbonate rocks that crop out in different climatic regions. Rocks such as limestone and marble, which consist almost entirely of carbonate minerals, are highly susceptible to chemical weathering in a moist climate and commonly form low, gentle landscapes. In a dry climate, however, the same rocks form bold cliffs because, with scant rainfall and only patchy vegetation, little carbonic acid is present to dissolve carbonate minerals.

Burrowing Animals

Large and small burrowing animals (for example, rodents and ants) bring partly decayed rock particles to

A.

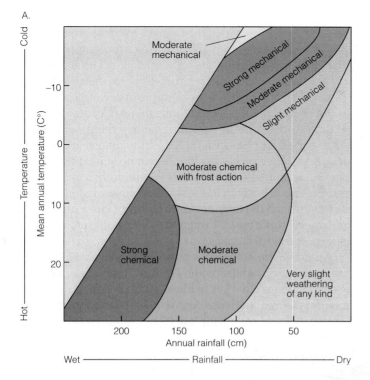

Figure 5.15 Climate and Weathering Climate (a combination of temperature and rainfall) plays a major role in controlling the type and effectiveness of weathering processes. Mechanical weathering is dominant where rainfall and temperature are both low. High temperature and precipitation favor chemical weathering.

the land surface, where they are exposed more fully to chemical action. More than 100 years ago, Charles Darwin made careful observations in his English garden and calculated that every year earthworms bring particles to the surface at the rate of more than 2.5 kg/m² (25 tons per hectare). After a study in the Amazon River basin, geologist J. C. Branner wrote that the soil there "looks as if it had been literally turned inside out by the burrowing of ants and termites." Although burrowing animals do not break down rock directly, the amount of disaggregated rock they move during many millions of years must be enormous.

Time
Studies of the decomposition of the stone in ancient buildings and monuments show that hundreds or even thousands of years are required for hard rock to decompose to depths of only a few millimeters. Granite and other hard bedrock surfaces in New England, Scandinavia, the Alps, and elsewhere still display polish and fine grooves made by ice-age glaciers before they disappeared about 10,000 years ago. In such regions, where cool climate and successive glaciations have significantly reduced the rate and effectiveness of chemical

weathering, it takes many tens of thousands of years to create weathered regolith like that shown in Figure 5.1. However, in regions that were not repeatedly glaciated and have been continuously exposed to weathering, the zone of weathering often extends to great depths. In some tropical areas, where weathering has continued for millions of years, mining operations have exposed bedrock that has been thoroughly decomposed to depths of 100 m or more.

The rates at which rocks weather have been estimated in several ways. First, experiments have been designed in which the length of the experiment provides time control and the processes were speeded up by increasing temperature and available water, and decreasing particle size. Second, studies have been made of the degree of weathering of ancient architectural structures of known age. Third, the thickness of weathering rinds on stones that have been exposed to weathering since prehistoric times can be used to estimate average weathering rates over long periods, if geologic information such as radiometric ages allow us to determine when the weathering began. The results of such investigations suggest that on geologic time scales, weathering rates that are initially rapid tend to decrease steadily with time as the weathering profile approaches a steady-state condition (Fig. 5.16).

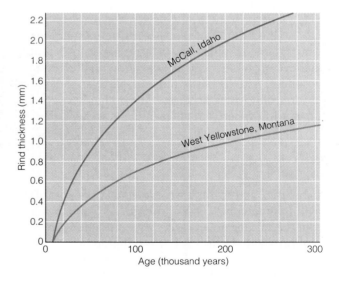

Figure 5.16 Weathering Rates Graph showing change in weathering rates through time for two localities in the northwestern United States. The thickness of weathering rinds on basaltic stones is plotted as a function of their estimated or known age. Weathering rates are initially rapid during the first few tens of thousands of years (steep parts of curves), but then decline steadily with increasing age; after several hundred thousand years, the rates are extremely low. The curve for McCall, Idaho, is steeper than the one for West Yellowstone. This indicates a higher rate of weathering, probably the result of a different climate at the two sites.

SOILS

Soils are one of our most important natural resources. Soils support the plants that are the basic source of our nourishment and provide food for domesticated animals. Soils also help maintain the Earth's surface environment as it is. They accomplish this by supporting vegetation, which, as it releases water to the atmosphere, is an integral part of the hydrologic cycle. In addition, soils store organic matter, thereby influencing the quantity of carbon that is cycled to the atmosphere as carbon dioxide, and they also trap pollutants.

Origin

The physical and chemical breakdown of solid rock by weathering processes is the initial step in the formation of soil. However, soil also contains organic matter mixed in with the mineral component. This organic fraction is essential in the definition of **soil**: the part of the regolith that can support rooted plants.

The organic matter in soil is derived from the decay of dead plants and animals. Living plants are nourished by the nutrients released from decaying organisms, as well as by those released during weathering of mineral matter. Plants draw these nutrients upward, in water solution, through their roots. Therefore, through their life cycle, plants are directly involved in the manufacture of the fertilizer that will nourish future generations of plants. These activities are an integral part of a continuous cycling of nutrients between the regolith and the biosphere. With its partly mineral, partly organic composition, soil forms an important bridge between the Earth's lithosphere and its teeming biosphere.

Soil Profile

As bedrock and regolith weather, the upper part of the regolith gradually evolves into soil. As a soil develops from the surface downward, an identifiable succession of approximately horizontal weathered zones, called **soil horizons**, forms. Each horizon has distinctive physical, chemical, and biological characteristics. Although soil horizons may resemble a sequence of deposits, or layers, they are not strata. Instead, they represent physical and chemical changes in the upper part of the regolith because of its closeness to the land surface. Taken together, the soil horizons constitute a **soil profile**, which consists of the succession of soil horizons that have developed within the regolith over time. The regolith within which the soil horizons have formed was the **parent material** for the soil; The parent material may be represented by still deeper regolith below the soil horizons (Fig. 5.17), and below the regolith there may be fresh rock from which the regolith itself was formed.

Soil profiles generally display two or more horizons. The uppermost horizon may be a surface accumulation of organic matter (*O horizon*) that overlies mineral soil. The organic matter is in various stages of decomposition.

An **A horizon** may either underlie an O horizon or lie directly beneath the surface. Typically, the A horizon is dark grayish or blackish (at least near its top) because of the presence of *humus*, the decomposed residue of plant and animal tissues, which is mixed with mineral matter. The A horizon has lost some of its original substance through the downward transport of clay particles and, more important, through the chemical leaching of more soluble substances.

An *E horizon*, sometimes present beneath the A horizon, has a gray or whitish color. The light color is due mainly to the lack of a darker oxide coating on light-colored mineral grains. E horizons commonly are found in acidic soils that develop beneath forests of evergreen trees.

The **B horizon** underlies the higher horizons and commonly is brownish or reddish. This horizon is enriched in clay and/or iron and aluminum hydroxides produced by the weathering of minerals within the horizon and also transported downward from overlying A and E horizons. Because the B horizon consists mostly of very fine particles, it tends to break into blocks or prisms when it dries. Where clay migration is an important process, each block or prism may be coated with clay. Although the B horizon generally is penetrated by plant roots, it contains less organic matter than the humus-rich A horizon.

A *K horizon*, present in some arid-zone soils beneath the B horizon, is densely impregnated with calcium carbonate which coats other mineral grains and constitutes up to 50 percent of the volume of the horizon.

The **C horizon** is the deepest horizon and constitutes rock in various stages of weathering, but it lacks the distinctive properties of the A and B horizons. Oxidation of iron from the original rock material in the C horizon generally imparts a light yellowish-brown color.

Soil Types

An astute observer traveling across the landscape will note that soils are not everywhere the same. Different soils result from the influence of six soil-forming factors: climate, vegetation cover, soil organisms, composition of parent material, topography, and time. The soil forming under prairie grassland differs from soil in a boreal forest or that of a tropical rainforest. The character of a soil may change abruptly as we move from basalt to limestone or from a gentle slope to a steep slope, and it also will change with the passage of time.

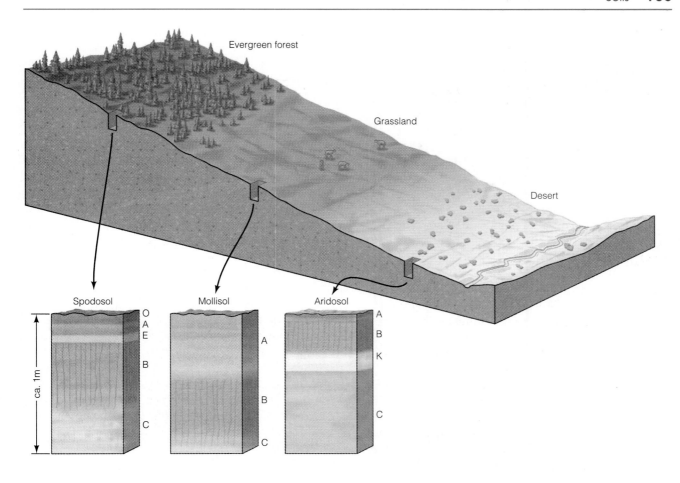

Figure 5.17 Climate and Soils Examples of soil horizons developed under different climatic and vegetation conditions (see Table 5.2).

Soil scientists classify soils according to their physical and chemical properties in much the same way that geologists classify rocks. Such classification makes it easier to map, study, and understand how soils are distributed on the landscape and how their properties reflect the six soil-forming factors (Fig. 5.18). In the soil classification scheme now in standard use in the United States, soils are classified into 11 orders, distinguished on the basis of easily recognizable characteristics (Table 5.2). The eleventh order—the *Andisols*—is restricted to soils developed on tephra.

Polar Soils

In cold, high-latitude deserts, like those found in Greenland, northern Canada, and Antarctica, soils generally are dry and lack well-developed horizons (and are therefore classified as *Entisols*). Weakly oxidized parent material may underlie a layer of coarse frost-churned stones. In wetter high-latitude and high-altitude environments, matlike tundra vegetation overlies perennially frozen ground that prevents water from percolating downward. This leads to water-logged soils that are rich

in organic matter *(Histosols)*. Only on well-drained sites do soils develop recognizable A and B horizons *(Inceptisols)*. Because the cold climate retards chemical processes, well-drained soils generally do not develop the thick, clay-enriched B horizon typical of highly developed temperate-latitude soils.

Temperate-Latitude Soils

Soils of the temperate zone vary largely in response to differences in climate and in the resulting vegetation cover. *Alfisols*, characteristic of deciduous woodlands, typically have a clay-rich B horizon beneath a light-gray E horizon. Acidic *Spodosols* developed in cool, moist evergreen forests have an organic-rich A horizon, an ashlike E horizon, and an iron-rich B horizon. Mountainous terrains, where cool climates cause low rates of soil formation and eroding slopes continually lose the uppermost soil, frequently have minimally developed profiles *(Entisols)* or display weakly developed B horizons lacking clay enrichment *(Inceptisols)*. Grasslands and prairies typically develop *Mollisols* having thick dark-colored, organic-rich A horizons. Soils formed in

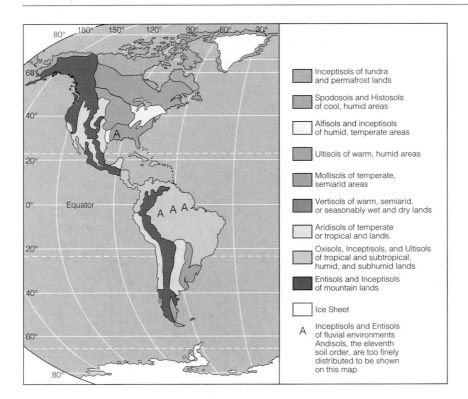

Figure 5.18 Distribution of Soil Types Map of the Western Hemisphere showing distribution of the major soil types.

Legend for map:

- Inceptisols of tundra and permafrost lands
- Spodosois and Histosols of cool, humid areas
- Alfisols and inceptisols of humid, temperate areas
- Ultisols of warm, humid areas
- Mollisols of temperate, semiarid areas
- Vertisols of warm, semiarid, or seasonably wet and dry lands
- Aridisols of temperate or tropical and lands.
- Oxisols, Inceptisols, and Ultisols of tropical and subtropical, humid, and subhumid lands
- Entisols and Inceptisols of mountain lands
- Ice Sheet
- A Inceptisols and Entisols of fluvial environments Andisols, the eleventh soil order, are too finely distributed to be shown on this map

TABLE 5.2	Orders of the Soil Classification System Used in the United States
Soil Order (Meaning of Name)	Main Characteristics
Alfisol (pedalfer[a] soil)	Thin A horizon over clay-rich B horizon, in places separated by light-gray E horizon. Typical of humid middle latitudes.
Aridosol (arid soil)	Thin A horizon above relatively thin B horizon and often with carbonate accumulation in K horizon. Typical of dry climates.
Entisol (recent soil)	Soil lacking well-developed horizons. Only a thin, incipient A horizon may be present.
Histosol (organic soil)	Peaty soil, rich in organic matter. Typical of cool, moist climates.
Inceptisol (young soil)	Weakly developed soil, but with recognizable A horizon and incipient B horizon lacking clay or iron enrichment. Generally occurs under moist conditions.
Mollisol (soft soil)	Grassland soil with thick dark A horizon, rich in organic matter. B horizon may be enriched in clay. E and K horizons may be present.
Oxisol (oxide soil)	Relatively infertile soil with A horizon over oxidized and often thick B horizon.
Spodosol (ashy soil)	Acidic soil marked by highly organic O and A horizons, an E horizon, and iron/aluminum-rich B horizon. Occurs in cool forest zones.
Ultisol (ultimate soil)	Strongly weathered soil characterized by A and E horizons over clay-rich B horizon. Characteristic of tropical and subtropical climates.
Vertisol (inverted soil)	Organic-rich soil having very high content of clays that shrink and expand as moisture varies seasonally.
Andisol (dark soil)	Soil developed on pyroclastic deposits and characterized by low bulk density and high content of amorphous minerals.

[a] Soils rich in iron and aluminum.

subtropical climates commonly display a strongly weathered B horizon *(Ultisols)*.

Desert Soils

In dry climates, where lack of moisture reduces leaching, carbonates often accumulate in the profile during the development of *Aridosols*. These strongly alkaline soils contrast with the more acid soils of humid regions that lack a carbonate-rich K horizon. An important part of the carbonate accumulation results from evaporation of water that rises in the ground, bringing dissolved salts from below. Recently, soil scientists have found that windblown dust also contributes to the accumulation of salts in arid-land soils. Over extensive arid regions of the southwestern United States, carbonates have in this way built up in the soil profile a solid, almost impervious layer of whitish calcium carbonate known as **caliche** (Fig. 5.19).

Tropical Soils

Soils that form where rainfall is high and the climate is warm are characterized by extreme chemical alteration of the parent material. These soils *(Oxisols)* are intensely weathered and infertile because essential nutrients have been leached away. *Vertisols* of tropical regions, where the climate alternates between a wet and a dry season, have a high clay content that causes the soil alternately to swell and shrink in response to seasonal wetting and

Figure 5.20 Temple Made of Laterite Reddish-brown blocks of laterite were used by the ancient Khymer people to construct a temple wall at Angkor Wat in the Cambodian jungle. The color and texture of the laterite contrast with the smooth, grayish sandstone used to construct the adjacent ornately carved temple structure.

Figure 5.19 Caliche in New Mexico A soil profile in semiarid central New Mexico includes whitish caliche forming a prominent K horizon between a yellowish-brown C horizon beneath and a reddish-brown B horizon above.

drying. During the dry season, open cracks extend from the surface down into the soil profile.

Of the many minerals formed during chemical weathering, goethite and hematite are among the least soluble. In tropical regions where the climate is very wet and warm, rock-forming minerals are slowly leached away, leaving a soft, mottled, reddish-gray residue, rich in iron. Geologists generally refer to this product of deep weathering as **laterite**. One product of lateritic weathering is ferric hydroxide [$Fe(OH)_3$]. As a result of climatic change or deforestation, the upper part of a laterite may dry out and become hardened because the ferric hydroxide dehydrates to goethite [$Fe(OH)_3 \rightarrow FeO.OH + H_2O$]. The resulting stonelike material, called *lateritic crust* or *ironstone*, is so hard that blocks of it can be used in construction (Fig. 5.20). The name laterite, which comes from the Latin word for brick *(latere)*, recognizes this property.

Rate of Soil Formation

Although chemical weathering is part of the complex process of soil development, soil formation and weathering are not the same thing. Chemical weathering chiefly concerns the decomposition of bedrock, a process that takes a very long time. The time required to form a soil profile in regolith can be much shorter.

A soil profile can form rapidly in some environments. A study in the Glacier Bay area of southern Alaska shows us how soils develop when retreating glaciers leave unweathered parent material that is finely ground rock flour exposed at the land surface. Despite the cold climate of Glacier Bay, within a few years after a glacier begins to recede, an A horizon develops on the newly exposed and revegetated landscape (Fig. 5.21). In this area, moderate temperatures and high rainfall promote rapid leaching of the finely ground parent material, which is a carbonate-rich glacial sediment. As the plant cover becomes denser, carbonic and organic acids acidify the soil, and leaching becomes more effective. After about 50 years, a B horizon appears and the com-

bined thickness of the A and B horizons reaches about 10 cm. The B horizon is not present in the youngest soils; it appears after about 30 years, and by 100 years the combined thickness of the A and B horizons is about 10 cm. Over the next 150 years, a mature forest develops on the landscape, the O horizon continues to become thicker, and iron oxides accumulate in the developing B horizon (but the A and B horizons do not increase in thickness). This entire succession of vegetation and soils is clearly visible on the deglaciated landscape in front of the retreating glaciers.

In less humid climates, soil forms more slowly, and it may take thousands of years for a detectable B horizon to appear. In the midcontinental United States, B horizons that have developed during the last 10,000 years contain little clay and lack structure, whereas those dating back about 100,000 years generally are rich in clay and have a distinctive structure. As an extreme example of the effect of wetness on rate of soil formation, the glacier-free cold deserts of Antarctica are so dry and cold that sediments more than a million years old have only weakly developed soils.

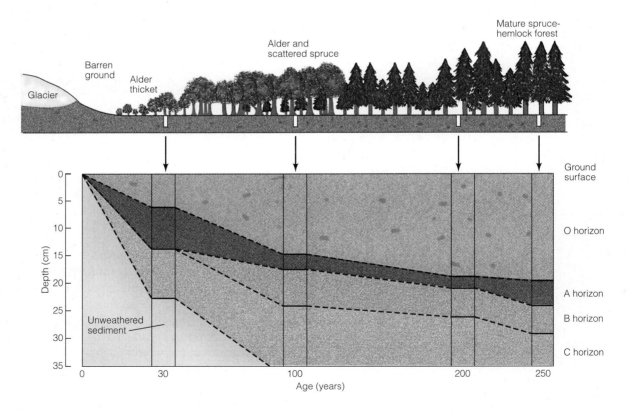

Figure 5.21 Soil Development Progressive soil development in Glacier Bay, Alaska, over the past 250 years can be seen as differences in soil profiles from sites of different age. With time, organic litter forming an O horizon increases in thickness and A and B horizons develop. For the first 30 years, the A horizon directly overlies the C horizon. The B horizon then begins to develop and after another 70 years has reached a thickness of 7 cm. As vegetation changes from alder thicket toward a mature spruce-hemlock forest, organic litter continues to pile up and reaches a thickness of 18 cm after 250 years. The age of each study site was determined by counting the annual growth rings of the oldest surrounding trees. (Based on data from F. C. Ugolini.)

Figure 5.22 Paleosols A thick reddish-brown paleosol (beside figure) developed on a pumice layer near Guatemala City, Guatemala, is overlain by another layer of pumice and pyroclastic flows that are separated by thinner brownish paleosols. A fault has displaced the layers about 2 m vertically.

Paleosols

If a soil is buried, it becomes part of the geologic record. The soil is now a **paleosol**, defined as a soil that formed at the ground surface and subsequently was buried and preserved (Fig. 5.22). Paleosols have been identified in rocks and sediments of many different ages; they are especially common in unconsolidated deposits of the Quaternary Period.

Distinctive and widespread paleosols have been used to subdivide, correlate, and date sedimentary sequences. They also can provide important clues regarding the nature of former landscapes, vegetation cover, and climate.

Soil Erosion

Global agricultural production increased dramatically during the first half of this century, when world popu-

lation was still less than 2.5 billion people. With world population now close to 6 billion, increasing deforestation and increasing agricultural use of land not well suited for agriculture are causing disastrous problems of soil erosion (Fig. 5.23). In many Third World nations, farmers have been forced beyond traditional farm and grazing lands onto steep, easily eroded slopes or into semiarid regions where periodic crop failure is a fact of life and plowed land is prone to severe wind erosion. At the same time, economic pressures in some of the more developed countries have increasingly led farmers to shift from ecologically favorable land-use practices to the planting of profitable row crops that often leave the land vulnerable to increased rates of erosion.

Although soil erosion is a natural process, with rates of erosion determined by topography, climate, or vegetation cover, the results of human activities have overwhelmed natural rates in many parts of the modern world. Widespread felling of trees has led to accelerated rates of surface runoff and destabilization of soils due to loss of anchoring roots. Soils in the humid tropics, when stripped of their natural vegetation cover and cultivated, quickly lose their fertility (Fig. 5.24). So widespread are the effects of soil erosion and degradation that the problem has been described as "epidemic" (see the boxed essay on *The Soil Erosion Crisis*).

Figure 5.23 Lost Soil Deforested hillslopes in Madagascar are deeply gullied by erosion. Streams draining this region turn reddish-brown as they carry away irreplaceable topsoil that is destabilized when anchoring tree roots disappear.

Understanding Our Environment

The Soil Erosion Crisis

The upper layers of a soil contain most of the organic matter and nutrients that support crops. When the O and A horizons are eroded away, not only the fertility but also the water-holding capacity of a soil diminishes. Because it generally takes between 80 and 400 years to form one centimeter of topsoil except under highly favorable circumstances such as the previously cited example of Glacier Bay, for all practical purposes, soil erosion is tantamount to mining the soil. Instead of remaining a continuously productive natural resource, it becomes an ever-decreasing and degraded resource. In the United States, the amount of farmland soil lost to erosion each year exceeds the amount of newly formed soil by more than 2 billion tons. It is estimated that farmers in the United States are now losing about 5 tons of soil for every ton of grain they produce. Soil loss in the former Soviet Union is at least as rapid, whereas in India the soil erosion rate is estimated to be more than twice this high. Worldwide, the estimated loss exceeds 25 billion tons a year. Expressed another way, at the current rate of loss, the world's most productive soils are being depleted at the rate of 7 percent each decade. One recent estimate projected that, as a result of excessive soil erosion and increasing population, only two-thirds as much topsoil is available to support each person at the end of the century as was available in 1984.

Soil erosion is a worldwide problem of massive proportions, and it affects each and every one of us, as the following numbers make clear. A person eats about 750 kg of food a year, and there now are more than 6 billion people on the Earth. Food production leads to an annual loss of 25 billion tons of topsoil, or 4.2 tons (4200 kg) per person. Thus, for every kilogram of food we eat, the land loses 5.6 kg of soil.

Figure 5.24 Deforestation Deforestation is destroying a luxurious rainforest in the Amazon basin near Maraba, Brazil. Stripped of their natural vegetation cover and planted with crops, soils on this landscape quickly lose their natural fertility.

Indirect Effects of Soil Erosion

Much of the topsoil eroded from agricultural lands is transported down rivers and deposited along valley floors, in marine deltas, or in reservoirs behind large dams. The resulting impact on society, often unanticipated, can be significant. For example, the designers of a major dam and reservoir in Pakistan projected a life expectancy for the reservoir of at least a century. However, increased population pressure on the region above the dam has resulted in greatly increased soil erosion, leading to such a high rate of sediment production that the reservoir is now expected to be filled with eroded soil within 75 years, making it unusable.

Control of Soil Erosion

Although soil erosion and degradation are severely impacting many countries, effective control measures can substantially reduce these adverse trends (Fig. 5.25).

In places where crops are grown, the surface is ordinarily bare during part of each year. On unvegetated, sloping fields, on pastures that are too closely grazed, and in areas planted with widely spaced crops such as corn, rates of erosion can be high. Wise farmers therefore reduce areas of bare soil to a minimum and prevent the grass cover on pastures from being weakened by overgrazing. If crops such as corn, tobacco, and cotton must be planted on a slope, strips of such crops are often alternated with strips of grass or similar plants that help resist soil erosion.

Another method of reducing soil loss involves crop rotation. A study in Missouri showed that land which lost 49 tons of soil per hectare when planted continually in corn lost only 7 tons per hectare when corn, wheat, and clover crops were rotated. In this case, the bare land

Figure 5.25 Fighting Soil Erosion Contour plowing is one effective method of reducing soil erosion. On this farm, crops have been planted in belts that follow, as much as practicable, the natural contours of the land, thereby inhibiting the development of rills and gullies that otherwise might form perpendicular to the contours.

exposed between rows of corn is far more susceptible to erosion than land planted with a more continuous cover of wheat or clover.

The most serious soil erosion problems occur on steep hillslopes. In Nigeria, for example, land planted with cassava (a staple food source) and having a gentle 1 percent slope lost an average of 3 tons of soil per hectare each year. On a 5 percent slope, however, the annual rate of soil loss increased to 87 tons per hectare. At this rate, a 15 cm thickness of topsoil would disappear in a single generation (about 20 years). On a 15 percent slope, the annual erosion rate increased to 221 tons per hectare, a rate that would remove all topsoil within a decade. Despite these grim statistics, steep slopes can be exploited through terracing.

In central China, where steep hillslopes are underlain by erodible deposits of wind-blown silt, terracing is a major factor in reducing soil loss (Fig. 5.26). In the arid regions of western China where strong winds and restricted vegetation cover can cause rapid soil loss, rows of trees planted as wind screens can substantially reduce erosion on the downwind side.

Soil Erosion and the World Economy
Because agriculture is the foundation of the world economy, progressive loss of soil signals a potential crisis that could undermine the economic stability of many societies. Halting the worldwide loss of soils is a formidable task that will require effective programs at the national and international levels. Soils can be viewed as a renewable resource only over geologically long intervals of time. Over the lifetime of individuals, or even of nations, they must be considered nonrenewable resources that must be carefully utilized and preserved if we are to have a sustainable global economy.

WEATHERING AND THE ROCK CYCLE

Weathering is an integral part of the rock cycle, and, as we saw in Chapter 2, the rock cycle is linked to the tectonic cycle through plate tectonics. If fresh rocks were not continually brought to the surface by tectonic uplift and volcanism, erosion would lower the land and a deep weathering profile would eventually develop on each continent. Therefore, the processes associated with plate tectonics provide the grist for the weathering mill.

The resistance of a silicate mineral to weathering is a function, principally, of three things: first, of the chemical composition of the mineral; second, of the extent to which the silicate tetrahedra in the mineral are polymerized (see Chapter 3); and, third, of the acidity of the waters with which the mineral reacts. Ferromagnesian minerals with no polymerization, such as olivine, or with chain polymers such as pyroxene, and amphibole, weathers most easily at the Earth's surface (Table 5.3).

Figure 5.26 Fighting Soil Erosion by Terracing A terraced hillside near Lanzhou, China, creates productive agricultural fields from steep hillslopes carved in deposits of windblown dust. Where unterraced, hillslopes are rapidly eroding and deeply gullied.

Calcium-rich plagioclase feldspar also weathers readily. Biotite, muscovite, and sodium-rich plagioclase are less easily weathered. Quartz, which has the highest degree of polymerization of all silicate minerals, is among the most resistant of rock-forming minerals and degrades chemically at an exceedingly slow rate. Nevertheless, given enough time even quartz can be dissolved.

A glance at the mineral-stability sequence (Table 5.3) shows that the order in which the minerals are ranked is approximately the reverse of that in which minerals crystallize in Bowen's reaction series (Fig. 4.40). This reversed order is not surprising because the least polymerized minerals crystallized earliest in Bowen's series.

PLATE TECTONICS AND GLOBAL WEATHERING RATES

We can estimate the rate at which a tombstone in a graveyard is being weathered chemically by measuring how much surface decomposition has occurred since the date was chiseled on the stone. Measuring the average global rate of chemical weathering is not quite as easy. The best approach is to measure the amount of dissolved substances delivered by rivers to the oceans, for the dissolved matter in a stream results from chemical weathering of rocks and sediments in the stream's drainage basin. Although a multitude of streams enter

TABLE 5.3	Order of Stability of Common Minerals Under Attack by Chemical Weathering		
MOST STABLE (Least susceptible to chemical weathering)	Ferric oxides and hydroxides Aluminum oxides and hydroxides Quartz Clay minerals Muscovite Potassium feldspar Biotite Sodium feldspar (albite-rich plagioclase) Amphibole	LEAST STABLE (Most susceptible to chemical weathering)	Pyroxene Calcium feldspar (anorthite-rich plagioclase) Olivine Calcite

the sea, the three largest contributors of dissolved substances are the Yangtze River which drains the high Tibetan plateau of China, the Amazon River which drains the northern half of the Andes in South America, and the Ganges-Brahmaputra river system that drains the Himalaya in India. Collectively, these three streams deliver about 20 percent of the water and the dissolved matter entering the oceans. If we add all the other streams draining these three highland regions, we can conclude that chemical weathering and erosion in Tibet, the Andes, and the Himalaya must provide a substantial part of the total dissolved load reaching the world's oceans. In other words, a direct relationship apparently exists between the occurrence of high-altitude landmasses and the global rate of chemical weathering. If true, then we might expect to see changes in weathering rates over geologic time, as mountain systems are uplifted and then worn away.

High rates of chemical weathering and high mountains are related for several reasons. First, high mountains are areas of rapid uplift. They occur where plates of lithosphere converge and cause mountain systems to form. Rapid uplift goes hand in hand with rapid mechanical breakdown and erosion of rock. This disintegration exposes large quantities of rock and mineral debris to chemical weathering.

Second, the amount of dissolved matter in streams is greatest in areas where easily eroded sedimentary rocks are exposed. If we examine the geology of high mountains like the Himalaya, the Andes, and the Alps, we find that these areas of active, rapid uplift are dominated by sedimentary rocks of the type that form along continental margins. Lowland regions in continental interiors tend to be dominated by ancient, less erodible metamorphic and plutonic rocks.

Third, high mountains force moisture-bearing winds upward and generally receive large amounts of precipitation. This means high rates of stream runoff and high erosion rates. This effect is especially pronounced in southern Asia where the intense monsoon rainfall on the southern flank of the Himalaya leads to unstable slopes, intense erosion, and a high discharge of dissolved matter and suspended sediment to the Indian Ocean (Fig. 5.27). Studies in South America have shown that more than three-quarters of the dissolved substances carried by the Amazon River to the sea come from the Andean highlands.

Past weathering rates apparently have not always been as high as they are today. Evidence from ocean sediments points to an increase in the amount of dissolved matter reaching the oceans during the past 5 million years. This implies an increase in global weathering rates during this time. Furthermore, the major mountain systems of the world have not always been as high as we find them today. Evidence points to increased uplift rates in the Andes, Himalaya, Tibetan plateau, and other tectonic regions like western North America and the Alps since the end of the Miocene Epoch (ca. 5.1 million years ago; Fig. 8.10). For example, sediments shed from the rising Himalaya coarsen upward, from silts of early Pliocene age (ca. 5.0 million years ago) to gravels of middle and late Pleistocene age (ca. 1.0 million years ago). This change implies that streams draining the mountains gained increasing energy with time as mountain slopes and stream channels steepened.

The changes we can detect in the rates of mountain uplift parallel changes in weathering rates that we can infer from the record of sediments and sedimentary rocks. Mountain uplift rates are believed to be closely related to rates of seafloor spreading. Thus, global rates of chemical weathering must be inextricably linked to plate tectonics.

Figure 5.27 Rapid Erosion The Indus River, which drains the high western Himalaya and Karakorum ranges, carries a high dissolved load to the Indian Ocean. This mountain region has one of the highest measured uplift rates in the world. High rates of erosion lead to high rates of chemical and physical weathering.

SUMMARY

1. Weathering extends to whatever depth air, water, and microscopic forms of life penetrate the Earth's crust. Water solutions, which enter the bedrock along joints and other openings, attack the rock chemically and physically, causing breakdown and decay.

2. Physical and chemical weathering, although involving very different processes, generally work together.

3. Growth of crystals, especially ice and salt, along fractures and other openings in bedrock is a major process of physical weathering. Others include intense fires that cause spalling and fracturing, the wedging action of plant roots, and the churning of rock debris by burrowing animals. Each of these processes can have large cumulative effects over time.

4. Chemical weathering involves the transformation of minerals that are stable deep in the Earth into forms that are stable at the Earth's surface. The principal processes are hydrolysis, leaching, oxidation, hydration, and dissolution.

5. Subdivision of large blocks into smaller particles increases surface area and thereby accelerates chemical weathering.

6. The effectiveness of weathering depends on rock type and structure, surface slope, local climate, and the time over which weathering processes operate.

7. Because heat and moisture speed chemical reactions, chemical weathering is far more active in moist, warm climates than in dry, cold climates.

8. Soils consist of regolith capable of supporting plants. Soils develop distinctive horizons whose character is a function of climate, vegetation cover, soil organisms, parent material, topography, and time.

9. The A horizon of a soil is rich in organic matter and has lost soluble minerals through leaching. Clay accumulates in the B horizon, together with substances leached from the A horizon. Both overlie the C horizon, which is less weathered parent material.

10. The classification system used in the United States places soils in 11 soil orders based on their physical characteristics.

11. Paleosols are buried soils that can provide clues about former landscapes, plant cover, and climate and are useful for subdividing, correlating, and dating the strata in which they are found.

12. Soil erosion and degradation are global problems that have been increasing as world population rises. Effective control measures include crop rotation, plowing along contours rather than downslope, terracing, and tree planting, but halting widespread loss of soils is a formidable challenge.

13. Chemical weathering leads to concentration of economically valuable mineral deposits that are primary sources of aluminum, nickel, iron, manganese, and copper.

14. Weathering is an integral part of the rock cycle. The silicate minerals least resistant to weathering are the ferromagnesian minerals with the least amount of ploymerization of silicate tetrahedra.

15. Global rates of weathering are linked to the presence or absence of high mountains, and thus to plate tectonics.

THE LANGUAGE OF GEOLOGY

A horizon (p. 134)

B horizon (p. 134)
bauxite (p. 128)

C horizon (p. 134)
caliche (p. 137)
carbonic acid (p. 125)
chemical weathering (p. 121)

dehydrate (p. 126)
dissolution (p. 126)

exfoliation (p. 130)

frost wedging (p. 123)

goethite (p. 128)

hematite (p. 126)
hydration (p. 126)
hydrolysis (p. 125)

joints (p. 121)

kaolinite (p. 125)

laterite (p. 137)
leaching (p. 125)

oxidation (p. 125)

paleosol (p. 139)
parent material (p. 134)
physical weathering (p. 121)

secondary enrichment (p. 128)
soil (p. 134)
soil horizon (p. 134)
soil profile (p. 134)
spall (p. 124)
spheroidal weathering (p. 130)

weathering (p. 121)
weathering rind (p. 130)

QUESTIONS FOR REVIEW

1. Explain the difference between decomposition and disintegration of rocks.

2. In what way do joints affect rock weathering?

3. How does acid rain in industrialized regions contribute to weathering?

4. Why is frost wedging likely to be most effective at temperatures between about –5° and –15° C?

5. What causes flakes of rock to spall off exposed boulders during a forest fire?

6. Explain why minerals such as gold or platinum can become concentrated in sediments.

7. Why does the physical breakup of a rock increase the effectiveness of chemical weathering?

8. Explain how rock type, rock structure, slope, climate, and time can influence weathering.

9. How does regolith formed on chemically weathered limestone differ from that formed on chemically weathered igneous rock, and why?

10. Describe the horizons in the profile of a well-developed soil in an evergreen forest; in an arid region.

11. If a buried soil is characterized by well-developed caliche in the top of the C horizon, what can you infer about the climate at the time the soil formed?

12. Why is the top of a paleosol an unconformity?

For an interactive case abstract, virtual tours, activities, and additional learning resources, go to
GEOSCIENCES TODAY: www.wiley.com/college/skinner

How Much Topsoil Is Left in the Corn Belt?
Land and Life on the North American Prairie

Each of the world's grassland biomes has its own distinct history, ecology, and local name. Like other biomes, grasslands share certain common ecological characteristics owing primarily to adaptation to soil, water, climate, fire, and other plant growth constraints. One of these is the Great Plains or Mid-Continent Prairie. The history, landscape, and agriculture of the Corn Belt illustrate the tight link that exists between landscape evolution, geology, ecology, climate, and natural resource availability.

OBJECTIVE: The primary objectives of this case are to introduce the underlying biophysical forces that characterize the global distribution of the grassland biome; consider how the Prairie/Great Plains ecoregion of mid-continent North America has evolved through time; and explore some of the complex issues of land degradation and ecosystem management.

The Human Dimension: In addition to the importance of managing resources for farmers and ranchers, how are the prairies becoming increasingly important recreational resources in providing access to open spaces and wilderness?

Questions to Explore:

1. What is the grassland biome, and where is it located around the world and in North America?

2. How have humans impacted the prairie landscape through time and particularly the subregion known as the Corn Belt?

3. How is the land degradation—particularly soil erosion, groundwater extraction, and biodiversity loss—monitored and mitigated using Earth Systems Science (ESS) tools and techniques.

6

Understanding
Our Environment

Using
Resources

GeoMedia

Anthropologists working in the Hadar region of Ethiopia. The skull of a male *Australopithecus afarensis* was discovered at this site and more bone fragments are being sought. Note the sedimentary layering visible in the hills in the distance.

Sediments and Sedimentation

Sedimentary Archives of Earth and Human History

The Hadar region of Ethiopia would seem to be a strange place to look for evidence of early human history, for the barren landscape looks quite inhospitable to life. Yet here, in a succession of ancient stream sediments, archaeologists have discovered skeletal remains of *Australopithecus*, a creature that many experts believe was an ancestral human form. In addition, they have found crude stone tools that may be the earliest evidence of human activity. The archaeological relicts and enclosing sediments constitute an invaluable archive detailing the environments when human beings first appeared on the Earth. Because this region lies along the great African rift-valley system, where volcanoes periodically have spread pyroclastic debris across the landscape, layers of volcanic ash interstratified with the stream deposits provide a means of dating both the sediments and their fossil content. Thus, we know from laboratory dating of the ash layers (Chapter 8) that the sediments were laid down between about 3.4 and 1 million years ago at a time when the landscape of this part of Africa provided more favorable habitats for life than it does today.

At this and other East African fossil sites, the enclosing sediments provide clues to ancient environments. Each area and each layer has its own story to tell, and collectively they weave a picture of changing landscapes, changing climate, and changing life. Each environment where sediments accumulate leaves its own distinctive imprint. Sediments and layered rocks exposed along a valley side may appear indistinguishable and uninteresting to the average person, but an experienced geologist can study the layers and determine how each unit was deposited, for example, whether by a stream, in a lake, or by the wind. The composition, shape, orientation, color, layering, and internal structure of the sedimentary particles, as well as the species and isotopic signatures of the remains of plants and animals, are all clues that can help us visualize the character of the land surface. From such clues, we can reconstruct an ancient geography that may bear little or no resemblance to that of the present. Part of the excitement a geologist gains from her or his work derives from using such sedimentary clues to puzzle out how the Earth's landscapes and life have evolved.

SEDIMENTS AND SEDIMENTATION

Like a perpetually restless housekeeper, nature is ceaselessly sweeping regolith off the solid rock beneath it, carrying the sweepings away, and depositing them as sediment in river valleys, lakes, and innumerable other places including, in particular, the seafloor. We can see sediment being transported by trickles of water after a rainfall and by every wind that carries dust. The mud on a lake bottom, the sand on a beach, even the dust on a windowsill is sediment.

Stratification and Bedding

Sedimentary **stratification** results from the arrangement of sedimentary particles in distinct layers (Fig. 6.1). Each sedimentary **stratum** (plural = **strata**) is a distinct layer of sediment that accumulated at the Earth's surface. Although layering is an obvious feature of most sediments and sedimentary rocks, it is also seen in the products of volcanic eruptions (lava flows and tephra deposits) and in many metamorphic rocks. A close look

Figure   6.1   Layers of Sedimentary Rock: Bedding   When you look at an outcrop of sedimentary rock, one of the first things you notice is layering, or bedding. Here, layered sedimentary rocks in Capital Reef National Park, Utah, have been exposed by erosion.

A. Sediment

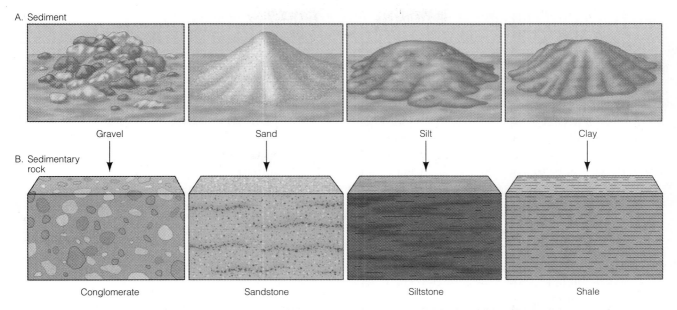

Gravel   Sand   Silt   Clay

B. Sedimentary rock

Conglomerate   Sandstone   Siltstone   Shale

Figure 6.2 Clastic Sediments and Clastic Sedimentary Rocks The main kinds of clastic sediments and the sedimentary rocks formed from them.

at sedimentary strata shows that they differ from one another because of differences in some characteristic of the particles or in the way in which they are arranged. For example, the average diameter of particles in one stratum may differ from the average diameter in another stratum.

The layered arrangement of strata in a body of sediment or sedimentary rock is referred to as **bedding**. Each **bed** within a succession of strata can be distinguished from adjacent beds by differences in thickness or character. The top or bottom surface of a bed is a **bedding plane**.

Clastic Sediment

A close look at sand and gravel beside a stream shows that pebbles and sand grains are simply bits of rock and minerals. The finer sedimentary particles are too small to see, but they are also derived from broken-up rock. The breaking up normally involves chemical as well as mechanical processes. Some of the finer particles are tiny pieces of the original rock, but most are minute mineral grains formed by chemical processes during weathering. For example, chemical decomposition of feldspar results in the formation of clay. The loose, fragmental debris produced by the mechanical breakdown of older rocks is called **detritus** (from the Latin for "worn down") or **detrital sediment**. Such sediment also is referred to as **clastic sediment** (from the Greek word *klastos*, meaning "broken") (Fig. 6.2). Any individual particle of clastic sediment is a *clast*. Particles of clastic sediment may

range in size from the largest boulders down to submicroscopic clay particles. This range of particle size is the primary basis for classifying clastic sediments and clastic sedimentary rocks. In Table 6.1 we can see that clastic sediment can be divided into four main classes, which, from coarsest to finest are gravel, sand, silt, and clay. Gravel is further subdivided on the basis of dominant clast size into boulder gravel, cobble gravel, and pebble gravel. If, for example, a gravel consists predominantly of clasts having diameters between 64 and 256 mm, we would call it a cobble gravel.

Mineral Composition of Clastic Sediment
Most coarse clastic sediments consist of mineral grains and rock fragments, with those least susceptible to chemical and physical breakdown predominating. Most sands, for example, have a high content of quartz, and some have a good bit of potassium feldspar; these are the common rock-forming minerals most resistant to weathering. As detrital rock fragments and mineral grains are transported, they are subjected to continuous chemical and physical breakdown. After several cycles of erosion and deposition, the result can be a sediment that consists almost entirely of quartz, the most resistant of the rock-forming minerals (Fig. 6.3).

Some Conspicuous Features of Clastic Sediments
The sizes of particles, the way they are packed together, and other distinctive features permit us to distinguish different types of sediment, as well as sedimentary rocks. Such characteristics also help us to infer the environment in which a sediment was deposited.

TABLE 6.1	Definition of Clastic Particles, Together with the Sediments and Sedimentary Rocks Formed from Them		
Name of Particle	Range Limits of Diameter (mm)[a]	Names of Loose Sediment	Name of Consolidated Rock
Boulder	More than 256	Boulder gravel	Boulder conglomerate[c]
Cobble	64 to 256	Cobble gravel	Cobble conglomerate[c]
Pebble	2 to 64	Pebble gravel	Pebble conglomerate[c]
Sand	1/16 to 2	Sand	Sandstone
Silt	1/256 to 1/16	Silt	Siltstone
Clay[b]	Less than 1/256	Clay	Mudstone and shale

Source: C. K. Wentworth, 1922, A scale of grade and class terms for clastic sediments: *Journal of Geology* 30, p. 377–392.

[a]Note that size limits of sediment classes are powers of 2, just as are memory limits in microcomputers (for example, 2K, 64K, 256K, 512K).

[b]Clay, used in the context of this table, refers to particle size. The term should not be confused with clay minerals, which are definite mineral species.

[c]If the clasts are angular, the rock is called a *breccia* rather than a conglomerate.

Figure 6.3 Well Rounded and Well Sorted Well-rounded grains of quartz sand from the St. Peter Sandstone of Wisconsin have been sorted by size and polished by constant shifting and abrasion in surf along an ancient shoreline.

Sorting. **Sorting** is a measure of the range of particle size of sediments. A sediment that has a wide range of particle size is said to be poorly sorted; if the range is small, the sediment is said to be well sorted (Figs. 6.3 and 6.4). In a clastic sediment, changes of grain size typically result from fluctuations in the speed of the transporting medium: the faster the speed, the larger are the particles that can be moved.

Sorting of particles also can be related to differences in their specific gravity. Particles of unusually dense minerals (e.g., gold, platinum, and magnetite) are deposited quickly on streambeds or on beaches, whereas lighter particles are carried onward.

Most of the particles transported by water or wind, however, are common rock-forming minerals, such as quartz and feldspar, that have similar specific gravities. Therefore, such particles typically are sorted not according to specific gravity but according to *size* (Fig. 6.3).

Long-continued movement of particles by turbulent water or air results in gradual destruction of the weaker particles, leaving behind the particles that can better survive in the turbulent environment. Very commonly, the survivor is quartz because it is hard and lacks cleavage. In this case, enrichment of a specific mineral in a sediment is based on *durability*.

Particle Shape. Mechanically weathered particles broken from bedrock tend to be angular because breakage typically occurs along grain boundaries, fractures, and surfaces separating rock layers. Nearly all such weathered particles become smooth and rounded as they are transported by water or air and are abraded by other rock fragments. *Roundness*, as measured by the sharpness of a particle's edges, is not the same as *sphericity*, which is a measure of how closely particle shape approaches that of a sphere (Fig. 6.4). A flat particle bounded by cleavage or fracture surfaces may have well-rounded edges, but it also may have a low degree of sphericity. In general, the greater the distance of travel, the greater is the degree of rounding.

Rhythmic Layering. Some sediments display a distinctive alternation of parallel layers having different properties. Such alternation suggests that some naturally occurring rhythm has influenced sedimentation. A pair of such sedimentary layers deposited over the cycle of a single year is termed a **varve** (Swedish for cycle). Varves are most commonly seen in deposits of high-latitude or high-altitude lakes, where there is a strong contrast in seasonal conditions. In spring, as a cover of winter ice melts away, the inflow of sediment-laden water increases and coarse

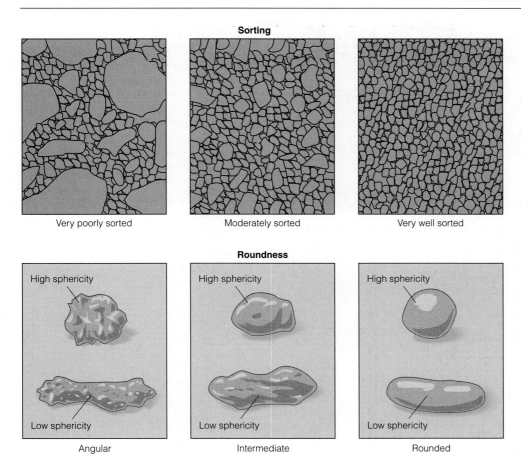

Sorting

Very poorly sorted   Moderately sorted   Very well sorted

Roundness

High sphericity   High sphericity   High sphericity

Low sphericity   Low sphericity   Low sphericity

Angular   Intermediate   Rounded

Figure 6.4 Sorting and Roundness Clastic sediments range from very poorly sorted to very well sorted, depending on the extent to which the constituent grains are of equal size. Particles also range from angular to rounded, depending on the degree to which sharp edges have been worn off, and they have a high or low degree of sphericity depending on how closely they approach a spherical shape.

sediment is then deposited throughout the summer. With the onset of colder conditions in the autumn, streamflow decreases and ice forms over the lake surface. During winter, very fine sediment that has remained suspended in the water column slowly settles to form a thinner, darker layer above the coarse lighter-colored summer layer. Varved lake sediments are common in Scandinavia and New England, where they formed beyond the retreating margins of ice-age glaciers (Fig. 6.5), and they are also seen in some ancient sedimentary rocks.

Cross Bedding. **Cross bedding** refers to beds that are inclined with respect to a thicker stratum within which they occur (Fig. 6.6). Cross beds consist of particles coarser than silt and are the work of turbulent flow in streams, wind, or ocean waves. As they move along, the particles tend to collect in ridges, mounds, or heaps in the form of ripples, waves, or dunes that migrate slowly forward in the direction of the current. Particles accumulate on the downcurrent slope of the pile to pro-

duce beds having inclinations as great as 30° to 35°. The direction in which cross bedding is inclined tells the direction in which the related current of water or air was flowing at the time of deposition.

Graded Bedding. The way particles are arranged within a layer provides important information about the conditions of sedimentation. If a mixture of small solid particles having different diameters and about the same specific gravity is placed in a glass of water, shaken vigorously, and then allowed to stand, the particles will settle out and form a deposit on the bottom of the glass. The largest particles settle first, followed by successively smaller ones. The finest may stay in suspension for hours or days before they finally settle out at the top of the deposit. In the resulting **graded bed**, the particles are sorted more or less according to size and grade upward from coarser to finer (Fig. 6.7). A graded bed also can form from a sediment-laden current. As the current slows down, the heaviest and largest particles settle out first, followed by lighter and smaller ones.

Figure 6.5 Sediment Cycles
Varves deposited in a glacial-age lake in southern Connecticut. Each pair of layers in a sequence of varves represents an annual deposit. Light-colored silty layers were deposited in summer, and the dark-colored clayey layers accumulated in winter. Each layer is well sorted.

Figure 6.6 Ancient Sand Dunes
Ancient cross-bedded sand dunes converted to sandstone are exposed widely in Zion National Park, Utah. The inclined bedding dips to the right, in the direction toward which the prevailing wind blew when the dunes were active.

Figure 6.7 Graded Bed A mudflow deposit that originated at Mount St. Helens volcano is graded. Conspicuous cobbles and pebbles in the lower part grade upward into finer pebbles and sand near the top. This graded bed resulted from a prehistoric eruption that sent a flood of muddy debris down a nearby valley.

Nonsorted Sediment. In some sedimentary rocks, the particles are a mixture of different sizes arranged chaotically, without obvious order. Such sediments are created, for example, by rockfalls, slow movement of debris down hillslopes, slumping of loose deposits on the seafloor, mudflows, and deposition of debris from glaciers or floating ice. Some nonsorted sediments are given specific names; for example, *till* is a nonsorted sediment of glacial origin, while the corresponding rock is a *tillite* (see Fig. 6.8).

CHEMICAL SEDIMENT

Some sediment contains no clastic particles, yet the material composing the sediment has been transported. In such sediment, the components were dissolved, transported in solution, and precipitated chemically. Sediment

Figure 6.8 Tillite The Dwyka Tillite, a nonsorted glacial deposit of boulders, cobbles, and pebbles in a fine-grained matrix, crops out in the western Transvaal, South Africa. It was deposited at a time when this part of Africa lay closer to the South Pole, allowing an ice sheet to exist.

formed by precipitation of minerals from solution in water is **chemical sediment**, and it forms in two principal ways: One way is through biochemical reactions resulting from the activities of plants and animals in the water. For example, tiny plants living in seawater can decrease the acidity of the surrounding water and thereby cause calcium carbonate to precipitate.

Chemical sediment also forms as a result of inorganic reactions in the water. When the water of a hot spring cools, it may precipitate opal or calcite. We can witness a similar effect when chemical sediment is deposited along the inside of a hot-water pipe, thereby constricting the flow of water and leading to expensive home repairs. Another common example is simple evaporation of seawater or lake water; as the water evaporates, dissolved matter is concentrated and salts begin to precipitate out as chemical sediment.

Biogenic Sediment

Many bodies of sediment contain **fossils**, the remains of plants and animals that died and were incorporated and preserved as the sediment accumulated. Sometimes the structure or skeleton of a plant or animal is preserved, but more commonly the remains are broken and scattered (Fig. 6.9). A sediment composed mainly of fossil remains, and therefore produced directly by the physiological activities of organisms, is called **biogenic sediment**. The solid parts generally end up as fragments, or clasts. If the sediment is composed largely of such fragments, we call

Figure 6.9 Fossil Dinosaur Fragmented remains of a hadrosaur being excavated from sedimentary strata approximately 70 million years old. The fossil was found in Coahuila State, Mexico.

it *bioclastic* sediment. The organic materials (soft tissues) of plants and animals often decompose, but in favorable environments they are transformed into new organic compounds that either are locally concentrated or widely dispersed in the enclosing sediment.

Calcareous and Siliceous Biogenic Sediment

One important kind of biogenic sediment is formed of calcium carbonate and is widespread in the oceans. Although calcium carbonate can be precipitated chemically from seawater, most carbonate sediments are biogenic and form mainly in warm surface ocean waters. There, carbonate-secreting organisms precipitate calcite or aragonite in building their hard parts. In doing so, the calcium and bicarbonate ions in the water are combined to form solid calcium carbonate:

$$\underset{\substack{\text{calcium} \\ \text{ion}}}{Ca^{2+}} + \underset{\substack{\text{bicarbonate} \\ \text{ion}}}{(HCO_3)^{1-}} = \underset{\substack{\text{calcium} \\ \text{carbonate}}}{CaCO_3} + \underset{\substack{\text{hydrogen} \\ \text{ion}}}{H^{1+}}$$

When floating microscopic marine organisms die, their remains settle and accumulate on the seafloor to form a muddy sediment called **deep-sea ooze** (Fig. 6.10). If such a sediment consists mainly of carbonate remains, it is *calcareous ooze*. Corals, algae, and other colonial organisms growing on the seafloor in shallow tropical waters also precipitate calcium carbonate, thereby creating extensive carbonate reefs.

Siliceous ooze resembles calcareous ooze but is composed mainly of the siliceous remains of tiny floating protozoa (radiolarians) and algae (diatoms). The hard parts in this ooze are often delicate and lacelike (Fig. 6.10). Some kinds of diatoms also live in lakes, where their remains accumulate as an important component of lake sediment.

Growth and Burial of Plant Matter

Nearly all living organisms derive their energy from the Sun. The chief energy-trapping mechanism is photosynthesis (see Figure B2.1), the process by which plants

Figure 6.10 Deep-Sea Ooze Skeletons of calcareous foraminifera (globular objects), siliceous radiolaria (delicate meshed objects), and siliceous rod-shaped sponge spicules from a deep-sea ooze, imaged by scanning electron microscope. The fossils are from a sediment core collected in the western Indian Ocean.

use the Sun's energy to combine water and carbon dioxide to make carbohydrates (organic compounds containing C, O, and H) and oxygen. Therefore, animals that consume plants are secondary consumers of trapped solar energy. When plants or animals die and decay, atmospheric oxygen combines with carbon and hydrogen in the organic compounds to form H_2O and CO_2 once again.

When organic matter is buried, some of the original entrapped solar energy becomes stored in sediments and ultimately in sedimentary rocks. The total amount of trapped organic matter is far less than 1 percent of the organic matter formed by growing plants and animals. Nevertheless, during the past 600 million years, the amount of trapped organic matter has grown to be very large. This trapped organic matter supplies the fossil fuel energy on which our modern society relies (see the boxed essay on *Fossil Fuels and the Politics of the Middle East*).

LITHIFICATION AND DIAGENESIS

Lithification (from the verb to *lithify*, meaning turn to stone) is the process whereby a newly deposited, unconsolidated sediment is slowly converted to sedimentary rock. During lithification, a number of changes commonly occur. Geologists refer collectively to all the chemical, physical, and biological changes that affect sediment after its initial deposition and during and after lithification as **diagenesis**.

The first and simplest diagenetic change is **compaction**, which occurs as the weight of an accumulating sediment forces the grains together. As the *pore space* (the space between grains) is reduced, water is forced out of the sediment. Substances dissolved in circulating pore water precipitate and cement the grains together, a process called **cementation**. Calcium carbonate is one of the most common cements (Fig. 6.11), but silica, a particularly hard cement, may also bond grains together.

As sediments accumulate, less stable minerals may recrystallize to more stable forms. The process of **recrystallization** is especially common in porous reef limestone. Over time, the mineral aragonite, which forms the skeletal structure of living corals, recrystallizes to its polymorph calcite.

Important chemical alterations also affect sediments. In the presence of oxygen (an *oxidizing environment*), organic remains are quickly converted to carbon dioxide and water. If oxygen is lacking (a *reducing environment*), the organic matter does not completely decay but instead may be slowly transformed to solid carbon.

The formation of peat, and its slow transformation to

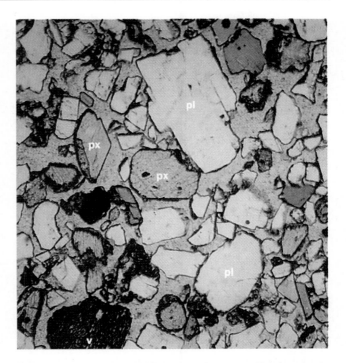

Figure 6.11 Cementation of a Sandstone Sand grains in a thin section of a sandstone from central Washington are bonded together by calcite cement. Light-colored grains are plagioclase (pl), brownish grains are pyroxene (px), and the large dark grain is a fragment of volcanic rock (v).

coal, provide an example of diagenesis under reducing conditions. Geologists recognized long ago that the plants responsible for deposits of peat and coal must have lived in ancient swamps because (1) a complete physical and chemical gradation exists from peat to coal, and peat now accumulates mainly in swamps; and (2) only under swamp conditions is the conversion of plant matter to peat chemically probable. On dry land and in running water, oxygen is abundant. Under these conditions, dead plant matter gradually rots away. Under stagnant or nearly stagnant swamp water, however, oxygen is used up and not replenished. Any plant matter lying in swamp water is attacked by anaerobic bacteria, which partly decompose it by splitting off some of the oxygen and hydrogen from the chemical compounds of which they are comprised. These two elements escape, combined in various gases, and the carbon gradually becomes concentrated in the residue. Although they work to destroy the plant matter, the bacteria themselves are destroyed before they can finish the job, because the poisonous acidic compounds they liberate from the dead plants kill them. In this manner, plant matter is gradually converted to peat. This could not happen in a stream because the flowing water would bring in new oxygen to decompose the plants and would also dilute the poisons and permit the bacteria to complete their destructive work.

Using Resources

Fossil Fuels and the Politics of Middle East Oil

Local large concentrations of organic substances in sediments and sedimentary rocks, in the form of coal, oil, and natural gas (collectively called the **fossil fuels**), provide most of the energy that runs our modern civilization. The nature and occurrence of each of these basic fuel resources depend on the kind of sediment, the kind of organic matter trapped, and the changes that have occurred during the long geological ages since the organic matter was buried.

Peat

On land, plants such as trees, bushes, and grasses contribute most of the trapped organic matter. These plants contain carbohydrates, but they are also rich in *resins, waxes,* and *lignins,* which tend to remain solid. In water-saturated environments, such as bogs or swamps, plant remains accumulate to form **peat,** an unconsolidated deposit of plant remains having a carbon content of about 60 percent (Fig. B6.1). Peat is the initial stage in the development of the combustible sedimentary rock we call coal, the most abundant of the fossil fuels.

Petroleum

In the ocean, microscopic phytoplankton (tiny floating plants) and bacteria (simple, single-celled organisms) are the principal sources of organic matter trapped in sediment. Most of the organic matter is trapped in clay that is slowly converted to shale. During this conversion, organic compounds are transformed into oil and natural gas. These two products are the main forms of **petroleum,** defined as gaseous, liquid, and semisolid naturally occurring substances that consist chiefly of **hydrocarbons** (chemical compounds of carbon and hydrogen).

Two kinds of evidence support the hypothesis that petroleum is a product of the decomposition of the remains of organisms: (1) oil possesses optical properties known only in hydrocarbons derived from biogenic organic matter, and (2) oil contains nitrogen and certain compounds believed to originate only in living matter.

Oil is nearly always found in marine sedimentary rocks. Sampling on the continental shelves and along the base of the continental slopes has shown that fine muds beneath the seafloor contain up to 8 percent organic matter. Geologists therefore conclude that oil originated primarily as organic matter deposited with marine sediment.

A long and complex chain of chemical reactions apparently is involved in the conversion of organic matter to crude petroleum and natural gas. In addition, chemical changes may occur in oil and gas even after they have

Figure B6.1 Cutting Peat A peat cutter harvests dark organic-rich peat from a bog in western Ireland. The peat has formed in a cool, moist climate that favors preservation of organic matter in wet environments. When dried, the peat provides fuel for heat and cooking.

accumulated. Such changes may explain, for example, why chemical differences exist between the oil in one body of petroleum and that in another.

Industrial societies are so dependent on petroleum that it constitutes a key component of their economies. In the best of all worlds, each nation would have its own reserve of petroleum suitable to meet its present and future needs. Unfortunately, however, geopolitical boundaries and geologic resources seldom coincide. A few of the major industrial countries, like Canada and Russia, have adequate supplies of petroleum for present needs. The United States can supply some, but not all, of the petroleum it uses. Other countries, such as Japan, France, and Germany, have either no petroleum or miniscule supplies and must import the oil and gas they require. The reason for this disparity lies in the unequal distribution of petroleum, which is a function of both sedimentary geology and tectonic history.

Sedimentary rocks that contain large quantities of oil are concentrated in a few relatively restricted geologic regions. Their distribution is a function of a long and often

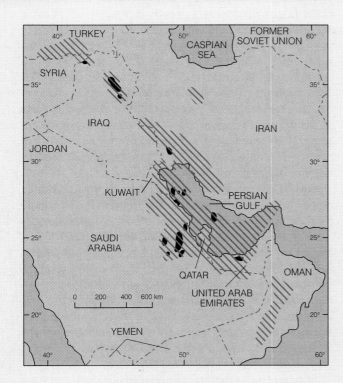

Figure B6.2 Where Middle East Oil Occurs Strata containing oil and gas underlie large areas of the Middle East. Areas outlined enclose the more than 400 separate oil fields that have been discovered so far. The outlines of the largest individual fields are highlighted.

complex history of sediment accumulation and tectonic events related to moving lithospheric plates. Countries like the United States and Russia cover such large geographic areas that the laws of geologic chance made it probable that they would contain at least some large petroleum deposits. For smaller countries, however, the chance is almost like a roll of the dice. Many small countries contain little or no petroleum and thus are losers, while others are big winners.

The largest known reserve of petroleum happens to lie in a limited geographic region centered around the Persian Gulf (Fig. B6.2). Here special sedimentary and tectonic circumstances have converged to produce huge underground reservoirs of fossil fuel in a region otherwise blessed with few natural resources. A few relatively small countries surrounding the gulf control more than half of the world's known petroleum resources. As a result, their economic and political significance has skyrocketed in the decades since oil was discovered in this region (Fig. B6.3).

Always a politically sensitive region, the Middle East lies at the center of the world political stage. International conflicts, the legacy of centuries of intertribal and religious disputes, now include petroleum as an added complicating factor, for control of oil resources is associated with both economic and political power. Although this oil-rich region may remain politically unstable well into the future, eventually its importance will decline as continued exploitation causes the nonrenewable petroleum resources to diminish and the world turns to alternative sources of power.

Figure B6.3 A Harsh Environment Hides Great Riches Geologic exploration teams working in the harsh desert environment of the Persian Gulf region have discovered vast reservoirs of petroleum that comprise more than half the world's known supplies. These discoveries quickly brought the oil-rich Middle East countries to economic and political prominence.

COMMON SEDIMENTARY ROCKS

Sedimentary rocks, like the sediments from which they are derived, fall into three categories: clastic, chemical, and biogenic.

Clastic Sedimentary Rocks

If a sedimentary rock is made up of mineral particles derived from the weathering and erosion of igneous rock, how can we tell it is sedimentary and not igneous? In addition to such obvious clues as sedimentary layering, rock texture also provides evidence that tells us whether a rock is sedimentary or igneous.

The mineral grains in most igneous rock are irregular and interlocked, but the mineral grains and rock fragments in sedimentary rock commonly are rounded and show signs of the abrasion they received during transport. Also, clastic sedimentary rock contains cement holding the particles together, whereas most igneous rocks consist of interlocking crystals. Fossils are another important feature for distinguishing between the two classes of rock. No organism can survive the high temperatures at which igneous rocks form, and so the presence of ancient shells or similar evidence of past life is an important clue to sedimentary origin.

Clastic sedimentary rocks are classified on the basis of predominant particle size, just as sediments are. The four basic classes are conglomerate, sandstone, siltstone, and mudstone or shale, which are the rock equivalents of gravel, sand, silt, and clay (Fig. 6.2).

A **conglomerate** is a lithified gravel. Because the clasts in gravels can vary greatly in size, we often use a modifying term to indicate the predominant size of the particles in a conglomerate, just as we do with the corresponding sediment (e.g., a boulder conglomerate, a cobble conglomerate, or a pebble conglomerate). The large clasts in a conglomerate generally are pieces of preexisting rock, whereas fine particles between the clasts (collectively known as the *matrix*) consist mainly of mineral fragments. In a conglomerate, the clasts are rounded in shape; if the clasts are equally large but angular, the rock is a **breccia**.

Sandstone consists mainly of grains of sand, although coarser or finer particles may be present (e.g., a pebbly sandstone or a silty sandstone). Different types of sandstone are recognized on the basis of composition. Quartz, being very resistant to weathering, is a common mineral in sandstones. A quartz sandstone consists predominantly of quartz grains. If feldspars are a major component, the sediment is called an *arkose* or *arkosic sandstone*. A dark-colored sandstone containing quartz, feldspar, and a large amount of tiny rock fragments (*lithic* particles) is a *lithic sandstone* (also called a *greywacke*).

Siltstone is composed mainly of silt-size mineral fragments, predominantly quartz and feldspar. Clastic rocks of still finer grain size include *mudstone*, which breaks down into blocky fragments, and **shale**, which cleaves into sheetlike fragments when it weathers. In shales, the clay-size particles are so small that composition

A

B

Figure 6.12 Chemical Rocks A. High evaporation rates in the desert basin of Searles Lake in eastern California have led to precipitation of salts that are interbedded with organic muds. The most common salts found in cores drilled through the lake sediments are halite (NaCl) and trona ($Na_2CO_3 \cdot NaHCO_3 \cdot 2H_2O$). B. Layers of black chert are interstratified with limestone in this marine sedimentary deposit that crops out along the shore of Fidalgo Island in western Washington State.

generally must be determined by X-ray methods. The most common components of shale are clay minerals, quartz, feldspar, and calcite.

Chemical Sedimentary Rocks

Chemical sedimentary rocks result from lithification of organic or inorganic chemical precipitates. Most of these rocks contain only one important mineral, which forms the basis for classification.

Common Rock Types

Among the most common chemical rocks are *rock salt* (halite = NaCl), *gypsum* ($CaSO_4 \cdot 2H_2O$), and *trona* ($Na_2CO_3 \cdot NaHCO_3 \cdot 2H_2O$) each of which is formed by evaporation of seawater or lake water and has economic value (Fig. 6.12A).

Chert is a hard, very compact sedimentary rock composed almost entirely of very fine-grained, interlocking crystals of quartz. Its typical splintery to conchoidal fracture made it useful to primitive people for tool production. (A conchoidal fracture has a curved, ribbed fracture like that of a clam shell.) It occurs as extensive continuous layers (bedded chert) or as nodules in carbonate rocks (Fig. 6.12B).

Mineral Deposits in Chemical Sedimentary Rocks
Sedimentary rocks provide employment for most of the world's geologists, for in these rocks are trapped vast accumulations of the sedimentary minerals, petroleum, and coal that help drive our modern civilization. For as long as society requires these mineral resources, there will be a need for geologists who have learned how these earthly treasures form and where to find them. Petroleum and coal are biogenic sedimentary materials, but several mineral resources are also found in sedimentary rocks that are not the remains of plants or animals.

Banded Iron Deposits. Modern civilization is more dependent on iron than on any other industrial metal. Some of the world's most important concentrations of iron are found in ancient sedimentary rocks that are billions of years old (e.g., in Brazil, Canada, the former USSR, South Africa, and Australia). These deposits are among the most unusual kinds of chemical sedimentary rocks known (Fig. 6.13). They consist of sedi-

Figure 6.13 Banded Iron Formation Iron-rich chemical sediment (dark) is interbedded with siliceous chert layers (white) in the Brockman Iron Formation, a 2-billion-year-old banded iron deposit in the Hamersley Range of Western Australia. The woman in the foreground taking the photograph is Dr. Janet Watson, a distinguished English geologist.

ment wholly chemical, or possibly biochemical, in origin and are free of detritus, although they are commonly interbedded with clastic sedimentary rocks. Every aspect of the banded iron deposits indicates chemical precipitation, but the cause of precipitation remains a problem. The precipitation may be related to the chemistry of the ancient atmosphere. Very likely, the most important difference between modern seawater and the seawater in which the iron deposits accumulated was the oxygen content of the water. If the amount of oxygen in the surface waters was very low, large amounts of dissolved Fe^{2+} could have been transported and precipitated. This would further imply that the atmosphere then contained much less oxygen than it does today.

Phosphorus Deposits. Sedimentary deposits of phosphorus minerals, which are a primary source of fertilizer, form through the precipitation of apatite $[Ca_5(PO_4)_3(OH,F)]$ from seawater. The surface waters of the ocean are depleted in phosphorus because fish and other marine animals extract phosphorus to make bone, scales, and other body parts. When the animals die and sink to the seafloor, their bodies slowly decay and release phosphorus to the deep ocean water. If such phosphorus-rich waters are brought to the surface by upwelling currents, precipitation of apatite can occur. Phosphorus-rich sediments are forming today off the western coasts of Africa and South America, but the process was much more common in the past, especially when shallow seas inundated broad areas of the continents.

Evaporite Deposits. Economically important concentrations of sedimentary minerals can result when lake or ocean water evaporates to form a deposit called an **evaporite** (Fig. 6.14). Examples of salts that precipitate from lake waters are sodium carbonate (Na_2CO_3), sodium sulfate (Na_2SO_4), and borax ($Na_2B_4O_7 \cdot 10\ H_2O$). These and other salts have many uses, including the production of paper, soap, detergents, antiseptics, and chemicals for tanning and dyeing. The most important salts that precipitate from seawater are gypsum ($CaSO_4 \cdot 2H_2O$), halite (rock salt; $NaCl$), and carnallite ($KCl \cdot MgCl_2 \cdot 6H_2O$). Much of the common table salt we use daily, the gypsum used for plaster and construction materials, and the potassium used in fertilizers are recovered from marine evaporites.

Biogenic Sedimentary Rocks

Biogenic rocks result from the lithification of biogenic sediment or of sediment having a high organic component.

Limestone and Dolostone

Limestone is the most important of the biogenic rocks and accounts for a major proportion of the carbon dioxide stored in the Earth's crust. If this CO_2 were somehow released, it would significantly change the composition of the atmosphere and cause the surface temperature of the planet to heat up dramatically (Chapter 19).

Limestone is formed chiefly of the mineral calcite. When calcite is replaced by dolomite, the resulting rock is *dolostone*. Limestone and dolostone are not easily classified because they can be either clastic or chemical in character. Bioclastic limestones consist of lithified shells or fragments of marine organisms that have carbonate hard parts; one coarse-grained type composed of shelly debris is called *coquina*. Other limestones consist of cemented reef organisms (*reef limestone*), the compacted carbonate shells of minute floating organisms (*chalk*), and accumulations of tiny, round, calcareous accretionary bodies (ooliths) that are 0.5 to 1 mm in diameter (*oolitic limestone*). Some lime muds are transformed into *lithographic limestone*,

Figure 6.14 Evaporites in Death Valley Evaporite salts encrust the surface of a desert playa on the floor of Death Valley, California. A shallow lake forms during rainy periods. As the water then evaporates and the playa lake dries up, salts crystallize out of the brine, and polygonal fractures form as the drying sediment contracts.

Common Sedimentary Rocks **161**

a very compact, fine-grained rock formerly used in lithography for engraving and the production of color plates.

Diatomite

Siliceous ooze consisting of radiolarian or diatom remains can become lithified to form a relatively soft, light-colored rock called **diatomite**. Although some diatomites formed in lakes, most are of marine origin. The rock is useful commercially as an abrasive and also, because of its very fine grain size and degree of porosity, as a filter.

Figure 6.15 Rich Coal Seam A thick seam of coal is mined from a pit near Douglas, Wyoming. Once the overlying rocks are stripped away, the exposed coal is easily and efficiently removed.

Coal

Coal occurs in strata (miners call them *seams*) along with other sedimentary rocks, mainly shale and sandstone. Most coal seams are 0.5 to 3 m thick, although some reach more than 30 m (Fig. 6.15), and they tend to occur in groups. A look through a magnifying glass at a piece of coal may reveal the shapes of bits of fossil wood, bark, leaves, roots, and other parts of land plants, chemically altered but still identifiable. Accordingly, **coal** can be defined as a usually black, combustible sedimentary rock, more than 50 percent of which consists of decomposed plant matter.

Diagenetic alteration of peat brings about a series of progressive changes. The peat is compressed and water is squeezed out, but there are also chemical changes that cause loss of water and organic gases such as methane (CH_4), leaving an increased proportion of carbon. The peat is thereby converted into *lignite* and eventually into *bituminous coal*, both of which are sedimentary rocks (Fig. 6.16).

Although peat can form even under subarctic conditions, the luxuriant plant growth needed to form thick and extensive coal seams develops most readily in a tropical or subtropical climate. This implies either that the global climate was warmer when the plant matter of coal accumulated, or else that the wet, swampy environments in which most of the world's coal seams formed existed in the tropics, within about 20° of the equator. Both conditions were probably involved. Coal deposits that must have formed in warm low-latitude environments and now lie in frigid polar lands (e.g., northern Alaska and Antarctica) provide some of the most compelling evidence we have for the slow drift of continents over great distances.

Oil Shales

Although shale is a clastic sedimentary rock, some shales have an unusually high content of organic matter and are of economic value. The organic oils and fats contained in dead organisms that are buried in marine or lake muds may be converted to hydrocarbon residues. While these residues may ultimately form petroleum, in some shales burial temperatures never have reached the levels required to break down the organic molecules completely. Instead, an alteration process occurs in which waxlike substances are formed. If such a rock (called *oil shale*) is heated, the solid organic matter is converted to liquid and gaseous hydrocarbons similar to those in petroleum. Extensive oil shales in Colorado, Utah, and Wyoming can produce as much as 240 liters of oil per ton of rock (Fig. 6.17).

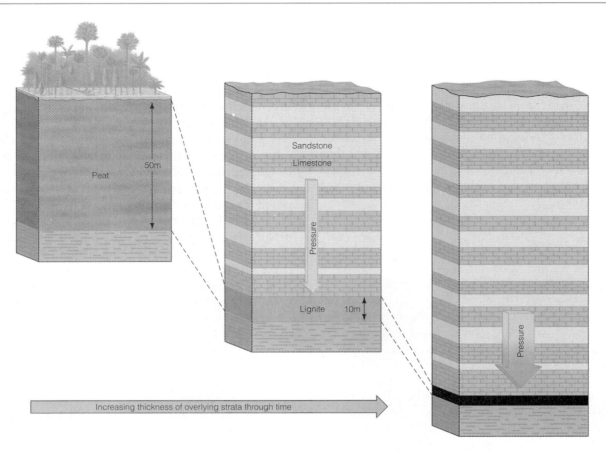

Figure 6.16   From Peat to Coal   Plant matter in peat (a biogenic sediment) is converted to coal (a biogenic sedimentary rock) by decomposition and increasing pressure as overlying sediments increase in thickness. By the time a layer of peat 50 m thick is converted to bituminous coal, its thickness has been reduced by 90 percent. In the process, the proportion of carbon has increased from 60 to 80 percent.

A

B

Figure 6.17   Oil Shale   A. Bedded shales in this Colorado canyon wall belong to the Green River Oil Shale, a sedimentary rock that formed as organic-rich sediment accumulated in ancient fresh-water lakes, was buried, then compacted and cemented.   B. A polished slab of laminated oil shale clearly shows the thin dark-colored layers in which the organic matter is concentrated.

SEDIMENTARY FACIES AND DEPOSITIONAL ENVIRONMENTS

In much the same way that history records the changing patterns and progress of civilization, sedimentary rocks record the history of our planet. Layers of sediment, like pages in a book, can reveal the changing environmental patterns of the Earth's surface and the progress of life over more than 3 billion years. If we can learn to read them, the sediments allow us to journey back through the ages and visualize how the world has evolved during its long, dynamic history.

Environmental Clues in Sedimentary Rocks

We have already seen that the size, shape, and arrangement of particles in sediments, as well as the geometry of sedimentary strata, provide us with evidence about the geological environment in which sediments accumulate. These and other clues, several of which are mentioned below, enable us to demonstrate the existence of ancient oceans, coasts, lakes, streams, glaciers, swamps, and other agents of transport and places of deposition of sediment that were important in the past.

Features on Bedding Planes

Wavelike irregularities formed by currents moving across a sediment, together with cracks, grooves, and other minor depressions, can be preserved on the bedding surfaces of sandstone or siltstone. Such features, collectively called *sole marks*, are useful in reconstructing past current directions and bottom conditions.

Bodies of sand that are being moved by wind, streams, or coastal waves are often rippled, and such ripples may be preserved in sandstones and siltstones as *ripple marks* (Fig. 6.18). Some mudstones and siltstones contain layers that are cut by polygonal markings. By comparing them with similar features in modern sediments, we infer that these are *mud cracks*, caused by shrinkage and cracking of wet mud as its surface dries (Fig. 6.19). Mud cracks imply former tidal flats, exposed stream beds, desert lake floors, and similar environments.

Footprints and *trails* of animals are often found with ripple marks and mud cracks (Fig. 6.20). Even *raindrop impressions* made during brief, intense showers may be preserved in strata. All provide evidence of moist surface conditions at the time of formation.

Fossils

Fossils provide significant clues about former environments. Some animals and plants are restricted to warm, moist climates, whereas others are associated only with

A

B

Figure 6.18 Modern and Ancient Ripples Modern ripples and ancient ripple marks A. Ripples forming in shallow water near the shore of Ocracoke Island, North Carolina. B. Fossil ripple marks on the top surface of a sandstone bed exposed at Artist's Point in Colorado National Park, Colorado.

cold, dry climates. By using the climatic ranges of modern plants and animals as guides and invoking the Principle of Uniformitarianism, we can infer the general character of the climate in which similar ancestral forms

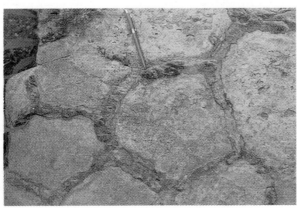

B

Figure 6.19 Modern and Ancient Mudcracks Modern and ancient mud cracks. A. Mud cracks formed on the surface of a dry lake floor. B. Ancient mud cracks preserved on the surface of a mudstone bed exposed at Ausable Chasm, New York.

lived. For example, plant fossils can provide estimates of past precipitation and temperature for sites on land, whereas fossils of tiny floating organisms can tell us about former surface temperatures and salinity conditions in the oceans. Fossils are also the chief basis for relating strata in different parts of the world in respect

A

B

Figure 6.20 Footprints in the Sand A. Seagull footprints are imprinted in sandbars along the Alsek River, Glacier Bay National Park, Alaska. In the background is the Brabazon Range of the St. Elias Mountains. B. Fossil footprints of a three-toed dinosaur are preserved in sandstone in the Painted Desert near Cameron, Arizona.

to age; they have therefore been very important in reconstructing the past 600 million years of Earth history (Chapter 8).

Color

The color of fresh sedimentary rock is determined by the colors of the minerals, rock fragments, and organic matter that compose it. Iron sulfides and organic detritus, buried with sediment, are responsible for most of the dark colors in sedimentary rocks and imply deposition in a reducing environment. Reddish and brownish colors result mainly from the presence of iron oxides, occurring either as powdery coatings on mineral grains or as very fine particles mixed with clay, and point to oxidizing conditions.

The weathered surface of a sedimentary rock may have a different color than a fresh, unweathered surface. For example, a sandstone that is pale gray on freshly broken surfaces may have a surface coating of iron oxide in weathered outcrop that gives it a yellowish-brown color. In this case, the surface color is derived from the chemical breakdown of iron-bearing minerals in the rock.

Sedimentary Facies

If we examine a sequence of exposed sedimentary rocks, we will likely see differences as we move up from one layer to the next. These differences reflect changes over time in depositional conditions at a particular place. If any group of layers in the sequence is traced away from the initial outcrop, they may change laterally. Most sedimentary strata change character laterally as a result of changes in the conditions under which the sediments accumulated.

A diversity of environments would be encountered if we were to travel across the edge of a continent and into the adjacent ocean basin (Fig. 6.21). Distinctive sediments and associated organisms serve to identify each. For each environment, we can make a list of the distinctive physical, chemical, and biological characteristics that permit us to distinguish sediment accumulating there from sediment being deposited in another environment. The change in sediment character that takes place as we move from one depositional environment to another is referred to as a change of **facies** (pronounced *fay-seez*). A *sedimentary facies*, therefore, can be thought of as any sediment that can be distinguished from another, contemporary sediment that accumulated in a different depositional environment. A facies may be characterized, for example, by distinctive grain size, grain shape, stratification, color, chemical composition, depositional structures, or fossils. Adjacent facies can merge into one another either gradually or abruptly (Fig. 6.22). Coarse gravel and sand of a beach may pass very gradually offshore into finer sand, silt, and clay on the floor of the sea or a lake. Coarse, bouldery glacial sediment, on the other hand, may end abruptly against stream sediments at the margin of a glacier.

By determining the distinctive characteristics of different bodies of sediment or sedimentary rock, studying the relationships of different facies, and using these characteristics to identify original depositional settings, we can reconstruct a picture of the varied environments of a region during past geologic intervals.

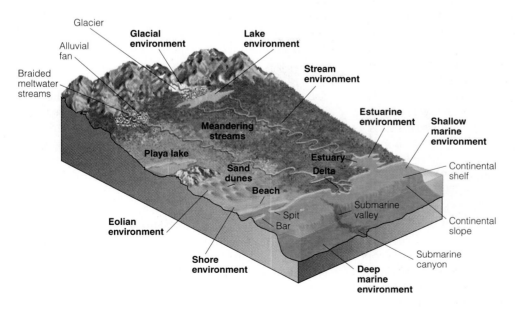

Figure 6.21 Depositional Environments Various depositional environments are seen while traveling from the crest of a mountain range across the edge of a continent to the adjacent margin of a nearby ocean basin.

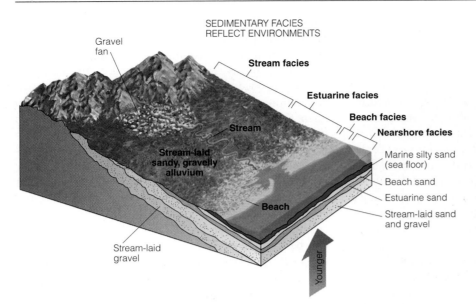

SEDIMENTARY FACIES
REFLECT ENVIRONMENTS

Gravel fan

Stream facies

Estuarine facies

Beach facies

Nearshore facies

Stream

Marine silty sand (sea floor)

Stream-laid sandy, gravelly alluvium

Beach sand

Estuarine sand

Beach

Stream-laid sand and gravel

Stream-laid gravel

Younger

Figure 6.22 Sedimentary Facies
A section from a mountain valley across a gravelly plain and estuary to the adjacent ocean shows a variety of depositional environments in which distinctive facies are deposited. In this example, facies adjoin or merge into one another on the land surface, whereas in a vertical section offshore, they lie one above another. The boundaries between adjacent facies dip seaward, indicating that the boundaries have shifted progressively in the landward direction through time. This implies that a rise of sea level relative to the land has taken place.

Common Sediments of Nonmarine Environments

Sediment derived from the mechanical and chemical breakdown of rocks is moved toward the sea or toward the lowest level of an inland basin. En route it is moved by water, ice, wind, or gravity. The sediment may be temporarily stored and then reworked repeatedly by one or several of these agencies before reaching its final resting place, where it is slowly converted to sedimentary rock.

Stream Sediments
Streams constitute the principal agency for transporting sediment across the land (Figs. 6.21 and 6.22). Their deposits are widespread. Stream-deposited sediment differs from place to place depending on the type of stream, the energy available for doing work, and the nature of the sedimentary load. A typical large, smoothly flowing stream may deposit well-sorted layers of coarse and fine particles as it swings back and forth across its valley. Silt and clay are deposited on the valley floor adjacent to the stream during floods; clays and organic sediments may accumulate in abandoned sections of the stream channel that no longer carry flowing water. By contrast, a stream flowing from the front of a mountain glacier may divide into an intricate system of interconnected channels that change in size and direction as the volume of water fluctuates and as the stream copes with an abundance of detritus. The resulting sediments consist largely of coarse-grained stream-channel deposits. A large mountain stream flowing down a steep valley can transport an abundant load of sediment. When the stream reaches the mountain front and is no longer constrained by valley walls, it is free to shift lat-

erally back and forth across the more gentle terrain as its load is dropped. The result is a fan-shaped deposit in which the sediments may range from coarse, poorly sorted gravels to well-sorted, cross-bedded sands.

Lake Sediments
Sediments deposited in a lake (Fig. 6.21) accumulate chiefly on the lakeshore and on the lake floor. Lakeshore deposits of generally well-sorted sand and gravel form beaches, bars, and spits across the mouths of bays. The sediment load of a stream entering a lake will be dropped as the velocity and transporting ability of the stream suddenly decrease. The resulting deposit, built outward into the lake, is a *delta* (Fig. 6.21; Chapter 10). Inclined, generally well-sorted layers on the front of a delta pass downward and outward into thinner, finer, and evenly laminated layers on the lake floor (Fig. 6.5).

Glacial Sediments
Sedimentary debris eroded and transported by a glacier is either deposited along the glacier's base or released at the glacier margin as melting occurs. The sediment may then be subjected to reworking by running water. Debris deposited directly from ice commonly forms a random mixture of particles that range in size from clay to boulders and consist of all the rock types over which the ice has passed (Fig. 6.8). Such sediment characteristically is neither sorted nor stratified, in contrast to most other nonmarine sediments. Stones in glacially deposited sediment often are angular, and sand grains typically show distinctive fractures due to crushing and abrasion (Fig. 6.23A).

Eolian Sediments
Both wind activity and the geologic results of it are referred to as **eolian**, after Aeolus, the Greek god of

A

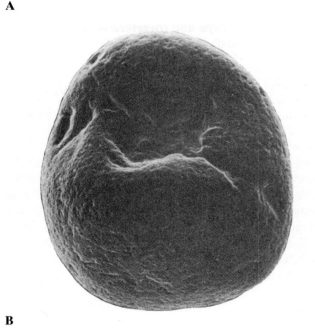

B

Figure 6.23 Shapes of Sand Grains Surface features of sand grains, seen on enlarged pictures taken with a scanning electron microscope, aid in differentiating among transporting agencies. A. The surface of a quartz gain (0.1 mm across) that has been crushed and abraded during transport at the bed of a Swiss glacier displays distinctive conchoidal fractures. B. The surface of a wind-transported quartz grain (0.5 mm diameter) from south-central Libya has a distinctive pitted appearance, thought to result from the long-continued action of dew that slowly dissolves mineral substance.

wind. Sediment carried by the wind tends to be finer than that moved by other erosional agents. Sand grains are easily moved where strong winds blow and where vegetation is too discontinuous to stabilize the land surface, as along seacoasts and in deserts (Fig. 6.21). In such places, the sand may pile up to form dunes composed of well-sorted sand grains and with bedding inclined in the downwind direction (the direction toward which the air is flowing). Individual grains may have a frosted appearance (Fig. 6.23B), thought to result from the long-continued action of dew which dissolves mineral substance on a grain, thereby creating a finely pitted surface texture. Using these characteristics, geologists can rather easily identify ancient dune sands in the rock record (Fig. 6.6).

Powdery dust picked up and moved by the wind is deposited as a blanket of sediment across the landscape (Chapter 13). Such sediment is thickest and coarsest near its source and becomes progressively thinner and finer with increasing distance downwind. Although common as a sediment in many parts of the world, wind-blown silt is virtually unknown as sedimentary rock, probably because it is easily eroded from the landscape and therefore is unlikely to be widely preserved.

Sediments of the Continental Shelves

The world's rivers continuously transport detritus to the edges of the sea where it can accumulate near the mouths of streams, be moved laterally along the coast by currents, or be carried seaward to accumulate on the continental shelves, sometimes to great thicknesses. In part spurred by the search for large undersea reservoirs of oil and gas, geologists have learned a great deal about the sediments accumulating on the shelves.

Estuarine Sediments

Much of the load transported by a large river may be trapped in an **estuary**, a semi-enclosed body of coastal water within which seawater is diluted with fresh water (Figs. 6.21 and 6.22). Coarse sediment tends to settle close to land, while fine sediment is carried in the seaward direction. Tiny individual particles of clay carried in suspension settle very slowly to the seafloor. However, when fresh water meets seawater, the clay particles tend to aggregate into clumps which, because of their greater mass, settle more rapidly to the bottom. If the rate of sedimentation is high and the land is slowly subsiding, a thick body of estuarine sediment can form.

Deltaic Sediments

Marine deltas are built outward in the sea at places where streams reach the shore and deposit their sediment load (Fig. 6.21). Large deltas are complex deposits consisting of coarse stream-channel sediments, fine sediments deposited between channels, and still finer sediments deposited on the seafloor (Chapter 10).

Beach Sediments

Quartz, the most durable of common minerals in continental rocks, is a typical component of beach sands. However, not all ocean beaches are sandy. Any beach consists of the coarsest rock particles contributed to it by the erosion of adjacent sea cliffs, or carried to it by rivers or by currents moving along the shore. Beach sediments tend to be better sorted than stream sediments of comparable coarseness and typically display cross stratification. Dragged back and forth by the surf and turned over and over, particles of beach sediment become rounded by abrasion. Although beach gravel and gravelly stream sediment may be similar in appearance, the pebbles and cobbles on many beaches acquire a distinctive flattened shape.

Offshore Sediments

Fresh water flowing through an estuary or past a river mouth may continue seaward across the submerged continental shelf as a distinct layer overlying denser, salty marine water. Some fine-grained sediment, carried in suspension, thereby reaches the outer shelf. The sediment then either settles slowly to the seafloor or is ingested by floating organisms that excrete it as small pellets that fall to the bottom. On the continental shelf of eastern North America, up to 14 km of fine sediment has accumulated over the last 70 to 100 million years. To build the whole pile, an average of less than a millimeter of sediment need have been deposited each year.

Most coarse marine sediment is deposited within 5 to 6 km of the land after being dispersed by currents that flow parallel to the shore. Coarse sediment is also found as far offshore as the seaward limits of the shelves. Its observed patchy distribution is mostly the result of changing sea level. At times when the sea fell below its present level, the shoreline migrated seaward across the shelves, exposing new land. Bodies of coarse sediment deposited near shore or on the land at such times were submerged as sea level again rose across the shelves. As much as 70 percent of the sediment cover on the continental shelves is probably a relict of such past conditions.

Only about 10 percent of the sediment reaching the continental shelves remains in suspension long enough to arrive in the deep sea, so it is clear that the great bulk of the Earth's sedimentary strata is shelf strata whose sediment originated on the continents. In effect, the shelves conserve continental crust which is continually recycled within the continental realm.

Carbonate Shelves

Carbonate sediments of biogenic origin accumulate on the continental shelves wherever the influx of land-derived sediment is minimal and the climate and sea-surface temperature are warm enough to promote the abundant growth of carbonate-secreting organisms. Carbonate sediments accumulate mainly on broad, flat carbonate shelves that border a continent or rise as platforms off the seafloor (Fig. 6.24). Most of the sediment consists of sand-sized skeletal debris, together with inorganic precipitates that form fine carbonate muds. Coarser debris is found mainly near coral and algal reefs or in areas of turbulence and strong currents.

Marine Evaporite Basins

Ocean water occupying a basin with restricted circulation that lies in a region of very warm climate will evaporate. This leads to the precipitation of soluble substances and the accumulation of marine evaporite deposits. Such deposits are widespread. In North America, for example, marine evaporite strata underlie as much as 30 percent of the entire land area.

The Mediterranean Sea is an example of an evaporite basin (Fig. 6.25). Were it not for continuous inflow of Atlantic water at its western end, the Mediterranean would gradually decrease in volume because of evapo-

Figure 6.24 Carbonate Shelf Carbonate sediments consisting of fine skeletal debris and inorganic precipitates accumulate in the warm, shallow marine waters of a broad, flat carbonate shelf surrounding the numerous islands of the Bahamas. The sediments show up as white sand bars.

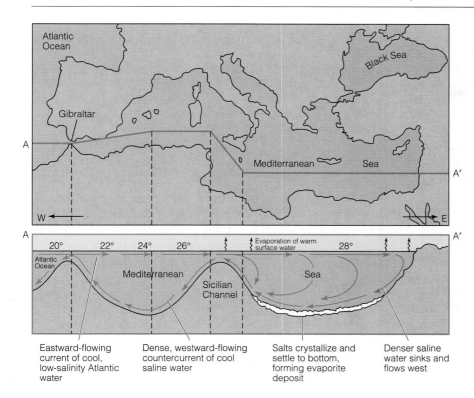

Figure 6.25 Water Flow in the Mediterranean Sea In the Mediterranean Sea, the inflow of fresh water from rivers in Europe and Africa is too small to offset the water evaporated from the sea surface in the hot eastern part of the basin. An inflowing current of Atlantic Ocean water therefore balances the loss. As the relatively cool ocean water (summer temperature 20°C) enters the basin and moves eastward, it warms (to 28°C) and evaporates. The salinity of the remaining water rises, increasing its density and setting up a westward-flowing current of cool saline water at depth. Salts crystallizing from the salty water settle to the bottom to form evaporite-rich sediments.

ration. It is estimated that if the Mediterranean were deprived of new water, evaporation would cause this landlocked sea to dry up completely in about 1000 years. In the process, a layer of salt about 70 m thick would be precipitated. Far thicker evaporites that underlie the Mediterranean basin are regarded as evidence of former periods when a high evaporation rate, together with a continuous inflow of Atlantic water to supply the necessary salt, allowed evaporites 2 to 3 km thick to accumulate.

Sediments of the Continental Slope and Rise

Along most of their length, the shallowly submerged continental shelves pass abruptly into continental slopes that descend to depths of several kilometers. Sediments that reach the shelf edge are poised for further transport down the slope and onto the adjacent continental rise.

Turbidity Currents and Turbidites
Thick bodies of coarse sediment of continental origin lie at the foot of the continental slope at depths as great as 5 km. The origin of these coarse accumulations at great depths in the oceans was difficult to explain until marine geologists demonstrated that the sediments could be deposited by **turbidity currents**. These are gravity-driven currents consisting of dilute mixtures of sediment and water having a density greater

than the surrounding water. Such currents have been produced in the laboratory in water-filled tanks into which a dense mixture of water, silt, and clay has been introduced. They have also been documented moving across the floors of lakes and reservoirs (Fig. 6.26). In the oceans, turbidity currents have been set off by earthquakes, landslides, and major coastal storms. Off the mouths of rivers, they can be set in motion by large floods (see the boxed essay on *Breaks in the Transatlantic Cables*).

The accumulated evidence leads us to believe that turbidity currents are effective geologic agents on continental slopes, where they can reach velocities greater than those of the swiftest streams on land. Some achieve a velocity of more than 90 km/h and transport up to 3 kg/m³ of sediment, spreading it as far as 1000 km from the source.

A turbidity current typically deposits a graded layer of sediment called a **turbidite** (Fig. 6.27). Such a graded layer is the result of rapid, continuous loss of energy in the moving current. As a rapidly flowing turbidity current slows down, successively finer sediment is deposited. At any site on the continental rise or an adjacent abyssal plain, a turbidite is deposited very infrequently, perhaps only once every few thousand years. In these places, far distant from the sediment source, the deposits are mainly thin layers a few millimeters to 30 cm thick. Although deposition is infrequent, over millions of years turbidites can slowly accumulate to form vast deposits beyond the continental realm.

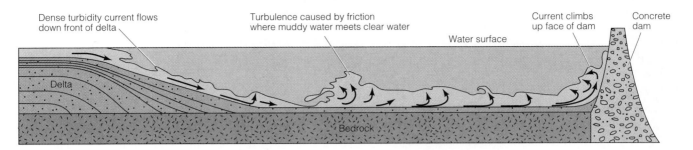

Figure 6.26 Turbidity Current A turbidity current generated by a surge of sediment-laden water enters the quiet water of a reservoir behind a dam. Moving rapidly down the face of a delta, the current passes along the lake floor and climbs up the face of the dam before subsiding. As the sediment settles to the bottom, it forms a graded turbidite layer.

Understanding Our Environment

Breaks in the Transatlantic Cables

In 1935, soon after the completion of the Hoover Dam on the Colorado River, engineers were surprised to find that from time to time the normally clear water flowing through the outlet pipes suddenly became muddy. Inquiry soon revealed that muddy water entering Lake Mead, far upstream from the dam, formed dense sediment-laden currents that swept along the floor of the lake and had enough momentum to rise up the side of the dam and flow through discharge pipes. The engineers had discovered *turbidity currents*, which are gravity-driven currents consisting of dilute mixtures of sediment and water having a density greater than that of the surrounding water.

Although it was not realized at the time, evidence of turbidity currents in the ocean, in the North Atlantic, was present in the record of broken phone cables between North America and Europe. The evidence was missed at the time because the existence of turbidity currents was not known when 13 cables were broken in 28 places on November 18, 1929 following a severe earthquake on the continental slope off Nova Scotia. At the time the breaks were thought to be the result of movements of the seafloor due to the earthquake. How it eventually became clear that turbidity currents caused the breaks is a nice example of scientific sleuthing.

The 1929 event was all but forgotten when, many years later, a thick file of measurements made by a repair ship was reviewed by scientists. There were two odd things about the cable breaks. First, although all the cables on the continental slope and deep-sea floor were broken, not one of the cables that crossed the continental shelf was damaged. Second, the breaks occurred in sequence over a period of 13 hours, in order of increasing depth, and through a distance of 480 km from the earthquake center. The repair ship found that each damaged cable had been broken at two or three points more than 160 km apart. The detached segments of cable between the breaks had been carried partway down the continental slope or buried beneath sediment beyond its base.

The whole event was like a vast laboratory experiment

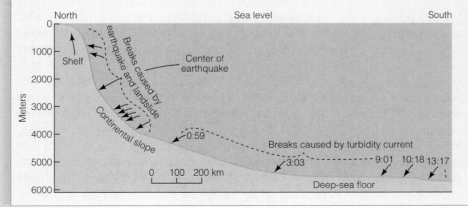

Figure B6.4 Record of Turbidity Currents Profile of seafloor off Nova Scotia showing events of the earthquake of November 18, 1929. Short arrows point to the loctions of breaks in transatlantic cables. The numbers show the times of breaks in hours and minutes after the earthquake. The vertical scale is exaggerated greatly. (*Source:* After Heezen and Ewing, 1952.)

Deep-Sea Fans

Some large submarine canyons on the continental slopes are aligned with the mouths of major rivers, like the Amazon, Congo, Ganges, and Indus. At the base of many such canyons is a huge **deep-sea fan**, a fan-shaped body of sediment that spreads downward and outward to the deep seafloor (Fig. 6.28). The sediments, which are derived mostly from the land, include fragments of land plants, as well as fossils of shallow- and deep-water marine organisms. Also present are many graded layers that are interpreted as turbidites.

Deep-sea fans are a major exception to the generalization that the final deposition of land-derived sediment in the ocean is largely confined to the continental shelves. When shelves are exposed at times of lowered sea level and rivers extend across them nearly to the continental slope, the stage is set for the rapid building of deep-sea fans.

Sediment Drifts

Discrete bodies of sediment up to hundreds of kilometers long, tens of kilometers wide, and two kilometers high have been discovered along the continental margins bordering the North Atlantic Ocean. Called *sediment drifts*, these huge deposits are associated with deep-ocean currents that flow near the base of the continental margins. Rippled sands along the axis of a sediment

Figure 6.27 Deep Sea Turbidities Deep-sea turbidite beds that have been tilted, uplifted, and exposed in a wave-eroded bench along the coast of the Olympic Peninsula, Washington.

with times and distances controlled by measurement. Each break was timed by the equipment that automatically records the messages transmitted through the cables, and was accurately located after the quake by the electric-resistance measurements which were made to enable repair ships to locate breaks and repair damage.

The only hypothesis that could reasonably explain all these facts was that the quake triggered a large submarine landslide on the continental slope. As the landslide developed, a series of slumps started turbidity currents that flowed downslope, breaking each cable as it was reached. Eight cables on the slope were broken instantaneously at the moment the quake occurred, and five others were broken at times ranging from 59 minutes to 13 hours 17 minutes after the quake (Fig. B6.4). The area affected was about 320 km wide, and the currents traveled well over 720 km from their source.

Based on the times of the breaks and the distances between them, the velocities of the turbidity currents could be calculated. Velocities on the continental slope are believed to have been at least 30 km/h and may have reached 40–55 km/h, or about six to eight times the average velocity of the Mississippi River.

Similar events to the 1929 turbidity currents are now known to have occurred at many localities around the world. All involved the slumping of unconsolidated sediments. Some were related to earthquakes. Others occurred off the mouths of large rivers, indicating that turbidity currents can apparently be set off by large floods. Still others resulted from major storms along coastlines. This evidence is impressive, and it leads to the conclusion that turbidity currents are very effective geologic agents in determining the environment of the continental slope.

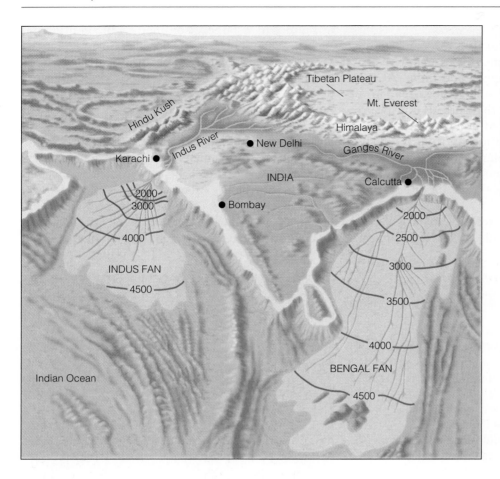

Figure 6.28 Deep Sea Fans The Indus and Ganges– Bramaphutra rivers have built the vast Indus and Bengal deep-sea fans on the seafloor adjoining the Indian subcontinent. Most of the sediment in the fans was transported from the high Himalaya in the north as the mountain system was uplifted over tens of millions of years. Contours are water depths in meters.

drift, where current velocities are high, grade laterally into muds deposited where currents decrease in velocity. Factors that lead to the formation of sediment drifts appear to include a deep current system in a stable location, a prevailing current velocity (<15 cm/s) that permits deposition, and a supply of sediment. At typical accumulation rates of tens of meters per million years, the formation of sediment drifts requires many millions of years.

Sediments of the Deep Sea

Analyses of samples from sediment cores allow geologists to sort out the various sources of seafloor sediment. Study of a great number of samples indicates that all the sediments are mixtures. Although a large portion is produced by biologic activity in the surface waters, some sediment is transported over great distances from continental interiors and eventually reaches the deep sea.

Deep-Sea Oozes

Calcareous ooze occurs over wide areas of ocean floor at low to middle latitudes where warm sea-surface temperatures favor the growth of carbonate-secreting organisms in the surface waters. However, calcareous ooze is not found in these same warm latitudes where the water is deeper than about 4 km. Cold deep-ocean waters are under high pressure and contain more dissolved carbon dioxide than shallower waters. As a result, they can easily dissolve any carbonate particles that settle to their level.

Other parts of the deep-ocean floor are mantled with siliceous ooze, most notably in the equatorial Pacific and Indian oceans, and in a belt encircling the Antarctic region where siliceous organisms predominate. These are areas where surface waters have a high biological productivity, in part related to the rise of deep-ocean water rich in nutrients.

Land-Derived Sediment

In addition to biogenic oozes, deep-sea strata include sediments carried to the oceans by rivers, eroded from coasts by wave action, transported by wind (fine desert dust [Fig. 13.22] and volcanic ash), and released from floating ice. Over much of the North Pacific and part of the central South Pacific the ocean floor lies at depths of more than 4 km and lacks calcareous ooze. In these regions, the seafloor is covered by oxidized reddish or brownish clay composed mainly of extremely fine clay minerals, quartz grains, and micas that originated on land.

7

**Measuring
Our Earth**

**Understanding
Our Environment**

**Using
Resources**

GeoMedia

The famous marble quarries of Carrara, Italy are high in the mountains above the town of Carrara. The view is looking west across Carrara to the Tyrrhenian Sea and the Gulf of Genoa.

IMPORTANT ROCK NAMES TO REMEMBER

breccia (p. 158)

coal (p. 161)
conglomerate (p. 158)

diatomite (p. 161)

limestone (p. 160)

sandstone (p. 158)
shale (p. 158)
siltstone (p. 158)

QUESTIONS FOR REVIEW

1. On what bases are sediments and sedimentary rocks classified?

2. What obvious clues can be used to tell a clastic sedimentary rock from an igneous rock?

3. What chemical reactions can lead to precipitation of chemical sediments?

4. What features in a sediment or sedimentary rock are responsible for stratification?

5. Describe the processes involved in the conversion of sediment to sedimentary rock.

6. Explain why petroleum is thought to originate in marine sediments.

7. Describe cross bedding and graded layers, and explain what they tell us about conditions of deposition.

8. What clues regarding depositional environment are provided by features on sedimentary bedding surfaces?

9. How would you tell sediments deposited by running water from sediment deposited by wind or in a lake?

10. If estuaries are generally shallow bodies of water, what hypothesis can you suggest to explain the occurrence of thick estuarine accumulations in the sedimentary record?

11. What explanation can be given for the occurrence of widespread relict sediments on the continental shelves?

12. If the Mediterranean Sea, which is very nearly an enclosed marine basin, were to evaporate completely, less than 100 m of evaporite sediments would be deposited. How, then, might *continuous* evaporite deposits under the Mediterranean that are more than 2 km thick have formed?

13. What distinctive features of the sediments in deep-sea fans provide clues about the way the sediments were transported?

14. What factors explain the distribution of calcareous and siliceous oozes on the ocean floors?

15. Explain where and why you would expect to find exceptionally thick accumulations of sedimentary rock.

Virtual Internship: You Be the Geologist

Your CD-ROM contains a virtual internship through which you will become a geologist as you assess sedimentary basin processes.

New Rocks from Old: Metamorphism and Metamorphic Rocks

Michelangelo's Pietà

In the year 1498, the young sculptor Michelangelo Buonarroti was living in Rome at the home of his benefactor, Jacopo Galli, a banker. Michelangelo was just finishing his first life-sized marble statue, a figure of the Greek god Bacchus, when he was commissioned by a Benedictine monk, Cardinal Groslaye of San Dionigi, to produce a sacred work in marble for the Basilica of St. Peter's Cathedral. The theme of the Pietà—Mary holding the lifeless body of Jesus—was quickly agreed upon.

However, it remained to secure an appropriate stone for the work. A perfect marble block 2 m across, and even wider and taller than 2 m, would not be quarried just on the chance of a sale; Michelangelo resigned himself to having to visit the quarries at Carrara at his own expense. Then, suddenly, a messenger came with word that the perfect block had arrived. It had been cut on order but had never been paid for. Michelangelo went quickly to the docks, as described in Irving

Stone's biography of Michelangelo, *The Agony and the Ecstasy:*

There it stood, gleaming pure and white in the summer sun, beautifully cut by the quarrymen high in the mountains of Carrara. It tested out perfect against the hammer, against water, its crystals soft and compacted with fine graining. He came back before dawn the next morning, watched the rays of the rising sun strike the block and make it as transparent as pink alabaster, with not a hole or hollow or crack or knot to be seen in all its massive white weight. His Pietá had come home.

Michelangelo's contract with the cardinal's monastery had stated specifically that "the work will be more beautiful than any work in marble to be seen in Rome today, and such that no master of our own time will be able to produce better." At the age of twenty-four, Michelangelo completed what many believe to be the most perfect, most deeply spiritual work of his career. He had infused the faultless marble with a light and life of its own. Although the cardinal died before the statue was installed in St. Peter's, during his last viewing of the incomplete work he declared that Michelangelo had indeed fulfilled the requirements of his contract.

The qualities of the stone so eagerly sought by Michelangelo and other sculptors—the fine, uniform texture and seamless white purity—are characteristic of the famous statuary marble quarried at Carrara, Italy. Marble forms when limestone is metamorphosed. A true white marble starts out as a sediment composed of materials such as shells, corals, or calcareous ooze, all of which are made of calcium carbonate. When subjected to heat and pressure, the shells and particles of calcium carbonate are recrystallized into uniformly sized, interlocking grains of calcite. All traces of fossils are eventually destroyed and, if it is pure enough, the rock will become a uniformly crystalline white marble like that found in Carrara.

METAMORPHISM

The term **metamorphism** is used to describe the changes in mineral assemblage and rock texture that take place in rocks within the Earth's crust as a result of changes in temperature and pressure.

Metamorphic rocks are of particular interest because the changes happen in the solid state. As tectonic plates move and crustal fragments collide, rocks are squeezed, stretched, bent, heated, and changed in complex ways. But even if a rock has been altered two or more times, vestiges of its earlier forms are usually preserved because the changes occurred in the solid state. Solids, unlike liquids and gases, tend to retain a memory of the events that changed them. In many ways, therefore, metamorphic rocks are the most complex, but they are also the

most interesting of the rock families. In them is preserved the story of many of the things that have happened to the crust. Deciphering the record is an exceptional challenge for geologists. For example, when tectonic plates collide, distinctive kinds of metamorphic rocks form along the plate edges. Geologists are attempting to determine where the boundaries of ancient continents once were by studying metamorphic rocks. Geologists are also trying to use evidence derived from metamorphic rocks to determine how long plate tectonics has been active on the Earth. So far the evidence suggests that plate tectonics has been operating for at least 2 billion years!

THE LIMITS OF METAMORPHISM

Before we discuss metamorphism in detail, it is important to define the limits of the process. Metamorphism describes changes in mineral assemblage and texture in sedimentary and igneous rocks subjected to temperatures above about 100°C and usually under pressures of hundreds of MPa caused by the weight of a few thousand meters of overlying rock. Metamorphism does not refer to changes caused by weathering (Chapter 2) or by diagenesis (Chapter 6), because both weathering and diagenesis take place at temperatures below 100°C and low pressures.

There is, of course, an upper limit to metamorphism, because at sufficiently high temperatures rock will melt. Remember then, metamorphism refers only to changes in solid rock, not to changes caused by melting. Changes due to melting involve igneous phenomena, as we discussed in Chapter 4.

Because at least a small amount of H_2O is present in most rocks, the upper limit of metamorphism in the crust is determined by the onset of wet partial melting, as discussed in Chapter 4 and shown in Figure 7.1. The H_2O present controls the temperature at which wet partial melting commences and the amount of magma that can form from a metamorphic rock. The upper limit of metamorphism is therefore a temperature range that depends on the amount of H_2O present. As shown in Figure 7.1, the upper limits of metamorphism overlap with the region of temperature and pressure where magma formation commences. When a tiny amount of H_2O is present, only a small amount of melting occurs and the melt stays trapped as small pockets in the metamorphic host. In many places it is possible to find evidence that melting started in H_2O-rich layers of metamorphic rock, even though adjacent, drier layers do not show any sign of melting. Composite volumes of rock containing an igneous component formed by a small amount of melting plus a metamorphic portion are

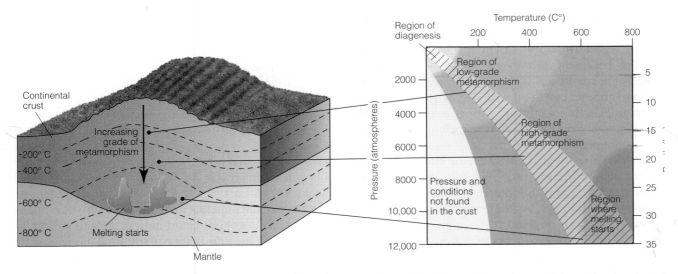

Figure 7.1 Temperature and Pressure Conditions of Metamorphism This diagram shows the conditions of pressure and temperature in which diagenesis, metamorphism, and melting occur. On the left is a sketch showing where in the crust metamorphism and melting occur. On the right is a plot of pressure versus temperature, with the pressure-temperature regimes of metamorphism indicated. The broad, curved band shows how pressure and temperature increase with depth through much of the continental crust. Pressure is shown along the lefthand side of the diagram, equivalent to pressure in the crust, is shown on the righthand side.

called **migmatites**. When large volumes of magma develop by partial melting, they will rise upward and intrude the metamorphic rock above. Eventually, a large volume of rising magma formed by wet partial melting will solidify as an intrusive igneous rock. As a result, we observe that batholiths of granitic rock and large volumes of metamorphic rock tend to be closely associated. The geological setting of this igneous-metamorphic rock association is along subduction and collision margins of tectonic plates.

Low-grade metamorphism refers to metamorphic processes that occur at temperatures from about 100°C to 500°C and at relatively low pressures (Fig. 7.1). The **high-grade metamorphism** refers to metamorphic processes at high temperature (above about 500°C) and high pressure.

CONTROLLING FACTORS IN METAMORPHISM

In a simplistic way you can think of metamorphism as cooking. When you cook, what you get to eat depends on what you start with and on the cooking conditions. So too with rocks; the end product is controlled by the initial composition of the rock and by the metamorphic (or cooking) conditions. The chemical composition of a rock undergoing metamorphism plays a controlling role in the new mineral assemblage; so do changes in temperature and pressure. The controls of temperature and pressure on metamorphism are not entirely straightforward, however, because they are strongly influenced by such things as the presence or absence of fluids, how

long a rock is subjected to high pressure or high temperature, and whether it is simply compressed or is twisted and broken during metamorphism.

Chemical Reactivity Induced by Fluids

The innumerable open spaces between the grains in a sedimentary rock and the tiny fractures in many igneous rocks are called pores, and most pores are filled by a watery fluid. The fluid is never pure water, for it always has dissolved within it small amounts of gases and salts plus traces of all the mineral constituents that are present in the enclosing rock. At the temperatures of high-grade metamorphism, the water is technically a vapor, but high pressure makes the vapor virtually as dense as liquid water would be, so it is just as effective a solvent as liquid water. Thus, the **intergranular fluid**, for that is its best designation, plays a vital role in metamorphism.

When the temperature and pressure of a rock undergoing metamorphism change, so does the composition of the intergranular fluid. Some of the dissolved constituents move from the fluid to the new minerals growing in the metamorphic rock. Other constituents move in the other direction, from the minerals to the fluid. In this way the intergranular fluid serves as a transporting medium that speeds up chemical reactions in much the same way that water in a stew pot speeds up the cooking of a tough piece of meat.

When intergranular fluids are either absent or else present only in trace amounts, metamorphic reactions are very slow. When a dry rock is heated, few changes occur because the growth of new minerals means that atoms must move by diffusing through the solid minerals. Diffusion through solids is an exceedingly slow process. If somehow an intergranular fluid is introduced, perhaps because pores are created as a result of the rock being crushed, diffusion of the atoms from one place to another can take place through the intergranular fluid. This is a vastly faster process, and as a result, new minerals grow rapidly and the metamorphic effects are pronounced.

As pressure increases because of burial of a rock and as metamorphism proceeds, the amount of pore space decreases and the intergranular fluid is slowly driven out of the rock. As the temperature of the rock increases, hydrous minerals recrystallize to anhydrous minerals and in the process water is released. The released water joins the intergranular fluid and is also slowly driven out of the metamorphic rock. For this reason, rock subjected to high-grade metamorphism contains fewer hydrous minerals, and less intergranular fluid, than does low-grade metamorphic rock.

Because intergranular fluid is a solution, any fluid that escapes during metamorphism will carry with it small amounts of dissolved mineral matter. If such a fluid flows through fractures formed as a result of tectonic forces, some of the dissolved minerals may be precipitated; the result is a vein (Fig. 7.2).

Figure 7.2 Vein in a Metamorphic Rock Quartz vein in a metamorphic rock. Vein is thought to have formed by deposition from escaping intergranular fluid during metamorphism.

The metamorphic changes that occur while temperatures and pressures are rising, and usually while abundant intergranular fluid is present, are termed **prograde metamorphic effects**. Those that occur as temperature and pressure are declining, and usually after much of the intergranular fluid has been expelled, are called **retrograde metamorphic effects**. Not surprisingly, because of the lack of fluid, retrograde metamorphic effects happen less rapidly and are less pronounced than prograde effects. Indeed, it is only because retrograde reactions happen so slowly that we see high-grade metamorphic rocks at the Earth's surface. If retrograde reactions were rapid, the minerals that are characteristic of high-grade metamorphism could not persist and we would lose much of the information about what has happened to rocks under intense metamorphism.

Pressure and Temperature

When a mixture of flour, salt, sugar, yeast, and water is baked in an oven, the high temperature causes a series of chemical reactions—new compounds grow and the final result is a loaf of bread. When rocks are heated, some minerals recrystallize and others react to form new minerals, and the final result is a metamorphic rock. In the case of the rocks, the source of heat is the Earth's internal heat. Rock can be heated simply by burial or by a nearby igneous intrusion. But burial is always accompanied by an increase in pressure, and intrusion usually occurs under high pressure. Therefore, whatever the cause of the heating, metamorphism can rarely be considered to be entirely due to the rise in temperature. The effects due to changing temperature and pressure must be considered together as discussed in the essay on *Pressure–Temperature–Time Paths: The History of Metamorphic Rocks*.

The combined influence of temperature and pressure on the melting properties of rocks and minerals was discussed in the boxed essay on rock melting in Chapter 4. Metamorphic transformations of mineral assemblages are also controlled by the dual effects of temperature and pressure. By measuring the effects of temperature and pressure on assemblage transformations in the laboratory, we can delineate the ranges of pressure and temperature conditions under which metamorphism has occurred in the crust. We will return to the important issue of the controls exerted by temperature and pressure on mineral assemblages later in the chapter.

When talking about deformation in a solid and development of new textures, we often use the term *stress* instead of pressure. We do so because stress has the connotation of direction. Rocks are solids, and solids can be squeezed more strongly in one direction than another; that is, stress in a solid, unlike a liquid, can be different in different directions. The textures in many metamor-

Measuring Our Earth

Pressure–Temperature–Time Paths: The History of Metamorphic Rock

Rock subjected to metamorphism develops a distinctive mineral assemblage characteristic of the grade of metamorphism reached. In order to reach the temperature and stress at which the highest grade of metamorphism occurs in a given rock, that rock must first have been subjected to lower temperatures and stresses, and thus it previously must have contained other mineral assemblages. Prograde metamorphism is a dynamic process, and mineral assemblages replace one another in succession.

One of the triumphs of modern scientific instrumentation has been the development of techniques by which tiny mineral fragments can be examined and analyzed. In many metamorphic rocks, microscopic relicts of those earlier mineral assemblages remain (Fig. B7.1); by analyzing these relicts, scientists can decipher the way stress (here called pressure, P, for simplicity) and temperature (T) changed with time (t) during metamorphism. In the language of the scientists who carry out such research, it is possible to determine the P–T–t path of a metamorphic rock. Surprising as it may seem, metamorphic rocks are rarely subjected simultaneously to the highest pressures and highest temperatures. This can only mean that, with time, geothermal gradients change in regions where metamorphism is occurring. For example, the P–T path shown in Figure B7.2 might apply to rock that is buried rapidly as a result of subduction, if the downward movement ceases and

the rock at depth continues to warm up over a long period of time. When cool rock moves downward relatively rapidly because of subduction, pressure increases in proportion to depth, but temperature will not increase in accord with a normal geothermal gradient because the warming of the subducted slab is a very slow process. If the downward movement slows and then ceases, because of a tectonic change from subduction to continental collision, for example, the rock will experience a maximum pressure (at point B of the figure). Because the rock will still be cooler than normal for its depth, temperature will continue to increase even as uplift starts and for some time afterward until point C is reached. Thus, the highest temperature and the highest-grade mineral assemblages will be produced at point C. If uplift continues until the rock is eventually exposed at the surface, the remainder of the P–T path will be a curve like $C \rightarrow A$.

By a further advance of modern scientific instrumentation, it is sometimes possible to obtain radiometric ages for the metamorphic mineral assemblages (Chapter 8). Knowing the P–T path of metamorphism and two or more time points on the path, we can calculate rates of burial and uplift and thereby compare the subduction rate of former tectonic plates with present subduction rates. They turn out to be very similar. Bodies of metamorphic rock are sensitive monitors of large-scale tectonic processes.

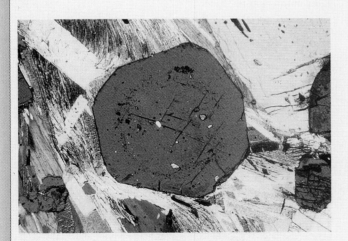

Figure B 7.1 Record of Changing Conditions A grain of garnet in a garnet–biotite gneiss. The small inclusions are relicts from an earlier mineral assemblage formed under P–T conditions different from those that formed the garnet–biotite gneiss. The garnet grain is 2 mm across.

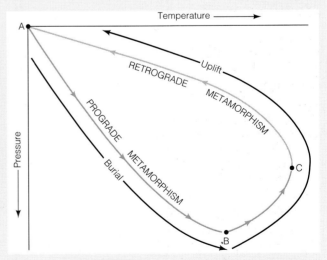

Figure B7.2 A Record of Changing P and T P–T path for a body of rock undergoing metamorphism along a subduction zone.

A

B

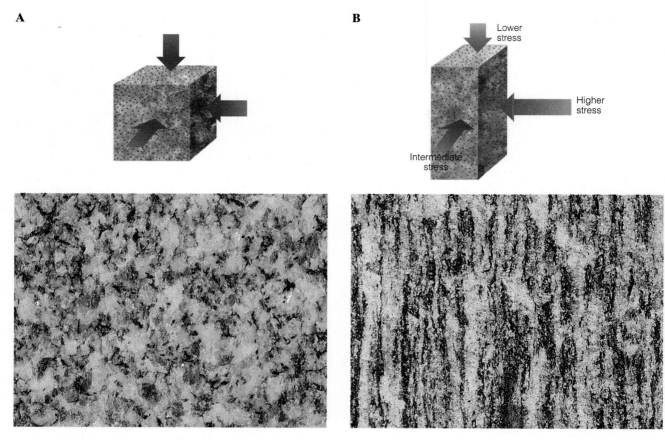

Figure 7.3 Uniform and Differential Stress Comparison of textures developed in rocks of the same composition under uniform and differential stress.   A. Granite, consisting of quartz, feldspar, and biotite, that crystallized in a uniform stress field. Note that biotite grains are randomly oriented.

B. High-grade metamorphic rock, also consisting of quartz, feldspar, and biotite, that crystallized in a differential stress field. Biotite grains are roughly parallel, giving the rock a distinct foliation. This kind of rock, called a *gneiss,* is discussed later in the chapter.

phic rocks record **differential stress** (meaning not equal in all directions) during metamorphism. Most igneous rocks, by contrast, have textures formed in **uniform stress** (meaning equal in all directions) because igneous rocks crystallize from liquids.

The most visible effect of metamorphism in a differential stress field involves the orientation of silicate minerals, such as micas and chlorites, that contain polymerized $(Si_4O_{10})^{4-}$ sheets. Compare Figure 7.3A and B. Figure 7.3A is a granite that has a typical texture of randomly oriented mineral grains that grew in a uniform stress field. Figure 7.3B, on the other hand, is a high-grade metamorphic rock containing the same minerals that formed in a differential stress field. Note that in Figure 7.3B the mica grains are roughly parallel, giving the rock a distinctive texture. Another example of a metamorphism in a differential stress field is shown in Figure 7.4. A conglomerate that originally consisted of uniformly rounded pebbles has been squeezed so that the pebbles are now flattened discs.

Note that it is the texture, not the mineral assemblage of a metamorphic rock, that is controlled by differential

Figure 7.4 Flattened Conglomerate Deformation of a conglomerate during metamorphism. Sandstone pebbles, originally round, have been flattened.

versus uniform stress. For this reason, geologists often use the terms *stress* and *pressure* interchangeably when describing mineral assemblages and metamorphic grades.

Time

Chemical reactions involve energy, and two compounds will react only if a lower state of energy is reached in the process. When the lowest energy state possible under the prevailing temperature and pressure has been reached, change due to chemical reaction ceases, and we say a state of equilibrium has been reached. A certain amount of time is needed for any chemical reaction to reach equilibrium. Some reactions, such as the burning of methane gas (CH_4) to yield carbon dioxide and water, can happen so rapidly that they create explosions. At the other end of the scale are reactions that require millions of years to proceed to completion.

Many of the chemical reactions that occur in rocks undergoing metamorphism are of the latter kind. No reliable ways have yet been developed to determine exactly how long a given metamorphic rock has remained at a given temperature and pressure. However, it can be demonstrated in the laboratory that high temperature, high pressure, and long reaction times produce large mineral grains. Thus, it is possible to draw the interesting general conclusion that coarse-grained rocks are the products of long-sustained metamorphic conditions (possibly over millions of years) at high temperatures and pressures. Fine-grained rocks are products of lower temperatures, lower pressures, or, in some cases, short reaction times.

METAMORPHIC RESPONSES TO CHANGES IN TEMPERATURE AND PRESSURE

Textural Responses

Most metamorphic rocks form in a differential stress field, and as a result they develop conspicuous, directional textures. As metamorphism proceeds and the sheet-structure minerals, such as mica and chlorite, start to grow, the minerals are oriented so that the sheets are perpendicular to the direction of maximum stress as shown in Figure 7.3B. The new, parallel flakes of mica produce a planar texture called **foliation** (Fig. 7.5), named from the Latin word *folium*, meaning leaf because many foliated rocks tend to split into thin, leaflike flakes. Foliation

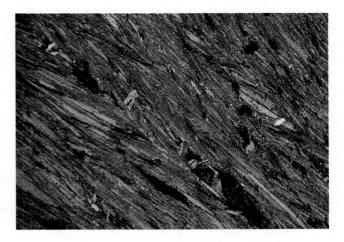

Figure 7.5 Foliation Seen Through a Microscope Microscopic thin section of a metamorphic rock showing pronounced foliation due to the roughly parallel arrangement of mica grains. The dark-colored grains are quartz. Direction of maximum stress is indicated by arrows. The sample is 1 cm wide.

may be pronounced or subtle, but when present it provides strong evidence of metamorphism.

Slaty Cleavage

During the earliest stages of low-grade metamorphism, the weight of the overlying rock is the only cause of stress. The new sheet-structure minerals, and therefore the foliation, tend to be parallel to the bedding planes of the sedimentary rock being metamorphosed. But deeper burial along a continental margin is typically followed by compression from a plate collision which deforms the flat sedimentary layers into folds and causes the sheet-structure minerals and the foliation to no longer be parallel to the bedding planes (Fig. 7.6). Low-grade metamorphic rocks tend to be so fine grained that the new mineral grains can be seen only with the microscope. The foliation is then called **slaty cleavage**, which is defined as the property by which a low-grade metamorphic rock breaks into platelike fragments along planes (Fig. 7.7).

Schistosity

At intermediate and high grades of metamorphism, grain sizes increase and individual mineral grains can be seen with the naked eye. Foliation in coarse-grained metamorphic rocks is called schistosity and is not necessarily planar. **Schistosity** is derived from *schistos*, a Greek word meaning "cleaves easily" and referring to the parallel arrangement of coarse grains of the sheet-structure minerals. Schistosity differs from slaty cleavage mainly in grain size. Intermediate and high-grade metamorphic rocks tend to break along wavy, or slightly distorted, surfaces, reflecting the presence and orientation of grains of quartz, feldspar, and other minerals.

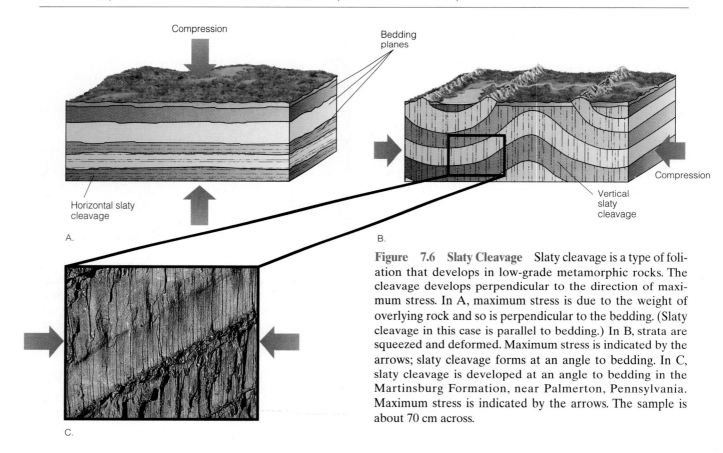

Figure 7.6 Slaty Cleavage Slaty cleavage is a type of foliation that develops in low-grade metamorphic rocks. The cleavage develops perpendicular to the direction of maximum stress. In A, maximum stress is due to the weight of overlying rock and so is perpendicular to the bedding. (Slaty cleavage in this case is parallel to bedding.) In B, strata are squeezed and deformed. Maximum stress is indicated by the arrows; slaty cleavage forms at an angle to bedding. In C, slaty cleavage is developed at an angle to bedding in the Martinsburg Formation, near Palmerton, Pennsylvania. Maximum stress is indicated by the arrows. The sample is about 70 cm across.

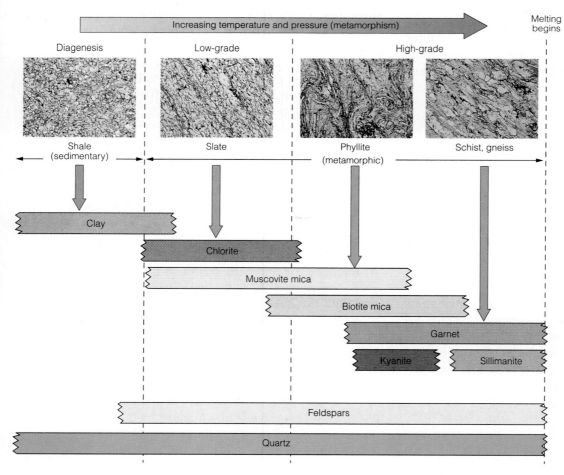

Figure 7.7 From Shale to Schist The bars in this diagram show the changes in mineral assemblage that occur as a shale is metamorphosed from low to high grade. Metamorphism progresses from left to right in the diagram. Before metamorphism occurs, the shale is a sedimentary rock made of clay particles and quartz grains. The first metamorphic rock to develop is a low-grade slate, then a phyllite, and finally a high-grade schist and gneiss. The photographs show the rocks under a microscope; each photo shows a field of view approximately 3 mm (0.12 in.). The minerals kyanite and sillimanite have the same composition (Al_2SiO_5) but different crystal structures —they are found only in metamorphic rocks.

Assemblage Responses

Metamorphism produces new mineral assemblages as well as new textures. As temperature and pressure rise, one new mineral assemblage follows another. For any given rock composition, each assemblage is characteristic of a given range of temperature and pressure. Some of the minerals in these assemblages are found rarely (or not at all) in igneous and sedimentary rocks. Their presence in a rock is usually evidence enough that the rock has been metamorphosed. Examples of these metamorphic minerals are chlorite, serpentine, epidote, talc, and the three polymorphs of Al_2SiO_5: kyanite, sillimanite, and andalusite. Figure 7.7 illustrates how mineral assemblages change with grade of metamorphism as a shale is metamorphosed.

KINDS OF METAMORPHIC ROCK

The naming of metamorphic rocks is based partly on texture and partly on mineral assemblage. The most widely used names are those applied to metamorphic derivatives of shales, sandstones, limestones, and basalts. This is so because shales, sandstones, and limestones are the most abundant sedimentary rock types, whereas basalt is by far the most abundant kind of igneous rock.

Metamorphism of Shale and Mudstone

Slate
The low-grade metamorphic product of either shale or mudstone is **slate**. The minerals usually present in both shale and mudstone include quartz, clays of various kinds, calcite, and possibly feldspar. Under conditions of low-grade metamorphism, muscovite and/or chlorite crystallize. Although the rock may still look a lot like shale or mudstone, the tiny new mineral grains produce slaty cleavage (Fig. 7.8A). The presence of slaty cleavage is clear proof that a rock has gone from being a sedimentary rock to a metamorphic rock.

Phyllite
Continued metamorphism of a slate to intermediate grade produces both larger grains of mica and a changing mineral assemblage; the rock develops a pronounced

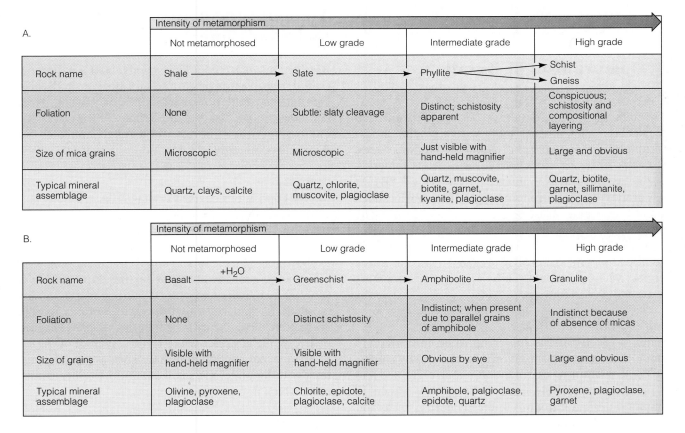

Figure 7.8 Progressive Metamorphism Progressive metamorphism of shale and basalt. Mineral assemblage and foliation change as a result of increasing temperature and differential stress.   A. Shale.   B. Basalt.

foliation and is called **phyllite** (from the Greek, *phyllon*, meaning a leaf). In a slate it is not possible to see the new grains of mica with the unaided eye, but in a phyllite they are just large enough to be visible (Fig. 7.8A).

Schist and Gneiss

Still further metamorphism beyond that which produces a phyllite leads to a coarse-grained rock with pronounced schistosity, called **schist**. The most obvious differences between slate, phyllite, and schist are in grain size, but, as can be seen in Figure 7.8A, increasing grain size is only one of a number of changes. At the high grades of metamorphism characteristic of schists, minerals may start to segregate into separate bands. A high-grade rock with coarse grains and pronounced foliation, but with layers of micaceous minerals segregated from layers of minerals such as quartz and feldspar, is called a **gneiss** (pronounced nice, from a word in the German, *gneisto*, meaning to sparkle) (Fig. 7.3B).

The names *slate* and *phyllite* describe textures and are commonly used without adding mineral names as adjectives. The names of the coarse-grained rocks, schists and gneisses, are also derived from textures, but in these cases mineral names are commonly added as adjectives; for example, we might refer to a quartz–plagioclase–biotite–garnet gneiss. The difference arises because minerals in coarse-grained rocks are large enough to be seen and readily identified.

Metamorphism of Basalt

Greenschist

The main minerals in basalt are olivine, pyroxene, and plagioclase, each of which is anhydrous. When a basalt is subjected to metamorphism under conditions where H_2O can enter the rock and form hydrous minerals, distinctive mineral assemblages develop (Fig. 7.8B). At low grades of metamorphism, mineral assemblages such as chlorite + plagioclase + epidote + calcite form. The resulting rock is equivalent in metamorphic grade to a slate but has a very different appearance. It has pronounced foliation as a phyllite does, but it also has a very distinctive green color because of its chlorite content; it is termed *greenschist*.

Amphibolite and Granulite

When a greenschist is subjected to an intermediate grade of metamorphism, chlorite is replaced by amphibole; the resulting rock is generally coarse-grained and is called an **amphibolite** (Fig. 7.9). Foliation is present in amphibolites but is not pronounced because micas and chlorites are usually absent. At highest grade metamorphism, amphibole is replaced by pyroxene, and an indistinctly foliated rock called a *granulite* develops.

Figure 7.9 Metamorphosed Basalt Amphibolite resulting from metamorphism of a pillow basalt. Compare this with Figure 4.21. The pillow structure was compressed during metamorphism but can be discerned by the borders of pale-yellow epidote formed from the original glassy rims of the pillows. The outcrop is in Namibia.

Metamorphism of Limestone and Sandstone

Marble and *quartzite* are, respectively, the metamorphic derivatives of limestone and quartz sandstone. Neither limestone nor quartz sandstone (when pure) contains the necessary ingredients to form sheet- or chain-structure minerals. As a result, marble and quartzite commonly lack foliation.

Marble

Marble consists of a coarsely crystalline, interlocking network of calcite grains. During recrystallization of a limestone, the bedding planes, fossils, and other features of sedimentary rocks are largely obliterated. The end result, as shown in Figure 7.10A, is an even-grained rock with a distinctive, somewhat sugary texture. Pure marble is snow white in color and consists entirely of pure grains of calcite. Such marbles are favored for statuary, marble gravestones, and statues in cemeteries, perhaps because many people consider white to be a symbol of purity. Many marbles contain impurities such as organic matter, pyrite, limonite, and small quantities of silicate minerals, which impart various colors.

Quartzite

Quartzite is derived from sandstone by the filling in of the space between the original grains with silica and by

A

B

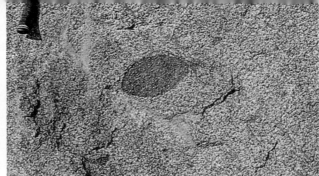

A

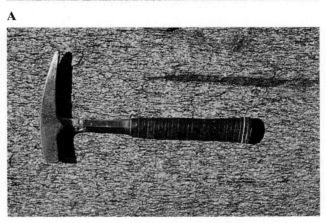

B

Figure 7.10 Nonfoliated Metamorphic Rocks Textures of nonfoliated metamorphic rocks seen in hand-sized specimens and in thin sections and viewed in polarized light. Notice the interlocking grain structure produced by recrystallization during metamorphism. Each specimen is 2 cm across.   A. Marble, composed entirely of calcite. All vestiges of sedimentary structure have disappeared.   B. Quartzite. Arrows point to faint traces of the original rounded quartz grains in some of the grains.

Figure 7.11 Foliation in a Metamorphosed Granite Development of foliation in a granite by cataclastic metamorphism. From Groothoek, South Africa.   A. Undeformed granite consisting of quartz, feldspar, and biotite. The dark patch in the center of the field of view is a xenolith of amphibolite. Foliation is not present in the granite.   B. The original granitic texture has been completely changed, and the granite has been transformed to a gneiss with a distinct foliation. Amphibolite xenoliths have been flattened and elongated.

recrystallization of the entire mass (Fig. 7.10B). Sometimes, the ghostlike outlines of the original sedimentary grains can still be seen, even though sometimes recrystallization may have completely rearranged the original grain structure.

KINDS OF METAMORPHISM

The processes that result from changing temperature and stress, and that cause the metamorphic changes observed in rocks, can be grouped under the terms *mechanical deformation* and *chemical recrystallization*. The two processes usually proceed together, but sometimes the change is due more to mechanical deformation than to chemical recrystallization, and at other times the reverse is true. Mechanical deformation includes grinding, crushing, and the development of foliation. The deformed conglomerate shown in Figure 7.4 is an example in which mechanical deformation probably played the dominant role in the changes that occurred. Concrete, as discussed in the essay *"Concrete: The Artificial Metamorphic Rock*

That Changed Our Environment," provides an example of change induced solely by chemical recrystallization. Chemical recrystallization includes all the changes in mineral composition, in growth of new crystals, and in the additions or losses of H_2O and CO_2 that occur as rock is heated. Different kinds of metamorphism reflect the different levels of importance of the two processes.

Cataclastic Metamorphism

Mechanical deformation of a rock can sometimes occur with only minor chemical recrystallization. Such deformation is usually localized and usually seen in igneous rocks. For example, when a coarse-grained rock such as granite is subjected to intense differential stresses, individual mineral grains may be shattered and pulverized. This sort of deformation is called **cataclastic metamorphism**. As cataclastic metamorphism proceeds, grain and rock fragments become elongated and a foliation develops. The two photographs shown in Figure 7.11 illustrate the change that can be produced in a granite by cataclastic metamorphism.

Understanding Our Environment

Concrete: The Artificial Metamorphic Rock that Changed Our Environment

The word *cement* comes from the Latin, *caementum*, meaning a substance that binds rock particles together. When cement is used to bind an aggregate of crushed rock, gravel, or oyster shells, the result is *concrete*—a term derived from another Latin word, *concrescere*, meaning to grow together.

The forerunner of modern cement was discovered by Roman engineers more than 2000 years ago. They found that water added to a mixture of quicklime (CaO, made by heating limestone) and glass-rich volcanic ash from Pozzuoli, near Naples, produced chemical reactions that caused new compounds to form, resulting in a hardening of the mixture. The hardened mass was stable in air and water, and the Roman engineers soon discovered that by mixing their Pozzuolan cement with crushed rock they could make a concrete that was physically strong and chemically stable.

The Romans eventually discovered that cement could be made from materials other than glassy volcanic ash. They found that ordinary shale, when heated to a sufficiently high temperature, would partially melt, forming a product similar to volcanic glass. When mixed with quicklime and water, the heated shale produced a tough, stable cement.

The Roman formula for cement seems to have been forgotten during Europe's Dark Ages. It was rediscovered in 1756 by John Smeaton, a British engineer charged with building a lighthouse on the treacherous Eddystone Rocks off the mouth of England's Plymouth Harbor. While researching ancient Latin documents for ideas about how to build underwater foundations for the lighthouse, Smeaton discovered the formula for Roman cement.

Following Smeaton's discovery and use of cement, concrete quickly became a popular building material. In 1824 another English engineer, Joseph Aspidin, patented a formula for *portland cement*, so-called because its color resembles Portland stone, a limestone widely used in England for building.

Today, concrete made from portland cement and rock aggregate is the most widely used building material around the world. Concrete has changed the environment of our cities, our motorways, harbors, and airports in ways that are unmatched by any other material (Fig. B7.3). In fact, it is difficult to imagine a concrete-free environment.

Limestone and shale, the starting materials for cement, are common and widely available so concrete can be, and is, made in essentially every country in the world. When an appropriate mixture of limestone and shale are heated to about 1480°C, carbon dioxide and water vapor are expelled, the limestone becomes quicklime, and the shale partially melts to glass. When water is added to the mixture, the quicklime reacts with the glass to form an interlocking mesh of calcium-aluminum silicate compounds—in short, an instant metamorphic rock.

Figure B7.3 A Concrete Landscape La Grande Arche in the Place de La Defense, Paris. The building was erected between 1982 and 1990. It is made of reinforced concrete and faced with marble.

Contact Metamorphism

Contact metamorphism occurs adjacent to bodies of hot magma that are intruded into cool rocks of the crust. Such metamorphism involves chemical recrystallization and happens in response to a pronounced increase in temperature. Mechanical deformation is minor or absent because the stress around a mass of magma tends to be uniform. Rock adjacent to the intrusion becomes heated and metamorphosed, developing a well-defined shell, or **metamorphic aureole**, of altered rock (Fig. 7.12).

The width of an aureole of contact metamorphosed rock depends on the size of the intrusive body and on the amount of H_2O in the rock being metamorphosed, as well as on fluids released by the solidifying magma. With a small intrusion, such as a dike or sill a few meters thick, and in the absence of a fluid, the aureole may be

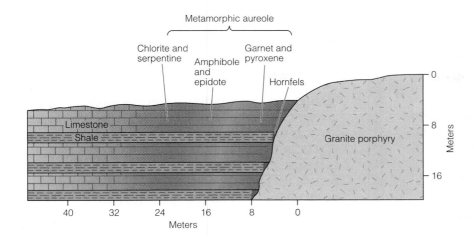

Figure 7.12 Contact Metamorphism Contact metamorphism around an intrusion of granite porphyry near Breckenridge, Colorado. Limestone composition was changed by fluids released by the cooling granite, and as a result a metamorphic aureole up to 20 m wide developed. New minerals form a series of concentric shells in the aureole, each with a distinct mineral assemblage. Shale between limestone layers was impervious to fluids and except for a narrow band of hornfels immediately adjacent to the granite was not affected.

only a few centimeters wide. The metamorphic rock is usually a hard, fine-grained rock composed of an interlocking mass of uniformly sized mineral grains, which is called a **hornfels**. A large intrusion contains much more heat energy than a small one, and it may also give off a lot of H_2O vapor. Around some large intrusions, more than a kilometer in diameter, aureoles reach more than 100 m in width and the metamorphic rocks are typically coarse-grained.

Within a large metamorphic aureole through which fluid has passed, several different, and roughly concentric, zones of mineral assemblages can usually be identified, as shown in Figure 7.12. Each zone is characteristic of a certain temperature range. Immediately adjacent to the intrusion, where temperatures were very high, we find anhydrous minerals such as garnet and pyroxene. Beyond them are found hydrous minerals such as epidote and amphibole, and beyond them micas and chlorites. The exact assemblage of minerals in each zone depends, of course, on the chemical composition of the intruded rock and on that of the invading fluid, as well as on temperature and pressure.

Burial Metamorphism

Sediments, together with interlayered pyroclastics, may attain temperatures of a few hundred degrees Celsius when buried deeply in a sedimentary basin. Abundant pore water is present in buried sediment, and this water speeds up chemical recrystallization and helps new minerals to grow. But water-filled sediment is weak and acts more like a liquid than a solid. The stress during **burial metamorphism** tends, therefore, to be uniform. As a result, little mechanical deformation is involved in burial metamorphism, and the texture of the metamorphic rock that results looks like that of an essentially unaltered sedimentary rock, even though there has been a substantial change in the mineral assemblage.

The family of minerals that particularly characterize the conditions of burial metamorphism are *zeolites*, a group of minerals with fully polymerized silicate structures containing the same chemical elements as feldspars but also containing water.

Burial metamorphism, which is the first stage of metamorphism following diagenesis, occurs in deep sedimentary basins, such as trenches on the margins of tectonic plates. As temperatures and pressures increase, burial metamorphism grades into regional metamorphism.

Regional Metamorphism

The most common metamorphic rocks of the continental crust occur throughout areas of tens of thousands of square kilometers, and the process that forms them is called **regional metamorphism**. Unlike contact or burial metamorphism, regional metamorphism involves pronounced differential stresses and a considerable amount of mechanical deformation in addition to chemical recrystallization. As a result, regionally metamorphosed rocks tend to be distinctly foliated.

Slate, phyllite, schist, and gneiss are the most common varieties of regionally metamorphosed rocks, and they are usually found in mountain ranges or the eroded remnants of former mountain ranges. Mountain ranges that contain rocks formed by regional metamorphism form as a result of subduction or collision between fragments of continental crust. During a collision between continents, sedimentary rock along the margin of a continent is subjected to intense differential stresses. The foliation that is so characteristic of slates, schists, and gneisses is a consequence of those intense stresses. Regional metamorphism is therefore a consequence of plate tectonics. Greenschists and amphibolites are also products of regional metamorphism. They tend to be found where segments of ancient oceanic crust of basaltic composition have been incorporated into the continental crust and metamorphosed.

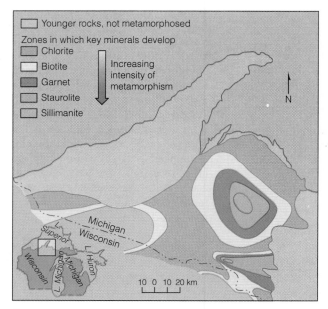

A

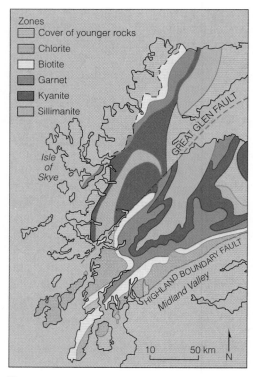

B

Figure 7.13 Regional Metamorphism and Metamorphic Zones Metamorphic zones resulting from regional metamorphism. A. Michigan. B. Scottish Highlands.

conductors of heat; the heating-up process can therefore be very slow. The temperatures reached depend both on depth and on how long a rock is buried in the thickened pile. If the folding and thickening are very slow, heating of the pile keeps pace with the temperature of adjacent parts of the crust and mantle (i.e., a normal continental geothermal gradient is maintained). However, if burial is very fast, as it is with sediment dragged down in a subduction zone, the pile has insufficient time to heat up and conditions of high pressure but rather low temperature prevail. Depending on the rate of burial, therefore, the same starting rock can yield two quite different metamorphic rocks because different pressures and temperatures are reached.

Metamorphic Zones

The first geologists to make a systematic study of a regionally metamorphosed terrain did so in the Scottish Highlands. They observed that rocks having the same overall chemical composition (that of a shale) could be subdivided into a sequence of zones, each zone having a distinctive mineral assemblage. Each assemblage was characterized by the appearance of new minerals. They selected characteristic *index minerals*, which, proceeding from low-grade rocks to high-grade rocks, marked the appearance of each new mineral assemblage. Their index minerals were, in order of appearance, chlorite, biotite, garnet, kyanite, and sillimanite. By plotting on maps the places where each of the index minerals first appeared in rocks having the chemical composition of shale, the workers in the Scottish Highlands defined a series of isograds. An **isograd** is a line on a map connecting points of first occurrence of a given mineral in metamorphic rocks. The concept of isograds is now widely used to study metamorphic rocks of all kinds; it is just as applicable to burial and contact metamorphism as it is to regional metamorphism.

The regions on a map between isograds are known as **metamorphic zones**. We speak of the chlorite zone, the biotite zone, and so forth, and it is the zones that are commonly depicted on maps showing the relationships among metamorphic rocks (Fig. 7.13).

METAMORPHIC FACIES

Careful study of metamorphic rocks around the world has demonstrated that the chemical compositions of most rocks are little changed by metamorphism. The main changes that do occur are the addition or loss of

In order to see what happens during regional metamorphism, consider a segment of the crust that is subjected to a horizontal compression. Rock in the crust becomes folded and buckled. The folding and buckling cause the crust to become locally thickened, as shown in Figure 7.1. The bottom of the thickened mass is pushed deeper where temperatures become higher. As a result, the rocks near the bottom of the thickened pile are subjected to both elevated stress and higher temperature. New minerals start to grow. However, rocks are poor

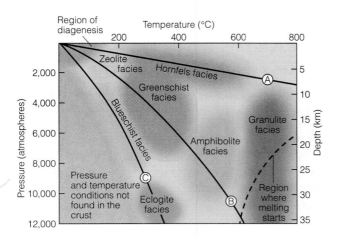

Figure 7.14 Metamorphic Facies Metamorphic facies plotted with respect to temperature and depth. Curve A is a typical thermal gradient around an intrusive igneous rock that is causing contact metamorphism. Curve B is a normal continental geothermal gradient. Curve C is the geothermal gradient developed in a subduction zone. Note that the zeolite facies overlaps in temperature and pressure with the conditions of digenesis.

volatiles such as H_2O and CO_2, but the principal constituents of rocks, such as SiO_2, Al_2O_3, and CaO, remain fixed or approximately so. The principal changes brought about during metamorphism, then, are changes in the mineral assemblage, not changes in the overall chemical composition of the rocks. The conclusion to be drawn from this observation is that the mineral assemblages in the metamorphic derivatives of common sedimentary and igneous rocks must be determined by the temperatures and stress to which the rocks are subjected during metamorphism. Based on this conclusion, the famous Finnish geologist Pennti Eskola proposed, in 1915, the concept of **metamorphic facies**. According to this concept, for given conditions of metamorphism, the equilibrium assemblages of minerals that form during metamorphism in rocks of different composition belong to the same metamorphic facies. Eskola drew his conclusions from studies of metamorphosed basalts that were interlayered with rocks of entirely different composition.

Metamorphic facies were originally described in terms of recurring mineral assemblages; for each assemblage there was assumed to be a specific set of temperature and pressure conditions. The realization that temperature, pressure, and rock composition each play a role in determining the mineral assemblage provided the link geologists needed; they could now determine the conditions under which rocks were metamorphosed by comparing their mineral assemblages to assemblages produced in laboratory experiments. It is now possible to prepare a detailed *petrogenetic grid* in which the mineral assemblage of almost any rock composition

can be plotted in terms of temperature and pressure. The concept is now applied to a wide range of temperatures and pressures. The principal metamorphic facies, together with geothermal gradients to be expected in the continental crust under three different geological conditions, are shown in Figure 7.14.

Because Eskola was studying metamorphosed basalts when he proposed the metamorphic facies concept, most of the names he gave to metamorphic facies reflect the mineral assemblages developed in rocks of basaltic composition. It is important to remember, however, as shown in Table 7.1, that mineral assemblages are just as much a result of original rock composition as they are of the temperature and pressure of metamorphism. When comparing mineral assemblages of rocks subjected to different grades of metamorphism, therefore, one must be certain that the rocks have the same overall chemical composition.

METASOMATISM

The metamorphic processes we have discussed so far involve essentially fixed compositions and relatively small amounts of fluid. The amount of fluid is small because the pore volume in rocks undergoing metamorphism tends to be small and because the release of H_2O and CO_2 from minerals involved in metamorphic reactions happens slowly rather than all at once. Geologists use the jargon expression "a small water–rock ratio" to describe the metamorphic environment. What they mean by the expression is that the weight ratio of fluid (mainly water) to rock is about 1:10 or smaller. This is enough fluid to serve as a metamorphic juice, but not enough to dissolve a lot of the rock and so change the rock composition noticeably.

Under a few circumstances, however, large water–rock ratios occur. When an open rock fracture occurs, for example, through which a lot of fluid flows, a water–rock ratio can be 10:1 or even 100:1, and the rocks adjoining the fracture can be drastically altered by the addition of new ions, removal of material in solution, or both. The term **metasomatism** (from *meta*, change, and *soma*, derived from the Latin word for juice) is applied to the process whereby the chemical compositions of rocks are distinctively altered by exchange with ions in solution.

Metasomatism is commonly associated with contact metamorphism, especially where the rocks being metamorphosed are limestones, as in the example shown in Figure 7.12. Metasomatic fluids released by a cooling magma pass outward through the volume of rock undergoing contact metamorphism. Because the fluids may carry constituents such as silica, iron, and magnesium in solution, the composition of the limestone close to the

TABLE 7.1	Characteristic Minerals of Differing Metamorphic Facies for Selected Rocks

	Precursor Rock Type	
Facies Name	Basalt	Shale
Granulite	Pyroxene, plagioclase, garnet	Biotite, K-feldspar, quartz, sillimanite
Amphibolite	Amphibole, plagioclase, garnet, quartz	Garnet, biotite, muscovite, kyanite or sillimanite, quartz
Epidote— Amphibolite	Amphibole, epidote, plagioclase, garnet, quartz	Garnet, chlorite, muscovite, biotite, quartz
Greenschist	Chlorite, amphibole, plagioclase, epidote	Chlorite, muscovite, plagioclase, quartz
Blueschist	Glaucophane,[b] chlorite, Ca-rich silicates	Glaucophane,[b] chlorite quartz, muscovite, lawsonite[c]
Eclogite	Pyroxene (variety jadeite), garnet, kyanite	Not yet observed
Hornfels	Pyroxene, plagioclase	Andalusite,[d] biotite, K-feldspar, quartz
Zeolite	Calcite, chlorite, zeolite (variety laumontite)	Zeolite, pyrophyllite,[e] Paragonite[f]

[a] For temperature and pressure conditions of each facies, refer to Figure 7.16.
[b] Glaucophane is a bluish-colored amphibole, $Na_2(Mg,Fe)_3Al_2Si_8O_{22}(OH)_2$.
[c] Lawsonite is a high pressure mineral similar in composition to anorthite; its formula is $CaAl_2Si_2O_7(OH)_2H_2O$.
[d] Andalusite is a polymorph of kyanite and sillimanite.
[e] Pyrophyllite is $AlSi_2O_5(OH)$.
[f] Paragonite is a sodium mica, $NaAl_2(Al_1Si_3)O_{10}(OH)_2$.

Figure 7.15 Metasomatism Metasomatically altered marble. The white mineral is calcite, brown is garnet, green is a pyroxene, and purple is fluorite. Materials needed to form the garnet, pyroxene, and fluorite were added to the marble by metasomatic fluids. The sample is 8 cm wide and comes from King Island, Australia.

cooling magma can be drastically changed, even though the limestone distant from the magma, beyond reach of invading fluids, remains unchanged. Figure 7.15 is a photograph of a contact metamorphic rock that was originally a limestone. Without the addition of new material, the limestone would have become a marble, but through metasomatism it was changed to an assemblage of garnet, a green pyroxene called diopside, fluorite, and calcite. Metasomatic fluids may also carry valuable metals and form mineral deposits, as discussed in the essay *"Valuable Ores from Subterranean Waters."*

PLATE TECTONICS AND METAMORPHISM

The revolution in geology brought about by plate-tectonics theory provides, for the first time, an explanation for the distribution of metamorphic zones in regionally metamorphosed rocks.

Burial metamorphism is believed to occur in the lower portions of the thick piles of sediment that accumulate on the continental shelf and continental slope.

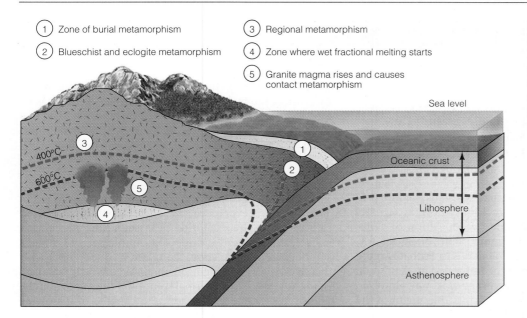

① Zone of burial metamorphism

② Blueschist and eclogite metamorphism

③ Regional metamorphism

④ Zone where wet fractional melting starts

⑤ Granite magma rises and causes contact metamorphism

Figure 7.16 Where Metamorphism Occurs Diagram of a convergent plate boundary showing the different regions of metamorphism. Dashed lines indicate temperature contours.

Such metamorphism is known to be happening today in the great pile of sediments accumulated in the Gulf of Mexico, off the mouth of the Mississippi River.

Regional metamorphism occurs at the subduction boundary of a plate, as shown in Figure 7.16. The temperatures and pressures characteristic of *blueschist* and *eclogite facies* metamorphism are reached when crustal rocks are dragged down by a rapidly subducting plate. Under such conditions, pressure increases more rapidly than temperature, and as a result rock is subjected to high pressure and relatively low temperature. Rocks subjected to blueschist and eclogite facies metamorphism are widespread in the Coast Ranges of California. Blueschist metamorphism is probably happening today along the subducting margin of the Pacific Plate where it plunges under the coast of Alaska and the Aleutian Islands.

The metamorphic conditions characteristic of *greenschist* and *amphibolite facies* metamorphism occur where continental crust is thickened by plate convergence and heated by rising magma. Continental collision is the most common setting for regional metamorphism, and broad areas of regionally metamorphosed rocks can be observed throughout the Appalachians and the Alps. Such metamorphism is no doubt occurring today beneath the Himalaya, where the continental crust has been thickened by collision, and beneath the Andes, where it has been both thickened and heated by rising magma. If the crust is sufficiently thick, rocks subjected to amphibolite facies or higher grade metamorphism can reach temperatures at which wet partial melting commences, initially with the formation of migmatites and eventually, of magmas.

Metasomatism and the generation of hydrothermal solutions can also be linked to plate tectonics because metasomatism is closely related to regional metamor-

phism and magmatic activity. A striking example can be seen in the distribution of chalcopyrite-rich copper deposits in North and South America. As shown in Figure 7.17, a pronounced belt of deposits formed in, or

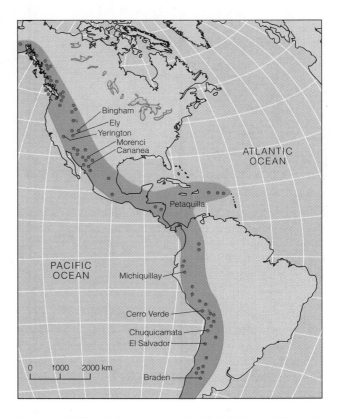

Figure 7.17 Subduction and Rich Mineral Deposits Chalcopyrite-rich copper deposits formed by hydrothermal solutions create a well-defined belt parallel to the subduction edges of the South and North American plates. Subduction is still active in South America and northern North America but has ceased in the southern part of North America.

Using Resources

Valuable Ores from Subterranean Water

The fluids that cause metasomatism are rich in H_2O, and they are hot—250°C or higher. Such fluids are called **hydrothermal solutions** (*hydro* being the Greek word for water, thermal derived from *therme*, the Greek word for heat).

Hydrothermal solutions form veins by depositing dissolved substances as minerals in cracks they flow through. As shown in Figure B7.4, hydrothermal solutions can also produce distinctive changes in the rocks they flow through. By alteration of existing rocks and by deposition of veins, hydrothermal solutions play an important role in the formation of many valuable mineral deposits. Probably more valuable mineral deposits have been formed by deposition from hydrothermal solutions than in any other way. However, despite the importance of such deposits, the origins of the solutions are often difficult to decipher. Whatever the origin of a given hydrothermal solution, deposition of materials carried in solution generally occurs deep underground where

Figure B7.4 Alteration by Hydrothermal Solutions Dark-colored metasomatic alteration produced in a calcareous siltstone by hydrothermal solutions. Hot solutions flowed through fracture-controlled openings introducing iron and other chemical elements and allowing a mineral assemblage of amphiboles, pyroxenes, and garnets to grow.

it cannot be seen. By the time a deposit is finally uncovered by erosion, the hydrothermal solution that formed it is no longer present. Nevertheless, clues have been found, so that many details of the process of deposition are now understood. Even so, a great deal of research remains to be done.

Composition of the Solutions

The principal ingredient of hydrothermal solutions is water. The water is never pure and always contains dissolved salts such as sodium chloride, potassium chloride, calcium sulfate, and calcium chloride. The amounts vary, but most solutions range from about the saltiness of seawater (3.5% dissolved solids by weight) to about ten times the saltiness of seawater. A hydrothermal solution is therefore a brine, and brines, unlike pure water, are capable of dissolving minute amounts of seemingly insoluble minerals such as gold, chalcopyrite, galena, and sphalerite.

Origins of the Solutions

Hydrothermal solutions have many sources. One way they form is through the cooling and crystallization of magma formed by wet partial melting (Chapter 4). Most of the water that causes the wet partial melting is released when such a magma solidifies. Instead of being pure water, however, it carries in solution both the most soluble constituents in the magmas, such as NaCl, and is also relatively enriched in chemical elements such as gold, silver, copper, lead, zinc, mercury, and molybdenum.

High temperatures increase the effectiveness of brines in forming hydrothermal minerals deposits. Volcanism and high temperatures go together. It is not surprising, therefore, that many mineral deposits are associated with hot volcanic rocks that were invaded by deep circulating water that started as rainwater or seawater. Nor is it surprising that a great many valuable mineral deposits are found in the upper portions of volcanic piles, where they were deposited when upward-moving hydrothermal solutions cooled and precipitated the ore minerals. Figure B7.5 is a simplified diagram of hydrothermal solutions being generated by seawater penetrating oceanic crust, being heated, and rising

associated with, old stratovolcanoes that lie along the western margin of the Americas. The magmas that produced the stratovolcanoes were formed by wet partial melting of mantle rocks that lie above subducted oceanic crust. The magmas served as the heat sources for the hydrothermal solutions that formed the mineral deposits. The magmas may also have carried up from the mantle the metals now found in the mineral deposits. The actual

source of the metals, whether from the mantle-generated magmas or from the continental crust and gathered by the magmas as they rose upward, is a matter of ongoing research. Regardless of the exact source of the valuable metals, the abundance of mineral resources we enjoy on the Earth is due to the combined effects of magmatic, metamorphic, and metasomatic processes, all of which occur because of plate tectonics.

upward by convection. Heated seawater reacts with the rocks it is in contact with, causing changes in the chemical and mineral composition. As the minerals are changed, trace metals such as copper and zinc present in the rocks become concentrated in the hot seawater.

Hydrothermal solutions formed beneath the sea can become so hot that they rise rapidly through fractures and form jetlike eruptions of hot hydrothermal solutions into cold seawater. Figure B7.6 is a photograph of such a submarine hot spring at 21°N latitude on the East Pacific Rise, the plate spreading center that runs through the Pacific Ocean. Many such eruptions have been observed along midocean ridges. As a jetting hydrothermal solution cools, it deposits its minerals in a massive blanket around the erupting vent.

A.

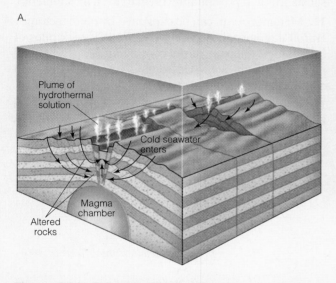

Figure B7.5 Submarine Hot Springs Seawater penetrates volcanic rocks on the seafloor. Heated by a magma chamber and by the hot rocks that surround the magma chamber, seawater becomes a hydrothermal solution, produces metasomatic alterations, and rises at a midocean ridge as a hydrothermal solution.

Figure B7.6 Hydrothermal Vent Hydrothermal solution forming an ore deposit on the seafloor. A so-called black smoker photographed at a depth of 2500 m below sea level on the East Pacific Rise at 21° N latitude. The "smoker" has a temperature of 320° C. The rising hydrothermal solution is actually clear; the black color is due to fine particles of iron sulfide and other minerals precipitated from solution as the plume is cooled through contact with cold seawater. The chimneylike structure is composed of pyrite, chalcopyrite, and other ore minerals deposited by the hydrothermal solution.

Source: "1998: Warmest Year of Past Millennium," *Science News* (March 20, 1999), p. 191.

SUMMARY

1.   Metamorphism involves changes in mineral assemblage and rock texture and occurs in the solid state as a result of changes in temperature and pressure.

2.   Mechanical deformation and chemical recrystallization are the processes that affect rock during metamorphism.

3.   The presence of intergranular fluid greatly speeds up metamorphic reactions.

4.   Foliation, as expressed by directional textures such as slaty cleavage and schistosity, arises from parallel growth of minerals formed during metamorphism and from the mechanical deformation of materials under differential stress.

5.   Cataclastic metamorphism involves mechanical deformation together with chemical recrystallization, but mechanical deformation is the predominant effect.

6.   Heat given off by bodies of intrusive igneous rock causes contact metamorphism and creates contact metamorphic aureoles. Contact metamorphism involves chemical recrystallization but little mechanical deformation.

7.   Regional metamorphism, which involves both mechanical deformation and chemical recrystallization, is a result of plate tectonics. Regionally metamorphosed rocks are produced along subduction and collision edges of plates.

8.   Rocks of the same chemical composition that are subjected to identical metamorphic environments react to form the same mineral assemblages. For given conditions of metamorphism, the equilibrium assemblages of minerals that form during the metamorphism of rocks of different composition belong to the same metamorphic facies.

9.   Metasomatism involves the changes in rock composition that occur when material in solution is added to the rock, or material is taken away, as the result of fluids flowing through a rock.

10.   Hydrothermal solutions are naturally formed hot brines that are capable of dissolving and transporting substances and precipitating them to form new minerals.

11.   Metamorphism can be explained by plate tectonics. Burial metamorphism occurs within the thick piles of sediment at the base of the continental slope; regional metamorphism occurs in regions of subduction and continental collision.

12.   Metasomatism due to hydrothermal solutions is linked to plate tectonics because the solutions tend to form in, or be associated with, stratovolcanoes. Stratovolcanoes are formed above subduction zones.

THE LANGUAGE OF GEOLOGY

amphibolite (p. 186)

burial metamorphism (p. 189)

cataclastic metamorphism (p. 187)
contact metamorphism (p. 188)

differential stress (p. 182)

foliation (p. 183)

gneiss (p. 186)

high-grade metamorphism (p. 179)
hornfels (p. 189)

hydrothermal solution (p. 194)

intergranular fluid (p. 179)
isograd (p. 190)

low-grade metamorphism (p. 179)

marble (p. 186)
metamorphic aureole (p. 188)
metamorphic facies (p. 191)
metamorphic zones (p. 190)
metamorphism (p. 178)
metasomatism (p. 191)
migmatite (p. 179)

phyllite (p. 186)
prograde metamorphic effects (p. 180)

quartzite (p. 186)

regional metamorphism (p. 189)
retrograde metamorphic effects (p. 180)

schist (p. 186)
schistosity (p. 183)
slate (p. 185)
slaty cleavage (p. 183)

uniform stress (p. 182)

QUESTIONS FOR REVIEW

1. Briefly discuss the factors that control metamorphism.

2. How and why does slaty cleavage form?

3. What is schistosity? How does it differ from slaty cleavage?

4. What is the difference between a schist and a gneiss?

5. How does a quartzite differ from a sandstone?

6. What is a metamorphic aureole?

7. What is regional metamorphism?

8. What is the geological setting of regional metamorphism? Can you name two places in the world where regional metamorphism is probably happening today?

9. What is burial metamorphism? Suggest some place on the Earth where it is probably happening today.

10. Geologists have found that the concept of metamorphic zones is very helpful in studying regional metamorphism. Suggest a reason why.

11. What is the metamorphic facies concept, and how does it help in the study of metamorphic rocks?

12. Under what conditions of pressure and temperature does blueschist facies metamorphism occur? What is the geologic environment where such temperatures and pressures are found? Suggest some place on the Earth where blueschist metamorphism is probably happening today.

13. Name three minerals that are found only in metamorphic rocks.

14. What are the two main volatiles that are added to, or lost from, rocks undergoing metamorphism? What roles do volatiles play in metamorphism?

15. What is cataclastic metamorphism? How would you distinguish it from contact metamorphism of an impure limestone?

16. Discuss the importance of metasomatism and describe how metasomatic processes can form valuable mineral deposits.

8

Measuring
Our Earth

Understanding
Our Environment

GeoMedia

The little coastal village of Étretat, Normandy, France, is built on
Cretaceous chalk.

Geologic Time

What's in a Name?

Chalk is a soft limestone, usually white or buff in color, composed chiefly of the shells of tiny aquatic creatures called foraminifera—forams for short. Forams proliferate in warm marine waters, and their accumulated remains have, at times in the geological past, built great thicknesses of chalk. The white cliffs of Dover, in southern England, and in Normandy, in France, are striking examples.

In the nineteenth century, when European geologists started to arrange fossils in the sequence in which they had lived, they designated type areas where strata containing fossils of a specific age could be found. A Belgian geologist named Omalius d'Halloy selected the fossiliferous chalk strata in the Paris Basin as one of the type examples. In the style of the day, d'Halloy gave the name Cretaceous, from the Latin *cretaceus*, meaning chalk, or, literally, Cretan earth, to the strata.

The Cretaceous Period is the time during which Cretaceous strata were deposited. Today we recognize the Cretaceous as the last period of the Mesozoic ("middle life") Era of the

Earth's history. The Mesozoic is called the Age of the Dinosaurs; the time in the Earth's history when dinosaurs ruled the land. At the very end of the Cretaceous Period, dinosaur fossils abruptly disappear.

The period that follows the Cretaceous is the Tertiary, a name left over from a mistaken eighteenth century idea that there had only been four time intervals in the Earth's history—the primary, secondary, tertiary, and quaternary intervals. The terms primary and secondary have fallen into disuse, but the Tertiary and Quaternary have been retained and are the two periods of the Cenozoic ("recent life") Era.

When geologists indicate the ages of strata on maps, they use the first letter of the name rather than writing out the full name. The letter C poses a problem for this system of abbreviation because there are two other periods, besides the Cretaceous, with names that start with the letter C, the Cambrian and the Carboniferous. To avoid confusion C is used for Carboniferous, Cambrian is abbreviated €, and the letter K, from the German word for chalk, *kreide*, is used for the Cretaceous. The letter for Tertiary is T. The boundary between the Cretaceous and the Tertiary, then, is the K/T boundary. The scientific shorthand notation—the K/T boundary—is now in general use. It appears in popular magazines, newspapers, television programs, and radio news reports. This has come about because it is now thought to be highly likely—though not absolutely proven—that dinosaur fossils disappear at the K/T boundary because a great meteorite impact killed off the last of the animals.

STRATIGRAPHY

The historical information that geologists have to work with is largely in the form of layered rocks that crop out at the Earth's surface or that can be penetrated by drilling. If we examine the rocks exposed in the upper walls of the Grand Canyon (Fig. 8.1A), where the Colorado River has cut nearly 2 km into the Earth's crust, we can see many nearly horizontal layers. These strata formed one atop the other as sediment accumulated on the floor of a shallow sea and, at certain times when the sea withdrew, on the land. Such rocks contain important clues about past environments at and near the Earth's surface. Their sequence and relative ages provide the basis for reconstructing much of Earth history.

The study of strata is called **stratigraphy**. Knowledge of stratigraphic principles and the relative ages of rock sequences makes it possible to work out many of the fundamental principles of physical geology. Two straightforward and simple, but nevertheless very powerful, laws govern stratigraphy: the law of original horizontality and the principle of stratigraphic superposition.

Original Horizontality

Most sediments were laid down in the sea, generally in relatively shallow waters, or by streams on the land. Under such conditions, each new layer was laid down horizontally over older ones. These observations are consistent with the **law of original horizontality**, which states that water-laid sediments are deposited in strata that are horizontal or nearly horizontal. From this generalization we can infer that layers of sedimentary rock, now inclined, or even buckled and folded, must have been disturbed since the time when they were deposited as sediments.

Stratigraphic Superposition and the Relative Ages of Strata

Toward the end of a cold winter, it is often possible to see layers of old snow that are compact and perhaps also dirty, overlain by fresh, looser, clean snow deposited during the latest snowstorm. Here are layers, or strata, that were deposited in sequence, one above the other. The simple principle involved here also applies to layers of sediment and sedimentary rock. Known as the **principle of stratigraphic superposition**, it states that in any sequence of sedimentary strata, the order in which the strata were deposited is from bottom to top. The relative ages of any two strata can therefore be determined according to whether one of the layers lies above or below the other. Figures 8.1A and 8.1B offer examples of the principle of stratigraphic superposition. The horizontal strata at the top of the Grand Canyon are younger than the horizontal strata below.

The principle of stratigraphic superposition must be employed with a certain amount of care. Observe in Figure 8.2 that tilting and buckling of strata, such as happens during continental collisions, can sometimes be so severe that overturning can bring older strata to overlie younger strata. Observations made on overturned strata would obviously lead to incorrect conclusions concerning the relative timing of deposition unless the overturning were recognized. Evidence such as ripple marks, graded beds, and cross-stratified beds, as discussed in Chapter 6, can be used to determine whether strata are correctly oriented or overturned.

Breaks in the Stratigraphic Record

Lyell and other geologists of the nineteenth century speculated that it might be possible to determine numerical ages by using the stratigraphic record. If one measures the rate of sedimentation in the sea, they argued,

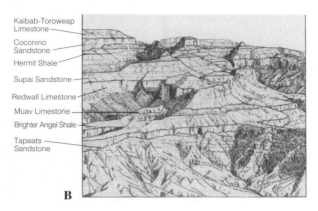

Kaibab-Toroweap
Limestone

Coconino
Sandstone

Hermit Shale

Supai Sandstone

Redwall Limestone

Muav Limestone

Brighter Angel Shale

Tapeats
Sandstone

A   **B**

Figure 8.1 The Principle of Stratigraphic Superposition
The Grand Canyon of the Colorado River. A. Flat-lying strata, nearly 2000 m thick and accumulated over 300 million years, were laid down on older strata, now tilted and tectonically deformed.

B. Sequence of horizontal sedimentary strata laid down on the tilted and deformed basement rocks of the Grand Canyon, from the Tapeats Sandstone (oldest) to the Kaibab-Toroweap Limestone (youngest), illustrates both the principle of stratigraphic superposition and the law of original horizontality.

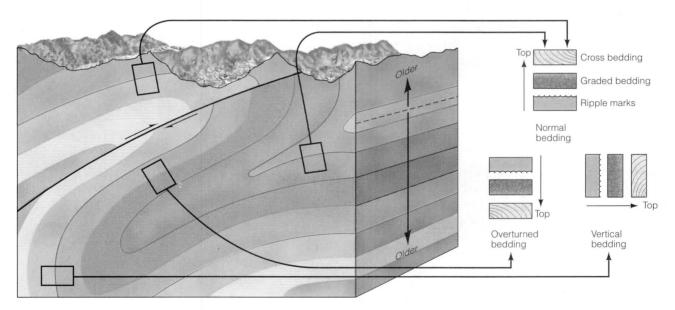

Top — Cross bedding

Graded bedding

Ripple marks

Normal
bedding

Older

Older

Top

Overturned
bedding

Top

Vertical
bedding

Figure 8.2 Orientation of Strata Sketch illustrating how some sedimentary features can indicate whether strata are normal (right side up), vertical, or overturned as a consequence of tectonic deformation.

and if one determines the thickness of all strata in a given body of rock, it should be possible to calculate how long it took for all the sediments represented by the strata in question to accumulate. Two assumptions must be correct for the method to work. First, it must be assumed that the rate of sedimentation was constant throughout the time of sediment accumulation. Second, it must be assumed that all strata exhibit **conformity**, meaning they have been deposited layer after layer without interruption. If there are any gaps in the stratigraphic record due to erosion or nondeposition, the elapsed time must somewhere be represented by strata

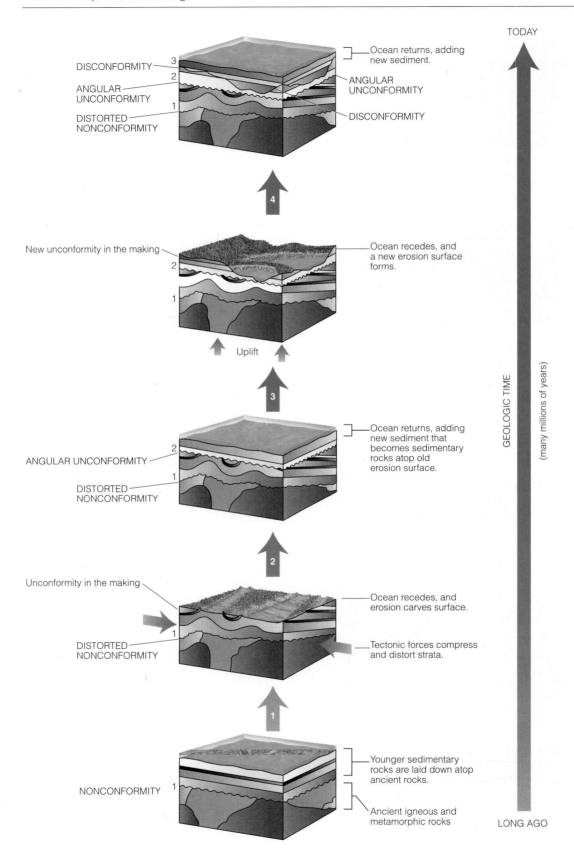

TODAY

Ocean returns, adding new sediment.

DISCONFORMITY 3
2 ANGULAR UNCONFORMITY

ANGULAR UNCONFORMITY 1

DISTORTED NONCONFORMITY DISCONFORMITY

4

New unconformity in the making 2 Ocean recedes, and a new erosion surface forms.

1

Uplift

3

GEOLOGIC TIME

(many millions of years)

ANGULAR UNCONFORMITY 2 Ocean returns, adding new sediment that becomes sedimentary rocks atop old erosion surface.

DISTORTED NONCONFORMITY 1

2

Unconformity in the making Ocean recedes, and erosion carves surface.

DISTORTED NONCONFORMITY 1 Tectonic forces compress and distort strata.

1

Younger sedimentary rocks are laid down atop ancient rocks.

NONCONFORMITY 1

Ancient igneous and metamorphic rocks

LONG AGO

Figure 8.3 Gaps in the Rock Record Sequence of geologic events leading to the three kinds of unconformity: (1) nonconformity; (2) angular unconformity; and (3) disconformity.

that can be placed in correct sequence by correlation. Estimates of hundreds of millions of years for all strata that have been deposited throughout geologic time were made, but they are wrong because both assumptions are false. The first assumption is false because it can be observed today that sedimentation rates vary widely from place to place and time to time, and there is a lot of evidence to suggest that rates have varied even more widely throughout geologic history. The second and even more important assumption is false because sedimentation can be disrupted periodically by major environmental changes, such as sea level changes and tectonic activity, that lead to intervals of erosion or nondeposition. In many cases there may not be correlatable strata elsewhere that can be used to fill the missing time gap. Numerous such breaks are found in the sedimentation record, with no way of knowing how much time the breaks represent. Because erosion destroys some of the record, the part that is preserved is necessarily incomplete and marked by discontinuities where intervals of geologic time, some brief and others very long, are not represented by deposits.

An **unconformity** is a substantial break or gap in a stratigraphic sequence. It records a change in environmental conditions that caused deposition to cease for a considerable time.

Kinds of Unconformities

Three important kinds of unconformities are found in sedimentary rocks (Fig. 8.3). The most obvious is the **angular unconformity**, which is marked by angular discontinuity between older and younger strata. It is labeled (2) in Figure 8.3. An angular unconformity implies that the older strata were deformed and then truncated by erosion before the younger layers were deposited across them. The outcrop at Siccar Point (Fig. 1.3) observed by James Hutton is obviously an angular unconformity. The second kind of unconformity is called a **disconformity**; it is an irregular surface of erosion between parallel strata. The surface numbered (3) in Figure 8.3 is a disconformity. A disconformity implies a cessation of sedimentation, plus erosion, but no tilting. Disconformities can be hard to recognize because the strata above and below are parallel. The way disconformities are usually discovered is through recognition that fossils of very different ages are present in adjacent strata.

The third kind of unconformity, labeled (1) in Figure 8.3, is a **nonconformity**, in which strata overlie igneous or metamorphic rocks.

The three types of unconformity can be seen in the Grand Canyon, as shown in Figure 8.4. At the base of the sedimentary sections is a nonconformity, and some distance above it is an obvious angular unconformity; still higher are three disconformities. Some of the same unconformities can be seen in Figures 8.1A and 8.1B; see if you can pick them out.

The Significance of Unconformities

A study of unconformities brings out the close relationship between tectonics, erosion, and sedimentation. All of the Earth's land surface is a potential surface of unconformity. Some of today's surface will be destroyed

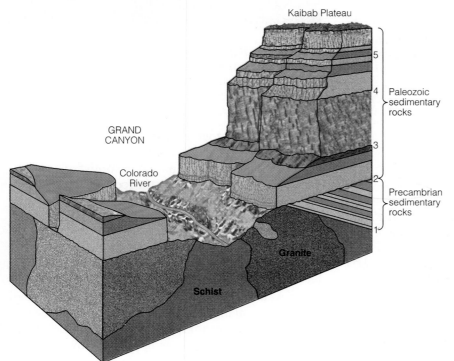

Kaibab Plateau

GRAND CANYON

Colorado River

5

4 } Paleozoic sedimentary rocks

3

2

Precambrian sedimentary rocks

1

Granite

Schist

Figure 8.4 Unconformities in the Grand Canyon Geologic section through rocks exposed in the Grand Canyon. The lowest unconformity (1), separating tilted sedimentary strata from older crystalline rocks, is a nonconformity. An angular unconformity (2) separates the tilted strata from horizontally layered strata above, while three disconformities (3, 4, and 5) are seen still higher in the section. Unconformities 2, 3, 4, and 5 are visible in Figure 8.1A.

by erosion, but other parts will be covered by sediment and preserved as a record of the present landscape. For example, the Swiss Alps, which were elevated by plate-tectonic movements, are being rapidly eroded away. Meanwhile, the eroded material is being carried away by streams and deposited in the Mediterranean Sea. The Mediterranean seafloor was once dry land, but tectonic forces depressed it, just as tectonic forces elevated the Alps. A surface of unconformity separates the young, river-transported sediments and the older rocks of the seafloor on which the sediments are being piled. In a sense, accumulation in one place compensates for destruction in another. The many surfaces of unconformity exposed in rocks of the Earth's crust are evidence that former seafloors were uplifted by tectonic forces and exposed to erosion. Preservation of a surface of erosion occurs when later tectonic forces depress the surface, so it, in turn, becomes a site of deposition of sediment. Unconformities testify that interactions between the internal and external processes have been going on throughout the Earth's long history.

STRATIGRAPHIC CLASSIFICATION

Every rock stratum can tell us something about the physical and biological character of a part of the Earth at some time in the geologic past. Anyone counting strata would quickly realize that the rock record is like a vast library consisting of thousands upon thousands of volumes. Like a library too, the rock record is a complex catalogue that is employed by geologists who work on the record of past events. Only the most important cataloging terms used in stratigraphic classification are introduced here.

Three related concepts are employed in stratigraphic classification. The first two are based on the rock units, while the third is somewhat abstract in that it concerns intervals of geologic time.

The first kind of unit employed in stratigraphic classification uses any distinctive stratum that differs from the strata above and below. An example of a *rock-stratigraphic unit* is the Navajo Sandstone in Zion National Park, Utah, seen in Figure 8.5. This striking sandstone is very different from the strata immediately above and below and can be easily mapped in the field.

The basis of rock stratigraphy is the **formation**, which is a group of similar strata that are sufficiently different from adjacent groups of strata so that on the basis of physical properties they constitute a distinctive, recognizable unit that can be used for geologic mapping over a wide area. The Navajo Sandstone is a formation. (Each formation is given a name: in North America it typically

Figure 8.5 Rock-Stratigraphic Unit Spectacular view of the Navajo Sandstone, Zion National Park, Utah. The Navajo Sandstone is an example of a rock-stratigraphic unit.

is the name of a geographic locality near which the unit is best exposed.)

The second kind of unit used in stratigraphic classification is one representing all the rocks that formed during a specific interval of geologic time. Each of the boundaries of a *time-stratigraphic unit*, upper and lower, is everywhere the same age.

A formation is defined only on the basis of its material characteristics, and its upper and lower boundaries lie where a recognizable change in physical properties occur. As illustrated in Figure 8.6, the ages of the boundaries of a formation can differ from place to place. In contrast, a time-stratigraphic unit may include more than one rock type, and its upper and lower boundaries may not necessarily coincide with a formational

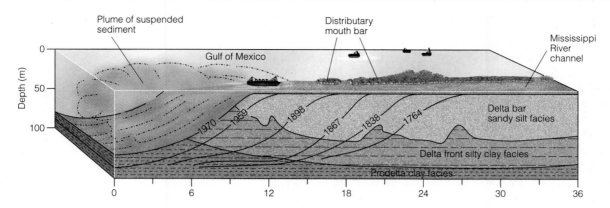

Figure 8.6 Time-Stratigraphic Unit Section through the Southwest Pass lobe of the Mississippi River delta, which is expanding outward into the Gulf of Mexico. Dated time lines show successive positions of the delta front between 1764 and 1979. The boundaries separating three sedimentary facies intersect each time line. Should this deposit be preserved and eventually converted to sedimentary rock, the resulting formations, corresponding to the three facies, will each be younger in the direction of delta growth.

boundary, but as demonstrated in Figure 8.6, each boundary is everywhere the same age.

The primary time-stratigraphic unit is a **system**, which is chosen to represent a time interval sufficiently great so that such units can be used all over the world. Names for systems arose from nineteenth-century studies of strata, mainly in Europe, and often derive from geographic localities. Most systems are now known to encompass numerical time intervals of tens of millions of years.

The final units used in stratigraphic classification are the intervals of geologic time during which strata in time-stratigraphic units accumulated. Units of geologic time are nonmaterial, while time-stratigraphic units are material. The primary unit of geologic time is a geologic *period*; it is the time during which a geologic system accumulated.

The geologic record is incomplete and punctuated by numerous unconformities; a given area may not contain a complete depositional record for the particular geologic time interval (Fig. 8.3). However, many of the gaps can be bridged by piecing together sequences of strata from different geographic areas, thereby providing us with a more complete picture of Earth history. The piecing together involves correlation.

CORRELATION OF ROCK UNITS

William Smith was an English land surveyor who was active around the beginning of the nineteenth century. His profession gave him an ideal opportunity to observe not only the landscape but also the rocks that underlie it. While surveying for the construction of new canals in western England, he observed many sedimentary strata and soon realized that they lay, as he put it, "like slices of bread and butter" in a definite, unvarying sequence. He became familiar with the physical characteristics of each layer, with the fossils each contained, and with the sequence of the layers. By looking at a specimen of sedimentary rock collected from anywhere in southern England, he could name the layer from which it had come and, of course, the position of the layer in the sequence.

Smith did not believe that his discovery reflected any particular scientific principle; it was purely practical. Nevertheless, it opened the door to the correlation of sedimentary strata over increasingly wide areas. **Correlation** means the determination of equivalence in time-stratigraphic units of the succession of strata found in two or more different areas. Smith correlated strata on the dual basis of physical similarity and fossil content initially over distances of several kilometers, and later over tens of kilometers. By means of fossils alone, it ultimately became possible to correlate through hundreds and then thousands of kilometers.

Correlation involves two main tasks. One is to determine the relative ages, one to another, of units exposed in local sections within an area being studied. Then the ages of the units relative to a standard scale of geologic time must be found. To accomplish these goals, a geologist employs various physical and biological criteria; one is not necessarily more dependable or precise than the others.

Continuous exposures are not common, and so we are often faced with correlating between widely spaced outcrops. A formation may be eroded so that only parts remain. The physical matching of the remnants generally involves the use of rock characteristics such as grain size, color, and sedimentary structures that permit the unit to be distinguished from others. In some instances quite different formations look almost identical, so correlation must be done with care. The Navajo Sandstone in Zion National Park shown in Figure 8.5, for example,

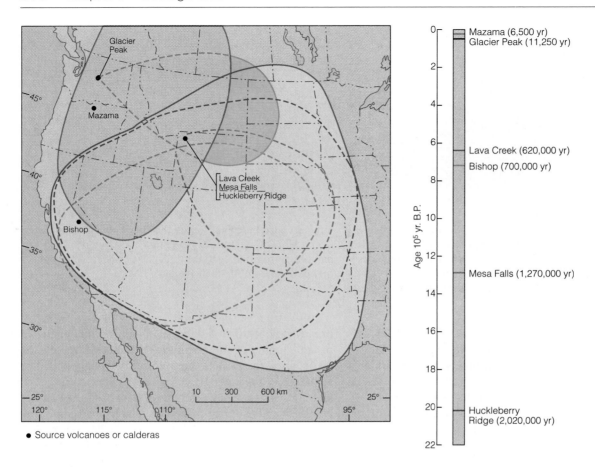

• Source volcanoes or calderas

Figure 8.7 Key Beds Map showing distribution of widespread tephra layers in western North America which erupted during the last 2 million years. Each tephra layer is a key bed.

is very similar to the Coconino Sandstone in the Grand Canyon (Fig. 8.1B), 120 km to the south. However, correlation of the two formations based on physical criteria would be incorrect because careful mapping reveals that the Navajo Sandstone sits much higher in the stratigraphic section than the Coconino Sandstone.

A thin and generally widespread sedimentary bed with characteristics so distinctive that it can be easily recognized but not confused with any other bed is called a *key bed*. Such beds are exceedingly useful in correlating major rock sections. In areas of volcanic activity, ash layers can serve as distinctive key beds for purposes of regional correlation as shown in Figure 8.7.

We call a fossil that can be used to identify and date the stratum in which it is found an *index fossil*. To be most useful, an index fossil should have common occurrence, a wide geographic distribution, and a very restricted age range. The best examples are swimming or floating organisms that underwent rapid evolution and quickly became widely distributed (Fig. 8.8). If a distinctive index fossil is recognizable at an outcrop, a rapid and reliable means of correlation is available (Fig. 8.9). Although some genera and species permit long-range correlation of rocks in different geographic

Figure 8.8 Index Fossil *Homotelus bromidensis,* a trilobite that lived about 500 million years ago, during the middle of the Ordovician Period, is a distinctive index fossil. These specimens came from the Bromide Formation in Oklahoma.

areas or even on different continents, more often close dating and correlation involve using assemblages of fossils of as many different types as possible.

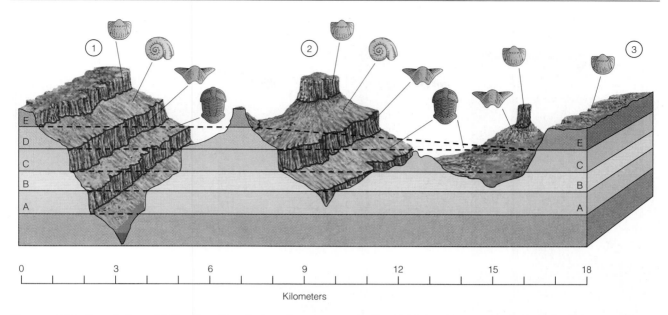

Figure 8.9 Correlating with Fossils Correlation of strata exposed at three localities, many kilometers apart, on the basis of similarity of the fossils they contain. The fossils show that at Locality 3 stratum D is missing because E directly overlies C. Either D was never deposited there, or it was deposited and later removed by erosion before the deposition of layer E.

THE GEOLOGIC COLUMN AND THE GEOLOGIC TIME SCALE

One of the great successes of the nineteenth-century geologists was the demonstration, through stratigraphic correlation, that time-stratigraphic sequences are the same on all continents. Through worldwide correlation those nineteenth-century geologists assembled a **geologic column**, which is a composite columnar section containing in chronological order the succession of known strata, fitted together on the basis of their fossils or other evidence of relative age. This worldwide standard is still being added to and refined as more rock units are described and mapped.

Standard names have evolved for the subdivisions of the geologic time units corresponding to the rock units of the geologic column. The units of the geologic time scale, which, like the geologic column, can be used worldwide, are eons, eras, periods, and epochs as shown in Figure 8.10.

Eons

An **eon** is the largest interval into which geologic time is divided, and there are four of them. The term **Hadean** (Greek for beneath the Earth) is given to the oldest eon. This is the earliest part of the Earth's history, an interval for which no rock record is known. However, rocks of this age are present on other planets where the earliest crustal rocks have been little modified since they accumulated. The **Archean** (Greek word for ancient) Eon follows the Hadean. Archean rocks are the oldest rocks we know of on the Earth, and they contain microscopic life-forms of bacterial character. The **Proterozoic** (earlier life) Eon follows the Archean. Proterozoic rocks include evidence of multicelled organisms that lacked preservable hard parts. Understandably, the record of the ancient Archean and Proterozoic is not as well known as the record of younger rocks because many of these ancient rocks have been intensely deformed, metamorphosed, and eroded. The **Phanerozoic** (visible life) is the most recent of the four eons. Phanerozoic rocks often contain plentiful evidence of past life in the form of well-preserved hard parts. Most examples of fossils that we see displayed in museums or illustrated in books are from the Phanerozoic Eon.

Eras

Geologic **eras** encompass major spans of time that also are defined on the basis of the life-forms found in the corresponding rocks. No formal eras are yet widely recognized for the Archean and Proterozoic rocks, but the Phanerozoic Eon is divided into the **Paleozoic** (old life), **Mesozoic** (middle life), and **Cenozoic** (recent life) eras, each name reflecting the relative stage of development of the life of these intervals (Fig. 8.10). Paleozoic forms of life progress from marine invertebrates to fishes, amphibians, and reptiles. Early land plants also appeared,

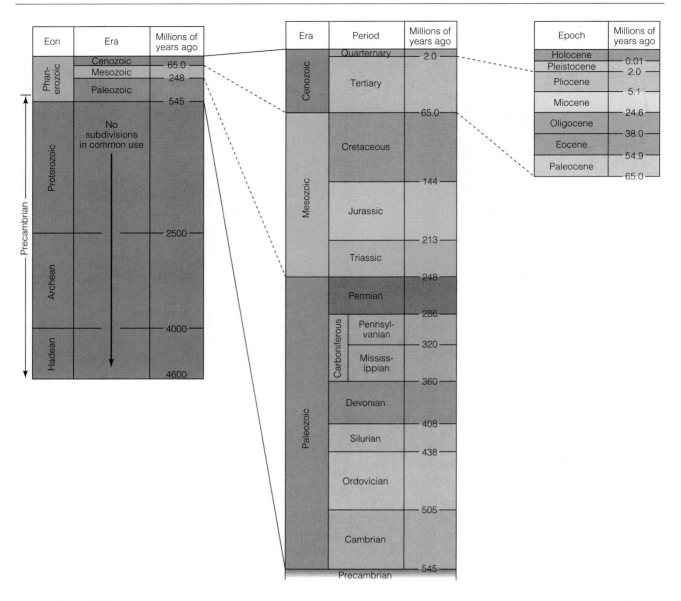

Figure 8.10 The Geologic Column and Time Scale The geologic time scale. Numerical ages obtained from radiometric dates. Note that the Pennsylvanian and Mississippian Periods are equivalent to the Carboniferous Period of Europe. The time boundary between the Archean and Hadean is a matter of definition because no rocks of the Hadean Eon are known on the Earth. Hadean rocks are known to exist on other planets in the solar system.

expanded, and evolved. The Mesozoic Era saw the rise of the dinosaurs, which became the dominant vertebrates on land. Toward the end of that era, mammals first appeared and later dominated the Cenozoic Era. The Mesozoic Era also witnessed the evolution of flowering plants, while during the Cenozoic Era grasses appeared and became an important food for grazing mammals.

Periods

The eras of the Phanerozoic eon are divided into **periods**. The geologic periods have a haphazard nomenclature. They were defined over an interval of nearly 100 years on the basis of strata that crop out in Britain, Germany, Switzerland, Russia, and the United States. The names are partly geographic in origin, but in some cases they are based on characteristics of the strata in the place of original study (Table 8.1).

The lowermost period of the Paleozoic Era, the Cambrian Period, is the time when animals with hard shells first appear in the geologic record. Prior to the Cambrian Period, all animals were soft bodied and the fossil evidence they left is very sparse. Rocks formed during the Archean and Proterozoic eons cannot be readily differentiated on the basis of the fossils they contain, so geologists often refer to the entire time preceding the Cambrian as simply Precambrian.

TABLE 8.1	Origin of Names for Periods of the Paleozoic, Mesozoic, and Cenozoic Eras, and the Epochs of the Quaternary and Tertiary Periods

Era	Period	Epoch	Origin of Name
Cenozoic	Quaternary[a]	Holocene	Greek for wholly recent
		Pleistocene	Greek for most recent
	Tertiary[a]	Pliocene	Greek for more recent
		Miocene	Greek for less recent
		Oligocene	Greek for slightly recent
		Eocene	Greek for dawn of the recent
		Paleocene	Greek for early dawn of the recent
Mesozoic	Cretaceous	↑	Latin for chalk, after chalk cliffs of southern England and France
	Jurassic	Epoch	Jura Mountains, Switzerland and France
	Triassic	Names	Threefold division of rocks in Germany
Paleozoic	Permian	Used	Province of Perm, Russia
	Pennsylvanian	Only	State of Pennsylvania
	Mississippian	By	Mississippi River
	Devonian	Specialists	Devonshire, County of Southwest England
	Silurian		Silures, ancient Celtic tribe of Wales
	Ordovician	↓	Ordovices, ancient Celtic tribe of Wales
	Cambrian		Cambria, Roman name for Wales

[a] Derived from eighteenth- and nineteenth-century geologic time scale that separated crustal rocks into a fourfold division of Primary, Secondary, Tertiary, and Quaternary, based largely on relative degree of lithification and deformation.

Epochs

The **epochs** of the Tertiary Period were defined in a piecemeal fashion. Studies of marine strata in sedimentary basins of France and Italy led Charles Lyell to subdivide the rocks into groupings based on the percentage of their fossils that are represented by still living species (Fig. 8.10). Each of the various periods of the Paleozoic and Mesozoic eras is also subdivided into epochs, the names of which are primarily geographic in origin. They are used mainly by specialists concerned with detailed studies of these strata and their contained fossils.

The names of the geologic time scale constitute the standard time language of geologists the world over. Through their use one can begin to comprehend numerous details of Earth history that have led to the discovery of many of the important principles of physical geology discussed in this book.

MEASURING GEOLOGIC TIME NUMERICALLY

The scientists who worked out the geologic column and time scale were challenged by the question of numerical time. They knew the relative time order in which the different systems had formed, but they also wished to know whether the sediments in each system had accumulated during the same length of time. They sought answers to questions such as these: "How much time elapsed between the end of the Cambrian Period and the beginning of the Permian Period?" "How long was the Tertiary Period?" The question of numerical time is as important as the geologic column and time scale. Numerical ages must be determined in order to answer such challenging and important questions as the age of the Earth, the age of the ocean, how fast mountain ranges rise, and how long humans have inhabited the Earth.

Early Attempts to Measure Geologic Time Numerically

During the nineteenth century, many attempts were made to develop a scale of years for the geologic time scale. One widely used method was discussed earlier in this chapter. It consisted of estimates of the time during which the rock cycle has been at work based on rates of sedimentation and the thickness of sedimentary strata. Estimates for the age of the Earth, which was presumed to be the same as the duration of the rock cycle, ranged from 3 million to 1.5 billion years!

A clever suggestion for estimating the age of the ocean concerned the saltiness of seawater. Because sea salts come from the erosion of common rocks and reach the sea dissolved in river water, why not measure the salts in modern river water and calculate the time needed to transport all the salts now in the sea? The first person to suggest that sea salt might be used to date the ocean was Edmund Halley in 1715. There is no record that Halley (for whom the comet is named) ever made the necessary measurements. John Joly finally made the necessary measurements and calculations in 1889. His answer for the ocean's age, 90 million years, does not represent the age of the ocean at all. The composition of the ocean is essentially constant because all the salts in the sea, like all other chemical constituents, are cyclic. Salts are added both by erosion and by submarine volcanism, but salts are also removed from solution. Some are removed as evaporite minerals (Chapter 6), whereas others are removed by reaction with hot volcanic rocks on the seafloor.

Perhaps the most interesting estimates were those made by Lord Kelvin, a physicist, who attempted to calculate the time the Earth has been a solid body. The Earth started as a very hot object, he argued. Once it had cooled sufficiently to form a solid outer crust, it could continue to cool only by the conduction of heat through solid rock. By measuring the thermal properties of rock and estimating the present temperature of the Earth's interior, he calculated the time for the Earth to cool to its present state. Kelvin's logic was faultless, and his mathematical calculations were correct. However, his estimate of 100 million years for the maximum age of the Earth is incorrect because he made an incorrect assumption. Kelvin assumed that no additional heat had been added since the Earth was formed. When Kelvin made his calculations, radioactivity was not known. Radioactivity continuously supplies heat to the Earth's interior, so that instead of cooling rapidly, the Earth's interior is cooling so slowly it has a nearly constant temperature over periods as long as hundreds of millions of years.

Resolving the dilemma of numerical time required finding a way to measure geologic time by some process that runs continuously, that is not reversible, that is not influenced by other processes and other cycles, and that leaves a continuous record without gaps in it. In 1896, the discovery of radioactivity provided the needed method.

Radioactivity and the Measurement of Numerical Time

Natural Radioactivity

We learned in Chapter 3 that the atomic number of a given element—that is, the number of protons in the atomic nucleus of the element—is constant and characteristic of that element. However, an atomic nucleus also contains neutrons, and the number of neutrons can vary without changing the number of protons. For example, all carbon atoms contain six protons, but the protons can be joined by six, seven, or eight neutrons. Different kinds of atoms of an element that contain different numbers of neutrons are called **isotopes** (Fig. 8.11). An isotope is identified by its **mass number**, which is the sum of the neutrons plus the number of protons. Carbon, therefore, has three isotopes with mass numbers 12, 13, and 14, which are written ^{12}C, ^{13}C, and ^{14}C, respectively. Most of the common chemical elements are mixtures of several isotopes.

Most of the isotopes of the chemical elements found in the Earth are stable and not subject to change. However, a few, such as ^{14}C, are radioactive because there are limits within which the ratio of the number of neutrons (n) to the number of protons (p) gives rise to a stable nucleus. If the ratio of n/p is too high or too low, the atomic nucleus of a radioactive isotope will transform spontaneously to a nucleus of a more stable isotope of a different chemical element. The rate of transformation is different for each isotope. Even though the process is one of transformation—from an unstable nucleus to the nucleus of an atom of a different kind—it has become common practice to call the process radioactive decay. An atomic nucleus undergoing radioactive decay is said to be a **parent**; the product arising from radioactive decay is called a **daughter**. ^{14}C decays to ^{14}N and ^{238}U decays to ^{206}Pb; ^{14}C and ^{238}U are parents, ^{14}N and ^{206}Pb daughters.

Radioactive decay is a phenomenon of the atomic nucleus, not of the electrons that orbit the nucleus. To understand radioactivity, we need to expand the previous discussion of the nucleus in Chapter 3.

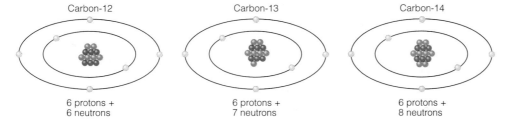

Carbon-12 Carbon-13 Carbon-14

6 protons + 6 neutrons   6 protons + 7 neutrons   6 protons + 8 neutrons

Figure 8.11 Naturally Occurring Isotopes of Carbon The three isotopes of carbon. Note that in each case there are six protons in the nucleus and six electrons in the energy-level shells. The differences lie in the numbers of neutrons in the nucleus.

Radioactive decay can happen in five ways:

1. By emission of an electron from the nucleus; such an electron is referred to as a β^- (beta) particle, and emission occurs because a neutron becomes a proton:

$$n \rightarrow p + \beta^-$$

Thus, by a β^- emission, the number of neutrons in a nucleus is decreased by one and the number of protons is increased by one, thus reducing the ratio n/p.

2. By emission of a particle with the same mass as an electron but with a positive charge; such a particle is called a *positron* and is written β^+. Positron emission decreases the number of protons in a nucleus by one and increases the number of neutrons by one, according to reaction:

$$p \rightarrow n + \beta^+$$

Thus, by positron emission the ratio of n/p is increased.

3. By capture into the nucleus of one of the orbital electrons, a process that decreases the number of protons in the nucleus by one, according to the reaction:

$$p + \text{electron} \rightarrow n$$

Note that the term β^- particle is only used for electrons emitted from the nucleus. Electron capture, like positron emission, increases the ratio of n/p.

4. By emission from the nucleus of a heavy atomic particle consisting of two neutrons and two protons ($2p + 2n$), called an α (alpha) particle. Loss of an α particle reduces the mass number by four and the atomic number by two. Alpha decay is a process that happens with isotopes that have large atomic numbers. Because isotopes with large atomic numbers tend to have more neutrons than protons in the nucleus, emission of a α particle causes an increase in n/p.

5. By emission of γ rays (gamma rays), which are very short-wavelength, high-energy electromagnetic rays. γ rays have no mass, so γ-ray emission does not affect either the atomic number or the mass number of an isotope. Emission of a γ ray happens when decay by one of the four previous methods produces an isotope with a high-energy state. By emitting a γ ray, the newly formed isotope moves to a more stable, lower energy state.

All of the radioactive decay schemes discussed above occur in isotopes found in the Earth, but three of the schemes, β^- decay, electron capture, and α decay, illustrated in Fig. 8.12, are especially important for determining numerical ages of geologic samples.

Rates of Decay

Many of the radioactive isotopes that were once in the Earth have decayed away and are no longer present. This is so because their rates of spontaneous decay are fast. A few radioactive isotopes that transform very slowly are still present, however. Careful study of radioactive isotopes in the laboratory has shown that decay rates are unaffected by changes in the chemical and physical environment. Thus, the decay rate of a given isotope is the same in the mantle, or a magma, as it is in a sedimentary rock. This is a particularly important point because it leads to the conclusion that rates of radioactive decay are not influenced by geologic processes, like erosion, metamorphism, or the melting of rock to form magma.

In the least complicated process of radioactive decay, the number of radioactive parent atoms continuously decreases, while the number of nonradioactive daughter atoms continuously increases. More complicated processes of radioactive decay involve daughter atoms that are also radioactive and so undergo further decay until a stable daughter is finally produced. Regardless of the complications involved in the decay scheme, all decay timetables follow the same basic law: The *proportion* of parent atoms that decay during each unit of time is always the same (see Fig. 8.13).* Proportion means a fraction, a percentage, and not a whole number. The rate of radioactive decay is determined by the **half-life**, which is the amount of time needed for the number of parent atoms to be reduced by one-half. For example, let's say that the half-life of a radioactive isotope is 1 hour. If we started an experiment with 1,000,000 parent atoms and the daughter atoms are not radioactive, only 500,000 parent atoms would remain at the end of an hour, and 500,000 daughter atoms would have formed. At the end of the second hour, another half of the parent atoms would be gone, so there would be 250,000 parent atoms and 750,000 daughter atoms. After the third hour, another half of the parent atoms would have decayed, leaving 125,000 parent atoms and 875,000 daughter atoms. The proportion of parent atoms that decay during each time interval is 50 percent. (The same kind of law governs compound interest paid on a bank account.)

In the graphic illustration of radioactive decay in Figure 8.13, the time units marked are half-lives. Of course, the time units are of equal length, but at the end of each unit the number of parent atoms, and therefore the radioactivity of the sample, has decreased by exactly one-half. Figure 8.13, which is drawn for the case where a radioactive isotope decays directly to a stable daughter isotope, also shows that the growth of daughter atoms just matches the decline of parent atoms. When the number of remaining parent atoms (N_p) is added to the num-

---

*The application of this law is valid only when the number of atoms is large enough (at least tens of thousands) so that statistical variations in the counting procedure are negligible.

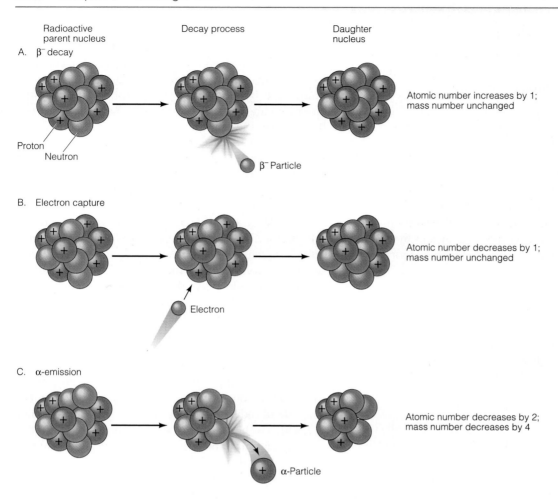

Radioactive
parent nucleus   Decay process   Daughter
                                               nucleus
A. β⁻ decay

Atomic number increases by 1;
mass number unchanged

Proton
    Neutron
                                        β⁻ Particle

B. Electron capture

Atomic number decreases by 1;
mass number unchanged

Electron

C. α-emission

Atomic number decreases by 2;
mass number decreases by 4

α-Particle

Figure 8.12   Radioactive Decay   Three of the ways radioactive isotopes decay. Note that in each case the atomic number (number of protons) of the daughter isotope differs from the atomic number of the parent isotope.

ber of daughter atoms (N_d), the result is N_o, the number of parent atoms that a mineral sample started with. That fact is the key to the use of radioactivity as a means of measuring geologic time and determining ages.

Potassium-Argon ($^{40}K/^{40}Ar$) Dating
We have selected one of the naturally radioactive isotopes, potassium-40 (^{40}K), to illustrate how minerals can be dated to determine when they formed. Potassium has three natural isotopes: ^{39}K, ^{40}K, and ^{41}K. Only one, ^{40}K, is radioactive and its half-life is 1.3 billion years. The decay of ^{40}K is interesting because two different decay schemes occur. Twelve percent of the ^{40}K atoms decay by electron capture to ^{40}Ar, an isotope of the gas argon. The remaining 88 percent of the ^{40}K atoms change by β⁻ decay to ^{40}Ca. The appropriate equations are:

$$^{40}K + \text{electron} \rightarrow {}^{40}Ar$$

and

$$^{40}K \rightarrow {}^{40}Ca + \beta^-$$

It is important to know that the fraction of ^{40}K atoms decaying to ^{40}Ar is always 12 percent; the percentage is not affected by changes in physical or chemical conditions.

When a potassium-bearing mineral crystallizes from a magma, or grows within a metamorphic rock, it includes some ^{40}K in its crystal structure. As soon as the mineral is formed, ^{40}Ar and ^{40}Ca daughter atoms start accumulating in the mineral because, like the parent ^{40}K atoms, they are trapped in the crystal structure. Because the ratio of ^{40}Ar to ^{40}Ca daughter atoms is always the same, it is only necessary to measure either ^{40}Ar or ^{40}Ca daughter atoms in order to know how many ^{40}K atoms have decayed. It is more accurate to use ^{40}Ar because argon is an element that has unusual atomic properties.

There is another reason, too, why ^{40}Ar is measured rather than ^{40}Ca. The electron energy-level shells of argon are filled, so atoms of argon do not readily form chemical bonds. Thus, ^{40}Ar atoms are not chemically bound in minerals; they are trapped by the crystal lattice but do not form bonds. The trapping is effective only at

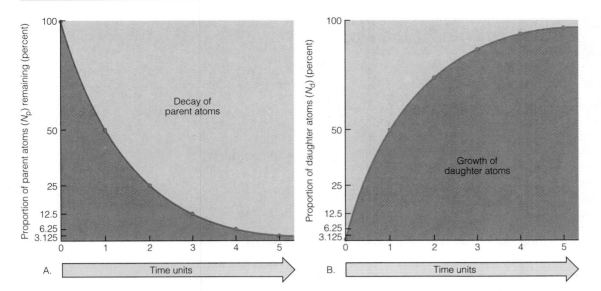

Figure 8.13 Radioactivity and Time Curves illustrating the basic law of radioactivity. The curves are drawn for the circumstance where a radioactive parent isotope decays to a stable daughter isotope. A. At time zero, a sample consists of 100 percent radioactive parent atoms. During each time unit, half the atoms remaining decay to daughter atoms. B. At time zero, no daughter atoms are present. After one time unit corresponding to a half-life of the parent atoms, 50 percent of the sample has been converted to daughter atoms. After two time units, 75 percent of the sample is daughter atoms, 25 percent parent atoms. After three time units, the percentages are 87.5 and 12.5, respectively. Note that at any given instant N_p, the number of parent atoms remaining, plus N_d, the number of daughter atoms, equals N_o, the number of parent atoms at time zero.

low temperatures, so at high temperatures argon rapidly diffuses out of a mineral. At lower temperatures argon cannot diffuse out and stays trapped. When the argon content of a mineral is measured, therefore, what is determined is the ^{40}Ar accumulated during the time since a mineral started trapping and retaining argon. Even if some ^{40}Ar is present in a magma, when a mineral crystallizes, therefore, a potassium-bearing mineral will not retain any initial argon because magmatic temperatures are far above trapping temperatures. All the ^{40}Ar atoms in a potassium-bearing mineral in an extrusive igneous rock such as a rhyolite or andesite must have come from decay of ^{40}K and must have accumulated since the temperature fell below the trapping temperature. Because extrusive igneous rocks cool very quickly, the trapping time and eruption time are essentially the same.

All that has to be done to determine the numerical time of eruption of an extrusive igneous rock is to select a potassium-bearing mineral and measure the amount of parent ^{40}K that remains and the amount of trapped ^{40}Ar. With the half-life of ^{40}K being known, it is a straightforward matter to calculate the radiometric age—the length of time a mineral has contained its built-in radioactivity clock. For an example of a calculation, see the boxed essay: *Calculations of a $^{40}K/^{40}Ar$ Age.*

Flexibility of $^{40}K/^{40}Ar$ Dating

Dating by ^{40}K is not limited to minerals that contain potassium as a major element. Even minerals that contain small amounts of potassium will serve the purpose. Thus, hornblende, a calcium–iron–magnesium silicate, can be used for $^{40}K/^{40}Ar$ dating because it generally contains a small quantity of potassium present by ionic substitution.

$^{40}K/^{40}Ar$ dating is most successfully applied to volcanic rocks because their solidification and cooling are rapid. As a result, they have solidification ages that are essentially coincident with their trapping ages. Because argon analyses can be performed with great sensitivity and because contamination by initial argon at the time of crystallization is generally not a problem, the method can be used for volcanic rocks as young as 20,000 years. For this reason, $^{40}K/^{40}Ar$ dating has proved very useful in studies of archaeology as well as geology.

Other Radiometric Dating Methods

Many naturally radioactive isotopes can be used for radiometric dating, but six predominate in geologic studies. These are two radioactive isotopes of uranium plus radioactive isotopes of thorium, potassium, rubidium, and carbon. These isotopes occur widely in different minerals and rock types, and they have a wide range

Calculation of a $^{40}K/^{40}Ar$ Age

Chemical analysis of a potassium feldspar sample from a pyroclastic rock shows that for every 20,000 parent atoms of ^{40}K present there are 1200 atoms of ^{40}Ar. We know that the ratio $^{40}Ar/^{40}Ca$ is constant, so 1200 atoms of ^{40}Ar means that 8800 atoms of ^{40}Ca are also present. This in turn means that N_d, the number of daughter atoms, is 1200 + 8800, or 10,000. The equation for radioactive decay (i.e., the equation for the curve in Fig. 8.13), is:

$$\frac{N_p}{N_o} = (1 - \lambda)^y$$

where N_p is the number of parent atoms now, in our example 20,000.

N_o is the number of parent atoms when the mineral formed. Because each parent atom that decays produces only one daughter atom,
$N_o = N_p + N_d = 20,000 + 10,000 = 30,000$
λ is the decay constant, which is the fraction of parent atoms that decays per unit time.

y is the number of time units.

We can simplify the calculation in two ways. First, we select the unit of time to be equal to the half-life, which means $\lambda = 0.5$. This in turn makes y equal to the number of half-lives since the mineral formed. The second simplification is to put the equation in a logarithmic form.

$$\frac{N_p}{N_o} = (1 - 0.5)^y = 0.5^y$$

or, in logarithmic form,

$$\log \frac{N_p}{N_o} = -0.3y$$

or

$$\log N_o - \log N_p = 0.3y$$

Thus

$$\log 30,000 - \log 20,000 = 0.3y$$

or

$$y = \frac{4.477 - 4.301}{0.3} = 0.587$$

The feldspar therefore formed 0.587 half-lives ago. The half-life of ^{40}K has been measured in the laboratory to be 1300 million years. Thus, the age of the mineral is $1300 \times 0.587 = 760$ million years

---

of half-lives, so that many geologic materials can be dated radiometrically (Table 8.2).

Radiocarbon Dating

Among the radiometric dating methods listed in Table 8.2, the one based on ^{14}C (also known as radiocarbon) is unique for two reasons. The first is that the half-life of ^{14}C is short by comparison with the half-lives of the long-lived isotopes, ^{40}K, ^{87}Rb, ^{232}Th, ^{235}U, and ^{238}U. The second reason is that daughter atoms of ^{14}N formed by radioactive decay of ^{14}C cannot be distinguished from other ^{14}N atoms in the specimen being dated.

Radiocarbon is continuously created in the atmosphere through bombardment of ^{14}N by neutrons created by cosmic radiation (Fig. 8.14). ^{14}C, with a half-life of 5730 years, decays back to ^{14}N by β^- decay. The ^{14}C mixes with ^{12}C and ^{13}C and diffuses rapidly through the atmosphere, hydrosphere, and biosphere. Because the rates of mixing and exchange are rapid compared with the half-life, the proportion of ^{14}C is nearly constant throughout the atmosphere. As long as the production rate remains constant, the radioactivity of natural car-

bon in the atmosphere remains constant because the rate of production balances the rate of decay.

While an organism is alive, it will continuously take in carbon from the atmosphere and so will contain the balanced proportion of ^{14}C. However, at death the balance is upset, because replenishment by life processes such as feeding, breathing, and photosynthesis ceases. The ^{14}C in dead tissues continuously decreases by radioactive decay. Analysis for the radiocarbon date of a sample involves only a determination of the radioactivity level of the carbon it contains. This is done by measuring the particles emitted as a result of radioactive decay. As previously mentioned, the daughter isotope, ^{14}N, cannot be measured successfully because it leaks away and because of atmospheric contamination.

Because of its application to organisms (by dating fossil wood, charcoal, peat, bone, and shell material) and its short half-life, radiocarbon has proved to be enormously valuable in establishing dates for prehistoric human remains and for recently extinct animals. In this way it is of extreme importance in archaeology. It is also of great value in dating the most recent part of geologic his-

| TABLE 8.2 | Some of the Principal Isotopes Used in Radiometric Dating |

Parent	Isotopes Decay System	Daughter	Half-Life of Parent (years)	Effective Dating Range (years)	Minerals and Other Materials That Can Be Dated
Uranium-238	$\alpha + \beta^-$ decay	Lead-206	4.5 billion	10 million–4.6 billion	Zircon and uraninite
Uranium-235	$\alpha + \beta^-$ decay	Lead-207	710 million	10 million–4.6 billion	Zircon and uraninite
Thorium-232	$\alpha + \beta^-$ decay	Lead-208	14 billion	10 million–4.6 billion	Zircon and uraninite
Potassium-40	Electron capture β^- decay	Argon-40 Calcium-40	1.3 billion	50,000–4.6 billion	Muscovite Biotite Hornblende Whole volcanic rock
Rubidium-87	β^- decay	Strontium-87	47 billion	10 million–4.6 billion	Muscovite Biotite Potassium feldspar Whole metamorphic or igneous rock
Carbon-14	β^- decay	Nitrogen-14	$5,730 \pm 30$	100–70,000	Wood, charcoal, peat, grain, and other plant material Bone, tissue, and other animal material Cloth Shell Stalactites Groundwater Ocean water Glacier ice

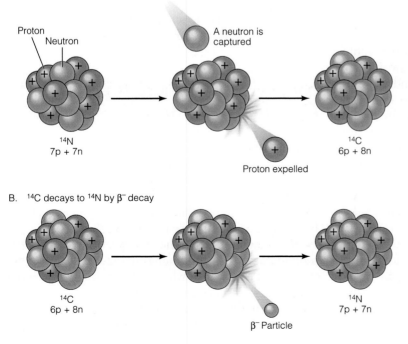

A. ^{14}C created by neutron capture

Proton
Neutron

A neutron is captured

^{14}N
7p + 7n

Proton expelled

^{14}C
6p + 8n

B. ^{14}C decays to ^{14}N by β^- decay

^{14}C
6p + 8n

β^- Particle

^{14}N
7p + 7n

Figure 8.14 Radiocarbon Production and decay of radiocarbon. A. The nucleus of a ^{14}N atom captures a neutron that was created by cosmic ray bombardment of the atmosphere and expels a proton, thereby changing ^{14}N to ^{14}C. B. The nucleus of a ^{14}C atom is radioactive. By β^- decay, ^{14}C reverts back to ^{14}N.

When Were There Glaciers on Hawaii?

The ability to attach dates to events in the geologic record has been an enormous boon for geologists. It is now possible to obtain quantitative answers to many questions that a few years ago could be approached only in a descriptive way. Two examples demonstrate how radiometric dating can be used.

During the recent ice age, ice caps and glaciers existed in many places that are ice-free today. The problem is to determine whether the age of the ice was everywhere the same. The age problem is especially severe on oceanic islands. On Mauna Kea, a 4200 m high shield volcano on Hawaii, an ice cap formed, then expanded and contracted several times. Mauna Kea is now dormant, but it was active when the ice cap existed. On the slopes of the volcano, flows of potassium-rich lavas are interlayered with sediment deposited by the ice as shown in Figure B8.1. The lavas have

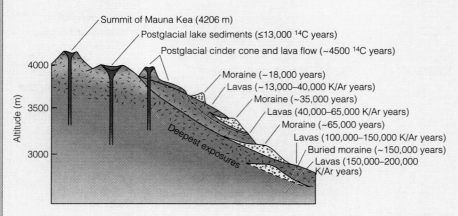

Figure B8.1 Radiometric Dating An example of radiometric dating. Glacial deposits on Mauna Kea Volcano, Hawaii, are interlayered with basaltic lava flows that have been dated by the $^{40}K/^{40}AR$ method. See text for discussion.

tory, particularly the latest glacial age. For example, the dates of many samples of wood taken from trees overrun by the advance of the latest great ice sheet and buried in the rock debris thus deposited show that the ice reached its greatest extent in the Ohio–Indiana–Illinois region about 18,000 to 21,000 years ago. It is even possible to date young ice, such as that in the Greenland Ice Sheet, directly. As the ice forms, bubbles of air are trapped in it. The carbon dioxide in the air bubbles can be liberated in the laboratory and dated, providing a date for ice formation.

Similarly, radiocarbon dates afford the means for determining rates of geologic processes, such as the rate of advance of the last ice sheet across Ohio; the rate of rise of the sea against the land while glaciers melted throughout the world; average rates of circulation of water in the deep ocean; the rates of local uplift of the crust that raised beaches above sea level; and even the frequency of volcanism. For an example of the way radiocarbon dating can be combined with other techniques to solve a geologic problem, see the essay, *When Were There Glaciers on Hawaii?*

Numerical Time and the Geologic Time Scale

Through the various methods of radiometric dating, geologists have determined the dates of solidification of many bodies of igneous rock. Many such bodies have identifiable positions in the geologic column, and as a result, it becomes possible to date, approximately, a number of the sedimentary layers in the column.

The standard units of the geologic column consist of sedimentary strata containing characteristic fossils, but the typical rocks from which radiometric dates (other than radiocarbon dates) are determined are igneous rocks. It is necessary, therefore, to be sure of the relative time relations between an igneous body that is datable

been dated by the ^{40}K/^{40}Ar method; organic matter in sediment from a postglacial lake has been dated by the ^{14}C method.

The deepest exposed glacial sediments, which are part of a buried moraine (Chapter 12), are close to 150,000 years old based on K/Ar ages of underlying and overlying lavas. A younger moraine lying partly exposed at the surface is overlain by lavas with ages close to 65,000 years, while the youngest moraines are associated with lavas and lake sediments that are between about 40,000 and 13,000 years old. These ages indicate that the glaciers on Mauna Kea advanced at essentially the same time as glaciers in the mountains of western North America.

A second dating example concerns the Haddar region of northern Ethiopia, one of the most productive places in the world for finding fossils of ancestral human beings. The sediments give good magnetic signals (Fig. B8.2). The problem is to know where in the magnetic polarity time scale the Haddar reversals fall. ^{40}K/^{40}Ar dates were obtained on a tuff and a basalt flow. The two radiometric dates determine the magnetic reversal ages unambiguously and indicate that early hominids lived in the region between 3 and 4 million years ago; that is, they lived there during the Pliocene Epoch.

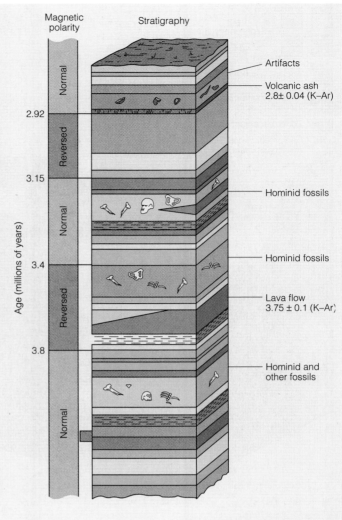

Figure B8.2 Dating Hominoid Fossils Example of two different dating techniques brought to bear on a geologic problem.

and a sedimentary layer whose fossils closely indicate its position in the column.

Figure 8.15 shows how the ages of sedimentary strata are approximated from the ages of igneous bodies. A sequence of sedimentary strata containing fossils of known ages are separated by an unconformity and two disconformities. Intrusive stock A cuts strata 1 and 2 but is truncated by the disconformity at the top of stratum 2. Thus, A must be younger than strata 1 and 2 but older than stratum 3, which was laid down on the erosion surface at the top of stratum 2 and contains weathered fragments of A among the sedimentary particles. Similarly, the combination of dikes and sills that make up the intrusive igneous complex B is truncated by the disconformity at the top of stratum 3, and they must be younger than stratum 3 but older than stratum 4. Lava flow C above the disconformity at the top of stratum 3 must also be younger than stratum 3 and younger than the dike–sill complex B. Lava flow C must be older than stratum 4, however, because it is covered by stra-

tum 4, and lava flow D must be even younger because it overlies stratum 4.

From the radiometric dates of the igneous bodies and the relative ages of the rock units shown in Figure 8.15, we can draw inferences concerning the ages of the sedimentary strata, as shown in the table that accompanies the figure.

Through a combination of geologic relations and radiometric dating methods, twentieth-century scientists have been able to fit a scale of numerical time to the geologic column worked out in the nineteenth century. The scale is being continuously refined, and so the numbers given in Figure 8.10 should be considered the best available now. Further work will make them more accurate.

It is a great tribute to the work of geologists during the nineteenth century that radiometric dating has fully confirmed the geologic column they established by the ordering of strata into relative ages. It is interesting too, to see just how wrong Lord Kelvin was in his estimate of 100 million years for the age of the Earth.

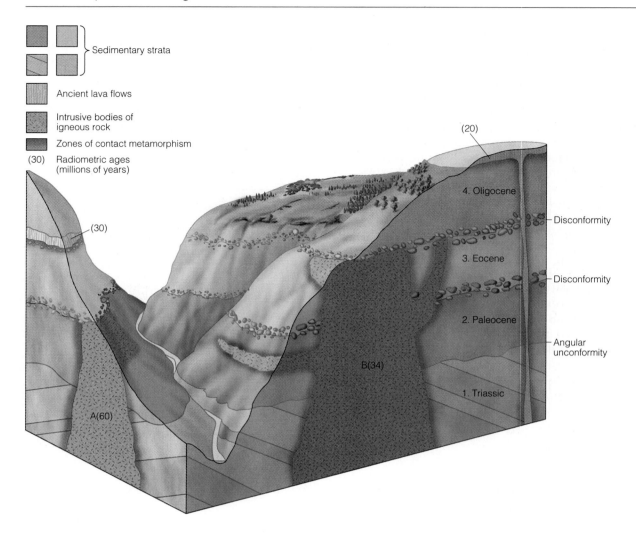

Sedimentary strata

Ancient lava flows

Intrusive bodies of igneous rock

Zones of contact metamorphism

(30) Radiometric ages (millions of years)

(20)

4. Oligocene

Disconformity

(30)

3. Eocene

Disconformity

2. Paleocene

B(34)

Angular unconformity

1. Triassic

A(60)

Stratum	Age (Millions of years)	Interpretation
4	<34 (younger than B) <30 (younger than C) >20 (older than D)	Age lies between 20 and 30 million years
3	<60 (younger than A) >34 (older than B) >30 (older than C)	Age lies between 34 and 60 million years
2	>60 (older than A)	Age of both is greater
1	>60 (older than A)	than 60 million years

Figure 8.15 Radiometric Dating and the Geologic Column Idealized section illustrating the application of radiometric dating to the geologic column. For method, see the text discussion.

The earliest record, as indicated in Figure 8.10, comes from the great assemblage of metamorphic and igneous rocks formed during the Precambrian that all lack fossils with hard parts. Of the many radiometric dates obtained from them, the youngest are around 600 million years and the oldest about 4.0 billion years. The Precambrian unit of the geologic column, then, existed during a *minimum* time equal to 4.0 billion minus 600 million years, or 3.4 billion years—a span about five times as long as the entire Phanerozoic Eon.

Given that some Precambrian rocks are about 4 billion years old, the beginning of the Earth's history must be still further back in time. The oldest radiometric dates, 4.1 billion years, have been obtained on individual mineral grains in clastic sedimentary rocks from Australia. Dates that are almost as old—4.0 billion years—have been obtained from granitic rocks from Canada. The existence of ancient granite proves that continental crust was present 4.0 billion years ago, whereas the 4.1 billion-year sedimentary grains prove that the rock cycle was operating then. Further confirmation of the ancient age of continental crust comes from another body of very ancient Precambrian rock, a 3.6 billion-year-old granite in South Africa. Although itself an igneous rock, this ancient granite contains xenoliths of quartzite. At an earlier time, before it became enveloped by the granite magma, the quartzite must have been part of a layer of sandstone. Before that, it must have been part of a layer of loose sand. Even earlier still, an igneous rock must have been subjected to weathering and erosion to produce the grains of quartz sand. Clearly, therefore, the rock cycle must have been operating in its present manner well before the granite magma solidified. Hence, as far back as we can see through the geologic column, we find evidence of the rock cycle, and, because we see ancient sediment that must have been transported by water, we know that when that sediment was deposited there must have been a hydrosphere.

The ancient rocks we have been discussing are all Archean. How long did the Hadean Eon last, and therefore how much older might our planet be? Strong evidence suggests that the Earth formed at the same time as the Moon, the other planets, and meteorites (small independent bodies that have "fallen" onto the Earth). Through various methods of radiometric dating and, in particular, the Rb/Sr and U/Pb systems, it has been possible to determine the ages of some meteorites and of some rock fragments in "Moon dust" (brought back by astronauts) as about 4.6 billion years. By inference, the time of formation of the Earth, and indeed of all the other planets and meteorites in the solar system, is believed to be approximately 4.6 billion years ago.

THE MAGNETIC POLARITY TIME SCALE

The Earth is like a gigantic magnet. It has an invisible magnetic field that permeates everything. Because of an unusual characteristic of the Earth's magnetism, it is possible to use the magnetic properties of rocks to make highly precise correlations and, in conjunction with other data, as an accurate dating technique. If a small magnet is allowed to swing freely in the Earth's magnetic field, the magnet will become oriented so that its axis is parallel to the Earth's magnetic lines of force and therefore the axis will point to the direction of the Earth's magnetic poles (Fig. 8.16). This is true for all places on the Earth. All free-swinging magnets will point to the

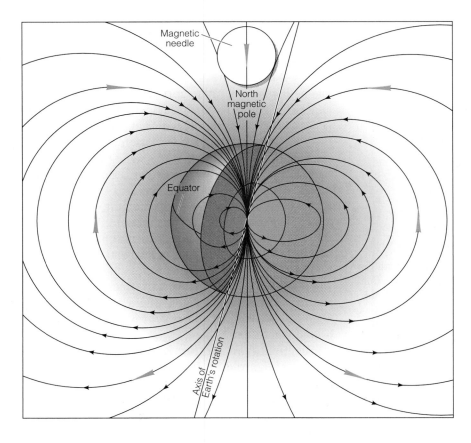

Figure 8.16 The Earth's Magnetic Field The Earth is surrounded by a magnetic field generated by the magnetism of the core. A small magnet, if allowed to swing freely, will always line up parallel to the magnetic field and point to the magnetic north pole.

magnetic poles. Certain rocks become permanent magnets as a result of the way they form, and like free-swinging magnets, the direction of their magnetic axis points to the magnetic poles. That property leads them to be useful as dating tools in the following manner.

Magnetism in Rocks

Magnetite and certain other iron-bearing minerals can become permanently magnetized. This property arises because orbital electrons spinning around a nucleus create a tiny atomic magnetic field. In minerals that can become permanent magnets, the atomic magnets line up in parallel arrays and reinforce each other. In nonmagnetic iron minerals, the atomic magnets are oriented in random directions.

Above a temperature called the **Curie point**, the thermal agitation of atoms is such that permanent magnetism is impossible. The Curie point for magnetite is about 580°C. Above that temperature, the magnetic fields of all the iron atoms are randomly oriented and cancel each other out. Below the Curie point, the magnetic fields of adjacent iron atoms reinforce each other (Fig. 8.17). When an external magnetic field is present, all magnetic domains (regions) parallel to the external magnetic field become larger and expand at the expense of adjacent, nonparallel domains. Quickly, the parallel domains become predominant, and a permanent magnet is the result.

Consider what happens when lava cools and solidifies. All the minerals crystallize at temperatures above 700°C—well above the Curie points of any magnetic minerals present. As the solidifed lava continues to cool, the temperature will drop below 580°C, the Curie point for magnetite. When the temperature drops below the Curie point, all the magnetite grains in the rock become tiny permanent magnets with the same polarity as the Earth's field. Grains of magnetite locked in a lava cannot move and reorient themselves the way a freely swinging bar magnet can. Therefore, as long as that solidified lava lasts (until it is destroyed by weathering or metamorphism), it will carry a record of the Earth's magnetic field at the moment it passed through the Curie point.

Sedimentary rocks can also acquire weak but permanent magnetism through the orientation of magnetic grains during sedimentation. As clastic sedimentary grains settle through ocean or lake water, or even as dust particles settle through the air, any magnetite particles present will act as freely swinging magnets and orient themselves parallel to the magnetic lines of force caused by the Earth's magnetic field. Once locked into a sediment, the grains make the rock a weak permanent magnet.

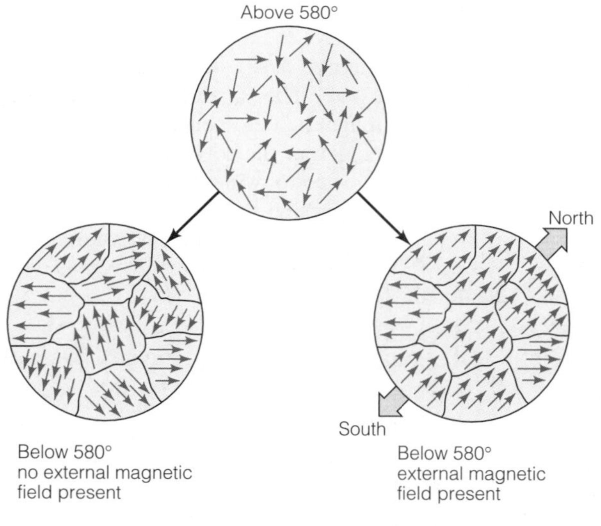

Figure 8.17 Magnetization of Magnetite Magnetization of magnetite. Above 580°C (the Curie point), the thermal motion of atoms is so great that the magnetic poles of individual atoms, shown as arrows, point in random directions. Below 580°C, atoms in small domains influence one another and form tiny magnets. In the absence of an external field, the domains are randomly oriented. In the presence of a magnetic field (lower right), most pole directions tend to be parallel to that of the external field and magnetite becomes permanently magnetized. The example in the lower right is that for a magnetite grain in a crystallizing lava.

The Polarity-Reversal Time Scale

From a study of magnetism in lavas, it was discovered early in the twentieth century that some rocks contain a record of reversed polarity. That is, when their magnetism was measured, some lavas indicated a south magnetic pole where the north magnetic pole is today, and vice versa (Fig. 8.18). Just why the Earth's magnetic field reverses polarity is not yet understood, but the fact that it does provides a very useful correlation tool because all lavas formed at the same time record the same magnetic polarity information, regardless of where on the Earth they solidified. The ages of lavas can be accurately determined using radiometric dating techniques, especially the $^{40}K/^{40}Ar$ method. Through combined radiometric dating and magnetic polarity measurements in thick piles of lava extruded over several million years, it has been possible to determine when magnetic polarity reversals occurred. A detailed record of all changes back to the Jurassic Period has now been assembled, and still earlier reversals are the topic of ongoing research.

The polarity record for the past 20 million years is shown in Figure 8.19. Periods of predominantly normal polarity (as at present), or predominantly reversed polarity, are called **magnetic chrons**. The four most recent chrons have been named for scientists who made

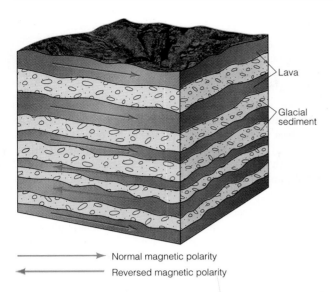

Normal magnetic polarity

Reversed magnetic polarity

Figure 8.18 Lavas and the Record of the Earth's Magnetic Field Lavas retain a record of the polarity of the Earth's magnetic field at the instant they cool through the Curie point. At Tjornes, in Iceland, a series of lava flows, separated by glacial sediment, demonstrate the reversibility of the magnetic field.

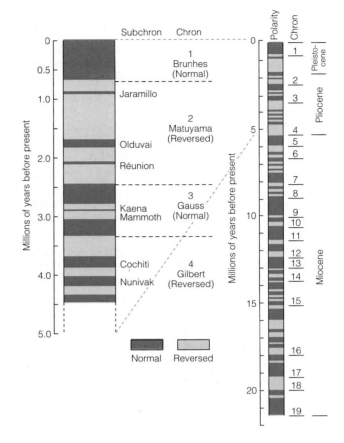

Figure 8.19 Magnetic Reversal Record Polarity reversals back 20 million years. Periods of normal polarity, as today, and periods of primarily reversed polarity are called magnetic chrons. Nineteen magnetic chrons have been identified to the beginning of the Miocene. Within each chron, one or more subchrons may occur. During a normal chron, subchrons are times of reversed polarity. Polarity reversals have been dated back to the mid-Jurassic, approximately 162 million years ago.

great contributions to studies of magnetism: Brunhes, Matuyama, Gauss, and Gilbert.

Use of magnetic reversals for geological dating differs from other dating methods. One magnetic reversal looks like any other in the rock record. When evidence of a magnetic reversal is found in a sequence of rocks, the problem is to know which of the many reversals it actually is. Additional information is needed. When a continuous record of reversals can be found, starting with the present, it is simply a matter of counting backward. This is the technique used in the dating of oceanic crust discussed in Chapter 17.

Magnetism in sedimentary rocks has proved to be a very sensitive and important correlation and dating technique. When fossils are present, an approximate age can be given to sedimentary rocks. Knowing an approximate age, geologists can determine the exact age from the magnetic reversals. Sediment cores recovered from the seafloor can be correlated very accurately using a combination of fossils and magnetic reversals. The measurements are so good that magnetic

reversals can even provide an accurate way to measure rates of sedimentation in the world ocean.

In the later chapters of this book, many examples of actual rates of geological processes are mentioned. Few, if any, examples would be possible without the numerical dates obtained through magnetic and radiometric dating. The ability to determine numerical dates has, more than any other contribution by geologists, changed the way we humans think about the world.

SUMMARY

1. Strata provide a basis for reconstructing Earth history and past surface environments. Most strata were horizontal when deposited (law of original horizontality), and all strata accumulated in sequence from bottom to top (principle of stratigraphic superposition).

2. Stratigraphic superposition concerns relative time. The relative ages of two strata can be fixed according to whether one of the layers lies above or below the other.

3. Unconformities are physical breaks in a stratigraphic sequence marking a period of time when sedimentation ceased and erosion removed some of the previously laid strata. Angular unconformity results when rocks are disturbed by tectonic activity prior to deposition of overlying strata.

4. A formation is a fundamental rock unit for field mapping distinguished on the basis of its distinctive physical characteristics and usually named for a geographic locality.

5. Systems are rock sequences and are the primary time-stratigraphic units used to construct the geologic column.

6. Geologic time units are based on time-stratigraphic units and represent the time intervals during which the corresponding systems accumulated.

7. Correlation of strata from place to place is based on physical and biological criteria that permit demonstration of time equivalence. Reliability of correlation is greatest if several criteria are used.

8. The geologic column is a composite section of all known strata, arranged on the basis of their contained fossils or other age criteria.

9. The geologic time scale is a hierarchy of time units established on the basis of corresponding time-stratigraphic units. Systems (time-stratigraphic units) and periods (geologic-time units) are based on type sections or type areas in Europe and North America. The geologic time scale constitutes the global standard to which geologists correlate local sequences of strata.

10. Decay of radioactive isotopes of various chemical elements is the basis for radiometric dating. The main radioactive isotopes and their daughters are $^{40}K/^{40}Ar$, $^{238}U/^{206}PPb$, $^{235}U/^{207}Pb$, $^{232}Th/^{208}Pb$, $^{87}Rb/^{87}Sr$, and $^{14}C/^{14}N$.

11. A sedimentary rock layer can be dated when it is bracketed between two bodies of igneous rock to which a radiometric dating method can be applied.

12. Radiocarbon dating is effective only for relatively young carbon-bearing materials (less than 70,000 years).

13. The age of the Earth is 4.6 billion years.

14. Magnetism in rocks and the polarity-reversal time scale are useful for dating oceanic crust, lavas, and young sedimentary rocks.

THE LANGUAGE OF GEOLOGY

angular unconformity (p. 203)
Archean Eon (p. 207)

Cenozoic Era (p. 207)
conformity (p. 201)
correlation (of strata) (p. 205)
Curie point (p. 220)

daughter (from radioactive decay)
(p. 210)
disconformity (p. 203)

eon (p. 207)
epoch (p. 209)
era (p. 207)

formation (p. 204)

geologic column (p. 207)

Hadean Eon (p. 207)
half-life (p. 211)

isotope (p. 210)

magnetic chron (p. 220)
mass number (p. 210)
Mesozoic Era (p. 207)

nonconformity (p. 203)

original horizontality (law of) (p. 200)

Paleozoic Era (p. 207)
parent (radioactive) (p. 210)
period (p. 208)
Phanerozoic Eon (p. 207)
Proterozoic Eon (p. 207)

stratigraphic superposition (principle
of) (p. 200)
stratigraphy (p. 200)
system (p. 205)

unconformity (p. 203)

QUESTIONS FOR REVIEW

1. How do the law of original horizontality and the principle of stratigraphic superposition help geologists unravel the history of deformed belts of sedimentary rock? Is the history so determined known in numerical or relative time?

2. What geologic events are implied by an angular unconformity? by a disconformity?

3. How does a rock-stratigraphic unit differ from a time-stratigraphic unit?

4. How can strata be correlated from place to place? Can you identify an important geological advance that came about through correlation?

5. What is the geologic column? Is the column in North America the same as the column in Australia? How could you prove that your answer is correct?

6. How is a newly discovered formation placed in its correct position in the geologic column?

7. The geologic time scale is divided into four eons. Name them in the correct order, starting with the most ancient.

8. What are the three eras of the Phanerozoic Eon, and what do their names mean?

9. What features make radioactivity an ideal way to measure geologic time? Would the radioactivity ages of a rock on the Moon and one on the Earth, formed at the same instant, be the same? Why?

10. What radiometric dating method would be suitable for obtaining the age of (a) a rhyolite thought to be about 100 million years old; (b) an Archean granite containing uraninite; and (c) charcoal from an archaeological site thought to be about 10,000 years old?

11. How has ^{14}C been used to date the advance of the last great ice sheet in central North America?

12. What is the Curie point, and why is it important for magnetic dating?

13. Polarity reversals of the Earth's magnetic field can become recorded in rocks in two different ways. What are they?

14. How can magnetic polarity reversals be used to determine the timing of past geologic events?

15. Can you explain why the vast time span known as the Precambrian does not have a detailed geologic time scale of periods and epochs? Which dating schemes would you consider using in Precambrian igneous rocks?

16. The radiometric age for a sheetlike mass of igneous rock is 20 million years. The sheet of igneous rock is parallel to the layering of a sandstone below, a shale above. How could you tell if the sheet is a lava flow or a sill? What could you say about the age of the sandstone if the sheet is a sill? a lava flow?

17. Uranium has an atomic number of 92. When ^{238}U decays to lead, it does so in a series of steps, emitting in the process 8 α-particles and 6 β-particles. What are the atomic and mass numbers of the daughter isotope of lead?

18. The half-life of ^{235}U decaying to ^{207}Pb is 710 million years. What is the age of a grain of zircon that contains ^{235}U/^{207}Pb atoms in the ratio of 3:1? What would the age be if the ratio were 1:3?

9

Understanding
Our Environment

GeoMedia

Devastated remains of Armero, Colombia, which was destroyed by a volcanic mudflow that swept down from the volcano Nevado del Ruiz at 11 p.m., November 13, 1984. More than 20,000 people died in the disaster.

Mass-Wasting

Andean Mudflow Hazards

The high Andes of South America include numerous active volcanoes and rugged peaks thrust up along converging lithospheric plates. The steep, unstable slopes rise above densely populated valleys, conditions spelling potential disaster in a landscape where major earthquakes and volcanic eruptions can bring sudden death and destruction.

In Colombia, the Andes culminate in a group of lofty active volcanoes lying west of Bogotá, the capital. Nevado del Ruiz (5400 m) has a history of volcanic activity extending back to at least 1595, when thunderous eruptions of tephra occurred and volcanic mudflows generated by melting ice and snow rushed down several valleys. In late 1984, the dormant volcano awakened and began belching clouds of steam and ash that continued through the autumn of 1985. People in the city of Armero, far downvalley from the volcano, grew alarmed. The local authorities seemed unconcerned and reassured them, even though recent geologic studies of the volcano had disclosed a history of repeated large volcanic mudflows. In early November, when the volcano showed signs of increasing activity, geologists warned that such mudflows could pose a danger for Armero in the event of an eruption. At 3 P.M. on the 13th, a technical emergency committee urged that Armero be evacuated, but the warning went unheeded. That night, as the local radio station played cheerful music and urged people to be calm, the volcano erupted. Torrents of water released from rapidly melting ice and snow near the summit sent huge waves of muddy debris surging downslope into surrounding valleys. The largest of several mudflows moved rapidly in the direction of Armero. Just after 11 P.M., as most of the population was sleeping soundly, a turbulent wall of mud came rushing out of a mountain canyon and inundated the city. More than 20,000 citizens of Armero perished, buried in a tomb of sulfurous volcanic mud. The geologists' prediction, based on a careful analysis of the geologic record, proved correct. Had their warning been heeded, the resulting human tragedy might have been avoided.

MASS MOVEMENT AND SLOPES

The landscapes we see about us may outwardly appear fixed and unchanging, but if we could make a time-lapse motion picture of almost any hillslope, the slope would seem almost alive and constantly changing. Much of the recorded motion would be the result of **mass-wasting**, the movement of regolith and masses of rock downslope under the pull of gravity. In contrast to other erosional processes, the debris is not carried by a transporting medium, such as water, wind, or ice. Mass-wasting involves a wide range of velocities, from imperceptibly slow to extremely rapid.

Because of gravity, a loose particle of regolith will always tend to move downslope. The particle's journey downslope can be very slow or very fast, but in either case the movement is caused primarily by gravity. Under

most conditions, a slope evolves toward an angle that allows the quantity of regolith reaching any point from upslope to be balanced by the quantity that is moving downslope. Such a slope is said to be in a balanced, or steady-state, condition.

Mass-wasting is not confined to the land. It occurs in lakes, and evidence of it is seen over vast areas of the seafloor. As on land, mass-wasting in the oceans is controlled by gravity and takes place wherever slopes exist.

ROLE OF GRAVITY

A smooth, vegetated slope may appear outwardly stable and show little obvious evidence of geologic activity. Yet if we examine the regolith beneath the surface, it is likely we will find some rock particles derived from bedrock farther upslope. We can deduce, therefore, that the particles have moved downslope.

In any body of rock or rock debris located on a slope, two opposing forces determine whether the body will remain stationary or will move. The first of these is **shear stress**, which is the force per unit area acting tangentially to any surface. If a body of rock is held by friction on a sloping surface, then a shear stress acts on the bottom surface of the rock, and that stress is *directed upslope*. The primary factor opposing a shear stress is the pull of gravity, and this pull is related to slope steepness. On a horizontal surface, gravity holds objects in place by pulling on them in a direction perpendicular to the surface (Fig. 9.1). On any slope, however, gravity can be resolved into two component forces. The perpendicular component of gravity (g_p in Fig. 9.1) acts at right angles to the slope and tends to hold objects in place. The *tangential component of gravity* (g_t in Fig. 9.1) acts along and down the slope in the opposite direction to the shear stress, and it is this force that causes objects to move downhill. As a slope becomes steeper, the tangential component increases relative to the perpendicular component, and the friction giving rise to the shear stress is more readily overcome.

The second force involved with downslope movement is **shear strength**, which is the internal resistance of the body to such movement. Shear strength concerns aggregates of material and is governed by factors inherent in the aggregates of rock or regolith. These factors include frictional resistance and cohesion between particles, and the binding action of plant roots.

As long as shear strength exceeds shear stress, an aggregate of rock or debris will not move. However, as these two forces approach a balance, the likelihood of movement increases. Another way to express this relationship is as a ratio of shear strength to shear stress, a ratio known as the *safety factor (Fs)*:

$$Fs = \text{Shear Strength/Shear Stress}$$

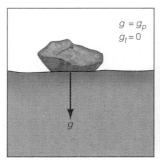
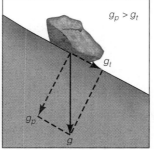

Figure 9.1 Gravity Effects of gravity on a rock lying on a hillslope. Gravity acts vertically and can be resolved into two components. One is perpendicular (g_p) and one tangential (g_t).

When this factor is less than 1, failure of a slope will occur. Steepening a slope by erosion, jolting it by an earthquake, or shaking it by blasting can cause an increase in shear stress. Shear strength might be reduced by weathering, by the decay of plant roots, or by saturation during a heavy rain.

The relationship between shear stress and slope angle (the steeper the slope, the greater the shear stress) means that conditions favoring mass movement tend to increase as slope angle increases. Steep slopes, of course, are most common in mountainous areas, so it is not surprising that mass-wasting is especially important in high mountains. Because the distribution and altitude of mountains is related to the interaction of lithospheric plates, mass-wasting on mountain slopes is a natural consequence of plate tectonics. It also is a basic part of the rock cycle: weathering, mass-wasting, and other aspects of erosion constitute a continuum of interacting processes, the end result of which is the gradual breakdown of solid bedrock and the redistribution of its weathered components.

ROLE OF WATER

Water is almost always present within rocks and regolith near the Earth's surface. Although, by definition, water is not a transporting agent in mass-wasting, it nevertheless has a very important role to play. Unconsolidated sediments behave in different ways depending on whether they are dry or wet, as anyone knows who has constructed a sand castle at the beach. Dry sand is unstable and difficult or impossible to mold, but when some water is added, the sand gains strength and can be shaped into vertical castle walls. The water and sand grains are drawn together by a force called *capillary attraction*. The attraction results from surface tension, a property of liquids that causes the exposed surface to contract to the smallest possible area. This force tends to hold the wet sand together as a cohesive mass. However, the addition of too much water saturates the sand and turns it into a slurry that easily flows away, as the sand-castle builder sees with dismay when the rising tide on the beach destroys the elaborate work of an afternoon.

Moist or weakly cemented fine-grained sediments, such as fine silt and clay, may be so cohesive that they can stand in near-vertical cliffs, like the walls of a sand castle at the beach. However, if the silt or clay becomes saturated with water and the internal fluid pressure rises above a critical limit, this fine-grained sediment will lose strength and begin to flow.

The movement of some large masses of rock has been attributed to the effects of increased water pressure in voids in the rock. If the voids along a contact between two rock masses of low permeability are filled with water, and the water is under pressure, a supporting effect may result. In other words, the water pressure bears part of the weight of the overlying rock mass, thereby reducing friction along the contact. The result can be sudden *failure*, the collapse of a rock mass due to reduced friction. An analogous situation, in which water pressure buoys up a heavy object, can make driving in a heavy rainstorm extremely dangerous. When water is compressed beneath the wheels of a moving car, the increasing fluid pressure can cause the tires to "float" off the roadway. The driver quickly loses control of the vehicle, a condition known as hydroplaning.

An experiment illustrating this same principle was described by geologists M. King Hubbert and William Rubey in 1959. An empty beverage can is placed in an upright position on the wetted surface of a sheet of glass (Fig. 9.2A). If the glass is then slowly tilted, the can will not begin to slide until a critical angle (approximately 17°) is reached. This angle is characteristic for the

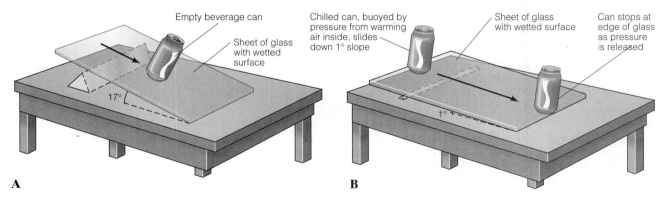

A   **B**

Figure 9.2 Interesting Experiment An experiment illustrating how compressed air reduces friction at the base of a mass resting on a slope. A. An empty beverage can placed on a wet sheet of glass will begin to slide down the surface when the angle reaches about 17°. B. A chilled empty can, placed upside down on wet glass, will slide when the glass has a slope of only 1° because friction at the base is reduced as warming air in the can expands. Sliding ceases when the can reaches the edge of the glass and pressure at the base is suddenly released.

Slope Failures

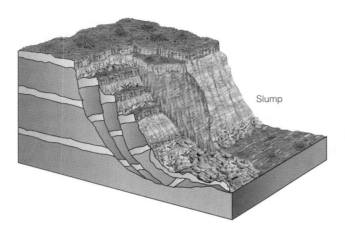

Slump

FALLS

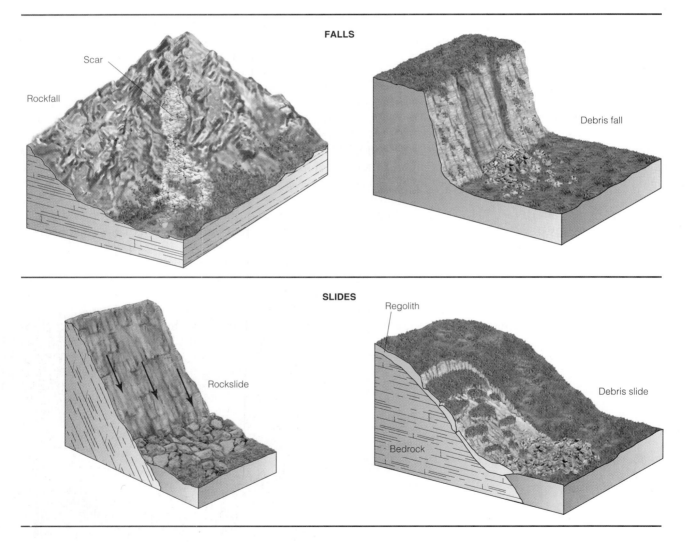

Scar

Rockfall

Debris fall

SLIDES

Regolith

Rockslide

Debris slide

Bedrock

Figure 9.3 Slope Failures Examples of slope failures giving rise to slumps, falls, and slides.

substances used in this demonstration (metal and wet glass).

Next, the can is chilled in a freezer, and the experiment is repeated. With the can placed on the wetted glass with its open end upward, the angle at which sliding begins is seen to be the same. Finally, the cold can is placed with its open end downward on the wetted glass which is tilted so it has a 1° slope (Fig. 9.2B). The can will slide the length of the glass and then stop abruptly at the edge. The reason it now moves on such a gentle slope is that the conditions at the base of the can have changed. As the can warms, the air inside expands, increasing the pressure. The increased air pressure at the base buoys up the can, reducing the friction between metal and glass, and allows the can to glide down the gentle slope. The can stops at the edge of the glass as the pressure is suddenly released. This simple experiment constitutes an analogy for natural geologic conditions in which high water pressure at the base of a large rock mass can promote downslope movement of the rock.

As seen from these examples, water can be instrumental in reducing friction and shear strength and thereby promoting movement of rock and sediment downslope under the pull of gravity. It does so in two important ways: (1) by reducing the natural cohesiveness between grains and (2) by reducing friction at the base of a rock mass through increased fluid pressure.

MASS-WASTING PROCESSES

Mass-wasting processes all share one characteristic: they take place on slopes. Any perceptible downslope movement of a mass of bedrock, regolith, or both is commonly referred to as a **landslide**. However, we can recognize many different kinds of slope movements, and because they often grade into one another, no simple and ideal classification exists. The composition and texture of the material involved, the amount of water and air mixed with the solid material, and the steepness of slope, all influence the type and velocity of movement. There is a broad array of mass-wasting processes in which a progressive transition exists from those in which water plays no significant role to those in which water plays an increasingly important role. The transition continues beyond mass-wasting to the flow of sediment-laden stream water and on to clear stream water.

The approach we will take here is to separate mass-wasting processes into those involving (1) the sudden failure of a slope that results in the downslope transfer of relatively coherent masses of rock or rock debris by slumping, falling, or sliding, and (2) the downslope flow of mixtures of solid material, water, and air. In the latter group, which involves internal motion of flowing masses of solid particles, processes are distinguished

on the basis of their velocity and the concentration in the flowing mixture. We also will examine some processes and deposits representative of mass-wasting in cold-climate regions and on the ocean floors.

Slope Failures

The constant pull of gravity makes all hillslopes and mountain cliffs susceptible to failure. When failure does occur, rock debris is transferred downslope and a more stable slope condition is reestablished.

Slumps

A **slump** is a type of slope failure in which a downward and outward rotational movement of rock or regolith occurs along a curved concave-up surface (Fig. 9.3). The top of the displaced block usually is tilted backward, producing a reversed slope. Slumps may occur singly or in groups, and they can range in size from small displacements only a meter or two in dimension to large slump complexes that cover hundreds or even thousands of square meters.

Slumps are one of the types of mass movement that we are most likely to see, for many result from artificial modification of the landscape. They are numerous along roads and highways where bordering slopes have been oversteepened by construction activity (Fig. 9.4). We can also see them along river banks or seacoasts where currents or waves have undercut the base of a slope.

Slumps frequently are associated with heavy rains or sudden shocks, such as earthquakes. Distinct episodes of

Figure 9.4 Slump A large slump in a high gravel terrace beside the Yakima River in central Washington has broken up a major highway and displaced it more than 100 m laterally into the river channel.

Figure 9.5 Landslide A jumble of boulders, many the size of houses, forms the deposit of a massive landslide that descended a steeply sloping bedrock surface on a high Andean mountain in Argentina. Other large masses of jointed rock are poised at the top of this slope, ready to produce future rockslides.

Figure 9.6 Talus A talus at the base of a steep cliff in the Brooks Range, northern Alaska. When most rockfall debris moves downslope via a gully, the resulting deposit at the base of the gully is a talus cone.

slumping may be related to changing climatic conditions. Slumping may recur seasonally and be associated with seepage of water into the ground during the rainy season. In parts of the western United States, increased slumping during recent decades appears to be correlated with an overall rise in average rainfall. A similar increase about 5000 years ago apparently was related to a shift from a warm, dry climate in the middle Holocene to cooler and wetter conditions since.

Falls and Slides

Ask mountain climbers about the greatest dangers associated with their sport and it is likely they will place falling rock near the top of the list. **Rockfall**, the free falling of detached bodies of bedrock from a cliff or steep slope, is common in precipitous mountainous terrain, where rockfall debris forms conspicuous deposits at the base of steep slopes (Fig. 9.3). As a rock falls, its speed increases. Knowing the distance of fall *(H)*, we can calculate the velocity *(v)* on impact as

$$v = \sqrt{2gH}$$

where *g* is the acceleration due to gravity. This formula tells us that a rock of a given size will be traveling at a much higher velocity if it falls free from a high cliff than from a low cliff.

A rockfall may involve the dislodgment and fall of a single rock fragment, or it may involve the sudden collapse of a huge mass of rock that plunges hundreds of meters, gathering speed until it breaks on impact into a vast number of smaller pieces. These pieces continue to bounce, roll, and slide downslope before friction and decreasing slope angle bring them to a halt.

When a mountain slope collapses, not only rock but overlying regolith and plants are generally involved. The resulting **debris fall** is similar to a rockfall, but it consists of a mixture of rock and weathered regolith, as well as vegetation (Fig. 9.3).

Slides, like falls, often involve the rapid displacement of masses of rock or sediment. A **rockslide** is the sudden downslope movement of detached masses of bedrock (or of debris, in the case of a **debris slide**) along an inclined surface, such as a bedding plane (Fig. 9.3). Like falling rock and debris, rock slides and debris slides are common in high mountains where steep slopes abound. When large rockslides occur, the resulting deposit generally is a chaotic, jumbled mass of rock, with individual boulders sometimes measuring tens of meters across (Fig. 9.5).

Accumulations of angular rock fragments are a common sight at the bases of steep cliffs. The rock debris commonly ranges in size from sand grains to large boulders. Such a body of debris sloping outward from the cliff that supplies it is a **talus** (Fig. 9.6). From cliff to talus, the movement is chiefly by falling, sliding, bounding, and

Figure 9.7 Angle of Repose Coarse, angular limestone blocks stand at the angle of repose (about 30°) in a talus below steep cliffs in the central Brooks Range, Alaska.

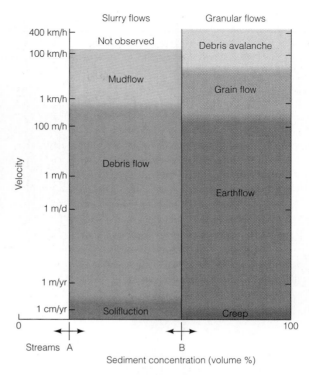

Figure 9.8 Sediment Flows Classification of sediment flows on the basis of their velocity and sediment concentration. The transition from a sediment-laden stream to a slurry flow occurs when the sediment concentration becomes so high that the stream no longer acts as a transporting agent; instead, the direct action of gravity becomes the primary force causing the saturated sediment to flow. As the percentage of water decreases further, a transition from slurry flow to granular flow takes place. Now the sediment may contain water and/or air. The boundaries between muddy streams and slurry flows (A) and between slurry and granular flows (B) are not assigned sediment-concentration percentages because the position of the boundaries can shift to the left or right depending on the physical and compositional characteristics of the sediment+water+air mixture.

rolling. The rock fragments come to rest at the steepest angle (measured from the horizontal) at which the debris remains stable (Fig. 9.7). This angle, referred to as the **angle of repose**, typically lies between 30° and 37°. Fine particles falling onto a talus tend to settle into open voids between coarser fragments. Large falling rocks have more momentum than small particles and tend to move farther down the slope. Some may bound beyond the toe of the talus and form a scattered array of isolated boulders.

Sediment Flows

There are many mass-wasting processes in which solid particles move in such a way that the overall motion of the entire mass may be described as a flowing motion. The material that flows is usually a mixture of solid particles, water, and sometimes air, but the presence of water is not required for flow to occur (consider for example, the flow of dry sand in an hourglass). Mass-wasting processes that involve the flow of such mixtures are called **sediment flows**. Note that this name is used broadly to include, in many cases, flowing rock fragments and regolith that were not sediments when movement began.

Factors Controlling Flow

At above-freezing temperatures, the way a sediment flows depends on (1) the relative proportion of solids, water, and air, and (2) the physical and chemical properties of the sediment.

All streams carry at least some sediment, but if the sediment becomes so concentrated that the water no longer can transport it, a sediment-laden stream will change into a very fluid sediment flow. Because the water in a sediment flow is not a transporting medium, a sediment flow, by definition, is a mass-wasting process. The entrained water helps promote flow, but the pull

of gravity on the solid particles remains the primary reason for their movement (whereas in a stream particles are moved largely by the force of flowing water, which flows because of the force of gravity on the water).

In Figure 9.8, sediment flows are subdivided into two classes based on sediment concentration. A **slurry flow** is a moving mass of water-saturated sediment. A **granular flow** is a mixture of sediment, air, and water, but unlike a slurry flow it is not saturated with water; instead, the full weight of the flowing sediment is supported by grain-to-grain contact or collision between grains. Each of these two classes is further subdivided into several processes on the basis of flow velocity (e.g., creep is a very slow type of granular flow, measured in millimeters or centimeters per year, whereas a debris avalanche is measured in kilometers per hour). In this classification of sediment flows, the boundaries between processes are

Sediment Flows

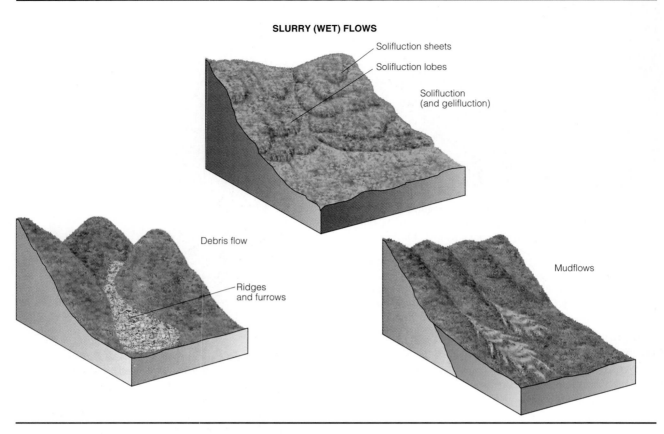

SLURRY (WET) FLOWS

Solifluction sheets

Solifluction lobes

Solifluction (and gelifluction)

Debris flow

Ridges and furrows

Mudflows

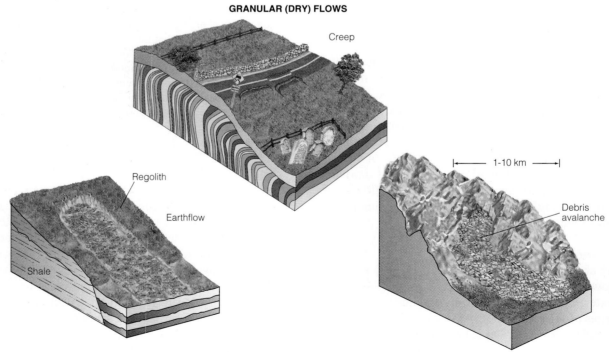

GRANULAR (DRY) FLOWS

Creep

Regolith

Earthflow

Shale

1-10 km

Debris avalanche

Figure 9.9 Wet and Dry Flows Examples of slurry flows and granular flows.

Figure 9.10 Solifluction A meter-thick solifluction lobe has slowly moved downslope and covers glacial deposits on the floor of the Orgière Valley in the Italian Alps.

placed only approximately and depend on the grain-size distribution, sediment concentration, and other factors.

Slurry Flows

In slurry flows, the sediment mixture is often so dense that large boulders can be suspended in it. Boulders too large to remain in suspension may be rolled along by the flow. When flow ceases, fine and coarse particles remain mixed, resulting in an unsorted or very poorly sorted sediment.

Solifluction. The very slow downslope movement of saturated soil and regolith is known as **solifluction**. As can be seen in Figure 9.8, this process lies at the lower end of the velocity scale for flowing sediment–water mixtures. Rates of movement are less than about 30 cm/yr, generally so slow as to be detectable only by measurements made over several seasons. The slow movement results in distinctive surface features, including lobes and sheets of debris that sometimes override one another (Figs. 9.9 and 9.10). Solifluction occurs on hillslopes in temperate and tropical latitudes, where regolith remains saturated with water for long intervals.

Debris Flows. A **debris flow** involves the downslope movement of unconsolidated regolith, the greater part being coarser than sand, at rates ranging from only about 1 m/yr to as much as 100 km/h (Fig. 9.8). In some cases a debris flow begins with a slump or debris slide, the lower part of which then continues to flow downslope (Figs. 9.9 and 9.11). A typical debris flow, once mobilized, may flow far enough to enter a stream channel and then spread across the surface of an alluvial fan, where it forms a poorly sorted deposit.

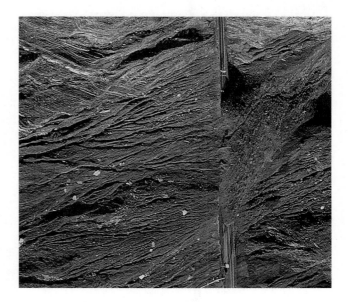

Figure 9.11 Debris Slide A debris slide from rapidly eroding cliffs along the coast of California has blocked this stretch of highway.

Debris flow deposits commonly have a tonguelike front. They also have a very irregular surface, often with concentric ridges and depressions that resemble the surfaces of deposits left by mountain glaciers. Debris flows are frequently associated with intervals of extremely heavy rainfall that lead to saturation of the ground.

Mudflows. A debris flow that has water content sufficient to make it highly fluid and in which particles no coarser than sand predominate is commonly called a

Figure 9.12 Debris Flow Passage of a muddy debris flow along a canyon bottom near Farmington, Utah, in June 1983. A. The bouldery front of a muddy debris flow advances from left to right along a stream channel in the wake of an earlier surge of muddy debris. B. The steep bouldery front, about 2 m high and advancing at 1.3 m/s, acts as a moving dam, holding back the flow of muddy sediment upstream. C. The main slurry, having a sediment concentration of about 80 percent and now moving at about 3 m/s, is viscous enough to carry cobbles and boulders in suspension.

mudflow (i.e., the term *mudflow* is a synonym for a rapidly moving debris flow). In Figure 9.8, the velocity range of mudflows lies at the upper range of debris flows (more than about 1 km/h). Most mudflows are highly mobile and tend to travel rapidly along valley floors (Fig. 9.9).

If you were to scoop up a handful of moving mudflow sediment, its consistency could range from that of freshly poured concrete to a souplike mixture only slightly denser than very muddy water. After heavy rain in a mountain canyon, a mudflow can start as a muddy stream that continues to pick up loose sediment until a moving dam of mud and rubble forms that extends to each wall of the canyon and is urged along by the force of the muddy sediment behind it (Fig. 9.12). On reaching open

Figure 9.13 Volcanic Mudflow A roadcut in the Chilean Andes near Santiago exposes a poorly sorted prehistoric volcanic mudflow in which large boulders cluster near the base of the deposit and finer clasts are scattered near the top.

country at the mountain front, the moving dam collapses, muddy sediment pours around and over it, and mud mixed with boulders is spread as a wide, thin sheet. Sediment fans at the base of mountain slopes in such arid regions as the central Andes, the Hindu Kush, and the eastern slope of the Sierra Nevada in California consist largely of superposed sheets of mudflow sediments interstratified with stream sediments (Fig. 9.13).

On active volcanoes in wet climates, layers of tephra and other volcanic debris commonly cover the surface and are easily mobilized as mudflows. As we saw in the case of Armero, Colombia, highly fluid mudflows can travel great distances and at such high velocities that they constitute one of the major hazards associated with volcanic eruptions. Mudflow sediments in valleys surrounding many of the stratovolcanoes in the Cascade Range contain much of the total volume of tephra and volcanic rocks that had earlier accumulated closer to the vents. A particularly large mudflow that originated on the slopes of Mount Rainier about 5700 years ago traveled at least 72 km. The sediment spread out beyond the mountain front as a broad lobe as much as 25 m thick. Its volume is estimated to be well over a billion cubic meters. Mount St. Helens, an unusually active volcano, has produced mudflows throughout much of its history. The most recent occurred during the huge eruption of May 1980 (Fig. 9.14).

Granular Flows
The sediment of granular flows may be largely dry, with air filling the pores, or it may be initially saturated with water but have a range of grain sizes and shapes that allows water to escape easily.

Creep and Colluvium. Most of us have seen old fences, telephone poles, or gravestones leaning at an angle on

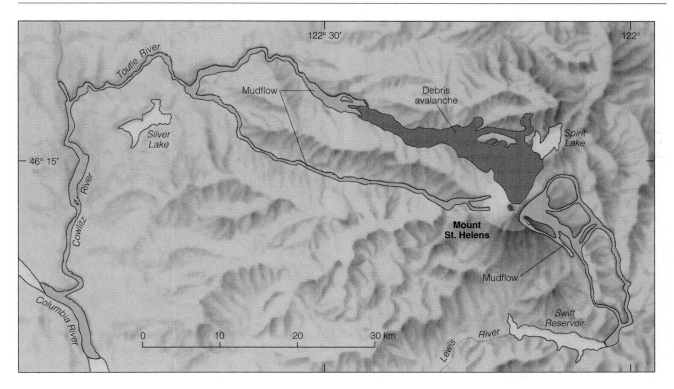

Figure 9.14 Mount St. Helens Mudflows During the 1980 eruption of Mount St. Helens in Washington, volcanic mudflows were channeled down valleys west and east of the mountain. Some mudflows reached the Columbia River, having traveled more than 90 km. Flow velocities were as high as 40 m/s and averaged 7 m/s.

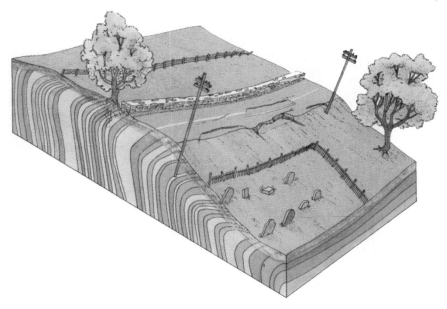

Figure 9.15 Creep Effects of creep on surface features and on bedrock. Steeply inclined strata have been deformed near the surface by differential creep, so they appear folded. Telephone poles and fence posts affected by creep are tilted, stone fences are deformed, roadbeds are locally displaced, tree trunks may be bent, and gravestones are tilted or fallen.

hillslopes, or have seen evidence of the downslope displacement of fractured road surfaces (Fig. 9.15). All are common evidence of **creep**, an imperceptibly slow downslope movement of regolith. Steeply inclined rock strata are sometimes bent over in the downslope direc-

tion just below the ground surface, showing further evidence of creep.

A number of factors contribute to creep (Table 9.1). Although creep occurs at a rate too slow to be seen, careful measurements of the downslope displacement

TABLE 9.1	Factors Contributing to Creep of Regolith
Frost Heaving	Freezing and thawing, without necessarily saturating the regolith, causing lifting and subsidence of particles
Wetting and Drying	Causes expansion and contraction of clay minerals
Heating and Cooling Without Freezing	Causes volume changes in mineral particles
Growth and Decay of Plants	Causes wedging, moving particles of downslope; cavities formed when roots decay are filled from upslope
Activities of Animals	Worms, insects, and other burrowing animals displace particles, as do animals trampling the surface
Dissolution	Mineral matter taken into solution creates voids in bedrock that tend to be filled from upslope
Activity of Snow	A seasonal snow cover tends to creep downslope and drag with it particles from the underlying ground surface

A

B

Figure 9.16 Creep on a Greenland Hillslope Colored targets placed in a straight line across a hillslope in Greenland: A, had been moved differentially downslope by creep when photographed a year later; B, the maximum recorded movement along the slope averaged 12 cm/yr.

of objects at the surface record the rates involved (Fig. 9.16). As might be expected, rates tend to be higher on steep slopes than on gentle slopes. Measurements in Colorado, for example, document a creep rate of 9.5 mm/yr on a slope of 39° but only 1.5 mm/yr on a 19° slope. Rates also tend to increase as soil moisture increases. However, in wet climates vegetation density also increases and the roots of plants, which bind the soil together, tend to inhibit creep. The creep rate measured on one grassy hillside in England having a slope of 33° is only 0.02 mm/yr. Despite a slow rate of movement, creep affects all hillslopes covered with regolith, and its cumulative effect is therefore very great.

Loose, incoherent deposits on slopes that are moving mainly by creep are termed **colluvium**. The particles in colluvium tend to be angular and lack obvious sorting. These characteristics generally make it possible to distinguish colluvium from sediments deposited by flowing water or air, which tend to consist of rounded particles, sorted and deposited in layers.

Earthflows. An **earthflow**, among the more common mass-wasting features on the landscape, is a granular flow having a velocity that falls in the range of about 1 cm/day to several hundred m/h (Fig. 9.8). Earthflows may remain active for several days, months, or even years.

Even after initial motion ceases, they may be highly susceptible to renewed movement. Like debris flows, earthflows are often made up of weak regolith, predominantly of silt and clay-sized particles, and occur on gentle to moderately steep slopes (2° to 35°). Earthflows occur where the ground is saturated, at least intermittently, and they frequently are associated with intervals of excessive rainfall.

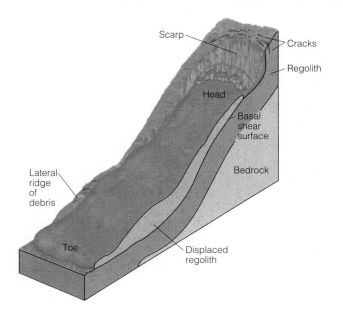

Figure 9.17 Earthflow In this section through an idealized earthflow, regolith moves downslope across a basal shear surface, leaving a scarp at the head where the regoliths separated from the slope above. A bulging toe protrudes beyond ridges of sediment that have piled up along the lower margins of the earthflow.

An earthflow typically heads in a steep *scarp*, which is a cliff formed where slide material has moved away from undisturbed ground upslope (Fig. 9.17). An earthflow generally has a narrow tonguelike shape and a rounded, bulging front. Earthflows range in size from features several meters long and wide and less than a meter deep, to features as much as several hundred meters wide, more than 1 km long, and more than 10 m deep. Many exist as parts of large earthflow complexes. In a longitudinal profile from head (top) to toe (leading edge), an earthflow is concave upward near the head and convex upward near the toe. This profile implies thinning of the earthflow sediment in the upper part of the displaced mass and thickening in the lower part during movement.

Field studies show that earthflows become mobilized when water pressure along shear surfaces rises. Pressures can reach such high values that movement will occur on slopes of only a few degrees.

A special type of earthflow occurs in wet, highly porous sediment consisting of clay- to sand-size particles. Such units may weaken if shaken suddenly, as by an earthquake. An abrupt shock increases shear stress and may cause a momentary buildup of water pressure in pore spaces, which decreases the shear strength. The result is rapid fluidization of the sediment and abrupt failure, a process known as **liquefaction**. Any structure built on such sediments, or in their path, may be quickly demolished (Fig. 9.18).

Grain Flows. If you have ever walked along the crest of a sand dune and stepped too close to the steep slope that faces away from the wind, your footstep likely started a cascade of sand flowing down the dune face. This example illustrates a type of mass-wasting called *grain flow* that involves movement of a dry or nearly dry granular sediment with air filling the pore spaces. Such grain flow occurs naturally when accumulating sand grains produce a slope that exceeds the angle of repose, leading to failure. During flow, moving grains collide frequently. Velocities of the moving sediments typically range between 0.1 and 35 m/s.

Debris Avalanches. A large, rapidly moving **debris avalanche** is a rare but spectacular event. This type of granular flow travels at high velocity (tens to hundreds of km/h) and can be extremely destructive (Fig. 9.9 and Table 9.2). A debris avalanche frequently involves huge

Figure 9.18 Liquefaction Chaotically tilted trees and houses in suburban Anchorage, Alaska, show how violent shaking of the ground during the great 1964 earthquake caused sudden liquefaction of underlying clays and widespread slumping.

Understanding Our Environment

Landslides, Floods, and Plate Tectonics

If we were to plot on a map the location of the world's major historic and prehistoric landslides, we would find that most tend to cluster along belts that lie close to the boundaries between converging lithospheric plates. They do so for two main reasons.

First, the world's highest mountain chains lie at or near plate boundaries, and on steep mountain slopes the safety factor often lies close to 1. The rocks of many mountain ranges consist of well-jointed strata that have been strongly fractured and deformed as they were uplifted. Both the joint planes and the bedding surfaces are potential zones of failure. Furthermore, along some of these belts lie the world's high stratovolcanoes, the slopes of which also tend to lie at steep angles.

Second, it is along the boundaries between plates, where plate margins slide past or over one another, that most large earthquakes occur. Earthquakes also are associated with upward-moving magma that feeds volcanic eruptions at the Earth's surface. The major landslides listed in Tables 9.2 and 9.3 occurred in active tectonic zones near plate margins, and several are known to have been directly related to major earthquakes.

Stream valleys carved deeply into the lofty Himalaya of Pakistan, India, and Nepal are the site of frequent landslides (Fig. B9.1). The collision that resulted when the northward-moving Australian-Indian Plate encountered the vast Eurasian Plate has led to uplift of the mountain range. Measured uplift rates here are the highest in the world. At Nanga Parbat (8125 m), which lies at the boundary between the converging plates, uplift rates are as high as 5 mm/yr. At this rate, the mountain should increase in altitude by 5000 m every million years. However, high mountains also mean high erosion rates, and erosion is tearing down the mountain about as rapidly as it is rising. Much of this destruction is the result of mass-wasting.

At the base of Nanga Parbat, only 21 km from the summit, lies the deep gorge of the Indus River, one of the two largest streams draining the Himalaya. From the river to the top of the mountain, the *relief* (the difference in altitude between highest and lowest points on a landscape) measures nearly 7000 m. High relief, steep slopes, fractured rocks, persistent downcutting by active streams and glaciers, and frequent large earthquakes all make Nanga Parbat an obvious place for landsliding. In fact, landsliding is probably the most effective of the agents involved in tearing down the mountain. One such landslide, and its resulting human impact, have been vividly recorded.

In early June 1841, a Sikh army was camped upstream from the town of Attock, which lies beyond the mouth of the steep gorge where the Indus flows out of the Himalaya. Suddenly, in midafternoon, a huge wall of muddy debris came rushing out of the gorge and overwhelmed the army. An eyewitness described the terrible scene: "It was a horrible mess of foul water, carcasses of soldiers, peasants, war-steeds, camels, prostitutes, tents, mules, asses, trees, and household furniture, in short every item of existence jumbled together in one flood of ruin....As a woman with a wet towel sweeps away a legion of ants, so the river blotted out the army of the Raja."

The cause of the disaster lay 400 km up the river canyon where, in January of that same year, an earthquake caused a spur of Nanga Parbat to collapse, triggering a massive landslide that rushed downslope toward the river. The slide debris quickly dammed the Indus, forming a lake that steadily grew until it was 150 m deep and more than 30 km long. As the rising water reached the top of the landslide dam, it overflowed and rapidly cut through the unconsolidated debris. The gigantic flood of water that was released rushed swiftly downstream, sweeping everything before it, toward the unsuspecting army on the plains below.

TABLE 9.2	Characteristics of Some Large Debris Avalanches				
Locality	Date	Volume (million m³)	Vertical Movement (m)	Horizontal Movement (km)	Estimated Velocity (km/h)
Huascaran, Peru	1971	10	4000	14.5	400
Sherman Glacier, Alaska	1964	30	600	5.0	185
Mount Rainier, Washington	1963	11	1890	6.9	150
Madison, Wyoming	1959	30	400	1.6	175
Elm, Switzerland	1881	10	560	2.0	160
Triolet Glacier, Italy	1717	20	1860	7.2	≥125
Black Hawk, California	prehistoric	280	1220	8.0	120
Saidmarreh, Iran	prehistoric	2000	1650	14.5	340

Figure B9.1 Landslide in Nepal A recent landslide in the Anapurna region of western Nepal has carried away a narrow trail on this steep mountainside. Steep slopes and frequent earthquakes make this region particularly susceptible to large-scale landsliding.

When Mount St. Helens erupted in 1980 and the north side of the volcano slid quickly downward into the adjacent valley, a flood similar to that associated with the Nanga Parbat landslide seemed possible. The landslide debris blocked the North Fork of the Toutle River near its head, causing the level of Spirit Lake to rise (Fig. 9.14). Left unchecked, the lake would have continued to rise until it reached the lowest portion of the landslide dam. Overflowing and cutting down rapidly into the unconsolidated debris, the escaping water would likely have caused a major flood, creating havoc downvalley. In this case, however, engineering equipment was quickly moved to stabilize the lake outlet artificially, thereby eliminating the potential threat.

masses of falling rock and debris that break up, pulverize on impact, and then continue to travel downslope, often for great distances. Such avalanches can lead to major disasters, as discussed in the boxed essay *Landslides, Floods, and Plate Tectonics*.

Large rockslides and rockfalls that give rise to debris avalanches have had the greatest human impact in populated mountain regions like the Alps and the Andes. In September 1717, a large mass of rock and ice fell onto Triolet Glacier from a mountain crest near Mont Blanc along the French-Italian border (Fig. 9.19). Pulverizing on impact, the fragmented debris moved rapidly downvalley before its leading edge came to rest about 7 km from, and 1860 m lower than, the site of detachment. The estimated velocity of the mass on impact was close to 320 km/h. As the sheet of debris rushed downward to the floor of the main valley, its momentum carried it up the opposite valley wall to a height of at least 60 m. At this point, just as the debris overwhelmed two small mountain villages, killing all the inhabitants and their livestock, its velocity is estimated to have been at least 125 km/h. From the estimated velocities and distance of travel, it is clear that the total travel time over the entire 7 km was between 2 and 4 minutes. Such rapid travel times mean that escape from large and destructive debris avalanches is seldom possible.

Because large debris avalanches are infrequent and extremely difficult to study while they are moving, few observational data about the process are available. Their extreme mobility has been attributed to the debris rid-

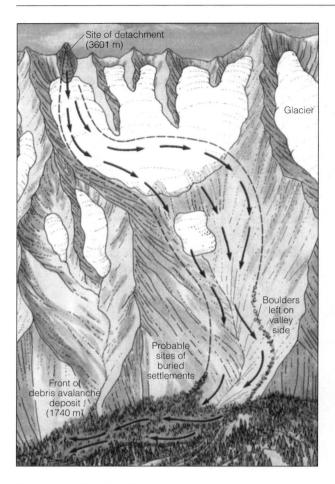

Figure 9.19 Val d'Aosta Avalanche A large debris avalanche in upper Val d'Aosta, Italy, traveled 7 km from its source high on a mountain spur. Within only a few minutes the debris buried two settlements on the valley floor, killing all the inhabitants and livestock. The reconstructed trajectory of the debris avalanche (arrows), which occurred in 1717, is based on deposits left along the valley sides.

ing on a layer of compressed air. If true, debris avalanches behave somewhat like a commercial hovercraft that travels across land or water on air compressed by a large propeller. Alternatively, air trapped and compressed within the moving debris may reduce friction between particles and cause the mass to behave in a highly fluid manner.

The flanks of steep stratovolcanoes are especially susceptible to collapse that can lead to the production of debris avalanches. The deposits of such debris avalanches can be difficult to recognize because of their huge dimensions. A broad valley that extends some 40 km north of towering Mount Shasta in northern California contains a complex of hills and mounds of volcanic rock that many geologists had interpreted as relicts of numerous minor eruptions from isolated vents. Recognition of the close similarity between these deposits and those of a huge debris avalanche associated with the

1980 Mount St. Helens eruption prompted a reassessment. A new study led to the remarkable conclusion that the entire array of features resulted from a similar, but far larger, collapse of a flank of the Shasta volcano about 300,000 years ago. In this case, the volume of rock involved was at least 26 km^3—almost ten times the volume of the landslide on Mount St. Helens. Whereas the St. Helens deposits cover an area of 60 km^2, the Shasta debris avalanche overwhelmed at least 450 km^2 (Fig. 9.20).

Mass-Wasting in Cold Climates

Mass-wasting is especially active at high latitudes and high altitudes, where average temperatures are very low. These are regions where much of the landscape is underlain by perennially frozen ground and frost action is an important geologic process.

Frost Heaving and Creep

When water freezes, it increases in volume. Ice forming in saturated regolith therefore pushes the ground surface up. This lifting of regolith by the freezing of contained water is called **frost heaving**.

Frost heaving strongly influences downslope creep of regolith in cold climates. When water freezes within the pores of regolith, the expansion of the water upon freezing increases the sizes of the pores and the ground surface is forced upward. As the ground thaws, the regolith returns to a more natural state as particles move to reduce the size of the oversized pores that had been created by the ice. During such settling of the regolith under the influence of gravity, some net movement takes place downslope. One might say this is because an oversized pore is more likely to be reduced by movement of particles from upslope than by movement of particles from downslope. Thus, a particle's net motion during each freeze–thaw cycle is a very short distance downslope (Fig. 9.21). The net result of repeated episodes of freezing and thawing, during which a particle experiences a succession of up and down movements, is slow but progressive downslope creep.

Gelifluction

In cold regions underlain year-round by frozen ground, a thin surface layer thaws in summer and then refreezes in winter. During the summer, this thawed layer becomes saturated with meltwater and is very unstable, especially on hillsides. As gravity pulls the thawed sediment slowly downslope, distinctive lobes and sheets of debris are produced. Similar to solifluction in temperate and tropical climates, this process is known as **gelifluction**. Although measured rates of movement are low, generally less than 10 cm/yr, gelifluction is so widespread on high-latitude landscapes that it constitutes a highly

Figure 9.20 Debris Avalanche, Mount Shasta A massive prehistoric debris avalanche from the northwestern flank of Mount Shasta volcano in northern California left a chaotic deposit (hills in middle distance) that extends 34 km from the volcano and covers at least 450 km^2.

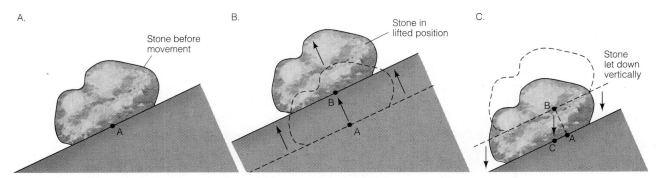

Figure 9.21 Freeze–Thaw Cycle Stone moved downslope by alternate freezing and thawing of the ground. As freezing occurs. (B) the stone is raised perpendicular to the ground surface, which also rises: when the ground thaws and settles. (C) gravity pulls the stone down approximately vertically, giving it a small but significant component of movement downslope. (A → C).

important agent of mass transport. Hillslopes in arctic Alaska and Canada, for example, commonly are mantled by superimposed sheets or lobes of geliflucted regolith.

Rock Glaciers

A **rock glacier**, another characteristic feature of many cold, relatively dry mountain regions, is a tongue or

Figure 9.22 Rock Glacier A jumbled mass of angular rock debris, supplied from a steep cliff on Sourdough Mountain in the Wrangell Mountains of southern Alaska, moves slowly downslope as a rock glacier.

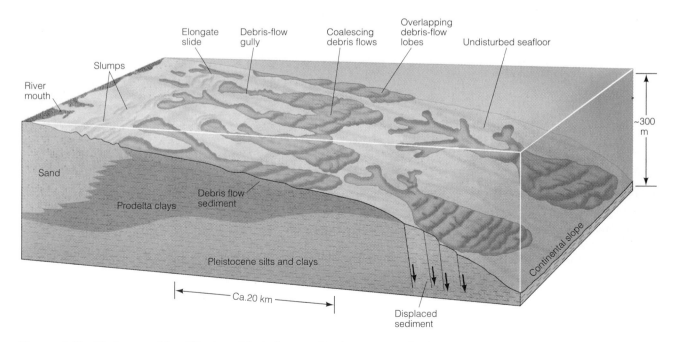

Figure 9.23 Underwater Mass-Wasting Block diagram showing mass-wasting features on the submarine surface of the Mississippi delta.

lobe of ice-cemented rock debris that moves slowly downslope in a manner similar to glaciers (Fig. 9.22). Rock glaciers generally originate below steep cliffs, which provide a source of rock debris. Active rock glaciers may reach a thickness of 50 m or more and advance at rates of up to about 5 m/yr. They are especially common in high interior mountain ranges like the Swiss Alps, the Argentine Andes, and the Rocky Mountains.

Subaqueous Mass-Wasting

As geologists have extended their search for petroleum to the continental shelves and slopes, their explorations have shown that mass-wasting is an extremely common and widespread means of sediment transport on the seafloor. Mass-wasting also has been documented in lakes. As on land, so also in subaqueous (underwater) environments, the potential for gravity-induced movement of rock and sediment exists.

Submarine slope failures can give rise to turbidity currents (Chapter 6), a type of subaqueous sediment flow, that travel down submarine canyons and deposit turbidites on the continental rise. In some regions subaqueous mass-wasting also accounts for a substantial portion of shelf deposits, especially in areas seaward of large rivers.

Marine Deltas
Major marine deltas commonly display surface features and sediments attributable to slope failures. In such subaqueous environments, failure can occur even on slopes as low as 1°. The slope failures generally display three distinct zones: a source region where subsidence (sinking or downward settling) and slumping take place, a central channel where sediment is transported, and a zone where sediment is deposited, often in the form of overlapping lobes of sediment. These and other features are common on the submarine slopes of the Mississippi delta, which is among the best-studied deltas in the world (Fig. 9.23). Slides and sediment flows are extremely active on the delta front. In places, sediment flows have deposited more than 30 m of sediment in the last 100 years.

Mass-Wasting in the Western North Atlantic
Extensive studies of the continental slope of eastern North America using a variety of modern techniques (deep-ocean drilling, piston coring, side-scan sonar, echograms, submersible vessels) have shown that vast areas of the seafloor are disrupted by submarine slumps, slides, and flows (Fig. 9.24). Some large slide complexes cover areas of more than 40,000 km^2 and reach depths as great as 5400 m. Series of slumps heading in abrupt scarps up to 10 m high and 50 km long lie upslope from slide complexes. The slides generally have affected the

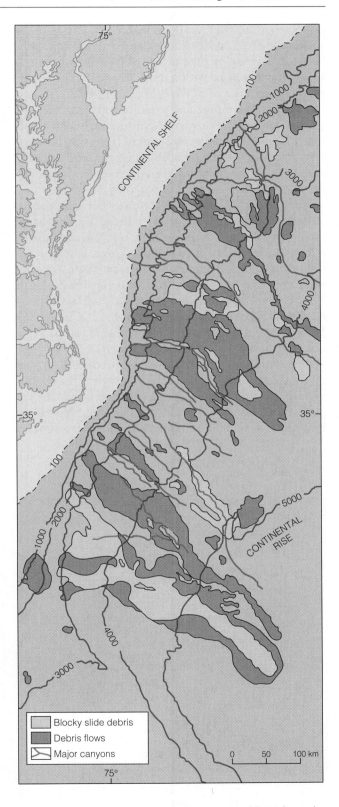

Figure 9.24 Submarine Mass Movement Map of a region off the eastern coast of the United States showing the distribution of large blocky landslide and debris-flow deposits on the continental slope and rise.

uppermost 50 m of seafloor sediment. Cores taken within the slide masses show disturbed and contorted layering, overlain by silty turbidite beds. Buried deposits have been identified within slide complexes, pointing to multiple episodes of movement.

Localized slumping and small-scale slides mark the walls of submarine canyons. Some geologists postulate that large-scale mass-wasting may have started these canyons, which were then enlarged through continuous slumping on canyon walls and the repeated movement of turbidity currents down the canyons.

Submarine slope failures often occur when shear strength of the sediments is reduced by an abrupt increase in the pore pressure. One way this can happen is by a sudden earthquake shock. Several major turbidity currents and slumps off eastern North America were contemporaneous with strong earthquakes, e.g., 1929: Grand Banks, Newfoundland (see the boxed essay on *Breaks in the Transatlantic Cables* in Chapter 6); 1886: Charleston, South Carolina; 1775: Cape Ann, Massachusetts.

The dating of numerous sediment cores from submarine slump masses and debris flows has shown that most of the mass-wasting events occurred during the last glacial age, when world sea level was at least 100 m lower than now and during the subsequent rise of world sea level that resulted from the melting of continental ice sheets. At such times, large quantities of sediment were dumped in the ocean by rivers crossing the emergent continental shelf and were then transported by mass-wasting into the deep sea. Because there have been numerous periods of low sea level during the ice ages of the last several million years, such processes have slowly built a thick wedge of sediment along the base of the Atlantic continental slope.

Hawaiian Submarine Landslides

The largest volcanoes on the Earth rise 7000 m or more from the floor of the Pacific Ocean. Like their smaller counterparts on the continents, they are composed of huge piles of well-jointed lava flows and volcanic rubble. Chaotic topography along the submerged lower margins of the Hawaiian volcanoes has been interpreted as evidence of repeated massive landslides on the volcano flanks (Fig. 9.25). Irregular terrain on the seafloor adjacent to Molokai suggests that much of the northern half of this volcano has slid downslope from a steep landslide scarp visible on submarine surveys. Anomalous coral-bearing gravels found to altitudes of 326 m on Lanai and nearby islands have recently been attributed to a giant wave that deposited the coral fragments high above sea level. The wave is believed to

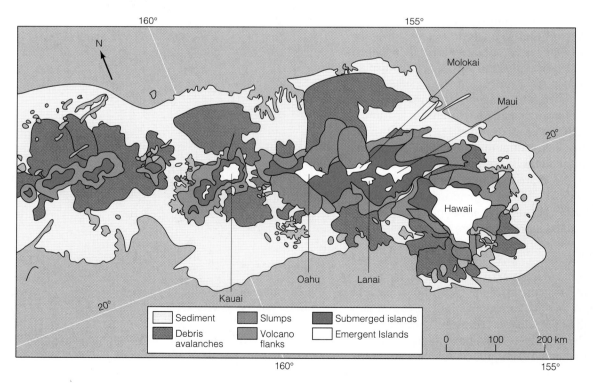

Figure 9.25 Chaotic Submarine Terrain A broad belt of chaotic terrain extending hundreds of kilometers across the seafloor adjacent to the Hawaii Islands is interpreted as deposits of massive landslides that originated on the steep, unstable submarine flanks of the volcanoes.

have resulted from a huge submarine landslide off the western coast of the island of Hawaii that traveled nearly 100 km from its source near sea level to depths of 4800 m. Based on dating of the corals, the landslide occurred about 105,000 years ago.

TRIGGERING OF MASS-WASTING EVENTS

Mass-wasting events sometimes seem to occur at random, with no apparent reason. However, most events, particularly the largest and most disastrous, are related to some extraordinary activity or occurrence.

Shocks

A sudden shock, such as an earthquake, may release so much energy that slope failures of many types and sizes are triggered simultaneously. In 1929 a major earthquake in northwestern South Island, New Zealand, triggered at least 1850 landslides larger than 2500 m^2 within an area of 1200 km^2 near the quake's center. An estimated 210,000 m^3 of debris was displaced, on average, in each 1 km^2 of land. Landslides were reported to be most numerous on well-bedded and well-jointed mudstones and fine sandstones. The Alaska earthquake of 1964 triggered many rockfalls, one of which became a huge rock avalanche that swept across the surface of Sherman Glacier, burying it with up to several meters of coarse, angular debris.

Slope Modification

Landslides often result when natural slopes are modified by human activities. Slides often occur, for example, where roads have been cut into the ground, creating artificially steep slopes that are much less stable than the more gentle original slopes (Fig. 9.26). Such landslides are especially common along the coastal cliffs of California where roads have been carved into deformed sedimentary rocks. Retaining walls may reduce the likelihood of landslides in thick colluvium, but unless such barriers are very strong, persistent downslope creep of the colluvial debris may ultimately cause them to fail.

Undercutting

Slumps and other types of landslides can be triggered by the undercutting action of a stream along its bank or by surf action along a coast. Coastal landslides are often associated with major storms that direct their energy against rocky headlands or the bases of cliffs of unconsolidated sediment. Windward coasts of the Hawaiian Islands retreat as pounding surf removes jointed lava from the base of steep seacliffs, causing them to collapse (Fig. 9.27).

Exceptional Precipitation

Landslides are often associated with heavy or persistent rains that saturate the ground and make it unstable. Such was the case in 1925 when prolonged rains, coupled

A

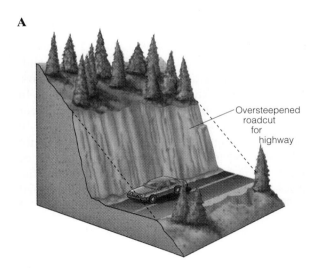

B

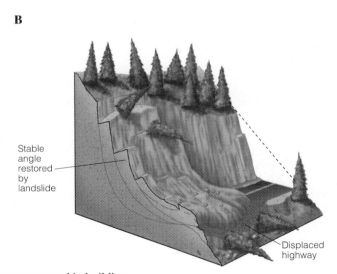

Figure 9.26 Manmade Landslide When a natural slope is oversteepened in building a road, failure can result. A. A highway cut exceeds the natural angle of the slope, producing an unstable situation. B. The oversteepened slope fails, and a landslide buries the road. The slope angle of the resulting deposit now is similar to the original natural one.

Figure 9.27 Coastal Landslide Steep seacliffs of jointed basalt along the windward coast of Hawaii are undercut by pounding surf. When a cliff collapses, the resulting landslide debris is rapidly reworked by surf and currents, and the process begins anew.

with melting snow, started a large rockfall in the Gros Ventre River basin of western Wyoming. The water saturated a porous sandstone that overlies impermeable shale and dips toward the valley floor. This saturated condition was an ideal trigger for slope failure. The estimated 37 million m^3 of rock, regolith, and organic debris started as a rockfall moved rapidly downslope, and grew into an avalanche. The mass of material created a natural dam that ponded the river. Two years later the dam failed, causing a flood that led to several deaths. Today, more than 75 years after the debris flow began, the scar at the head of the slide is still quite obvious, as is the distinctive chaotic topography downslope.

Volcanic Eruptions

Animation

Lahar

Volcanic eruptions are another means of initiating mass-wasting events. Large stratovolcanoes consist of interstratified lava flows, rubble, and pyroclastic layers that form steep slopes. On high, ice-clad volcanoes, slopes may be further steepened by glacial erosion. During eruptions, slope failure is often widespread. Large volumes of water, released when summit glaciers and snowfields melt during eruption of hot lavas or pyroclastic debris, can combine with unconsolidated deposits to form mudflows or debris flows (which geologists call *lahars*) that move rapidly downslope and often continue for many kilometers downvalley.

Submarine Slope Failures

Factors leading to unstable conditions on continental

slopes and delta fronts include (1) high internal water pressures resulting from rapid deposition of sediment and the inability of water trapped in accumulating sediment to escape; (2) generation of methane gas from organic matter deposited with sediments, which increases pressure in pores between grains; (3) local oversteepening of slopes, a result of high rates of sedimentation; (4) displacement along vertical fractures in seafloor sediments or rocks, which leads to oversteepening of slopes; and (5) shocks induced by earthquakes.

MASS-WASTING HAZARDS

As the human population increases and cities and roads expand across the landscape, the likelihood that mass-wasting processes will affect people increases. Landslides occur worldwide, and their impact, in terms of loss of lives and property, can be devastating (Table 9.3). In the United States alone, landslides in a typical year cause more than $1 billion in economic losses and 25 to 50 deaths, and the figures are rising. Although it may not always be possible to predict accurately the occurrence of significant mass-wasting events, knowledge of the processes and their relationship to local geology can lead to intelligent planning that will help reduce the loss of lives and property.

Assessments of Hazards

Assessments of potential hazards resulting from major mass-wasting events are based mainly on reconstructions of similar past events in order to evaluate their magni-

TABLE 9.3	Fatalaties Resulting from Some Major Landslides During This Century	
Year	Location	Fatalities
1916	Italy, Austria	10,000
1920	China[e]	200,000
1945	Japan[f]	1200
1949	Former USSR[e]	12,000-20,000
1954	Austria	200
1962	Peru	4000-5000
1963	Italy	2000
1970	Peru[e]	70,000
1985	Colombia[v]	23,000
1987	Ecuador[e]	1000

Source: National Research Council (1987).
Landslides related to earthquakes ([e]), floods ([f]), and volcanic eruptions ([v])

tude and frequency. From such information, it is possible to calculate how often an event of a certain magnitude is likely to recur.

Maps showing potential areas of impact of mass-wasting events are important tools for land-use planners.

For example, large debris avalanches and small rockfalls are ever-present hazards in the Italian Alps (Fig. 9.28). Field studies have shown that large debris avalanches similar to that of 1717 (Fig. 9.19) have repeatedly blanketed valley floors with rocky debris during the last 3000 years. From this evidence, a map has been constructed showing areas that could be affected by future rockfalls having various trajectories and distribution patterns. A number of small communities are at hazard from small- to intermediate-size rock avalanches (traveling 3 to 5 km), while several large communities, including one village with a population of several thousand, could be affected by large debris avalanches (traveling up to 7 km) like some recorded in deposits on the valley floors.

Valleys in the Cascade Range of Washington and Oregon contain deposits of large mudflows that spread from high volcanoes repeatedly during the last 10,000 years. Based on the number and extent of such deposits, hazards maps have been prepared, like that shown for Mount Rainier and vicinity in Figure 9.29. This map shows that risk from mudflows is high within about 25 km of the volcano's slopes, and that risk exists even at distances of 100 km or more along densely populated valley floors. A similar map prepared prior to the 1980 eruptions of Mount St. Helens proved prophetic, for mudflows generated during that series of eruptions had distributions closely similar to those predicted on the basis of geologic studies of past events.

Figure 9.28 Lucky Escape A new apartment building at the base of a steep mountain slope in the Italian Alps was struck by a large boulder falling from the cliffs above. This relatively small rockfall, which occurred just one day before the new occupants were to move in, demolished the bedroom and most of the living room.

Mitigation of Hazards

The impact of mass-wasting processes on human environments often can be reduced or eliminated by careful advanced planning. Slopes subject to creep can be stabilized by draining or pumping water from saturated sediment, while oversteepened hillslopes can be prevented from slumping if they are regraded to angles equal to or less than the natural angle of repose. In some mountain valleys subject to mudflows from active volcanoes, water-filled reservoirs can be quickly emptied so that dams will pond potentially destructive mudflows before they reach population centers (Fig. 9.29). Although we generally have no way of anticipating or preventing large rockfalls and debris avalanches, eliminating or restricting human activities in possible impact zones offers the best means of mitigating such hazards.

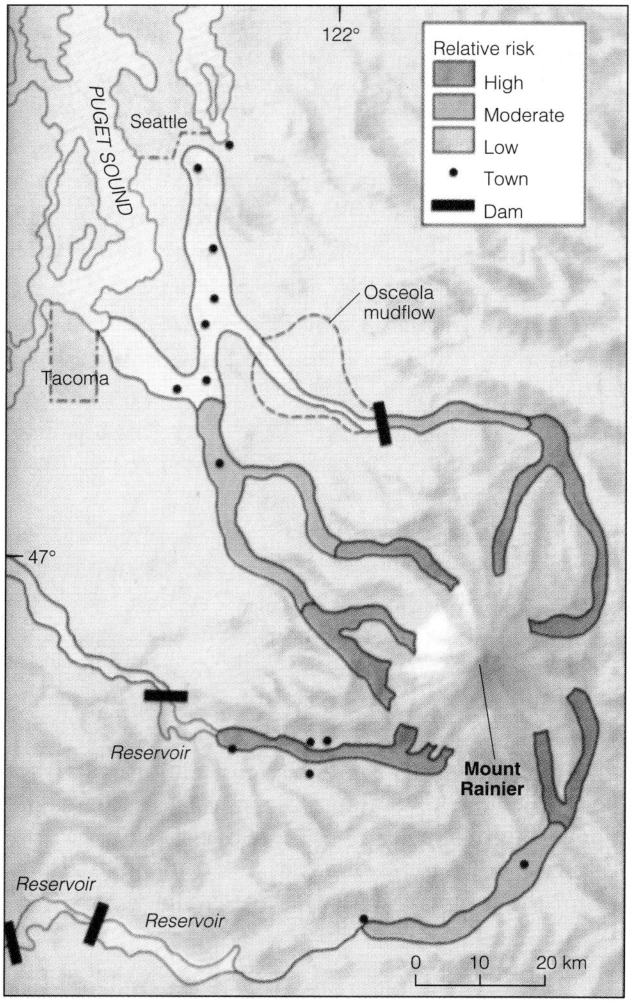

Figure 9.29 Mount Rainier Mudflows Map of the southeastern Puget Lowland, Washington, showing areas of low, moderate, and high risk from mudflows and floods originating at Mount Rainier volcano. Also shown is the extent of the huge prehistoric Osceola mudflow that was associated with a summit eruption about 5700 years ago.

SUMMARY

1. Mass-wasting is the downslope movement of rock debris under the pull of gravity without a transporting medium. Mass-wasting occurs both on land and beneath the sea.

2. The composition and texture of debris, the amount of air and water mixed with it, and the steepness of slope influence the type and velocity of slope movements.

3. Mass-wasting processes include sudden slope failures (slumps, falls, and slides) and downslope flow of mixtures of regolith, water, and air.

4. Failures occur when shear stress reaches or exceeds the shear strength of slope materials. High water pressure in rock voids or sediment reduces shear strength and increases the likelihood of failure.

5. Slumps involve a rotational movement along a concave-up surface that results in backward-tilted blocks of rock or regolith.

6. Falling and sliding masses of rock and debris are common in mountains where steep slopes abound.

7. Rockfall debris accumulates at the base of a cliff to produce a talus with slopes that stand at the angle of repose.

8. Slurry flows involve dense moving masses of water-saturated sediment that form nonsorted deposits when flow ceases. Flow velocities range from very slow (solifluction) to rapid (debris flows).

9. In granular flows, sediment is in grain-to-grain contact or grains constantly collide. The sediment may be largely dry, or it may be saturated with water that can escape easily.

10. Although creep is imperceptibly slow, it is widespread and therefore quantitatively important in the downslope transfer of debris.

11. Large, rapidly moving debris avalanches are relatively infrequent but potentially hazardous to humans.

12. In regions of perennially frozen ground, frost heaving, creep, and gelifluction are important mass-wasting processes.

13. Large areas of seafloor on the continental slopes show evidence of widespread slumps, slides, and flows. Mass-wasting on submarine slopes was especially active during glacial ages, when sea level was lower and large quantities of stream sediment were transported to the edge of the continental shelves.

14. Slope failures can be triggered by earthquakes, undercutting by streams, heavy or prolonged rains, or volcanic erup- tions. Subaqueous slope failures are frequently related to earthquake shocks and to oversteepening of slopes caused by rapid deposition of sediments.

15. Loss of life and property from mass-wasting events can be prevented or mitigated by adequate assessment and planning based on geologic studies of previous occurrences.

THE LANGUAGE OF GEOLOGY

angle of repose (p. 231)

colluvium (p. 236)
creep (p. 235)

debris avalanche (p. 237)
debris fall (p. 230)
debris flow (p. 233)
debris slide (p. 230)

earthflow (p. 236)

frost heaving (p. 240)

gelifluction (p. 240)
granular flow (p. 231)

landslide (p. 229)
liquefaction (p. 237)

mass-wasting (p. 226)
mudflow (p. 234)

rockfall (p. 230)
rock glacier (p. 241)
rockslide (p. 230)

sediment flows (p. 231)
shear strength (p. 226)
shear stress (p. 226)
slump (p. 229)
slurry flow (p. 231)
solifluction (p. 233)

talus (p. 230)

QUESTIONS FOR REVIEW

1. How does mass-wasting differ from weathering and from stream erosion?

2. What primary factors influence shear stress and shear strength on hillslopes? How can the shear strength of sediment or a rock mass on a hillslope suddenly change?

3. Why can the presence of water in rock or regolith promote downslope movement?

4. What distinctive landscape features might enable you to identify an area where numerous slumps have occurred?

5. What conspicuous type of deposit generally is found at the base of a cliff subject to frequent rockfalls? How would you expect the particles in the deposit to be sorted?

6. Why is a regolith that has been artificially excavated so that the slope of its surface exceeds the natural slope angle likely to fail?

7. How might one prove that creep is occurring on a slope, and how can its rate be measured?

8. Why are lava flows erupted from stratovolcanoes largely restricted to the volcanic cones, while mudflows originating on the same mountains are often distributed for many tens of kilometers beyond the volcanoes along adjacent valleys?

9. Explain why large-scale mass-wasting on the continental slopes apparently was far more active during the last glacial age than it is today.

10. Why are large-debris avalanches and volcanic mudflows generally far more dangerous to people than are lava flows?

11. How might prolonged and heavy rainfall affect the shear strength of a body of regolith and make it susceptible to failure?

12. What geologic conditions make high mountains especially prone to landslide activity?

10

Understanding Our Environment

Using Resources

GeoMedia

When the Mississippi River reached a height of 2 meters above flood level on July 3, 1993, the downtown part of Davenport, Iowa, was inundated. An anxious citizen tried to prevent water from flooding into a local restaurant.

Streams and Drainage Systems

The Great Flood of 1993

Like a great interstate highway system, the Mississippi River and its major tributaries span the United States from north to south and, even more so, from west to east, directing water from this vast drainage basin toward the Gulf of Mexico. In a typical year, the river rises during the spring, as melting snows and early rains swell its volume. Before cities appeared along the river, and farms spread over its fertile floodplain, the river wandered back and forth across its valley and, during times of flood, could spread over hundreds of square kilometers of adjacent land. Today the river is ordinarily confined to its immediate channel by artificial levees, built to keep the river from invading valuable farmlands and communities during seasonal floods. Now the river has no place to go when a major flood occurs, and so it rises higher in its confined channel and flows faster.

In the spring of 1993, record rains fell across the upper Mississippi Valley. Nearly 1 meter of rain fell in

Iowa during the three spring months, an amount equal to the average annual precipitation. The heavy, persistent rain resulted when the jet stream in the upper atmosphere dipped south of its usual track and cool, dry northern air encountered warm moist air moving inland from the Gulf of Mexico. High pressure stalled over the eastern coast of the country and forced the Gulf air to move northward into the midwestern United States, leading to the wettest summer on record in a broad belt from the Dakotas to Iowa and Illinois. The land, already saturated by spring rains, could not absorb the additional water falling on it, and this led to increased runoff that quickly caused streams to rise. By July, parts of nine states were under water as floodwaters overtopped levees and inundated floodplains. The main period of flooding continued for 79 days, beating the previous record of 77 days set by the flood of 1973. Fifty people were killed in the flood, at least 55,000 homes were partially or completely destroyed, and damage was estimated at $12 billion, much of it in lost crops. At the height of flooding, more than 40,000 km^2 bordering the Missouri and Mississippi rivers were under water.

The impact of the flood of 1993 on people's lives and property will not soon be forgotten, but will a basic lesson in geology be learned anew? People who choose to live on river floodplains must understand that floodplains are so named because they are periodically subjected to floods, and floods are a natural and an expected part of any river's behavior. An old Mississippi folksaying states that "When God made the world, He had a large amount of surplus water which he turned loose and told to go where it pleased; it has been going where it pleased ever since and that is the Mississippi River." The surplus water will likely always be with us, and no doubt will continue to prove very difficult to control.

STREAMS IN THE LANDSCAPE

Almost anywhere we travel over the land surface, we can see evidence of the work of running water. Even in places where no rivers flow today, we are likely to find deposits and landforms that tell us water has been instrumental in shaping the landscape. Most of these features can be related to the activity of streams that are part of complex drainage systems.

A **stream** is a body of water that flows downslope along a clearly defined natural passageway, in the process transporting detrital particles and dissolved substances. The passageway is called the stream's **channel**, and for most streams the detritus constitutes the bulk of its **load**, which is the sediment and dissolved matter the stream transports. The quantity of water passing by a point on the streambank in a given interval of time is the

stream's **discharge**. As a stream moves sediment from place to place, its channel is continually being altered. A stream and its channel are closely related and form an ever-changing, interrelated system.

Streams play important roles in our lives. Large streams, like the Amazon, the Rhine, and the Mississippi, are important avenues of transportation. Many of the world's great cities are built in stream valleys; New Orleans, Cairo, London, Paris, Rome, and Moscow are examples (Fig. 10.1). People choose to live near streams because valley floors are flat and easy to build on, soils tend to be deep and fertile, and water is available. But stream valleys have drawbacks as well. They can be threatened by floods, and as cities grow, human and industrial wastes begin to pollute the water. How to achieve an acceptable balance between human needs and the capacities of streams to maintain a safe, clean water supply is one of the major issues facing society.

Figure 10.1 Life on the Water's Edge Paris, like many of the world's other great cities, was founded along the banks of a major river. The Seine provides water for human and industrial use, is an avenue of transportation, and has great aesthetic and recreational value. However, under the stress of a growing population, the Seine, like other urban rivers throughout the world, is increasingly susceptible to pollution.

In addition to their immediate practical and aesthetic importance, streams are vital geologic agents:

- Streams carry most of the water that goes from land to sea and so are an essential part of the hydrologic cycle.

- Streams transport billions of tons of sediment to the oceans each year; there, the sediment is deposited and eventually becomes part of the rock record.

- Streams also carry billions of tons of soluble salts, released by weathering, to the oceans each year. These salts play an essential role in maintaining the saltiness of seawater.

- Streams shape the surface of the Earth. Most landscapes consist of stream valleys separated by higher ground and are the result of weathering, mass-wasting, and stream erosion working in combination.

STREAM CHANNELS

A stream's channel is an efficient conduit for running water. The discharge varies both along the channel and through time, mainly because of changes in precipitation. In response to varying discharge and load, the channel continuously adjusts its shape and orientation. Therefore, a stream and its channel are dynamic elements of the landscape.

Cross-Sectional Shape

The size and shape of any particular channel cross section reflect the typical stream conditions at that place. Very small streams may flow in channels that are as deep as they are wide, whereas very large stream channels usually have widths many times greater than their depths (Fig. 10.2). Because the volume of water moving through a channel generally increases downstream, it follows that the ratio of channel width to channel depth is likely to change downstream as the volume of water increases.

Long Profile

If we measure the vertical distance that a stream channel falls between two points a known distance apart along its course, we will have obtained a measure of the stream's **gradient**. The gradients of steep mountain streams, such as the Sacramento River in the Trinity Mountains of California, are typically greater than 30 m/km, whereas near the mouth of a large stream like the Missouri River, the gradient is typically about 0.1 m/km (Fig. 10.3).

Usually, the gradient of a river decreases downstream, and so the stream's **long profile** (a line drawn along the surface of a stream from its source to its mouth) is a curve that decreases in gradient downstream (Fig. 10.3). However, a long profile is not a perfectly smooth curve because irregularities in the gradient occur along the channel. For example, a local change in gradient may occur where a channel passes from a bed of resistant rock into one that is more erodible, or where a landslide or lava flow forms a temporary dam across the channel. Abrupt increases in the gradient cause water to flow rapidly and turbulently through a stretch of rapids or to plunge over a steep drop as a waterfall. A hydroelectric dam also introduces an irregularity in the long profile of a stream channel and may create an extensive reservoir upstream.

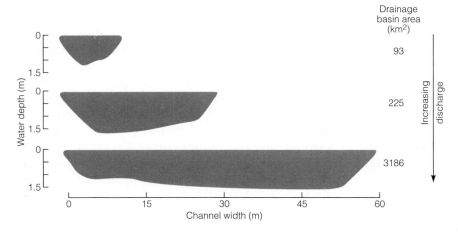

Figure 10.2 Stream Channels Cross sections of three natural streams in the upper Green River drainage basin demonstrating that the ratio of width to depth increases with increasing drainage basin area and discharge. In each cross section, the vertical scale is five times the horizontal scale.

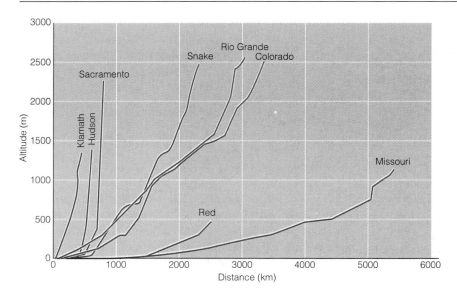

Figure 10.3 Stream Profiles Long profiles of some streams in the United States. The Klamath, Hudson, and Sacramento are relatively short, steep-gradient streams, whereas the Missouri is a long stream with a low average gradient.

DYNAMICS OF STREAMFLOW

The average annual rainfall on the area of the United States is equivalent to a layer of water 76 cm thick covering this same land surface. An amount equivalent to 45 cm returns to the atmosphere by evaporation and transpiration (Fig. 2.18), and 1 cm infiltrates the ground; the remaining 30 cm forms **runoff**, the portion of precipitation that flows over the land surface. By standing outside during a heavy rain, you can see that water initially tends to move down slopes in broad, thin sheets, a process called **overland flow**. You will also notice, however, that after traveling a short distance overland flow begins to concentrate into well-defined channels, thereby becoming streamflow. Runoff is a combination of overland flow and **streamflow**.

Factors in Streamflow

Several basic factors control the way a stream behaves: (1) *gradient*, expressed in meters per kilometer; (2) *stream cross-sectional area* (width × average depth), expressed in square meters; (3) *average velocity* of water flow, expressed in meters per second; (4) *discharge*, expressed in cubic meters per second; and (5) *load*, expressed in kilograms per cubic meter. Unlike the sediment of a stream's load, dissolved matter generally does not affect stream behavior.

Because gravity pulls on water, just as it pulls on rock and regolith, the steepness of a stream's gradient is a factor in stream behavior. A stream plunging over a waterfall or cascading down a series of steep rapids obviously is behaving differently from the same stream where it reaches level terrain.

The Discharge Equation for Streams
The relationship among discharge, velocity, and channel shape can be expressed by the equation

$$Q = A \times V$$

Q	A	V
Discharge (m^3/s)	Cross-sectional area of stream (width × average depth) (m^2)	Average velocity (m/s)

This equation tells us that when discharge changes, one or more of the factors on the right side of the equation also must change. For example, if discharge increases, both velocity and cross-sectional area are likely to increase in order to accommodate the added flow. Conversely, a decrease in discharge can lead to a corresponding decrease in stream dimensions and velocity.

A dramatic example of changes in stream factors can be seen when floods occur. During 1956, the Colorado River at Lees Ferry, Arizona, experienced a major change in dimensions as discharge increased and then fell (Fig. 10.4). Prior to the flood, the river averaged about 2 m deep and 100 m wide. As discharge increased in late spring, the water rose in the channel and erosion scoured the bed until at peak flow the river was about 7 m deep and 125 m wide. Together with an increase in velocity, the enlarged river was now able to accommodate the increased flood discharge and carry a greater load. As discharge fell, the stream was unable to transport as much sediment, and the excess load was dropped in the channel, causing its floor to rise. At the same time, decreasing discharge caused the water level to fall, thereby returning the cross-sectional area to its preflood dimensions.

As we can see from this example, a stream and its channel are intimately related. The channel is always responsive to changes in discharge, so that the system,

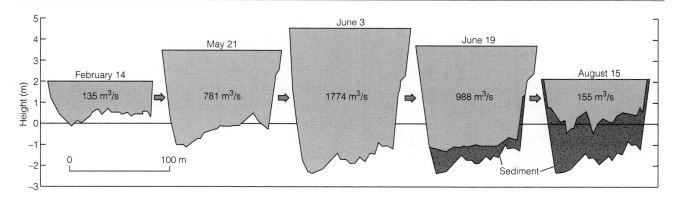

Figure 10.4 Changes in the Colorado River Changes in the cross-sectional area of the Colorado River at Lees Ferry, Arizona, during 1956. As discharge increased from February to June, the channel floor was scoured and deepened and the water level rose higher against the banks. During the falling-water phase, the river level fell and sediment was deposited in the channel, decreasing its depth.

at any point along the stream, moves to a more balanced condition.

Changes Downstream

Traveling down a stream from its head to its mouth, we can see that orderly changes occur along the channel: (1) discharge increases; (2) stream cross-sectional area increases; (3) velocity increases slightly; and (4) gradient decreases (Fig. 10.5).

The fact that velocity increases downstream seems to contradict the common observation that water rushes turbulently down steep mountain slopes and flows smoothly over nearly flat lowlands. However, the physical appearance of a stream is not a true indication of its velocity. Most of the time, discharge is low in the headward reaches of a stream, so the flowing water is shallow. The stream bed causes much more resistance to the flow of shallow water than it would for deep water, so much so that the velocity of shallow mountain streams is low even though their gradients are large. Discharge increases downstream as each **tributary** (a stream joining a larger stream) and inflow of groundwater introduce

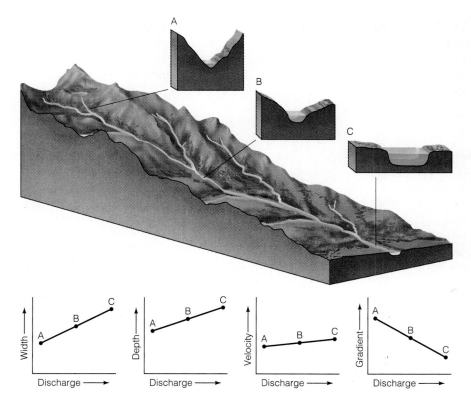

Figure 10.5 Downstream Changes Changes in the downstream direction along a stream system. Discharge increases as new tributaries join the main stream and as groundwater seeps in. Width and depth of the stream are shown by cross sections A, B, and C. Graphs show the relationship of discharge to stream width and depth, to velocity, and to stream gradient at the same three cross sections.

more water. To accommodate the greater volume of water, velocity increases accordingly, together with the cross-sectional area of the stream.

Floods

Animation

Floods

The uneven distribution of rainfall through the year causes many streams to rise seasonally in flood. A *flood* occurs when a stream's discharge becomes so great that it exceeds the capacity of the channel, therefore causing the stream to overflow its banks (Fig. 10.6). People affected by floods are frequently surprised and even outraged at what a rampaging stream has done to them. Geologists, however, tend to view floods as normal and expected events, and they hypothesize that floods have been occurring as long as rain has been falling on the Earth.

Unusually large discharges associated with floods appear as major peaks on a *hydrograph*, a graph that plots stream discharge against time. In the example shown in Figure 10.7, a passing storm generated a brief interval of intense rainfall. As the runoff moved into the stream channel, the discharge rose quickly. The crest of the resulting flood, when peak flow was reached, passed the point where the discharge was being measured about 2 hours after the storm passed. It took an addi-

tional 9.5 hours before all of the storm runoff passed through the channel at that point and discharge decreased to the normal nonflood level.

The example cited above is for a small catchment basin, so the times are short. For large catchment basins, times are much longer. Regardless of the size of the catchment basin, however, as discharge increases during a flood, so does velocity. This velocity increase has the double effect of enabling a stream to carry not only a greater load, but also larger particles. The collapse of the large St. Francis Dam in southern California in 1928 provides an extreme example of the exceptional force of floodwaters. When the dam gave way, the water behind it rushed down the valley as a spectacular flood, moving blocks of concrete weighing as much as 9000 metric tons through distances of more than 750 m. Because natural floods are also capable of moving very large objects as well as great volumes of sediment, they are able to accomplish considerable geologic work.

Flood Prediction

Major floods can be disastrous events, causing both loss of life and extensive property damage (Table 10.1), and so it is highly desirable to be able to predict their occurrence. By plotting the occurrence of past maximum discharges, including floods of different sizes, on a probability graph, a *flood-frequency curve* can be pro-

Figure 10.6 Mississippi River in Flood During a major flood in 1993, the Mississippi River (upper right corner of view) overtopped its levee and inundated adjacent farmland on the river's floodplain in Illinois.

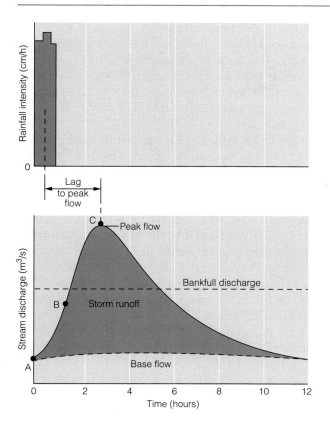

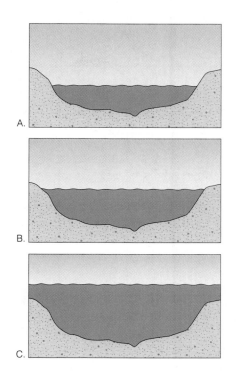

Figure 10.7 Discharge After a Storm Hydrograph of a stream following a brief storm. An interval of intense rainfall causes a rise in stream discharge as the runoff passes through the stream. Peak discharge lags about 2 hours behind the peak rainfall. It takes an additional 9.5 hours for all of the storm runoff to pass and the discharge to fall to its prestorm value (the stream's base flow).

duced (Fig. 10.8). The measure of how often a flood of a given magnitude is likely to occur is called the *recurrence interval*.

We can use a flood-frequency curve to estimate the probability that a flood of a certain magnitude will occur in any one year. In the case of the Skykomish River at Gold Bar, Washington, a flood having a discharge of 1750 m^3/s has a 1 in 10 (10%) probability of occurring in any given year, whereas a larger flood of 2500 m^3/s has a 1 in 50 (2%) probability. We refer to a flood having a recurrence interval of 10 years as a 10-year flood, whereas if the interval is 50 years, it is a 50-year flood, and so forth.

One potential problem with using flood-frequency curves to estimate the probability of future floods is that the curves are based on floods of the recent past. If

TABLE 10.1	Fatalities from Some Disastrous Floods		
River	Date	Fatalities	Remarks
Huang He, China	1887	ca. 900,000	Flood inundated 130,000 km^2 and swept many villages away.
Johnstown, Pennsylvania	1889	2200	Dam failed. Wave 10 – 12 m high rushed downvalley.
Yangtze, China	1911	ca. 100,000	Formed lake 130 km long and 50 km wide.
Yangtze, China	1931	ca. 200,000	Flood extended from Hankow to Shanghai (>800 km), leaving tens of millions homeless.
Huang He, China	1938	ca. 900,000	Chinese troops destroyed dikes to divert river as a means of blocking advance of Japanese army.
Vaiont, Italy	1963	2000	Landslide into lake caused wave that overtopped dam and inundated villages below.

Source: *Encyclopedia Americana* (1983).

A.

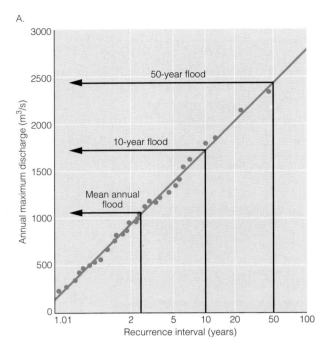

B.

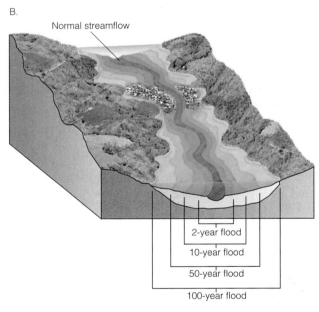

Figure 10.8 Predicting Floods A. The graph shows how often floods of different sizes occur on the Skykomish River at Gold Bar, Washington, plotted on a probability graph. A flood with a discharge of 1750 m³/s has a recurrence interval of 10 years and, thus, a 1-in-10 chance of occurring in any given year (a 10-year flood). B. The landscape diagram shows the normal, nonflood discharge for a hypothetical stream in the darkest blue. In lighter blues, the diagram shows what larger floods might be like—the 2-year, 10-year, 50-year, and 100-year floods for this stream.

the Earth's climate changes during the next several decades (Chapter 19), present flood-frequency curves may be of little value in predicting future floods.

Catastrophic Floods

Exceptional floods—those well outside a stream's normal range—occur very infrequently, perhaps only once in several centuries. Even greater floods, evidence for which we can find in the geologic record, can be viewed as catastrophic events that occur very rarely even on geologic time scales.

In the 1920s, geologist J Harlen Bretz began a study of a curious landscape in eastern Washington State called the Channeled Scabland. An array of dark, channel-like features mark places where bare lava flows lie exposed at the surface, stripped of the fertile topsoil that makes this a prime wheat-producing region (Fig. 10.9A). Bretz carefully documented the character and distribution of a variety of landforms that provide evidence of the Scabland's origin: dry coulees (canyons) with abrupt cliffs marking sites of former huge waterfalls, deep rock basins carved in the basalt, massive piles of gravel containing enormous boulders, linear deposits of gravel in the form of huge current ripples, and upper limits of water-eroded land that lie hundreds of meters above valley floors (Fig. 10.9B).

Bretz considered different hypotheses to explain this array of features, but he was led inescapably to conclude that they could be accounted for only by a catastrophic event—a truly gigantic flood, far larger than any historic flood. The source of the enormous volume of floodwater was resolved with the discovery that the continental ice sheet covering western Canada during the last glaciation had advanced across the Clark Fork River, damming it to create a huge lake in the vicinity of Missoula, Montana. The glacier-blocked lake contained between 2000 and 2500 km³ of water when it was filled and remained in existence only as long as the ice dam was stable. When the glacier retreated or its front began to float in the rising lake water, the dam failed, and water was released rapidly from the basin, as though a plug had been pulled from a giant bathtub. The main exit route lay across the Channeled Scabland region and down the Columbia River to the sea. Recent geologic studies have shown that the lake formed and drained repeatedly, creating numerous floods. The array of features scattered throughout the Scabland region thus provides us with dramatic evidence that the geologic work accomplished by catastrophic floods can be prodigious.

Base Level

As a stream flows downslope, its potential energy decreases and finally falls to zero as it reaches the sea. At this level the river no longer has the ability to deepen its channel. The limiting level below which a stream cannot erode the land is called the **base level** of the stream. The base level for most streams is global sea level (Fig. 10.10). Exceptions are streams that drain into

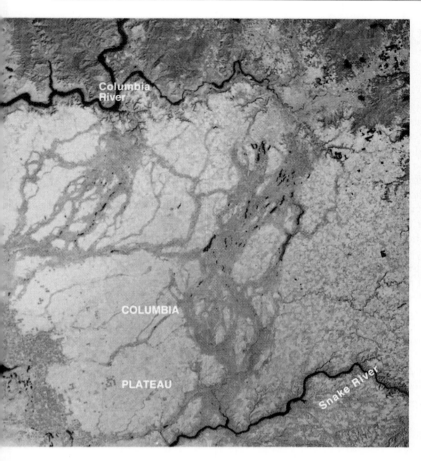

B

Figure 10.9 Channeled Scabland A. Satellite view of the Columbia plateau showing dark channels of Channeled Scabland where floodwaters stripped away a deposit of light-colored wind-blown dust from underlying dark-gray basaltic lava flows. The floodwaters traveled from the upper right to lower left. The area shown measures approximately 100 × 100 km. B. Giant ripple marks formed by raging floodwaters as they swept around a bend of the Columbia River. Composed of coarse gravel, the ripples are up to several meters high, and their crests are as much as 100 m apart.

A

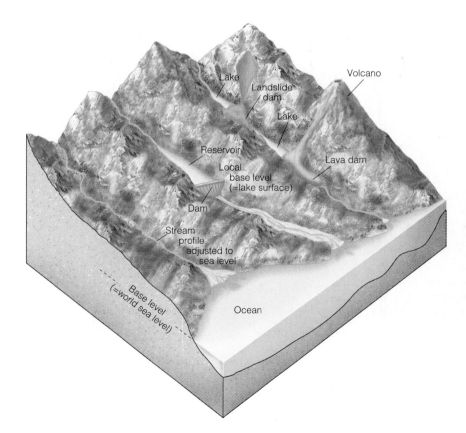

Figure 10.10 Base Levels Relationship of streams to base level (the world ocean) and to local base levels. The stream on the left flows directly to the ocean, which is its base level. The dammed middle stream flows into a reservoir, which acts as a local base level; the base level of the stream segment below the dam is the ocean. The stream to the right is dammed in two places, by landslide sediment and by a lava flow, each of which forms a lake upstream that acts as a local base level; the stream below the lava flow has the ocean as its base level.

Understanding Our Environment

Tampering with the Nile

For more than seven millennia people have lived along the banks of the lower Nile River, their fields forming a ribbon of green that crosses the vast desert of Egypt. Long before the pyramids were built, people learned to live with the Nile's annual cycle. During part of each year the river flowed peacefully within its banks, but by late summer it began to rise during the annual period of flooding. Although floods could devastate those living along the stream's margins, at the same time the floods were beneficial, for they carried an abundant load of silt that added annual nourishment to agricultural fields.

In the early nineteenth century, when a French mission sent by Napoleon conducted a census, the estimated population of Egypt was less than 2.5 million. By the 1920s the number had risen to 14 million, and since then the population has tripled. This exploding population placed a severe strain on the ability of the Nile to supply basic water needs.

To provide an adequate water supply and stabilize flow through the lower, densely populated Nile basin, a high dam was constructed at Aswan in the 1960s (Figs. B10.1 and 10.22). The Aswan Dam was designed to reduce downstream the immense difference in discharge between the flood and low-flow seasons, and to permit full utilization of the Nile water. It was intended to provide water not only for human consumption, but also for irrigation, hydroelectric power, and inland navigation. These primary goals have been realized, for the dam led to a marked change in the river's annual discharge pattern. No longer is there a large seasonal flood below the dam; instead, the annual hydrograph reflects the more uniform, controlled discharge of water from the reservoir (Fig. B10.2). Despite the success of the project, some important geologic consequences resulted from tampering with the Nile's hydrology.

The Nile, like all large other streams, is a complex natural system, and its behavior reflects a balance between discharge and sediment load. Ninety-eight percent of the Nile's load is suspended sediment. Prior to construction of the Aswan Dam, an average of 125 million metric tons of sediment passed downstream each year, but the dam

Figure   B10.1   Aswan Dam   View of Aswan High Dam, which impounds the Nile River (right) to form Lake Nasser (to south behind dam). Sediment formerly carried northward to the Mediterranean Sea and now settling out in the lake will eventually fill the reservoir and make it unusable.

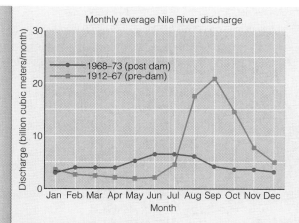

Figure B10.2 Nile Discharge Prior to construction of the Aswan Dam, discharge varied seasonally, with peak discharge coming during the late summer and early fall interval of flooding. Controlled release of water after the dam was built greatly reduced the seasonal variability in discharge.

reduced this value to only 2.5 million metric tons. Nearly 98 percent of the suspended sediment is now deposited in the reservoir behind the dam; within a few hundred years, the reservoir will be filled with silt and no longer be usable. Under natural conditions, this sediment was carried downstream by floodwaters, where much of it was deposited over the floodplain and delta, thus adding to the rich agricultural soils at a rate of 6 to 15 cm/century. With this natural source of nourishment eliminated, farmers must now resort to artificial fertilizers and soil additives to keep the land productive.

Before the Aswan Dam was constructed, the annual Nile flood transported at least 90 million metric tons of sediment to the Mediterranean Sea, adding it to the front of the delta. The shoreline at that time reflected a balance between sediment supply and the attack of waves and currents that redistributed the sediment along the coast. Because the annual discharge of sediment has now been cut off, the coast has become increasingly vulnerable to erosion. Over the long term, we can expect to see continuing changes to the delta as the shoreline adjusts to the reduction in Nile sediment and a new balance is reached.

closed interior basins having no outlet to the sea. Where the floor of a tectonically formed basin lies below sea level (e.g., Death Valley, California), the base level coincides with the basin floor. A stream flowing into such a basin can erode its channel below the level of the world ocean.

In relating base level of most streams to sea level, we must recognize that world sea level fluctuates over geologically long periods of time because of changes in the shape and capacity of the ocean basins and the growth and shrinkage of glaciers on the continents. Thus, base level is always slowly changing.

When a stream flows into a lake, the surface of the lake acts as a local base level (Fig. 10.10). However, the lake outlet may be lowered by erosion, causing the water to drain away. Once the local base level is destroyed, the stream will adjust its long profile to the changed conditions.

Natural and Artificial Dams

Not all streams flow steadily from their headwaters to the sea. The courses of many are interrupted by lakes that have formed behind natural dams consisting, for example, of landslide sediments, glacial deposits, lava flows, or even glacier ice (Fig. 10.10). Such a dam acts as a local base level and creates an irregularity in a stream's long profile. However, any natural dam is a temporary feature, for eventually it will be breached and eroded away.

Large artificial dams also disrupt the normal flow of water in a stream. They are being constructed in ever-

increasing numbers to provide water storage, flood control, and hydroelectric power.

An artificial dam built across a stream creates a reservoir that traps nearly all the sediment that the stream formerly carried to the ocean. Therein lies one of the major long-term problems in the generation of hydroelectric power: accumulating sediment will eventually fill a reservoir, often within 50 to 200 years, making it useless. Lake Nasser, ponded behind the Aswan Dam on the Nile River, will be half filled with silt in the next century. Thus, although water power is continuous, the reservoirs needed to convert this power to electricity have limited lifetimes. (See the boxed essay *Tampering with the Nile*.)

Hydroelectric Power

Hydroelectric power is recovered from the potential energy of water in streams as they flow downslope to the sea. Water ("hydro") power is just another expression of solar power because it is the Sun's heat energy that drives the water cycle. That cycle is continuous, and so energy obtained from flowing water is also continuously available.

Water power has been used in small ways for thousands of years, but only in the twentieth century has it been widely used for generating electricity. All the water flowing in the streams of the world has a total recoverable energy estimated as 9.2×10^{19} J/yr. This is an amount of energy equivalent to burning 15 billion barrels of oil per year. Unlike coal and oil, however, hydropower cannot be used up; it is a renewable resource. Nevertheless, hydropower cannot solve all the world's energy problems, for streams powerful enough to provide large amounts of power are few and do not always flow near places where power is needed. Furthermore, reservoir siltation limits the useful life of power dams and will eventually give rise to the problem of dealing with huge obsolete concrete structures and the vast accumulation of sediment behind them. Sometimes the sites of power dams are in places with high population densities. As the essay on *Taming the Yangtze River* makes clear, its consequences are not easy to predict.

CHANNEL PATTERNS

From an airplane, it is easy to see that no two streams are alike: they vary in size and shape. The variety of channel patterns on the landscape can be explained if we understand the relationships among stream gradient, discharge, and sediment load.

Straight Channels

Straight channel segments are rare. Generally, they occur for only brief stretches before the channel begins to curve. If a stream channel has many curves, we refer to the channel pattern as a *sinuous* one. Close examination of a straight segment of natural channel shows that it has some of the features of sinuous channels. A line connecting the deepest parts of the channel typically does not follow a straight path equidistant from the banks but wanders back and forth across the channel (Fig. 10.11A). This pattern may result from random variations in channel depth. At places where the deepest water lies at one side of a channel, a deposit of sediment (a *bar*) tends to accumulate on the opposite side, where velocity is lower. The sinuous flow of water within a channel causes a succession of bars to form on alternate sides of the channel.

Meandering Channels

In many streams, the channel forms a series of smooth bends that are similar in size and resemble in shape the switchbacks of a mountain road (Fig. 10.12). Such a bend in a stream channel is called a **meander**, after the Menderes River (in Latin, *Meander*) in southwestern Turkey, which is noted for its winding course. Meanders are not accidental. They occur most commonly in channels that lie in fine-grained stream sediments and have gentle gradients. The meandering pattern reflects the way

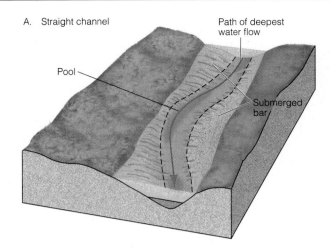

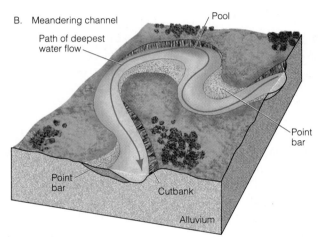

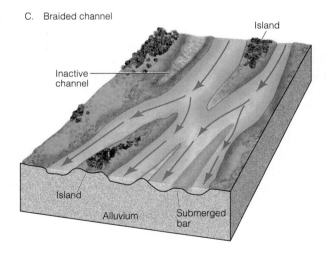

Figure 10.11 Few Stream Channels Are Straight Features associated with A. straight, B. meandering, and C. braided channels. Pools are places along a channel where water is deepest. Arrows indicate the direction of streamflow and trace the path of the deepest water.

in which a river responds to resistance to flow and the way energy is dissipated as uniformly as possible along its course.

Try to wade or swim across a meandering stream

Using Resources

Taming the Yangtze River

When completed, it will be the largest and costliest dam ever built. It will benefit millions of consumers by generating at least 18,000 megawatts of electricity, the equivalent of 10 major coal-fired power plants. It will help control one of the world's most flood-prone rivers, which has killed more than 300,000 people during the twentieth century alone.

The Three Gorges Dam on China's 6300-km long Yangtze River is a monumental undertaking that will cost at least $25 billion and take more than a decade to build. When completed in 2009, it will change the map of central China. The 185-m-high dam will back up a reservoir 600 km long and as much as 175 m deep. It will be more visible from an orbiting spacecraft than China's famous Great Wall.

However, the water that the dam impounds will submerge more than 450 villages and towns, and displace as many as 1.5 million people, thereby creating a massive resettlement problem. Half these people will be urban dwellers and the other half rural residents, all of whom will need new jobs or new farmland to begin their lives anew. Resettlement is not the only problem caused by the dam. Despite its strong promotion by the government, the project has generated a storm of adverse criticism, in large part focused on environmental issues.

According to some critics, the Three Gorges project may become one of the world's largest environmental disasters. Not only will the vast reservoir submerge one of China's most scenic and historically important areas, it will threaten rare plant and animal species along the river, including the giant sturgeon, giant salamander, fresh-water jelly fish, alligators, monkeys, and various birds. In addition, the reservoir will receive large amounts of industrial wastes and as much as 1 billion tons of sewage each year, creating a toxic soup of human waste, industrial chemicals, and heavy metals. The dam will slow the flow of water, reducing the natural self-cleaning process that now flushes waste products downstream toward the East China Sea. Silting of the reservoir, some critics claim, could block drainage outlets in Chongqing, home to 30 million people, causing sewage to back up and slosh through the streets. Some geologists fear that the weight of the huge lake on the Earth's crust could trigger an earthquake. In a worst-case scenario, this could cause the dam to fail, inundating millions of people downstream in a flood far larger and more disastrous than the natural floods of historical times.

Whether or not the benefits and environmental problems are as great as the opposing sides profess, it is clear that China has embarked on a great environmental and social experiment, the consequences of which will not be known until well into the next century.

---

and it quickly becomes apparent that the velocity of the flowing water is not uniform. Velocity is lowest along the bed and walls of the channel because here the water encounters maximum frictional resistance to flow. The highest velocity along a straight channel segment usually is found near the surface in midchannel (Fig. 10.13). However, wherever the water rounds a bend, the zone of highest velocity swings toward the outside of the bend.

Over time, meanders migrate slowly down a valley. As water sweeps around a meander bend, the zone of highest velocity swings toward the outer streambank. Strong turbulence causes undercutting and slumping of sediment where the fast-moving water meets the steep

Figure 10.12 Meandering Stream A meandering stream near Phnom Penh, Cambodia, flows past agricultural fields that cover the river's floodplain. Light-colored point bars, composed of gravelly alluvium, lie opposite cutbanks on the outsides of meander bends. Two oxbow lakes, the product of past meander cutoffs, lie adjacent to the present channel.

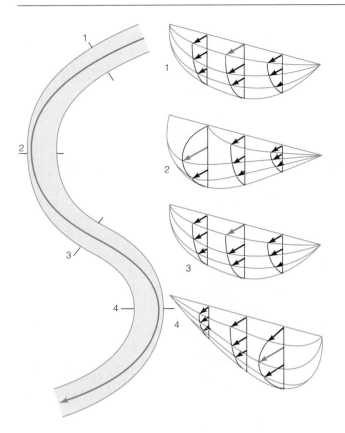

Figure 10.13 Stream Velocity Velocity distribution in cross sections through a sinuous channel (lengths of arrows indicate relative flow velocities). The zone of highest velocity (red arrow) lies near the surface and toward the middle of the stream where the channel is reactively straight (sections 1 and 4). At bends, the maximum velocity swings toward the outer bank and lies below the surface (sections 2 and 5).

bank. Meanwhile, along the inner side of each meander loop, where water is shallow and velocity is low, coarse sediment accumulates to form a *point bar* (Figs. 10.11B and 10.12). As a result, meanders slowly change shape and shift position along a valley as sediment is removed from and added to their banks.

The behavior of artificial streams and their channels has been studied in laboratory experiments using large sedimentation tanks. The experiments show that if the streambank sediment has a uniform grain size, meanders are symmetrical and all tend to migrate downstream at the same rate. In nature, however, bank sediment generally is not uniform, as shown in Figure 10.14, which is an example from the lower Mississippi River. Wherever the downstream part of a meander that is cutting into sandy sediment encounters a less erodible sediment, such as clay, its migration can be slowed. Meanwhile, the next meander upstream, migrating more rapidly, may intersect the slower moving meander. The water in the channel now can take a shorter route down-

stream over a steeper gradient, so the stream bypasses the meander that has been cut off. As sediment is deposited along the margin of the new channel route, the cutoff meander is blocked and converted into a curved *oxbow lake* (Figs. 10.12 and 10.14).

Nearly 600 km of the Mississippi River channel has been abandoned through cutoffs since 1776. In 1883, former steamboat pilot Mark Twain, then a best-selling humorist and an observant amateur geologist, speculated about the future history of the Mississippi River in his book *Life on the Mississippi* (1883):

The Mississippi between Cairo and New Orleans was 1215 miles long 176 years ago. It was 1180 after the cutoff of 1722. It was 1040 after the American Bend cutoff. It has lost 67 miles since. Consequently, its length is only 973 miles at present.

Now, if I wanted to be one of those ponderous scientific people, and "let on" to prove what had occurred in the remote past by what had occurred in a given time in the recent past, or what will occur in the far future by what has occurred in late years, what an opportunity here! Geology never had such a chance, nor such exact data to argue from! Please observe:

In the space of 176 years the lower Mississippi has shortened itself 242 miles. That is an average of a trifle over one mile and a third per year. Therefore, any calm person, who is not blind or idiotic, can see that in the Old Oolitic Silurian Period, just over a million years ago next November, the Lower Mississippi River was upwards of 1,300,000 miles long, and stuck out over the Gulf of Mexico like a fishing rod. And by the same token any person can see that 742 years from now the Lower Mississippi will be only a mile and three-quarters long, and Cairo and New Orleans will have joined their streets together, and be plodding comfortably along under a single mayor and a mutual board of aldermen. There is something fascinating about science. One gets such wholesale returns of conjecture out of such a trifling investment of fact.

Despite Mark Twain's perceptive analysis, the river's length has not changed appreciably over the last two centuries because the loss of channel due to cutoffs has been balanced by lengthening of the channel as other meanders have enlarged.

Braided Channels

The intricate geometry of a **braided stream** resembles the pattern of braided hair, for the water repeatedly divides and reunites as it flows through two or more adjacent but interconnected channels separated by bars or islands (Fig. 10.11C). The cause of braiding is related

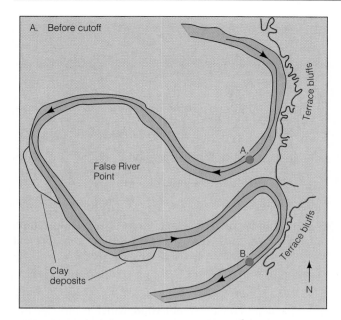

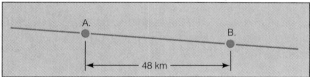

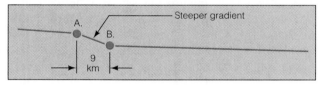

Figure 10.14 Cutoffs Cutoff of a meander loop of the Mississippi River in Louisiana. A. The downvalley migration of a meander loop was halted when the channel encountered a body of clay in the floodplain sediments. This allowed the next meander loop to advance and finally cut off the river segment surrounding False River Point. B. The new, shorter channel had a steeper gradient than the abandoned course, and a braided pattern developed.

to the stream's ability to transport sediment. If a stream is unable to move all the available load, it tends to deposit the coarsest sediment as a bar that locally divides the flow and concentrates it in the deeper segments of channel to either side. As the bar builds up, it may emerge above the surface as an island and become stabilized by vegetation that anchors the sediment and inhibits erosion.

A braided pattern tends to form in streams having highly variable discharge and easily erodible banks that can supply abundant sediment load to the channel system. Streams of meltwater issuing from glaciers generally have a braided pattern because the discharge varies both daily and seasonally, and the glacier supplies the stream with large quantities of sediment. The braided pattern, therefore, seems to represent an adjustment by which a stream increases its efficiency in transporting sediment.

Large braided rivers typically have numerous constantly shifting shallow channels (Fig. 10.15). Although at any moment the active channels may cover no more than 10 percent of the width of the entire channel system, within a single season all or most of the surface sediment may be reworked by the laterally shifting channels.

Figure 10.15 Braided River Intricate braided pattern of Brahmaputra River where it flows out of the Himalaya en route to the Ganges delta. Noted for its huge sediment load, the river can be 8 km wide during the rainy monsoon season.

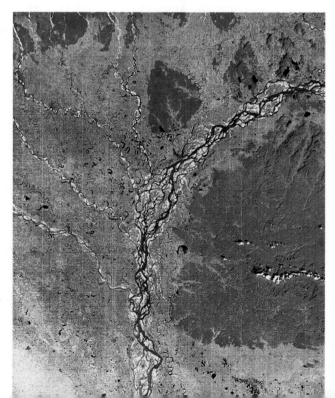

EROSION BY RUNNING WATER

Erosion by water begins even before a distinct stream has formed. It occurs in two ways: by impact as raindrops hit the ground and by overland flow during heavy rains, a process known as **sheet erosion**. As raindrops strike bare ground, they dislodge small particles of loose soil, spattering them in all directions. On a slope the result is net displacement downhill. One raindrop has little effect, but the number of raindrops is so great that together they can accomplish a large amount of erosion.

The effectiveness of raindrops and overland flow in eroding the land is greatly diminished by a protective cover of vegetation. The leaves and branches of trees break the force of falling raindrops and cushion their impact on the ground. More important, the intricate network of roots forms a tight mesh that holds soil in place, greatly reducing erosion. The root network also holds water, letting it percolate slowly down through the soil. As a consequence, in vegetated areas there is less overland flow than in areas of bare ground.

The ability of streams to erode is related to the way water moves through a stream channel. If the velocity is very slow, the water particles travel in parallel layers, a motion called **laminar flow**. With increasing velocity, the movement becomes more erratic and complex, giving rise to the swirls and eddies that characterize **turbulent flow**. The velocity in stream channels is sufficiently high for turbulent flow to dominate. Only in a very thin zone along the bed and channel walls, where frictional drag is high, is velocity low enough for laminar flow to occur.

The ability of a stream to pick up particles of sediment from its channel and move them along depends largely on the turbulence and velocity of the water. Figure 10.16, based on experimental data, shows the velocities required to erode particles of different size from a stream bed, the range of velocity in which particles can be transported, and the velocities at which particles can no longer be moved and will settle to the bottom. In general, as velocity increases, so does the ability of the turbulent water to lift ever larger particles of sediment. Silt and clay are an exception, for they tend to be cohesive and difficult to erode except under conditions of high velocity.

THE STREAM'S LOAD

The solid portion of a stream's load consists of two parts. The first part is the coarse particles that move along the stream bed (the **bed load**), while the second is the fine particles that are suspended in the water (the **suspended load**). Wherever they are dropped, these

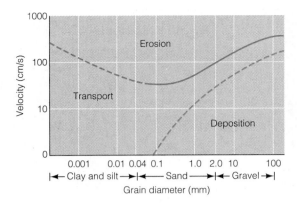

Figure 10.16 Velocity and Transport of Grains Graph showing how stream velocity controls erosion, transport, and deposition of sediment grains of different sizes.

solid particles constitute **alluvium**, which is any detrital sediment deposited by a stream.

Streams also carry dissolved substances (the **dissolved load**) that are chiefly a product of chemical weathering.

Bed Load

The bed load generally amounts to between 5 and 50 percent of the total load of most streams. Bed-load particles move at a slower velocity than the stream water, for the particles are not in constant motion. Instead, they move discontinuously by rolling or sliding. Where forces are sufficient to lift a particle, it may move short distances by saltation, a motion that is intermediate between suspension and rolling or sliding. **Saltation** involves the progressive forward movement of a particle in a series of short intermittent jumps along arcuate paths (Fig. 10.17). Saltation continues as long as currents are turbulent enough to lift particles from the bed.

The distribution of bed-load sediment in a stream channel is related to the velocity distribution (Fig. 10.18). Coarse-grained sediment is concentrated where the velocity is high, whereas finer-grained sediment is relegated to zones of progressively lower velocity.

Placer Deposits

The famous California gold rush of 1849 followed the discovery that the sand and gravel in the bed of a small stream contained bits of gold. Similar gold-bearing gravels are found in many other parts of the world. The gravels themselves are sometimes rich enough to be mined, but even when they are too lean, the gold is a clue that a source must lie upstream. In fact, many mining districts have been discovered by following trails of gold

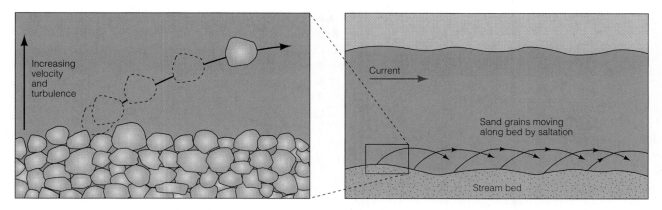

Figure 10.17 Saltation A sandy bed load moves by saltation when sand grains are carried up into a stream at places where turbulence locally reaches the bottom or where suspended grains impact other grains on the bed. Once raised into the flowing water, the grains are transported along arc-shaped trajectories as gravity pulls them toward the stream bed where they impact other particles which, in turn, are set in motion.

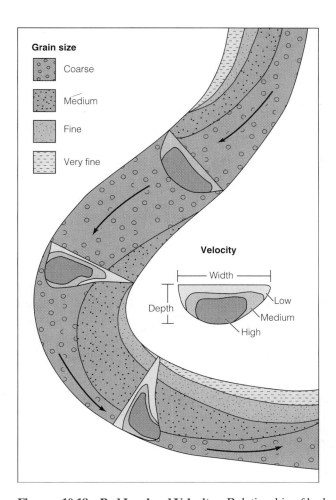

Figure 10.18 Bed Load and Velocity Relationship of bed-load grain size to velocity in a section of meandering channel. The coarsest sediment is associated with the zone of highest velocity; both lie on the outside of a bend adjacent to the cutbank but in the center of the channel between bends. The finest sediment is associated with the zone of lowest velocity and lies on the inside of a meander bend, opposite the cutbank.

and other minerals upstream to their sources in veins in bedrock.

Because pure gold is dense (specific gravity = 19), it is deposited from the bed load of a stream very quickly, while quartz, with a specific gravity of only 2.65, is washed away. As most silicate minerals have specific gravities that are low by comparison with gold, grains of gold become mechanically concentrated in places where the velocity of streamflow is high enough to remove the low-density particles but not high enough to remove the higher-density ones. Such concentration occurs, for example, behind rock bars or in bedrock holes along the channel, below waterfalls, on the inside of a meander bend, and downstream from the point where a tributary enters a main stream. A deposit of heavy minerals concentrated mechanically is a **placer**.

Many heavy, durable minerals other than gold also form placers. These include minerals that occur as pure metals, such as platinum and copper, as well as tinstone (cassiterite, SnO_2) and nonmetallic minerals such as diamond, ruby, and sapphire. Even if a vein contains a low percentage of a mineral, the placer it yields may be quite rich. In order to become concentrated in placers, the minerals must be not only dense but also resistant to chemical weathering and not readily susceptible to cleaving as the mineral grains are tumbled in the stream.

Every phase of the conversion of gold in a vein to placer gold has been traced. Chemical weathering of the exposed vein releases the gold, which then moves slowly downslope by mass-wasting. In some places mass-wasting alone concentrates the gold sufficiently to justify mining. More commonly, however, the mineral particles get into a stream, which concentrates them more effectively than mass-wasting.

Most placer gold occurs as grains the size of silt particles, the "gold dust" of miners. Some of it is coarser. A

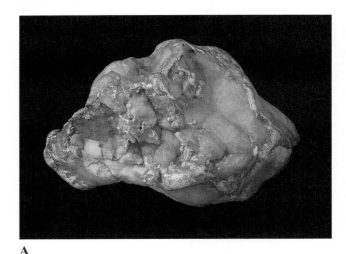

A

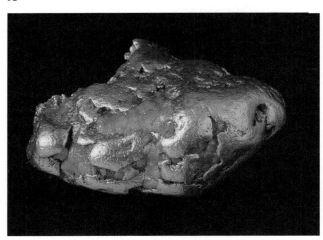

B

C

Figure 10.19 Nuggets Formation of a nugget. A. A vein of metallic gold cutting through a pebble of vein quartz. Stream abrasion causes the brittle quartz to chip and be reduced in size, while the malleable gold deforms but is not reduced. B. The ratio of gold to quartz increases as the quartz is abraded away. Eventually a nugget of almost solid gold forms. C. A nugget of metallic gold from California. No quartz remains. Each of the specimens has a diameter of about 4 cm.

lump of pebble or cobble size is a nugget (Fig. 10.19). The largest nugget ever recorded weighed 80.9 kg.

In following placers along stream channels, prospectors have learned that rounding and flattening (by pounding) of nuggets increase downstream. Therefore, when they find rough, angular nuggets, prospectors know the primary source is close.

Suspended Load

The muddy character of many streams is due to the presence of fine particles of silt and clay moving in suspension (Fig. 10.20). Most of the suspended load is derived from fine-grained regolith washed from areas unprotected by vegetation and from sediment eroded and reworked by the stream from its own banks. The Yellow River (or Huang He) of China is yellow because of the great load of yellowish silt it erodes and transports seaward from widespread deposits of eolian sediment that underlie much of its basin. Prior to construction of several major dams along its course, the Colorado River was also extremely muddy; those who lived along it sometimes remarked that the river was "too thin to plow, but too thick to drink."

Because upward-moving currents within a turbulent stream exceed the velocity at which particles of silt and clay can settle toward the bed under the pull of gravity, such particles tend to remain in suspension longer than they would in nonturbulent waters. They settle and are deposited only where velocity decreases and turbulence ceases, as in a lake or in the sea.

Dissolved Load

All stream water contains dissolved chemical substances that constitute part of its load. The bulk of the dissolved content of most rivers consists of seven ionic species: bicarbonate (HCO_3^{1-}), calcium (Ca^{2+}), sulfate (SO_4^{2-}), chloride (Cl^{1-}), sodium (Na^{1+}), magnesium (Mg^{2+}), potassium (K^{1+}), plus dissolved silica as $Si(OH)_4$.

Although in some streams the dissolved load may represent only a small percentage of the total load, in others it amounts to more than half. Streams that receive large contributions of underground water (Chapter 11) generally have higher dissolved loads than those whose water comes mainly from surface runoff.

Downstream Changes in Grain Size

The size of the particles a stream can transport is related mainly to the flow velocity. Therefore, we might expect the maximum size of sediment to increase in the down-

Figure 10.26 Alluvial Fan A symmetrical alluvial fan has formed at the margin of Death Valley, California, at the mouth of a steep mountain canyon.

Deltas

When a stream enters the standing water of the sea or a lake, its speed drops rapidly, its ability to transport sediment decreases markedly, and it deposits its solid load. As we learned in Chapter 6, a sedimentary deposit that forms where a stream flows into standing water is a **delta**, so named because the Nile delta has a crudely triangular shape that resembles the Greek letter delta (Δ) (Fig. 10.27), and all other deltas take their name from the Nile delta.

Deltas built by streams transporting coarse sediments are of two types. A gravel-rich delta that is formed where an alluvial fan is building outward into a standing body of water (a *fan delta*, Fig. 10.28A) typically is built adjacent to a mountain front. Stream-channel, sheet-flood, and debris-flow sediments characteristic of alluvial fans form the upper parts of such deltas. The sediments in a fan delta show evidence of highly variable currents and abrupt changes of facies.

A *braid delta* is a coarse-grained delta constructed by a braided stream that builds outward into a standing body of water (Fig. 10.28A). Its upper part displays features characteristic of braided streams. Braid deltas are especially conspicuous where braided glacial meltwater streams flow into lakes or the sea.

As a stream enters standing water, particles of the bed load are deposited first, in order of decreasing weight. Then the suspended sediments settle out. A layer representing one depositional event (such as a single flood) therefore grades from coarse sediment at the stream mouth to finer sediment offshore. The accumulation of many successive layers creates an embankment that grows progressively outward (Fig. 10.28A). The coarse, thick, steeply sloping part of a depositional layer in a delta is a *foreset layer*. Traced away from the shore, the same layer changes facies, becoming rapidly thinner and finer, and covering the bottom over a wide area. This part of a depositional layer in a delta is called a *bottomset layer*.

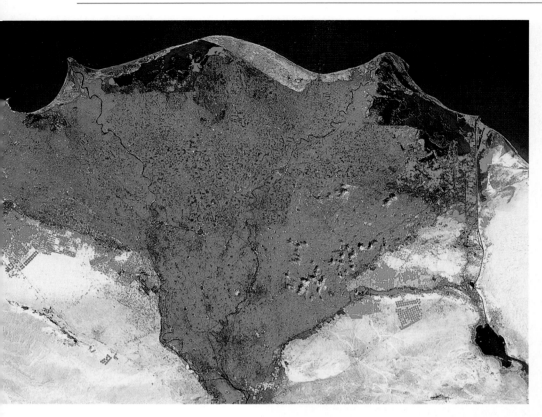

Figure 10.27 Nile Delta The delta of the Nile River, as seen from an orbiting spacecraft. Crops covering the well-watered agricultural land of the delta surface appear red in this false-color image and contrast sharply with the barren desert landscape beyond.

As deposition proceeds and the delta builds outward, the coarse foreset layers progressively overlap the bottomset layers. Thus, the stream gradually extends its channel outward over the growing delta. Coarse channel deposits and finer sediment deposited between channels, together called topset layers, overlie the foreset layers in a delta (Fig. 10.28A).

Many of the world's largest streams, among them the Ganges-Brahmaputra, the Huang He, the Amazon, and the Mississippi, have built massive deltas at their mouths. Each delta has its own peculiarities, determined by such factors as the stream's discharge, the character and volume of its load, the shape of the bedrock coastline near the delta, the offshore submarine topography, and the intensities and directions of currents and waves.

Most major streams transport large quantities of fine suspended sediment, the bulk of which is carried seaward as the fresh stream water overrides denser saltwater at the coast. The fine sediment then settles out to form a gently sloping delta front.

Where strong currents and wave action redistribute sediment as quickly as it reaches the coast, delta formation may be inhibited. However, if the rate of sediment supply exceeds the rate of coastal erosion, then a delta will be built seaward. The Mississippi River delivers a huge sediment load to the margin of the Gulf of Mexico each year. Much of the load is deposited along and around numerous *distributaries*, which are long fingerlike channels that branch from the main channel. The coarsest sediment is found along the channels. Finer sediment reaches the front of the delta and also accumulates between distributary channels during floods. The result is a complex intertonguing of facies and an intricate delta margin (Fig. 10.28B).

DRAINAGE SYSTEMS

To city dwellers used to a structured, artificial environment, nature sometimes seems to lack any obvious organization or pattern. But organization does exist, even though we may have to look for it. Streams are not distributed randomly across the landscape but are organized into intricate drainage systems, the geometry of which can provide clues about the underlying geology and the evolution of continents.

Drainage Basins and Divides

Every stream is surrounded by its **drainage basin**, the total area that contributes water to the stream. The line that separates adjacent drainage basins is a **divide**. Drainage basins range in size from less than a square kilometer to vast areas of near-continental dimensions (Fig. 10.23). The huge drainage basin of the Mississippi River encompasses an area that exceeds 40 percent of the area of the conterminous United States (Fig. 10.29). Not surprisingly, the area of any drainage basin is related to both the length and the mean annual discharge of the stream that drains the basin.

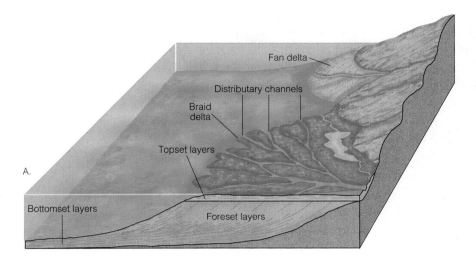

A.

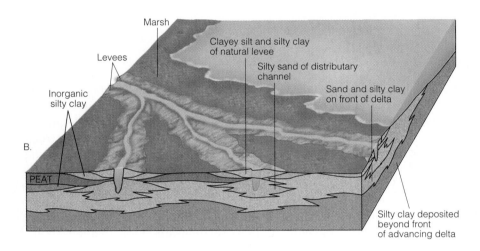

B.

Figure 10.28 Features of Deltas Main features of deltas. A. A braid delta built into a lake displays topset, foreset, and bottomset layers. A nearby fan delta is an alluvial fan that is building out into the body of water. B. Part of a large delta built into the sea shows the intertonguing relationship of coarse channel deposits and finer sediments deposited on the delta front and elsewhere.

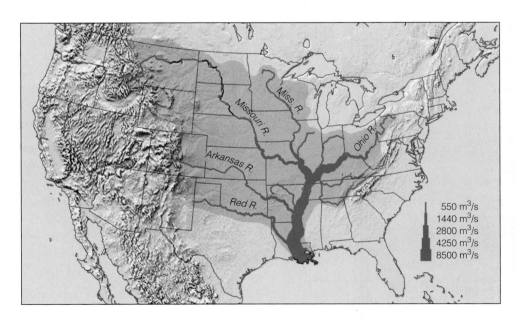

Figure 10.29 Drainage Basin of the Mississippi River The drainage basin of the Mississippi River encompasses a major portion of the central United States. In this diagram, the width of the river and its major tributaries reflect discharge values.

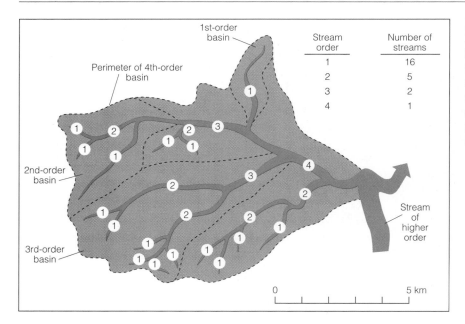

Stream order	Number of streams
1	16
2	5
3	2
4	1

Figure 10.30 Stream Orders Drainage basin of a fourth-order stream in the Appalachian region showing tributary channels numbered according to stream order. Both the number of tributaries and their length are related to stream order. Basins are classified according to the order of the largest stream they contain.

Stream Order

The arrangement and dimensions of streams in a drainage basin tend to be orderly. This can be verified by examining a stream system on a map and numbering the observed stream segments according to their position, or order, in the system. The smallest segments lack tributaries and are classified as first-order streams. Where two first-order streams join they form a second-order stream, which has only first-order tributaries. Third-order streams are formed by the joining of two second-order streams and can have first-order and second-order tributaries, and so forth for successively higher stream orders (Fig. 10.30).

If for any drainage basin, the number of streams assigned to each order is counted, we quickly see that the sums increase with decreasing stream order. In its pattern of a main stream joined by increasingly greater numbers of successively smaller tributaries, a stream system resembles a tree with its trunk and increasing numbers of successively smaller branches. This orderliness is like that inherent in a stream's long profile, in which gradient decreases systematically from head to mouth, while discharge, velocity, and channel dimensions increase. All these relationships imply that in response to a given quantity of runoff, stream systems develop with just the size and spacing required to move the water off each part of the land with greatest efficiency.

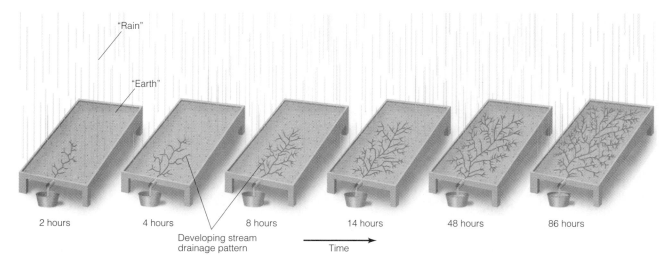

Figure 10.31 Evolution of a Stream Evolution of a drainage network in an experimental rainfall-erosion container. The initial channel, which directed runoff toward the lower end of the container, grew headward and developed new tributaries as it spread to encompass the drainage basin.

Evolution of Drainage

A stream system can develop quickly, as indicated by the following example. In August 1959, an earthquake raised and tilted the bed of Hebgen Lake, near West Yellowstone, Montana, exposing a large area of silt and sand. With the first rain, small stream systems began to develop on the newly exposed lakebed. Sample areas were surveyed and mapped one and two years after the earthquake. The results showed the same basic geometry that characterizes much larger and older stream systems. The small, newly formed valleys, together with the areas between them, were disposing of the available runoff in a highly systematic way, and all within a period of two years after the surface had emerged from beneath the lake.

A similar result has been obtained experimentally using sprinkler systems that subject large sediment-filled containers to artificial rainfall (Fig. 10.31). As erosion proceeds, the stream network spreads upslope, eventually encompassing the entire basin. In the course of drainage evolution, both the length of each channel and the number of tributaries increase.

As a system of drainage develops, details of its pattern change. Streams acquire new tributaries, and some old tributaries are lost as a result of *stream capture*, which is the interception and diversion of one stream by another stream that is expanding its basin by erosion in the headward direction. When stream capture occurs, some stream segments are lengthened and others are shortened. Just as the hydraulic factors within a stream are constantly adjusting to changes, so too is the drainage system constantly changing and adjusting as it grows. Like a stream channel, a drainage system is a dynamic system tending toward a condition of equilibrium.

Drainage Patterns, Rock Structure, and Stream History

One of the best ways to view stream systems on the landscape is from the window of an orbiting spacecraft; next best is from an airplane. From an altitude of 8 or 9 km, stream patterns can tell us a great deal about geologic structure and landscape history.

The ease with which a formation is eroded by streams depends chiefly on its composition and structure. The course a stream takes across the land therefore bears a close relationship to these factors. Thus, drainage patterns we can see from an airplane or those that can be traced on a topographic map give us information about underlying rock type and structure. Figure 10.32 shows some of the most common drainage patterns and the geologic factors that control them. An experienced geologist can use these drainage patterns to infer such things as rock type, the direction and degree of slope of an inclined rock

Dendritic — *Branching of channels ("treelike") in many directions.* Common in massive rock and in flat-lying strata. In such situations, differences in rock resistance are so slight that their control of the directions in which valleys grow headward is negligible.

Parallel — *Parallel or subparallel channels that have formed on sloping surfaces underlain by homogeneous rocks.* Parallel rills, gullies, or channels are often seen on freshly exposed highway cuts or excavations having gentle slopes.

Radial — *Channels radiate out, like the spokes of a wheel, from a topographically high area,* such as a dome or a volcanic cone.

Rectangular — *Channel systems marked by right-angle bends.* Generally results from the presence of joints and fractures in massive rocks, or foliation and fractures in metamorphic rocks. Such structures, with their cross-cutting patterns, have guided the directions of valleys.

Trellised — *Rectangular arrangement of channels in which principal tributary streams are parallel and very long,* like vines trained on a trellis. This pattern is common in areas where the outcropping edges of folded sedimentary rocks, both weak and resistant, form long, nearly parallel belts.

Annular — *Streams follow paths that are segments of circles that ring a dissected dome or basin where erosion has exposed successive belts of rock of varying degrees of erodibility.*

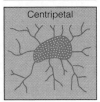
Centripetal — *Streams converge toward a central depression,* such as a volcanic crater or caldera, a structural basin, a breached dome, or a basin created by dissolution of carbonate rock.

Figure 10.32 Stream Patterns Some common stream patterns and their relationship to rock type and structure.

unit, the manner in which the rocks are folded or offset, and the orientation and spacing of joints.

A close relationship between streams and the rock units across which they flow can provide important insights regarding the structure and geologic history of an area. Geologists classify such relationships into several categories that reflect distinctive stream histories (Fig. 10.33).

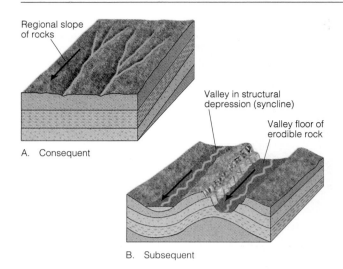

A. Consequent

B. Subsequent

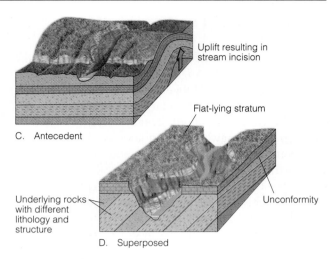

C. Antecedent

D. Superposed

Figure 10.33 Relationship Between Streams and Geology The relationship of streams to geology provides information about the structural history of an area. A. The course of a *consequent stream* is determined by the slope of land surface. B. A *subsequent stream* occupies belts of weak rock having taken a course determined by geologic structure.

C. An *antecedent stream* maintains its course across rocks that have been raised across its path; typically, it occupies a gorge that crosses a structural ridge rather than taking a course around the obstruction. D. A *superposed stream* has cut down through strata until its channel lies in underlying rock of different lithology or structure; the initial course of the stream was not determined by the rocks across which it is now flowing.

SUMMARY

1. Streams are part of the hydrologic cycle and the chief means by which water returns from the land to the sea. They help shape the Earth's surface and transport sediment to the oceans.

2. Usually, a stream's long profile decreases in gradient downstream.

3. The discharge of a stream at any place along its course is equal to the product of its cross-sectional area and its average velocity there. When a stream deepens because of an increase in discharge, its velocity also increases.

4. As discharge increases downstream, stream width and depth increase, and velocity increases slightly.

5. Streams experiencing large floods are capable of transporting large loads and moving large boulders. Exceptional floods can do a great deal of geologic work, but they have a low recurrence interval.

6. World sea level constitutes the base level for most stream systems. A local base level, such as a lake, may temporarily halt downward erosion upstream.

7. Straight channels are rare. Meandering channels form where gradients are low and the load is fine-grained. Braided patterns develop in streams with highly variable discharge and a large load to transport.

8. Stream load is the sum of bed load, suspended load, and dissolved load. Bed load is usually a small fraction of the total load of a stream. Most suspended load is derived from erosion of fine-grained regolith or from streambanks. Streams that receive large contributions of underground water commonly have higher dissolved loads than those deriving their discharge principally from surface runoff.

9. Sediment size decreases downstream because of sorting and abrasion of particles. The composition of a stream's load

changes downstream as sediments of different compositions are introduced.

10. Sediment yield is influenced by rock type and structure, climate, and topography. The greatest sediment yields are recorded in mountainous terrain with steep slopes and abundant runoff, and in small basins that are transitional from grassland to desert conditions. In moist climates, vegetation anchors the surface, thereby inhibiting erosion.

11. During floods, streams overflow their banks and construct natural levees, which grade laterally into silt and clay deposited on the floodplain. Terraces are due to the abandonment of a floodplain as a stream erodes downward.

12. Alluvial fans are constructed where a stream experiences a sudden decrease in gradient. The area of a fan is closely related to the size of the area upstream that supplies sediment to the fan.

13. A delta forms where a stream enters a body of standing water and loses its ability to transport sediment. The shape of a delta reflects the balance between sedimentation and erosion along the shore.

14. A drainage basin encompasses the area supplying water to the stream system that drains the basin. Its area is related to the stream's length and annual discharge.

15. Stream systems possess an inherent orderliness, with the number of stream segments increasing with decreasing stream order.

16. Drainage patterns are related to underlying rock type and structure, and often can reveal information about a stream's history.

THE LANGUAGE OF GEOLOGY

alluvial fan (p. 272)
alluvium (p. 266)

base level (p. 258)
bed load (p. 266)
braided stream (p. 264)

channel (p. 252)

delta (p. 273)
discharge (p. 252)

dissolved load (p. 266)
divide (p. 274)
drainage basin (p. 274)

floodplain (p. 271)

gradient (p. 253)

laminar flow (p. 266)
load (p. 252)
long profile (p. 253)

meander (p. 262)

natural levee (p. 271)

overland flow (p. 254)

placer (p. 267)

runoff (p. 254)

saltation (p. 266)

sheet erosion (p. 266)
stream (p. 252)
streamflow (p. 254)
suspended load (p. 266)

terrace (p. 272)
tributary (p. 255)
turbulent flow (p. 266)

QUESTIONS FOR REVIEW

1. What evidence leads us to think that streams must be a major force in shaping the Earth's landscapes?

2. Describe how overland flow differs from streamflow.

3. Why does vegetation decrease the effectiveness of sheet erosion?

4. How do a stream's dimensions (depth, width) and velocity adjust in response to changes in discharge?

5. What controlling factors would have to change in order for a braided channel system to change to a meandering system?

6. Why is it that stream velocity generally increases downstream, despite a decrease in stream gradient?

7. When a large dam is built across a stream to create a reservoir upstream, the channel immediately below the dam commonly experiences erosion. Why should this erosion occur?

8. What is meant by a "200-year flood"? What is the likely effect of such a flood on landscape evolution compared to the cumulative effect of annual floods?

9. What factors in a stream would have to change to cause fine gravel being moved as bed load to be transported as suspended load? In what way would the rate of forward movement of such particles change?

10. Why, if velocity increases downstream, does the maximum size of bed load particles typically decrease in size in that direction?

11. How does increasing urbanization affect the amount of sediment eroded from a drainage basin, and why?

12. How might internal stratification and sedimentary characteristics permit you to distinguish between a delta and an alluvial fan that are preserved in the stratigraphic record?

For an interactive case abstract, virtual tours, activities, and additional learning resources, go to
GEOSCIENCES TODAY: **www.wiley.com/college/skinner**

The 1993 Floods on the Mississippi and Missouri Rivers

During the spring and summer months of 1993, record rains fell across the Upper Mississippi River Basin, and parts of nine Midwestern and Great Plains states were under water! A total of 532 counties in the Midwest (including all of the counties in Iowa) were considered federal disaster areas as a result of the flood.

OBJECTIVE: The primary goals of the case are to determine what lessons have been learned from the 1993 Midwest floods; how U.S. government agencies provide disaster support; and how new technologies can aid flood forecasters and planners in anticipating future floods.

The Human Dimension: What role did the human-altered landscape play in exacerbating the flood, and what lessons were learned after the floodwaters receded?

Questions to Explore:

1. How do scientists define and characterize floods?

2. What were the climate conditions in the U.S. Midwest during the spring and summer months of 1993?

3. How can geographic computer technologies such as remote sensing, GPS, GIS, and GVIS aid in modeling flood hydrology?

11

Measuring
Our Earth

Understanding
Our Environment

GeoMedia

Downtown Tucson, Arizona, with the Santa Catalina Mountains in the rear. Most of Tucson is supplied by groundwater that is pumped from the alluvial sediments on which the city is built. The demand for water now exceeds the capacity of the groundwater reservoir.

Groundwater

Water in Tucson

Tucson is home to 750,000 people and is growing larger every year. People flock to Tucson to enjoy southern Arizona's warm, dry climate, to admire the rugged scenery of the surrounding mountains, and to become involved in its pleasant, laid-back lifestyle. But Tucson has a nasty little problem, a troublesome cloud on the future horizon. Tucson is short of water.

Tucson is located in the Basin and Range Province, a physiographic region that stretches from Idaho to northern Mexico. The province is characterized by steep-sided mountain ranges that run on a roughly north-south orientation, interspersed with broad, flat valleys. The valleys are basins filled with sediment derived from the erosional debris shed from the adjacent mountains.

Tucson is located in a valley that is surrounded by four mountain ranges, one on each side—the Santa Catalinas to the north, the Rincons to the east, the Tucsons to the west, and the Santa Ritas to the south.

The main river draining the region is the Santa Cruz River, and downtown Tucson is built astride the Santa Cruz.

The first people who lived in the Tucson area got their water from the Santa Cruz River or from springs along the edge of the mountains. When Europeans arrived—first Spanish missionaries, then, in the nineteenth century, a few farmers and army personnel—the Santa Cruz was a flowing river, fish were present, and water birds nested along its banks. By the late years of the nineteenth century, surface water flow was insufficient and people were sinking wells in order to tap the groundwater. The great thickness of sediments that lie beneath Tucson are like a sponge—a huge aquifer full of potable water. What has now become clear, however, is that the water under Tucson is mostly old water; water that is thousands of years old and that flowed in and saturated the basin when the climate was wetter than today. Now the wells under Tucson are pumping this old water at a rate greatly in excess of the replacement rate by today's streams and snowmelt from the surrounding mountains, The water table has dropped, springs have dried up, the Santa Cruz rarely flows, and the fish are gone. If pumping continues unabated, the potable groundwater will eventually be gone too.

Tucsonians have another option. They can bring water from the Colorado River that runs across the northern border of the state. The plan to do so is called the Central Arizona Project, or CAP. But CAP water is a bit turbid and doesn't have the sweet taste of local well water, so Tucsonians are reluctant to use it. Perhaps there are intermediate options such as using CAP water to recharge the Tucson aquifer, but so far a solution that a majority find acceptable has not been found.

The Tucson water problem is monitored closely by city, state, and federal geologists. How it will play out is still not clear. But the underlying message is clear: the warm, dry climate that draws people to Tucson is the very reason that water is scarce—or, to use an old saying, Tucsonians can't have their cake and eat it too.

WATER IN THE GROUND

Access to water, whether from streams, lakes, springs, and direct rainfall or from underground, is a vital human need. Most early cities and towns were founded close to streams that would provide a reliable source of water. As the population of these towns and cities grew, the streams often became insufficient. People then either resorted to bringing water from a more distant source through canals, or they dug wells to obtain water from underground

As society has become increasingly more populous and industrialized, communities have generated ever larger amounts of human and industrial wastes, a good deal of which has inevitably found its way into the very water that people must rely on for their existence. In many places water is dwindling in both quantity and quality, creating important questions for the communities involved: Will there be enough clean water to sustain future needs? Is the quality adequate for the uses to which we put this water? Is the water being used with a minimum of waste?

Origin of Groundwater

Less than 1 percent of the water on the Earth is **groundwater**, defined as all the water in the ground (below the unsaturated zone) occupying the pore spaces within bedrock and regolith. Although the volume of groundwater sounds small, it is 40 times larger than the volume of all the water in fresh-water lakes or flowing in streams and nearly a third as large as the water contained in all the world's glaciers and polar ice.

Most groundwater originates as rainfall. Rainwater that soaks into the ground and moves down to the saturated zone becomes part of the groundwater system and moves slowly toward the ocean, either directly through the ground or by flowing out onto the surface and joining streams (Fig. 2.17B).

That groundwater comes from rain was finally established on a quantitative basis in the seventeenth century, when Pierre Perrault, a French physicist, measured the mean annual rainfall for part of the drainage basin of the Seine River in eastern France and the mean annual stream runoff in the same basin area. After estimating the loss by evaporation, Perrault concluded that the difference between the amounts of rainfall and runoff was ample enough, over a period of years, to account for the amount of water in the ground.

Depth of Groundwater

Water is present everywhere beneath the land surface, but more than half of all groundwater, including most of what is usable, occurs within about 750 m of the Earth's surface. The volume of water in this zone is estimated to be equivalent to a layer of water approximately 55 m deep spread over the world's land areas. Below a depth of about 750 m, the amount of groundwater gradually, though irregularly, diminishes. Holes drilled for oil have found water as deep as 9.4 km, and one deep experimental hole drilled on the Kola Peninsula by Russian scientists encountered water at more than 11 km. However, even though water may be present in crustal rocks at such depths, the pressure exerted by overlying rocks is so high and openings in rocks are so small that it is unlikely that much water is present.

The Water Table

Much of what we know about the occurrence of groundwater has been learned from the accumulated experience of generations of people who have dug or drilled millions of wells. This experience tells us that a hole penetrating the ground ordinarily passes first through a layer of moist soil and then into a zone in which open spaces in regolith or bedrock are filled mainly with air (Fig. 11.1). This is the **zone of aeration** (also called the *unsaturated zone*, for although water may be present, it does not saturate the ground). The hole then enters the saturated zone, a zone in which all openings are filled with water. We call the upper surface of the **saturated zone** the **water table**. Normally, the water table slopes toward the nearest stream or lake, but in deserts, for example, it may lie far underground. Just as water is present at some depth everywhere beneath the land surface, so too is the water table always present.

In fine-grained sediment, a narrow fringe as much as 60 cm thick immediately above the water table is kept wet by capillary attraction that temporarily holds water above the water table a small distance into the zone of aeration (the capillary fringe shown in Fig. 11.1). *Capillary attraction* is the adhesive force between a liquid and a solid that causes water to be drawn into small openings. This is the same force that draws ink through blotting paper and kerosene through the wick of a lamp.

In humid regions, the water table is a subdued imitation of the land surface above it (Fig. 11.1). It is high beneath hills and low beneath valleys because water tends to move toward low points in the topography. If all rainfall were to cease, the water table would slowly flatten and gradually approach the levels of the valleys; water seepage into the ground would diminish and then cease, and streams would dry up as the water table fell beneath valleys. In times of drought, when rain may not fall for several weeks or even months, we can sense the flattening of the water table in the drying up of wells. When a well becomes dry, we know that the water table has dropped to a level below the bottom of the well. It is repeated rainfall, dousing the ground with fresh supplies of water, that maintains the water table at a normal level.

Whatever its depth, the water table is a significant surface, for it represents the upper limit of all readily usable groundwater. For this reason, a major aim of groundwater geologists and well drillers alike is to determine the depth and shape of the water table.

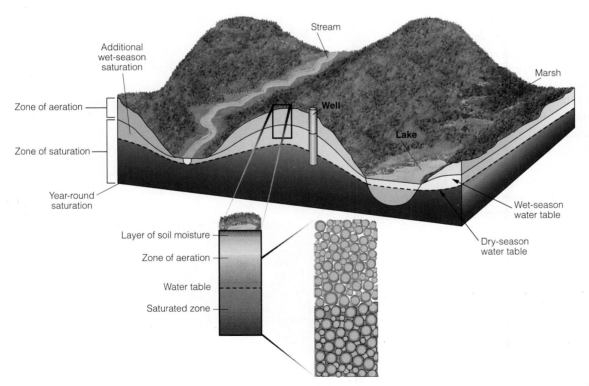

Figure 11.1 Water In the Ground In a typical groundwater system, the water table separates the zone of aeration from the saturated zone and fluctuates in level with seasonal changes in precipitation. Corresponding fluctuations are seen in the water level in wells that penetrate the water table. Lakes, marshes, and streams occur where the water table intersects the land surface. A layer of moisture coincides with the surface soil, and a thin capillary fringe lies immediately above the water table. In shape, the water table is a subdued imitation of the overlying land surface.

HOW GROUNDWATER MOVES

Groundwater operates continuously as a small but integral part of the hydrologic cycle (Fig. 2.17B). Water, evaporated mainly from the oceans and falling on the land as rain, seeps into the ground and enters the groundwater reservoir. Some of this slowly moving underground water reaches stream channels and contributes to the water they carry to the ocean, and the cycle continues.

Most of the groundwater within a few hundred meters of the surface is in motion. Unlike the swift flow of rivers, however, which is measurable in kilometers per hour, groundwater moves so slowly that velocities are expressed in centimeters per day or meters per year. The reason for this contrast is simple. Whereas the water of a stream flows unimpeded through an open channel, groundwater must move through small, constricted passages, often along a tortuous route. Therefore, the flow of groundwater to a large degree depends on the nature of the rock or sediment through which the water moves.

Porosity and Permeability

Porosity is the percentage of the total volume of a body of regolith or bedrock that consists of open spaces, called *pores*. It is porosity that determines the amount of water that a given volume of regolith or bedrock can contain.

The porosity of sediments is affected by the sizes and shapes of the rock particles, as well as by the compactness of their arrangement and by the weight of any overlying material (Fig. 11.2A and B). In some well-sorted sands and gravels that are not deeply buried, the porosity may reach 20 percent, while some very porous clays have porosities as high as 60 percent.

The porosity of a sedimentary rock is affected not only by the sorting and arrangement of the particles, but also by the extent to which the pores become filled with the cement that holds the particles together (Fig. 11.2C). The porosity of igneous and metamorphic rocks, on the other hand, generally is low. However, if such rocks have many joints and fractures, or, if a lava is vesicular, the porosity will be higher.

Permeability is a measure of how easily a solid allows fluids to pass through it. A rock of very low porosity is also likely to have low permeability. However, a high porosity does not necessarily mean a correspondingly high permeability. The sizes of pores, how well they are connected, how crooked or straight a path water must follow as it travels through porous material, all determine the permeability of a rock or sediment.

An example of a sediment with high porosity and low permeability is clay, the particles of which have diameters of less than 0.005 mm (Table 6.1). Although clay may

have a very high porosity, because the pores are very small the permeability is low.

By contrast, in a sediment with grains at least as large as sand (grain diameters of 0.06 to 2 mm), the pores commonly are wide and the water in the pores is free to move. Such sediment is permeable. As the diam-

A.

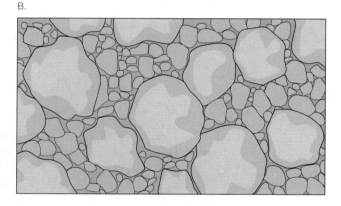

B.

C.

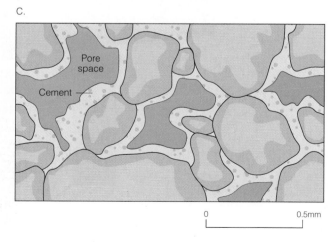

Pore space

Cement

0   0.5mm

Figure 11.2 Porosity in Sediments Porosity in different sediments. A. A porosity of 30 percent in a reasonably well-sorted sediment. B. A porosity of 15 percent in a poorly sorted sediment in which fine grains fill spaces between larger grains. C. Reduction in porosity in an otherwise very porous sediment due to cement that binds grains together.

eters of the pores increase, permeability increases. Gravel, with very large pores, is more permeable than sand and can yield large volumes of water to wells.

Movement in the Zone of Aeration

Water from a rain shower soaks into the soil, which usually contains clay resulting from the chemical weathering of bedrock. Because of these fine clay particles, the soil is generally less permeable than underlying coarser regolith or rock. The low permeability and the fine clay particles cause part of the water to be retained in the soil by forces of molecular attraction. This is the layer of soil moisture shown in Figure 11.1. Some of this moisture evaporates directly into the air, but much of it is absorbed by the roots of plants that later return it to the atmosphere through transpiration (Fig. 2.18).

Because of the pull of gravity, water that cannot be held in the soil by molecular attraction seeps downward until it reaches the water table. With every rainfall, more water enters the ground, but apart from soil moisture and the capillary fringe, the zone of aeration is likely to be nearly dry between rains.

Movement in the Saturated Zone

The movement of groundwater in the saturated zone, termed **percolation**, is similar to the flow of water when a saturated sponge is squeezed gently. Water moves slowly by percolation through very small pores along parallel, threadlike paths. Movement is easiest through the central parts of the spaces but diminishes to zero immediately adjacent to the sides of each space because there the water is retarded by the tendency of water to wet the surfaces of mineral grains.

Responding to gravity, water percolates from areas where the water table is high toward areas where it is lowest. In other words, it generally percolates toward surface streams or lakes (Fig. 11.3). Much of it flows along innumerable long, curving paths that go deep through the ground. Some of the deeper paths turn upward and enter the stream or lake from beneath. This upward flow is possible because water tends to flow toward points where pressure is least. However, most of the groundwater entering a stream travels along shallow paths not far beneath the water table.

Recharge and Discharge Areas

Replenishment, or **recharge**, of groundwater occurs as rainfall and snowmelt enter the ground in **recharge areas**, which are areas of the landscape where precipitation seeps downward beneath the surface and reaches the saturated zone (Fig. 11.4). The water continues to move slowly toward **discharge areas**, which are areas where subsurface water is discharged to streams or to lakes, ponds, or swamps. The surface extent of recharge areas is invariably larger than that of discharge areas.

The time water takes to move through the ground from a recharge area to the nearest discharge area depends on rates of movement and on the travel distance. It may take only a few days, or possibly thousands of years in cases where water moves through the deeper parts of a groundwater body (Fig. 11.4).

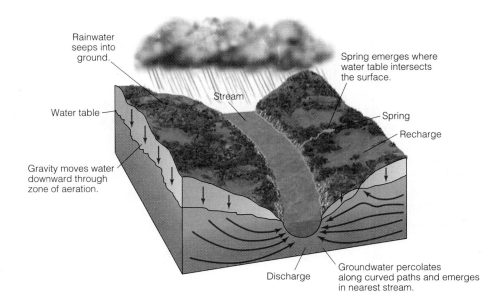

Rainwater seeps into ground.

Spring emerges where water table intersects the surface.

Stream

Water table

Spring

Recharge

Gravity moves water downward through zone of aeration.

Discharge

Groundwater percolates along curved paths and emerges in nearest stream.

Figure 11.3 Groundwater Flow Paths of groundwater flow in a humid region in uniformly permeable rock or sediment. Long, curved arrows represent only a few of many possible paths. Springs are located where the water table intersects the land surface.

Groundwater

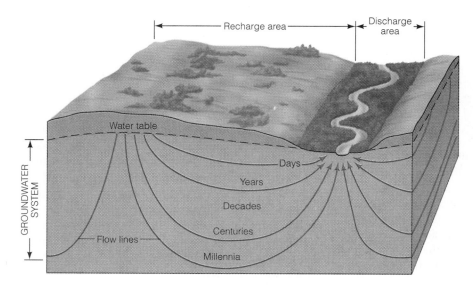

Figure 11.4 Recharge and Discharge Distribution of recharge and discharge areas in a humid landscape. The time required for groundwater to reach the discharge area from the recharge area depends on the path and distance of travel. Downward and upward percolation is faster and more direct in the most permeable pathways.

Measuring Our Earth

How Fast Does Groundwater Flow?

As shown in Figure B11.1, the slope of the water table can be determined by measuring the difference in altitude of two points (h_1 and h_2) on the water table and dividing this figure by the horizontal distance (l) between the points. The resulting slope value is generally referred to as the **hydraulic gradient**, and the velocity of groundwater (V) is proportional to the hydraulic gradient:

$$V \propto \frac{h_1 - h_2}{l}$$

In 1856, Henri Darcy, a French engineer, concluded that the velocity of groundwater must be related not only to the slope of the water table (the hydraulic gradient), but also to the permeability of the rock or sediment through which the water is flowing. He proposed an equation in which permeability, together with the acceleration due to gravity and the density and viscosity of water, is expressed as a coefficient (K). This coefficient, referred to as the *coefficient of permeability* or the *hydraulic conductivity*, is simply a measure of the ease with which water moves through a rock or sediment. The equation Darcy proposed can thus be expressed as:

$$V = \frac{K(h_1 - h_2)}{l}$$

In Chapter 10 we learned that discharge (Q) in streams varies as a function of both stream velocity (V) and cross-sectional area (A). The discharge of groundwater through a rock or sediment also depends on the velocity of flow and cross-sectional area of flow. In this case, however, the cross-sectional area is not that of an open channel but rather that

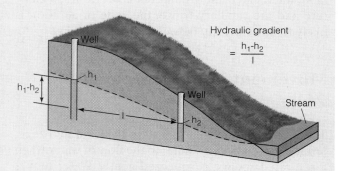

Figure B11.1 Hydraulic Gradient The hydraulic gradient is the slope of the water table. In the example illustrated, the hydraulic gradient is found by subtracting the altitude of h_2, the level of water in the downslope well, from h_1, the level of the water in the upslope well, and dividing the difference by the distance (l) between the two wells.

of an interconnected system of pores. We can express discharge by using an equation we learned in Chapter 10: $Q = AV$. If we substitute the value for V given in equation 1 in this equation from Chapter 10, we arrive at a new equation,

$$Q = \frac{AK(h_1 - h_2)}{l}$$

that is known as **Darcy's Law**. If we take the cross-sectional area (A) as constant for any given situation, by measuring any two of the remaining three variables [discharge (Q), coefficient of permeability (K), and hydraulic gradient ($h_1 - h_2$)/l], we can calculate the value of the third.

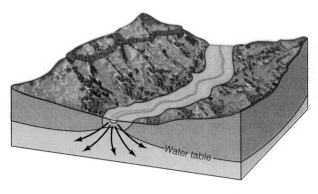

Figure 11.5 Water Loss from Streams In arid regions, direct recharge is minimal and the water table lies below the bed of a river. During times of low flow, large throughflowing streams and intermittent streams lose water, which seeps downward to resupply groundwater in the saturated zone.

In humid regions, recharge areas encompass nearly all the landscape beyond streams and their adjacent floodplains (Fig. 11.4). In more arid regions, recharge occurs mainly in mountains and in the alluvial fans that border them. In such regions, recharge also occurs along the channels of major streams that are underlain by permeable alluvium; the water leaks downward through the alluvium and recharges the groundwater (Fig. 11.5).

Discharge and Velocity

Groundwater does not move everywhere at a constant rate. We can demonstrate this by injecting colored dye at some point in a recharge area and measuring the time it takes for the dye to appear in nearby springs or wells. Experiments conducted with materials of uniform permeability have shown that the velocity of groundwater flow increases as the slope of the water table increases. In other words, the steeper the slope, the faster the water moves.

Because percolating groundwater encounters a large amount of frictional resistance, flow rates tend to be very slow. Normally, velocities range between half a meter a day and several meters a year. The highest rate yet measured in the United States, in exceptionally permeable material, was only about 250 m/yr. For a discussion of its factors that control the rate of groundwater flow, see the essay *How Fast Does Groundwater Flow?*

SPRINGS AND WELLS

People generally obtain supplies of groundwater either from springs or by excavating wells that reach the saturated zone underground.

Springs

A **spring** is a flow of groundwater emerging naturally at the ground surface. The simplest kind of spring is one that issues from a place where the land surface intersects the water table (Fig. 11.3). Small springs are found in all kinds of rocks, but almost all large springs issue from lava, limestone, or gravel.

A vertical or horizontal change in permeability is a common reason for the localization of springs (Fig. 11.6). Often this change involves the presence of an **aquiclude**, a body of impermeable or distinctly less permeable

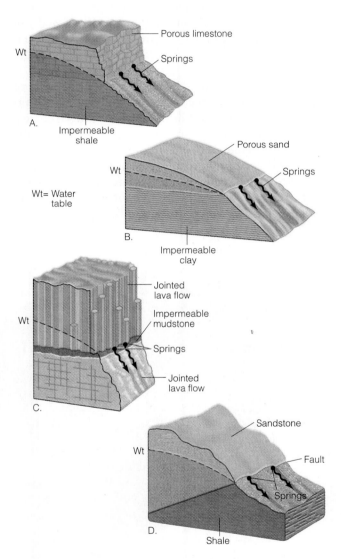

Figure 11.6 What Causes Springs? Examples of springs formed in different geologic conditions. A. Two springs discharge water at the contact between a porous limestone and an underlying impermeable shale. B. Springs lie at the contact between a porous sandy unit and an underlying impermeable clay. C. Springs issue along the contact between a highly jointed lava flow and an underlying impermeable mudstone. D. Springs issue along the trace of a fault where it intersects the land surface.

rock adjacent to a permeable one (Fig. 11.6A). If sand overlies a relatively impermeable clay aquiclude, water percolating through the sand will flow laterally and will emerge as a spring where the stratigraphic boundary between the sand and the aquiclude intersects the land surface, as along the side of a valley or a coastal cliff (Fig. 11.6B). Springs may also issue from lava flows, especially where a jointed lava bed overlies an aquiclude (Fig. 11.6C), or along the trace of a fault (Fig. 11.6D).

Wells

A well will supply water if it intersects the water table (Fig. 11.1). Figure 11.7 shows that a shallow well can become dry at times when the water table is low, whereas a nearby deeper well may yield water throughout the year.

When water is pumped from a new well, the rate of withdrawal initially exceeds the rate of local groundwater flow. This imbalance in flow rates creates a conical depression in the water table immediately surrounding the well called a **cone of depression** (Fig. 11.7). The locally steepened slope of the water table increases the flow of water to the well, consistent with Darcy's Law. Once the rate of inflow balances the rate of withdrawal, the hydraulic gradient stabilizes, but it will change if either the rate of pumping or the rate of recharge changes. In most small domestic wells the cone of depression is hardly discernible. Wells pumped for irrigation and industrial uses, however, withdraw so much water that the cone can become very wide and deep and can lower the water table in all wells of a district.

If the source of a groundwater supply is inhomogeneous rock or sediment, water yields from wells may vary considerably within short distances. For example, igneous and metamorphic rocks generally are not very permeable because the spaces between their mineral grains are extremely small and constricted. However, what is true for small samples does not necessarily apply to large masses of the same rock. Many massive igneous and metamorphic bodies contain numerous fissures, joints, and other openings that permit free circulation of groundwater. A well reaching such a conduit may produce water, whereas one that does not intersect any openings may be dry (Fig. 11.8A).

Discontinuous bodies of permeable and impermeable rock or sediment (Fig. 11.8B) result in very different water yields from wells. An impermeable layer of rock or sediment in the zone of aeration can produce a *perched water body* (a water body perched atop an aquiclude that lies above the main water table). The impermeable layer catches and holds the water reaching it from above.

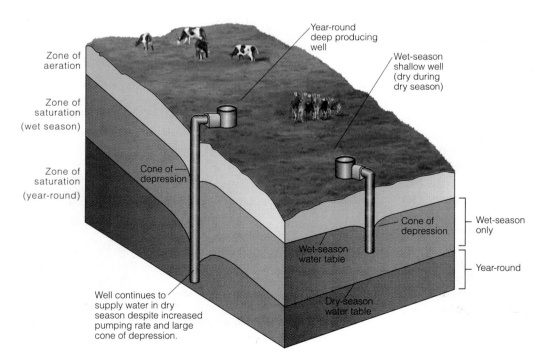

Figure 11.7 Wells: Year-Round and Seasonal Effect of seasonal changes in precipitation on the position of the water table. During the wet season, recharge is high and the water table is high, so water is present both in a shallow well and in a deeper well upslope. During the dry season, the water table falls, the hydraulic gradient decreases, and the shallow well is dry. The deeper well continues to supply water, but increased pumping during the dry season enlarges the cone of depression.

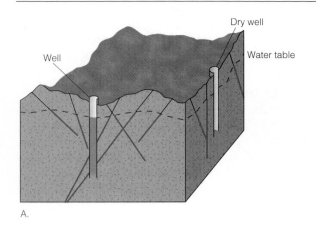

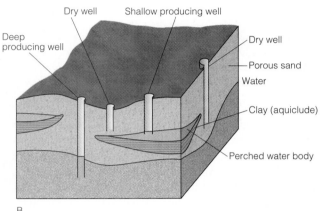

Figure 11.8 Successful and Unsuccessful Wells Yield to wells from nonhomogeneous rocks can be highly variable. A. Wells that penetrate fractures in metamorphic and igneous rocks produce water. Dry wells result if no water-bearing fractures are encountered. B. Perched water bodies above the main water table are held up by aquicludes and provide shallow sources of groundwater. Wells that miss the perched water body and do not reach the deeper water table are dry.

AQUIFERS

When we wish to find a reliable supply of groundwater, we search for an **aquifer** (Latin for water carrier), which is a body of highly permeable rock or regolith that can store water and yield sufficient quantities to supply wells. Bodies of gravel and sand generally are good aquifers, for they tend to be highly permeable and often have large dimensions. Many sandstones are also good aquifers. However, the presence of a cementing agent between grains of a sandstone reduces the diameter of the openings and thus reduces the permeability.

Unconfined and Confined Aquifers

An aquifer that is not overlain and confined by an aquiclude is called an **unconfined aquifer**. An aquifer that is bounded by aquicludes is a **confined aquifer**.

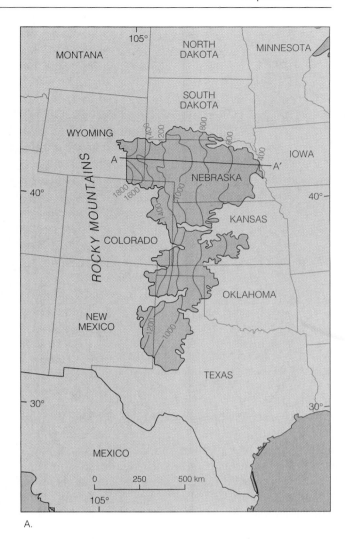

A.

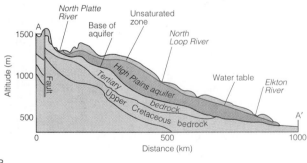

B.

Figure 11.9 High Plains Aquifer High Plains aquifer, an example of an unconfined aquifer. A. Regional extent of the aquifer and contours (in meters) on the water table. Water flow is generally east, perpendicular to the contour lines. B. Cross section along profile A-A' showing the slope of the water table and the relation of the High Plains aquifer to underlying bedrock units.

About 30 percent of the groundwater used for irrigation in the United States is obtained from the High Plains aquifer, an unconfined aquifer that lies at shallow depths beneath the High Plains (Fig. 11.9). The aquifer,

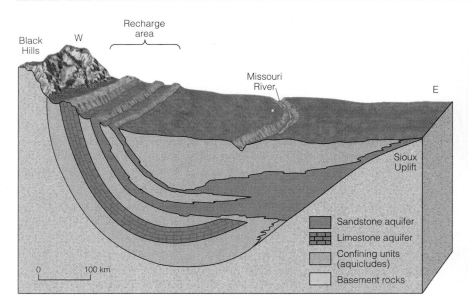

Figure 11.10 Dakota Aquifer Dakota aquifer system in South Dakota, an example of a confined aquifer. The aquifer units consist of porous sandstones and a permeable limestone that reach the surface in the recharge area as linear ridges lying beyond the central core of the Black Hills. Intervening confining layers are aquicludes. In this diagram, the vertical scale is greatly exaggerated.

which averages about 65 m thick and is tapped by about 170,000 wells, consists of a number of sandy and gravelly rock units of late Tertiary and Quaternary age. The water table at the top of this aquifer slopes gently from west to east, and water flows through the aquifer at an average rate of about 30 cm/day. Recharge comes directly from precipitation and seepage from streams.

The use of groundwater for irrigation in the High Plains was spurred by severe regional droughts in the 1930s and again in the 1950s. Annual recharge of the High Plains aquifer from precipitation is much less than the amount of water being withdrawn; the inevitable result is a long-term fall in the level of the water table. In parts of Kansas, New Mexico, and Texas, the water table has dropped so much over the past half century that the thickness of the saturated zone has declined by more than 50 percent. The resulting decreased water yield and increased pumping costs have led to major concern about the future of irrigated farming on the High Plains.

The Dakota aquifer system in South Dakota provides a good example of a confined aquifer (Fig. 11.10). It lies to the east of the Black Hills, an elongate dome of uplifted rocks around which permeable strata and bounding aquicludes form the land surface. Rain falling on the permeable units where they reach the surface recharges the groundwater, which flows down the inclined rock layers toward the east.

Artesian Systems

Water that percolates into a confined aquifer flows downward under the pull of gravity. As it flows to greater depths, the water is subjected to increasing hydrostatic pressure. If a well is drilled to the aquifer, the difference in pressure between the water table in the recharge area and the level of the well intake will cause water to rise in the well. Potentially, the water could rise to the same height as the water table in the recharge area. If the top of the well is lower in altitude than the recharge area, the water will flow out of the well without pumping. Such an aquifer is called an **artesian aquifer**, and the well is called an *artesian well* (Fig. 11.11). Similarly, a freely flowing spring supplied by an artesian aquifer is an *artesian spring*. The term *artesian* comes from a French town, Artois (called Artesium by the Romans), where artesian flow was first studied. Under unusually favorable conditions, artesian water pressure can be great enough to create fountains that rise as much as 60 m above ground level. Such wells and springs are naturally attractive, for the costs of pumping are avoided as long as the amount of recharge to the system is sufficient to maintain the necessary water pressure.

The Floridan Aquifer

A vertical sequence of highly permeable carbonate rocks along the peninsula of Florida provides an example of a complex regional aquifer system in which both confined and unconfined units are present and in which water locally reaches the surface by artesian flow (Fig. 11.12). The aquifer system is restricted mainly to middle and late Tertiary limestones, in the upper part of which are numerous caves and smaller openings that have been dissolved in the rock. The permeable beds are interconnected to varying degrees, and their permeability is at least ten times greater than that of strata above and below.

The Floridan aquifer gives rise to numerous springs. Their concentration and discharge probably exceed

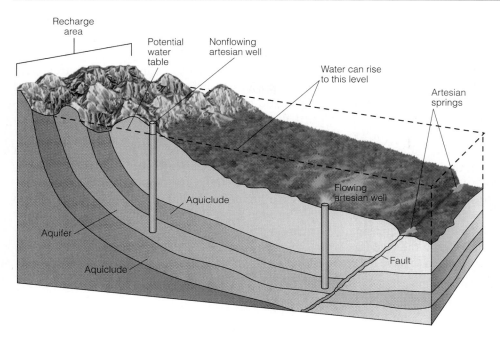

Figure 11.11 Artesian Wells Two conditions are necessary for an artesian system: a confined aquifer and sufficient water pressure in the aquifer to make the water in a well rise above the aquifer. The water in a well drilled into the aquifer rises to the height of the water table in the recharge zone, indicated by the dashed line. If this height is above the ground surface, the water will flow out of the well without being pumped.

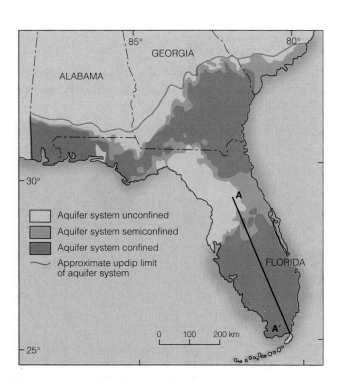

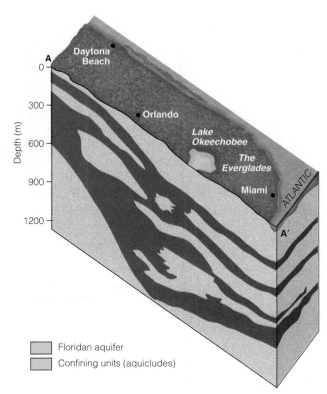

Figure 11.12 Floridan Aquifer Floridan aquifer system. The map shows the distribution of the system and areas where it is unconfined, semiconfined (overlying confining unit is less than 30 m thick), and confined (overlying confining unit is more than 30 m thick).

The cross section along line A-A' shows the relation of aquifer to overlying, underlying, and intervening aquicludes.

those of any country in the world. Most of the major springs are artesian and occur where the overlying impermeable beds have been breached and the water pressure is high enough to allow water to rise above the land surface.

The age of groundwater in the Floridan aquifer system has been determined by radiocarbon dating of carbonate molecules dissolved in the water. Most of the radiocarbon enters the ground with rain falling on the recharge area and moves through the aquifer with the groundwater. Samples were collected from wells penetrating the aquifer system along a line 133 km long and approximately parallel to the slope of the aquifer. The age of the water was found to increase with increasing distance from the recharge area. Water in the well farthest from the recharge area is calculated to have been in the ground for at least 19,000 years.

MINING GROUNDWATER

In the dry regions of western North America, where streams are few and average discharge is low, groundwater is a major source of water for human consumption. In many of these dry regions, withdrawal exceeds natural recharge. Thus, the volume of stored water is steadily diminishing. In the same way that petroleum is being steadily withdrawn from the most accessible oil pools and minerals are being mined from the most accessible rocks of the upper crust, groundwater also is being mined from the best aquifers. We regard fossil fuels and minerals as nonrenewable resources, for they form only over geologically long intervals of time. We don't often stop to think that groundwater can also be a nonrenewable resource. In some regions, natural recharge would take so long to replenish a depleted aquifer that formerly vast underground water supplies have essentially been lost to future generations. Even where the problem has been recognized and measures have been taken to stem the loss, centuries or millennia of natural recharge will be required to return aquifers to their original state.

Lowering of the Water Table

As we have seen in the case of the High Plains aquifer, when groundwater withdrawal exceeds recharge, the water table falls. This fall in level can lead to the drying up of springs and streams if the water table no longer intersects the land surface. It can also cause shallow wells to run dry and necessitate the drilling of still deeper wells. Under these conditions, the groundwater reservoir is steadily depleted while the cost of pumping water from an ever-deepening water table continues to increase.

To halt the fall of the water table, groundwater sometimes can be artificially recharged. Artificial recharge might involve, for example, spraying biodegradable liquid wastes from a food-processing or sewage-treatment plant over the land surface. The pollutants are removed by biologic processes as the liquid percolates downward through the soil, and the purified water then recharges the groundwater system. Runoff from rainstorms in urban areas can be channeled and collected in basins where it will seep into permeable strata below, raising the water table. In addition, groundwater withdrawn for nonpolluting industrial use may be pumped back into the ground through injection wells, thereby recharging the saturated zone.

Land Subsidence

The water pressure in the pores of an aquifer helps support the weight of the overlying rocks or sediments. When groundwater is withdrawn, the pressure is reduced, and the particles of the aquifer shift and settle slightly. As a result, the land surface subsides (Fig. 11.13). The amount of subsidence depends on how much the water pressure is reduced and on the thickness and compressibility of the aquifer. Such land subsidence is widespread in the southwestern United States, where withdrawal of groundwater has caused disruptions of the land surface; structural damage to buildings, roads, and

Figure 11.13 Subsidence Fissure located near Chandler Heights, Arizona, caused by subsidence of the ground due to removal of large quantities of underground water.

bridges; damage to buried cables, pipes, and drains; and an increase in areas subject to flooding.

Land subsidence can be especially damaging where water is pumped from beneath cities. A well-known example is Mexico City, built on the site of the ancient Aztec capital of Tenochtitlán, which lay in the middle of a shallow lake. As groundwater was exploited, the porous lake sediments slowly compressed, and many buildings began to shift and tilt as the land subsided. Another example is Pisa's famous Leaning Tower, which leans because of subsidence. The tower was built on unstable fine-grained floodplain sediments and began to tilt when construction began in 1174 (Fig. 11.14). The tilting increased rapidly during the present century as groundwater was withdrawn from deep aquifers. Recent strengthening of the foundation is designed to keep the tower stable in the future, but it will do so only if groundwater withdrawal is strictly controlled.

WATER QUALITY AND GROUNDWATER CONTAMINATION

Groundwater
Contamination

Citizens of modern industrialized nations take it for granted that when they turn on a faucet, water that is safe and drinkable will flow from the tap. Throughout much of the world, however, drinking water is barely adequate for human consumption. Not only do natural dissolved substances make some water unpalatable, but many water supplies have become severely contaminated by human and industrial waste products.

Chemistry of Groundwater

Analyses of many wells and springs show that the compounds dissolved in groundwater are mainly chlorides, sulfates, and bicarbonates of calcium, magnesium, sodium, potassium, and iron. We can trace these substances to the common minerals in the rocks from which they were weathered. As might be expected, the composition of groundwater varies from place to place according to the kind of rock in which the water occurs. In much of the central United States, for instance, the water is rich in calcium and magnesium bicarbonates dissolved from the local limestone and dolostone bedrock. Taking a bath in such water, termed *hard water*, can be frustrating because soap does not lather easily and a crustlike ring forms in the tub. Hard water also leads to deposition of scaly crusts in water pipes, eventually restricting water flow. By contrast, water that contains little dissolved matter and no appreciable

Figure 11.14 The Learning Tower Continues to Lean The Leaning Tower of Pisa, Italy, the tilting of which accelerated as groundwater was withdrawn from aquifers to supply the growing city.

calcium is called *soft water*. Such water is found, for example, in parts of the northwestern United States where volcanic rocks and greywacke sandstones are common. With soft water, we can easily get a nice soapy lather in the shower.

Groundwater can dissolve noxious elements from rocks it flows through, making the water unsuitable for consumption. Water circulating through sulfur-rich rocks may contain dissolved hydrogen sulfide (H_2S) that, though harmless to drink, has the disagreeable odor of rotten eggs. In some arid regions, the concentration of dissolved sulfates and chlorides is so great that the groundwater is unusually noxious. In very dry regions, groundwater moving through porous sedimentary rocks will dissolve salts that then are deposited as the water evaporates in the zone of aeration. The resulting saline soils are thereby rendered unsuitable for agriculture (see the boxed essay, *Toxic Groundwater in the San Joaquin Valley*).

Toxic Groundwater in the San Joaquin Valley

The only time most people ever hear of selenium is in their high school chemistry course when they learn that it is thirty-fourth in the periodic list of the elements. Selenium is a naturally occurring element and a necessary trace nutrient in our diet, as well as in the diet of livestock and many wild animals. However, if excessive amounts are ingested, selenium can prove toxic.

In 1983, selenium attracted national attention when the U.S. Fish and Wildlife Service reported fish kills and high incidences of mortality, birth defects, and decreased hatching rates in nesting waterfowl at the Kesterson National Wildlife Refuge in California's San Joaquin Valley (Fig. B11.2). Laboratory studies showed high concentrations of selenium in fish from Kesterson Reservoir in the wildlife refuge, while birds using the reservoir were found to have high concentrations of selenium and obvious symptoms of selenium poisoning. Federal agencies quickly began inves-

tigations to learn how these toxic levels of selenium were entering the ecological system.

Geologists have now shown that the poisoning of wildlife at Kesterson resulted from a combination of geological, hydrological, and agricultural factors. The western San Joaquin Valley is a prime agricultural area, but because it lies in a zone of arid climate, the land is irrigated. The irrigation artificially raised the water table to such a degree that a system of subsurface drains was established to remove excess water. The drainage system carries subsurface water to a surface canal that funnels the water northward along the valley to Kesterson Reservoir in the wildlife refuge.

A natural source of selenium lies in the Coast Ranges immediately to the west of the San Joaquin Valley. Rainwater falling in these mountains dissolves selenium-bearing salts from marine sedimentary strata. Surface runoff then carries the dissolved salts to broad alluvial fans in the

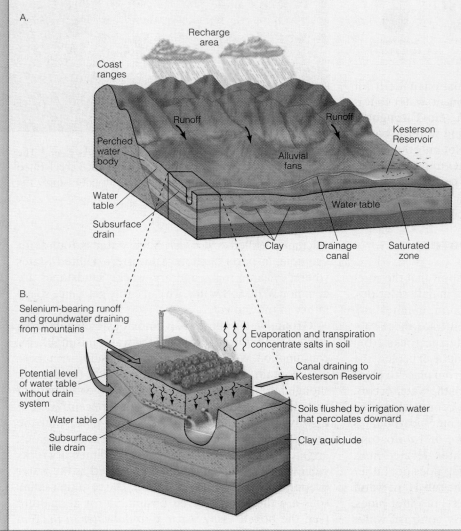

Figure B11.2 Contamination of a Reservoir Geologic setting of Kesterson Reservoir in western San Joaquin Valley, California. Runoff carries dissolved selenium to alluvial fans where irrigation water flushes it into a drainage system that concentrates it in the Kesterson Reservoir. Subsurface drains lower the water table, which otherwise would lie at shallow depths because the groundwater is perched above a clay aquiclude.

valley where the water seeps into the ground and recharges a shallow regional aquifer.

Evaporation in this arid region, where an annual rainfall of less than 250 mm is greatly exceeded by an annual evaporation rate of about 2300 mm, concentrates the salts in the soil. An irrigation system on the fan surfaces is designed to supply water to crops and also to flush salts out of the soil and into the drainage canal that leads toward Kesterson Reservoir. Because the reservoir has no outlet, the selenium is concentrated there and now has reached toxic levels.

Groundwater conditions in the western San Joaquin Valley contribute to the Kesterson Reservoir problem. Most of the shallow groundwater in this area is alkaline and slightly to highly saline. Under these conditions, selenium is very soluble and is carried with the flowing groundwater. This mobility greatly increases the ease with which selenium moves from the soils into the artificial drainage system.

A further condition is the presence of a clay layer 3 to 23 m beneath the land surface. Being impermeable, the clay layer restricts the downward percolation of selenium-bearing groundwater and produces a perched water body. It is this perched water body above the clay aquiclude that necessitates the drain system, which in turn creates the environmental hazard at Kesterson Reservoir.

Although we are constantly alerted to the environmental impact of pesticides and other manufactured poisons that are introduced into natural ecosystems, the Kesterson saga illustrates how natural substances that pose no special hazard under normal conditions reach toxic levels through human intervention. The solution to such problems, which are increasing in number as an expanding human population places greater demands on limited natural resources, rests on an understanding of the complex interrelationship of the factors involved. In seeking these solutions, geologists have an increasingly important role to play.

Pollution by Sewage

The most common source of water pollution in wells and springs is sewage. Drainage from septic tanks, broken sewers, privies, and barnyards contaminates groundwater. If water contaminated with sewage bacteria passes through sediment or rock with large pores, such as coarse gravel or cavernous limestone, it can travel long distances and remain polluted (Fig. 11.15). On the other hand, if the contaminated water percolates through sand or permeable sandstone, it can become purified within short distances, in some cases less than about 30 m from where the pollution occurred (Fig. 11.15). Sand is an especially suitable cleansing agent because it promotes purification by (1) mechanically filtering out bacteria (water gets through, but most of the bacteria do not); (2) oxidizing bacteria so they are rendered harmless; and (3) placing bacteria in contact with other organisms that consume them. For this reason, purification plants that treat municipal water supplies and sewage percolate these fluids through sand.

Contamination by Seawater

Along coasts, fresh groundwater is separated from seawater by a thin transition zone of brackish water (Fig. 11.16A). Any pumping from an aquifer near the coast will reduce the flow of fresh groundwater toward the sea. This reduced fresh-water flow may then allow saltwater to move landward through permeable strata. Excessive pumping that exceeds the natural flow of fresh groundwater toward the sea may eventually permit saline water to encroach far inland and reach major

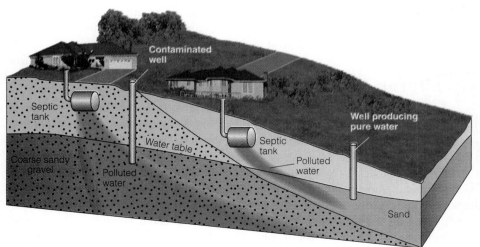

Figure 11.15 Contaminated Groundwater Purification of groundwater contaminated by sewage. Pollutants percolating through a highly permeable sandy gravel contaminate the groundwater and enter a well downslope from the source of contamination. Similar pollutants moving through permeable fine sand higher in the stratigraphic section are removed after traveling a relatively short distance and do not reach a well downslope.

A.

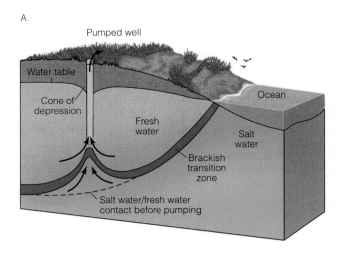

B.

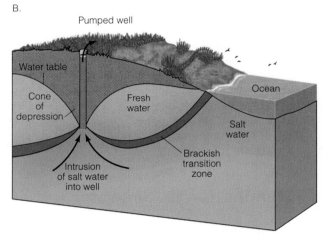

Figure 11.16 Contamination by Seawater Seawater contaminates a pumping well in a coastal area. A. Near the coast, a body of fresh groundwater overlies salty marine water. If pumping is not excessive, the well will draw only fresh water from the aquifer. B. Heavy pumping of groundwater forms a pronounced cone of depression both at the top and at the base of the groundwater body, eventually permitting salty water to enter and contaminate the well.

pumping centers. Such seawater intrusion can then contaminate water supplies (Fig. 11.16B). Once intrusion occurs, it is very difficult to reverse.

Toxic Wastes and Agricultural Poisons

Vast quantities of human garbage and industrial wastes are deposited each year in open basins or excavations at the land surface. When such a landfill site reaches its capacity, it generally is covered with dirt and then revegetated. Many of the waste products, now underground, are mobilized when rainwater seeps downward through the site and carries away soluble substances. In this way, harmful chemicals slowly leach into groundwater

reservoirs and contaminate them, making them unfit for human use. The pollutants travel from landfill sites as plumes of contaminated water in directions that depend on the regional groundwater flow pattern and are dispersed at the same rates as the percolating water (Fig. 11.17). The pollutants often are toxic to humans as well as to plants and animals.

In the United States, the pollution problems associated with landfill wastes have become so severe that the government has begun a major long-term program to clean up such sites and render them environmentally safe. However, the identified sites number in the tens of thousands, and it is difficult to judge how much time and money will be required to accomplish this formidable task.

Each year pesticides and herbicides are sprayed over agricultural fields and suburban gardens to help improve quality and productivity. Some of these chemicals have been linked with cancers and birth defects in humans, and some have led to disastrous population declines of wild animals. For example, a dramatic drop in the bald eagle population in the United States has been linked to the introduction of pesticides (primarily DDT) into the natural food chain. Because of the manner in which they are spread, such toxic chemicals invade the groundwater over wide areas as precipitation flushes them into the soil.

Underground Storage of Hazardous Wastes

One of the leading environmental concerns of industrialized countries is the necessity of dealing with highly toxic industrial wastes. Experience has demonstrated that surface dumping quickly leads to contamination of surface and subsurface water supplies and thereby to the possibility of serious and potentially fatal health problems (Fig. 11.18). In addition, countries with nuclear capacity have the special problem of disposing of high-level radioactive waste products. Some of the isotopes involved (e.g., ^{90}Sr and ^{137}Ce) are so highly radioactive that even minute quantities can prove fatal to people if released to the surface environment.

Most studies concerning disposal of hazardous wastes—both toxic and radioactive—have concluded that underground storage is appropriate, provided safe sites can be found. In the case of high-level nuclear wastes, which can remain dangerous for tens or hundreds of thousands of years because of the long half-lives of some of the radioactive isotopes, a primary requirement is that a site will be stable over a very long time interval. Therefore, the only completely safe sites for disposing of radioactive wastes and their containers are those that will not be affected chemically by groundwater, physically by earthquakes or other disruptive events, or accidentally by people.

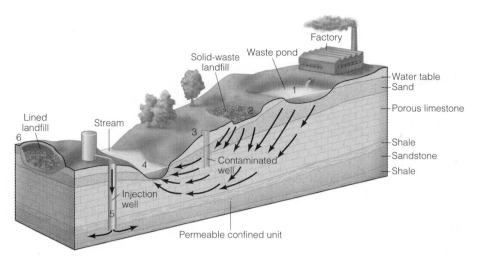

Figure 11.17 Toxic Wastes A groundwater system contaminated by toxic wastes. Toxic chemicals in an open waste pond (1) and an unlined landfill (2) percolate downward and contaminate an underlying aquifer. Also contaminated are a well downslope (3) and a stream (4) at the base of the hill. Safer, alternative approaches to waste management include injection into a deep confined rock unit (5) that lies well below aquifers used for water supplies, and a carefully engineered surface landfill (6) that is fully lined to prevent downward seepage of wastes. Because neither of the latter approaches is completely foolproof, constant monitoring at both sites would be required.

The placement of hazardous wastes underground, even far underground, immediately raises concerns about groundwater. Water is a nearly universal solvent, and the weakly acidic character of most groundwater means that any container of toxic or radioactive substances eventually is likely to corrode, so that the contents will then dissolve and be transported away from the storage site. Water is present in crustal rocks to depths of many kilometers, and in many of these rocks it is circulating at rates of 1 to 50 m/yr. Over tens or hundreds of thousands

Figure 11.18 Wastes from a Landfill Toxic wastes leak from rusting containers and soak into the groundwater system beneath this unsupervised waste disposal dump in the Czech Republic.

of years, even such slow rates can move dissolved substances over great distances and introduce them to more rapidly flowing parts of the hydrologic system.

Geologists generally agree that the ideal underground storage site for radioactive wastes should possess the following characteristics:

- The enclosing rock should have few fractures and low permeability.
- The enclosing rock should have no present or future economic mineral potential.
- Local groundwater flow should be away from plant and animal life.
- Only very long paths of groundwater flow should be directed toward places accessible to humans.
- The area should have low rainfall.
- The zone of aeration should be thick.
- The rate of erosion should be very low.
- The probability of earthquakes or volcanic activity should be very low.
- Future change of climate in the region should be unlikely to affect groundwater conditions substantially.

The safe long-term storage of toxic and nuclear wastes at underground sites provides a major challenge for geologists. Historically, geologists have studied past events, but now they are being asked to predict possible future events. To do so with any confidence requires considerable knowledge of local and regional groundwater conditions. It also demands a much better understanding of how complex groundwater systems might respond to crustal movements, local and global climatic change, and other natural factors that can affect the stability of a storage site.

GEOLOGIC ACTIVITY OF GROUNDWATER

In regions underlain by rocks that are highly susceptible to chemical weathering, groundwater creates distinctive landscapes that are among the most interesting and picturesque on our planet.

Dissolution

As soon as rainwater reaches the Earth's surface, it begins to react with minerals in regolith and bedrock and weathers them chemically. An important part of chemical weathering involves minerals and rock materials passing directly into solution through dissolution and hydrolysis. Among the rocks of the Earth's crust, the carbonate rocks are some of the most readily attacked by this process (Fig. 11.19).

Figure 11.19 Marble Dissolves in Rainwater A marble balustrade of the Forbidden City in Beijing, China, shows the effects of more than 300 years of dissolution. The original sharply carved design has become smooth and indistinct as chemical weathering, enhanced by acid rainfall, has dissolved the stone.

Limestone, dolostone, and marble are the most common carbonate rocks and underlie millions of square kilometers of the Earth's surface. Although carbonate minerals are nearly insoluble in pure water, they are readily dissolved by carbonic acid (Table 5.1, reaction 6) in the form of downward-percolating rainwater. As a result, groundwater becomes charged with calcium cations and bicarbonate anions.

The weathering attack occurs mainly along joints and other partings in the carbonate bedrock. The result is impressive. When granite is weathered chemically, quartz and other resistant minerals are little affected and remain as part of the regolith. However, when limestone weathers, nearly all its volume can be dissolved away in slowly moving groundwater.

By measuring over a period of time the amount of dissolution observed on small, precisely weighed limestone tablets placed at open sites in various areas, geologists have obtained estimates of the average rate at which limestone landscapes are being lowered by dissolution. In temperate regions with high rainfall, a high water table, and a nearly continuous cover of vegetation, carbonate landscapes are being lowered at average rates of up to 10 cm/1000 yr. In dry regions with scanty rainfall, low water tables, and discontinuous vegetation, rates are far lower. Measured rates of dissolution by groundwater in carbonate terrains of the United States show that the dissolution rate can exceed the average erosional reduction of the surface by mass-wasting, sheet erosion, and streams.

Chemical Cementation and Replacement

The conversion of sediment into sedimentary rock is primarily the work of groundwater. A body of sediment lying beneath the sea is generally saturated with water, as is sediment lying in the saturated zone beneath the land. Substances in solution in the water are precipitated as cement in the spaces between rock and mineral particles of the sediment. As we learned in Chapter 6, this diagenetic process transforms the loose sediment into firm rock. Calcite, quartz, and iron compounds (mainly hydroxides such as limonite) are, in that order, the chief cementing substances.

Less common than the deposition of cement between the grains of a sediment is **replacement**, the process by which a fluid dissolves matter already present and at the same time deposits from solution an equal volume of a different substance. Evidently, replacement takes place on an approximately volume-for-volume basis because the new material preserves the most minute textures of the material replaced. Both mineral and organic substances can be replaced. *Petrified wood* is a common

Figure 11.20 Petrified Wood Logs of petrified wood weather out of mudstone layers in Petrified Forest National Park, Arizona.

example of the replacement of organic matter (Fig. 11.20).

Carbonate Caves and Caverns

People have long been interested in caves. The earliest evidence we have of human dwellings comes from limestone caves in Europe and Asia that provided shelter for paleolithic peoples during the Pleistocene glacial ages. The walls of these caves were the rocky canvases of prehistoric artists, whose polychrome paintings provide us with superb renditions of the prey of ice-age big-game hunters.

Caves come in many sizes and shapes. Although most caves are small, some are of exceptional size. A very large cave or system of interconnected cave chambers is often called a *cavern*. The Carlsbad Caverns in southeastern New Mexico include one chamber 1200 m long, 190 m wide, and 100 m high. Mammoth Cave, in Kentucky, consists of interconnected caverns with an aggregate length of at least 48 km. The recently discovered Good Luck Cave on the tropical island of Borneo has one chamber so large that it could accommodate not only the world's largest previously known chamber (in Carlsbad Caverns), but also the largest chamber in Europe (in Gouffre St. Pierre Martin, France) and the largest chamber in Britain (Gaping Ghyll).

Cave formation is mainly a chemical process involving the dissolution of carbonate rock by circulating groundwater. The usual sequence of development is thought to involve (1) initial dissolution along a system of interconnected open joints and bedding planes by percolating groundwater, (2) enlargement of a cave passage along the most favorable flow route by water that fully occupies the opening, (3) deposition of carbonate formations on the cave walls while a stream occupies the cave floor, and (4) continued deposition of carbonate on the walls and floor of the cave after the stream has stopped flowing. Although geologists have argued for years as to whether caves form in the zone of aeration or in the saturated zone, available evidence favors the idea that most caves are excavated in the shallowest part of the saturated zone, along a seasonally fluctuating water table.

The rate of cave formation is related to the rate of dissolution. Where the water is acidic, the rate of dissolution increases with increasing flow velocity. Therefore, as a passage increases in size and the flow changes from very slow laminar flow to more rapid turbulent flow, the rate of dissolution will rise. The development of a continuous passage by slowly percolating waters has been estimated to take up to 10,000 years, while the further enlargement of the passage by more rapidly flowing water into a fully developed cave system may take an additional 10,000 to 1 million years.

Although limestone caves are generally believed to involve dissolution by carbonic acid, chemical evidence suggests that at least some caves, including Carlsbad Caverns, may involve dissolution by sulfuric acid. The proposed agent is hydrogen-sulfide-bearing solutions derived from petroleum-rich sediments. The solutions rise along joints, where they meet and interact with oxygenated water to form sulfuric acid, which then dissolves the limestone.

Cave Deposits

Some caves have been partly filled with insoluble clay and silt, originally present as impurities in limestone and gradually concentrated as the limestone was dissolved. Others contain partial fillings of **dripstone** and **flowstone**, deposits chemically precipitated from dripping and flowing water, respectively, in an air-filled cavity. Both kinds of deposits are commonly composed of calcium carbonate. The carbonate pre-

Figure 11.21 Dripstones Spectacular dripstone and flowstone formations ornament Lehman Caves in Great Basin National Park, Nevada.

cipitates take on many curious forms, which are among the chief attractions for cave visitors. Among the most common shapes are *stalactites* (iciclelike forms of dripstone hanging from ceilings), *stalagmites* (blunt "icicles" of dripstone projecting upward from cave floors), *columns* (stalactites joined with stalagmites, forming connections between the floor and roof of a cave), and crenulated or curtainlike formations of flowstone (Fig. 11.21).

As its name implies, dripstone is deposited by successive drops of water. As each drop forms on the ceiling of a cave, it loses a tiny amount of carbon dioxide gas and precipitates a particle of calcium carbonate (Fig. 11.22). This chemical reaction is the reverse of the one by which calcium carbonate is dissolved by carbonic acid.

Dripstone and flowstone can be deposited only in caves that are at least partially filled with air and therefore lie at or above the water table. Yet many, perhaps most, caves are believed to have formed below the water table, as is suggested by their shapes and by the fact that some caves are lined with crystals, which can form only in an aqueous environment. How can we reconcile these apparently conflicting observations? The answer lies partly in the observation that both dissolution and deposition are known to take place in the zone of aeration. In some cases, the answer probably also lies in a change in the level of the water table. This change in level can result when uplift of the land causes

Figure 11.22 How Stalactites Form A drop of water collects at the end of a growing stalactite in Carlsbad Caverns, New Mexico. As the water loses carbon dioxide, a tiny amount of calcium carbonate precipitates from solution and is added to the end or sides of the stalactite.

a stream to cut downward into the landscape or when a change of climate causes a lowering of the regional water table (Fig. 11.23). Caves formed in the upper part of the saturated zone when the water table is high would shift into the zone of aeration as the water table falls. Dissolution could then give way to deposition of dripstone and flowstone.

Sinkholes

In contrast to a cave, a **sinkhole** is a large dissolution cavity that is open to sky. Some sinkholes are caves whose roofs have collapsed; others are formed at the surface, where rainwater is freshly charged with carbon dioxide and is most effective as a solvent. Many sinkholes located at the intersections of joints, where downward movement of water is most rapid, are funnel shaped.

Sinkholes of the Yucatan Peninsula in Mexico, which are locally called *cenotes* (a word of Mayan origin), have high, vertical sides and contain water because their floors lie below the water table (Fig. 11.24). The cenotes were the primary source of water for the ancient Maya and formerly supported a considerable popu-

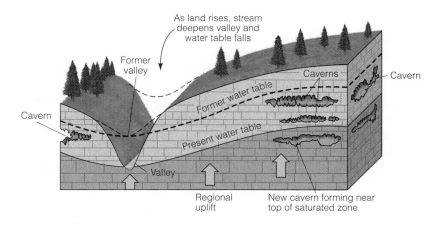

Figure 11.23 Formation of Caverns One possible history of a cavern containing carbonate deposits. The cavern formed in the saturated zone when the water table lay at a former, higher level. Uplift of the region caused streams to deepen their valleys. The water table then lowered in response to valley deepening, leaving the cavern above the lowered, present water table (level 2). Stalactites and stalagmites could then form in the air-filled cavern.

Figure 11.24 Cenote The sacred well at Chichen Itza, a ruined Mayan city on the Yucatan Peninsula of Mexico. This cenote, formed in flat-lying limestone strata, contained a rich store of archaeological treasures that were cast into the water with human sacrifices centuries ago.

Figure   11.25   Sinkhole in Florida   Most of a city block in Winter Park, Florida disappeared into a widening crater as this sinkhole formed in underlying carbonate bedrock.

lation in Yucatan. A large cenote at the ruined city of Chichen Itza was sacred and dedicated to the rain gods. Remains of more than 40 human sacrifices, mostly young children, have been recovered from the cenote, together with huge quantities of jade, gold, and copper offerings.

In the carbonate landscape of Florida, new sinkholes are forming constantly (Fig. 11.25). In one small area of about 25 km^2, more than 1000 collapses have occurred in recent years. In this case, lowering of the water table brought on by drought and excessive pumping of local wells has led to extensive collapse of cave roofs.

Some sinkholes form catastrophically. An account of one such event in rural Alabama describes how a resident was startled by a rumble that shook his house. He then distinctly heard the sound of trees snapping and breaking. A short time later, hunters walking through nearby woods discovered a huge 50 m deep sinkhole that measured 140 m long and 115 m wide.

Karst Topography

Video

Karst

In some regions of exceptionally soluble rocks, caves and sinkholes are so numerous that they form a peculiar topography characterized by many small, closed basins and a disrupted drainage pattern. Streams disappear into the ground and eventually reappear elsewhere as large springs. Such terrain is called **karst topography**

after the classic karst regions of former Yugoslavia (extending from Slovenia to Montenegro), a remarkable limestone landscape of closely spaced sinkholes. The term *karst* is derived from the German form of *krs*, a Slav word meaning rock. Although most typical of carbonate landscapes, karst can also develop in areas underlain by gypsum and salt.

Several factors control the development of karst landscapes, First, the topography must produce a steep enough hydraulic gradient to permit the flow of groundwater through soluble rock under the pull of gravity. In addition, precipitation must be adequate to supply the groundwater system, soil and plant cover must supply an adequate amount of carbon dioxide, and temperature must be high enough to promote dissolution. Although karst terrain is found throughout a wide range of latitudes and at varied altitudes, it often is best developed in moist temperate to tropical regions underlain by thick and widespread soluble rocks.

Several distinctive kinds of karst landscape are recognized. The most common is *sinkhole karst*, a landscape dotted with sinkholes of various sizes and shapes. Such landscapes are seen in southern Indiana, south-central Kentucky, central Tennessee, and Jamaica, among other places.

Cone karst and *tower karst* occur in thick, well-jointed limestone that separates into isolated blocks as it weathers. Cone karst consists of many closely spaced conical- or pinnacle-shaped hills separated by deep sinkholes

Figure 11.26 Astronomy and Karst Cone karst near Arecibo in northwestern Puerto Rico, The Arecibo radiotelescope (lower right) occupies a circular depression in the midst of a vast limestone landscape of closely spaced conical hills separated by deep sinkholes.

Figure 11.28 Karst in Ireland The Burren region in western Ireland is an extensive pavement karst developed on limestone. Crevices have developed by solution along prominent joints, giving the terrain a regular, geometric texture.

(Fig. 11.26). Tower karst, by contrast, consists of isolated towerlike limestone hills separated by expanses of alluvium. If sediment being deposited in the bottom of sinkholes is not removed by the local drainage system, the depressions slowly fill, and areas between towers merge to form a flat alluvial surface. Cone and tower karst landscapes are found in parts of Mexico, Central America, and the Caribbean islands of Cuba and Puerto Rico, as well as in the South Pacific.

One of the most famous and distinctive of the world's tower karst regions lies near Guilin in southeastern China. There, vertical-sided peaks of limestone rise up to 200 m high. The dramatic landscape of this region has inspired both classical Chinese painters and present-day photographers (Fig. 11.27).

Pavement karst consists of broad areas of bare limestone in which joints and bedding planes have been etched and widened by dissolution, creating a distinctive land surface. Pavement karsts are especially common in high-latitude regions where continental glaciation has stripped away regolith and left carbonate bedrock exposed to weathering. These rocky, intricately etched landscapes, nearly devoid of vegetation, are found in such places as Spitzbergen, Greenland, and the Burren region of western Ireland (Fig. 11.28).

Figure 11.27 Tower Karst in China Steep limestone pinnacles up to 200 m high, surrounded by flat expanses of alluvium, form a spectacular karst landscape around the Li River near Guilin, China.

SUMMARY

1. Groundwater is derived mostly from rainfall and occurs everywhere beneath the land surface.

2. The water table is the top of the saturated zone. In humid regions, the form of the water table is a subdued imitation of the overlying land surface.

3. Groundwater moves chiefly by percolation, at rates far slower than those of surface streams. In rock or sediment of constant permeability, the velocity of groundwater increases as the slope of the water table increases.

4. In moist regions, groundwater in recharge areas percolates downward under the pull of gravity. It moves away from hills toward valleys, where it may emerge to supply streams (groundwater discharge areas). In dry regions, the groundwater is recharged by water percolating downward beneath surface streams.

5. According to Darcy's Law, the discharge of water in a groundwater system is equal to the product of the cross-sectional area of flow, the coefficient of permeability, and the hydraulic gradient.

6. Springs often occur at places where either the water table or an aquiclude intersects the land surface.

7. Groundwater flows into most wells directly by gravity. Pumping of water from wells creates cones of depression in the water table.

8. Major supplies of groundwater are found in aquifers, among the most productive of which are porous sand, gravel, and sandstone.

9. If the top of a well that penetrates an artesian aquifer lies below the altitude of the water table in the recharge area, hydrostatic pressure will allow water to rise in the well and flow out at the surface without pumping.

10. An unconfined aquifer is one that is not constrained above by an aquiclude, whereas a confined aquifer is bounded by aquicludes.

11. Excessive withdrawal of groundwater can lead to lowering of the water table and to land subsidence.

12. Water quality is influenced by the content of natural dissolved substances, seawater intrusion, and pollution by human and industrial wastes that percolate into groundwater reservoirs.

13. Hazardous (toxic and radioactive) wastes should be stored underground only if geologic conditions imply little or no change in groundwater systems over geologically long intervals of time.

14. Groundwater dissolves mineral matter from rock. It also deposits substances as cement between grains of sediment, thereby reducing porosity and converting the sediments to sedimentary rock.

15. In carbonate rocks, groundwater not only creates caves and sinkholes by dissolution but also deposits calcium carbonate as dripstone and flowstone.

16. Karst topography forms in areas of porous carbonate or other soluble rocks where the relief is great enough to permit gravitational flow of groundwater.

THE LANGUAGE OF GEOLOGY

aquiclude (p. 287)
aquifer (p. 289)
artesian aquifer (p. 290)

cone of depression (p. 288)
confined aquifer (p. 289)

Darcy's Law (p. 286)
discharge area (p. 285)
dripstone (p. 300)

flowstone (p. 300)

groundwater (p. 282)

hydraulic gradient (p. 286)

karst topography (p. 302)

percolation (p. 285)
permeability (p. 284)
porosity (p. 284)

recharge (p. 285)
recharge area (p. 285)
replacement (p. 299)

saturated zone (p. 283)
sinkhole (p. 301)
spring (p. 287)

unconfined aquifer (p. 289)

water table (p. 283)

zone of aeration (p. 283)

Virtual Internship: You Be The Geologist

In the Groundwater Contamination internship on your CD-ROM, you are an environmental geologist. Your task is to discover the source of contaminated groundwater found in the basement of a small-town school.

QUESTIONS FOR REVIEW

1. What is the ultimate source of groundwater?

2. Why does a thin zone immediately above the water table remain continuously moist?

3. Why do the flow paths of groundwater moving beneath a hill tend to turn upward toward a stream in an adjacent valley?

4. What variables determine how long it takes water to move from a recharge area to a discharge area?

5. Explain why it is possible to determine the discharge, velocity, or coefficient of permeability of groundwater passing through an aquifer provided two of these three factors are known.

6. What is the hydraulic gradient, and what importance does it have in determining the rate of flow of groundwater?

7. Why are sandstones generally better aquifers than siltstones or shales?

8. What features in igneous and metamorphic rocks promote the flow of groundwater through them?

9. How are springs related to the water table?

10. What causes a cone of depression to form around a producing well?

11. What causes water to rise to or above the ground surface in an artesian well?

12. Why is sand especially effective in purifying water flowing through it?

13. What is the origin of "hard" water in regions of carbonate bedrock?

14. Why do dripstone and flowstone formations not form in a cave that is in the saturated zone and therefore completely filled with water?

15. Why are karst pavements common in carbonate terrain at high latitudes?

16. A large area on a hillside has been suggested as a landfill site for garbage generated by a small nearby city. You are asked for a geologic appraisal of the site to determine if local subsurface water supplies might be affected. What geologic factors would you investigate and why?

17. What geologic and biologic factors are likely to disqualify a site from being selected for underground storage of high-level radioactive waste?

For an interactive case abstract, virtual tours, activities, and additional learning resources, go to
GEOSCIENCES TODAY: www.wiley.com/college/skinner

Environmental Hazards on the U.S.–Mexican Border: The Case of Ambos Nogales

Nogales, Arizona and Nogales, Sonora, together known as Ambos Nogales, are binational urban regions or twin cities. It is important to understand the human–environment interaction of this area and its inhabitants' vulnerability to a potential chemical emergency that might occur on the Mexican side and the potential impact on the U.S. side. It is equally important to understand and assess issues of transboundary and transnational environmental contamination.

OBJECTIVE: The primary objective is to demonstrate the interaction of the dispersion of a contaminant gas with local patterns of topography and human settlement, as well as illustrate how GIS (Geographical Information Systems) might be used to model and simulate chemical emergencies so that urban planners and emergency personnel can be better prepared to respond to or prevent such occurrences.

The Human Dimension: How does the concept of vulnerability relate to the study of natural or human-induced hazards within the Earth System Sciences?

Questions to Explore:

1. What are some of the most important geographical, socioeconomic, and historical patterns that characterize binational urban regions (twin cities) along the U.S.–Mexican border?

2. How does the geographic nexus involving topography, relief, human settlement, and gas or liquid contaminant dispersion patterns create a potentially hazardous situation for humans living in Ambos Nogales?

3. How do human factors exacerbate or mitigate vulnerability to natural or human-induced hazards such as risk of exposure to a chemical hazard?

12

**Measuring
Our Earth**

**Understanding
Our Environment**

**Using
Resources**

GeoMedia

Emperor penguins gather near the Dawson-Lambton glacier, Antartica,
at point where the glacier meets the sea ice of the Weddell Sea.

Glaciers and Glaciation

The Icy Record of Past Events

In most temperate mountain glaciers of average size, the oldest ice is only a few hundred years old. By contrast, in the great polar ice sheets, and in some high-altitude glaciers at lower latitudes having temperatures well below freezing, the oldest ice spans the last glacial age or even the last several glacial ages. For more than four decades, these latter glaciers have been the focus of a unique endeavor: obtaining from ice cores a high-resolution record of changing climate and atmospheric composition extending far into the past.

When snow accumulates on a glacier, it compacts and the air between snow crystals becomes trapped in bubbles as the snow slowly changes to glacier ice. The trapped air is a sample of the atmosphere over the glacier at the time the ice forms. By sampling the ice and analyzing its trapped gases and solid particles, an array of information can be obtained. Most of what we now know about the past composition of the atmosphere during glacial times has been obtained from such studies.

Measurements of the stable isotopes of oxygen in ice cores provide a measure of air temperature when the ice formed. In much the same way that annual rings enable us to date the age of trees, cyclic shifts in the isotopic signal, related to the annual solar cycle, permit ages within the upper part of ice cores to be dated to the nearest year. The deeper part of ice cores loses the annual signal owing to internal deformation and flow, and so the oldest ice is dated by modeling ice flow at the coring site. Among the remarkable discoveries is a high-amplitude, high-frequency variation in oxygen isotopes in the Greenland Ice Sheet that characterized glacial times. These significant and rapid changes of temperature likely were related to oceanographic changes in the adjacent North Atlantic.

In addition to past temperatures, glaciologists can measure variations in atmospheric carbon dioxide and methane, and thereby show that major changes in the atmospheric content of these greenhouse gases occur during a glacial-interglacial cycle. Similar variations are seen in the content of microparticles in the ice (i.e., wind-blown dust) which demonstrate that the glacial atmosphere was very dusty compared to that of inter-glacial times. By analyzing the chemistry of the dust, source areas of the sediment can be determined, as well as major wind trajectories. Thus, we now know that most of the dust in Greenland ice cores originated in the desert basins of central Asia and that dust in the Antarctic Ice Sheet originated in Patagonia.

Because glaciers suitable for obtaining ice core records occur not only in the two polar regions (Greenland and Antarctica), but also at the crests of many high plateaus and mountain ranges (for example, in Tibet, the high Andes, and the Himalaya), changes in atmospheric conditions through time can be reconstructed in a variety of geographic settings and then compared to identify both regional and global patterns.

GLACIERS

At any place where more snow falls on the land during each winter than is melted during the following summer, the snow will gradually grow thicker. As the snow accumulates, the increasing weight of overlying snow causes the basal layers to recrystallize, forming a solid mass of ice. This process is analogous to the way sedimentary rocks, buried under a deep pile of strata within the Earth's crust, recrystallize to form metamorphic rocks. When the accumulating snow and ice become so thick that the pull of gravity causes the frozen mass to move, a glacier is born. Accordingly, we define a **glacier** as a permanent body of ice, consisting largely of recrystallized snow, that shows evidence of downslope or outward movement due to the pull of gravity.

Glaciers are found in areas where average temperature is so low that water can exist year round in the frozen state. As we might expect, therefore, most glaciers are found in high latitudes, the coldest parts of our planet. However, because low temperatures also occur at high altitudes, many small glaciers are found in middle and low latitudes on high mountains.

Because glaciers vary considerably in their physical characteristics, we can distinguish several kinds based on their shape and size (Table 12.1), as well as on their internal temperature.

Mountain Glaciers and Ice Caps

The shape and direction of movement of most mountain glaciers are determined by the surrounding bedrock topography (Fig. 12.1). A *cirque glacier* occupies a **cirque**, a protected bowl-shaped depression on a mountain-

TABLE 12.1	Principal Types of Glaciers, Classified According to Form
Glacier Type	Characteristics
Cirque glacier	Occupies a cirque on a mountainside
Valley glacier	Flows from cirque(s) onto and along the floor of a valley
Fjord glacier	Valley glacier that occupies a fjord. Base lies below sea level. May have a steep front that recedes rapidly as icebergs break off and float away
Piedmont glacier	Broad lobe of ice that terminates on open slopes beyond a mountain front. Fed by one or more large valley glaciers
Ice cap	Dome-shaped body of ice and snow that covers mountain highlands (or lower-lying lands at high latitudes) and displays generally radial outward flow
Ice sheet	Continent-sized mass of ice that overwhelms nearly all land within its margins
Ice shelf	Thick slablike glacier that floats on the sea and is fed by one or more glaciers on land. Commonly located in large bays

Figure 12.1 Kinds of Glaciers
Common types of mountain glaciers, classified according to shape and size. A small cirque glacier occupies a bowl-shaped hollow at the head of a valley, while in the adjacent valley two cirque glaciers have expanded and merged to form a valley glacier. In the next valley a fjord glacier sheds icebergs into a long fjord. Nearby, several tributary glaciers merge into a single ice stream that terminates in a broad piedmont glacier. In the distance, a high mountain plateau is mantled by a broad ice cap.

side, open downslope and bounded upslope by a steep cliff called a *headwall*. A growing cirque glacier that spreads outward and downward along a valley becomes a *valley glacier* (Fig. 12.2). Many of the Earth's high

Figure 12.2 Ice Streams Dark bands of rock debris mark the boundaries between adjacent tributary ice streams that have merged to form Kaskawulsh Glacier, a large valley glacier in Yukon Territory, Canada.

mountain ranges (e.g., the Alaska Range and the Himalaya) contain glacier systems that include valley glaciers tens of kilometers long (Fig. 12.3).

Valley glaciers in some coastal mountain ranges at middle to high latitudes occupy deep glacier-carved valleys that are filled by an arm of the sea. Such a valley is called a **fjord**, and a glacier that occupies it is a *fjord glacier* (Fig. 12.1).

A very large mountain glacier may spread out onto a gentle piedmont slope beyond the mountain front. It then becomes a *piedmont glacier*, forming a broad lobe of ice that resembles an inverted spoon (Fig. 12.1).

An *ice cap* covers a mountain highland or lower-lying land at high altitude and displays generally radial outward flow (Figs. 12.1 and 12.4).

Ice Sheets and Ice Shelves

An *ice sheet* is the largest type of glacier on the Earth. These continent-sized masses of ice overwhelm nearly all the land surface within their margins. Modern ice sheets, which occur only on Greenland and Antarctica, include about 95 percent of the ice in existing glaciers. If all the ice in these vast ice sheets were to melt, their

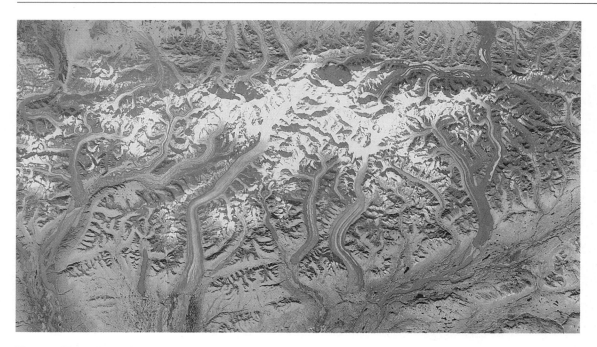

Figure 12.3 Valley Glaciers A vertical satellite view of the valley-glacier complex that covers much of Denali National Park in south-central Alaska. Mount Denali (formerly McKinley), the highest peak in North America, lies near the center of the glacier-covered region.

Figure 12.4 Ice Caps Several ice caps over areas of high land on Iceland. Vatnajökull, in the southeastern part of the island, is the largest, covers 8300 km², and overlies an active volcano.

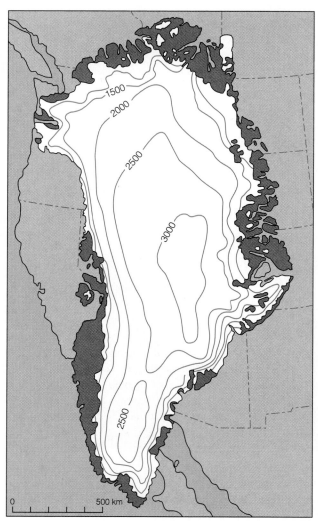

Figure 12.5 Greenland Ice Sheet Map of Greenland superimposed on a map of western United States showing that the Greenland Ice Sheet covers an area approximating that of the United States west of the Rocky Mountains. Contours are the height of the ice surface above sea level, in meters. At the crest of the ice sheet, the ice is more than 3000 m above sea level.

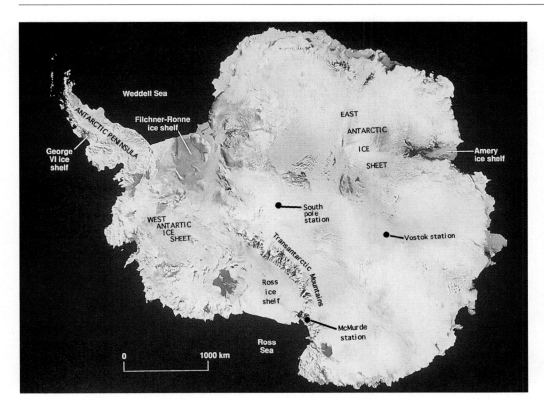

Figure 12.6 Antarctic Ice Sheets Satellite view of Antarctica. The East Antarctic Ice Sheet overlies the continent of Antarctica, whereas the much smaller West Antarctic Ice Sheet overlies a volcanic island arc and surrounding seafloor. Three major ice shelves occupy large bays. The ice-covered regions of Antarctica nearly equal the combined area of Canada and the conterminous United States.

combined volume—close to 24 million km³—would be sufficient to raise world sea level by nearly 66 m.

The Greenland Ice Sheet has about the same area as the United States west of the Rocky Mountains (Fig. 12.5). It is so thick (about 3000 m) that its weight is enough to cause the land surface beneath much of it to be depressed below sea level.

Antarctica is covered by two large ice sheets that meet along the lofty Transantarctic Mountains (Fig. 12.6). The East Antarctic Ice Sheet is the larger of the two and covers the continent of Antarctica. Because of its ice sheet, Antarctica has the highest average altitude and the lowest average temperature of all the continents. The smaller West Antarctic Ice Sheet overlies numerous islands of the Antarctic archipelago. Parts of it rest on land that rises above sea level, but extensive portions cover land lying below sea level.

An *ice shelf* is a thick, nearly flat sheet of floating ice that is fed by one or more glaciers on land and terminates seaward in a steep ice cliff as much as 50 m high. Ice shelves are found at several places along the margins of the Antarctic ice sheets, where they are confined in large bays (Fig. 12.6), and also in the Canadian Arctic islands. The largest Antarctic ice shelves extend hundreds of kilometers seaward from the coastline and reach a thickness of at least 1000 m near their landward margins.

Temperate and Polar Glaciers

Glaciers can be classified not only according to their size and shape, but also their internal temperature. Temperature is an important criterion for classification because it helps determine how glaciers move and how they shape the landscape.

Ice in a **temperate glacier** is at the **pressure melting point**, which is the temperature at which ice melts at a particular pressure (Fig. 12.7A). In such glaciers, which are restricted mainly to low and middle latitudes, meltwater and ice exist together in equilibrium. At high latitudes and high altitudes, where the mean annual air temperature lies below freezing, the temperature in a glacier remains below the pressure melting point and little or no seasonal melting occurs. Such a glacier, in which the ice remains below the pressure melting point, is called a **polar glacier** (Fig. 12.7B).

Since the temperature of snow crystals falling to the surface of a temperate glacier is below freezing, how can

A.

TEMPERATE GLACIER

$-10°$ $-5°$ $0°$ $5°C$

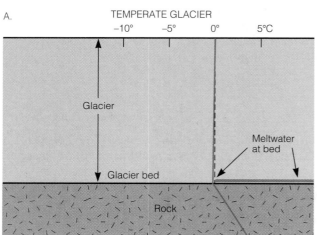

B.

POLAR GLACIER

$-20°$ $-15°$ $-10°$ $-5°$ $0°$

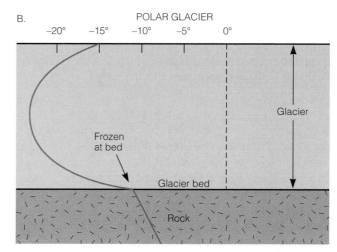

Figure 12.7 Temperate and Polar Glaciers Temperature profiles (red lines) through temperate and polar glaciers.   A. Ice in a temperate glacier is at the pressure melting point from surface to bed. The terminus is rounded, as shown in this view of Pré de Bar Glacier in the Italian Alps, because melting occurs at the surface.   B. Ice in a polar glacier remains below freezing, and the ice is frozen to its bed. Subfreezing temperatures inhibit melting at the terminus, which forms a steep cliff of ice, as shown in this view of Commonwealth Glacier in Antarctica.

ice throughout the glacier reach the pressure melting point? The answer lies in the seasonal fluctuation of air temperature and in what happens when water freezes to form ice. In summer, when air temperature rises above freezing, solar radiation melts the glacier's surface snow and ice. The meltwater percolates downward, where it encounters freezing temperatures and therefore freezes. When changing state from liquid to solid, each gram of water releases 335 J of heat. This released heat warms the surrounding ice and, together with heat flowing upward from the solid Earth beneath the glacier, keeps the temperature of the ice at the pressure melting point.

Glaciers and the Snowline

Glaciers can form only at or above the **snowline**, which is the lower limit of perennial snow. The snowline is sensitive to local climate, especially temperature and precipitation, and rises in altitude from near sea level in

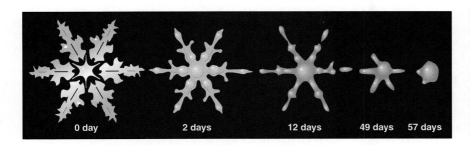

Figure 12.8 From Snow to Ice Conversion of a snowflake into a granule of old snow. The delicate points of a snowflake disappear as melting and evaporation occur. The resulting meltwater refreezes, and water vapor formed by evaporation at the margin condenses near the center of the crystal, making it more compact.

polar latitudes to as much as 6000 m in the tropics. It also rises inland from moist coastal regions toward the drier interiors of large islands and continents.

A glacier potentially can form at any place above the snowline where the topography will permit a glacier to exist. This explains why glaciers are found not only at sea level in the polar regions, where the snowline is low, but also near the equator, where some high peaks in New Guinea, East Africa, and the Andes are high enough for perennial snow to exist.

Conversion of Snow to Glacier Ice

Glacier ice is a metamorphic rock that consists of interlocking crystals of the mineral ice. It owes its characteristics to deformation under the weight of overlying snow and ice.

Newly fallen snow is very porous and has a density less than a tenth that of water. The delicate points of snowflakes gradually disappear by evaporation. The resulting water vapor condenses, mainly in constricted places near the centers of snowflakes. In this way, the fragile ice crystals slowly become smaller and rounder (Fig. 12.8), which allows the material to settle and become more compact as the amount of pore space is reduced.

Snow that survives a year or more gradually increases in density until it is no longer permeable to air, at which point it becomes glacier ice. Although now a rock, such ice has a far lower melting temperature than any other naturally occurring rock, and its density of about 0.9 g/cm^3 means that it will float in water.

Further changes in ice take place with increasing depth in a glacier. Figure 12.9 shows a core obtained by Russian glaciologists who drilled through the East Antarctic Ice Sheet at Vostok Station (Fig. 12.6). As snowfall adds to the glacier's thickness, the increasing pressure causes initially small grains of glacier ice to grow in size until, near the base of the ice sheet, they reach a diameter of 1 cm or more. This increase in grain size is analogous to what happens in a fine-grained rock that is carried deep within the Earth's crust: as with metamorphic rocks, high pressure and high temperature

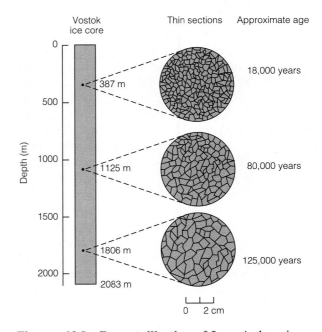

Figure 12.9 Recrystallization of Ice A deep ice core drilled at Vostok Station penetrates through the East Antarctic Ice Sheet to a depth of 2083 m. Thin-section samples of ice taken from different depths in the core show a progressive increase in the size of ice crystals, a result of slow recrystallization as the weight of overlying ice increases with time.

over a long time cause large mineral grains to slowly develop (Chapter 7).

Why Glaciers Change in Size

The pioneer Scottish glaciologist James D. Forbes visited the Alps in the mid-nineteenth century and carefully mapped the extent of long valley glaciers flowing from Mont Blanc, the highest peak in Europe. Since Forbes's day, these glaciers have shrunk greatly, exposing extensive areas of valley floor that only a century ago were buried beneath thick ice. Over this same period, changes in glacier size have been recorded worldwide. Like those in the Alps, most other glaciers have receded dramatically since the end of the last century, although some have remained relatively unchanged and others have expanded. To understand why glaciers advance and retreat, and why different glaciers in any region may behave in different ways, we must examine how a glacier responds to a gain or loss of mass.

Mass Balance

The mass of a glacier is constantly changing as the weather varies from season to season and, on longer time scales, as local and global climates change. These ongoing environmental changes cause fluctuations in the amount of snow added to the glacier surface and in the amount of snow and ice lost by melting.

We can think of a glacier as being like a checking account at the bank. The amount of money in a bank account at the end of the year is the difference between the amount of money added to the initial account and the amount removed during the year. The glacier's account is measured in terms of the amount of snow added to the existing ice, mainly in the winter, and the amount of snow (and ice) lost, mainly during the summer. Additions to the glacier's account are collectively called **accumulation**, and losses are termed **ablation**. The total in the account at the end of a year—in other words, the difference between accumulation and ablation—is a measure of the glacier's **mass balance** (Fig. 12.10). The account may have a surplus (a positive balance) or a deficit (a negative balance), or it may be in exact balance (accumulation = ablation).

If a glacier is viewed at the end of the summer ablation season, two zones are generally visible on its surface (Fig. 12.11). An upper zone, the *accumulation area*, is the part of the glacier covered by remnants of the previous

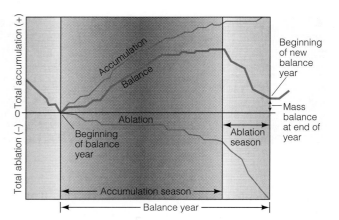

Figure 12.10 Mass Balance in a Glacier Graph showing how accumulation and ablation determine glacier mass balance (heavy line) over the course of a year. The balance curve, obtained by summing values of accumulation (positive values) and ablation (negative values), rises during the accumulation season as mass is added to the glacier, then falls during the ablation season as mass is lost. The mass balance at the end of the balance year reflects the difference between mass gain and mass loss.

winter's snowfall and is an area of net gain in mass. Below it lies the *ablation area*, a region of net loss where bare ice and old snow are exposed because the previous winter's snow cover has melted away.

The **equilibrium line** marks the boundary between the accumulation area and the ablation area (Fig. 12.11). It lies at the level on the glacier where net mass loss equals

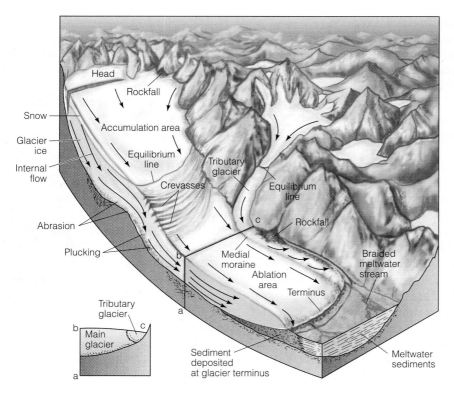

Figure 12.11 Features of a Valley Glacier Main features of a valley glacier. The glacier has been cut away along its center line; only half is shown. Crevasses form where the glacier bed has a steeper slope. Arrows show directions of ice flow. A band of rock debris marks the boundary between the main glacier and a tributary glacier that joins it from an adjacent valley.

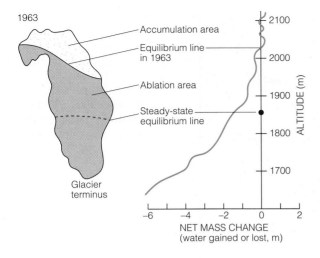

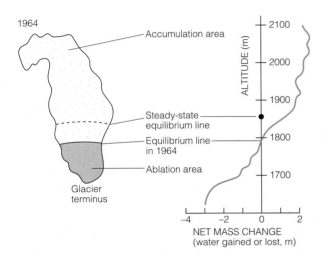

a succession of years in which negative mass balance predominates will lead to retreat of the terminus. If no net change in mass occurs, the glacier is in a balanced state. If this condition persists, the terminus is likely to remain relatively stationary.

Response Lags. Advance or retreat of a glacier terminus does not necessarily give us an accurate picture of changing climate because a lag occurs between climatic change and the response of the glacier terminus to that change. The lag reflects the time it takes for the effects of an increase or a decrease in accumulation above the equilibrium line to be transferred through the slowly moving ice to the glacier terminus. The length of the lag depends both on the size of a glacier and the way the ice flows; the lag will be longer for large glaciers than for small ones and longer for polar glaciers than for temperate ones. Temperate glaciers of modest size (like those in the European Alps) have response lags that range from several years to a decade or more. This lag time can explain why, in any area with glaciers of different sizes, some glaciers may be advancing while others are stationary or retreating.

Calving. During the last century and a half, many Alaskan fjord glaciers have receded at rates far in excess of typical glacier retreat rates on land. Their dramatic recession is due to frontal **calving**, which can be defined as the progressive breaking off of icebergs from the front of a glacier that terminates in deep water (Fig. 12.13). Although the base of a fjord glacier may lie

Figure 12.12 Changes to a Glacier Maps of South Cascade Glacier in Washington State at the ends of the 1963 and 1964 balance years showing the position of the equilibrium line relative to the position it would have under a balanced (steady-state) condition. The curves show values of mass balance as a function of altitude. During 1963, a negative balance year, the glacier lost mass and the equilibrium line was high (2025 m). In 1964, a positive balance year, the glacier gained mass and the equilibrium line was low (1800 m). Net change is measured in water lost or gained per unit area of the glacier.

net mass gain. The equilibrium line on temperate glaciers coincides with the local snowline, which marks the lower limit of fresh snow at the end of the ablation season. Being very sensitive to weather, the equilibrium line fluctuates in altitude from year to year and is higher in warm, dry years than in cold, wet years (Fig. 12.12).

Fluctuations of the Glacier Terminus
If, over a period of years, a glacier's mass balance is positive more often than negative, the mass of the glacier increases. The front, or *terminus*, of the glacier is then likely to advance as the glacier grows. Conversely,

Figure 12.13 Calving from a Fjord Glacier Icebergs break away from the calving front of a fjord glacier in southwestern Greenland. This fjord was filled by an arm of the sea as the glacier retreated back rapidly by frontal calving.

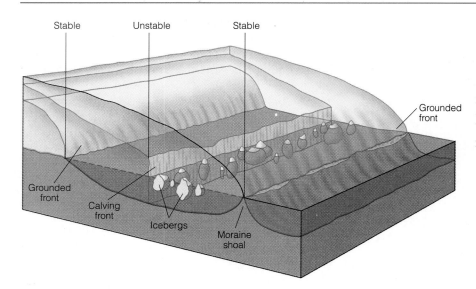

Stable   Unstable   Stable

Grounded front

Grounded front

Calving front

Icebergs

Moraine shoal

Figure 12.14 Fjord Glacier The terminus of a fjord glacier remains stable if it is grounded against a shoal, but if it retreats into deeper water, calving will begin. The unstable terminus then retreats at a rate that depends on the water depth. Once it becomes grounded farther up the fjord, the terminus is stable once again.

far below sea level along much of its length, its terminus can remain stable as long as it is resting (or "grounded") against a shoal (Fig. 12.14). However, if the terminus retreats off the shoal, water will replace the space that had been occupied by ice. With the glacier now terminating in water, conditions are right for calving. Because a fjord is a deep basin, the water becomes progressively deeper as the calving terminus continues to retreat upfjord. The deepening water leads to faster retreat, because the greater the water depth, the faster the rate of calving. Once started, calving will continue rapidly and irreversibly until the glacier front recedes into water too shallow for much calving to occur, generally near the head of the fjord.

Icebergs produced by calving glaciers constitute an ever-present hazard to ships in subpolar seas. When the S.S. *Titanic* sank after striking a berg in the North Atlantic in 1912, the detection of approaching icebergs relied on sailors' vision. Today, with sophisticated electronic equipment, large bergs can generally be identified well before an encounter. Nevertheless, ice has a density of 0.9, so that 90 percent of an iceberg lies under water, making it difficult to detect. In coastal Alaska, where calving glaciers are commonplace, icebergs pose a potential threat to huge oil tankers. For this reason, Columbia Glacier, which lies adjacent to the main shipping lanes from Valdez at the southern end of the Alaska Pipeline, is being closely monitored as its terminus pulls steadily back and multitudes of bergs are released.

Calving is the principal means of ablation for the large ice sheets of Antarctica. Large tabular icebergs breaking off the vast floating ice shelves that fill embayments between the East and West Antarctic ice sheets reach thicknesses of hundreds of meters. One exceptionally large berg about the size of Rhode Island drifted 2000 km in three years before it broke apart. Large bergs and a myriad of smaller ones that enter the ocean

every year constitute a vast potential resource of fresh water (see the boxed essay, *Ice from Antarctica for Arid Lands*).

How Glaciers Move

We can easily prove to ourselves that glaciers move. One way is to visit a glacier near the end of the summer and carefully measure the position of a boulder lying on the ice surface with respect to some fixed point beyond the glacier margin. If we return a year later, we will find that the boulder has moved at least several meters in the downglacier direction. Actually, it is the ice that has moved, carrying the boulder along for the ride.

What causes a glacier to move may not be immediately obvious, but we can find clues by examining the ice and the terrain over which it lies. These clues tell us that ice moves in two ways: by internal flow and by sliding of the basal ice over underlying rock or sediment.

Internal Flow
When an accumulating mass of snow and ice on a mountainside reaches a critical thickness, the mass will begin to deform and flow downslope under the pull of gravity. The flow takes place mainly through movement within individual ice crystals, which are subjected to higher and higher stress as the weight of the overlying snow and ice increases. Under this stress, ice crystals are deformed by slow displacement (termed *creep*) along internal crystal planes in much the same way that cards in a deck of playing cards slide past one another if the deck is pushed from one end (Fig. 12.15). As the compacted, frozen mass begins to move, stresses between adjacent ice crystals cause some to grow at the expense of others, and the resulting larger crystals end up with their internal planes oriented in the same direction.

A.

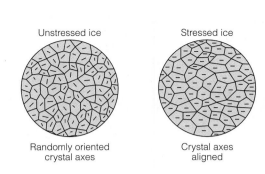

Unstressed ice   Stressed ice

Randomly oriented   Crystal axes
crystal axes   aligned

Figure 12.15 Deforming Ice Internal creep in ice crystals of a glacier. A. Randomly oriented crystals of ice in the upper layers of a glacier are transformed under stress so that their axes are aligned. B. When a stress is applied to an ice crystal, creep along internal planes results in slow deformation of the crystal.

B.

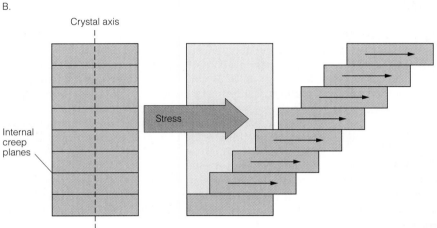

Crystal axis

Internal
creep
planes

Stress

This alignment of crystals leads to increased efficiency of flow, for the internal creep planes of all crystals now are parallel.

In contrast to deeper parts of a glacier where the ice flows as a result of internal creep, the surface portion of a glacier has relatively little weight on it and is brittle. Where a glacier passes over an abrupt change in slope, such as a bedrock cliff, the surface ice cracks as tension pulls it apart. The cracks open up and form crevasses. A **crevasse** is a deep, gaping fissure in the upper surface of a glacier, generally less than 50 m deep (Fig. 12.11). At depths greater than about 50 m, continuous flow of ice prevents crevasses from forming.

Basal Sliding

Ice temperature is very important in controlling the way a glacier moves and its rate of movement. Meltwater at the base of a temperate glacier acts as a lubricant and permits the ice to slide across its bed (the rocks or sediments on which the glacier rests). In some temperate glaciers, such sliding accounts for up to 90 percent of the total observed movement (Fig. 12.16). By contrast, polar glaciers are so cold that they are frozen to their bed. Their motion largely involves internal deformation rather than basal sliding, and so their rate of movement is greatly reduced.

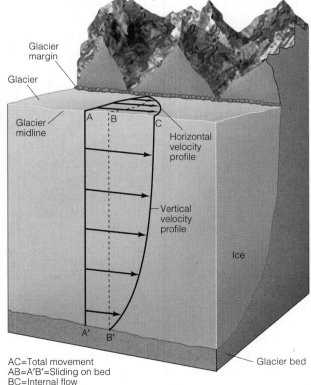

Glacier
margin

Glacier

Glacier
midline

A   B   C

Horizontal
velocity
profile

Vertical
velocity
profile

Ice

A'   B'

AC=Total movement
AB=A'B'=Sliding on bed
BC=Internal flow

Glacier bed

Figure 12.16 Glacial Flow and Sliding Three-dimensional view through half of a glacier showing horizontal and vertical velocity profiles. A portion of the total observed movement is due to internal flow within the ice, whereas part is due to sliding of the glacier along its bed, lubricated by a film of meltwater.

Using Resources

Ice from Antarctica for Arid Lands

As world population increases and our insatiable demand for fresh water continues to rise, a critical problem we face is how to meet this demand. Some imaginative people think the answer may lie in a vast unutilized resource in the frigid polar seas: Antarctic icebergs.

Captain James Cook was among the first to recognize the potential for using fresh water locked up in polar icebergs. In 1773, while his ship lay off the coast of Antarctica, his sailors hoisted 15 tons of berg ice on board, and Cook noted in his log that "this is the most expeditious way of watering I ever met with."

Today, scientists visualize towing large icebergs far to the north where they could supply water to the arid coasts of South America, Africa, Australia, and the southwestern United States. Although icebergs are shed from arctic glaciers as well, these icebergs are fewer and on average much smaller than those of the Antarctic region. A French engineering firm, commissioned by Saudi Arabia, has already studied the feasibility of towing Antarctic icebergs weighing as much as 85 million tons northward more than 8000 km to the Red Sea where they could supply drinking and agricultural water.

As impressive as such proposals sound, the technological problems are formidable. How can blocks of ice hundreds of meters thick and several square kilometers in area be towed or pushed thousands of kilometers through warm and often stormy seas and still have enough mass remaining to make the venture economically feasible? Large icebergs are plentiful around Antarctica, where an estimated 1000 km^3 of glacier ice breaks off as icebergs each year (Fig. B12.1). Those measuring 1 to 2 km long may be optimal for towing. A huge berg towed 3000 km northward through relatively cool waters to Australia at an average speed of 0.5 m/s (1 knot) would require 70 days to make the transit, but about 60 m of ice would melt from its sides. An unprotected iceberg of this size is unlikely to survive the long trip (up to a year) through much warmer waters to southern California or the Middle East. To avoid excessive loss, the ice would have to be insulated against melting and evaporation and protected against collapse and disintegration. In addition, the towing vessels or propulsion systems large enough to move huge icebergs over great distances have yet to be developed. Although these technological challenges appear immense, the day may come when ice that has been stored in Antarctic glaciers for tens of thousands of years will bring life-sustaining waters to the Earth's low-latitude arid lands.

Figure B12.1 Water for Deserts? A small tabular iceberg near Antarctica has calved off the margin of a nearby ice shelf. Scientists visualize towing bergs many times this size to supply fresh water to arid lands in Australia, the American Southwest, and the Middle East.

Velocities and Directions of Flow

Measurements of the surface velocity across a valley glacier show that the uppermost ice in the central part of the glacier moves faster than ice at the sides, similar to the velocity distribution in a river (Figs. 10.13 and 12.16). The reduced rates of flow toward the margins are due to frictional drag of the ice against the valley walls. A similar reduction in flow rate toward the bed is observed in a vertical profile of velocity (Fig. 12.16).

Snow continues to pile up in the accumulation area each year, while melting removes snow and ice from the ablation area. The surface profile of a glacier does not change much, however, because ice is transferred from the accumulation area to the ablation area. In the accumulation area, the mass of accumulating snow and ice is pulled downward by gravity, and so the dominant flow direction is toward the glacier bed. However, the ice does not build up to ever greater thickness because a downglacier component of flow is also present. Ice flowing downglacier replaces ice being lost from the glacier's surface in the ablation area, and so in this area the flow is upward toward the surface (Fig. 12.11). Water molecules that fall as snowflakes on the glacier near its head therefore have a long path to follow before they emerge near the terminus. Those falling close to the equilibrium line, however, travel only a short distance through the glacier before reaching the surface again.

Even if the mass balance of a glacier is negative and the terminus is retreating, the downglacier flow of ice is maintained. Retreat does not mean that the ice-flow direction reverses; instead, it means that the rate of flow downglacier is insufficient to offset the loss of ice at the terminus.

In most glaciers, flow velocities range from only a few centimeters to a few meters a day, or about as fast as the rate at which groundwater percolates through crustal rocks. Hundreds or thousands of years have elapsed since ice now exposed at the terminus of a very long glacier fell as snow near the top of its accumulation area.

Glacier Surges

Although most glaciers slowly grow or shrink in size as the climate fluctuates, some glaciers experience episodes of very unusual behavior marked by rapid movement and dramatic changes in size and form. Such an event, called a **surge**, is unrelated, or only secondarily related, to a change in climate. When a surge occurs, a glacier seems to go berserk. Ice in the accumulation area begins to move rapidly downglacier, producing a chaos of crevasses and broken pinnacles of ice in the ablation area. Medial moraines, which are bands of rocky debris marking the boundaries between adjacent tributary glaciers

(Fig. 12.11), are deformed into intricate patterns (Fig. 12.17). The termini of some glaciers have advanced up to several kilometers during surges. Rates of movement as great as 100 times those of nonsurging glaciers and reaching as much as 6 km a year have been measured.

The cause of surges is still imperfectly understood, but available evidence points to a reasonable hypothesis. We know that if water from melting ice is confined at the base of a glacier, the weight of the glacier will be supported by the water. Over a period of years, steadily increasing amounts of water trapped beneath the ice may lead to widespread separation of the glacier from its bed. The resulting effect is similar to the displacement of a beverage can on a sheet of glass in the experiment described in Chapter 9. According to this hypothesis, as the ice is floated off its bed, its forward mobility is greatly increased and it moves rapidly forward before the escape of water brings the surge to a halt.

GLACIATION

Glaciation, the modification of the land surface by the action of glacier ice, has occurred so recently over large areas of Europe, Asia, and North and South America that weathering, mass-wasting, and erosion by running water have not yet had time to alter the landscape appreciably. Except for a cover of vegetation, the appearance of these glaciated landscapes has remained nearly unchanged since they emerged from beneath the ice.

Like the geologic work of other surface processes, glaciation involves erosion and the transport and deposition of sediment.

Figure 12.17 Surging Glaciers Surging tributary glaciers flowing from the mountains on the right of this view have deformed the medial moraines of Alaska's Yanert Glacier into a series of complex folds.

Glacial Erosion and Sculpture

In changing the surface of the land over which it moves, a glacier acts collectively like a plow, a file, and a sled. As a plow, it scrapes up weathered rock and soil and plucks out blocks of bedrock; as a file, it rasps away firm rock; and as a sled, it carries away the load of sediment acquired by plowing and filing, along with additional rock debris that falls onto the glacier from adjacent slopes.

Small-Scale Features of Glacial Erosion

The base of a temperate glacier is studded with rock fragments of various sizes that are all carried along with the moving ice. Small fragments of rock embedded in the basal ice scrape away at the underlying bedrock and produce long, nearly parallel scratches called *glacial striations* (Fig. 12.18). Larger rock fragments that the ice drags across the bedrock abrade *glacial grooves* aligned in the direction of glacier flow. Grains of silt in the basal ice act like fine sandpaper and polish the rock until it has a smooth, reflective surface.

Because striations and grooves are aligned with the direction of ice flow, geologists use these and other aligned erosional features to reconstruct the flow paths of former glaciers. Debris-laden ice striates and polishes the upglacier sides of small bedrock knobs or hills over which it moves, and plucks blocks of rock from the downglacier sides. The resulting asymmetrical landforms, smooth and gently sloping upglacier and steeper and angular downglacier, provide clear evidence of the direction in which the glacier was moving (Fig. 12.19).

Landforms of Glaciated Mountains

Skiers racing down the steep slopes at Alta, Mammoth, or Whistler, and rock climbers inching their way up the cliffs of Yosemite Valley, the granite spires of Mont Blanc, or the icy monoliths of the southern Andes owe a debt to the ancient glaciers that carved these mountain playgrounds. The scenic splendor of most of the world's high mountains is the direct result of glacial sculpturing that has produced a distinctive assemblage of alpine landforms.

Cirques. Cirques are among the most common and distinctive landforms of glaciated mountains. The characteristic bowl-like shape of a cirque is the combined result of frost-wedging, glacial plucking, and abrasion (Fig. 12.20). Many cirques are bounded on their down-valley side by a bedrock threshold that impounds a small lake (a *tarn*).

A cirque probably begins to form beneath a large snowbank or snowfield at or just above the snowline. As meltwater infiltrates rock openings under the snow, it refreezes and expands, disrupting the rock and dis-

Figure 12.18 Glacial Markings A deglaciated bedrock surface beyond Findelen Glacier in the Swiss Alps displays grooves and striations etched by rocky debris in the base of the moving glacier when it overlay this site. In the background rises the Matterhorn, a glacial horn sculpted by glaciers that surround its flanks.

Figure 12.19 Glacial Sculpturing Asymmetrical glacially sculptured rock surface beyond the terminus of Franz Josef Glacier in New Zealand's Southern Alps. The glacier flowed from right to left. Slopes facing toward the glacier are smooth and polished. Steep slopes facing downvalley result from the plucking of bedrock blocks by flowing ice.

Figure 12.20 Aretes, Cirques, and Horns Sharp-crested arêtes flank the Western Cum, a deep cirque on the west side of Mount Everest in the central Himalaya. Sharp-crested peaks in the distance are horns that have resulted from headward growth of flanking cirque and valley glaciers.

lodging fragments. Small rock particles are carried away by snowmelt runoff during periods of thaw. A shallow depression in the land is thereby created, and the depression gradually becomes larger and larger. As snow continues to accumulate in the deepening hollow, the snowbank thickens and becomes a glacier. Plucking then helps to enlarge the cirque still more, and abrasion at the bed further deepens it.

As cirques on opposite sides of a mountain grow larger and larger, their headwalls intersect to produce a sharp-crested ridge called an *arête*. Where three or more cirques have sculptured a mountain mass, the result

can be a high, sharp-pointed peak (a *horn*), an example of which is the Matterhorn in the Swiss/Italian Alps (Fig. 12.18).

Glacial Valleys. Valleys that were shaped by former glaciers differ from ordinary stream valleys in several ways. The chief characteristics of glacial valleys include a U-shaped cross profile and a floor that lies below the floors of smaller tributary valleys (Fig. 12.21). Streams commonly descend as waterfalls, or cascades, as they flow from the tributary valleys into the main valley. The glaciated Cascade Range of western United States

Figure 12.21 Glaciation in Yosemite Repeated glaciations by thick valley glaciers carved the deep, U-shaped Yosemite Valley in California's Sierra Nevada. The valley glacier was nourished by an extensive mountain ice cap that covered the undulating upland surface of the range, seen in the distance.

derives its name from such streams. The long profile of a glaciated valley floor may possess steplike irregularities and shallow basins. These usually are related to the spacing of joints in the rock, which influenced the ease of glacial plucking, or to changes in rock type along the valley. Finally, the valley typically heads in a cirque or a group of cirques.

Fjords. Fjords are common features along the mountainous, west-facing coasts of Norway, Alaska, British Columbia, Chile, and New Zealand, as well as in northern Canada. Fjords typically are shallow at their seaward end and deepen inland. This is a result of deep glacial erosion upstream of a glacial terminus at an earlier time, which created elongate basins in glacial valley floors, sometimes to depths of more than 300 m below modern sea level (Fig. 12.22). Sognefjord in Norway reaches a depth of 1300 m, yet near its seaward end the water depth is only about 150 m.

Landforms Produced by Ice Caps and Ice Sheets
Abrasional Features. Landscapes that were shaped by overriding ice sheets display the same small-scale erosional features typical of valley glaciation. Striations have been especially helpful to geologists in recon-

Figure 12.22 Canadian Fjord A deep fjord indents the northeast coast of Baffin Island in northeastern Canada. If the fjord were drained of water, its cross-section would resemble that of Yosemite Valley (Fig. 12.21).

structing the flow lines of long-vanished ice sheets. Striations also demonstrate that basal ice in the thick central zones of former ice sheets was at the pressure melting point, a necessary condition for the sliding action required to produce these features. In peripheral zones, evidence of glacial erosion is sometimes less obvious, leading to the conclusion that the thinner ice there was cold and largely frozen to its bed.

Wherever ice sheets overwhelmed mountainous terrain, as in the high ranges of northwestern North America during the most recent glacial age, the highest evidence of glaciation can frequently be seen as the level where smooth, ice-abraded slopes pass abruptly upward into rugged, frost-shattered peaks and mountain crests that stood above the glacier surface. Some divides between adjacent drainages are broad and smooth and show evidence of glacial abrasion and plucking where they were overridden by ice.

Streamlined Forms. In many areas inside the limits of former ice sheets, the land surface has been molded into smooth, nearly parallel ridges, some of which are several kilometers long. Among the most distinctive of these landforms is the **drumlin**, a streamlined hill consisting largely of glacially deposited sediment and elongated parallel to the direction of ice flow (Fig. 12.23). Glacially molded drumlin-shaped hills of bedrock (called *rock drumlins*) also owe their shape to erosion by flowing ice. Drumlins and rock drumlins, like the streamlined bodies of supersonic airplanes that are designed to reduce air resistance, offered minimum resistance to glacier ice flowing over and around them.

Transport of Sediment by Glaciers

One way a glacier differs from a stream is the way in which it carries its load of rock particles. Unlike a stream, part of a glacier's coarse load can be carried at its sides and even on its surface. A glacier can carry far larger pieces of rock, and it can transport large and small pieces side by side without segregating them according to size and density into a bed load and a suspended load. Because of these differences, sediments deposited directly by a glacier are neither sorted nor stratified.

The load of a glacier typically is concentrated at its base and sides because these are the areas where glacier and bedrock are in contact and where abrasion and plucking are effective. Much of the rock debris on the surfaces of valley glaciers arrived there by rockfalls from adjacent cliffs. If a rockfall reaches the accumulation area, the flow paths of the ice (Fig. 12.11) will carry the debris downward through the glacier and then

Figure 12.23 Drumlins A field of drumlins in Dodge County, Wisconsin, each shaped like the inverted hull of a ship, are aligned parallel to the flow direction of the continental ice sheet that shaped them during the last glaciation.

upward to the surface in the ablation area. If rocks fall onto the ablation area, the debris will remain at the surface and be carried along by the moving ice. Where two glaciers join, rocky debris at their margins merges to form a distinctive, dark-colored medial moraine (Figs. 12.2 and 12.11).

Much of the load in the basal ice of a glacier consists of very fine sand and silt grains informally called *rock flour*. These particles have sharp, angular surfaces that are produced by crushing and grinding (Fig. 6.23A).

Glacial Deposits

Glaciers are efficient agents of erosion and transport. Many of the sediments and landforms they produce are distinctive, making it relatively easy for geologists to interpret the record of past glacier variations.

Sediments deposited by a glacier or by streams produced by melting glacier ice are collectively called

glacial drift, or simply **drift**. The term *drift* dates from the early nineteenth century when it was vaguely conjectured that all such deposits had been "drifted" to their resting places during the biblical flood of Noah or in some other ancient body of water. Glacial drift includes sediments associated both with moving ice and with stagnant ice. Several types of sediment are recognized. They form a gradational series ranging from nonsorted to sorted deposits.

Ice-laid Deposits

Till and Erratics. At one end of the range is **till**, which is nonsorted drift deposited directly from ice. The term was used by Scottish farmers long before the origin of the sediment was understood. The rock particles in a body of till lie just as they were released from the ice (Fig. 12.24). Most tills are a random mixture of rock fragments in which a matrix of fine-grained sediment surrounds larger stones of various sizes. The till matrix consists largely of sand and silt derived by abrasion of

Figure 12.24 Till Till deposited by the Laurentide Ice Sheet in the James Bay Lowland of southern Canada contains numerous igneous and metamorphic clasts resting in a fine-grained matrix. The well-developed jointing probably is related to the release of stress as the 2-km-thick glacier thinned and disappeared at the end of the last glaciation.

the glacier bed and from reworking of preexisting fine-grained sediments. Pebbles and larger rock fragments in till often have smoothed and abraded surfaces, and some are striated. Both the stones and the coarser matrix grains in till tend to lie with their longest axis aligned with the direction of ice flow.

As we learned in Chapter 6, a *tillite* is an ancient till that has been converted to rock. Tillites constitute a

Figure 12.25 Erratic Boulder Large erratic boulder of granite embedded in an end moraine of the last glaciation on Tierra del Fuego in southernmost Chile. The local bedrock is sedimentary. The nearest possible source area for the boulder lies in the high Cordillera Darwin to the south, on the opposite side of a deep fjord system.

primary line of evidence for pre-Pleistocene glacial ages.

In many cases, not all the boulders and smaller rock fragments in a till are the same kind of rock as the underlying bedrock, indicating that these components of the till were carried to their present site from somewhere else. A glacially deposited rock or rock fragment with a lithology different from that of the underlying bedrock is called an **erratic** (Latin for wanderer) (Fig. 12.25). Some huge erratics weigh many tons and are found tens or even hundreds of kilometers from their sources. In areas of ice-sheet glaciation, erratics derived from distinctive bedrock sources may have a fanlike distribution, spreading out from the area of outcrop and reflecting the diverging pattern of ice flow.

Glacialmarine Drift. Closely resembling till, **glacialmarine drift** is sediment deposited on the seafloor from ice shelves or bergs. As an iceberg or the base of an ice shelf slowly melts, the contained sediment is released and settles to the seafloor, where it forms a nonsorted deposit. Individual stones released from icebergs that have drifted into the open sea fall to the seafloor and plunge into unconsolidated marine sediments. The impact causes any laminated structure in the uppermost sediment layers to be deformed. Such *dropstones* are also common in the sediments of lakes that are ponded along glacier margins.

Moraines. A moving glacier carries with it rock debris eroded from the land over which it is passing or dropped on the glacier surface from adjacent cliffs. As the debris is transported past the equilibrium line and ablation reduces ice thickness, the debris begins to be deposited. Some of the basal debris is plastered directly onto the ground as till. Some also reaches the glacier margin where it is released by the melting ice and either accumulates there or is reworked by meltwater that transports it beyond the terminus.

An accumulation of till having a surface form that is unrelated to the underlying bedrock is called a **moraine**. If a body of drift is widespread, has a relatively smooth-surface topography, and consists of gently undulating knolls and shallow, closed depressions, we call it *ground moraine*; commonly, it is a blanket of till 10 m or more thick that was deposited beneath a glacier. By contrast, an *end moraine* is a ridgelike accumulation of drift deposited along the margin of a glacier (Fig. 12.26). An end moraine deposited at the glacier terminus is a *terminal moraine*, whereas a similar deposit along the side of a valley glacier is a *lateral moraine*. Normally, terminal and lateral moraines are segments of a single, continuous landform deposited below the equilibrium line.

End moraines can form in several ways: as sediment is bulldozed by a glacier advancing across the land, as loose surface debris on a glacier slides off and piles up

Figure   12.26   End Moraine   Tumbling glacier on Mount Robson in British Columbia has retreated upslope from a sharp-crested terminal moraine that it constructed during the nineteenth and twentieth centuries.

along a glacier margin, or as debris melts out of ice and accumulates at the margin of a glacier. End moraines range in height from a few meters to hundreds of meters. The great thickness of some lateral moraines results from the repeated accretion of sediment from debris-covered glaciers during successive ice advances.

Stratified Drift

In contrast to nonsorted till and glacialmarine drift, some glacial drift is both sorted and stratified. This kind of drift is not deposited directly by glacier ice but rather by meltwater flowing from the ice. **Stratified drift** ranges from coarse, very poorly sorted sandy gravels deposited by turbulent streams of meltwater to fine-grained, well-sorted silts and clays deposited in quiet water.

Outwash.   Stratified sediment deposited by meltwater streams as they flow away from a glacier margin is called **outwash** (because the sediment is washed "out" beyond the ice). Such streams typically have a braided pattern because of the large sediment load they are moving. If the streams are free to swing back and forth widely beyond the glacier terminus, they deposit outwash to form a broad *outwash plain*. Meltwater streams confined by valley walls build an outwash body called a *valley train* (Figs. 12.11 and 12.27A).

Following glacier retreat, a stream's sediment load is reduced and the underloaded stream cuts down into its outwash deposits to produce *outwash terraces* (Fig. 12.27B). A succession of terraces commonly is found in valleys that have experienced repeated glaciations (Fig. 12.28). Generally, each prominent terrace can be traced upstream to an end moraine or to the limit of a former glacier.

Deposits Associated with Stagnant Ice.   When rapid ablation greatly reduces a glacier's thickness in its terminal zone, ice flow may virtually cease. Sediment carried by meltwater flowing over or beside such stagnant ice is deposited as stratified drift that slumps and collapses as the supporting ice slowly melts away. Such sediment, called **ice-contact stratified drift**, is recognized by abrupt changes in grain size; distorted, offset, and irregular stratification; and extremely uneven surface form. Bodies of ice-contact stratified drift have many distinctive forms and are classified according to their shape (Fig. 12.27). Among the landforms most likely to be seen are a *kame*, a small hill of ice-contact stratified drift; a *kettle*, a basin in drift created by the melting away of a mass of underlying glacier ice; and an *esker*, a long, sinuous ridge of sand and gravel deposited by a meltwater stream flowing under or within stagnant glacier ice. Extremely uneven terrain underlain by ice-contact stratified drift and marked by numerous kettles and kames is clear evidence of former stagnant-ice conditions.

A

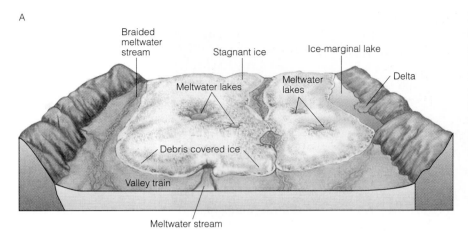

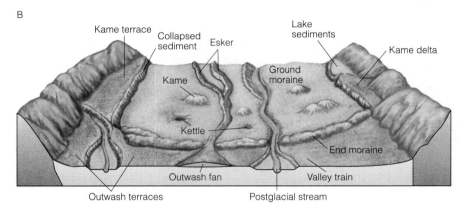

Figure 12.27 Features Formed by Stagnant Ice Origin of ice-contact stratified drift and related landforms associated with stagnant ice. A. Ablating stagnant ice furnishes temporary retaining walls for bodies of stratified sediment deposited by meltwater streams and in meltwater lakes. B. As ice melts, bodies of sediment slump, creating kettle-and-kame topography.

Figure 12.28 Outwash Terraces Flat-topped outwash terraces related to several phases of glaciation border meandering Cave Stream on South Island, New Zealand.

PERIGLACIAL LANDSCAPES AND PERMAFROST

Land areas beyond the limit of glaciers where low temperature and frost action are important factors in determining landscape characteristics are called **periglacial** zones. Periglacial conditions are found over more than 25 percent of the Earth's land areas, primarily in the circumpolar zones of each hemisphere and at high altitudes.

Permafrost

A common feature of periglacial regions is perennially frozen ground, also known as **permafrost**—sediment, soil, or even bedrock that remains continuously below freezing for an extended time (from two years to tens of thousands of years). The largest areas of permafrost occur in North America, northern Asia, and the high, cold Tibetan plateau (Fig. 12.29). It has also been found on many high mountain ranges, even including some lofty summits in tropical and subtropical latitudes. The southern limit of continuous permafrost in the northern hemisphere generally lies where the annual air temperature is between −5 and −10°C (23 and 14°F).

Most permafrost is believed to have originated during either the last glacial age or earlier glacial ages. Remains of woolly mammoth and other extinct ice-age animals found well preserved in frozen ground indicate that permafrost existed at the time of their death.

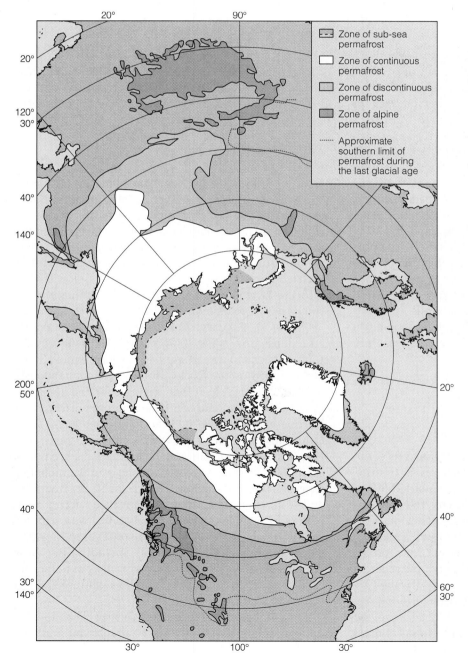

	Zone of sub-sea permafrost
	Zone of continuous permafrost
	Zone of discontinuous permafrost
	Zone of alpine permafrost
----	Approximate southern limit of permafrost during the last glacial age

Figure 12.29 Extent of Permafrost Distribution of permafrost in the northern hemisphere. Continuous permafrost lies mainly north of the 60th parallel and is most widespread in Siberia and Arctic Canada. Extensive alpine permafrost underlies the high, cold plateau region of central Asia. Smaller, isolated bodies occur in the high mountains of the western United States and Canada.

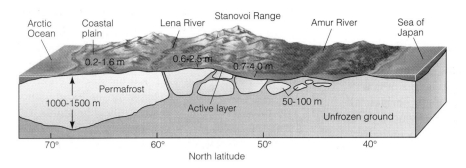

Figure 12.30 Permafrost in Siberia Diagrammatic transect across northeastern Siberia (vertical scale is greatly exaggerated) showing distribution and thickness of permafrost and thickness of the active layer. Thick, continuous permafrost under the Arctic coastal plain thins southward where it becomes discontinuous in response to warmer mean annual temperature. The active layer increases in thickness southward due to warmer summer temperatures.

The depth to which permafrost extends depends not only on the average air temperature but also on the rate at which heat flows upward from the Earth's interior and on how long the ground has remained continuously frozen. The maximum reported depth of permafrost is about 1500 m (4900 ft) in Siberia (Fig. 12.30). Thickness of about 1000 m (3300 ft) in the Canadian Arctic and at least 600 m (2000 ft) in northern Alaska have been measured. These areas of very thick permafrost all occur in high latitudes outside the limits of former ice sheets. The ice sheets would have insulated the ground surface and, where thick enough, caused ground temperatures beneath them to rise to the pressure melting point of ice. On the other hand, open ground unprotected from subfreezing air temperatures by an overlying ice sheet could have become frozen to great depths during prolonged cold periods.

Permafrost presents a real challenge to people who live and work in permafrost regions; see the essay *Living with Permafrost.*

THE GLACIAL AGES

As early as 1821, European scientists began to recognize features characteristic of glaciation in places far from any existing glaciers. They drew the then remarkable conclusion that glaciers must once have covered extensive regions. The concept of a glacial age with widespread effects was first proposed in 1837 by Louis Agassiz, a Swiss scientist who achieved considerable fame through his hypothesis. Although at first many people regarded Agassiz's idea as outrageous, gradually, through the work of many geologists, the concept gained widespread acceptance. Today, the study of the glacial ages provides us with dramatic evidence of rapid global climatic changes on the Earth and with clues about how natural

physical and biological systems responded to these changes. It also gives us important information about how glaciers behave and helps us understand some of the basic physical processes of the crust and upper mantle.

Ice-Age Glaciers

Over tens of millions of years, the climate slowly grew cooler as the Earth moved into a late Cenozoic glacial era. During the last few million years, the planet has experienced numerous glacial-interglacial cycles superimposed on the long-term cooling trend. At present, we find ourselves near a time of maximum warmth in such a cycle and possibly poised to begin a slow decline into the next glacial age which will culminate many thousands of years in the future.

About 30,000 years ago, late in the Pleistocene Epoch, an extensive ice sheet that had formed over eastern Canada began to spread south toward the United States and west toward the Rocky Mountains. Simultaneously, another great ice sheet that originated in the highlands of Scandinavia spread southward across northwestern Europe and overwhelmed the landscape (Fig. 12.31). Other large ice sheets grew over northern regions of North America and Eurasia, including some areas now submerged by shallow polar seas, and over the mountain ranges of western Canada. The ice sheets in Greenland and Antarctica grew larger and advanced across areas of the surrounding continental shelves that were exposed by falling sea level. Glaciers also developed in the world's major mountain ranges, including the Alps, Andes, Himalaya, and Rockies, as well as in numerous smaller ranges and on isolated peaks scattered widely through all latitudes.

On a global scale, the areas of former glaciation add up to an impressive total of more than 44 million km², which is about 29 percent of the Earth's present land

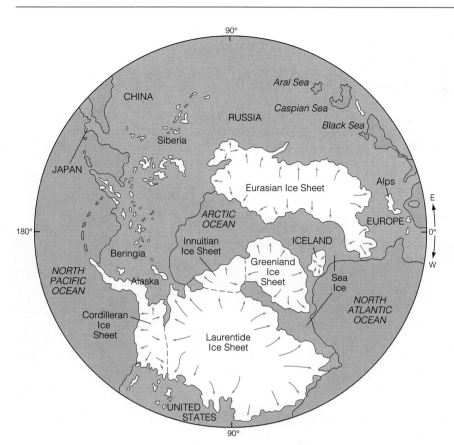

Figure 12.31 Northern Hemisphere Ice Sheets Areas of the northern hemisphere that were covered by glaciers during the last glacial age. Arrows show the general directions of ice flow. Coastlines are shown as they were at that time, when world sea level was at least 100 m lower than present. Sea ice, shown covering the Arctic Ocean, extended south into the North Atlantic. Some scientists postulate that thick ice shelves, rather than sea ice, covered these portions of the ocean. The extent of former glacier ice over shallow continental shelves of northern Eurasia, as well as in parts of northern North America, is controversial.

area. Today, by comparison, only about 10 percent of the world's land area is covered with glacier ice; of this area, 84 percent lies in the Antarctic region.

Drainage Diversions and Glacial Lakes

The growth of ice sheets over the continents caused disruption of major stream systems. In North America, repeated encroachment of Pleistocene ice sheets displaced the Missouri and Ohio rivers into new courses beyond the ice margin. When glaciers blocked preglacial drainage paths, water was ponded to form ice-dammed lakes. Vast ice-marginal lakes that developed beyond the edges of the expanding ice sheet in eastern North America shifted in location and changed size as the glacier receded. Major disruption of drainage also produced large ice-marginal lakes in northern Asia as ice expanding southward in western Siberia blocked the courses of major north-flowing rivers.

Lowering of Sea Level

Whenever large glaciers formed on the land, the moisture needed to produce and sustain them was derived primarily from the oceans. As a result, sea level was lowered in proportion to the volume of ice on land. During the most recent glacial age, world sea level fell at least 100 m, thereby causing large expanses of the shallow continental shelves to emerge as dry land. At that time, the Atlantic coast of the United States south of New York lay as much as 150 km east of its present position. At the same time, lowering of sea level joined Britain to France where the English Channel now lies, and North America and Asia formed a continuous landmass across what is now the Bering Strait (Fig. 12.31). These and other land connections allowed plants and animals, as well as humans, to pass freely between land areas that now are separated by ocean waters.

Deformation of the Crust

The weight of the massive ice sheets caused the crust of the Earth to subside beneath them, a process described further in Chapter 16. Because ice (density 0.9 g/cm^3) is one-third as dense as average crusted rock (2.7 g/cm^3), an ice sheet 3 km thick could cause the crust to subside by as much as 1 km. The Hudson Bay region of Canada, which 20,000 years ago lay near the center of the vast Laurentide Ice Sheet (Fig. 12.31), is still rising as the lithosphere and asthenosphere adjust to the removal of this ice load. Using glacial-geologic evidence, we can measure accurately the rates at which the crustal rocks

Understanding Our Environment

Living with Permafrost

In permafrost terrain, a thin surface layer of ground that thaws in summer and refreezes in winter is known as the *active layer*. In summer this thawed layer tends to become very unstable. The permafrost beneath, however, is capable of supporting large loads without deforming. Many of the landscape features we associate with periglacial regions reflect movement of regolith within the active layer during annual freeze and thaw cycles.

Permafrost presents unique problems for people living on it. If a building were constructed directly on the surface, the warm temperature developed when the building was heated would be likely to thaw the underlying per-

Figure B12.2 Cabin Sinks on Permafrost Ground This cabin in central Alaska settled more than a meter in eight years as permafrost beneath its foundation thawed.

have risen over many thousands of years. Such measurements provide us with important information about how the lithosphere and asthenosphere behave when subjected to changing loads.

Earlier Glaciations

Until recently, it was thought that the Earth had experienced four glacial ages during the Pleistocene Epoch. This assumption was based on studies of ice-sheet and mountain-glacier deposits and had its roots in early studies of the Alps, where geologists mapped moraines and outwash terraces and interpreted them as evidence of four Pleistocene glaciations. This traditional view had to be modified, however, when studies of deep-sea sediments disclosed a long succession of glaciations, the most recent of which was shown by radiocarbon dating to correspond to the youngest extensive glacial drift on land. Paleomagnetic dating of these marine sediments showed that during the last 800,000 years the lengths of the glacial-interglacial cycles averaged about 100,000 years. For the Pleistocene Epoch as a whole, more than 20 glacial ages are recorded, rather than the traditional four. Geologists now realize that the glacial record on land is incomplete and marked by numerous unconformities, whereas many areas of the deep sea contain a record of continuous sedimentation.

Seafloor Evidence

Deep-sea sediments provide some of the best evidence we have of the cycling of the climate from glacial to interglacial conditions. When we study the history of progressively earlier times by examining seafloor sedi-

mafrost, making the ground unstable (Fig. B12.2). Arctic inhabitants learned long ago that they must place the floors of their buildings above the land surface on pilings or open foundations so that cold air can circulate freely, thereby keeping the ground beneath the building frozen.

Wherever a continuous cover of low vegetation on a permafrost landscape is ruptured, melting can begin. As the permafrost melts, the ground collapses to form impermeable basins containing ponds and lakes. Thawing can also be caused by human activity, and the results can be environmentally disastrous. Large wheeled or tracked vehicles crossing the Arctic tundra can quickly rupture it. The water-filled linear depressions that result from thawing can remain as features of the landscape for many decades.

The discovery of a commercial oil field on the North Slope of Alaska in the 1960s generated the need to transport the oil southward by pipeline to an ice-free port. The company formed to construct the pipeline was faced with some unique problems. In order for the sticky oil to flow through a pipeline in the frigid Arctic environment, the oil had to be heated. However, an uninsulated, heated pipe in the frozen ground could melt the surrounding permafrost. Even if the pipe were insulated before placing it underground, the surface vegetation cover would be disrupted, likely leading to melting and instability. For these reasons, along much of its course, the pipeline was constructed in piers above the ground, thereby greatly reducing the possibility of ground collapse (Fig. B12.3).

Figure B12.3 Alaska Pipeline The Alaska Pipeline carries petroleum from the North Slope fields near Prudoe Bay southward across two mountain ranges en route to the port of Valdez. To increase the ease of flow through the pipe, the oil is heated. Because much of the pipeline route lies along permafrost terrain, in many sectors the huge pipe is suspended above ground to keep it from melting the frozen ground beneath.

ments that are sampled by drilling, we find that the biologic component of the sediments records repeated shifts in the composition of surface-water animal and plant populations—from warm interglacial forms to cold glacial forms and back to warm interglacial forms. The ratio of the amounts of the isotopes ^{18}O to ^{16}O in layers of calcareous ooze in these cores also fluctuates with a similar pattern. These $^{18}O/^{16}O$ variations in Pleistocene marine sediments are thought primarily to represent changes in global ice volume. When water evaporates, water containing the light isotope ^{16}O evaporates more easily than water containing the heavier ^{18}O. Consequently, when water evaporates from the ocean and is precipitated on land to form glaciers, the ice is somewhat enriched in the lighter isotope relative to the ocean water that remains behind. As a result, Pleistocene glaciers contained more of the light isotope,

while the oceans became enriched in the heavy isotope. Isotope curves derived from the seafloor sediments therefore provide a continuous reading of changing ice volume on the planet (Fig. 12.32). Because glaciers wax and wane in response to climatic changes, the isotope curves also give a generalized view of global climatic change.

Pre-Pleistocene Glaciations

Ancient glaciations, identified mainly by tillites and striated rock surfaces, are known from older parts of the geologic column (Fig. 6.9). The earliest recorded glaciation dates to about 2.3 billion years ago, in the early Proterozoic. Evidence of other glacial episodes has been found in rocks of late Proterozoic, early Paleozoic, and late Paleozoic age. During the late Paleozoic, 50 or more glaciations are believed to have occurred.

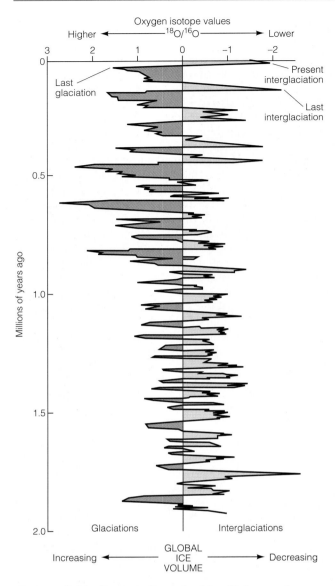

Figure 12.32 Isotopic Record of Ice Volume Average curve of oxygen-isotope variations in deep-sea cores representing global changes in ice volume during the last 2 million years. During most of the last million years, each glacial-interglacial cycle was about 100,000 years long; earlier cycles were about 40,000 years long.

Little Ice Ages

Old documents, lithographs, and paintings of Alpine valleys dating to the sixteenth and seventeenth centuries describe and depict glaciers that advanced over small villages and became larger than at any time in human memory (Fig. 12.33). Similar ice advances took place in other parts of the world and in many cases led to the greatest expansion of glaciers since the end of the last glacial age 10,000 years ago. During this recent interval of generally cool climate, which started in the mid-thirteenth century and lasted until the mid-nine-

teenth century, mountain glaciers expanded worldwide. This Little Ice Age, as it is commonly known, was similar to other brief episodes of glacier expansion that were superimposed on the much longer glacial-interglacial cycles.

WHAT CAUSES GLACIAL AGES?

Ever since geologists first became convinced that the Earth had experienced a succession of ice ages, the problem of their cause has been a subject of continuing debate and investigation. We are now much closer to resolving this problem than ever before, but many answers still elude us, for climatic change is a complex phenomenon that involves not only the atmosphere, but also the solid earth, the oceans, and the biosphere, as well as extraterrestrial factors. The ultimate solution to the problem of the ice ages, therefore, will involve the cooperative efforts of specialists from many different fields of science.

Glacial Eras and Shifting Continents

Several different successions of glacial and interglacial ages, each succession lasting tens of millions of years, can be identified in the geologic record. What seems to be the only reasonable explanation for their pattern is suggested by slow but important geographic changes that affect the crust of the planet. These changes include (1) the movement of continents as they are carried along with shifting plates of lithosphere; (2) the large-scale uplift of continental crust where continents collide; (3) the creation of mountain chains where one plate overrides another; and (4) the opening or closing of ocean basins and seaways between moving landmasses.

The effect of such earth movements on climate is illustrated by the fact that low temperatures are found, and glaciers tend to form and persist, in two kinds of situations: (1) on landmasses at high latitudes and (2) at high altitudes. Furthermore, glaciers are particularly common in places where winds can supply abundant moisture evaporated from a nearby ocean. Today, 84 percent of the Earth's glacier ice lies in Antarctica, where temperatures are constantly below freezing and the land is surrounded by ocean. The only glaciers found at or close to the equator lie at extremely high altitudes.

Abundant evidence leads us to conclude that the positions, shapes, and altitudes of landmasses have changed with time (Chapter 17), in the process having altered the paths of ocean currents and atmospheric circulation. Where evidence of ancient ice-sheet glacia-

A

B

Figure 12.33 Historic Changes in Glaciers   A. A lithograph made in the 1850s shows Rhone Glacier in the Swiss Alps close to its maximum extent during the Little Ice Age. The glacier terminates in a broad lobe that crosses the floor of the upper Rhone Valley

B. A photograph of the same area taken 110 years later shows the terminus of Rhone Glacier perched high up in the headward part of the valley. Extensive thinning and recession of this glacier and other glaciers worldwide marked the end of the Little Ice Age.

Measuring Our Earth

Understanding Milankovitch

The concept that slow, cyclic changes in the path the Earth takes around the Sun, and in the tilt of its axis, are related to changing climate on the scale of glacial and interglacial ages is generally referred to as the Milankovitch Theory. Three movements are involved:

First, the axis of rotation, which now points in the direction of the North Star, wobbles like a spinning top (Fig. B12.4A). The wobbling movement causes the Earth's axis to trace a cone in space, completing one full revolution every 26,000 years. If we were to look down on the Earth from a point above the Earth's axis, this movement would be clockwise. At the same time, the axis of the Earth's elliptical orbit is also rotating, but much more slowly, in the opposite direction. These two motions together cause a progressive shift in the position of the four cardinal points of the Earth's orbit (spring and autumn equinoxes and winter and summer solstices). As the equinoxes move slowly around the orbital path, a motion called *precession of the equinoxes,* they complete one full cycle in about 23,000 years.

Second, the *tilt* of the axis, which is now 23.5°, shifts over a range of about 3° during a span of about 41,000 years (Fig. B12.4B).

Finally, the *eccentricity* of the orbit, which is a measure of the departure of the orbit from circularity, changes over periods of 100,000 and 400,000 years. About 50,000 years ago, the orbit was more circular (lower eccentricity) than it has been for the last 10,000 years (Fig B12.4C).

A. Precession of the equinoxes (period = 23,000 years)

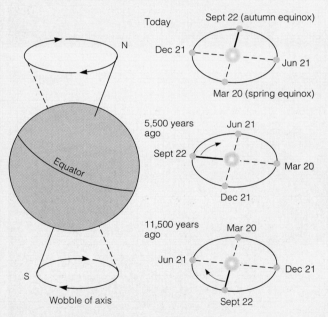

B. Tilt of the axis (period = 41,000 years)

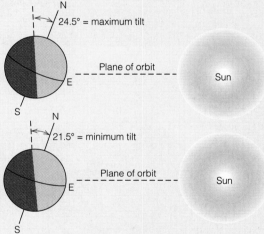

C. Eccentricity (dominant period =100,000 years)

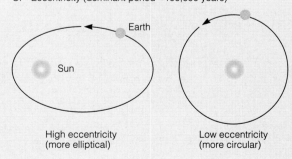

High eccentricity (more elliptical)   Low eccentricity (more circular)

Figure   B12.4   Geometry of the Orbit   Geometry of the Earth's orbit and axial tilt.   A. *Precession.* The Earth wobbles on its axis like a spinning top, making one revolution every 26,000 years. The axis of the Earth's elliptical orbit also rotates, though more slowly, in the opposite direction. These motions together cause a progressive shift, or precession, of the spring and autumn equinoxes, with each cycle lasting about 23,000 years.   B. *Tilt.* The tilt of the Earth's axis, which now is about 23.5°, ranges from 21.5 to 24.5°, with each cycle lasting about 41,000 years. Increasing tilt means a greater difference, for each hemisphere, between the amount of solar radiation received in summer and that received in winter.C. *Eccentricity.* The Earth's orbit is an ellipse with the Sun at one focus. The shape of the orbit changes from almost circular (low eccen-tricity) to more elliptical (high eccentricity) over periods of 100,000 and 400,000 years. The higher the eccentricity, the greater is the seasonal variation in radiation received at the higher latitudes of the hemisphere tilted toward the Sun when the Earth is in the part of the orbit that brings it closest to the Sun.

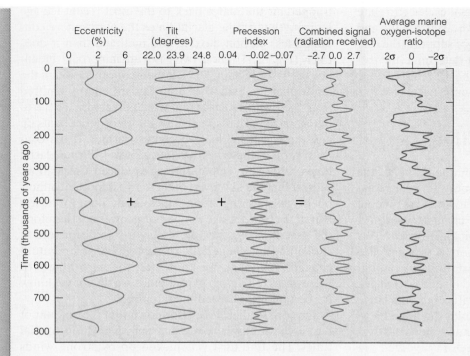

Figure B12.5 Orbital Changes Curves showing variations in orbital eccentricity, tilt, and precession index during the last 800,000 years. Summing the effects of these factors produces a combined signal that shows the amount of radiation received on the Earth at a particular latitude through time.

The slow but predictable changes in orbital path and tilt cause long-term variations of as much as 10 percent in the amount of radiant energy that reaches any particular place on the Earth's surface in a given season. Such variations are most pronounced at high latitudes. The effects of the three variables can be summed, and we can derive a combined signal that depicts the amount of radiation received at any specified latitude and at any specified time in the past (Fig. B12.5). Because the astronomical variables are predictable, we can also use this theory to extrapolate the signal into the future and thereby forecast the timing of future glaciations and interglaciations.

tion is now found in low latitudes, we infer that such lands were formerly located in higher latitudes where large glaciers could be sustained. The most dramatic case concerns glacial deposits of late Paleozoic age that are exposed in South America, southern Africa, India, Australia, and Antarctica. Tillites and related rocks have been interpreted as deposits of continental glaciers that repeatedly covered large portions of a vast southern continent (called Gondwanaland) which was located near the south pole of the Earth. As a result of the breakup and subsequent northward movement of large fragments of Gondwanaland, many of the ancient glacial rocks are now found in low latitudes (Fig. 17.1).

The absence of widespread glacial deposits in rocks of Mesozoic age implies that during this era most of the world's landmasses had moved away from polar latitudes and that climates were mild. By the early Cenozoic, slowly shifting landmasses had once again moved into polar latitudes, and tectonic movements were beginning to uplift large areas of the western United States and central Asia to high altitudes. By middle Cenozoic time, the Earth was again poised to enter another lengthy glacial era.

Ice Ages and the Astronomical Theory

As initially discovered through studies of glacial deposits and later verified by studies of deep-sea cores, glacial and interglacial ages have alternated for almost 3 million years (Fig. 12.32). Determining their cause has long been a fundamental challenge to the development of a comprehensive theory of climate. A preliminary answer was provided by Scottish geologist John Croll, in the mid-nineteenth century, and later elaborated by Milutin Milankovitch, a Serbian astronomer of the early twentieth century.

Croll and Milankovitch recognized that minor variations in the Earth's orbit around the Sun and in the tilt of the Earth's axis cause slight but important variations in the amount of radiant energy reaching any given latitude on the Earth's surface (see *Measuring Our Earth: Understanding Milankovitch*). By reconstructing and dating the history of climatic variations during the Quaternary Period, geologists have shown that fluctuations of climate on the time scale of glacial cycles correlate strongly with cyclic variations in the Earth's tilt and orbital path. This persuasive evidence supports the theory that astronomical changes that determine the distribution of radiation reaching the Earth's surface control the *timing* of the glacial-interglacial cycles.

Atmospheric Composition

Although orbital factors can explain the timing of the glacial-interglacial cycles, the variations in solar radiation reaching the Earth's surface are too small to account for the average global temperature changes of 4 to 10° C that are implied by geologic and biologic evidence. We therefore must conclude that other factors are also involved. Somehow, the slight variations in radiant energy received from the Sun, caused by orbital changes, must be translated into a temperature change sufficiently large to generate and maintain the huge Pleistocene ice sheets. We do not yet know how this was accomplished, but some of the factors involved are likely to be changes in the chemical composition and dustiness of the atmosphere, and changes in the reflectivity of the Earth's surface.

Air bubbles in glacier ice of the present Antarctic and Greenland ice sheets are samples of the Earth's ancient atmosphere. Studies of the chemical composition of trapped air that dates back to the most recent ice age indicate that during glacial times the atmosphere contained less carbon dioxide and methane than it does today (Fig. 12.34). These two gases are important "greenhouse" gases (Chapter 19). If their concentration in the atmosphere is high, they trap radiant energy emitted from the Earth's surface that would otherwise escape to outer space. As a result, the lower atmosphere heats up and the Earth's climate becomes warmer. If the concentration of these gases is low, as it was during glacial times, surface air temperatures are reduced. Calculations suggest that the low levels of these two important atmospheric gases during glacial times can account for nearly half of the total ice-age temperature lowering. Therefore, the greenhouse gases likely play a significant role in explaining the *magnitude* of past global temperature changes. Although we know that the percentages of these gases fell during glacial times, we do not yet know for certain what caused them to drop.

Ice core studies have also shown that the amount of dust in the atmosphere was unusually high during glacial times. The fine dust was picked up by strong winds blowing across outwash deposits and dry desert basins. So much dust was delivered to the atmosphere that the sky must have appeared hazy much of the time. The fine atmospheric dust scattered incoming radiation back into space, which would have further cooled the Earth's surface.

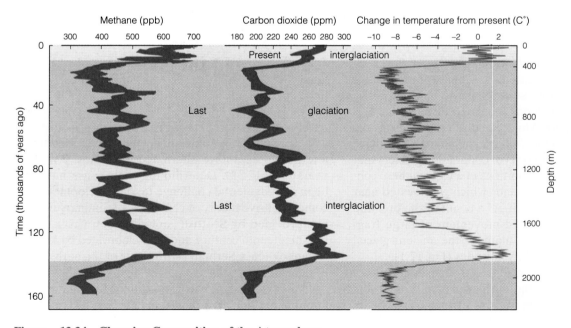

Figure 12.34 Changing Composition of the Atmosphere
Curves comparing changes in carbon dioxide and methane with temperature changes based on oxygen-isotope values in samples from a deep ice core drilled at Vostok Station, Antarctica. Concentrations of these greenhouse gases were high during the early part of the last interglaciation, just as they are now, but they were lower during glacial times. The curves are consistent with the hypothesis that these gases contributed to warm interglacial climates and cold glacial climates.

Whenever the world enters a glacial age, large areas of land are progressively covered by snow and glacier ice. The highly reflective surfaces of snow and ice scatter incoming radiation back into space, further cooling the lower atmosphere. Together with lower greenhouse gas concentrations and increased atmospheric dust, this additional cooling would favor the expansion of glaciers.

Changes in Ocean Circulation

The circulation of ocean waters also plays an important role in global climate. As warm surface water moving northward into the North Atlantic evaporates, the remaining water becomes saltier and cooler. The resulting cold, salty water is dense and sinks deep within the ocean. Heat released to the atmosphere as the water cools maintains a relatively mild climate in northwestern Europe. Consider, however, what would happen if this circulation ceased.

The rate of deep-ocean circulation is sensitive to surface salinity at sites where dense water forms. Studies have shown that during times of reduced salinity, the deep-ocean circulation is reduced. We therefore can postulate that as summer radiation decreased at the onset of a glaciation, the high-latitude ocean and atmosphere cooled, decreasing evaporation and leading to the expansion of sea ice. The resulting freshening of the high-latitude surface water would have halted the production of dense, salty water, thereby shutting down the vertical circulation system. Reduction of high-latitude evaporation, significantly reducing the release of heat to the atmosphere, would have maintained cold air masses moving eastward across the North Atlantic. Further cooled by an expanding sea-ice cover in the North Atlantic and by the growing ice sheets on the continents, the climate of Europe would have become increasingly colder, ultimately causing the formation of permafrost in a broad zone beyond the ice-sheet margin.

Thus, a change in the ocean's circulation system provides a means of amplifying the relatively small climatic effect attributable to astronomical changes. Furthermore, it may help explain why the Earth's climate system appears to fluctuate between two relatively stable modes—one in which the ocean circulation system is operational (during interglaciations) and one in which it has shut down (during glaciations).

Solar Variations, Volcanic Activity, and Little Ice Ages

Climatic fluctuations lasting for centuries or decades were responsible for the Little Ice Age and similar episodes of glacier expansion. However, such fluctuations are too brief to be caused either by movements of continents or variations in the Earth's orbit; they therefore require us to seek other explanations for their cause. Two have received special attention.

One hypothesis regarding the cause of glacial events like the Little Ice Age is based on the concept that the energy output of the Sun fluctuates over time. The idea is appealing because it might explain climatic variations on several different time scales. However, although correlations have been proposed between weather patterns and rhythmic fluctuations in the number of sunspots appearing on the surface of the Sun, as yet there has been no clear demonstration that solar variations are responsible for climatic changes on the scale of the Little Ice Age.

Large explosive volcanic eruptions can eject huge quantities of ash into the atmosphere to create a veil of fine dust that circles the globe. Like other types of dust, the fine ash particles tend to scatter incoming solar radiation, resulting in a slight cooling at the Earth's surface. Although the dust settles out rather quickly, generally within a few months to a year, tiny droplets of sulfuric acid, produced by the interaction of volcanically emitted SO_2 gas with oxygen and water vapor, also scatter the Sun's rays, and such droplets remain in the upper atmosphere for several years. After a large eruption, volcanic dust and gases in the atmosphere can lower average surface air temperature by 0.5 to 1.0 C°, sufficient to influence the mass balance of glaciers. A close association between intervals of glacier advance and periods of unusually strong volcanic activity during the last several centuries lends support to the hypothesis that volcanic emissions can produce detectable changes of climate on the decadal time scale.

SUMMARY

1.   Glaciers are permanent bodies of moving ice that consist largely of recrystallized snow.

2.   Based on geometry, glaciers are classified as cirque glaciers, valley glaciers, fjord glaciers, piedmont glaciers, ice caps, ice sheets, and ice shelves.

3.   Ice in a temperate glacier is at the pressure melt-

ing point, and liquid water exists at the base of the glacier; in a polar glacier, ice is below the pressure melting point and is frozen to the rock on which it rests.

4. Glaciers can form only at or above the snowline, which is close to sea level in polar regions and rises to high altitudes in the tropics.

5. The mass balance of a glacier is measured in terms of accumulation and ablation. The equilibrium line separates the accumulation area from the ablation area and marks the level on the glacier where gain is balanced by loss.

6. Temperate glaciers move as a result of internal flow and basal sliding. In polar glaciers, which are frozen to their bed, motion is much slower and involves only internal flow. Surges involve extremely rapid flow, probably related to excessive amounts of water at the base of a glacier.

7. Glaciers erode rock by plucking and abrasion. Rock debris, transported chiefly at the base and sides of a glacier, includes fragments of all sizes, from fine rock flour to large boulders.

8. Mountain glaciers erode stream valleys into U-shaped glacial valleys with cirques at their heads. Fjords are excavated far below sea level by glaciers in high-latitude coastal regions.

9. Glacial drift is sediment deposited by glaciers and glacial meltwater. Till is deposited directly by glaciers, while glacialmarine drift is deposited on the seafloor from floating glacier ice. Stratified drift includes outwash deposited by meltwater streams and ice-contact stratified drift deposited on or against stagnant ice.

10. Ground moraine is built up beneath a glacier, whereas end moraines (both terminal and lateral) form at a glacier margin.

11. Permafrost, a common feature of periglacial zones,

is confined mainly to areas where annual air temperature is at least $-5°C$. It reaches maximum thickness of at least 1500 m and is believed to have formed during glacial ages in subfreezing landscapes not covered by continental ice sheets.

12. Permafrost can present unique engineering problems, for thawing commences when the vegetation cover is broken, leading to collapse and extreme instability of the ground surface.

13. During glacial ages, huge ice sheets repeatedly covered northern North America and Eurasia, causing the crust beneath the ice to subside and world sea level to fall.

14. Glacial ages have alternated with interglacial ages in which temperatures approximated those of today. Studies of marine cores indicate that more than 20 glacial-interglacial cycles occurred during the Pleistocene Epoch.

15. Glacial eras in Earth history probably are related to the favorable positioning of continents and ocean basins, brought about by movements of lithospheric plates. The timing of glacial-interglacial cycles appears to be closely controlled by the Earth's precession as well as by changes in the eccentricity of the Earth's orbit and the tilt of the axis rotation, which affect the distribution of solar radiation received at the Earth's surface. Changes in the atmospheric concentration of carbon dioxide, methane, and dust may help explain the magnitude of global temperature lowering during glacial ages, while changes in ocean circulation may help explain the shifts between relatively stable glacial and interglacial modes of the climate system.

16. Climatic variations on the scale of centuries and decades have been ascribed to fluctuations in energy output from the Sun or to injections of volcanic dust and gases into the atmosphere.

THE LANGUAGE OF GEOLOGY

ablation (p. 314)
accumulation (p. 314)

calving (p. 315)
cirque (p. 308)
crevasse (p. 317)

drift (p. 323)
drumlin (p. 322)

equilibrium line (p. 314)
erratic (p. 324)

fjord (p. 309)

glacial drift (p. 323)
glacialmarine drift (p. 324)
glaciation (p. 319)
glacier (p. 308)

ice-contact stratified drift (p. 325)

mass balance (p. 314)
moraine (p. 324)

outwash (p. 325)

periglacial (p. 327)
permafrost (p. 327)
polar glacier (p. 311)
pressure melting point (p. 311)

snowline (p. 312)
stratified drift (p. 325)
surge (p. 319)

temperate glacier (p. 311)
till (p. 323)

QUESTIONS FOR REVIEW

1. What distinguishes temperate glaciers from polar glaciers?

2. What is the snowline and how are glaciers related to it?

3. Describe how snow is converted to glacier ice.

4. Why does the position of the equilibrium line provide a rough estimate of a glacier's mass balance?

5. Why is there a lag in time between a change of climate and the response of a glacier's terminus to the change?

6. In what ways does ice temperature influence the way a glacier moves?

7. Describe the unique motions of surging and calving glaciers.

8. Illustrate how small-scale and large-scale erosional features can be used to infer directions of flow of former glaciers.

9. What is the active layer in permafrost terrain, and how does it form?

10. Where, and why, would you expect to find permafrost at latitudes of less than 40°?

11. Describe what potential foundation problems a home builder might encounter in northern Alaska if the contractor were to clear the building site of vegetation and begin construction on the exposed ground surface.

12. How might you distinguish till from stratified drift in a roadside outcrop?

13. In what different ways are moraines built at a glacier margin?

14. What can you say about the condition of a glacier in association with which kettles and kames are developing?

15. Why, and by approximately how much, does world sea level fall and rise during glacial-interglacial cycles?

16. What evidence obtained from deep-sea cores indicates that glacial-interglacial cycles have occurred repeatedly during the Pleistocene Epoch?

17. What natural factors explain the recurrence of glacial events on time scales of tens of thousands of years?

18. How might large volcanic eruptions influence climate and cause glaciers to grow or shrink?

13

Measuring
Our Earth

Understanding
Our Environment

GeoMedia

Death of a date palm grove. Advancing sand dunes are overwhelming a grove of date palms growing around an oasis near Chinguetti, Mauritania.

Wind Action and Deserts

The Dust Bowl: Could It Happen Again?

During the mid-1930s, many farm families in the southern Great Plains region of the United States abandoned their homes and lands and trekked west, becoming part of a great migration described by John Steinbeck in his award-winning novel *The Grapes of Wrath*. A primary factor behind the exodus was the onset of a severe drought. The drought led to an increase in the frequency of major dust storms that devastated the land by destroying crops and burying formerly productive fields with drifting sand and dust. As a consequence, the southern plains came to be called the "dust bowl," and we refer to those times as the "dust-bowl years." In a particularly great storm on March 20, 1935, a cloud of suspended sediment extended 3.6 km above Wichita, Kansas. The dust load in the lowermost 1.6 km of this cloud was estimated to be 35 million kg in each cubic kilometer. Samples of sediment collected from flat roofs showed that, on the day of the storm, about 280,000 kg of rock particles, or about 0.5 percent of the load suspended in the lowermost layer of air, was deposited on each square kilometer of land. Enough sediment was carried eastward on March 21 to bring temporary twilight at mid-day over New York and New England.

The effectiveness of the wind in creating the dust bowl was aided by decades of poor land-use practices. Grasses growing on the prairies when the original settlers arrived protected the rich topsoil from wind erosion. However, the grasses were progressively replaced by plowed fields and seasonal grain crops that left the ground bare and vulnerable part of the year. Although today the land is still potentially vulnerable in drought years, improved farming practices should reduce the likelihood of similar catastrophes in the future.

WIND AS A GEOLOGIC AGENT

If we lived on Mars instead of the Earth, a substantial percentage of this book would likely be devoted to the theme of wind action and deserts, for Mars is an arid, windy, and dusty planet. When the Mariner 9 spacecraft approached Mars in 1971, for instance, a dust storm of major proportions enveloped much of the planet and continued unabated for several months. Photographs taken both during this Mariner mission and during subsequent Viking missions revealed a planetary surface that has been extensively modified by wind action.

Wind is also an important agent of erosion and sediment transport on the Earth, but its effects are visible mainly in desert regions, where few people live. Most of the world's population is concentrated in the relatively moist parts of the temperate and tropical latitudes (0–66° N and S), where a protective cover of vegetation makes wind an ineffective geologic agent. Nevertheless, even in these populated regions the occurrence of ancient sand dunes and widespread deposits of dust show us that wind has been important in shaping the landscape at times when the continents were drier and windier places than they are today.

Planetary Wind System

To help us understand why winds are effective geologic agents in some regions but not in others, we need to see how surface winds are related to the global circulation of the atmosphere.

Circulation of the Atmosphere
The atmosphere consists of a mixture of gases that together we call air. The atmosphere is always moving, a fact we are well aware of whenever we feel a gentle breeze or a strong wind blowing. The basic reason the atmosphere is always in motion is that more of the Sun's heat is received per unit of land surface near the equator than near the poles. This unequal heating gives rise to convection currents. The heated air near the equator expands, becomes lighter, and rises. High up, it spreads outward in the direction of both poles. As the upper air travels both northward and southward, it gradually cools, becomes heavier, and sinks. On reaching the Earth's surface, this cool, descending air then flows back toward the equator, warms up, and rises, thereby completing a cycle of convection.

The Coriolis Effect
If the Earth did not rotate, the convection currents in the atmosphere would simply flow from the equator to the poles and back again. But of course the Earth does rotate, and its rotation complicates the convection currents in the atmosphere (and in the ocean). The **Coriolis Effect** is named for the nineteenth-century French mathematician, Gaspard-Gustave de Coriolis, who first analyzed the effect. To the observer on the Earth, the Coriolis effect causes anything that moves freely with respect to the rotating Earth (such as a plane, a missile, or the wind) to veer off course. It is like trying to throw a ball to a friend when you are both on spinning merry-go-rounds.

In the northern hemisphere the Coriolis effect causes a moving mass to veer to the right of the direction in which it is moving. and in the southern hemisphere to the left. To understand how the Coriolis effect works, consider what happens to a small mass of air that flows northward from the equator. Because the Earth and the atmosphere are rotating eastward, the northward-flowing air is also moving eastward. The eastward speed due to the Earth's rotation is 1670 km/h at the equator, but at successively higher latitudes the eastward speed due to rotation is less and less, until at the north pole it is zero.

A northward mass of flowing air will keep the eastward velocity with which it started because it is not fixed to the Earth's surface. Thus, as the northward-flowing-mass of air moves further and further north, its eastward velocity will be more rapid than the eastward velocity of the Earth's surface immediately below. To an astronaut in a space vessel, the air mass would flow in a straight line, but to an observer on the ground who would have an eastward velocity due to the Earth's rotation, the air mass would appear to have been deflected eastward, which is to say, to the right. The amount of the deflection is a function of the speed of the moving air mass and the latitude.

The Coriolis effect breaks up the simple flow of air between the equator and the poles into belts (Fig. 13.1). The result, in both the northern and southern hemispheres, is a large cell of circulating air lying between the equator and about 30° latitude. In these low-latitude cells, the prevailing winds are northeasterly in the north-

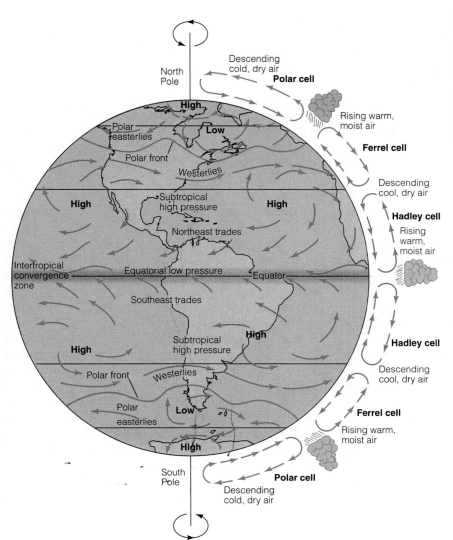

Figure 13.1 Global Wind Patterns
The Earth's planetary wind system, shown schematically. Moist air, heated in the warm equatorial zone, rises convectively and forms clouds that produce abundant rain. Cool, dry air descending at latitudes 20–30° produces a belt of subtropical high pressure in which lie many of the world's great deserts.

ern hemisphere (that is, they flow *from* the northeast toward the southwest), while in the southern hemisphere they are southeasterly. These wind systems are called the *tradewinds*, for their direction and consistent flow carried trade ships across the tropical oceans at a time when winds were the chief source of power.

In each hemisphere, a second cell of circulating air lies poleward of the low-latitude cell. In these second, middle-latitude cells, westerly winds prevail (blowing *from* the west). In these cells, cold equatorward-flowing upper air descends near 20–30° latitude in both hemispheres, and northward-moving surface air rises in higher latitudes where it meets dense, cold air flowing from the polar regions.

A third cell of circulating air lies over the polar regions. In each cell, cold, dry, upper air descends near the pole and moves equatorward in a wind system called the polar easterlies. As this air slowly warms and encounters the belt of westerlies, it rises along the *polar front* and returns toward the pole.

The cold upper air converging where the low- and mid-latitude cells meet cannot hold as much moisture as warm air, so where this cold air descends, dry conditions are created at the land surface. As a result, much of the arid land in both the northern and southern hemispheres is centered between latitudes 20° and 30°. By contrast, abundant moisture in the warm air surrounding the equator condenses as the air rises and becomes cooler and denser, creating clouds that release their moisture as tropical rains.

Climate

The global pattern of air flow, which is determined by the nonuniform heating of the Earth's surface, the Coriolis effect, the distribution of land and sea, and the topography of the land, ultimately controls the variety and pattern of the Earth's climates. **Climate** is the average weather of a place, together with the degree of variability of that weather, over a period of years. It is measured by such factors as temperature, precipitation, cloudiness, and windiness. If the Earth had no mountains and no oceans to affect the moving atmosphere, the major climatic zones would all lie parallel to the equator. However, the pattern of climatic zones is distorted by the distribution of oceans, continents, high mountains, and plateaus. As a result, average temperature, precipitation, cloudiness, and windiness vary greatly from one place to another and give rise to an array of distinct climatic regions.

Movement of Sediment by Wind

We have all seen pictures of the tremendous destruction wreaked by hurricanes and typhoons when winds achieve speeds of at least 130 km/h and sometimes reach 300 km/h. The force of such a wind is so strong that trees are uprooted, houses are ripped apart, and large objects are thrown substantial distances. Fortunately, hurricane winds are the exception, but they provide a glimpse of the potential power of wind as a geologic agent. (see *Measuring the Earth's Hum.*)

Because the density of air at sea level (1.2 kg/m^3) is far less than that of water (1000 kg/m^3), air cannot move as large a particle as water flowing at the same velocity can. In extraordinary wind storms, when wind speeds locally reach or exceed 300 km/h, coarse rock particles up to several centimeters in diameter can be lifted to heights of a meter or more. Pebbles swept aloft by exceptional winds have been found lodged in buildings, trees, and cracks in telephone poles. In most regions, however, wind speed rarely exceeds 50 km/h, a velocity described as a strong wind. In a strong wind, the largest particles of sediment that can be suspended in the air stream are grains of sand. Larger particles settle out too quickly to remain aloft. At lower wind speeds, sand moves along close to the ground surface, and only finer grains of dust move in suspension.

Wind-Blown Sand

If a wind blows across a bed of sand, the grains begin to move when the wind speed reaches about 4.5 m/s (16 km/h). The resulting forward rolling motion of the sand is called *surface creep* (Fig. 13.2). With increasing wind speed, turbulence near the ground lifts moving sand grains into the air, where they travel along arcuate paths, landing a short distance downwind. This is the same process (*saltation*) we see in a stream, where grains of sand move close to the bottom, also following arcuate paths (see Chapter 10).

Saltation. Saltation accounts for at least three-quarters of the sand transport in areas covered by sand dunes. Measurements of the rate of sand movement in deserts of the Middle East indicate a rapid increase in sand movement with increasing wind speed. For example, a strong wind blowing at 58 km/h will move as much sediment in one day as it would take a wind blowing at 29 km/h to move in three weeks.

If a wind is strong enough, it can start a grain rolling along the ground where it may impact another grain and knock it into the air. As this second grain falls to the ground, it will impact other grains, some of which are thrown upward into the air stream. Within a very short time the air close to the ground may contain a very large number of saltating sand grains, all moving along with the wind and moving in arclike paths somewhat similar to those of a Ping-Pong ball bouncing across a table (Fig. 13.3). However, even in strong winds, saltating sand grains seldom rise far off the ground, as shown by abrasion marks on utility poles and fence posts that are sandblasted up to a height of about a meter.

Sand Ripples. Sheets of well-sorted sand that have accumulated on the land surface are inherently unstable, even under gentle winds. As the wind passes across

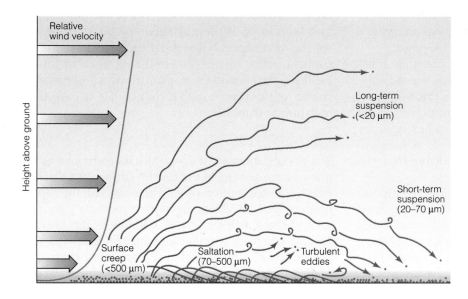

Figure 13.2 Suspension of Dust Under conditions of moderate wind, sand grains larger than 500 mm (0.5 mm) in diameter move by surface creep, while smaller grains (70–500 μm) saltate across the ground surface. Still finer particles (20–70 μm) are carried aloft in turbulent eddies and encounter faster moving air that transports them downwind as they slowly settle to the ground. The finest dust (less than 20 μm) reaches greater heights and is swept along in suspension as long as the wind is blowing.

Relative wind velocity

Height above ground

Long-term suspension (<20 μm)

Short-term suspension (20–70 μm)

Surface creep (<500 μm)

Saltation (70–500 μm)

Turbulent eddies

Measuring Our Earth

Measuring the Earth's Hum

Sometimes the Earth seems to clang like a bell. High-magnitude earthquakes, such as the great Chilean quake of 1960, cause the entire planet to vibrate for days; the oscillations move the ground up and down by as much as a centimeter. However, seismologists have recently identified a very different type of whole-earth vibration: a deep, soft hum that can be picked up only by the most sensitive seismographs.

At first, researchers speculated that the hum might result from the combined effects of all the earthquakes that occur throughout the world. But this would require a continual series of earthquakes with magnitudes of 5.8 or more. On the average such earthquakes occur every few days, yet the Earth's low hum is constant, with periods ranging from three to eight minutes. All the world's known earthquakes, taken together, would not be enough to produce this result.

At the annual meeting of the American Geophysical Union in the fall of 1998, a group of seismologists led by Naoki Suda of Japan proposed another idea: that the Earth hums along with the wind. They examined seismic records for periods of 50 to 80 days from especially quiet sites in Europe, South Africa, and central Asia. After removing background noise, they found that at each location the strength of the hum increased and decreased at predictable times each day—it was strongest from noon to 8 P.M. and weakest from midnight to 6 A.M. The same pattern of activity can be seen in the world's thunderstorms. The researchers concluded that the seismic hum may be a result of turbulent winds striking the Earth's surface.

Not everyone agrees with this proposal, and Suda's team has stated that it is preliminary. Other suggested explanations of the Earth's hum include small undetected earthquakes, ocean currents, and the movement of tectonic plates. More daily seismic records from more sites will need to be analyzed before seismologists can reach a firm conclusion about why the Earth hums.

Source: Richard A. Kerr, "New Data Hint at Why Earth Hums and Mountains Rise," *Science*, January 15, 1999, pp. 320–321.

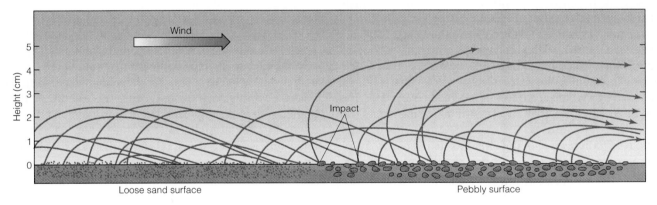

Figure 13.3 Saltation Strong wind causes movement of sand grains by saltation. Impacted grains bounce into the air and are carried along by the wind as gravity pulls them back to the land surface where they impact other particles, repeating the process.

such an accumulation, saltation moves the smaller, most easily transported grains. Sand grains too large to be moved are left behind. As the saltating finer grains impact the surface at some average distance downwind, additional fine particles are set in motion, and another accumulation of coarse grains develops as the fine sand moves onward. By this process, the coarse grains form a series of small, linear ridges of sand called **sand ripples**. Sand ripples tend to be aligned in a regular pattern with their crests oriented perpendicular to the wind direction (Fig. 13.4). Under very strong winds, the ripples disappear because then all grains can be moved and sorting is less likely to occur.

Figure 13.4 Sand Ripples Sand ripples cross the surface of a desert sand sheet on the floor of Death Valley, California.

Wind-Blown Dust

Sand grains blown across the land surface travel slowly and are deposited quickly when wind velocity subsides. Fine particles of dust (silt- and clay-size sediment), on the other hand, travel faster, longer, and much farther before settling to the ground.

As you might guess, the dustiest places on Earth tend to coincide with some of the world's major desert regions. Estimates of annual world dust production, which are imprecise at best, range as high as 5 billion tons. Among the many types of terrain that can give rise to large quantities of dust, the following are especially important: dry lake and stream beds, alluvial fans, outwash plains of glacial streams, and regions underlain by deposits of wind-blown dust that have lost their vegetation cover as a result of either climatic change or human disturbance.

Mobilization and Transport of Dust. As a result of frictional drag, the velocity of moving air decreases sharply near the ground surface. Right at the surface lies a layer of relatively quiet air less than 0.5 mm thick (Fig. 13.5). Sand grains that protrude above this layer of quiet air can be swept aloft by rising turbulent eddies. Dust particles, however, are so small and often so closely packed that they form a very smooth surface that does not protrude above the quiet air. Even a strong wind blowing over such a surface may not disturb the dust. Mobilization of the fine sediment may require the impact of saltating sand grains or other physical disruption of the smooth surface. We can see how such dust is set in motion by looking at a dusty desert road covered by dry, compact silt on a windy day. The wind blowing across the road generates little or no dust, but a vehicle driving over the road creates a choking cloud, which is blown away downwind before settling once more to the ground. The passing wheels have broken up the surface of the powdery dust that was too smooth to be disturbed by the wind, lifted dust particles into the air, and created turbulence in the atmosphere that keeps the particles aloft.

Once in the air, dust constitutes the wind's suspended load. The grains of dust are continually tossed about by eddies, like particles in a stream of turbulent water, while gravity tends to pull them toward the ground (Fig. 13.2). Meanwhile, the wind carries the dust forward. Although in most cases suspended sediment is deposited fairly near its place of origin, strong winds associated with large dust storms are known to carry very fine dust into the upper atmosphere, where it can be transported thousands of kilometers.

Dust Storms. Dust storms are the chief events leading to large-scale transport of dust. In a dust storm, the visibility at eye level is reduced to 1000 m or less by dust raised from the ground surface. Such storms are most frequent in the vast arid and semiarid regions of central Australia, western China, Russian Central Asia, the Middle East, and North Africa, as shown in Figure 13.6. In the United States, blowing dust is especially common in the southern Great Plains and in the desert regions of California and Arizona.

Deposition of Dust. Dust can be deposited if (1) wind velocity and air turbulence decrease so that particles can no longer remain in suspension, (2) the particles collide with rough or moist surfaces that trap them, or with surfaces having a weak electrical charge that attracts them, (3) the particles accumulate to form aggregates, which then settle out because of their greater mass, or (4) the particles are washed out of the air by rain.

Vegetation acts as a trap for descending dust particles because wind velocity is reduced over vegetated landscapes. Forest is more efficient at trapping dust than is low-lying vegetation because of the greater effect trees have on reducing wind velocity in the critical zone above the ground.

Deposition also occurs where a topographic obstacle causes a divergence of air flow, thereby leading to reduced wind velocity behind the obstruction (Fig. 13.7). This explains why deposits of dust are generally thick on the

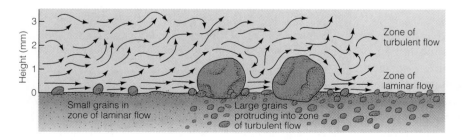

Figure 13.5 Laminar and Turbulent Flow Particles of fine sand and silt at the ground surface lie within a zone of laminar air flow less than 0.5 mm thick where wind velocity is extremely low. As a result, it is difficult for the wind to dislodge and erode these small grains. Larger grains protrude into a zone of faster moving, turbulent air. The turbulence, which exerts a greater push on the top of the grains than does the laminar flow at their base, makes it easy to start the grains moving.

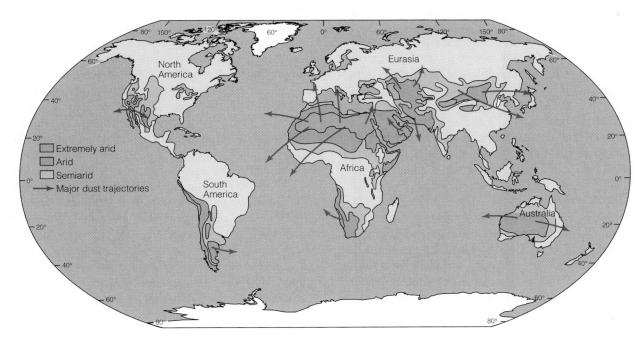

Figure 13.6   Dust Storms   Major dust storms are most frequent in arid and semiarid regions that are concentrated in the belts of subtropical high pressure north and south of the equatorial zone. Arrows show the most common trajectories of dust transported during major storms.

lee, or downwind, side of obstacles (the side away from the wind) and either thin or absent on the *windward*, or upwind, side (the side from which the wind is blowing).

Coarse and medium-grained dust particles, which are carried at relatively low altitudes, are likely to settle out first. Finer particles are carried higher in the atmosphere and therefore may remain suspended for long periods.

Detrimental Effects of Wind-Blown Sediment. Wind action can create many problems for people. Each year the material losses can range into the billions of dollars, and the direct and indirect effects of wind can lead to a significant loss in human lives.

Blowing sand and dust can severely damage crops and other vegetation. Large wind storms are reported to have caused major damage to wheat fields, citrus orchards, and vineyards, as well as to root crops. In one severe California storm in 1977, cattle were asphyxiated by dense dust, and hair and skin were sandblasted from their hind quarters. A strong, dense dust storm can quickly turn a car's clear windshield into a sheet of frosted glass, and blowing sand can pit and strip away the shiny paint of a new car.

The engines of vehicles operating in dusty areas are particularly susceptible to damage. Dust can contaminate fuel, clog air filters, abrade cylinders, and cause electrical short circuits. In the Second World War, higher-than-average engine failure among vehicles operating in

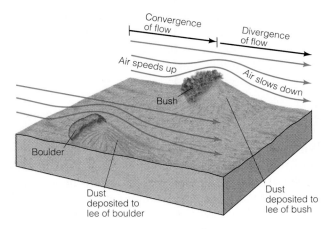

Figure 13.7   Effect of Obstacles on Wind   Obstacles in the path of wind produce a convergence of air flow, causing the air to move faster. As the air passes across the obstacle, the flow paths diverge, leading to lower velocity. The lower velocity reduces the wind's carrying capacity, causing the air to drop some of its dust load, which accumulates on the lee side of the obstruction.

desert areas was attributed to such abnormal cylinder wear. Dust similarly created a problem for military vehicles operating in the deserts of Saudi Arabia and Kuwait during the Gulf War of 1991.

Blowing dust can severely reduce visibility on roads and highways. Near Tucson, Arizona, clouds of dust caused so many accidents that a Dust Storm Alert system was introduced in 1976 to warn motorists of hazardous conditions.

Airborne dust also can pose a hazard to aircraft. In 1973 a Royal Jordanian Airlines plane crashed at an airport in northern Nigeria while flying through dense dust, killing 176 persons. An attempt to rescue U.S. hostages in Iran in 1979 was aborted after helicopter engine failure, attributed to dense airborne dust, caused a fatal crash in the desert staging area. Helicopter pilots operating in the Saudi Arabian desert in 1990–91 reported that dense dust raised while flying close to the ground produced a bright electrostatic glow around their rotor blades, thereby greatly reducing the effectiveness of light-sensitive night-vision goggles.

Inhalation of dust can lead to various medical problems. When too many fine mineral particles enter the lungs, tissue damage can cause emphysema. Quartz inhalation can lead to silicosis, a debilitating disease common among unprotected miners working in dusty conditions. In addition, disease-causing organisms may be carried in dust, where they may survive for long periods. Among the deadly germs that can be transported in wind-blown dust are anthrax and tetanus. In central China, a close correlation has been found between deaths due to cancer of the esophagus and the distribution of dust deposits, with the death rate increasing as the average grain size of the dust decreases. Outside China, the disease is mainly present in the dusty, arid regions of Iran, central Asia, Mongolia, and Siberia, suggesting that the cancer may somehow be related to persistent inhalation of fine dust.

Wind Erosion

Wind is an important agent of erosion wherever winds are strong and persistent and either the land is too dry to support vegetation or the influx of airborne sediment is so rapid that vegetation cannot gain a foothold and thereby stabilize the ground surface. Flowing air erodes in two ways. **Deflation** (from the Latin word meaning to blow away) is the picking up and removal of sand and dust by the wind. This process provides most of the wind's load. The second process, *abrasion*, results when rock is impacted by wind-driven grains of sediment.

Deflation

Deflation on a large scale takes place only where little or no vegetation exists and where loose rock particles are fine enough to be picked up by the wind. Areas of significant deflation are found mainly in deserts; nondesert sites where deflation occurs include ocean beaches, the shores of large lakes, and the floodplains of large glacial streams (Fig. 13.8). Of greatest economic importance, however, is the deflation of bare plowed fields in farmland. Although deflation may occur seasonally when land is

plowed, it is especially severe during times of drought, when no moisture is present to hold soil particles together.

Generally, deflation lowers the land surface slowly and irregularly, making measurements of wind erosion difficult. However, in the dry 1930s, deflation in parts of the western United States amounted to 1 m or more within only a few years. This is a tremendous rate compared with the long-term average rate of erosion for the region as a whole, which is only a few centimeters per thousand years.

Deflation Hollows and Basins. Small saucer- or trough-shaped hollows and larger basins created by wind erosion are among the most conspicuous evidence of deflation. Tens of thousands of these basins occur in the semiarid Great Plains regions of North America from Canada to Texas. Most are less than 2 km long and only a meter or two deep. In wet years, the larger ones are carpeted with grass and may contain shallow lakes. However, in dry years soil moisture evaporates, grass dies away, and the wind deflates the bare soil.

Where sediments are particularly susceptible to erosion by wind, deflation basins can be excavated to depths of 50 m or more. The immense Qattara Depression in the Libyan desert of western Egypt, the floor of which lies more than 100 m below sea level, has been attributed to intense deflation. In any basin, the depth to which deflation can reach is limited only by the water table. As deflation lowers the land to the level at which the ground is saturated, the surface soil becomes moist, thereby encouraging the growth of vegetation that inhibits further deflation.

Figure 13.8   Deflation of Glacial Sediments   Active deflation of meltwater sediments downstream from Tasman Glacier in the Southern Alps of New Zealand produces clouds of dust. Loess is accumulating on vegetation-covered glacial deposits in the foreground.

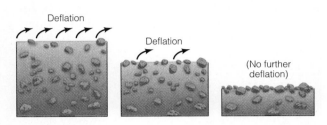

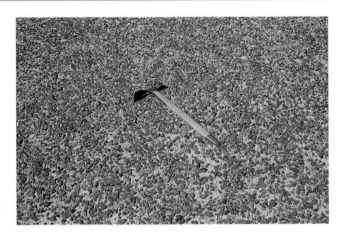

Figure 13.9 Deflation Progressive deflation of a poorly sorted sediment leads to the development of a desert pavement.

Figure 13.10 Desert Pavement A desert pavement on the floor of Searles Valley, California, consists of a layer of gravel, too coarse to be moved by the wind, that covers finer sediment and inhibits further deflation.

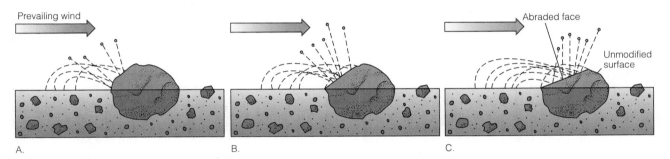

Figure 13.11 Ventifacts Stages in the formation of a ventifact. A. Saltating sand grains impact the upwind side of a stone that protrudes above a desert surface. B. As the sand grains chip away at the rock, a facet is cut across the upwind side of the stone. C. With continued sand blasting, the abraded face is reduced to a lower angle. Given time, the exposed face of the ventifact could be reduced to ground level.

Desert Pavement. When sand and dust are either blown away from a deposit of alluvium, or locally removed by sheet erosion (Chapter 10), stones too large to be moved become concentrated at the surface (Fig. 13.9). Eventually, a continuous cover of stones forms a **desert pavement**, so called because the stones fit together almost like the cobbles in a cobblestone pavement (Fig. 13.10).

Abrasion

Ventifacts. Where bedrock and loose stones are abraded by wind-driven sand and dust, they acquire a distinctive shape and surface polish. Any bedrock surface or stone that has been abraded and shaped by wind-blown sediment is a **ventifact** (Fig. 13.11). A ventifact can be identified by its having at least one smooth, abraded surface that faces upwind (Fig. 13.12). If erosion of surrounding sediment causes a stone to shift position, or if the wind direction changes, a new surface with a different orientation will be cut. Where this new abraded surface inter-

Figure 13.12 Field of Ventifacts Ventifacts litter the ground surface near Lake Vida in Victoria Valley, Antarctica. The most intensely abraded surfaces are inclined to the right, in the direction from which strong winds blow off the East Antarctic Ice Sheet.

sects the initial abraded surface, a sharp keel-like edge is produced. Because wind-abraded surfaces form facing upwind, ventifacts that have not been reoriented can be used to measure the prevailing wind direction.

Yardangs. Among the common eolian landforms of some desert regions is an elongate, streamlined, wind-eroded ridge called a **yardang** (from the Turkic word *yar*, meaning steep bank). Typically, yardangs are sharp-crested and carved from hard, compacted sediments or from highly weathered crystalline rocks (Fig. 13.13). Some have a shape that resembles an inverted ship's hull. Individual yardangs range up to a few tens of kilometers long and up to 100 m high. Generally, they occur in groups. Yardangs probably form initially by differential deflation along irregular depressions in the land surface that lie approximately parallel to the wind direction. Then they increase in size as the abrading action of wind-blown sand and dust further deepens and broadens the depressions, creating free-standing intervening ridges.

Eolian Deposits

Although wind action is now chiefly confined to arid and semiarid lands, deposits of wind-transported sediment are found not only in these regions but also in regions that are moist and covered by vegetation. The explanation for this distribution lies partly in changing climatic regimes and partly in the fact that wind-blown dust often settles out far from its place of origin.

Dunes

A **dune** is a hill or ridge of sand deposited by winds. Although little is known about how dunes initially form, it is likely that a dune develops where some minor surface irregularity or obstacle distorts the flow of air. Wind velocity within a meter or two of the ground varies with the slightest irregularity of the land surface. On encountering any small obstacle, wind sweeps over and around it but leaves a pocket of slower moving air immediately downwind. In such a pocket of low wind velocity, sand grains moving with the wind drop out and begin to build a mound. The mound, in turn, influences the flow of air and may continue to grow until it forms a dune.

Dune Form and Size. A typical isolated sand dune is asymmetrical. It has a gently sloping windward face, whose maximum slope is the angle of repose, generally 33–34° (Fig. 13.14). Sand grains move up the windward slope by saltation to reach the crest of the dune. As most saltation jumps are much shorter than the length of the lee face, grains making it past the crest of the dune generally fall onto the lee face near its top. The bulge thus created through grain-by-grain accumulation eventually reaches an unstable angle. The sand then

Figure 13.13 Yardang A sharp-crested yardang carved from compact lake sediments rises above the floor of Rogers Lake playa in southeastern California. The crest of the yardang is oriented parallel to the prevailing wind direction.

avalanches (slips) downward, spreading the grains of the bulge down the lee face. For this reason, the lee face of an active dune is also known as the *slip face*. The continual avalanching of sand grains keeps the slip face at the angle of repose and produces cross strata much like the foreset layers in a delta (Fig. 13.14).

The angle of the windward slope of a dune varies with wind velocity and grain size, but is always much less than that of the slip face. This asymmetry of form provides a means of telling the direction of the wind that shaped the dune, for the slip face always lies on the downwind side. Furthermore, the cross strata within the dune that represent former slip faces also slope in the direction toward which the wind was blowing when each stratum was deposited.

Many dunes grow to heights of 30 to 100 m, and some massive desert dunes in the western Alashan Plain of China reach heights of 500 m or more (Fig. 13.15). The height to which any dune can grow probably is determined by the maximum wind velocity, which increases above the land surface. At some height, the wind will reach a velocity great enough to carry the sand grains up into suspension off the top of a dune as fast as they move up the windward slope by saltation. Thus, the dune can grow higher only as long as the rate at which sand is supplied to the crest exceeds the rate of removal by the wind.

Dune Types. Five different dune types are shown in Table 13.1. Dune type is controlled by the amount of sand available, the variability of wind direction, and

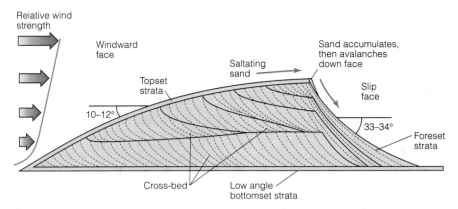

Figure 13.14 **Anatomy of a Dune** Cross section through a barchan dune showing typical gentle windward slope and steep slip face. Thin topset strata overlie sets of cross-bedded strata representing old slip faces. Sand grains saltating up the windward slope fall onto the top of the slip face where they accumulate before avalanching downward to produce foreset strata resting at the angle of repose.

the amount of vegetation cover. Where the amount of sand is limited and lack of moisture inhibits growth of vegetation, strong winds from one direction can build an isolated *barchan dune*. These crescent-shaped dunes, having horns that point downwind, are the best-studied dune type. Barchan dunes can migrate over great distances without much change in form. Where sand supply is greater, individual barchans may merge to form a *transverse dune* having a sinuous crest oriented perpendicular to the direction of the strongest wind. A *linear dune*, which tends to be long and relatively straight and to occur in rather regularly spaced groups, forms mainly in areas of limited sand supply and variable (often bidirectional) winds. Wind blowing from all directions can produce a large, pyramidal *star dune* with sinuous radiating arms. Although they range in general

from 50 to 150 m high, star dunes more than 300 m high are known. A *parabolic dune* is shaped like a U or a V and has two trailing arms, generally vegetated, pointing upwind. These dunes are common in coastal dune fields, where relatively constant wind off the ocean creates a moist environment that allows vegetation to grow. Parabolic dunes almost always develop where local disturbance of the vegetation allows deflation to build an accumulation of sand that grows to large size, and they may develop multiple crests and slip faces.

Dune Migration. Transfer of sand from the windward to the lee side of an active dune causes the whole dune to migrate slowly downwind. Measurements of barchan dunes show rates of migration as great as 25 m/yr. The migration of dunes, particularly along sandy coasts and

Figure 13.15 **Giant Dune** A huge sand dune in the Alashan Plain of western China towers over two horsemen, reaching a height of several hundred meters above the desert floor.

TABLE 13.1	Principal Types of Dunes Based on Form

Dune Type	Definition and Occurrence	
Barchan dune	A crescent-shaped dune with horns pointing downwind; occurs on hard, flat desert floors in areas of constant wind direction and limited sand supply; height 1 m to more than 30 m	
Transverse dune	A dune forming an asymmetrical ridge transverse to dominant wind direction; occurs in areas with abundant sand; can form by merging of individual barchans	
Linear dune	A long, relatively straight, ridge-shaped dune in deserts with limited sand supply and variable (bidirectional) winds; slip faces change orientation as wind shifts direction	
Star dune	An isolated hill of sand having a base that resembles a star in plan; sinuous arms of dune converge to form central peak as high as 300 m; tends to remain fixed in place in areas where wind blows from all directions	
Parabolic dune	A dune shaped like a U or V, with open end facing upwind; trailing arms, generally stabilized by vegetation, also point upwind; common in coastal dune fields; some form by piling of sand along lee and lateral margins of deflated areas in older dunes	

across desert oases, has been known to bury houses and farmers' fields (Fig. 13.16), fill in canals, and even threaten the existence of towns. In such places, sand encroachment is countered most effectively by planting vegetation that can survive in the very dry sandy soil of the dunes. Continuous plant cover inhibits dune migration for the same reason that it inhibits deflation: if the wind cannot move sand grains across it, a dune cannot migrate.

Sand Seas
Some large deserts contain vast tracks of shifting sand known as **sand seas**. Among the best examples are those found in northern and western Africa, the vast desert region of the Arabian Peninsula, and the large deserts of western China (Fig. 13.17). Sand seas contain a variety of dune forms, ranging from low mounds of sand to barchans, transverse dunes, and star dunes. In a typical sand sea, huge dune complexes form a seemingly endless and monotonous landscape.

Loess
Although most regolith contains a small proportion of wind-laid dust, the dust is thoroughly mixed with other fine sediments, making it indistinguishable from them. However, in some regions wind-laid dust is so thick and uniform that it constitutes a distinctive deposit and may

Figure 13.16 Field of Barchans Sand dunes advance from right to left across irrigated fields in the Danakil Depression, Egypt.

control the primary landscape characteristics. Known as **loess** (German for *loose*), this sediment is defined as wind-deposited dust consisting largely of silt but commonly accompanied by some fine sand and clay.

Loess is an important resource in countries where it is thick and widespread. Its importance lies in the productive soils developed on it. The rich agricultural lands of the upper Mississippi Valley, the Columbia Plateau of Washington State, the Loess Plateau region of central China, and much of eastern Europe are developed on rich loessial soils that provide food for millions. Loess deposits can also provide shelter: throughout the loess region of central China, caves carved in loess make dry homes for thousands of families (Fig. 13.18). Paleosols interstratified in the Chinese loess constitute a primary material for brick making.

Characteristics. Loess has two characteristics that indicate it was deposited by the wind rather than by streams, in marine water, or in lakes: (1) it forms a rather uniform

Figure 13.17 Sand Sea A vast sand sea stretches to the horizon in the center of China's vast Takla Makan desert, which lies thousands of kilometers from oceanic sources of precipitation.

Figure 13.18 Loess Cave A cave excavated by hand in compact loess provides a roomy and comfortable home for a Chinese family in the loess region near Xian.

blanket, mantling hills and valleys alike through a wide range of altitudes, and (2) it contains fossils of land plants and air-breathing animals. It typically is homogeneous, lacks stratification, and, where exposed, can stand at such a steep angle that it forms vertical cliffs (Fig. 13.19), just as though it were firmly cemented rock. This last property results from the fine grain size of loess, for molecular attraction among the particles is strong enough to make them very cohesive.

A vertical face of loess will stand for a long time without collapsing, but if shaken or disturbed, it may suddenly fail. The loess region of central China is prone to violent earthquakes. Although major earthquakes are infrequent, when one does occur, the results can be disastrous. The most devastating earthquake on record struck the loess country of Shaanxi Province in 1556. Of the estimated 830,000 people killed, many likely were buried when violent shaking of the ground caused steep cliffs of loess to fail and loess caves to collapse.

Origin. The distribution of loess shows that its principal sources are deserts and the floodplains of glacial meltwater streams. The loess that covers some 800,000 km^2 in central China, and in places reaches a thickness of more than 300 m, was blown there from the floors of the great desert basins of central Asia. The source of this loess has generally been ascribed to weathering of rocks in the deserts of northern China and Mongolia. However, it is difficult to account for the volume of the deposits and the rather high sedimentation rate by weathering alone. At least part of the sediment may have come from the breakdown of rocks by frost action and glacial processes in the high glaciated mountains of inner Asia and the subsequent deflation of resulting fine particles that were transported by streams onto large alluvial fans and dry lake floors in adjacent desert basins.

Glacial Loess of North America and Europe. Loess of glacial origin is widespread in the middle part of North America (especially Nebraska, South Dakota, Iowa, Missouri, and Illinois) and in east-central Europe (especially Austria, Hungary, the Czech Republic, and Slovakia). It has two distinctive features: first, the shapes and compositions of its particles resemble those of the fine sediment produced by the grinding action of glaciers. Second, glacial loess is thickest downwind from former large braided meltwater streams, such as the Mississippi and Missouri rivers in North America and the Rhine and Danube rivers in Europe, which were the primary source of the silt in loess deposits. We infer that areas just outside the margins of the large Pleistocene ice sheets were very cold and windy and that the floodplains of the constantly shifting meltwater streams were easily deflated because they remained largely bare of vegetation. In the central United States, the coarsest fraction of wind-blown sediment settled out adjacent to the source valleys, forming deposits 8 to 30 m thick (Fig. 13.20). Downwind, with increasing distance from the sediment source, the loess decreases progressively in thickness and average grain size.

Figure 13.19 Loess in Central China Loess deposited during the last glacial age forms the vertical wall of a valley adjacent to Huang He (Yellow River) in central China.

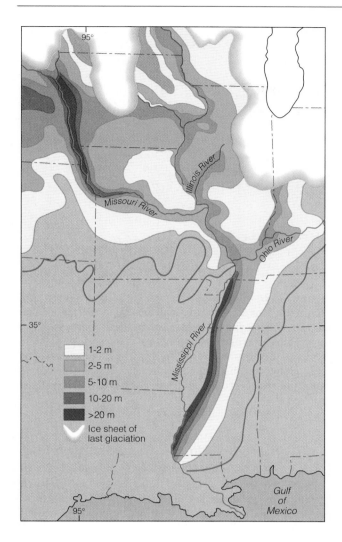

Figure 13.20 Loess in Central USA Loess deposits in the Mississippi Valley region of central United States. The loess is found largely beyond the limits of continental ice sheets that flowed into the northern United States during successive Pleistocene glaciations. The thickest loess lies adjacent to the courses of major streams that carried sediment-laden meltwater from the glacier margin. The loess thins downwind from the river source areas, showing that prevailing winds at times of loess deposition must have been predominantly from the west or northwest. Red line shows downwind limit of nearly continuous loess deposits.

Dust in Ocean Sediments and Glacier Ice

Dust blown over the oceans falls out and settles to the seafloor, where it forms an important component of deep-sea sediments. Plumelike deposits of eolian dust, identified by studying the mineral content and chemical composition of deep-sea muds, trend eastward across the North Pacific from China, westward across the subtropical North Atlantic from Africa, and westward into the Indian Ocean from Australia (Fig. 13.21). Fine particles of quartz deflated from Asian deserts have been found in soils of the Hawaiian Islands. Reddish dust

Figure 13.21 Wind-blown Dust on the Seafloor Fine grains of quartz and kaolinite identified in deep-sea cores define plumes of wind-blown dust derived from deserts in Asia, Australia, north Africa, and North and South America. The darker pattern within the plumes shows zones of highest concentration of these minerals, and therefore the primary paths of winds blowing from the desert regions.

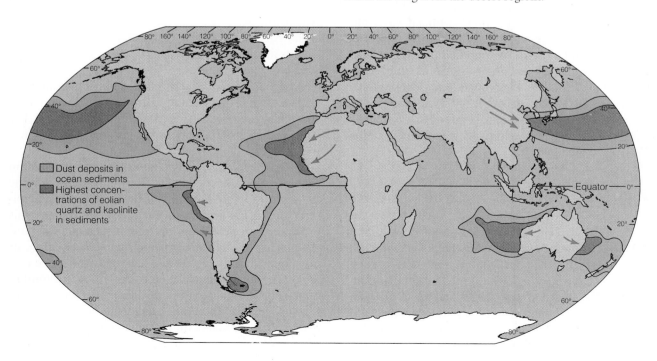

from the Sahara has settled on the decks of ships in the Atlantic Ocean, is deposited on glaciers in the Alps, and has been identified in soils of the Caribbean Islands.

Wind-blown dust is also found in cores drilled through polar ice sheets and low-latitude mountain glaciers. High rates of dust fall during the ice ages are correlated with dry conditions on the continents and strong winds that deflated fine sediment from desert basins and active outwash plains.

Volcanic Ash

Not all wind-transported sediment originates by deflation. Large quantities of tephra can be ejected into the atmosphere during explosive volcanic eruptions. Although coarse and dense particles fall out quickly downwind from a vent, small particles may be carried great distances. Fine ash that reaches the stratosphere may circle the Earth many times before it finally settles to the ground. The particles that fall out during an eruption commonly form an elongate plume of sediment that decreases in particle size and thickness downwind from the source volcano (Fig. 13.22). Such plumelike layers of prehistoric volcanic ash allow geologists to reconstruct the paths of winds that prevailed during ancient eruptions. Although deposits of volcanic ash may resemble loess, and even be interstratified with loess, a distinctive igneous mineralogy and tiny fragments of volcanic glass generally make tephra layers easy to recognize.

DESERTS

Although the word "desert" literally means a deserted (relatively uninhabited) region, nearly devoid of vegetation, the modern development of artificial water supplies has changed the meaning of this word by making many desert regions livable and suitable for agriculture. As a result, the term **desert** is now generally used as a synonym for land where annual rainfall is less than 250 mm or in which the potential evaporation rate exceeds the precipitation rate, regardless of whether the land is "deserted." Aridity, then, is a chief characteristic of any desert.

Origin of Deserts

Desert lands of various kinds total about 25 percent of the land area of the world outside the polar regions. In addition, a smaller though still large percentage of semiarid land exists in which the annual rainfall ranges between 250 and 500 mm. Together, these arid and semiarid regions form a distinctive pattern on the world map (Fig. 13.23). The desert regions are not randomly scattered across the globe, but instead are related to the

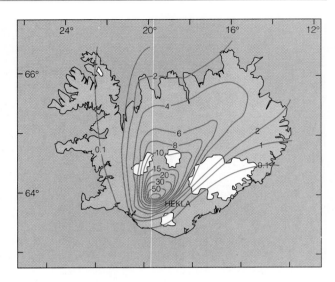

Figure 13.22 Wind-blown Tephra The distribution of a prehistoric tephra layer erupted from Hekla Volcano on Iceland about 4000 years ago is shown in this isopach map. The isopachs are lines of equal thickness of the deposit (in cm) and show that the tephra thins rapidly away from the volcano. The major plume of tephra extends to the north and northeast, indicating the direction of the prevailing winds at the time of the eruption.

Earth's geography and to atmospheric circulation. In all, five types of desert are recognized (Table 13.2).

When we compare Figure 13.23, showing the distribution of major deserts, with the general plan of atmospheric circulation shown in Figure 13.1, a relationship is immediately apparent. The most extensive deserts are associated with the two circumglobal belts of dry, descending air centered between latitudes 20° and 30°; examples include the Sahara and Kalahari deserts of Africa, the Rub-al-Khali Desert of Saudi Arabia, and the Great Australian Desert. These and other subtropical deserts at these latitudes comprise one of the five recognized types of deserts.

A second type of desert is found in continental interiors, far from sources of moisture, where hot summers and cold winters prevail (a continental-type climate). The Gobi and Takla Makan deserts of central Asia fall into this category.

A third kind of desert is found where a mountain range creates a barrier to the flow of moist air, forcing the air upward so that most of the moisture is precipitated out as the air crosses the mountain thereby this produces a zone of low precipitation, called a *rainshadow*, on the lee side of mountains. As marine air moving onshore rises against the windward slope of a mountain range, it cools, thereby lessening the amount of moisture it can hold. As a result, most of the moisture is lost as precipitation falls on the windward slope. Air reaching the lee side of the mountain range now contains little moisture, resulting in a dry climate over the coun-

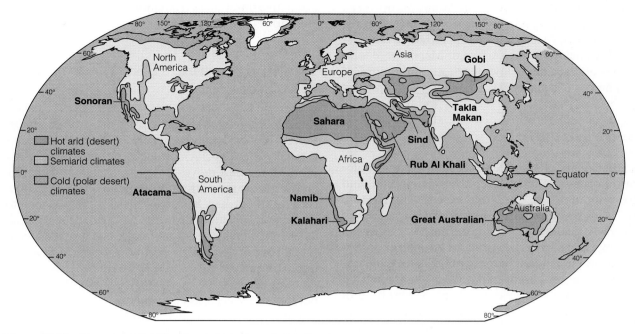

Figure 13.23 Deserts of the World   Arid and semiarid climates of the world and the major deserts associated with them. Very dry areas of the polar regions include areas known as polar deserts.

try beyond the mountains. The Cascade Range and Sierra Nevada of the western United States form such a barrier and are responsible for desert regions lying immediately east of these mountains.

Coastal deserts constitute a fourth category. They occur locally along the margins of continents, where cold, upwelling seawater cools maritime air flowing onshore. This decreases the air's ability to hold moisture, thereby creating coastal fogs. As the air encounters the warm land, it has a limited moisture content. In spite of the fog, the air contains too little moisture to generate much precipitation, so the coastal region remains a desert. In fact,

coastal deserts of this type in Peru and southwestern Africa are among the driest places on the Earth.

The four kinds of desert mentioned thus far are all hot deserts, where rainfall is low and summer temperatures are high. In the fifth category are vast deserts of the polar regions where precipitation is also extremely low due to the sinking of cold, dry air. However, cold deserts differ from hot deserts in one important respect: the surface of a polar desert, unlike those of warmer latitudes, is often underlain by abundant water, but nearly all in the form of ice. Even in midsummer, with the Sun above the horizon 24 hours a day, the air temperature may remain

TABLE 13.2	Main Types of Deserts and Their Origins	
Desert Type	**Origin**	**Examples**
Subtropical	Centered in belts of descending, dry air about 20–30° north and south latitude	Sahara, Sind, Kalahari, Great Australian
Continental	In continental interiors, far from moisture sources	Gobi, Takla Makan Desert
Rainshadow	To lee of mountain barriers that trap moist air flowing from ocean	Deserts to lee of Sierra Nevada, Cascades, and Andes
Coastal	Continental margins where cold, upwelling marine water cools maritime air flowing onshore	Coastal Peru and southwestern Africa
Polar	In regions where cold, dry air descends, creating very low precipitation	Northern Greenland, ice-free areas of Antarctica

below freezing. Polar deserts are found in northern Greenland, arctic Canada, and in the ice-free valleys of Antarctica. Such deserts are considered to be the closest earthly analogs to the surface of Mars, where temperatures also remain below freezing and the rarefied atmosphere is extremely dry.

Desert Climate

The arid climate of a hot desert results from the combination of high temperature, low precipitation, and high evaporation rate. The world's record temperature of 57.7°C (135.9°F) was measured in the Libyan desert of North Africa. In the Atacama Desert of northern Chile, intervals of a decade or more have passed without measurable rainfall. In these hot deserts, temperature and evaporation are interrelated: the higher the temperature, the greater the rate of evaporation. In parts of the desert region of the southwestern United States, evaporation from lakes and reservoirs is 10 to 20 times more than the annual precipitation.

During daylight hours, the air over hot desert regions is heated, expands, and rises. Strong winds are produced as surface air moves in rapidly to take the place of the rising hot air. Thus, in addition to being arid, hot deserts can also be quite windy.

SURFACE PROCESSES AND LANDFORMS IN DESERTS

No major geologic process is restricted entirely to desert regions. Rather, the same processes operate with different intensities in moist and arid landscapes. We have already explored the importance of wind in shaping arid landscapes. In a desert, surface sediments and landforms related to processes other than wind display some distinctive differences from those of the sediments and landforms of humid regions.

Weathering and Mass-Wasting in Deserts

In a moist region, regolith covers the ground almost universally and is comparatively fine textured because it usually contains clay, a product of chemical weathering. The regolith moves downslope, mainly by creep, and it is covered with almost continuous vegetation. As a result of creep, hillslopes in a moist region tend to be smooth and curvilinear. By contrast, the regolith in a desert is thinner, less continuous, and coarser in texture. Much of it is the product of mechanical weathering. Although some chemical weathering does take place in deserts, its intensity is greatly diminished because of reduced soil moisture.

Slope angles developed by downslope creep become adjusted to the average particle size of the regolith; the coarser the particles, the steeper the slope required to move them. Because the particles created by mechanical weathering tend to be coarse, desert slopes are generally steeper and more angular than those in a moist region.

Mechanically weathered fragments of rock tend to break off along joints, leaving steep, rugged cliffs. Among the most distinctive landforms we can see in deserts is a *butte* (French for *knoll* or *small hill*, and pronounced bewt), an isolated, steep-sided hill or precipitous pillar representing an erosional remnant carved from a resistant, flat-lying rock unit (Fig. 13.24). A flat-topped *mesa* (Spanish for *table*) is a wider landform of the same origin.

In many desert areas, the light hues of recently deposited sediments contrast with the darker hues of older deposits. The darker color can generally be attributed to **desert varnish**, a thin, dark, shiny coating (commonly manganese oxide) formed on the surface of stones and rock outcrops in desert regions after long

Figure 13.24 Butte in the Kalahari Desert The Finger of God, a pillarlike butte of sandstone in the Kalahari Desert of southwestern Africa, rests precariously on a pyramid of erodible shale. Erosion has separated the butte from a sandstone-capped mesa in the distance.

exposure (Fig. 13.25). The presence of manganese in the varnish is thought to be due either to release of this element from desert dust (which settles on the ground and is weathered by summer rainstorms) or, alternatively, to the manganese-concentrating activity of microorganisms that live on rock surfaces.

Desert Streams and Fluvial Landforms

Contrary to popular belief and to many Hollywood films, most deserts do not consist of endless expanses of sand dunes. Only a third of the Arabian Peninsula, the sandiest of all dry regions, and only a ninth of the Sahara are covered with sand. Scattered oases mark places where the water table locally reaches the surface, allowing vegetation to develop, but they occupy little space. Much of the nonsandy area of deserts is either crossed by systems of stream valleys or covered by alluvial fans and alluvial plains. Thus, in many deserts more geologic work apparently has been done by streams than by winds.

Most streams that flow into deserts from adjacent mountains never reach the sea, for they soon disappear as the water evaporates and some soaks into the ground. Exceptions are long rivers like the Nile, which originates in the moist highlands of Ethiopia and East Africa and then flows across the arid expanses of the Sudan and Egypt. Such a river carries so much water that it keeps flowing to the ocean despite great evaporative loss where it crosses a desert.

Flash Floods
The sparse vegetation cover in deserts presents no great impediment to surface runoff, which can readily erode

Figure 13.25 Drawings in the Desert Varnish Aboriginal drawings have been etched in a dark coating of desert varnish on a rock outcrop at Newspaper Rock, Utah. Removal of the oxide coating has exposed unweathered rock of lighter color.

loose, dry regolith. A major rainstorm is likely to be accompanied by a **flash flood**, a sudden, swift flood that can transport large quantities of sediment. The debris from such floods forms fans at the bases of mountain slopes and on the floors of wide valleys and basins.

Often streams in flood pass rapidly through desert canyons where they erode preexisting alluvium and undercut valley sideslopes, causing the slopes to cave in. As a flood subsides, its load is deposited rapidly, creating a flat alluvial surface (Fig. 13.26). The stratigraphy of such alluvial fills often discloses a complex history of cutting and filling.

Figure 13.26 Arroyo in Arizona A flash flood has just passed through this steep-walled arroyo on the Navajo reservation in northeastern Arizona. As the floodwater subsides, sediment is deposited across the flat alluvial floor of the canyon.

Fans and Bajadas

Alluvial fans develop under a wide range of climatic conditions, but they are especially common in arid and semiarid lands, where they typically are composed of both alluvium and debris-flow deposits. They are a characteristic landform of deserts and can be a major source of groundwater for irrigation. In some semiarid regions, entire cities have been built on alluvial fans or fan complexes (for example, San Bernardino, California, and Teheran, Iran). Alluvial fans in Iran, Afghanistan, and Pakistan are dotted with mounds of debris that mark the sites of deep artificial shafts passing downward from the surface to horizontal tunnel systems. The shafts were designed to collect water within the upper reaches of fans for use in surface irrigation. Some such systems date back a thousand years or more (Fig. 13.27).

In desert basins of the southwestern United States, the Middle East, and central Asia, alluvial fans form a prominent part of the landscape. In these regions, the fans border highlands, with the top of each fan lying at the mouth of a mountain canyon. Where a mountain front is straight and its canyons are widely spaced, each fan will encompass an arc of about 180° (Fig. 10.27). If canyons are closely spaced along the base of a mountain range, coalescing adjacent fans form a broad alluvial apron, or **bajada** (Spanish for *slope*), that has an undulating surface due to the convexities of the component fans (Fig. 13.28).

Desert Lakes and Playas

Runoff in arid regions is rarely abundant enough to sustain permanent lakes. Instead, the floor of a desert basin may contain a dry lakebed, called a **playa** (Spanish for *beach*) (Fig. 13.28). Following a major rainstorm,

Figure 13.27 Iranian quanat Line of mounds mark the course of a quanat, an underground tunnel for transporting water across the desert to Isphahan, Iran.

runoff may be sufficient to form a temporary playa lake that will last up to several weeks. White or grayish salts at the dry surface of a playa, left by the repeated formation and evaporation of temporary lakes, can accumulate to thicknesses of tens of meters and constitute an important source of industrial chemicals (see Chapter 4).

Figure 13.28 Death Valley playa and bajada A vast salt-encrusted playa occupies the floor of Death Valley in California. The playa is bordered by a bajada, composed of coalescing alluvial fans constructed beyond the mouths of adjacent mountain valleys.

Pediments

One of the most characteristic landforms of dry regions is the **pediment**, a broad, relatively flat surface, eroded across bedrock and thinly or discontinuously veneered with alluvium, which slopes away from the base of a highland (Fig. 13.29). Although from a distance it may resemble a bajada, a pediment is a bedrock surface rather than a thick alluvial fill. Rock debris scattered over a pediment is carried by running water from adjacent mountains and is also derived by weathering of the pediment surface. Downslope, the scattered rock debris gradually forms a continuous cover of alluvium that merges with the thick alluvial fill of an adjacent valley.

The long profile of a pediment, like that of an alluvial fan, is concave upward, becoming progressively steeper toward a mountain front. Such a profile is typically associated with the work of running water (see Chapter 10). Faint, shallow channelways on pediment surfaces show that water is involved in their formation, and it is generally agreed that pediments are slopes across which sediment is transported by mass-wasting and running water. Eyewitness accounts of lateral erosion by floodwaters associated with intense desert storms have led geologists to suggest that both these processes may be involved in pediment formation. However, the exact manner in which pediments form is still not firmly established.

The surface of a pediment rises toward a mountain that it meets at an abrupt angle. This suggests that desert mountain slopes do not become gentler with time, as they tend to do in moist regions, where chemical weathering and creep are dominant. Instead, the slopes apparently achieve an angle determined by the resistance of the bedrock and maintain that angle as they gradually retreat under the attack of weathering and mass-wasting (Fig. 13.30). In this way, as a mountain slope retreats,

Figure 13.29 Pediment in Mojave A pediment in the Mojave Desert of southeastern California has left only a few residual hills near the crest of a former mountain ridge. The flat bedrock surface cut across crystalline rocks passes downslope beneath a thin cover of alluvium.

a pediment will increase in size by expanding at its upslope edge. The growth of the pediment, at the expense of the mountain, may continue until the entire mountain has been consumed.

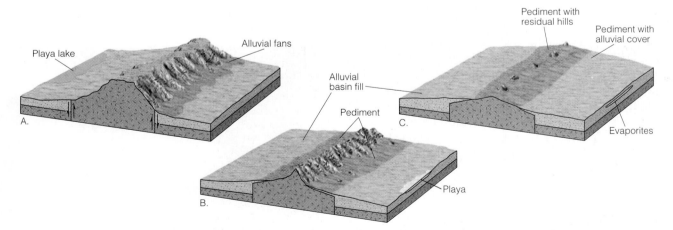

Figure 13.30 Formation of a Pediment Stages in the formation of a pediment. A. A mountain block, uplifted along bordering faults, is eroded by streams that contribute sediment to a growing alluvial basin fill. B. A pediment is cut across the margins of the uplifted block and grows headward into the mountains as sheetfloods and running water transport sediment across the planar rock surface toward the basin fill. C. Growing headward from both sides of the upland, the pediment slowly consumes the mountains, leaving only a few residual hills rising above the eroded bedrock surface.

Figure 13.31 Inselberg Ayers Rock, a massive inselberg, rises about 360 m above the surrounding flat plain in central Australia.

Inselbergs

Among the most impressive of the Earth's landforms are steep-sided mountains, ridges, or isolated hills that rise abruptly from adjoining plains like rocky islands standing above the surface of a broad, flat sea. Ayers Rock in central Australia is a famous example (Fig. 13.31). Called **inselbergs** (German for *island mountain*), these landforms appear in many environmental settings, ranging from coastal to interior and arid to humid. However, they are especially common and well developed in semiarid grasslands in the middle of tectonically stable continents. Numerous examples can be found in southern and central Africa, northwestern Brazil, and central Australia.

Field evidence suggests that inselbergs form in areas of relatively homogeneous, resistant rock (most commonly granite or gneiss, but also sedimentary rocks such as conglomerate and sandstone) that are surrounded by rocks more susceptible to weathering. Differential weathering over long time intervals lowers adjacent terrains, leaving these resistant rock masses standing high. Once formed, the bare rock hills tend to shed water, whereas surrounding debris-mantled plains absorb water, causing the underlying rocks to weather more rapidly. For this reason, inselbergs may remain as stable parts of a landscape and persist for tens of millions of years or more. Some may even date back to the Mesozoic Era. If that is true, they have remained prominent landscape features since the time of the dinosaurs.

DESERTIFICATION

In the region south of the Sahara lies a belt of dry grassland known as the Sahel (Arabic for *border*). There the annual rainfall is normally only 100 to 300 mm, most of it falling during a single brief rainy season. In the early 1970s, the drought-prone Sahel experienced the worst drought of this century (Fig. 13.32). For several years in succession the annual rains failed to appear, causing adjacent desert to spread southward—according to one estimate as much as 150 km. The drought extended from the Atlantic to the Indian Ocean, a distance of 6000 km, and affected a population of at least 20 million people, many of them seminomadic herders of cattle, camels, sheep, and goats. The results of the drought were intensified by the fact that between about 1935 and 1970 the human population had doubled, and the number of livestock had also increased dramatically. This increase in the number of people and animals led to severe overgrazing, so that with the coming of the

Figure 13.32 Drought in Mali Overgrazing during years of drought killed most of the vegetation around wells in the Azaouak Valley, Mali. Without vegetation, soil blows away and the desert advances.

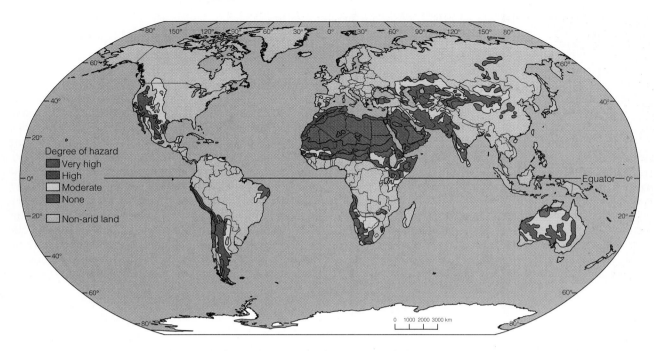

Figure 13.33 Possibility of Desertification Substantial portions of all the temperate-latitude continents face the hazard of moderate to very high desertication.

drought the grass cover almost completely failed. About 40 percent of the cattle—a great many millions—died. Millions of people suffered from thirst and starvation, and many died as vast numbers migrated southward in search of food and water. In the mid-1970s the rains returned briefly. Then, in the 1980s, drought conditions resumed. Ethiopia and the Sudan were especially hard hit and experienced widespread famine. Mass starvation was alleviated only by worldwide relief efforts.

Such invasion of desert into nondesert areas, referred to as **desertification**, can result from natural environmental changes as well as from human activities see *Understanding Our Environment: Tectonic Desertification.* The major symptoms are declining groundwater tables, increasing saltiness of water and topsoil, reduction in supplies of surface water, unnaturally high rates of soil erosion, and destruction of native vegetation. Although we can find evidence of natural desertification events in the geologic record, there is increasing concern that human activities, regardless of natural climatic trends, can in themselves help promote widespread desertification (Fig. 13.33).

Desertification in North America

The impact of desertification on human life in North America is less severe than in more densely populated regions of the world, but it nevertheless has important and far-reaching implications for the continent's food, water, and energy supplies, as well as its natural environment. Nearly 37 percent of the dry regions of the continent have experienced "severe" desertification. In the southwestern United States, about 10 percent of the land area—approximately the same size as the original 13 states—has been affected by desertification over the past century. Large shifting sand dunes have formed, erosion has virtually denuded the landscape of vegetation, numerous gullies have developed, and salt crusts have accumulated on nearly impermeable irrigated soils. This desertification has been brought about primarily by overgrazing, by excessive withdrawal of groundwater, and by unsound water-use practices, in part combined with a population increase and expanded agricultural production.

Countermeasures

How can the detrimental effects of desertification be halted and even reversed? The answer lies largely in understanding the geologic principles involved and in intelligently applying measures designed to reestablish a natural balance in the affected areas. Elimination of incentives to exploit arid lands beyond their natural capacity, coupled with long-range planning aimed at minimizing the negative effects of human activity, should help in reaching the desired goal. Because arid lands of the western United States supply about 20 percent of the nation's total agricultural output, the long-term benefits could be substantial.

Understanding Our Environment

Tectonic Desertification

Western China contains two of the world's major deserts. One, the Takla Makan, is a hot desert with vast regions of shifting sand dunes (Fig. 13.17), while the second encompasses the western Tibetan Plateau, a high, cold desert with sparse steppe vegetation underlain by perennially frozen ground (Fig. B13.1). Both are relatively young, geologically speaking, and their origin is tied to the tectonics of converging lithospheric plates.

Evidence of the long-term natural desertification of western China has been assembled by Chinese paleontologists who have studied the fossil remains of plants and animals in Tertiary strata. The fossil sites lie scattered about the high Tibetan Plateau and adjacent mountain ranges, mostly at altitudes of 4000 to 6000 m where the landscape now consists of cold alpine steppe and cold desert. Fossil plants and pollen of Eocene and Oligocene age disclose a flora that consisted of evergreen broadleaf forests and included eucalyptus, magnolia, and fig. Today, these warmth-loving trees are found in moist, tropical environments at low altitudes. Remains of a giant rhinoceros, a larger relative of the rhinoceros that lives in tropical southeastern Asia today, have also been found in the Oligocene strata. From this evidence, we can infer that western China during the early Tertiary must have had a relatively low altitude (perhaps 500 to 1000 m) and a warm climate, and that there were no mountain barriers to block the passage of moist oceanic air from the south. The widespread distribution of *Hipparian*

Figure B13.1 Himalayan barrier The snow- and ice-capped Himalaya rising high above the dense green vegetated plains of India (lower left) keep moisture-bearing winds from reaching the high Tibetan Plateau, which receives less than 250 mm of annual precipitation.

supports this view, for it implies that this primitive horse could migrate freely over the continent unimpeded by major mountain chains.

Changing plant and animal assemblages imply progressive uplift of the region during the Miocene and Pliocene, accompanied by cooling and drying of the climate. By the beginning of the Pleistocene, nearly 2 million years ago, the Tibetan Plateau had reached an altitude of 1000 to 2000 m, and alpine coniferous forests had replaced subtropical trees. During the last half of the Pleistocene, these forests gave way to dry alpine steppe as the rate of uplift accelerated, bringing the plateau and the adjacent Himalaya to high altitudes.

The progressive desertification of western China during the Tertiary coincides with the ongoing collision of India with Asia and resulting uplift of the Tibetan Plateau and associated mountain ranges (B13.2). Prior to this collision, the area of Tibet and the Takla Makan Desert stood at low altitude and lay close to an ancient seaway that separated India and Asia. As the two continents converged, the seaway narrowed and then disappeared, placing these future desert regions closer to the center of Asia and therefore in a more continental climatic environment. Uplift associated with continental collision further intensified the desertification process by raising Tibet into successively drier and colder climatic zones and simultaneously raising the lofty Himalaya across the path of the northward-flowing monsoonal air, thereby blocking off the primary source of precipitation. The cold desert of the lofty plateau and the adjacent hot Takla Makan therefore owe their existence both to their midcontinental position and the rainshadow formed by the high mountain barrier that separates the hot, humid plains of India from the frigid wastes of Tibet.

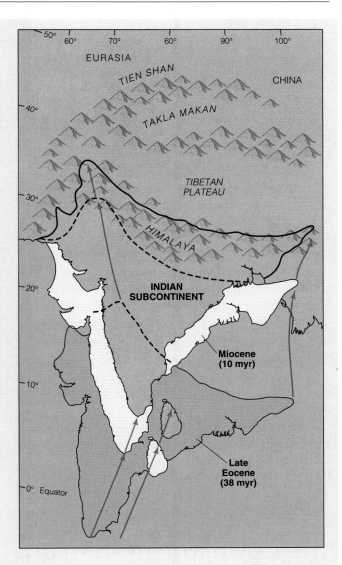

Figure B13.2 Asian Collision As India drifted northward during the Tertiary Period, the seaway separating it from Asia narrowed and disappeared. Successive positions of the continents during the late Eocene and the late Miocene are shown relative to the present continent of Asia. The collision of the two landmasses resulted in uplift of the Himalaya and associated high mountain ranges, and isolation of the Tibetan Plateau and Takla Makan Desert from moisture sources to the south.

SUMMARY

1. Unequal heating of the Earth by solar radiation sets up convective circulation in the atmosphere. The Coriolis effect breaks the equator-to-poleward atmospheric circulation in the northern and southern hemispheres into cells, that contain prevailing lower-level winds called the tradewinds, westerlies, and polar easterlies.

2. Climate is the average weather of a place over a period of years, together with the degree of variability of that weather. The Earth's climates are influenced by the distribution of land and oceans, and by surface topography.

3. Wind moves saltating sand grains close to the ground and suspended dust particles at higher levels. Sorting of sediment results.

4. Through deflation and abrasion, winds create deflation basins, desert pavement, ventifacts, and yardangs.

5. Dunes originate where obstacles distort the flow of air. Dunes have steep slip faces and gentler windward slopes. They migrate in the direction of wind flow, forming cross strata that dip downwind.

6. Loess is deposited chiefly downwind from deserts and from the floodplains of glacial meltwater streams. Once deposited, it is stable and is little affected by further wind action.

7. Airborne tephra deposited during explosive volcanic eruptions decreases in thickness and grain size away from the source vent.

8. Hot deserts constitute about a quarter of the world's nonpolar land area and are regions of slight rainfall, high temperature, excessive evaporation, relatively strong winds, and sparse vegetation. Polar deserts occur at high latitudes where descending cold, dry air creates arid conditions.

9. Mechanical weathering, flash floods, and winds are especially effective geologic agents in deserts.

10. Fans, bajadas, and pediments are conspicuous features of many deserts. Pediments are probably shaped by running water and are eroded surfaces across which sediment is transported.

11. Inselbergs form in relatively homogeneous, resistant rocks and may remain as persistent landforms for millions of years.

12. Recurring natural droughts can lower the water table, cause high rates of soil erosion, and destroy vegetation, thereby leading to the invasion of deserts into nondesert areas. Overgrazing, excessive withdrawal of groundwater, and other human activities can promote desertification. Desertification can be halted or reversed by measures that restore the natural balance.

THE LANGUAGE OF GEOLOGY

bajada (p. 360)

climate (p. 343)
Coriolis effect (p. 342)

deflation (p. 348)
desert (p. 356)
desert pavement (p. 349)
desert varnish (p. 358)
desertification (p. 363)

dune (p. 350)

flash flood (p. 359)

inselberg (p. 362)

loess (p. 353)

pediment (p. 361)
playa (p. 360)

sand ripples (p. 345)
sand sea (p. 352)

ventifact (p. 349)

yardang (p. 350)

QUESTIONS FOR REVIEW

1. What atmospheric factors cause most of the world's large hot deserts to be concentrated in belts lying between 20° and 30° from the equator?

2. Explain what controls the depth to which deflation is effective in arid regions.

3. Why are the erosional effects of blowing sand generally confined to a zone extending only about a meter above the ground surface?

4. Explain the origin of the internal stratification of a sand dune. How do sand dunes migrate downwind?

5. How might you tell the former direction of the prevailing wind from the form and internal stratification of an ancient, inactive sand dune? from a ventifact? from a tephra deposit?

6. What measures could you recommend that a farmer in the southwestern United States take to halt the migration of sand dunes now threatening his agricultural fields?

7. How might you tell a deposit of loess from an alluvial silt having a similar range of particle sizes?

8. Why do hillslopes in arid landscapes tend to be steeper and sharper than those in humid landscapes?

9. What evidence can you cite that points to streams being effective agencies of erosion and sediment transport in desert regions?

10. How would you tell a bajada from a pediment in the field? What process(es) are involved in the formation of each?

11. Why do playas often have a distinctive deposit of salts at their surface?

12. What factors influence the formation of inselbergs? Why are inselbergs likely to remain persistent features of a semiarid landscape?

13. What are some of the obvious symptoms of desertification? How might they be retarded or reversed by human intervention?

14

Measuring
Our Earth

Understanding
Our Environment

GeoMedia

Vacation homes on the Alantic coast, Cedar Island, Virginia, are endangered by the encroaching sea.

The Oceans and Their Margins

Our Changing Coastlines

Shoreline homes command premium prices in the real estate market. If built on solid rock, they can prove to be a lasting investment; however, coastlines are among the most dynamic places on the Earth's surface, and a house built on a coast composed of erodible sediment can prove to be a poor bargain.

A recent government assessment of coastal erosion problems in the United States showed that moderate to severe beach erosion is taking place along more than 80 percent of the shoreline. The most rapid erosion is occurring along sandy stretches of the Atlantic and Gulf coasts and the north coast of Alaska. Because these coasts consist of erodible sand, they are especially vulnerable to the attack of waves and currents.

A primary reason why erosion and other shoreline problems are receiving increasing attention is that coastal zones of the United States now have population densities that are five times the national average.

At the time of the 1990 census, 50 percent of all Americans lived within 75 km of a coast. Projections indicate that this number will increase to 75 percent by 2010. The nation's fragile shorelines, already severely stressed by the existing population, will come under increased pressure as continuing growth leads to demands for further development, additional fresh water, and adequate waste disposal.

Unlimited development, however, could well prove disastrous, for coastlines are places that change constantly under the attack of wind and sea. The natural risk of living in a coastal zone may increase still further if warming climate causes world sea level to rise. The combination of natural risk and increasing population pressure has led many well-informed scientists to conclude that we are rapidly approaching a coastal crisis. Like other environmental crises of our own making, the solution to coastal problems must rest on balancing human needs and expectations with a clear understanding of how natural processes control coastal environments.

THE WORLD OCEAN

Seawater covers 70.8 percent of the Earth's surface, and most of it is contained in three huge interconnected basins—the Pacific, Atlantic, and Indian oceans (Fig. 14.1). (The Arctic Ocean is generally considered an extension of the North Atlantic.) All three are connected with the Southern Ocean, a body of water south of 50° S latitude that completely encircles Antarctica. Collectively, these four vast interconnected bodies of water, together with a number of smaller ones, are often referred to as the *world ocean*. Smaller seas and gulfs vary considerably in shape and size; some are almost completely surrounded by land, whereas others are only partly enclosed. Each owes its distinctive geography to plate tectonics, which has led to the creation of numerous small basins both in and adjacent to the major ocean basins.

Depth and Volume of the Oceans

Over the past 70 years, the oceans have been crisscrossed many thousands of times by ships carrying acoustical instruments called echo sounders that measure ocean depth. As a result, the topography of the seafloor and the depth of the overlying water column are known in considerable detail for all but the most remote parts of the ocean basins. The greatest ocean depth yet measured (11,035 m) lies in the Mariana Trench near the

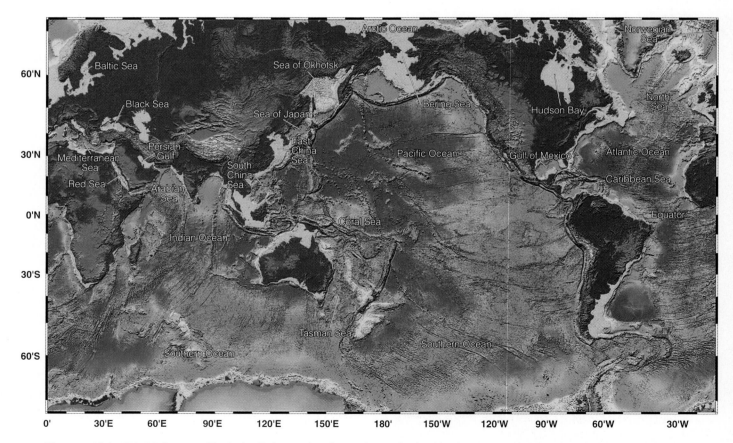

Figure 14.1 World Ocean Shaded relief map showing major and minor basins of the world ocean. The deepest basins are shown in purple and blue, shallower regions in shades of green and yellow, and the continental shelves in light blue.

A.

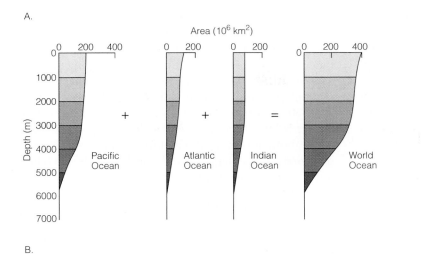

Figure 14.2 Area (A) and volume (B) of the oceans. The Pacific represents nearly half the volume of the oceans, with the Atlantic and Indian oceans being comparable to each other in both size and volume. Although the deepest known place in the oceans lies more than 11,000 m below sea level, nearly all the water lies at a depth of less than 6000 m.

island of Guam in the western Pacific. This is more than 2 km farther below sea level than Mount Everest rises above sea level. The average depth of the sea, however, is about 3.8 km, compared to an average height of the land of only 0.75 km.

If we measure the area of the sea and calculate its average depth, we then can calculate that the present volume of seawater is about 1.35 billion cubic kilometers (Fig. 14.2); more than half this volume resides in the Pacific Ocean. We say *present* volume because the amount of water in the ocean fluctuates somewhat over thousands of years, with the growth and melting of continental glaciers (Chapter 12).

Ocean Salinity

About 3.5 percent of average seawater, by weight, consists of dissolved salts, enough to make the water undrinkable. It is enough too, if these salts were precipitated, to form a layer about 56 m thick over the entire seafloor.

Salinity is the measure of the sea's saltiness, expressed in per mil (‰ = parts per thousand) rather than percent (% parts per hundred). The salinity of seawater normally ranges between 33 and 37 per mil. The principal elements

that contribute to this salinity are sodium and chlorine. Not surprisingly, when seawater is evaporated, more than three-quarters of the dissolved matter is precipitated as common salt (NaCl). However, seawater contains most of the other natural elements as well, many of them in such low concentrations that they can be detected only by extremely sensitive analytical instruments. More than 99.9 percent of the salinity reflects the presence of only eight ions: chloride, sodium, sulfate, magnesium, calcium, potassium, bicarbonate, and bromine.

Where do these ions come from? Each year streams carry 2.5 billion tons of dissolved substances to the sea. As exposed crustal rocks interact with the atmosphere and the hydrosphere (rainwater), cations are leached out and become part of the dissolved load of streams flowing to the sea. The principal anions found in seawater, on the other hand, are believed to have come from the mantle. Chemical analyses of gases released during volcanic eruptions show that the most important volatiles are water vapor (steam), carbon dioxide (CO_2), and the chloride (Cl^{1-}) and sulfate (SO_4^{2-}) anions. These two anions dissolve in atmospheric water and return to the Earth in precipitation, much of which falls directly into the ocean. Part of the remainder is carried to the sea dissolved in river waters. Volcanic gases are also released directly into the ocean from submarine eruptions. Other sources of ions include dust eroded from desert regions and blown out to sea, and gaseous, liquid, and solid pollutants released through human activity either directly into the oceans or carried there by streams or polluted air.

The quantity of dissolved ions added by rivers over the billions of years of Earth history far exceeds the amount now dissolved in the sea. Why, then, doesn't the sea have a higher salinity? The reason is that chemical substances are being removed at the same time they are being added. Some elements, such as silicon, calcium, and phosphorus, are withdrawn from seawater by aquatic plants and animals to build their shells or skeletons. Other elements, such as potassium and sodium, are absorbed and removed by clay particles and other minerals as they slowly settle to the seafloor. Still others, such as copper and lead, are precipitated as sulfide minerals in claystones and mudstones rich in organic matter. Because these and other processes of extraction are essentially equal to the combined inputs, the composition of seawater remains virtually unchanged.

The most important factors affecting salinity of surface waters are (1) evaporation (which removes water and leaves the remaining water saltier), (2) precipitation (which adds fresh water, thereby diluting the seawater and making it less salty), (3) inflow of fresh (river) water (which makes the seawater less salty), and (4) the freezing and melting of sea ice (when seawater freezes, salts are excluded from the ice, leaving the unfrozen seawater saltier). As one might expect, salinity is high in the latitudes where the Earth's great deserts lie, for in these zones evaporation exceeds precipitation, both on land and at sea (Fig. 14.3A). In a restricted sea, like the Mediterranean, where there is little inflow of fresh water, surface salinity exceeds the normal range; in the Red Sea, which is surrounded by desert, salinity reaches 41 percent. Salinity is lower near the equator because precipitation is high, and cool water, which rises from the deep sea and sweeps westward in the tropical eastern Pacific and eastern Atlantic oceans, reduces evaporation. It also is low at high latitudes that are rainy and cool. Up to 100 km offshore from the mouths of large rivers, the surface ocean water can be fresh enough to drink.

TEMPERATURE AND HEAT CAPACITY OF THE OCEAN

An unsuspecting tourist from Florida who decides to take a swim on the northern coast of Britain quickly learns how varied the surface temperature of the ocean can be. A map of global summer sea-surface temperature displays a pronounced east-west banding, with *isotherms* (lines connecting points of equal tempera-

ture) approximately paralleling the equator (Fig. 14.3B). The warmest waters during August (>28°C) occur in a discontinuous belt between about 30° N and 10°S latitude in the zone where solar radiation reaching the surface is at a maximum. In winter, when the belt of maximum incoming solar radiation shifts southward, the belt of warm water also moves south until it is largely below the equator. Waters become progressively cooler both north and south of this belt and reach temperatures of less than 10°C poleward of 50° N and S latitude. The average surface temperature of the oceans is about 17°C, while the highest temperatures (>30°C) have been recorded in restricted tropical seas, such as the Red Sea and the Persian Gulf.

The ocean differs from the land in the amount of heat it can store. For a given amount of heat absorbed, water has a lower rise in temperature than nearly all other substances; that is, it has a high *heat capacity*. Because of water's ability to absorb and release large amounts of heat with very little change in temperature, both the total range and the seasonal changes in ocean temperatures are much less than what we find on land. For example, the highest recorded land temperature is 58°C, measured in the Libyan Desert, and the lowest, measured at Vostok Station in central Antarctica, is –88°C; the range, therefore, is 146 C°. By contrast, the highest recorded ocean temperature is 36°C, measured in the Persian Gulf, and the coldest, measured in the polar seas, is –2°C, a range of only 38 C°.

The annual change in sea-surface temperatures is 0–2 C° in the tropics, 5–8 C° in the middle latitudes, and 2–4 C° in the polar regions. Corresponding seasonal temperature ranges on the continents can exceed 50 C°. Coastal inhabitants benefit from the mild climates resulting from this natural ocean thermostat. Along the Pacific coast of Washington and British Columbia, for example, winter air temperatures seldom drop to freezing, while east of the coastal mountain ranges they can plunge to –30°C or lower. In the interior of a continent, summer temperatures may exceed 40°C, whereas along the ocean margin they typically remain below 25°C. Here, then, is a good example of the interaction of the hydrosphere, atmosphere, land surface, and biosphere: ocean temperatures affect the climate, both over the ocean and over the land, and climate ultimately is a major factor in controlling the distribution of plants and animals.

VERTICAL STRATIFICATION

The physical properties of seawater vary with depth. To help understand why this is so, think about shaking up a bottle of oil-and-vinegar salad dressing and then setting it on a table. After a minute or two, the oil will rise

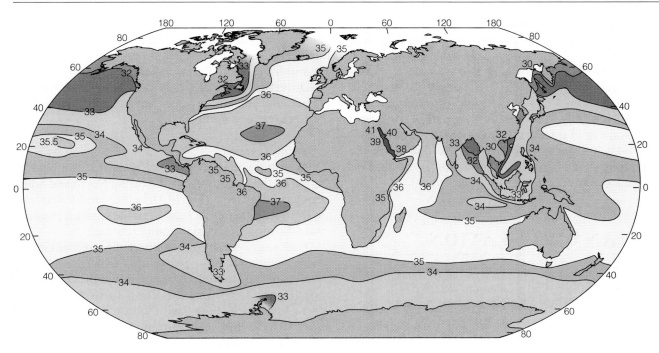

Figure 14.3A Salinity of the Ocean Average surface salinity of the oceans. High salinity values are found in tropical and subtropical waters where evaporation exceeds precipitation. The highest salinity has been measured in enclosed seas like the Persian Gulf, the Red Sea, and the Mediterranean Sea. Salinity values generally decrease poleward, both north and south of the equator, but low values also are found off the mouths of large rivers.

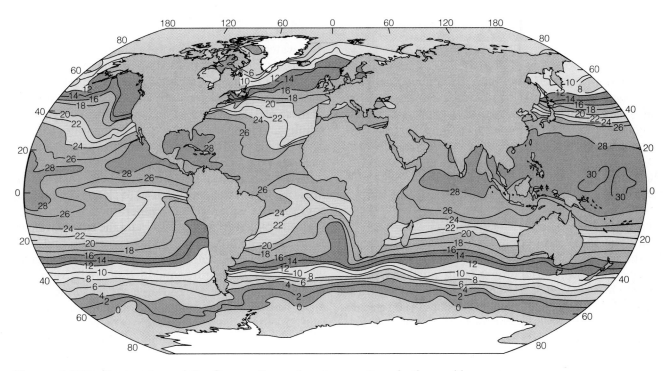

Figure 14.3B Temperature of the Ocean Sea-surface temperatures in the world ocean during August. The warmest temperatures (≥28°C) are found in the tropical Indian and Pacific oceans. Temperatures decrease poleward from this zone, reaching values close to freezing in the north and south polar seas.

to the top of the bottle and the vinegar will settle to the bottom. The two ingredients become stratified because they have different densities: the less-dense oil floats on the denser vinegar. The oceans also are vertically stratified, as a result of variations in the density of seawater. Seawater becomes denser as its temperature decreases and as its salinity increases. Gravity pulls dense water downward until it reaches a level where the surrounding water has the same density. These density-driven movements lead both to stratification of the oceans and to circulation in the deep ocean.

OCEAN CIRCULATION

When Christopher Columbus set sail from Spain in 1492 to cross the Atlantic Ocean in search of China, he took an indirect route. Instead of sailing due west, which would have made his voyage shorter, he took a longer route southwest toward the Canary Islands, then west on a course that carried him to the Caribbean Islands where he first sighted land. In choosing this course, he was following the path not only of the prevailing winds but also of surface ocean currents. Instead of fighting the westerly winds and currents at 40° N latitude, he drifted with the Canary Current and North Equatorial Current, as the northeast tradewinds filled the sails of his three small ships.

Surface Currents of the Open Ocean

Surface ocean currents, like those Columbus followed, are broad, slow drifts of surface water set in motion by the prevailing surface winds. Air that flows across the sea drags the water slowly forward, creating a current of water as broad as the current of air, but rarely more than 50 to 100 m deep. The ultimate source of this motion is the Sun, which heats the Earth unequally, thereby setting in motion the planetary wind system.

The Coriolis Effect
The direction taken by ocean currents is also influenced by the Coriolis effect, the phenomenon by which all moving bodies veer to the right in the northern hemisphere and to the left in the southern hemisphere (Chapter 13). If a freely floating object on the ocean in the northern hemisphere moves away from the pole, its angular velocity about the pole will be slower than that of the water. This causes the object to lag behind the rotation, and so it will be deflected in a clockwise direction (to the right). If such an object were moving toward the pole, its angular velocity about the pole would be faster than the water, resulting in a counterclockwise

deflection (again toward the right). Regardless of the direction of movement, an object in the northern hemisphere will be deflected to the right. In the southern hemisphere, the deflection is to the left, while at the equator the effect disappears. Although the Coriolis effect does not cause ocean currents, it deflects them once they are in motion.

Current Systems
Low-latitude regions in the tradewind belts are dominated by the warm, westward-flowing North and South Equatorial currents (Fig. 14.4). In their midst, and lying in the doldrums belt of light, variable winds, is the eastward-flowing Equatorial Countercurrent.

Each major ocean current is part of a large subcircular current system called a **gyre**. Within each gyre different names are used for different segments of the current system. Figure 14.4 shows the Earth's five major ocean gyres—two each in the Pacific and Atlantic oceans and one in the Indian Ocean. Currents in the northern hemisphere gyres circulate in a clockwise direction; those in the southern hemisphere circulate counterclockwise.

The northern Indian Ocean exhibits a unique circulation pattern in which the direction of flow changes seasonally with the changing pattern of monsoonal air flow. During the summer, strong and persistent monsoon winds blow the surface water eastward, whereas in winter, winds from Asia blow the water westward.

Major Water Masses

The water of the oceans is organized vertically into major water masses, stratified according to density. The density differences lead to large-scale circulation of water within the deep ocean. The identity and sources of these masses have been determined by studying the salinity and temperature structure of the water column at many places. The Atlantic Ocean provides a good example (Fig. 14.5).

In the Atlantic, water in the surface zone forms a *central water mass* extending to about 35° latitude north and south of the equator. The temperature of this water typically ranges from 6 to 19°C, and the salinity ranges from 34 to 36.5 per mil. Cooler subarctic and subantarctic surface-water masses are found at high latitudes where cool temperatures and high rainfall give rise to colder, less saline waters. The largest polar surface water mass, flowing as the Antarctic Circumpolar Current (ACC), moves clockwise around Antarctica. Its temperature is 0–2°C, and its salinity is 34.6–34.7 per mil.

The central water mass of the Atlantic overlies an *intermediate water mass* that extends to a depth of about 1500 m. The Antarctic Intermediate Water Mass (AAIW), the most extensive such body of water, originates as cold subantarctic surface water that sinks and

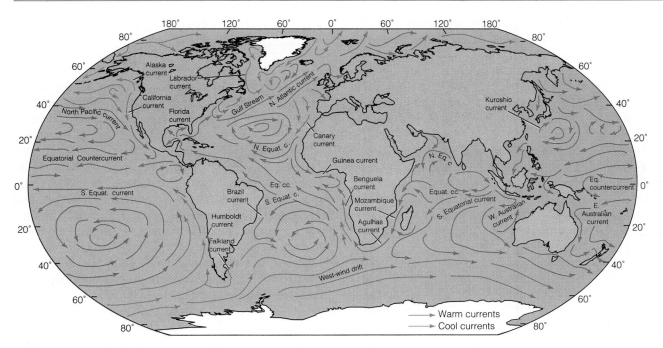

Figure 14.4 Ocean Currents Surface ocean currents form a distinctive pattern, curving to the right (clockwise) in the northern hemisphere and to the left (counterclockwise) in the southern hemisphere. The westward flow of tropical Atlantic and Pacific waters is interrupted by continents, which deflect the water poleward. The flow then turns away from the poles and becomes the eastward-moving currents that define the middle-latitude margins of the five great midocean gyres.

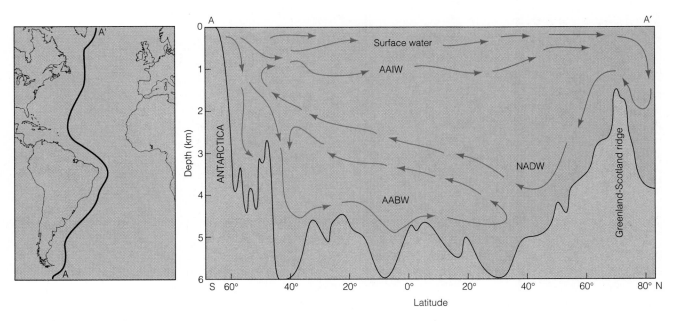

Figure 14.5 Circulation of the Atlantic Ocean Transect along the western Atlantic Ocean showing water masses and general circulation pattern. North Atlantic Deep Water (NADW) originates near the surface in the North Atlantic as northward-flowing surface water cools, becomes increasingly saline, and plunges to depths of several km. As NADW moves into the South Atlantic, it rises over denser Antarctic Bottom Water (AABW), which forms adjacent to the Antarctic continent and flows into the north Atlantic as Antarctic Intermediate Water (AAIW) at a mean depth of about 1 km.

spreads northward across the equator to about 20° N latitude. Its temperature ranges from 3 to 7°C, and its salinity lies within the range 33.8–34.7 per mil. Water entering the Atlantic from the Mediterranean Sea is so saline (37–38‰) that it flows over a shallow sill at Gibraltar (2400 m) and downward beneath intermediate water to spread laterally over much of the ocean basin.

In the North Atlantic, the deep ocean consists of a *deep-water mass* that extends from the intermediate water to the ocean floor. This dense, cold (2 to 4°C), saline (34.8 to 35.1‰) North Atlantic Deep Water (NADW) originates at several sites near the surface of the North Atlantic, flows downward, and spreads southward into the South Atlantic.

The deepest, densest, and coldest water in the Atlantic is the *bottom water mass* that forms off Antarctica and spreads far northward; in the Pacific it reaches as far as 30° N latitude. Because of its greater density, Antarctic Bottom Water (AABW) flows beneath North Atlantic Deep Water. It forms when dense brine, produced during the formation of winter sea ice in the Weddell Sea adjacent to Antarctica, mixes with cold circumpolar surface water and sinks into the deep ocean. This dense water has an average temperature of –0.4°C and a salinity of 34.7 per mil.

The Global Ocean Conveyor System

The sinking of dense cold and/or saline surface waters provides a link between the atmosphere and the deep ocean. It also propels a global **thermohaline circulation** system, so called because it involves both the temperature and salinity characteristics of the ocean waters. We can trace this circulation from the North Atlantic southward toward Antarctica and into the other ocean basins. The largest mass of NADW forms in the Greenland and Norwegian seas, where relatively warm and salty surface water entering from the western North Atlantic cools, becomes denser, and sinks into a confined basin north of a submarine ridge connecting Scotland and Greenland (Fig. 14.6A). The dense water then spills over low places along the ridge and plunges down into the deep ocean as NADW. Warm, salty surface and intermediate water is drawn toward the North Atlantic to compensate for the south-flowing deep water. It is the heat lost to the atmosphere by this warm surface water, together with heat from the warm Gulf Stream, that maintains a relatively mild climate in northwestern Europe.

In the South Atlantic, south-flowing NADW enters the Antarctic Circumpolar Current, which travels clock-

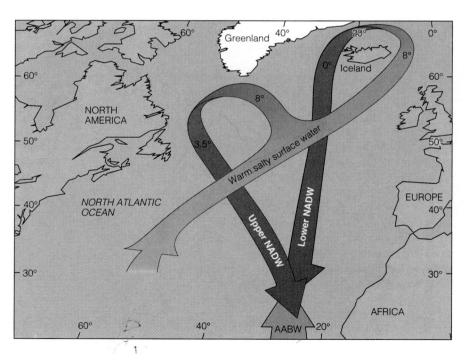

Figure 14.6A Deep Saline Water North Atlantic Deep Water (NADW) forms when the warm, salty water of the Gulf Stream/ North Atlantic Current cools, becomes increasingly saline due to evaporation and plunges downward to the ocean floor. The densest water then spills over the Greenland–Scotland ridge and flows southward as lower NADW. Less-dense water forming between Greenland and North America moves south and east as upper NADW, overriding the lower, denser water. Because both water masses are less dense than northward-flowing Antarctic Bottom Water (AABW), they pass over it on their southward journey.

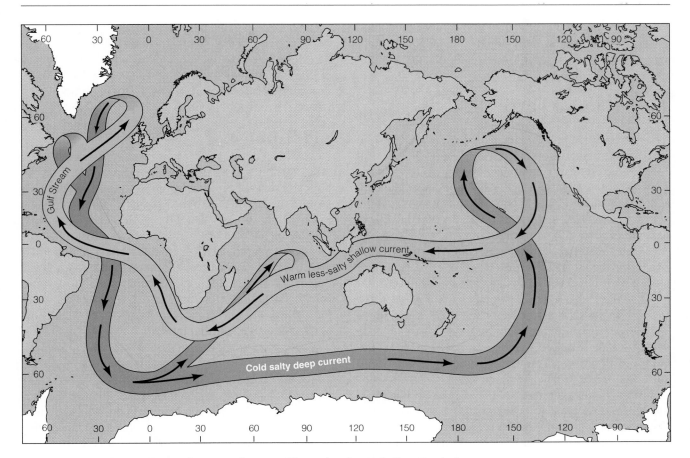

Figure 14.6B Global Ocean Conveyor System The major thermohaline circulation cells that make up the global ocean conveyor system are driven by exchange of heat and moisture between the atmosphere and ocean. Dense water forming at a number of sites in the North Atlantic spreads slowly along the ocean floor, eventually to enter both the Indian and Pacific oceans before slowly upwelling and entering shallower parts of the thermohaline circulation cells. Antarctic Bottom Water (AABW) forms adjacent to Antarctica and flows northward in fresher, colder circulation cells beneath warmer, more saline waters in the South Atlantic and South Pacific. It also flows along the Southern Ocean beneath the Antarctic Circumpolar Current to enter the southern Indian Ocean. Warm surface waters flowing into the western Atlantic and Pacific basins close the great global thermohaline cells.

wise around Antarctica. Surface and intermediate water flowing into the South Atlantic from the Pacific and from the Indian Ocean via the southern tip of Africa replenishes the NADW moving out of the basin (Fig. 14.6B). Meanwhile, Antarctic Bottom Water plunging down to the ocean floor in the southernmost Atlantic moves northward, slowly wells up, mixes with overlying NADW, and flows back toward Antarctica to join the Circumpolar Current (Fig. 14.6B). Thereby completing an important segment of the global system of ocean circulation, the Atlantic thermohaline circulation cell acts like a great conveyor belt.

Other circulation cells, linked to the Atlantic cell via the Antarctic Circumpolar Current, and also driven by density contrasts related to temperature and salinity, exist in the Pacific and Indian oceans. In concert, they move water along the global ocean conveyor system, slowly replenishing the waters of the deep ocean (Fig. 14.6B). NADW is estimated to form at a rate of 15 to 20 million m^3/s (equal to about 100 times the rate of outflow of the Amazon River), while Antarctic Bottom Water forms at a rate of about 20 to 30 million m^3/s. Together, these water masses could replace all the deep water of the world ocean in about 1000 years.

OCEAN TIDES

Tides, the rhythmic, twice-daily rise and fall of ocean waters, are caused by the gravitational attraction between the Moon (and to a lesser degree, the Sun) and the Earth. A sailor in the open sea may not detect tidal motion, but near coasts the effect of the tides is amplified and they become geologically important.

Tide-Raising Force

The gravitational pull that the Moon exerts on the solid Earth is balanced by an equal but opposite inertial force (which tends to maintain a body in uniform linear motion) created by the Earth's rotation about the center of mass of the Earth–Moon system (Fig. 14.7). At the center of the Earth, gravitational and inertial forces are balanced, but they are not balanced from place to place on the Earth's surface. A water particle in the ocean on the side facing the Moon is attracted more strongly by the Moon's gravitation than it would be if it were at the Earth's center, which lies at a greater distance. Although the attractive force is small, liquid water is easily deformed, and so each water particle on this side of the Earth is pulled toward a point located at the center of the Moon. This creates a bulge on the ocean surface.

Although the magnitude of the Moon's gravitational attraction on a particle of water at the surface of the ocean varies over the Earth's surface, the inertial force at any point on the surface is the same. On the side nearest the Moon, gravitational attraction and inertial force combine, and the excess inertial force (or *tide-*

raising force) is directed toward the Moon. On the opposite side of the Earth the inertial force exceeds the Moon's gravitational attraction, and the tide-raising force is directed away from the Earth (Fig. 14.7). These unbalanced forces generate the daily ocean tides.

Tidal Bulges

The tidal bulges created by the tide-raising force on opposite sides of the Earth appear to move continually around the Earth as it rotates. In fact, the bulges remain essentially stationary beneath the tide-producing body (the Moon) while the Earth rotates. At most places on the ocean margins, two high tides and two low tides are observed each day as a coast encounters both tidal bulges. In effect, at every high tide, a mass of water runs into the coastline, where it piles up. This water then flows back to the ocean basin as the coastline passes beyond each tidal bulge.

Earth–Sun gravitational forces also affect the tides, sometimes opposing the Moon by pulling at a right angle and sometimes aiding by pulling in the same direction. Twice during each lunar month, the Earth is directly aligned with the Sun and the Moon, whose gravitational effects are thereby reinforced, producing higher high tides and lower low tides (Fig. 14.8). At positions halfway between these extremes, the gravitational pull of the Sun partially cancels that of the Moon,

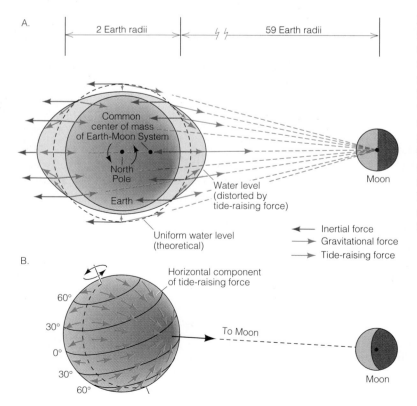

Figure 14.7 Origin of the Tides A. Tide-raising forces are produced by the Moon's gravitational attraction and by inertial force. On the side toward the Moon, both forces combine to distort the water level from that of a sphere, raising a tidal bulge. On the opposite side of the Earth, where inertial force is greater than the gravitational force of the Moon, the excess inertial force (called the tide-raising force) also creates a tidal bulge. B. The horizontal component of the tide-raising force is shown by arrows directed toward the point where a line connecting the Earth and Moon intersects the Earth's surface. This point shifts latitude with time as the relative position of the Earth and Moon changes.

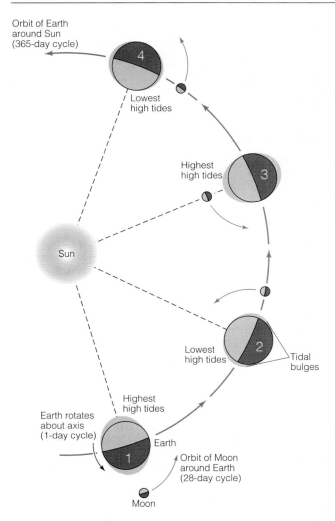

A

B

Figure 14.8 Highest and Lowest Tides When the Earth, Moon, and Sun are aligned (positions 1 and 3), tides of highest amplitude are observed. When the Moon and Sun are pulling at right angles to each other (positions 2 and 4), tides of lowest amplitude are experienced.

Video

Tides

Figure 14.9 Big Tidal Range The tidal range in the Bay of Fundy, eastern Canada, is one of the largest in the world. A. Coastal harbor of Alma, New Brunswick at high tide. B. Same view at low tide.

thus reducing the tidal range. However, the Sun is only 46 percent as effective as the Moon in producing tides, so the two tidal effects never entirely cancel each other.

In the open sea, the effect of the tides is small (≤1 m), and along most coasts the tidal range commonly is no more than 2 m. However, in bays, straits, estuaries, and other narrow places along coasts, tidal fluctuations are amplified and may reach 16 m or more (Fig. 14.9). Associated currents are often rapid and may approach 25 km/h. The incoming tide locally can create a wall of water a meter or more high (called a *tidal bore*) that moves up estuaries and the lower reaches of streams. Fast-moving *tidal currents*, though restricted in extent, constitute a potential source of renewable energy that is still largely untapped.

Tidal Power

Energy obtained from the tides is renewable energy, for it never can be used up. However, harnessing tidal power for human use has seen only limited success. Water in a restricted bay, retained behind a dam at high tide, can drive a generator the same way that river water can. One important difference between hydroelectric power from rivers and that from tidal power is that rivers flow continuously, whereas tides can be exploited only twice a day; therefore, electrical supply is erratic. Experts estimate that if all sites with suitable tidal ranges were developed to produce power, the total recoverable energy would be equivalent to only about a tenth that annually obtained from oil. Thus, although tidal power may prove important locally, it is unlikely ever to be a significant factor in global energy supply.

OCEAN WAVES

Ocean waves receive their energy from winds that blow across the water surface. The size of a wave depends on how fast, how far, and how long the wind blows. A gentle breeze blowing across a bay may ripple the water or form low waves less than a meter high. By contrast, storm waves produced by hurricane-force winds (>115 km/h) blowing for days across hundreds or thousands of kilometers of open water may become so high that they tower over ships unfortunate enough to be caught in them.

Wave Motion

Figure 14.10 shows the significant dimensions of a wave traveling in deep water, where it is unaffected by the bottom far below. Each small parcel of water in the wave revolves in a loop, returning, as the wave passes, very nearly to its former position. This looplike, or oscillating, motion of the water can be observed by injecting droplets of dye into a glass tank of water in which waves have been generated mechanically and then photographing the paths of the waves with a video camera.

Because wave form is created by a looplike motion of water parcels, the diameters of the loops at the water surface exactly equal wave height (*H* in Fig. 14.10). Downward from the surface, a progressive loss of energy occurs, expressed as a decrease in loop diameter. At a depth equal to only half the **wavelength** (the distance between successive wave crests or troughs; *L* in Fig. 14.10), the diameters of the loops have become so small that motion of the water is negligible.

Wave Base
The depth *L*/2 is the effective lower limit of wave motion (Fig. 14.10). Therefore, this depth must be the lower limit of erosion of the bottom by waves. This depth is generally referred to as the **wave base**. In the Pacific Ocean, wavelengths as long as 600 m have been measured. For them, *L*/2 equals 300 m, a depth half again as great as the average depth of the outer edge of the con-

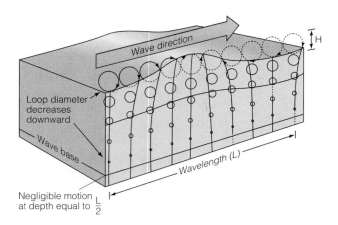

Figure 14.10 Motion of Water in a Wave Looplike motion of water parcels in a wave in deep water. To trace the motion of a water parcel at the surface, follow the arrows in the largest loops from right to left. The resultant motion is the same as watching the wave crest travel from left to right. Parcels of water in smaller loops beneath the surface have corresponding positions, marked by nearly vertical lines. Dashed lines represent waveform and parcel positions one-eighth of a period later.

tinental shelves (about 200 m). Although the wavelengths of most ocean waves are far shorter than 600 m, it nevertheless is possible for very large waves approaching these dimensions to affect even the outer parts of continental shelves.

Landward of depth *L*/2, the circular motion of the lowest water parcels is influenced by the increasingly shallow seafloor, which restricts movement in the vertical direction. As the water depth decreases, the orbits of the water parcels become flatter until the movement of water at the seafloor in the shallow water zone is limited to a back-and-forth motion (Fig. 14.11).

Breaking Waves
As a wave nears the shore, it undergoes a rapid transformation. When the wave reaches depth *L*/2, its base encounters frictional resistance exerted by the seafloor. This resistance interferes with wave motion and distorts the wave's shape, causing the height to increase and the wavelength to decrease. Now the front of the wave is in shallower water than the rear part and is also steeper than the rear. Eventually, the front becomes too steep to support the advancing wave, and as the rear part continues to move forward, the wave collapses, or *breaks* (Fig. 14.11).

The form of breaking waves differs from place to place. This is why surfers prefer some beaches more than others. Board surfing originated in Hawaii, where the gently sloping seafloor immediately offshore allows approaching waves to increase in height gradually before rolling over and breaking. An expert surfer can catch a wave and ride its arcing front for a long distance before turning beneath the breaking crest and exiting to safety.

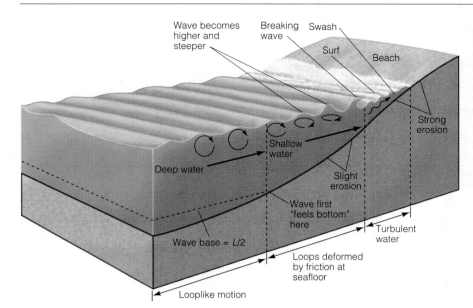

Wave becomes higher and steeper

Breaking wave

Swash

Surf

Beach

Deep water

Shallow water

Strong erosion

Slight erosion

Wave first "feels bottom" here

Wave base = *L*/2

Loops deformed by friction at seafloor

Turbulent water

Looplike motion

Figure 14.11 Waves Reach the Shore Waves change form as they travel from deep water through shallow water to shore. In the process, the circular motion of water parcels in deep water changes to elliptical motion as the water becomes shallow and the wave encounters frictional resistance to forward movement. Vertical scale is exaggerated, as is the size of loops relative to the scale of waves. (Compare with Figure 14.10.)

Surf

When a wave breaks, the motion of the water instantly becomes turbulent, like that of a swift river. Such "broken water," called **surf**, is wave activity between the line of breakers and the shore. In surf, each wave finally dashes against rock or rushes up a sloping beach until its energy is expended; then the water flows back toward the open sea. Water piled against the shore returns seaward in an irregular and complex way, partly as a broad sheet along the bottom and partly in localized narrow channels as *rip currents*, which are responsible for dangerous undertows that can sweep unwary swimmers out to sea.

The geologic work of waves is mainly accomplished by the direct action of surf. Surf is a powerful erosional agent because it possesses most of the original energy of each wave that created it. This energy is quickly consumed in turbulence, in friction at the bottom, and in movement of the sediment that is thrown violently into suspension from the bottom.

Wave Refraction

A wave approaching a coast generally does not encounter the bottom simultaneously all along its length. As any segment of the wave touches the seafloor, that part slows down, the wavelength begins to decrease, and the wave height increases. Gradually, the trend of the wave becomes realigned to parallel the bottom contours (Fig. 14.12). Known as **wave refraction**, this process changes the direction of a series of waves moving in

Figure 14.12 Wave Refraction Waves arriving obliquely onshore along a coast near Oceanside, California, change orientation as they encounter the bottom and begin to slow down. As a result, each wave front is refracted so that it more closely parallels the bottom contours. The arriving waves develop a longshore current that moves from left to right in this view.

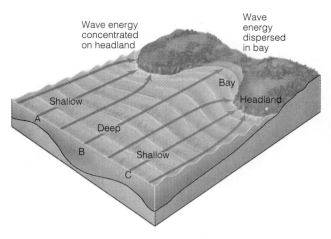

Figure 14.13 Erosion of Headlands Refraction of waves concentrates wave energy on headlands and disperses it along shores of bays. This oblique view shows how waves become progressively distorted as they approach the shore over a bottom that is deepest opposite the bay. The result is vigorous erosion on the exposed headland and sedimentation along the margin of the bay.

shallow water at an angle to the shoreline. Thus, waves approaching the margin of a deep-water bay at an angle of 40 or 50° may, after refraction, reach the shore at an angle of 5° or less.

Waves passing over a submerged ridge off a headland will be refracted and converge on the headland (Fig. 14.13). This convergence, as well as the increased wave height that accompanies it, concentrates wave energy on the headland, which is eroded vigorously. Conversely, refraction of waves approaching a bay will make them diverge, diffusing their energy at the shore. The net tendency of these contrasting effects, in the course of time, is to make irregular coasts smoother and less indented.

COASTAL EROSION AND SEDIMENT TRANSPORT

The world's ocean coastlines are dynamic zones of conflict where land and water meet. Erosional forces tear away at the land, and other forces move and deposit sediment, thereby adding to the land. Because the conflict is unending, few coasts achieve a condition of complete equilibrium.

Erosion by Waves

Most erosion along a seacoast is accomplished by waves moving onshore. Wave erosion takes place not only at sea level in the surf zone, but also below and above sea level.

Erosion Below Sea Level

Ocean waves typically break at depths that range between wave height and 1.5 times wave height. Because waves are seldom more than 6 m high, the depth of vigorous erosion by surf should be limited to 6 m times 1.5, or 9 m below sea level. This theoretical limit is confirmed by observation of breakwaters and other coastal structures, which are only rarely affected by surf to depths of more than 7 m.

Abrasion in the Surf Zone

An important kind of erosion in the surf zone is the wearing down of rock by wave-transported rock particles. By continuous rubbing and grinding with these tools, the surf wears down and deepens the bottom and eats into the land, at the same time smoothing, rounding, and making smaller the tools themselves. Because surf is the active agent, this activity is limited to a depth of only a few meters below sea level. In effect, the surf is like an erosional knife edge or saw cutting horizontally into the land.

Erosion Above Sea Level

During great storms, surf can strike effective blows well above sea level. During one great storm that struck the coast of Scotland, a solid mass of stone, iron, and concrete weighing 1200 metric tons was ripped from the end of a breakwater and moved inshore. The damage was repaired with a block weighing more than 2300 metric tons, but five years later storm waves broke off and moved that block too. The pressures involved in such erosion were about 27 metric tons/m^2. Even waves having much smaller force can break loose and move blocks of bedrock from sea cliffs. Waves pounding against a cliff compress the air trapped in fissures. So great is the force of the compressed air that blocks of rock can be dislodged. Nearly all the energy expended by waves in coastal erosion is confined to a zone that lies between 10 m above and 10 m below mean sea level.

Sediment Transport by Waves and Currents

Sediment produced by waves pounding against a coast, or brought to the sea by rivers, is redistributed by currents that build distinctive shoreline deposits or move sediment offshore onto the continental shelves.

Longshore Currents

Despite refraction, most waves reach the shore at an oblique angle (Fig. 14.14). The path of an incoming wave can be resolved into two directional components, one oriented perpendicular to the shore and the other parallel to the shore. Whereas the perpendicular component produces the crashing surf, the parallel compo-

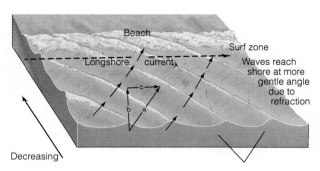

Figure 14.14 Longshore Current A longshore current develops offshore as waves approach a beach at an angle and are refracted. The line representing the front of each approaching wave can be resolved into two components: the component oriented perpendicular to the shore (b) produces surf, whereas that oriented parallel to the shore (c) is responsible for the longshore current. Such a current can transport considerable amounts of sediment along the coast.

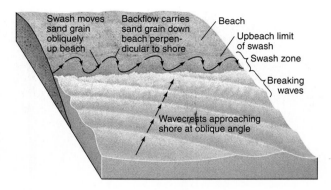

Figure 14.15 Beach Drift As surf rushes up a beach with each incoming wave, sand grains are picked up and carried shoreward. Arriving surf approaching the shore at an angle will travel obliquely up the beach. The return flow, pulled by gravity, flows back nearly perpendicular to the shoreline. A grain of sand will therefore move along a zigzag path as successive waves reach the shore. Net motion is down the beach and is called beach drift.

nent sets up a **longshore current** within the surf zone that flows parallel to the shore(i.e., along the shore) (Fig. 14.14). While surf erodes sediment at the shore, the longshore current moves the sediment along the beach. The direction of longshore currents may change seasonally if the prevailing wind directions change, thereby causing changes in the direction of the arriving waves.

Beach Drift

Meanwhile, on the exposed beach, incoming waves produce a second, irregular pattern of water movement directed along the shore. Because waves generally strike the beach at an angle, the swash (uprushing water) of each wave travels obliquely up the beach before gravity pulls the water back directly down the slope of the beach. This zigzag movement of water carries sand and pebbles first up, then down the beach slope. The net effect of successive movements of this type is the progressive transport of sediment along the shore, a process known as **beach drift** (Fig. 14.15). The greater the angle of waves to shore, the greater the rate of longshore movement. Marked pebbles have been observed to drift along a beach at a rate of more than 800 m/day. When the volume of sand moved by beach drift is added to that moved by longshore currents, the total can be very large.

Beach Placers

Gold, diamonds, and several other heavy minerals have been concentrated in beach sands by surf and longshore currents. Diamonds are being obtained in large quantities from gravelly beach placers, both above and below sea level, along a 350-km strip of coastal Namibia in southwestern Africa. Weathered from deposits in the

interior, the diamonds were transported by the Orange River to the coast and were spread southward by longshore drift. Later, some beaches were raised tectonically, and others were drowned by rising sea level.

Although Nome, Alaska, has become famous as the finish line of the grueling Iditarod sled-dog race, the town initially attracted world attention because of its six beaches, four above sea level and two below, which are rich with placer gold. Ilmenite, a primary source of titanium, is highly concentrated (50 to 70%) along several beaches in India. Magnetite-rich sands occur on the coasts of Oregon, California, Brazil, and New Zealand, while chrome-rich sands are mined in Japan.

Offshore Transport and Sorting

Seaward of the surf zone, bottom sediment is moved by currents and by unusually large waves during storms, with net movement seaward. Each particle of sediment is picked up again and again, whenever the energy of waves or currents is great enough to move it. As the particle gets into ever-deeper water, it is picked up less and less frequently. With increasing depth, energy related to wave motion decreases. Therefore, far from shore only fine grains can be moved. As a result, the sediment becomes sorted according to diameter, from coarse in the surf zone to finer offshore. This subaqueous sorting is analogous to sorting by wind, which transports fine sediment to greater distances than it transports coarse sediment.

As sediments accumulate on a continental shelf, they normally grade seaward from sand into mud. This gradation is true not only of the particles eroded from the shore by surf, but also of the particles contributed by rivers. Suspended sediment from the Columbia River in the northwestern United States moves obliquely across

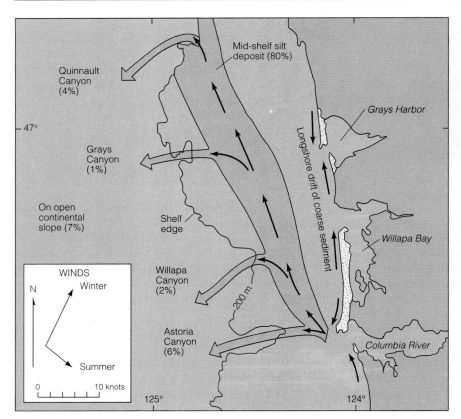

Figure 14.16 Sediment on the Continental Shelf Dispersal of sediment on the continental shelf off the state of Washington. Suspended sediment from the Columbia River moves northwest offshore, about 80 percent of it accumulating as a mid-shelf silt deposit. The remaining 20 percent either is deposited on the adjacent continental slope or moves down steep submarine canyons.

the continental shelf, where about 80 percent of it accumulates as a midshelf silt deposit (Fig. 14.16). Of the remaining 20 percent, about 13 percent is funneled down steep submarine canyons toward the continental rise, while the rest is deposited on the open continental slope.

A VARIETY OF COASTS

At a coast, waves that may have traveled unimpeded across thousands of kilometers of open ocean encounter an obstruction to further progress. They dash against firm rock, erode it, and move the eroded rock particles. Over the long term, the net effect is substantial.

Elements of the Shore Profile

To understand the changes along a coast, we must first examine the *shore profile*, a vertical section along a line drawn perpendicular to the shore. If we combine what we learn from such a profile with what we know about the forces that act parallel to the shore, we will have a three-dimensional picture of coastal activities.

Beaches

Beaches are characteristic features of many coasts, even those with steep, rocky cliffs. Along sandy coasts that lack cliffs, the beach constitutes the primary shore environment. A beach is regarded by most people as the sandy surface above the water along a shore. Actually, it is more than this. We define a **beach** as wave-washed sediment along a coast, including sediment in the surf zone. In this dynamic zone, sediment is continually in motion.

Part of the sediment of a beach may be derived from erosion of adjacent cliffs or cliffs elsewhere along the coast. However, along most coasts a much higher percentage of the sediment comes from alluvium brought to the shore by rivers.

On low, open shores an exposed beach typically has several distinct elements (Fig. 14.17). The first is a rather gently sloping *foreshore*, a zone extending from the level of lowest tide to the average high-tide level. Here is found a *berm*, which is a nearly horizontal or landward-sloping bench formed of sediment deposited by waves. Beyond lies the *backshore*, a zone extending inland from the berm to the farthest point reached by surf. On some coasts, beach sand is blown inland by onshore winds to form belts of coastal dunes (Chapter 13).

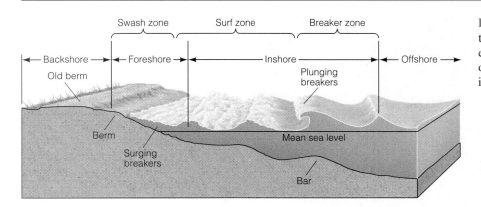

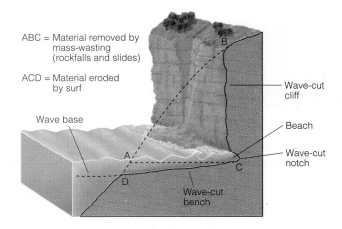

Figure 14.18 Erosion of a Cliffed Coast Principal features of a shore profile along a cliffed coast. Notching of the cliff by surf action undermines the rock, which then collapses and is reworked by surf. Note the large proportion of material removed by mass-wasting (ABC) relative to that eroded by surf (ACD).

Rocky (Cliffed) Coasts

Seen in profile, the usual elements of a cliffed coast (Fig. 14.18) are a wave-cut cliff and wave-cut bench, both the work of erosion, and a beach, which is the result of deposition. A *wave-cut cliff* is a coastal cliff cut by surf. Acting like a horizontal saw, the surf cuts most actively at the base of the cliff. As the upper part of the cliff is undermined, it collapses and the resulting debris is redistributed by surf (Fig. 9.27). An undercut cliff that has not yet collapsed may have a well-developed notch at its base.

Below a wave-cut cliff, you can often find a **wave-cut bench**, a platform cut across bedrock by surf. It slopes gently seaward and expands in the landward direction as the cliff retreats. Some benches are bare or partly bare, but most are thinly covered with sediment that is in transit from shore to deeper water. The shoreward parts of some benches are exposed at low tide (Fig. 14.19). If

Figure 14.19 Wave-Cut Bench A nearly horizontal wave-cut bench has formed along the coast at Bolinas Point, California, as the surf, acting like an erosional saw, has cut into and beveled the tilted sedimentary strata.

Figure 14.20 Sea Stacks and Sea Arches Stack and sea arch along the French shore of the English Channel near Étretat carved in horizontally bedded white chalk. The surf first hollows out a sea cave in the most erodible part of the bedrock. A cave excavated completely through a headland is then transformed into a sea arch. An isolated remnant of the cliff stands as a stack on a wave-cut bench offshore.

the coast has been uplifted, a wave-cut bench and its sediment cover can be completely exposed.

Other erosional features associated with cliffed coasts are *sea caves*, *sea arches*, and *stacks* (Fig. 14.20). Each is the result of differential erosion of rock as surf attacks a cliff, causing its gradual retreat.

Factors Affecting the Shore Profile

Through erosion and the creation, transport, and deposition of sediment, the form of a coast changes, often slowly but at some times very rapidly. At any given time, both the geometry of the shoreline and the shape of the shore profile represent a compromise among the constructive and destructive forces acting along the coast.

The compromise is reached in several ways. On a beach, for instance, a wave running up the beach as a thin sheet of water moves particles of sand and gravel upslope, while gravity pulls the particles back again (Fig. 14.15). More energy is needed to move pebbles downslope than to move sand grains. Therefore, the pebbles moved shoreward by the wave remain until the slope becomes steep enough for the returning flow of water (the backwash) to carry them back. This partly explains why gravel beaches are generally steeper than beaches built of sand.

Another factor determining the beach profile is permeability. Some of the water rushing up a beach easily flows into the underlying porous beach sediment, thereby reducing the volume and transporting capability of the backwash.

During storms, the increased energy in the surf erodes the exposed part of a beach and makes it narrower. In calm weather, the exposed beach is likely to receive more sediment than it loses and therefore becomes wider. Storminess may be seasonal, resulting in seasonal changes in beach profiles. Along parts of the Pacific coast of the United States, winter storm surf tends to carry away fine sediment, and the remaining coarse fraction assumes a steep profile. In calm summer weather, fine sediment drifts in and the beach assumes a gentler profile.

Major Coastal Deposits and Landforms

Up to this point we have been describing the erosional effects of surf and the related shaping of the shore profile. The deposits made by surf and by the currents it generates are equally important. These deposits are largely the result of longshore transport of sediment, and they produce some distinctive landforms.

Marine Deltas

In some places, constructional processes *prograde* (build out) the coastline more rapidly than it can be destroyed by surf. The extent to which a marine delta projects seaward from the land is a compromise between the rate at which a river delivers sediment at its mouth and the ability of currents and waves to erode sediment along the delta front and move it elsewhere along the coast. The great size of the Mississippi delta (Fig. 14.21) testifies to the huge volume of sediment carried by the river and to the relative ineffectiveness of waves in destroying it. The Columbia River formerly transported a large load of sediment to the Pacific coast (much of the sediment is now trapped behind numerous

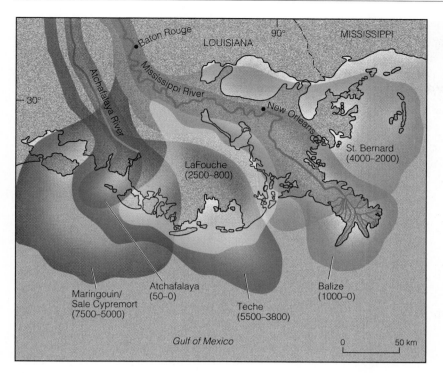

Figure 14.21 Mississippi Delta The Mississippi River has built a series of overlapping subdeltas as it has continually dumped sediment into the Gulf of Mexico. The ages of subdeltas are given in radiocarbon years before the present.

hydroelectric and irrigation dams built along the river's course), and yet no delta has been constructed at its mouth. In this case, strong winter storms and persistent wave action erode sediment as quickly as it arrives at the coast.

Many deltas are compound features that have a long and complicated sedimentary history. The Mississippi delta is really a complex of several coalescing subdeltas built successively over the last several thousand years (Fig. 14.21). The present active margin of the delta has a very irregular front (the so-called birdfoot delta, or Balize lobe). The shores of the adjacent older subdeltas have been extensively modified by coastal erosion and by gradual submergence as the crust slowly subsides under the weight of the accumulating pile of sediment.

Spits and Related Features

Among other conspicuous landforms on many coasts is the **spit**, an elongated ridge of sand or gravel that projects from land and ends in open water (Fig. 14.22).

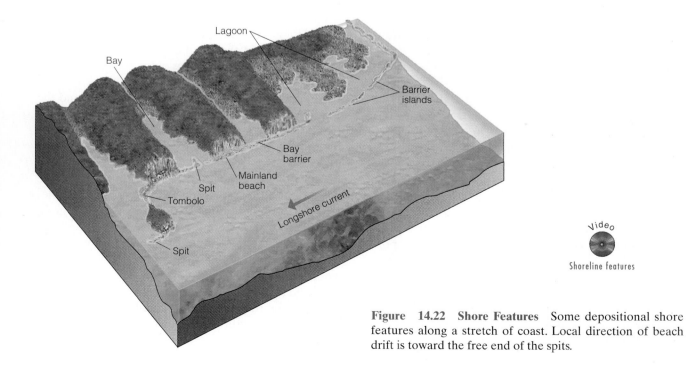

Figure 14.22 Shore Features Some depositional shore features along a stretch of coast. Local direction of beach drift is toward the free end of the spits.

Video

Shoreline features

Figure 14.23 Cape Cod The curved spit at the north tip of Cape Cod, Massachusetts, has been built by a north-flowing longshore current that reworks glacial deposits forming the peninsula east and south of Cape Cod Bay.

Most spits are merely continuations of beaches. Well-known examples are Sandy Hook on the southern side of the entrance to New York Harbor and Cape Cod, Massachusetts (Fig. 14.23). A spit is built of sediment moved by longshore drift and dropped at the mouth of a bay where the longshore current encounters deeper water and its velocity decreases. The free end curves landward in response to currents created by refraction as waves enter the bay.

A spitlike ridge of sand or gravel that connects an island to the mainland or to another island, called a *tombolo* (Fig. 14.22), forms in much the same way as a spit does. The shape of a spit normally reflects the quantity and source of sediment and the prevailing transport direction. For example, coarse sediment carried to the Pacific Ocean by the Columbia River, together with sediment eroded by waves pounding against sea cliffs, is moved by longshore currents and forms large spits across major embayments (Fig. 14.16). The spits are oriented in the directions of the prevailing summer

(southward) and winter (northward) longshore currents, which are controlled by seasonal wind patterns.

Along an embayed coast with abundant sediment supply, a ridge of sand or gravel may be built across the mouth of a bay to form a *bay barrier* (Fig. 14.22). A bay barrier develops as beach drift lengthens a spit across a bay in which tidal or river currents are too weak to scour away the spit as it is built.

Beach Ridges

Beaches on many sandy coasts lie seaward of, and nearly parallel to, a series of low, sandy *beach ridges*. Beach ridges are old berms, often built during major storms. Nearly all date to the last 5000 years, during which time sea level has been within several meters of its present position. Beach-ridge complexes in western Alaska, dated by prehistoric artifacts left by ancient maritime cultures, become progressively older inland from the present coast.

Barrier Islands

A **barrier island** (Figs. 14.22 and 14.24A) is a long, narrow sandy island lying offshore and parallel to a coast. An elongate bay lying inshore from a barrier island or other low, enclosing strip of land (such as a coral reef) is called a *lagoon*. Barrier islands are found along most of the world's lowland coasts. Well-known examples are Coney Island and Jones Beach (New York City's coastal playground areas), and the long chain of islands centered at Cape Hatteras on the North Carolina coast. Padre Island, one of a succession of barrier islands paralleling the coast of Texas, is 130 km long.

Many barrier islands off the southeastern United States probably were built as the rate of sea-level rise at the end of the last glacial age began to slow during the middle Holocene (Fig. 14.24B). Shells obtained at depths of 5 to 10 m from the basal deposits of some barrier islands have radiocarbon ages as old as 5000 years, an age that supports such a history. As the sea then rose more slowly across the very gentle continental shelf and the shoreline moved progressively inland, waves breaking in shallow water offshore eroded the bottom and piled up sand to form the long bars that ultimately became barrier islands.

A barrier island generally consists of one or more ridges of dune sand associated with successive shorelines that were occupied as the island formed. During great storms, surf washes across low places on the island and erodes it, cutting inlets that may remain permanently open. At such times, fine sediment is washed into the lagoon between barrier and mainland. In this way, the length and shape of barrier islands are always changing. Studies of the response of barrier islands to changing environmental conditions suggest that island development is closely related to sediment supply, the direction

A.

B.

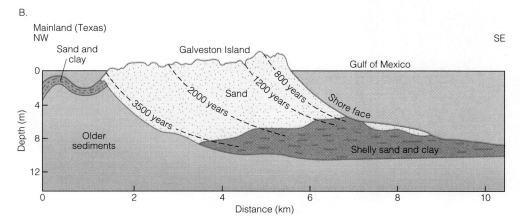

Figure 14.24 Barrier Islands A. This sandy barrier island off the coast of Mississippi lies so close to sea level that waves can surge across its surface during large storms, eroding and redistributing the sediment. B. Cross section through Galveston Island, one of a series of barrier islands off the coast of Texas. Dashed lines show the former position of the seaward side of the island, based on radiocarbon dating. Since 3500 years ago, the island has grown southeastward toward the Gulf of Mexico.

and intensity of waves and nearshore currents, the shape of the offshore profile, and the stability of sea level.

Organic Reefs and Atolls

Many of the world's tropical coastlines consist of organic reefs built by vast colonies of tiny organisms that secrete calcium carbonate to form corals. Such organisms require shallow, clear water in which the temperature remains above 18°C. Reefs therefore are built only at or close to sea level and are characteristic of low latitudes.

Three principal reef types are recognized. A **fringing reef** is either attached to or closely borders the adjacent land (Fig. 14.25A). It therefore lacks a lagoon. Typically, a fringing reef has a flat upper surface as much as 1 km wide, and its seaward edge plunges steeply into deeper water.

A **barrier reef** is separated from the land by a lagoon that may be of considerable length and width (Fig. 14.25B). Such a reef may surround an island or lie far off the coast of a continent, as in the case of the Great Barrier Reef off Queensland, Australia.

An **atoll**, a roughly circular coral reef enclosing a shallow lagoon (Fig. 14.25C), is formed when a tropical volcanic island with a fringing reef slowly subsides (Fig. 14.26). The subsidence causes the reef to grow upward, for its organisms can survive only near sea level. As the island continues to subside, the area of exposed volcanic land becomes progressively smaller, and the fringing reef is transformed into an offshore barrier reef. Eventually, the volcanic island disappears beneath the sea, and the surrounding reef is thereby transformed into an atoll. Atolls generally lie in deep

A

B

C

Figure 14.25 Coral Reefs Chief kinds of tropical coral reefs. A. Fringing reef on the island of Oahu in the Hawaiian Islands. B. Barrier reef enclosing the island of Moorea in the Society Islands. A narrow lagoon separates the high island, which is the eroded remnant of a formerly active volcano, from a shallow reef. C. The reef of a small atoll in the Society Islands is surmounted by low, vegetated sandy islands that lie inside a line of breakers along the reef margin.

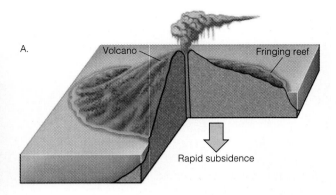

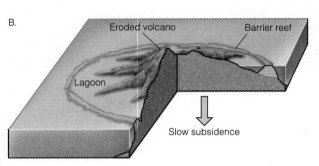

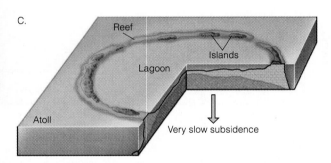

Figure 14.26 Origin of an Atoll Evolution of an atoll from a subsiding volcanic island. Rapid extrusion of lava to form an oceanic shield volcano causes the island to subside as the crust is loaded by the volcanic pile. A fringing reef grows upward, keeping pace with submergence, and becomes a barrier reef surrounding the eroding volcano. With continued subsidence and upward reef growth, the last remnants of the volcano disappear beneath sea level, and all that remains is an atoll reef surrounding a central lagoon.

water in the open ocean and range in diameter from as little as 1 km to as much as 130 km. Their origin has been proved by drill holes that reach volcanic rock after penetrating thick sections of ancient reef rocks.

COASTAL EVOLUTION

The world's coasts do not all fall into easily identifiable classes. Their variety is great because their configurations depend largely on the structure and erodibility of coastal rocks, on the active geologic processes at work, on the length of time over which these processes have operated, and on the history of world sea-level fluctuations.

Types of Coasts

Some coasts—for instance, most of the Pacific coast of North America—are steep and rocky and consist of mountains or hills separated by deep valleys. Others, such as the Atlantic and Gulf coasts from New York City to Florida and westward into northern Mexico, traverse a broad coastal plain that slopes gently seaward and are festooned with barrier islands.

The Pacific and Atlantic/Gulf coasts represent two extremes, between which are many intermediate types. Each type owes its general character to its structural setting. The rugged and mountainous Pacific coast lies along the margin of the North American Plate, which is continually being deformed where it interacts with adjacent plates to the west. Uplifted and faulted marine terraces are common features along parts of this coast (for example, in southern Alaska, Oregon, and California) and along similar coasts that are emerging from the sea (Japan, New Zealand, and northern Chile are examples). By contrast, the eastern continental margin lies within the same lithospheric plate as the adjacent ocean

floor and in a zone that is tectonically passive. The bedrock has low relief, and much of the coastal zone borders young sedimentary deposits of the Atlantic and Gulf coastal plains. Old shorelines inland from these coasts formed at times when glaciers were fewer, the ocean basins contained more water than they do now, and world sea level therefore was higher.

Much of the New England coast, like those of Sweden and Finland, is a coast of low relief that owes its special character to repeated glaciation and changing sea level. Each overriding ice sheet eroded bedrock in the coastal zone and depressed the land below sea level. With retreat of the ice, both the land and world sea level rose (although not always at the same rate). The result is an embayed, rocky coastline that shows the effects both of differential glacial erosion and drowning of the land by the most recent sea-level rise.

Where rocks of different erodibilites are exposed along a coast, marine erosion typically will produce a shoreline that is strongly controlled by rock type and structure. Such control is especially impressive in folded sedimentary or metamorphic belts that have been partially submerged, as along the coasts of Norway, Ireland, and Croatia (Fig. 14.27). It is also responsible for some of the world's deeply embayed fjord coasts, the pattern of which reflects deep glacial erosion along regional fracture systems and subsequent invasion by seawater

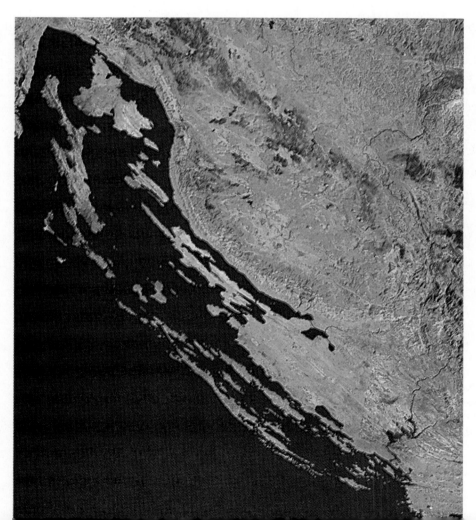

Figure 14.27 Islands in the Adriatic Sea Differential erosion of sedimentary rocks along the coast of Croatia on the Adriatic Sea. The tectonically deformed and eroded rocks, now partially submerged by the sea, form a series of linear islands along the trend of major structures.

as glaciers retreated to the heads of their fjords (Fig. 12.22).

Effectiveness of Coastal Processes

The effectiveness of erosional and depositional processes is not equal along all coasts. Coasts lying at latitudes between about 45 and 60° are subjected to higher than average storm waves generated by strong westerly winds, whereas subtropical east-facing coasts are subjected to infrequent but often disastrous hurricanes (called typhoons west of the 180th meridian). In the polar regions, sea ice becomes an effective agent of coastal erosion. These and other factors influence the amount of energy expended in erosion along the shore and, together with the structural and compositional properties of the exposed rocks, contribute to the variety of coastal landforms.

Changing Sea Level

We have seen already that sea level fluctuates daily as a result of tidal forces. It also fluctuates over much longer time scales as a result of (1) changes in the volume of water in the oceans as continental glaciers wax and wane

and (2) the motions of lithospheric plates that cause the volume of the ocean basins to change. Over the span of a human lifetime, these slower changes appear insignificant, but on geologic time scales they contribute importantly to the evolution of the world's coasts.

Submergence

Whatever their nature, nearly all coasts have experienced **submergence**, a rise of water level relative to the land, owing to the worldwide rise of sea level that accompanied the most recent deglaciation. (see *The Flooding of the Black Sea*). As water from melting glaciers returned to the oceans, rising sea level caused widespread submergence of coastal zones (Chapter 12; Fig. 14.28). Most large *estuaries*, for example, are former river valleys that were drowned by this recent sea-level rise (see *How to Modify an Estuary*). Evidence of lower glacial-age sea levels is almost universally found seaward of the present coastlines and to depths of 100 m or more. Former beaches, sand dunes, and other coastal landscape features on the inner continental shelves mark shorelines built by the rising sea at the end of the glacial age and later drowned.

Emergence

Evidence of past higher sea levels is related mainly to past interglacial ages. Inland from the Atlantic coast of the United States from Virginia to Florida are many

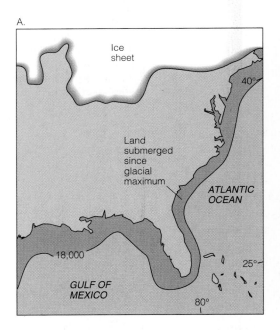

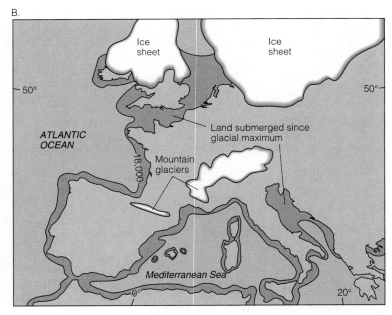

Figure 14.28 Effect of Rising Sea Level   Coastal submergence of eastern North America and western Europe resulted when meltwater from wasting ice sheets returned to the ocean basins at the close of the last glaciation. A. Area of northeastern North America covered by glacier ice during the last glacial maximum, the approximate position of the shoreline at the glacial maximum (18,000 years ago), and coastal areas submerged by the postglacial rise of sea level. B. Areas covered by ice sheets in western Europe at the last glacial maximum and land areas that have been submerged during the postglacial rise of sea level.

Measuring Our Earth

The Flooding of the Black Sea

The water spilling over the cliff created a "maelstrom of gigantic twisting eddies of water and wreckage" and "a huge crater of gigantic whirlpools." Beyond that rose "an immense fountain of water from which other waves rebounded," and above all this turmoil, "great bursts of spray atomized in a mist that steadily mushroomed into a swirling black cloud illuminated from within by continuous discharges of lightning."

This is how William Ryan and Walter Pitman of Columbia University's Lamont-Doherty Earth Observatory describe what took place when a natural dam was breached and a massive deluge poured from the Mediterranean Sea into the Black Sea. This huge flood occurred about 7500 years ago and was a result of the melting of the glaciers at the end of the last ice age 10,000 years ago.

As the glaciers retreated, meltwater raised the level of the world's oceans, including the Mediterranean Sea. But the water did not enter the Black Sea, despite its lower elevation, because the bottom of the Bosporus Strait was higher than the Mediterranean. In time, however, the rising waters entered the strait and began flowing into the Black Sea.

It may have begun as a trickle, but it eventually became a torrent: Ten cubic miles a day spilled from the Mediterranean into the Black Sea—130 times more than the daily flow over Niagara Falls. The water level in the Black Sea rose 15 centimeters a day, ending up 150 meters higher than it was before the inundation. In less than two years, a fresh-water lake was transformed into an inland sea one and a half times its original size.

The evidence for this massive flood comes from cores of sediment taken from the bottom of the Black Sea. The cores show that the marginal region of the sea was once dry land. In addition, the entire sea bottom is covered by a thin, uniform layer of sediment, which could only have resulted from a flood. Within that layer are Mediterranean saltwater mollusks dating from about 7600 years ago.

In an effort to gather further evidence to support their theory, Ryan and Pitman have undertaken an expedition to look for the remains of ancient settlements on the floor of the Black Sea. They believe that during the flood the sea's water rose so fast that coastal dwellers were driven inland at a rate of 1 to 2 kilometers a day. The population shift could have spread farming techniques into central Europe and may be the source of myths and legends like the biblical tale of Noah's ark and the ancient Greek myth of Deucalion's flood. These and other stories of a great flood in ancient times share a number of details that point to the possibility that they are different versions of the same original story.

Although it is impossible to prove that the overflow of the Mediterranean into the Black Sea was the flood described in the Bible, one can readily imagine the impact it would have had on the region's inhabitants. As the rising waters swallowed up their homes, they would have dispersed throughout eastern Europe and western Asia, bringing with them the story of an enormous flood that changed their world forever.

---

marine beaches, spits, and barriers, the highest of which reach an altitude of more than 30 m. These landforms owe their present altitude to a combination of broad upward arching of the crust and submergence during times when climates were warmer than now, glaciers were smaller, and sea level was therefore higher. The position of such features above present sea level points to **emergence**, a lowering of water level relative to the land.

Sea-level Cycles

Varied evidence demonstrates that many coastal and offshore features are relicts of times when sea level was either higher or lower than now. The youngest deposits along a coast often form a thin blanket over older, similar units that date to earlier episodes of submergence. Repeated emergence and submergence over many glacial-interglacial cycles, each accompanied by erosion

and redeposition of shoreline deposits, have resulted in complex coastal landform assemblages.

Relative Movements of Land and Sea

The rise and fall of sea level are global movements, affecting all parts of the world's oceans at the same time. By contrast, uplift and subsidence of the land, which cause emergence or submergence along a coast, are piecemeal movements, generally involving only parts of landmasses; nevertheless, such movements can cause rapid relative changes in sea level. Vertical tectonic movements at the boundary of converging lithospheric plates have uplifted beaches and tropical reefs to positions far above sea level (Fig. 14.29). Because movements of land and sea level may occur simultaneously, in either the same or opposite directions, unraveling the history of sea-level fluctuations along a coast can be a difficult and challenging exercise.

Understanding Our Environment

How to Modify an Estuary

San Francisco Bay is one of the world's best-known estuaries (Fig. B14.1). When first discovered by Europeans, its shores and productive waters supported as many as 20,000 Native Americans. The population boom that followed the discovery of gold in California in 1848 and continued unabated to the present day has led to major changes in the river systems that enter the bay and thereby has greatly affected the estuary environment.

The need for water is a continuing concern for Californians, who require it not only for human consumption and industry but also for support of huge agricultural enterprises. Diversion of water from natural streams has reduced fresh-water inflow to San Francisco Bay to less than half what it was in 1850; today, the flow is only about 30 percent of the post–1850 average. At the same time, increasing urbanization has led to a loss of 95 percent of the bordering tidal marshes as a result of filling and diking. Hydraulic mining for gold in the foothills of the Sierra Nevada following the initial discovery of gold produced vast quantities of fine sediment that choked streams, destroyed fish spawning grounds, obstructed navigation, and ultimately reached the bay, where it reduced the area and volume of the estuary and modified tidal circulation.

The unforeseen consequences of this human tampering with a natural stream and estuarine system have been many. Most commercial fisheries have disappeared.

San Francisco

Figure B14.1 Shrinking Size of San Francisco Bay Vertical satellite image of the shrinking San Francisco Bay estuary. Filling and diking of tidal marshes to create farmland, evaporation ponds, and residential and industrial developments have reduced 2200 km^2 of marsh land that existed before 1850 to less than 130 km^2 today.

Reduced fresh-water flushing of the bay has concentrated agricultural, domestic, and industrial waste products in estuarine sediments, contaminating both them and the organisms that feed on them and raising increasing concerns about human health. Natural habitats of migrating birds have been extensively destroyed, with presumed major effects on bird populations.

Such changes are not unique to San Francisco Bay. Many other large estuaries, which are favored sites of urban development and industrialization, have similar problems. The Rhine River in the Netherlands, the Thames River in England, and the Susquehanna and Potomac rivers in the United States are examples. All are sensitive to human-induced changes and susceptible to steady deterioration. The best hope for reversing their decline is through an improved understanding of how human activities affect the physical, chemical, and biological processes in river-estuarine systems, and of the measures that must be taken to curtail potentially destructive actions.

A.

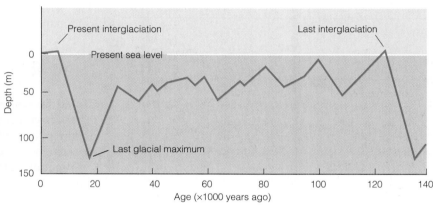

B.

Figure 14.29 New Guinea Sea-level Record Coastal emergence of eastern New Guinea. A. The emergent coast of the Huon Peninsula in eastern Papua New Guinea is flanked by a series of ancient coral reefs that form flat terracelike benches parallel to the shoreline. Each reef formed at sea level and was subsequently uplifted along this active plate margin. The highest reefs lie several hundred meters above sea level and are hundreds of thousands of years old. B. Curve of sea-level fluctuations constructed by uranium-isotope dating the uplifted reefs of the Huon Peninsula. Prior to a recent high stand, sea level was lower than at present ever since the last interglaciation, about 120,000 years ago.

COASTAL HAZARDS

People who choose to live along a coast, like those who live on the floodplains of large rivers, are often exposed to natural hazards that can prove devastating. During infrequent large events such as storms, the dynamic nature of the coastline environment becomes especially obvious.

Storms

The approximate equilibrium among the forces that operate along coasts is occasionally interrupted by exceptional storms that erode cliffs and beaches at rates far greater than the long-term average. During a single storm in 1944, for example, cliffs of compact sediment on Cape Cod retreated up to 5 m, which is more than 50 times the normal annual rate of retreat.

Such infrequent bursts of rapid erosion not only can be quantitatively important in the natural evolution of a coast but also can have a significant impact on coastal inhabitants. Thirty of the 50 states in the United States have coastlines on the Atlantic or Pacific oceans, the Gulf of Mexico, or the Great Lakes. These 30 states contain about 85 percent of the nation's population, and about half of these people live in the coastal zone. The concentration of such large numbers of people in coastal areas means that infrequent large storms not only can be hazardous to life but can also cause extraordinary damage to property. Alantic hurricanes that reach the eastern coast of the United States can be exceptionally devastating. The largest to strike the coast in the last decade have caused property damage running to tens of billions of dollars.

Tsunamis

A strong earthquake or other brief, large-scale disturbance of the ocean floor, such as a landslide or volcanic eruption, can generate a potentially dangerous seismic sea wave called a **tsunami** (a Japanese term meaning "harbor wave"). A tsunami is often erroneously referred to as a tidal wave, but it has nothing to do with the tides. Its threat lies in the great speed at which it travels (as much as 950 km/h), its long wavelength (up to 200 km), its low observable height in the open ocean, and its ability to pile up rapidly to heights of 30 m or more as it moves into shallow water along an exposed coast. The suddenness of the tsunami's arrival and consequent lack of warning time used to result in numerous fatalities and considerable damage when tsunamis moved into populated areas. Hawaii is especially susceptible to dangerous tsunamis approaching from the numerous earthquake regions surrounding the Pacific basin (Fig. 14.30 and Table 14.1). Today in Hawaii sirens and radio newscasts alert the population to arriving tsunamis, and people can refer to maps printed in their telephone books that show the coastal zones at greatest risk (Fig. 14.31); both measures have helped reduce the risk to humans.

Landslides

Cliffed shorelines are susceptible to frequent landsliding as erosion eats away at the base of a seacliff. Roads, buildings, and other structures built too close to such cliffs can become casualties when sliding occurs.

Sometimes landslides on cliffed shorelines give rise to giant waves that are even more destructive than the slides. We saw in Chapter 8 that large prehistoric submarine landslides on the submerged flanks of the Hawai-

TABLE 14.1	Historic Tsunamis Striking Hawaii and Resulting in Moderate to Severe Damage[a]		
Date	Location of Earthquake		Average Wave Speed (km/h)
November 7, 1837	South America		–
April 2, 1868	Hawaii		–
August 13, 1868	South America		–
July 25, 1869	South America (?)		–
May 10, 1877	South America		–
February 23, 1923	Kamchatka		695
April 1, 1946	Aleutian Islands		790
May 22, 1960	South America		710

[a] Data from Macdonald and Abbott, 1970, Table 16.

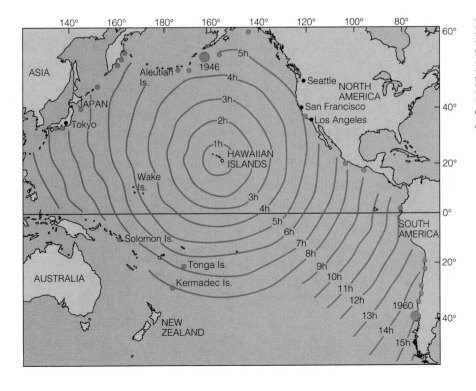

Figure 14.30 Tsunami Travel Times Map showing the time required for a tsunami to reach the island of Oahu, Hawaii. Small red dots mark origins of historic tsunamis that struck Hawaii. Large dots mark places where the disastrous tsunamis of 1946 and 1960 originated (Table 14.1).

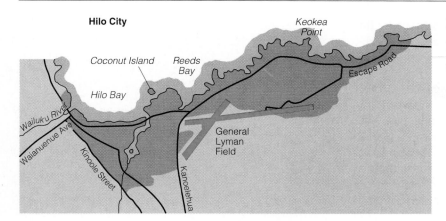

Hilo City

Coconut Island

Reeds Bay

Hilo Bay

Keokea Point

Escape Road

Wailuku River

Waianuenue Ave.

Kinoole Street

Kanoelehua

General Lyman Field

Figure 14.31 Tsunami Hazard Areas Areas (green) along coast of the island of Hawaii that are susceptible to inundation by tsunamis (from Hawaiian Telephone Book).

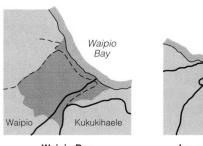

Waipio Bay

Waipio   Kukukihaele

Waipio Bay

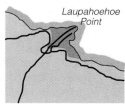

Laupahoehoe Point

Laupahoehoe Point

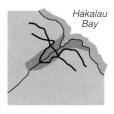

Hakalau Bay

Hakalau Bay

ian Islands apparently produced huge waves that moved rapidly landward and transported beach gravels to altitudes of more than 200 m. A repetition of such an event in the now densely populated coastal zones of the islands would wreak great havoc and loss of life.

Very large waves have also been produced by massive coastal landslides at Lituya Bay, which lies along the Fairweather Fault on the southern coast of Alaska. During the last such event, in 1958, a 7.9-magnitude earthquake caused a coastal rock cliff to collapse into an arm of the bay. The massive landslide created a wave that rose more than 500 m against the opposite wall of the fjord. The wave swept away vegetation, producing a sharp trimline in the forest, and moved rapidly down the bay toward the sea. A boat anchored in the bay was caught up by the onrushing wave, carried over the spit at the bay mouth, and transported out into the open ocean.

PROTECTION AGAINST SHORELINE EROSION

Oceanfront land is often considered prime real estate. People who build houses and towns along seacoasts are not eager to lose their investment by the actions of an unruly ocean and therefore take steps to protect their land and property.

Protection of Seacliffs

A strip of shoreline that consists of easily erodible rock or sediment is not easy to protect. A cliff can be clad with an armor consisting of tightly packed boulders so large that they can withstand the onslaught of storm waves. It can also be defended by a strong seawall built parallel to the shore on foundations deep enough to prevent the undermining by surf during storms. Both structures offer cliffs some protection against ordinary storms, but both are expensive and may not be effective against large storms or hurricanes.

Protection of Beaches

Because of their great recreational value, beaches in densely populated regions justify greater expense for maintenance than most headlands do. A beach, however, presents a special sort of problem. As a result of beach drift, what happens on one part of a beach affects all other parts that lie in the downdrift direction. This is true because any beach is in a state of delicate balance between forces of erosion and deposition. For example, a seawall, dock, or other structure built at the updrift end of a beach reduces the amount of sand available for beach drift. The surf becomes underloaded and makes good the loss by eroding sand from along the beach.

Small beaches have been completely destroyed by this process in only a few years.

Breakwaters and Groins

A **breakwater** is an offshore barrier designed to protect a beach or boat anchorages from incoming waves. However, a breakwater upsets the natural balance of the adjacent beach, leading to shoreline changes. Breakwaters constructed along the shore at Tel Aviv in Israel (Fig. 14.32) protect the beach from the onslaught of waves, but they have turned a straight coastline into a scalloped one. Arriving waves now are refracted around each breakwater. The currents created by the refracted waves transport sand toward the zone of protected water, causing the beach to expand seaward behind each barrier.

Beach erosion can be checked to some extent by building groins at short intervals along the beach. A **groin** is a low wall built out into the water at a right angle to the shoreline (Fig. 14.33). A groin acts as a check on the rate of beach drift because it traps sand carried to it along the shore. However, erosion tends to occur on the downdrift side of a groin, where the beach has been

Figure 14.32 Breakwaters Breakwaters constructed offshore from Tel Aviv, Israel, protect the beach zone from incoming waves. Wave refraction around the barriers has led to progradation of the beach behind each breakwater, producing a scalloped coastline.

Figure 14.33 Groins Groins have been built perpendicular to the shoreline of Miami Beach, Florida, to prevent excessive loss of sand by longshore drift at this popular resort area.

deprived of its constant supply of sand. The net effect, therefore, is to protect one part of a beach at the expense of another part.

Artificial Nourishment

Another way of protecting an eroding beach is to haul in sand and pile it on the beach at the updrift end. Surf then erodes the pile and drifts the new sand down the length of the beach. Sand for artificially nourishing a beach must be continuously replenished, however. As can be imagined, the feeding of a beach, like the construction and maintenance of groins, can be expensive.

Effects of Human Interference

Many beaches around the world are deteriorating because of human interference. In southern California, for example, most of the sand on beaches is supplied not by erosion of wave-cut cliffs but by alluvium carried to the sea at times of flood. However, because buildings and other structures in stream valleys are vulnerable to floods, dams have been built across the stream courses to control flooding. Of course, the dams also trap the sand and gravel carried by the streams, thus preventing the sediment from reaching the sea. Halting the through-flow of sand has upset the natural balance among the factors involved in the longshore transport of sediment. The result has been significant erosion of some beaches.

A further dramatic example of human interference can be seen along the Russian Coast of the Black Sea. Of the sand and pebbles that form the natural beaches there, 90 percent used to be supplied by rivers as they entered the sea. During the 1940s and 1950s, three things occurred: large resort developments were built at the beaches, large breakwaters were constructed so that two major harbors could be extended into the sea, and dams were built across some rivers inland from the coast. All this construction interfered with the steady state that had existed among the supply of sediment to the coast, longshore currents and beach drift, and deposition of sediment on beaches. By 1960, it was estimated that the combined area of all beaches along the coast had decreased by 50 percent. Then beachfront buildings began to sag or collapse as the surf ate away at their foundations. An ironic twist to the chain of events lies in the fact that large volumes of sand and gravel were removed from beaches and used as concrete aggregate, not only to construct the resort buildings but also to build the dams that cut off the supply of sediment to the coast.

SUMMARY

1. Seawater covers nearly 71 percent of the Earth's surface and is concentrated in the Pacific, Atlantic, and Indian oceans. Each is connected to the Southern ocean, which encircles Antarctica.

2. Although the greatest ocean depth is more than 11 km, the average depth is 3.8 km. More than half the ocean water resides in the Pacific basin.

3. More than 99.9 percent of the saltiness of seawater is due to eight ions that are derived through chemical weathering of rocks on land and then transported by streams to the sea. Other sources of ions include airborne dust and human pollutants.

4. Sea-surface temperatures are strongly related to latitude, with the warmest temperatures measured in equatorial latitudes. Surface salinity is also strongly latitude-dependent and related to both evaporation and precipitation.

5. Huge wind-driven surface ocean currents which, because of the Coriolis effect, circulate clockwise in the northern hemisphere and counterclockwise in the southern hemisphere carry warm equatorial water toward the polar regions.

6. Sinking of cold and/or saline high-latitude surface waters leads to oceanwide thermohaline circulation. Operating like a great conveyor belt and driven by density contrasts, this global circulation system replenishes the deep water of the world ocean, replacing it in about 1000 years.

7. Twice-daily ocean tides, resulting from the gravitational attraction of the Moon and Sun, are produced as the surface of the rotating Earth passes through tidal bulges on opposite sides of the planet. Tidal currents move sediment in bays, straits, estuaries, and other restricted places along coasts.

8. The motion of wind-driven surface waves terminates downward at the wave base, a distance equal to half the wavelength. A wave breaks in shallowing water as interference with the bottom causes the wave to grow higher and steeper. Waves approaching shore are refracted as the wave base reaches the bottom, realigning the wave so that it reaches the shore at gentler angle.

9. Longshore currents and beach drift can transport great quantities of sand along coasts.

10. On gentle, sandy coasts the beach typically consists of a foreshore, berm, and backshore. On rocky coasts the shore profile includes a wave-cut cliff, a wave-cut bench, and a beach.

11. Depositional shore features include beaches, marine deltas, spits, tombolos, bay barriers, and barrier islands. Barrier islands form offshore in areas where rising sea level causes a shoreline to advance across a gently sloping coastal plain.

12. Organic reefs form in tropical seas where water temperature averages at least 18°C. A fringing reef on a volcanic island can be transformed into a barrier reef as the island subsides, and eventually into an atoll.

13. The shape of coasts partly reflects the amount of energy available to erode and deposit sediment. Rock structure and degree of erodibility help dictate the form of rocky coasts.

14. Nearly all coasts have experienced recent submergence due to the postglacial rise of sea level. Some coasts have experienced more complicated histories of emergence and submergence due to crustal movements on which are superimposed the worldwide sea-level rise.

15. Infrequent but powerful storms, tsunamis, and large landslides can pose significant threats to people and structures in the coastal zone.

16. Erosional damage to a shore cliff can be minimized by a seawall or an armor of boulders. Beach erosion is a serious problem along many inhabited coasts, but beaches can be temporarily protected by breakwaters, groins, or artificial nourishment.

THE LANGUAGE OF GEOLOGY

atoll (p. 389)

barrier island (p. 388)
barrier reef (p. 389)
beach (p. 384)
beach drift (p. 383)
breakwater (p. 398)

emergence (of a coast) (p. 393)

fringing reef (p. 389)

groin (p. 398)
gyre (p. 374)

longshore current (p. 383)

salinity (p. 371)
spit (p. 387)
submergence (of a coast) (p. 392)
surf (p. 381)

thermohaline circulation (p. 376)

tsunami (p. 396)

wave base (p. 380)
wave-cut bench (p. 385)
wavelength (p. 380)
wave refraction (p. 381)

QUESTIONS FOR REVIEW

1. If the quantity of dissolved ions carried to the oceans by streams throughout geologic history far exceeds the known quantity of these substances in modern seawater, why is seawater not far more salty than it is?

2. How and why are the temperature and salinity of surface ocean water related to latitude?

3. Explain why the oceans are vertically stratified with respect to water density.

4. Explain how the ocean's thermohaline circulation system operates and how it affects the climate of western Europe.

5. Explain why at any place on a coast there are two high tides and two low tides each day.

6. Describe the motion of a parcel of water below a passing wave in the open ocean, and explain how and why the motion changes as the wave moves into shallow water.

7. How is wave base related to wavelength?

8. What causes a wave to "break" near the shore?

9. What is the effective depth of erosion by surf, and what determines this depth?

10. How does wave refraction explain why rocky headlands are more vigorously eroded by surf than bays are?

11. What causes longshore currents to develop and, in some cases, to shift direction seasonally?

12. Why are the sediments in a beach generally well sorted and well stratified?

13. Describe how you think an atoll would look during the peak of a glacial age, and why.

14. Why might a ship in midocean not detect the passage of a tsunami, while one anchored in a constricted ocean-facing harbor would not miss its arrival?

15. What geologic features would you look for to determine whether a coastal region had experienced emergence or submergence in the recent geologic past?

16. Describe measures that can be taken to reduce the impact of erosion (a) along a cliffed coast and (b) along a sandy beach. What negative effects might the measures have, even though they stop erosion?

17. How might the construction of a dam along the lower part of a large river affect the coast where the river enters the ocean?

15

Understanding
Our Environment

GeoMedia

The hills surrounding Monthermé on the River Meuse, in northern France, are examples of the natural ramparts that protected Paris from invasion by the German Army during World War I. The photograph looks in a north westerly direction. The hills run roughly north-south with the more gently sloping side facing west. The French army could easily haul armaments up the gentle slopes. The German army had to try to scale the much steeper slopes that faced east; they were unable to haul their guns up the slopes so had to try and fight their way through the narrow river valleys.

Deformation of Rock

Geology and World War I

Success or failure in military campaigns is often determined by geology. This was dramatically demonstrated in the great battles of the Western Front during World War I.

When the German Army advanced westward into France in 1915, it encountered terrain that favored the armies of France and France's allies. The German goal was to capture Paris. In resisting and ultimately repulsing the German attack, the Allies fought some of the most famous and deadly battles in history, battles with names like Verdun, St. Mihiel, Argonne, and Flanders that are now carved on monuments of memory around the world.

Paris, the goal of the German offensive, sits near the center of a large sedimentary basin, the Paris basin. Strata in the basin are sandstone, limestone, and shale. Beneath Paris the strata are horizontal, but to the east, along the margin of the basin, the strata are tilted upward so that they slope gently toward Paris at about 15° to the horizontal. Weathering has affected the rock types differently. Sandstones and limestones have weathered slowly and make high ridges. Shales have weathered rapidly and form broad valleys. The ridges run roughly north–south and form a sequence of ramparts that protect Paris.

The western slopes of the ridges, which are the upper surfaces of the resistant strata, slope gently toward Paris. The Allies could advance to the ridge crests by moving up the gentle slopes. The eastern slopes of the ridges are the broken edges of the strata; they are steep and difficult of access. These are the slopes that faced the Germans. To breach the natural defense of the cuestas, as asymmetric ridges formed by erosion of a sequence of hard and soft layers are called, the German Army had either to move down the few river valleys that cut through the ridges or avoid the cuestas by going north or south. Wherever they moved, the Allies were waiting with massed armies, and so, what came to be known (obviously incorrectly) as the "war-to-end-all-wars" ground to a hideous end in 1918, helped by geological activities that had started millions of years earlier.

HOW ROCK IS DEFORMED

Greece is well known for its earthquakes. One destroyed Sparta, the famous city of antiquity, in the fifth century B.C., and there have been many more since. Recently, a team of Greek and British scientists discovered that the tectonic forces that cause the earthquakes are also stretching Greece and slowly making it grow larger.

A century ago, the distances between a series of Greek survey monuments were measured very accurately. In 1988 a scientific team remeasured the distances and found that Greece is now a meter longer. They also discovered that Greece is being twisted so that the southern end, the Peloponnesus, is moving to the southwest relative to the rest of Greece. The reason for the stretching and twisting is plate tectonics. Africa is moving north and slowly forcing a slice of Mediterranean seafloor under Greece. The resulting forces are stretching and twisting the overlying continental crust.

Because Greece is being stretched, we can conclude that rock in the Greek crust is being deformed. There is nothing unique or unusual about such a conclusion. Evidence that rocks can be deformed is easy to find. If you look at a photograph of the Alps, the Rockies, the Appalachians, or any other mountain range, you will see strata, once horizontal, that are now tilted and bent. Enormous forces are needed to deform such huge masses of rock.

Tectonic forces continuously squeeze, stretch, bend, and break rock in the lithosphere. The source of energy for tectonic forces is the Earth's heat energy. As we discussed in Chapter 2, convection converts the Earth's heat energy to mechanical energy. The huge, slow, convective flows of hot rock in the mesosphere and asthenosphere continuously buckle and warp the lithosphere. It is those convective forces that are ultimately the cause of the rock deformation we observe in mountain ranges and that are stretching Greece.

In order to discuss deformation in rocks, such as bending, twisting, and fracture, it is helpful to review some of the elementary properties of solids. Knowledge of the factors controlling rock deformation comes largely from laboratory experiments in which cylinders or cubes of rock are squeezed and twisted under controlled conditions.

Stress and Strain

We learned in Chapter 7 that when rock deformation is being discussed the term *stress* is used rather than pressure. We also learned that *uniform stress* describes that situation where the stress is equal in all directions, such as the stress on a small body immersed in a liquid or gas. Uniform stress in rocks is also called **confining stress** because any body of rock in the lithosphere is confined by the rocks around it and is uniformly stressed by the weight of the overlying rocks. *Differential stress*, by contrast, is stress that is not equal in all directions. The stress that deforms rocks is usually differential stress. The three kinds of differential stress are shown in Figure 15.1. **Tensional stress** stretches rocks, **compressional stress** squeezes them, while **shear stress** causes slippage and translation. Differential stresses are caused by tectonic forces. An example of the difference between uniform and differential stress is illustrated in Figure 7.3.

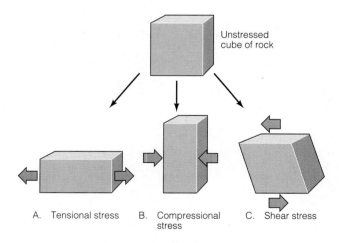

A. Tensional stress   B. Compressional stress   C. Shear stress

Figure 15.1 Differential Stress Shape of a cube of rock deformed by differential stress. Arrows indicate directions of maximum stress. A. Tensional stress. B. Compressional stress. C. Shear stress.

The term used to describe the deformation of a rock is **strain**, which is defined as the change in size or shape, or both, in a solid as a result of stress. Uniform stress causes a solid to change size but not shape. Differential stress causes a solid to change shape, but it may or may not also cause a change in size.

Stages of Deformation

When a rock is subjected to increasing stress, it passes through three stages of deformation in succession:

1. Elastic deformation is a reversible, or nonpermanent, change in the volume or shape of a stressed rock. When the stress is removed, the rock returns to its orig-inal size and shape. There is a limiting stress, called the **elastic limit**, beyond which a solid suffers permanent deformation and does not return to its original size or shape once the stress is removed.

A famous British scientist, Sir Robert Hooke (1635–1703), was the first to demonstrate that, for all materials, provided the elastic limit is not exceeded, a plot of stress versus strain is always a straight line. Hooke proved his point by using a spring, as in Figure 15.2A. However, Hooke's law is equally true for rocks, as illustrated in Figure 15.2B.

2. Ductile deformation is an irreversible change in shape and/or volume of a rock that has been stressed beyond the elastic limit. If a cylinder of rock is stressed by a compressional stress applied parallel to the long axis

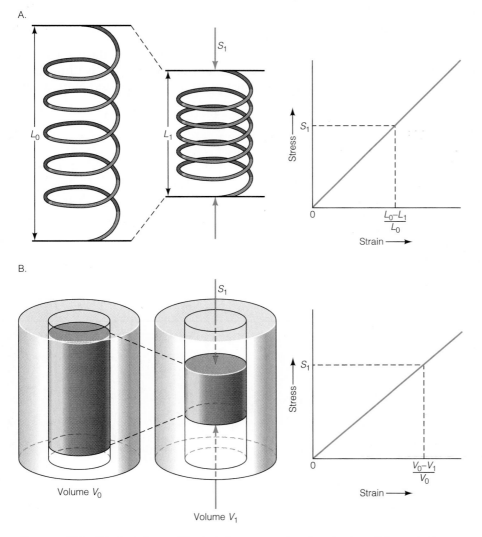

Figure 15.2 Hooke's Law Hooke's law states that for elastic solids, strain is pro-portional to stress. A. A spring, shortened by a compressional stress S_1, has its length reduced from L_0 to L_1. A plot of the strain $(L_0 - L_1)/L_0$ as a function of stress produces a straight line. B. A cylinder of rock subjected to a confining stress by a tight metal jacket, and shortened by a compressional stress S_1, has its volume reduced from V_0 to V_1. A plot of strain $(V_0 - V_1)/V_0$ as a function of stress produces a straight line.

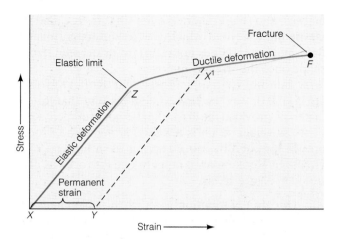

Figure 15.3 Stress and Strain Typical stress–strain curve for a cylinder of rock tested in the laboratory. Following elastic deformation (X to Z), the elastic limit Z marks the onset of ductile deformation. If, at point X', the stress is removed, the solid will return to an unstressed state at point Y by following the path X'Y. The distance XY is a measure of the permanent, irreversible strain produced by ductile deformation. If the stress is not released at X', but is increased and maintained, the strength of the solid is exceeded when rupture occurs at the point of fracture, F.

of the cylinder, an interesting result is obtained. As shown in Figure 15.3, the stress–strain curve for the cylinder rises first through the elastic region, and then, at the elastic limit (point Z), the curve flattens and additional stress causes ductile deformation. If the stress is removed at point X; the cylinder partially returns to its original shape—the strain decreases along the curve X'Y. A permanent strain, equal to XY, has been induced in the rock. The permanent strain XY is due to ductile deformation.

3. Fracture occurs in a solid when the limits of both elastic and ductile deformation are exceeded. Consider again the stress–strain curve in Figure 15.3. If, instead of releasing the stress at point X', we continued to increase the stress, the stress–strain curve would continue to point F, where the cylinder breaks by fracturing. Obviously, fracture, like ductile deformation, is an irreversible kind of deformation.

Ductile Deformation Versus Fracture

A brittle substance tends to deform by fracture, while a ductile substance deforms by a change of shape. Drop a piece of chalk on the floor and it will fracture, but if you drop a piece of butter it will bend but not break. Chalk is brittle; butter is ductile.

A typical stress–strain curve for a brittle substance is shown in Figure 15.4A. Note that Z, the elastic limit, is very close to F, the point of fracture. Thus, little ductile deformation occurs in a brittle substance. In contrast, in a ductile substance the elastic limit and fracture point are far apart, as shown in Figure 15.4B.

Examples of deformation of brittle and ductile rocks are shown in Figures 15.5A and B. The strata in Figure 15.5A have fractured cleanly with little or no evidence of bending due to ductile deformation. The strata in Figure 15.5B, by contrast, have been intensely twisted and bent by ductile deformation.

To evaluate deformation in rocks, it is necessary to estimate the relative importance of brittle properties versus ductile properties. The essential conditions controlling the relative importance of the two kinds of properties are (1) temperature; (2) confining stress; (3) time and strain rate; and (4) composition.

Temperature

The higher the temperature, the more ductile and less brittle a solid becomes. A rod of glass is difficult to bend at room temperature; if we try too hard, it will

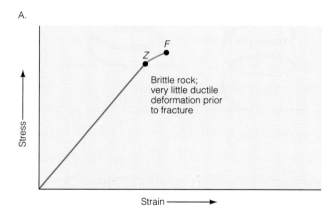

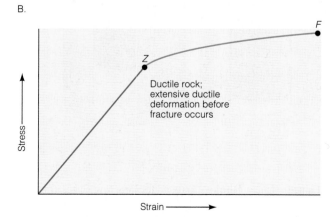

Figure 15.4 Stress–Strain Curves Comparison of stress–strain curves. A. Curve for a brittle substance. B. Curve for a ductile substance.

A   **B**

Figure 15.5 Brittle Versus Ductile Examples of rock deformation. A. Fracture of strata by brittle deformation; Triassic-aged Bunter Sandstone, Merseyside, U.K. B. Bending of strata by ductile deformation; limestones in Crete.

break because it is brittle. However, a glass rod becomes ductile and can be readily bent if it is heated to redness over a flame. Rocks are like glass rods. They are brittle at the Earth's surface, but at depth, where temperatures are high because of the geothermal gradient, rocks become ductile.

Confining Stress
The effect of confining stress on deformation is not familiar in common experience. Confining stress is a uniform squeezing of rock owing to the weight of all the overlying strata. High confining stress hinders the formation of fractures and so reduces brittle properties. At high confining stress, it is easier for a solid to bend and flow than to fracture. Reduction of brittleness by high confining stress is a second reason why solid rock can be bent and folded by ductile deformation.

Time and Strain Rate
The effect of time on rock deformation is vitally important, but as with confining stress, it is not obvious from common experience. Stress applied to a solid is transmitted by all the constituent atoms of the solid. If the stress exceeds the strength of the bonds between atoms, either the atoms must move to another place in the crystal lattice in order to relieve the stress or the bonds must break, which means fracture occurs. Atoms in solids cannot move rapidly. Nevertheless, if the stress builds up slowly and gradually and is maintained for a long period, the atoms have time to move, and the solid can slowly readjust and change shape by ductile deformation. The important point to appreciate is that the *rate*

at which a solid is strained is just as significant as how long a stress is active.

The term used for time-dependent deformation of a rock is **strain rate**, which is the rate at which a rock is forced to change its shape or volume. Strain rates are measured in terms of change of volume per unit volume per second. For example, a strain rate that is sometimes used in laboratory experiments is 10^{-6}/s, by which is meant a change in volume of one-millionth of a unit volume per unit volume per second. Strain rates in the Earth are much lower than this—about 10^{-14} to 10^{-15}/s. The lower the strain rate, the greater the tendency for ductile deformation to occur.

A comparison of the influences of temperature, confining pressure, and strain rate on rock properties can be seen in Figure 15.6. Low temperature, low confining stress, and high strain rate enhance brittle properties. These conditions are characteristic of the crust (especially the upper crust), and as a result failure by fracture is common in upper-crustal rocks. High temperature, high confining stress, and low strain rates, which are characteristic of the deeper crust and mantle, reduce brittle properties and enhance the ductile properties of rock.

Composition
The composition of a rock has a pronounced effect on its properties. Composition has two aspects. First, the kinds of minerals in a rock exert a strong influence on properties because some minerals (such as quartz, garnet, and olivine) are very brittle, while others (such as mica, clay, calcite, and gypsum) are ductile. Second, the

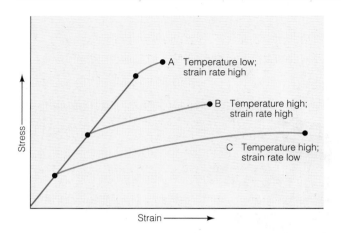

Figure 15.6 Controls on Rock Deformation The effects of temperature and strain rate on rock deformation. Curve A: low temperature, high strain rate. Curve B: high temperature, high strain rate. Curve C: high temperature, low strain rate.

presence of water in a rock reduces brittleness and enhances ductile properties. Water affects properties by weakening the chemical bonds in minerals and by forming films around mineral grains, thereby reducing the friction between grains. Thus, wet rocks have a greater tendency to be deformed in a ductile fashion than do dry rocks.

Rocks that readily deform by ductile deformation are limestone, marble, shale, slate, phyllite, and schist. Rocks that tend to be brittle rather than ductile are sandstone and quartzite, granite, granodiorite, and gneiss.

Brittle-Ductile Properties of the Lithosphere

Rock strength in the Earth does not change smoothly with depth. As is evident in Figure 15.7, there are two peaks in the plot of rock strength with depth. The reason is that strength is determined by composition, temperature, and pressure.

Rocks in the crust are quartz-rich, so the strength properties of quartz determine the strength properties of the crust. Rock strength increases down to a depth of about 15 km. Above 15 km rocks are strong; they fracture and fail by brittle deformation. Below 15 km fractures become less common as quartz weakens and rocks become increasingly ductile. The depth in the crust where ductile properties start to predominate over brittle properties is known as the *brittle-ductile transition.*

Quartz is absent from the mantle. Rocks in the mantle are olivine-rich. Olivine is stronger than quartz, and the brittle-ductile transition of olivine-rich rock is reached only at a depth of about 40 km. Thus, a second maximum in rock strength is reached at this depth (Fig.

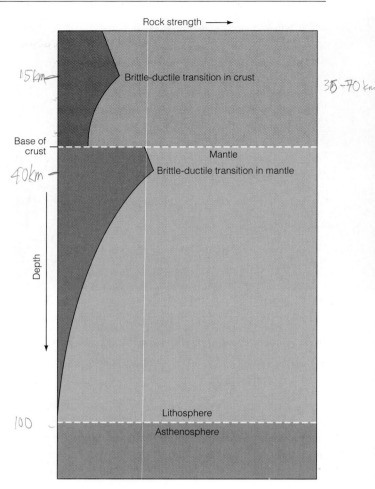

Figure 15.7 Rock Properties and Depth Schematic diagram showing the way rock strength changes with depth. Two brittle-ductile transitions are present. In the crust, the transition is controlled by the properties of quartz-rich rocks; in the mantle, olivine-rich rocks determine the properties. The depth at which failure by brittle deformation disappears is the lithosphere-asthenosphere boundary.

15.7). Below the brittle-ductile transition in the mantle, rock strength again declines. As discussed in Chapter 2 in the section on geothermal gradients, by about 1300°C, rock strength is very low; brittle deformation is no longer possible. The disappearance of all brittle deformation properties marks the lithosphere-asthenosphere boundary.

DEFORMATION IN PROGRESS

Most deformation in the crust is too slow or too deeply buried to be observed. Large movements happen so slowly, which means at such low strain rates, that they can be measured only over a hundred or more years. The deformation occurring in Greece, discussed at the beginning of this chapter, is an example. Nevertheless, deformation does sometimes happen fast enough to be detected and measured. For convenience we divide large-scale, observable deformation of the crust into two groups: abrupt movement that involves fracture, in which blocks of the crust suddenly move a few centimeters or a few meters in a matter of minutes or hours, and gradual movement involving ductile deformation, in which slow, steady motions occur without any abrupt jarring.

Abrupt Movement

Abrupt movement involves the fracture of brittle rocks and movement along the fractures. A fracture in a rock along which movement occurs is a **fault**. Once fracturing has started, friction inhibits continuous slippage. Instead, stress again builds up slowly until friction between the two sides of the fault is overcome. Then abrupt slippage occurs again. If the stresses persist, the whole cycle of slow buildup followed by an abrupt movement repeats itself many times. Although movement on a large fault may eventually total many kilometers, this distance is the sum of numerous small, sudden slips. Each sudden movement may cause an earthquake and, if the movement occurs near the Earth's surface, may disrupt and displace surface features (Fig. 15.8).

Figure 15.8 is an example of horizontal abrupt movement. Abrupt vertical movements are also well documented. The largest abrupt vertical displacement ever observed occurred in 1899 at Yakutat Bay, Alaska, during an earthquake. A stretch of the Alaskan shore (including the beach, barnacle-covered rocks, and other telltale features) was suddenly lifted as much as 15 m above sea level. This visible vertical displacement may be less than the total amount, because the fault is hidden offshore and the block of crust on the other side of it, entirely beneath the sea, may have moved downward, thus adding to the total displacement of the stretch of beach.

Abrupt movements in the lithosphere are commonly accompanied by earthquakes and can therefore be hazardous to people living nearby. Earthquakes and earthquake hazards are discussed in Chapter 16.

Gradual Movement

Movement along faults is usually, but not always, abrupt. Measurements along the San Andreas Fault in California reveal places where gradual slipping occurs, sometimes reaching a rate as high as 5 cm a year. Because the San Andreas Fault cuts through the entire lithosphere—that is, well below the brittle-ductile transition—it is probable that deformation of rocks on either side of the fault is ductile at depth but brittle near the surface.

Possibly no spot on the Earth is completely stationary. Measurements by U.S. government surveyors over 100 years, for example, reveal great areas of the United States where the land is slowly sinking and other places where it is slowly rising (Fig. 15.9). The causes of these vast, slow, vertical movements are not well understood, but the movements do prove that the solid Earth is not as rigid as it seems at first sight and that great internal forces are continuously deforming its crust.

Figure 15.8 Active Fault An orange grove in southern California planted across the San Andreas Fault. Movement on the fault displaced the originally straight rows of trees. The direction of motion is such that trees in the background moved from left to right relative to the trees in the foreground.

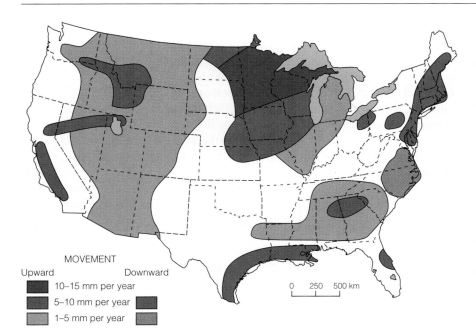

MOVEMENT

Upward   Downward

■ 10–15 mm per year ■

■ 5–10 mm per year ■

■ 1–5 mm per year ■

0   250   500 km

Figure 15.9 The Moving Land Surface Accurate measurements over a 100-year period show that in large areas of the United States, the surface is slowly moving either upward or downward. Subsidence along the coasts of California and the Gulf of Mexico is believed to have been caused in part by withdrawal of gas, oil, and water, which allows subsurface reservoirs to collapse. Uplift near the Great Lakes is a rebound effect following the melting of the last ice sheet. The causes of movements in other areas are not known with certainty. Those areas in which no movement is shown are not necessarily stationary. They are simply areas in which measurements are very few or movement is so slow it has not yet been detected.

EVIDENCE OF FORMER DEFORMATION

With such convincing evidence of present-day deformation of the Earth's crust, we might reasonably expect to find a great deal of evidence of former deformation. Studies of land and sea-bottom topography provide abundant evidence of vertical movements. In some areas the distribution of various kinds of rock provides clear evidence that horizontal movements have occurred through distances as great as several hundred kilometers.

Not all evidence of movement and deformation observed in bedrock is as obvious as the examples cited. But once we learn to recognize it, evidence of deformation is seen to be very widespread—so much so that a special branch of geology, *structural geology*, has the study of rock deformation as its primary focus.

Strike and Dip

The law of original horizontality (Chapter 6) tells us that sedimentary strata and lava flows were initially horizontal. Where we observe such rocks to be tilted, we can conclude that deformation has occurred. In order to decipher and explain this deformation, a geologist starts by measuring the angle and direction of tilting.

In order to measure the orientation of a tilted plane, we need to remember the two principles of geometry shown in Figure 15.10A: (1) the intersection of two planes defines a line and (2) in an inclined plane all horizontal lines are parallel. The line formed by the intersection of an inclined plane with a horizontal plane is always horizontal. Such a line can be visualized as the waterline on an inclined stratum along the shore of a lake, as shown in Figure 15.10B. The lake surface is a convenient horizontal plane. The waterline marks the **strike**, which is the compass direction of the horizontal line formed by the intersection of a horizontal plane and an inclined plane.

Once we know the strike, we need only one more measurement to fix the orientation of an inclined plane. That is the **dip**, the angle in degrees between a horizontal plane and the inclined plane, measured down from horizontal. The direction and angle of dip are indicated in Figures 15.10A and 15.10B.

Geologic Maps

Rarely is it possible for a geologist to see all the structural details of deformed rocks in a given area. Soil and vegetation usually cover much of the evidence. A geologist must therefore decipher the structures by making observations and measurements at a number of individual *outcrops* (places where bedrock is exposed at the surface). When the observations made at each outcrop are plotted on a map, and inferences are made as to what has happened beneath the cover of soil and vegetation, the result is a geologic map. An example of a simple geologic map is shown in Figure 15.11. Note the way the geologist recorded the strike and the angle and direction of dip of the contacts between strata at each outcrop.

Deformation by Fracture

Rock in the crust, especially rock close to the surface, tends to be brittle. As a result, rock at or near the Earth's surface tends to be cut by innumerable fractures called

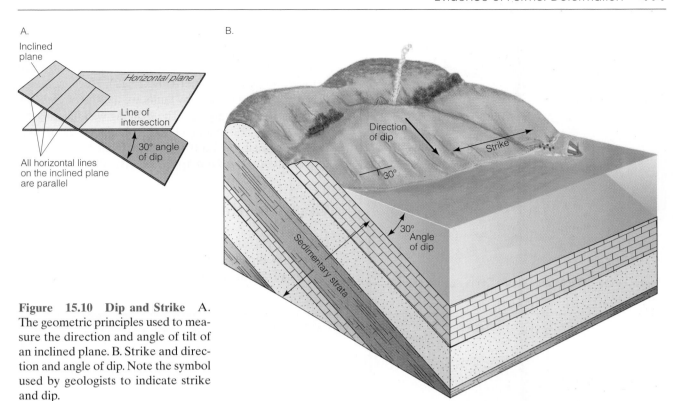

A.

Inclined plane

Horizontal plane

Line of intersection

All horizontal lines on the inclined plane are parallel

30° angle of dip

B.

Direction of dip

Strike

30°

Sedimentary strata

30° Angle of dip

Figure 15.10 Dip and Strike A. The geometric principles used to measure the direction and angle of tilt of an inclined plane. B. Strike and direction and angle of dip. Note the symbol used by geologists to indicate strike and dip.

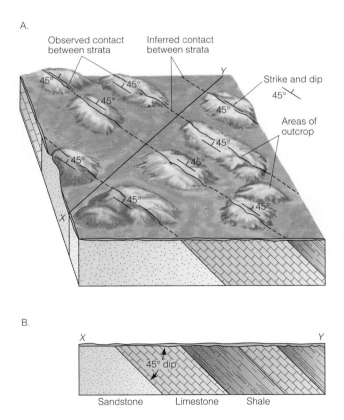

A.

Observed contact between strata

Inferred contact between strata

Strike and dip

45°

Areas of outcrop

45° 45° 45° 45° 45° 45° 45° 45° 45°

Y

X

B.

X   Y

45° dip

Sandstone   Limestone   Shale

Figure 15.11 Geologic Map A. Dotted lines denote areas of outcrop where the geologist identified rock types and strata. Dashed lines indicate inferred contacts. B. Section along the line *XY* in part A.

either *joints* or *faults*. As discussed in Chapter 5, a joint is a special kind of fracture because there has not been any movement along it. A fault, as we learned earlier, is a fracture along which visible displacement has occurred.

Relative Displacement

Generally, it is not possible to tell how much movement has occurred along a fault, nor which side of the fault has moved. In an ideal case—for example, if a single mineral grain or a single pebble in a conglomerate has been cut through by the fault and the halves have been carried apart a measurable distance—the amount of movement can be determined. Yet even then it is not possible to say whether one block stood still while the other moved past it, or whether both sides moved. In classifying fault movements, therefore, geologists can determine only relative displacements; that is, one side of a fault has moved in a given direction relative to the other side. For example, in Figure 15.5A it is apparent that there has been displacement on a fault, but all we can say about the movement is that the left-hand side moved down relative to the right-hand side. Similarly, in Figure 15.8 all we can say is that the orange trees in the rear moved to the right relative to those in the front.

Hanging Wall and Footwall

Most faults are inclined like the faults in Figure 15.5A. (That is, to use a geological term, they dip.) To describe

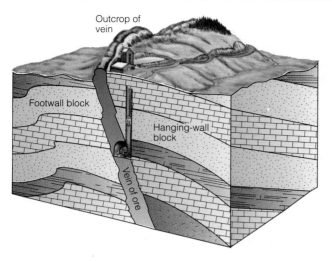

Figure 15.12 Old Mining Terms Hanging-wall and footwall.

the inclination, geologists have adopted two old mining terms. From a miner's viewpoint, the rocks above an inclined vein overhang him, while the rocks below the vein are beneath his feet (Fig. 15.12). Because veins occupy openings created by faults, we use the old miner's terms in the following way. The **hanging wall block** is the block of rock above an inclined fault; the block of rock below an inclined fault is the **footwall block**. These terms, of course, do not apply to vertical faults.

Classification of Faults

Faults are classified according to (1) the dip of the fault and (2) the direction of relative movement. The common classes of faults, together with the changes in local

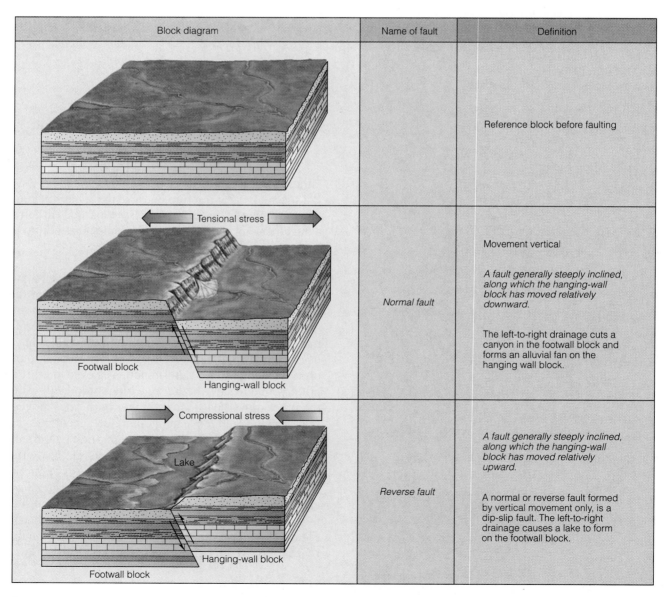

Block diagram	Name of fault	Definition
		Reference block before faulting
Tensional stress / Footwall block / Hanging-wall block	Normal fault	Movement vertical *A fault generally steeply inclined, along which the hanging-wall block has moved relatively downward.* The left-to-right drainage cuts a canyon in the footwall block and forms an alluvial fan on the hanging wall block.
Compressional stress / Lake / Hanging-wall block / Footwall block	Reverse fault	*A fault generally steeply inclined, along which the hanging-wall block has moved relatively upward.* A normal or reverse fault formed by vertical movement only, is a dip-slip fault. The left-to-right drainage causes a lake to form on the footwall block.

Figure 15.13 Kinds of Faults Kinds of faults, the directions of maximum stress that cause them, and some of the topographic changes they produce.

topography they sometimes create, are listed in Figure 15.13. The standard planes of reference in classifying faults are the vertical and the horizontal. Along many faults, movement is entirely vertical or entirely horizontal, but along some faults combined vertical and horizontal movements occur.

Normal Faults

Normal faults are caused by tensional stresses that tend to pull the crust apart, as well as by stresses created by a push from below that tend to stretch the crust. Movement on a normal fault is such that the hanging wall block moves down relative to the footwall block. The faults shown in Figure 15.5A are normal faults.

Commonly, two or more normal faults with parallel strikes but opposite dips enclose an upthrust or down-dropped segment of the crust. As shown in Figure 15.14, a down-dropped block is a **graben**, or **rift**, if it is bounded by two normal faults and a **half-graben** if subsidence occurs along a single fault. An upthrust block is a **horst**. The central, steep-walled valley that runs down the center of the mid-Atlantic Ridge and cuts through Iceland (Fig. 2.16) is a graben. Perhaps the world's most famous system of grabens and half-grabens is the African Rift Valley seen in Figure 1.6, which runs north–south through the countries of East Africa for more than 6000 km. Within parts of the Rift Valley magma has followed channels that lead upward along the fault surfaces, creating volcanoes.

Normal faults are innumerable. Therefore, horsts and grabens are also very common, although none is more spectacular than the African Rift Valley. The

Figure 15.13 *continued.*

A.

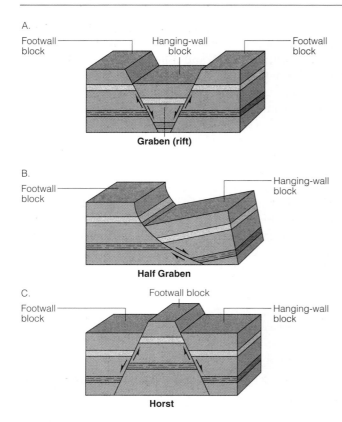

Figure 15.14 Horsts and Grabens Horsts and grabens form when tensional stresses produce normal faults.

north–south valley of the Rio Grande in New Mexico is a graben. The valley in which the Rhine River flows through Western Europe follows a series of grabens. A spectacular example of normal faulting is found in the Basin and Range Province in Utah, Nevada, and Idaho. There, movement on a series of parallel and nearly parallel north–south striking, normal faults has formed horsts and half-grabens. The horsts are now mountain ranges, and the half-grabens are sedimentary basins. As seen in Figure 15.15, the province, which is bounded in the east by the western edge of the Wasatch Range and continues westward to the eastern edge of the Sierra Nevada, contains some spectacularly beautiful scenery.

Reverse Faults and Thrust Faults

Reverse faults arise from compressional stresses. Movement on a reverse fault is such that a hanging wall block moves up relative to a footwall block. Reverse fault movement shortens and thickens the crust.

A special class of reverse faults, called **thrust faults**, are low-angle reverse faults with dips less than 15° (Fig. 15.16). Such faults, common in great mountain chains, are noteworthy because along some of them the hanging wall block has moved many kilometers over the footwall block. In most cases the hanging wall block, thousands of meters thick, consists of rocks much older than those adjacent to the thrust on the footwall block (Fig. 15.17).

Figure 15.15 Basin and Range Province Scenery in the northern part of the Basin and Range Province. Looking east toward the Lost River Range, Idaho, from Pioneer Mountain. Valley in the foreground is a graben; the Lost River Range is a horst.

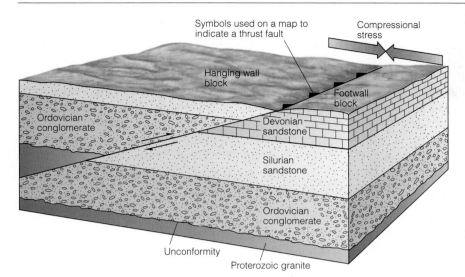

Figure 15.16 Thrust Fault—Older over Younger Example of a thrust fault. Note that thrusting causes older strata in the hanging-wall block to lie locally above younger units in the foot-wall block.

A.

B.

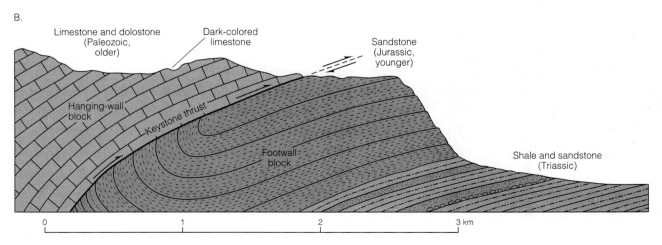

Figure 15.17 Keystone Thrust Keystone Thrust, west of Las Vegas, Nevada. A. Air view northward defines this thrust fault by a color contrast in the strata. Light-colored Jurassic sandstone, forming a cliff nearly 600 m high, lies below the fault and forms the footwall block; dark-colored Paleozoic lime-stones and dolostones lie above the fault and form the hanging wall block. B. Section drawn across area shown in the photograph.

Strike-Slip Faults

Strike-slip faults are those in which the principal movement is horizontal and therefore parallel to the strike of the fault (Fig. 15.13). Strike-slip faults arise from shear stresses. One strike-slip fault is so famous almost everyone has heard of it—the San Andreas Fault. In Figure 15.8 it is strike-slip movement on the San Andreas that is offsetting the rows of orange trees.

Horizontal fault movement is designated as follows; to an observer standing on either fault block, the movement of the other block is *left lateral* if it is to the left and *right lateral* if it is to the right. The sense of relative motion is the same regardless of which block the observer is standing on. The San Andreas is a right-lateral strike-slip fault. Apparently, movement has been occurring along it for at least 65 million years. The total movement is not known, but some evidence suggests that it now amounts to more than 600 km.

Many of the largest and most active faults are strike-slip faults. This is so because strike-slip faults, spreading centers, and subduction zones are the three kinds of margins that bound tectonic plates (Chapter 2). The three plate-margin components link together to form continuous networks encircling the Earth. Where one plate margin terminates, another commences; their junction point is called a *transform*. J. T. Wilson, the Canadian scientist who first recognized the network relation, proposed that the special class of strike-slip faults that form plate boundaries be called **transform faults**. The three possible configurations for transform faults are shown diagrammatically in Figure 15.18.

Evidence of Movement Along Faults

Often we find fractures in rock but cannot tell at first glance whether or not movement has occurred along them; in other words, we don't know whether we are looking at a fault or a joint. For example, in uniform, even-grained rock such as granite, or in a pile of thin-bedded strata no one of which is unique or distinctive, it may not be possible to see the displacement of any obvious features. However, examination of either the fault sur-

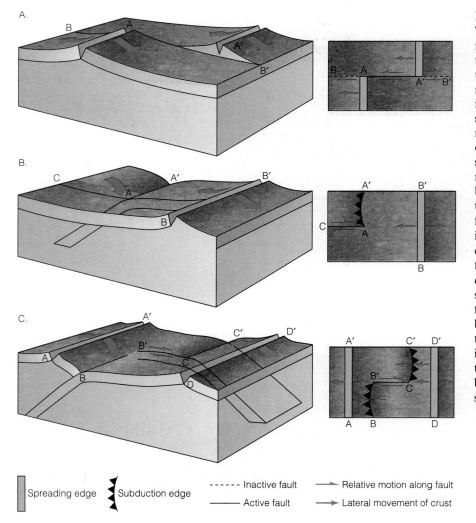

Figure 15.18 Transform Faults Transform faults as they appear on a map. A. A spreading center offset by a transform fault. Crust on both sides of the two spreading center segments moves laterally away from the spreading center. Between segments of the spreading center, along *A-A'*, movement on the two sides of the fault is in opposite directions. Beyond the spreading center, however, along segments *A-B* and *A'-B'*, movement on both sides of the projected fault is in the same direction. Therefore, a transform fault does not cause the spreading center segments to move continuously apart. B. A transform fault, a spreading center, and a subduction edge. New crust formed as a spreading center moves laterally away from the spreading center and plunges back into the mantle at the subduction edge. Note that a subduction edge is a thrust fault. Teeth marks on the thrust fault indicate the hanging-wall block. C. Transform fault joining two subduction edges.

Spreading edge Subduction edge ----- Inactive fault → Relative motion along fault
                                  ——— Active fault   ⇒ Lateral movement of crust

face or rock immediately adjacent to it sometimes reveals signs of local deformation, indicating that movement has occurred. Under special circumstances, even the relative direction of movement can be deciphered.

Movement of one mass of rock past another can cause the fault surfaces to be smoothed, striated, and grooved. Striated or highly polished surfaces on hard rocks, abraded by movement along a fault, are called *slickensides*. Parallel grooves and striations on such surfaces record the direction of the most recent movement (Fig. 15.19A).

Not all fault surfaces have slickensides. In many instances, fault movement crushes rock adjacent to the fault into a mass of irregular pieces, forming *fault breccia* (Fig. 15.19B). Intense grinding breaks the fragments into such tiny pieces that they may not be individually

A

Figure 15.19 Slickensides and Breccias The effects of faulting. A. Slickensides on a fault surface, Borrego, California. The hanging-wall rocks have been removed by erosion, thus exposing the fault on the footwall. B. Fault breccia, Titus Canyon, Death Valley. Angular gneiss fragments (dark) broken by faulting are set in a matrix of rock flour and calcite.

B

Understanding Our Environment

Rock Deformation and Oil Pools

Most of the world's largest oil pools are where they are because of the way rocks deform. "Oil pool" is a somewhat confusing term because it is not a pool in the usual sense of a "lake"; the term actually refers to a body of rock in which oil occupies all the pore spaces.

Oil and natural gas occur together, and for a pool to form, five essential requirements must be met:
1. A *source rock* rich in organic matter, such as shale, must provide the oil. Most shales contain some organic mat-

ter from dead plants and animals and can serve as source rocks. Conversion of organic matter to oil and gas happens spontaneously when shale is buried, and, in response to the geothermal gradient, the temperature of the source rock rises.

2. A permeable and porous *reservoir rock* must be present so that the oil can percolate in from the source rock.

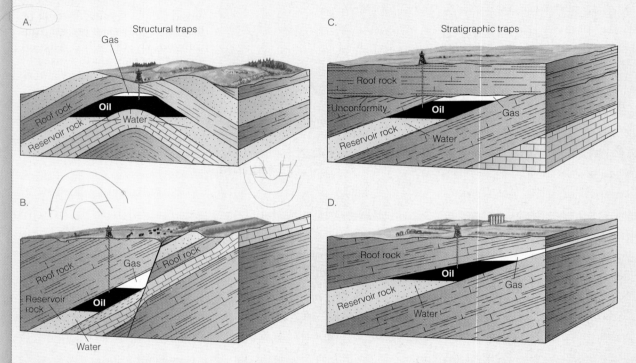

Figure B15.1 Oil Traps Four kinds of oil traps. A, B, structural traps; C, D, stratigraphic traps. Folds (part A) are traps formed by ductile deformation. They are the most important of all oil traps. Faults (part B) are traps formed by fracturing of brittle rocks. In C an unconformity marks the top of the reservoir; in D a porous stratum (reservoir) thins out and is overlain by an impermeable roof rock. Gas overlies oil, which floats on groundwater, saturates the reservoir rock, and is held down by a roof of shale. Oil fills only the pore spaces in the rock.

visible even under a microscope. Some contain such tiny fragments they resemble chert.

Deformation by Bending

Bending may consist of broad, gentle warping that extends over hundreds of kilometers, or it might be close, tight flexing of microscopic size, or anything in between. Regardless of the volume of rock involved or the degree of warping, the bending of rocks is referred to as *folding*. Folding is most easily recognized in layered rocks. An individual bend or warp in layered rock is called a **fold**.

Regardless of their size, folds are formed by ductile deformation as a result of compressional and shear stresses. At very low strain rates, even shallow rocks that are well above the brittle-ductile transition can be

3. Oil floats on water, and so oil in a reservoir rock will float upward to the water table unless the reservoir rock is covered by an impermeable *roof rock*.

4. Like groundwater in an aquifer, oil will move laterally in a reservoir rock and eventually escape. An important requirement for oil-pool formation is that the reservoir and roof rocks must form a trap that can hold the oil and prevent it from being flushed out by groundwater. As shown in Figure B15.1, traps are either structural—either anticlines or faults—or stratigraphic—either unconformities or changes in sedimentary facies. Structural traps are far more common than stratigraphic traps.

5. The final, and in some ways most important, requirement for pool formation is that the deformation that forms a trap must occur before all the oil and gas has escaped from the reservoir rock. This is important because conversion of organic matter to oil occurs soon after a shale is buried. If the trap were formed long after oil formation, no oil pool would form.

The great oil pools of Saudi Arabia, Kuwait, Iraq, and other parts of the Middle East are in structural traps. The deformation that produced the traps was caused by plate-tectonic movements. At the right moment for oil trapping, the sedimentary rocks in the Middle East were deformed by compressional stresses due to Africa's banging into Asia.

Iraq invaded Kuwait in 1990 in order to control the gigantic Kuwaiti oil pools; the United Nations went to war in 1991 to expel the Iraqi invaders. In their retreat the Iraqis set fire to the great Kuwaiti oil wells, thereby creating one of the worst environmental disasters of all times (Fig. B15.2). How sad it is that we humans find it necessary to kill each other and pollute the environment because of tectonic events that happened millions of years ago.

Figure B15.2 Oil Wells on Fire Oil wells left burning after the withdrawal of the Iraqis from Kuwait. Note the intense air pollution associated with the fires.

folded. That is apparently what is happening in Greece today, apparently too, it is what happened when the great oil fields of the Middle East were formed, as discussed in the essay *Rock Deformation and Oil Pools*. However, the very intense folding that we can observe so widely in mountain ranges probably occurred below the brittle-ductile transition when the rocks were deeply buried and thus were subjected to high temperature and high confining stress.

Types of Folds

The simplest fold is a **monocline**, a local steepening in an otherwise uniformly dipping pile of strata (Fig. 15.20A and B). An easy way to visualize a monocline is to lay a book on a table and then drape a handkerchief over one side of the book and out onto the table. So draped, the handkerchief forms a monocline. Most folds are more complicated than monoclines (Fig. 15.21). An upfold in the form of an arch is an **anticline**; a downfold with a

A.

B.

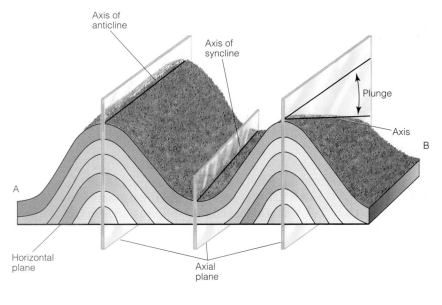

Approximate area
of photo in B

Figure 15.20 Monocline in Utah A. A monocline in southern Utah that interrupts the generally flat-lying sedimentary
strata of the wide Colorado plateau. In the area of maximum
bending, the strata are nearly vertical (right-hand side of photo).
The view is looking south. B. Sketch of a monocline.

Axis of
anticline

Axis of
syncline

Plunge

Axis

B

A

Horizontal
plane

Axial
plane

Figure 15.21 Geometry of a Fold Features of simple folds. Note that the strata
dip away from the axis of an anticline but
toward the axis of a syncline. A. Fold axis
horizontal. B. Fold axis plunging.

troughlike form is a **syncline**. Anticlines and synclines are usually paired.

Geometry of Folds

As shown in Figure 15.21A, the sides of a fold are the **limbs**, and the median line between the limbs, along the crest of an anticline or the trough of a syncline, is the **axis** of the fold. A fold with an inclined axis is said to be a *plunging fold*, and the angle between a fold axis and the horizontal is the **plunge** of a fold (Fig. 15.21B). An imaginary plane that divides a fold as symmetrically as possible, and that passes through the axis, is the *axial plane*.

Many folds, such as those shown in Figure 15.21, are nearly symmetrical. Others, however, are not symmetrical; intense stress can create complex shapes. The common forms of folds are shown in Figure 15.22.

An *open fold*, such as that depicted in Figure 15.22A, is one in which the two limbs dip gently and equally away from the axis. The more intense the compres-

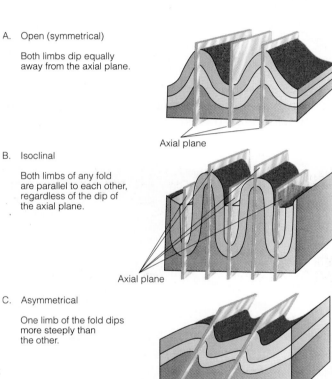

A. Open (symmetrical)

Both limbs dip equally away from the axial plane.

Axial plane

B. Isoclinal

Both limbs of any fold are parallel to each other, regardless of the dip of the axial plane.

Axial plane

C. Asymmetrical

One limb of the fold dips more steeply than the other.

Axial plane

D. Overturned

Strata in one limb have been tilted beyond the vertical. Both limbs dip in the same direction but not at the same angle.

Normal limb (gentle clip)

Axial plane

Overturned limb (steep clip)

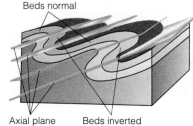

E. Recumbent

Axial planes are horizontal or nearly so. Strata on the lower limb of anticline and upper limb of syncline are upside down.

Beds normal

Axial plane   Beds inverted

Figure 15.22 Kinds of Folds Five kinds of folds.

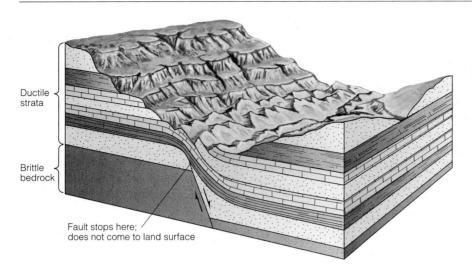

Ductile strata

Brittle bedrock

Fault stops here; does not come to land surface

Figure 15.23 Faulting Causes a Monocline to Form Monocline developed in soft, ductile strata due to movement on a fault in the underlying bedrock.

sional stress, the less open a fold will be. When stress is very intense, the fold closes up and the limbs become parallel to each other; such a fold is said to be isoclinal (Fig. 15.22B). Intense stress is also the reason why a fold becomes either asymmetric (Fig. 15.22C), in which limbs have unequal dips, or overturned, in which limbs dip in the same direction (Fig. 15.22D). Eventually, an overturned fold may become recumbent, meaning the two limbs are horizontal or nearly so (Fig. 15.22E). Recumbent folds are common in mountainous regions, such as the Alps and the Himalaya, that were produced by continental collisions.

If only fragmentary exposures of bedrock are available, it is apparent that difficulties might arise in deciding whether or not a given fold is overturned. As is apparent in Figures 15.22D and E, it is necessary to know whether a stratum is right-side up or upside down in order to decide which limb of a fold it is in. Figuring this out is not always possible, but in some cases sedimentary structures, such as mud cracks and graded layers, do record whether strata are in their original orientation or inverted (Fig. 8.2). In other examples, only careful, thorough mapping of all bedrock exposures can provide the answer.

Relationship Between Folds and Faults
Folds and faults do not continue forever. Faults tend to die out as folds. Note in Figure 15.13 that the hinge fault dies out as a small wrinkle—a fold. Folds die out by becoming smaller and smaller wrinkles, in much the same way as wrinkles in a bedsheet die out.

When two kinds of rock are subjected to the same stresses, one kind may be brittle and deform by fracture, and the other by ductile deformation. Certain monoclines form as a result of such differences. Monoclines commonly result from movement on a fault that causes flat-lying, ductile strata to bend as shown in Figure 15.23.

Some of the great thrust faults in the Alps probably started as recumbent folds. As depicted in Figure 15.24, when stress continues to build up and to deform a recumbent fold, the overturned limb may become so stretched and strained that it eventually breaks and

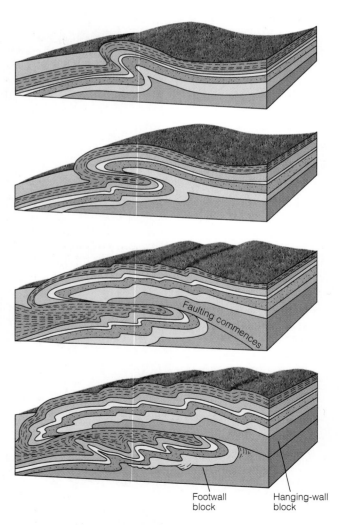

Faulting commences

Footwall block

Hanging-wall block

Figure 15.24 Recumbent Fold The evolution of a recumbent fold into a thrust fault. This kind of structure is important in the Alps and many other mountain ranges.

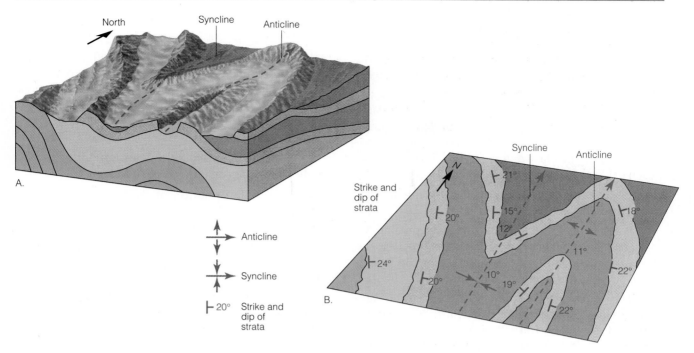

Figure 15.25 Folding Revealed by Topography Distinctive topographic forms and patterns resulting from differing resistance to erosion of different kinds of rock reveal the presence of plunging folds. A. Block diagram showing topographic effects. Note that resistant strata (layers 3 and 5) make topographic highs in both anticlines and synclines, while easily eroded strata (layers 2 and 4) make topographic lows in both anticlines and synclines. B. Geologic map of area shown in Part A. Compare with Figure 1.16.

becomes a thrust fault. Movement on some of the great recumbent-fold and thrust-fault structures in the Alps is in excess of 50 km.

Folds and Topography
Differences in erodibility of adjacent strata can lead to distinctive topographic forms, and commonly these topographic forms reveal the presence of folds. It is important to remember, however, that anticlines do not necessarily make ridges nor synclines valleys, as Figure 15.25 shows. A particularly striking example of the

relationship between plunging folds, erosion, and topography can be seen in the Valley and Ridge province of Pennsylvania (Fig. 1.16). There, a series of plunging anticlines and synclines were created during the Paleozoic Era by a continental collision between North America, Africa, and Europe. The Appalachian mountain chain was the result. Now deeply eroded, the folded rocks determine the pattern of the topography because soft, easily eroded strata (shales) underlie the valleys, while resistant strata (sandstones) form the ridges.

SUMMARY

1. Rocks can be deformed in three ways: by elastic deformation (no permanent change), by ductile deformation (folds), and by fracture (faults, joints).

2. High confining stress and high temperatures enhance ductile properties. Low temperatures and low confining stress enhance elastic properties and failure by fracture when the elastic limit is exceeded.

3. The rate at which a solid is deformed (strained) also controls the style of deformation. High strain rates lead to fractures; low strain rates cause folding.

4. Weak rocks (limestone, marble, slate, phyllite, and schist) enhance ductile properties. Strong rocks (sandstone, quartzite, and granite) enhance brittle properties. Dry rocks are stronger than wet rocks.

5. The orientations of contacts between strata, faults, joints, or any other inclined plane surfaces are determined by the strike (direction of intersection of the inclined plane and a horizontal plane) and the dip (angle between the inclined plane and the horizontal).

6. Fractures in rocks along which slippage occurs are called faults. Normal faults are caused by tensional stresses that tend to pull the crust apart, whereas thrust and other types of reverse faults arise from compressional stresses that squeeze, shorten, and thicken the crust. Strike-slip faults are caused by shear stresses; they are vertical fractures that have horizontal motion.

7. It is usually possible to determine only relative motions of rocks on either side of a fault.

8. Ductile deformation of strata causes bends or warps, which are called folds. Folding is due to compressional stress. An upward, arched fold is an anticline; a downward, troughlike fold is a syncline. The sides of a fold are called limbs.

9. Overturned folds (meaning both limbs dip in the same direction) are common in mountain ranges formed by continental collision. In some ranges, such as the Alps, overturned folds have nearly horizontal limbs, in which case the folds are said to be recumbent.

10. Faults die out by becoming folds, and folds die out by becoming smaller and smaller wrinkles.

THE LANGUAGE OF GEOLOGY

anticline (p. 419)
axis (of a fold) (p. 421)

compressional stress (p. 404)
confining stress (p. 404)

dip (p. 410)
ductile deformation (p. 405)

elastic deformation (p. 405)
elastic limit (p. 405)

fault (p. 409)
fold (p. 418)
footwall block (of a fault) (p. 412)

fracture (p. 406)

graben (p. 412)

half-graben (p. 412)
hanging wall block (of a fault) (p. 412)
horst (p. 412)

limbs (of a fold) (p. 421)

monocline (p. 419)

normal faults (p. 413)

plunge (of a fold) (p. 421)

reverse fault (p. 414)
rift (p. 412)

shear stress (p. 404)
strain (p. 405)
strain rate (p. 407)
strike (p. 410)
strike-slip fault (p. 416)
syncline (p. 421)

tensional stress (p. 404)
thrust fault (p. 414)
transform faults (p. 416)

QUESTIONS FOR REVIEW

1. What are stress and strain, and what is the relationship between them?

2. Discuss the ways a seemingly rigid solid, such as a rock, can be deformed. In what sequence do deformation properties come into play?

3. What properties determine whether a rock is brittle and fails by fracture or is deformed by ductile deformation?

4. Name three rock types that readily deform by ductile deformation and three that tend to deform by fracture.

5. Identify three prominent topographic features on the Earth made up of a system of grabens and horsts. What kind of stresses cause horsts and grabens to form?

6. Describe the way a transform fault works. Why are they called *transform* faults? Cite an example of a transform fault that is still active.

7. What kind of stresses produce folds? Draw a sketch of an anticline and mark the axis, axial plane, and limbs.

8. How do folds and faults die out?

9. What is a recumbent fold and where would you expect to find large recumbent folds? Describe the way a recumbent fold can become a thrust fault.

10. Draw a geologic map of a plunging anticline. Mark strike and direction of dip at several places around the fold. Draw the fold axis and indicate, if you can, the direction of plunge.

16

Measuring
Our Earth

Understanding
Our Environment

GeoMedia

Buildings in Los Angeles destroyed by the Whittier Narrows earthquake
of October 1987. The quake, which registered 6.0 on the Richter scale, is
thought to have been caused by movement on a concealed thrust fault
beneath the city.

Earthquakes and the Earth's Interior

Hidden Faults Beneath Los Angeles

Seismologists are understandably eager to pinpoint the locations of faults beneath major cities. In the case of Los Angeles, this resembles hunting for fish in a lake: You know there are lots of them but you can't exactly say where they are.

Recently, however, two scientists, John H. Shaw of Harvard University and Peter M. Shearer of the Scripps Institute of Oceanography, made a major catch: They discovered a large, active fault extending 40 kilometers from east to west under Los Angeles. It is one of many so-called blind thrusts, concealed cracks that are covered over by thick layers of surface sediments. Although blind thrusts are much smaller than the San Andreas Fault, they are very dangerous because of their proximity to the city.

The researchers located the fault, which they named Puente Hills thrust, by gaining access to data collected by oil companies. In the process of searching for petroleum, the companies have scanned the underground geology of Los Angeles by vibrating the ground with explosions. The resulting high-resolution subsurface images, along with seismic records for the area, enabled Shaw and Shearer to map the blind-thrust system lying directly beneath the metropolitan area.

The Puente Hills thrust appears to be broken into three segments, each of which could produce a magnitude 6.5 quake. Each segment would act like a giant ramp, with one wedge of crust riding up another. The researchers believe that one of these segments probably caused the magnitude 6.0 Whittier Narrows quake of 1987. They point out that the damage done by earthquakes of this size, combined with geologic evidence of even larger quakes in the past, make it more important than ever to focus attention on the hazards posed by hidden thrust faults.

EARTHQUAKES

When the Earth quakes, it is as if our world has been struck by a huge hammer. The reason for the quaking is that energy stored in elastically strained rocks is suddenly released. The more energy released, the stronger the quake. Just how the elastically stored energy is built up and how it is released continue to be subjects for intensive research.

Do an experiment yourself. Have a friend hit one end of a wooden plank or the top of a wooden table with her or his fist or a hammer while you press your hand on the other end. You will feel vibrations set up in the plank or tabletop by the energy of the blow. The harder the blow, the stronger the vibrations. You can feel those vibrations because some of the energy imparted by the hammer or fist is transferred to your hand by elastic vibrations through the whole length of the solid wood. Fortunately, fists and hammers don't hit the Earth, but a bomb blast or a violent volcanic explosion will serve as an energy source just as well. So, too, will the sudden slipping of rock masses along fault surfaces.

The most widely accepted theory concerning the origin of earthquakes involves slipping faults and the *elastic rebound theory*.

Origin of Earthquakes

Sudden movement along faults seems to be the cause of most earthquakes. But it cannot be that simple. Some earthquakes are millions of times stronger than others. Why? The same energy that in one case is released by thousands of tiny slips and tiny earthquakes is in another case stored and released in a single immense earthquake. The **elastic rebound theory** suggests that if fault surfaces do not slip easily past one another, energy can

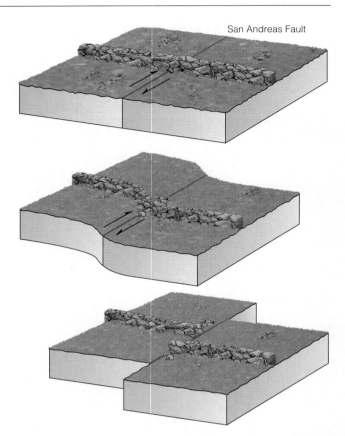

Figure 16.1 Cause of an Earthquake An earthquake, caused by sudden release of energy. Sketch based on detailed surveys near the San Andreas Fault, California, before and after the abrupt movement that caused the earthquake of 1906. A stone wall crosses the fault and is slowly bent as rock is elastically strained. After the earthquake, two segments of the wall are offset 7 m.

be stored in elastically deformed bodies of rock, just as in a steel spring that is compressed. Then when the fault finally does slip, the elastically strained bodies of rock rebound to their original shapes. The first evidence supporting the elastic rebound theory came from studies of the San Andreas Fault, which, as discussed in Chapter 15, is a right-lateral strike-slip fault. During long-term field observations in central California, beginning in 1874, scientists from the U.S. Coast and Geodetic Survey determined the precise positions of many points both adjacent to and distant from the fault (Fig. 16.1). As time passed, movement of the points revealed that the crust was slowly being bent. For some reason, in the area of measurement near San Francisco, the fault was locked and did not slip. On April 18, 1906, the two sides of this locked fault shifted abruptly. The elastically stored energy was released as the fault moved and the bent crust snapped back, thereby creating a violent earthquake. Repetition of the survey then revealed that the bending had disappeared.

Most earthquakes occur in the brittle rock of the lithosphere. As discussed in Chapter 15, brittleness is the tendency for a solid to fracture when the deforming stress exceeds the limits of elasticity. At great depth, temperatures and pressures are so high that ductile deformation occurs. Rocks can neither fracture nor store elastic energy under such conditions. Rather, they are like putty, and they undergo permanent changes of shape that remain even after the deforming forces have been removed. Earthquakes, then, are phenomena of the brittle, outer, cooler portion of the Earth.

How Earthquakes Are Studied

The name given to the study of earthquakes is **seismology**, a word that comes directly from the ancient Greek term for earthquakes, *seismos*.

Seismographs

The device used to record the shocks and vibrations caused by earthquakes is a **seismograph**. An ideal way to record the vibrations and motions of the Earth would be to put the seismograph on a stable platform that is not affected by the vibrations. But a seismograph must stand on the Earth's vibrating surface, and it will therefore vibrate with the surface. This makes the act of measurement difficult because there is no fixed frame of reference against which to make measurements. The

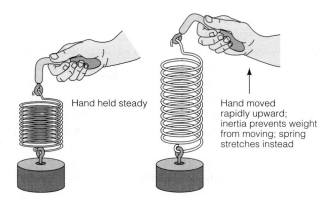

Figure 16.2 How a Seismograph Works The principle of the inertial seismograph.

problem is the same one that a sailor in a small boat faces when attempting to measure waves at sea.

To overcome the frame-of-reference problem, most seismographs make use of **inertia**, which is the resistance of a large stationary mass to sudden movement. If you suspend a heavy mass, such as a block of iron, from a light spring and then suddenly lift the spring, you will notice that because of inertia the block of iron remains almost stationary while the spring stretches (Fig. 16.2). This is the principle used in *inertial seismographs* (Fig. 16.3). Vertical motion is measured by the device

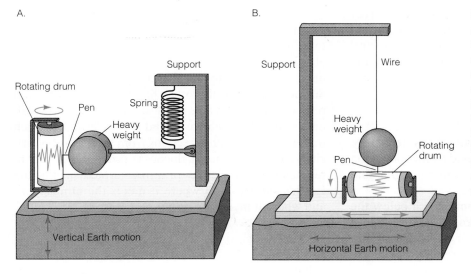

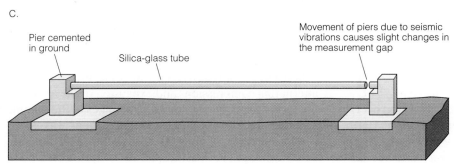

Figure 16.3 Recording Earthquake Activity Seismographs measure vibrations sent out by earthquakes. A. An inertial seismograph for measuring vertical motions. B. An inertial seismograph for measuring horizontal motions. C. A strain seismograph. A rigid silica-glass tube, as long as 35 m, is supported on a solid concrete pier anchored to the ground. A second pier carries a sensitive electronic measuring device to record any movement in the end of the tube. The distance between the two piers changes when a seismic vibration disturbs the Earth's surface.

shown in Figure 16.3A, in which a heavy mass is supported by a spring and the spring is connected to a support which in turn is connected to the ground. When the ground vibrates, the spring expands and contracts but the mass remains almost stationary. Then the distance between the ground and the mass can be used to sense vertical displacement of the ground surface.

Horizontal displacement can be similarly measured by suspending a heavy mass from a wire to make a pendulum (Fig. 16.3B). Because of its inertia, the mass does not keep up with the horizontal ground motion, and the difference between the pendulum and ground movement records the horizontal ground motion. Inertial seismographs are commonly used in groups so that motions can be measured simultaneously up-down, east-west, and north-south.

Another kind of device, called the *Benioff strain seismograph*, employs two concrete piers in the ground spaced at a distance of about 35 m (Fig. 16.3C). When the Earth vibrates, the two piers move independently of each other. Attached to one pier is a long, rigid, silica-glass tube. The other pier carries a very sensitive detector to measure even the slightest movement in the end of the silica-glass tube. Strain seismographs are frequently installed in mines, tunnels, and other places where temperature is relatively constant and wind disturbance minimal.

Modern seismographs are incredibly sensitive. Vibrational movements as tiny as one hundred millionth (10^{-8}) of a centimeter can be detected. Indeed, many instruments are so sensitive they can detect ground depression caused by a moving automobile several blocks away, and can even identify ocean waves and tides several kilometers from the seashore.

Earthquake Focus and Epicenter

The place where energy is first released to cause an earthquake is called the **earthquake focus**. In reality, because most earthquakes are caused by movement on a fault, the focus is not a point but rather a region that may extend for several kilometers.

An earthquake focus lies at some depth below the Earth's surface. It is more convenient to identify the site of an earthquake from the **epicenter**, which is the point on the Earth's surface that lies vertically above the focus of an earthquake (Fig. 16.4). A good way to describe the location of an earthquake focus is to state the location of its epicenter and its depth.

Seismic Waves

When an earthquake occurs, the elastically stored energy is transmitted from the focus to other parts of the Earth. As with any other vibrating body, waves (vibrations) spread outward from the focus. The waves, called **seis-**

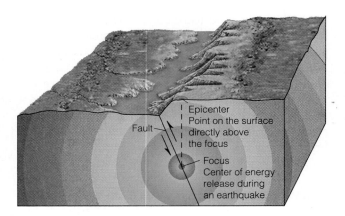

Figure 16.4 Location of an Earthquake   The focus of an earthquake is the site of first movement on a fault and the center of energy release. The epicenter of an earthquake is the point on the Earth's surface that lies vertically above the focus.

mic waves, spread out in all directions, just as sound waves spread in all directions when a gun is fired.

Seismic waves are elastic disturbances, and so unless the elastic limit is exceeded, the rocks through which they pass return to their original shapes after passage of the waves. Seismic waves must therefore be measured and recorded while the rock is still vibrating. For this reason, many continuously recording seismograph stations have been installed around the world.

There are several kinds of seismic waves, and they belong to two families. **Body waves**, of which there are two kinds, corresponding to the two ways a solid can be elastically deformed, travel outward from the focus and have the capacity to travel entirely through the Earth. **Surface waves**, on the other hand, are guided by and restricted to the Earth's surface. Body waves are analogous to either light or sound waves, both of which travel outward in all directions from their point of origin. Surface waves are analogous to ocean waves and are restricted to the vicinity of a free surface, such as the Earth's surface both where it meets the atmosphere and where it meets the ocean.

Body Waves

Rocks can be elastically deformed by either a change of volume or a change of shape. The first kind of body wave, *compressional waves*, deforms rocks by change of volume and consists of alternating pulses of compression and expansion acting in the direction of wave travel (Fig. 16.5A). Sound waves are also compressional waves. When a sound wave passes through the air, it does so by alternating compressions and expansion of the air. Compression and expansion produce changes in the volume and density of a medium. Compressional waves can pass through solids, liquids, or gases because all three can sustain changes in density. When a compressional wave

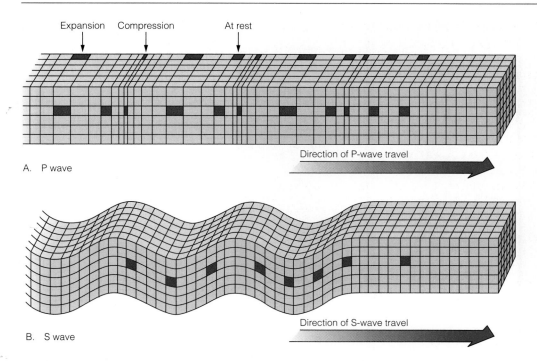

Expansion   Compression   At rest

Direction of P-wave travel

A. P wave

Direction of S-wave travel

B. S wave

Figure 16.5 Seismic Body Waves Seismic body waves of the P (compressional) and S (shear) types. A. P waves cause alternate compressions and expansions in the rock the wave is passing through. An individual point in a rock will move back and forth parallel to the direction of P-wave propagation. As wave after wave passes through, a square will repeatedly expand to a rectangle, return to a square, contract to a rectangle, return to a square, and so on. B. S waves cause a shearing motion. An individual point in a rock will move up and down, perpendicular to the direction of S-wave propagation. A square will repeatedly change to a parallelogram, then back to a square again.

passes through a medium, the compression pushes atoms closer together. Expansion, on the other hand, is an elastic response to compression, and it causes the distance between atoms to be increased. Movement in a solid subjected to compressional waves is back and forth in the line of the wave motion. Compressional waves have the greatest speed of all seismic waves—6 km/s is a typical value for the uppermost portion of the crust—and they are the first waves to be recorded by a seismograph after an earthquake. They are therefore called **P (for primary) waves**.

The second kind of body waves are *shear waves*. They deform materials by change of shape. Because gases and liquids do not have the elasticity to rebound to their original shape, shear waves can be transmitted only by solids. Shear waves consist of an alternating series of sidewise movements, each particle in the deformed solid being displaced perpendicular to the direction of wave travel (Fig. 16.5B). A typical speed for a shear wave in the upper crust is 3.5 km/s. Because shear waves are slower than P waves and reach a seismograph some time after a P wave arrives, they are called **S (for secondary) waves**.

Seismic body waves behave like light waves and sound waves, which is to say both can be transmitted

through a medium and also be *reflected* and *refracted*. Reflection is the familiar phenomenon of light bouncing off the surface of a mirror or a glass of water. Seismic body waves are reflected by numerous surfaces in the Earth. Refraction is a less familiar phenomenon. It occurs whenever a wave speed changes, and the effect is to cause a bending in the direction the wave is moving. Speed change and refraction can be either gradual or abrupt. An abrupt change is seen when a ray of light strikes a surface of water as shown in Figure 16.6. Some of the light is reflected, but some also crosses the surface and travels through the water. The speed of light is less in water than in air, and the wave path is sharply bent at the surface. The ray is said to have been refracted.

Seismic body wave speeds are a function of the density of the medium through which they are passing. If the Earth had a homogeneous composition and if density increased steadily with depth as a result of increasing pressure, refraction would cause seismic wave paths to be curved as shown in Figure 16.7A. Measurements show that wave paths are indeed curved, owing to gradual refraction, but measurements reveal, too, that the seismic waves are also refracted and reflected by several zones of sudden density change, such as the core-mantle boundary (Fig. 16.7B).

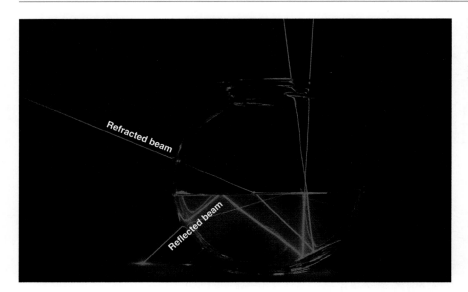

Figure 16.6 Reflection and Refraction Reflection and refraction demonstrated with laser beams. A beam of light is reflected from, or refracted across, a boundary between air and water.

A.

B.

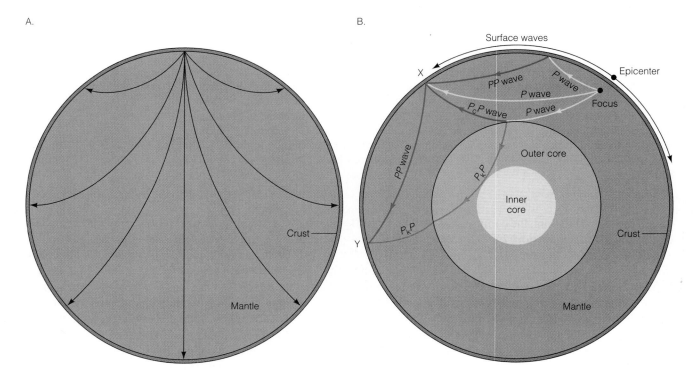

Figure 16.7 Passage of Seismic Body Waves Refraction and reflection of seismic body waves. A. Wave paths through a planet of homogeneous composition in which density increases with depth due to increasing pressure. Changing density causes wave paths to bend due to refraction. B. Various paths of P waves in the Earth, which is a compositionally layered planet. Seismographs at some places (locations X and Y, for example) receive both direct P waves (yellow) as well as reflected (blue) and refracted (red) P waves. A P wave reflected off the Earth's surface is called a PP wave; one reflected off the core-mantle boundary is a PcP wave; one refracted through the liquid outer core is a PkP wave.

Surface Waves

To an observer, surface waves appear very similar to P and S waves—one kind causes an up and down or rolling motion, rather like an S wave, and another kind causes a back and forth shaking. However, surface waves travel more slowly than P and S waves, and in addition they pass around the Earth rather than through it. Thus, surface waves are the last to be detected by a seismograph (Fig. 16.8A).

One important difference between body waves and surface waves concerns a property called dispersion. Body waves travel through the Earth at the same

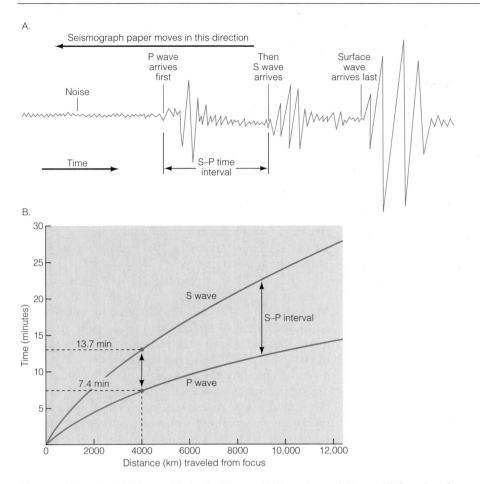

A.

Seismograph paper moves in this direction

Noise

P wave arrives first

Then S wave arrives

Surface wave arrives last

Time

S–P time interval

B.

Time (minutes)

S wave

S–P interval

13.7 min

7.4 min

P wave

Distance (km) traveled from focus

Figure 16.8 Travel Times of Seismic Waves Different travel times of P, S, and surface waves. A. Typical record made by a seismograph. The P and S waves leave the earthquake focus at the same instant. The fast-moving P waves reach the seismograph first, and some time later the slower-moving S waves arrive. The delay in arrival times is proportional to the distance traveled by the waves. The surface waves travel more slowly than either P or S waves. B. Average travel-time curves for P and S waves in the Earth, used to locate an epicenter. For example, when seismologists at a station measure the S-P time interval to be 13.7 min - 7.4 min = 6.3 min, they know the epicenter is 4000 km away from their station.

velocity regardless of wavelength, but this is not true for surface waves. The longer the wavelength of a surface wave, the deeper the wave motion penetrates the Earth. Because the Earth is compositionally layered, getting denser and denser as depth increases, long-wavelength surface waves penetrate to deeper (denser), higher velocity zones, while short-wavelength surface waves do not. In this way, surface waves of different wavelengths develop different velocities, and that property is called *dispersion*. Because some wavelengths reach a depth of several hundred kilometers, dispersion of surface waves is a very important way to determine the properties of the outer few hundred kilometers of the Earth. We will return to the importance of dispersion when we discuss the properties of the mantle later in this chapter.

Location of the Epicenter

The location of an earthquake's epicenter can be determined from the arrival times of the P and S waves at a seismograph. The farther a seismograph is away from an epicenter, the greater the time difference between the arrival of the P and S waves (Fig. 16.8). After using a graph like that shown in Figure 16.8B to determine how far an epicenter lies from a seismograph, the seismologist draws a circle on a map around the station with a radius equal to the calculated distance to the epicenter. The exact position of the epicenter can be determined when data from three or more seismographs are available—the center lies where the circles intersect (Fig. 16.9).

The depth of an earthquake focus below an epicenter can also be determined. If a local earthquake is

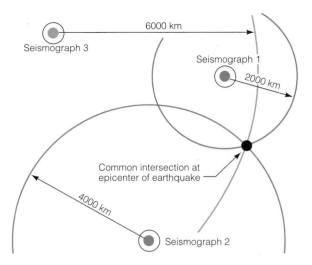

Figure 16.9 Locating Site of an Earthquake Locating an epicenter. The effects of an earthquake are felt at three different seismograph stations. The time differences between the first arrival of the P and S waves depend on the distance of a station from the epicenter. The following distances are calculated by using the curves in Figure 16.8B.

	Time Difference	Calculated Distance
Seismograph 1	8.8 min - 4.7 min = 4.1 min	2000 km
Seismograph 2	13.7 min - 7.4 min = 6.3 min	4000 km
Seismograph 3	17.5 min - 9.8 min = 7.7 min	6000 km

On a map, a circle of appropriate radius is drawn around each station. The epicenter is where the three circles interesect.

recorded by several nearby seismographs, the focal depth can be determined in the same way that the epicenter is determined, by using P–S time intervals. For distant earthquakes a different method is employed. Note in Figure 16.7 that a seismograph stationed at position X would record both a direct P wave and a P wave reflected from the Earth's surface (labeled *PP*). The direct P wave, having a shorter path length, would arrive before the *PP* wave. The travel-time difference between them is a measure of the focal depth of the earthquake.

Magnitudes of Earthquakes

Very large earthquakes (of the kind that destroyed San Francisco in 1906, T'ang Shan, China, in 1976, parts of Mexico City in 1985, and the Loma Prieta quake of 1989 that again destroyed parts of San Francisco) are, fortunately, relatively infrequent. In earthquake-prone regions, such as San Francisco and the surrounding area, very large earthquakes occur about once a century. They occur more frequently in some areas and less frequently in others, but a century is an approximate average. This means that the time needed to build up elastic energy to a point where the frictional locking of a fault is overcome is about 100 years. Small earthquakes may occur along a fault during this time as a result of local

slippage, but even so, elastic energy is accumulating because most of the fault remains locked. When the lock is broken and an earthquake occurs, the elastic energy is released during a few terrible minutes. By careful measurement of elastically strained rocks along the San Andreas Fault, seismologists have found that about 100 J of elastic energy can be accumulated in $1m^3$ of deformed rock. This is not very much—it is equivalent to only about 25 calories of heat energy—but when billions or trillions of cubic meters of rock are strained, the total amount of stored energy can be enormous. The amount of elastically stored energy released during the Loma Prieta earthquake was about 10^{15} J, and the 1906 San Francisco earthquake released at least 10^{17} J. The energy released by a hydrogen bomb blast is also about 10^{17} J!

Richter Magnitude Scale
Measurements of elastically deformed rocks before an earthquake, and of undeformed rocks after an earthquake, can provide an accurate measure of the amount of energy released. The task is very time-consuming, and all too frequently the pre-earthquake measurements are simply not available. Therefore, seismologists have developed a way to estimate the energy released by measuring the amplitudes of seismic waves. The **Richter magnitude scale**, named after the seismologist who developed it, is defined by the maximum amplitudes of the P and S waves (that is, the heights of the waves on a seismogram) 100 km from an epicenter. Because wave signals vary in strength by factors of a hundred million or more, the Richter scale is logarithmic, which means it is divided into steps called magnitudes, starting with magnitude 1 and increasing upward. Each unit increase in magnitude corresponds to a tenfold increase in the amplitude of the wave signal. Thus, a magnitude 2 signal has an amplitude that is ten times larger than a magnitude 1 signal, and a magnitude 3 is a hundred times larger than a 1 signal. To see how a Richter magnitude is calculated, see boxed essay: *Calculating a Richter Magnitude.*

Earthquake Risk

Most people in the United States think immediately of California when earthquakes are mentioned. However, the most intense earthquakes to jolt North America in the past 200 years were centered near New Madrid, Missouri. Three earthquakes of great size occurred on December 16, 1811, and January 23 and February 7, 1812. The exact magnitude of these earthquakes is unknown because instruments to record them did not exist at the time. However, judging from the local damage caused, as well as the fact that tremors were felt and minor damage occurred as far away as New York and Charleston, South Carolina, it is estimated that each was greater than mag-

Measuring Our Earth

Calculating a Richter Magnitude

We can see from Figure B16.1 how a Richter magnitude is calculated. The energy of a seismic wave is a function of both its amplitude and the duration of a single wave oscillation, T. Divide the maximum amplitude, X, measured in steps of 10^{-4} cm on a suitably adjusted seismograph, by T, measured in seconds. Then add a correction

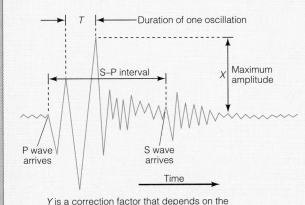

Y is a correction factor that depends on the distance of the seismograph from the epicenter. It is calculated from the S–P interval.

Figure B16.1 Measuring Magnitudes Measurements used for determining Richter magnitude (M) from a seismograph record.

factor, Y, determined from the S-P wave interval. The quantity X/T is a measure of the maximum energy reaching the seismograph. The formula is:

Richter Magnitude, $M = \log X/T + Y$

One Richter magnitude scale unit corresponds to a tenfold increase in X. However, the energy increase is proportional to X^2, which is to say a hundredfold. The duration of a single oscillation differs greatly from one earthquake to another. In particular, the most energetic earthquakes have a higher proportion of long-duration waves. As a result, the energy increase corresponding to one Richter scale unit increase, when summed over the whole range of waves in a wave record, is only a thirtyfold increase. Thus, the difference in energy released between an earthquake of magnitude 4 and one of magnitude 7 is 30 x 30 x 30 = 27,000 times!

How big can earthquakes get? The largest recorded to date have Richter magnitudes of about 8.6, which means they release about as much energy as 10,000 atom bombs of the kind that destroyed Hiroshima at the end of World War II. It is possible that earthquakes do not get any larger than this because rocks cannot store more elastic energy. Before they are deformed further, they fracture and so release the energy.

---

nitude 8 and the largest of the quakes was larger than the one that struck San Francisco in 1906.

Based on known geological structures (mainly faults) and on the location and intensity of past earthquakes, the National Oceanographic and Atmospheric Administration prepared the seismic-risk map shown in Figure 16.10. Although Figure 16.10 is very informative, it is not particularly useful to people who plan and

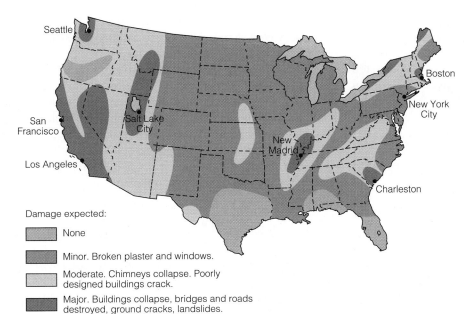

Damage expected:

☐ None

☐ Minor. Broken plaster and windows.

☐ Moderate. Chimneys collapse. Poorly designed buildings crack.

☐ Major. Buildings collapse, bridges and roads destroyed, ground cracks, landslides.

Figure 16.10 Where Quakes Are Most Intense Seismic-risk map of the United States based on quake intensity. Zones refer to maximum earthquake intensity and therefore to maximum possible destruction. The map does not indicate frequency of earthquakes. For example, frequency is high in southern California but low in eastern Massachusetts. Nevertheless, when earthquakes occur in eastern Massachusetts, they can be as severe as the more frequent quakes in southern California.

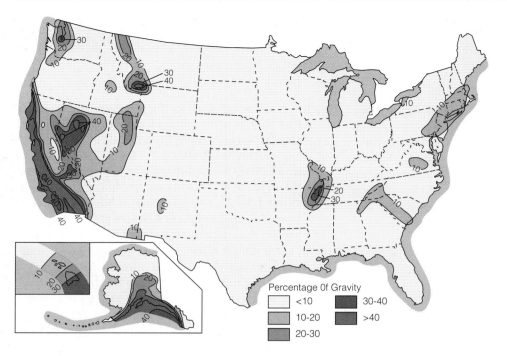

Figure 16.11 Seismic Risk and City Planning A seismic-risk map based on horizontal acceleration. Because acceleration during ground movement is the most important factor to be considered in the design of an earthquake-stable building, this type of risk map is preferred by builders and city planners. Numbers on contours refer to the maximum horizontal acceleration during an earthquake, expressed as a percentage of the acceleration due to gravity (980 cm/S^2). The probability that an acceleration of the amount indicated will occur in any given 50-year period is 1 chance in 10.

build roadways and public buildings. Builders need to know just how strongly the ground is likely to shake and what the frequency of large earthquakes might be. To answer this need, the U.S. Geological Survey publishes a seismic-risk map of the kind shown in Figure 16.11. The strength of a possible earthquake is compared to the acceleration due to gravity, which is 980 cm/s^2 (that is to say, 1.0G). The maximum acceleration is likely to occur with California quakes and to reach 80 percent of gravity (which would be 0.8G). Damage begins at about 0.1G.

Figure 16.11 includes some additional information. Scientists of the U.S. Geological Survey have calculated that there is only 1 chance in 10 that a given acceleration shown in Figure 16.11 will be exceeded in a 50-year period. That information is of great help to insurance companies. Could a great earthquake happen where none has been experienced in living memory? See the essay, *A Great Earthquake in the Pacific Northwest?*, for an answer.

Earthquake Disasters

Every year the Earth experiences many hundreds of thousands of earthquakes. Fortunately, only one or two are large enough, or close enough to major centers of population, to cause loss of life. Certain areas are known to be earthquake-prone, and special building codes in such places require structures to be as resistant as possible to earthquake damage. However, all too often an unexpected earthquake will devastate an area where buildings are not adequately constructed, as in the T'ang Shan earthquake in 1976 in China, where 240,000 people lost their lives. Other examples are the earthquake that destroyed parts of the center of Mexico City, killing 9500 people in 1985 (Fig. 16.12), and the earthquake that struck Soviet Armenia in 1988, killing an estimated 25,000 people (Fig. 16.13).

Eighteen earthquakes are known to have caused 50,000 or more deaths apiece (Table 16.1). The most disastrous one on record occurred in 1656, in Shaanxi Province, China, where an estimated 830,000 people died. Many of those people lived in cave dwellings excavated in loess (Chapter 13), which collapsed as a result of the quake. Since 1900, there have been 47 earthquakes worldwide in which 500 or more people have died.

Earthquake Damage

The dangers of earthquakes are profound, and the havoc they can cause is often catastrophic. Their effects are of six principal kinds. The first two, ground motion and

TABLE 16.1	Earthquakes During the Past 800 Years That Have Caused 50,000 or More Deaths	
Place	Year	Estimated Number of Deaths
Silicia, Turkey	1268	60,000
Chihli, China	1290	100,000
Naples, Italy	1456	60,000
Shaanxi, China	1556	830,000
Shemaka, USSR	1667	80,000
Naples, Italy	1693	93,000
Catania, Italy	1693	60,000
Beijing, China	1731	100,000
Calcutta, India	1737	300,000
Lisbon, Portugal	1755	60,000
Calabria, Italy	1783	50,000
Messina, Italy	1908	160,000
Gansu, China	1920	180,000
Tokyo and Yokohama, Japan	1923	143,000
Gansu, China	1932	70,000
Quetta, Pakistan	1935	60,000
T'ang Shan, China	1976	240,000
Iran	1990	52,000

Figure 16.12 Not Designed to Withstand an Earthquake A building that was not constructed to withstand expected earthquakes. A high-rise apartment house was one of the buildings that collapsed during the earthquake that struck Mexico City in 1985. Proper building design can minimize damage. Nearby buildings of sturdier construction withstood the shaking.

Figure 16.13 Collapse of City Buildings When a magnitude-6.8 earthquake struck Armenia on December 7, 1988, poorly constructed buildings with inadequate foundations collapsed like houses of cards. The principal cause of collapse was ground motion.

Understanding Our Environment

A Great Earthquake in the Pacific Northwest?

Subduction zones are places of intense seismic activity. Many of the greatest earthquakes ever recorded occurred in subduction zones: the Chilean earthquake of 1960, Richter magnitude 8.5, for instance, and the Alaskan earthquake of 1964, magnitude 8.6. Such high-magnitude earthquakes are totally destructive.

The foci of most earthquakes in subduction zones are within the lithosphere of either the sinking or over-riding plate. Such earthquakes tend to be no larger than about Richter magnitude 7.5 and to be caused by the bending and stretching of rock within the plates. The big quakes of magnitude 8 and larger have foci right at the interface between two plates. Presumably such quakes occur when the downgoing plate sticks to the bottom of the overriding plate. When the lock is broken, a tremendous earthquake results.

From northern California to central British Columbia, the Juan de Fuca Plate (named for an early Spanish explorer) has been slipping beneath the North American Plate at a rate of 3 to 4 cm/yr for the past million years (Fig. B16.2), but for the past 200 years no great earthquakes have occurred on the Cascadia subduction zone. Furthermore, seismographs do not record any current seismic activity along the interface between the two plates. Is the Pacific Northwest a likely spot for a gigantic quake sometime in the future? Is the Juan de Fuca Plate locked to the overriding North American Plate, or is it sliding smoothly downward?

Geologists of the U.S. Geological Survey and other institutions are attempting to answer these questions by looking at the stratigraphic record. In general, great earthquakes at subduction zones tend to cause pronounced elevation changes along nearby coastlines. Coastal subsidence up to 2 m or more can change well-vegetated coastal lowlands into intertidal flats. Careful mapping along the coastline of Washington reveals a

number of places (labled in Figure B16.2) where marshes and swamps suddenly became barren tidal flats in the geologically recent past (Figure B16.3).

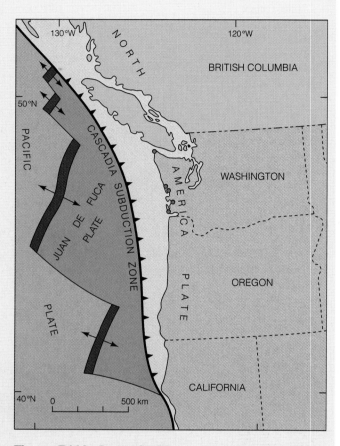

Figure B16.2 Interacting Plates The Cascadia subduction zone separates the downgoing Juan de Fuca Plate from the overriding North American Plate. Evidence of sudden, catastrophic sinking of near-coastal lowlands has been discovered along the Washington coast at the five sites marked.

faulting, are primary effects, and they cause damage directly. The other four effects are secondary and cause damage indirectly as a result of processes set in motion by the earthquake.

1. Ground motion results from the movement of seismic waves, especially surface waves, through surface-rock layers and regolith. The motions can damage and sometimes completely destroy buildings, especially buildings on regolith, because regolith responds more strongly than rock. Proper design of buildings (including such fea-

tures as steel framework and foundations tied to bedrock) can do much to prevent such damage, but in a very strong earthquake even the best buildings may suffer some damage. A lot of the damage from the 1988 Armenian and the 1989 Loma Prieta earthquakes was due to ground motion.

2. Where a fault breaks the ground surface, buildings can be split, roads disrupted, and any feature that crosses or sits on the fault broken apart.

3. A secondary effect, but one that is sometimes a

Radiocarbon dating of organic matter buried in sediment when lush coastal lowlands were suddenly flooded shows that a catastrophic change in the coastline occurred about 300 years ago in southwestern Washington. Evidence also suggests that catastrophic coastline changes occurred about 1700, 2700, 3100, and 3400 years ago in that area.

Geological research suggests that the Puget Sound area, which is lined with deep faults, could suffer a magnitude-9 subduction-zone earthquake. The region could also be subject to large earthquakes in the shallower part of the crust.

The new findings are part of a study known as SHIPS, for Seismic Hazards Investigation in Puget Sound. It is one of the most complex studies ever done to assess earthquake hazards in an urban area, and its goal is to produce a three-dimensional map of the geologic structure from Puget Sound through the Georgia Strait. The area covered includes the cities of Olympia, Tacoma, Seattle, Everett, and Bellingham in Washington State, as well as Victoria and Vancouver in British Columbia.

Once the map is complete, it will help identify where the amplitudes of earthquake waves would be highest and where their durations would be longest. This information can be used by business owners, city planners, and emergency managers to design a seismic hazard map for the area.

Figure   B16.3   Evidence of a Prehistoric Earthquake   Geologist Brian Atwater stands at the high-tide level on the surface of a brackish tidal marsh in Willapa Bay on the coast of Washington. The marsh plants overlie roots of a former spruce forest that was killed by sudden subsidence of the land during a major earthquake about 300 years ago.

greater hazard than moving ground, is fire (Fig. 16.14). Ground movement displaces stoves, breaks gas lines, and loosens electrical wires, thereby starting fires. Ground motion also breaks water mains, so that no water is available to put out fires. In the earthquakes that struck San Francisco in 1906, and Tokyo and Yokohama in 1923, more than 90 percent of the damage to buildings was caused by fire.

4. In regions of steep slopes, earthquake vibrations may cause regolith to slip, cliffs to collapse, and other rapid mass-wasting movements to start. This is particularly true in Alaska, parts of southern California, China, and hilly places such as Iran and Turkey. Houses, roads, and other structures are destroyed by rapidly moving regolith.

5. The sudden shaking and disturbance of water-saturated sediment and regolith can turn seemingly solid ground to a liquidlike mass of quicksand. This process is called liquefaction, and it was one of the major causes of damage during the earthquake that

Figure 16.14 Broken Gas Lines Fire caused by gas lines broken as a result of the Loma Prieta earthquake in 1989. This shot shows a fire in the Marina district of San Francisco.

destroyed much of Anchorage, Alaska, on March 27, 1964, and that caused apartment houses to sink and collapse in Niigata, Japan, that same year (Fig. 16.15).

6. Finally, there are **seismic sea waves**, called *tsunami* as discussed in Chapter 14, which have been particularly destructive in the Pacific Ocean. About 4.5 hours after a severe submarine earthquake near Unimak Island,

Figure 16.15 Shaking Causes Liquefaction Destruction of part of Anchorage, Alaska, caused liquefaction as a result of the earthquake of 1964. The building in the foreground was the home of the publisher of the *Anchorage Times*.

TABLE 16.2	Earthquake Magnitudes, Frequencies for the Entire Earth, and Damaging Effects			
Richter Magnitude	Number per Year	Modified Mercalli Intensity Scale[a]		Characteristic Effects of Shocks in Populated Areas
<3.4	800,000	I		Recorded only by seismographs
3.5–4.2	30,000	II and III		Felt by some people who are indoors
4.3–4.8	4,800	IV		Felt by many people; windows rattle
4.9–5.4	1,400	V		Felt by everyone; dishes break, doors swing
5.5–6.1	500	VI and VII		Slight building damage; plaster cracks, bricks fall
6.2–6.9	100	VII and IX		Much building damage; chimneys fall; houses move on foundations
7.0–7.3	15	X		Serious damage, bridges twisted, walls fractures; many masonry buildings collapse
7.4–7.9	4	XI		Great damage; most buildings collapse
>8.0	One every 5–10 yr	XII		Total damage, waves seen on ground surface, objects thrown in the air

[a] Mercalli numbers are determined by the amount of damage to structures and the degree to which ground motions are felt. These depend on the magnitude of the earthquake, the distance of the observer from the epicenter, and whether an observer is in or out of doors.

Alaska, in 1946, a tsunami struck Hawaii. The wave traveled at a velocity of 800 km/h. Although the amplitude of the wave in the open ocean was less than 1 m, the amplitude increased dramatically as the wave approached land. When it hit Hawaii, the wave had a crest 18 m higher than normal high tide. This destructive wave demolished nearly 500 houses, damaged a thousand more, and killed 159 people.

Modified Mercalli Scale

Because damage to the land surface and to human property is so important, the scale of earthquake-damage intensity (called the **Modified Mercalli Scale**) is based on the amount of vibration people feel during low-magnitude quakes and the extent of building damage during high-magnitude quakes. The correspondence between Mercalli intensity, Richter magnitude, and the estimated number of earthquakes is listed in Table 16.2. Note that an earthquake has one magnitude, but it can vary in intensity of damage (and hence the Mercalli Scale estimate) depending on distance from the epicenter.

WORLD DISTRIBUTION OF EARTHQUAKES

Although no part of the Earth's surface is exempt from earthquakes, several well-defined **seismic belts** are subject to frequent earthquake shocks (Fig. 16.16). Of these, the most obvious is the *circum-Pacific belt*, for it is here that about 80 percent of all recorded earthquakes originate. The belt follows the mountain chains in the western Americas from Cape Horn to Alaska, crosses to

Asia where it extends southward down the coast, through Japan, the Philippines, New Guinea, and Fiji, where it finally loops far southward to New Zealand. Next in prominence, giving rise to 15 percent of all earthquakes, is the Mediterranean-Himalayan belt, extending from Gibraltar to Southeast Asia. Lesser belts follow the midocean ridges.

Seismic belts are places where a lot of the Earth's internal energy is released. Therefore, it might be expected that other manifestations of internal energy would also appear in these belts, and indeed some of them do. Midocean ridges, deep-sea trenches, andesitic volcanoes, and many other features that outline the margins of plates of lithosphere either coincide with or else closely parallel these margins. Compare Figure 16.16 with Figure 1.15 to see that seismic belts outline the plate boundaries.

The depths of earthquake foci around plate edges are also informative. Most foci are no deeper than 100 km, because, as already mentioned, earthquakes occur in brittle rocks and around plate edges the brittle lithosphere is only 100 km thick. However, a few earthquakes do originate at depths as great as 700 km. It is noteworthy that these deep earthquakes are not associated with either oceanic ridges or transform faults. Rather, they are related to seafloor trenches. Those trenches mark the places where cool, brittle lithosphere sinks down into the mantle.

Benioff Zone

Detailed study of deep-earthquake foci beneath a seafloor trench (Fig. 16.17) shows that the foci follow a well-defined pathway called a **Benioff zone**, named after the scientist who first recognized this phenomenon.

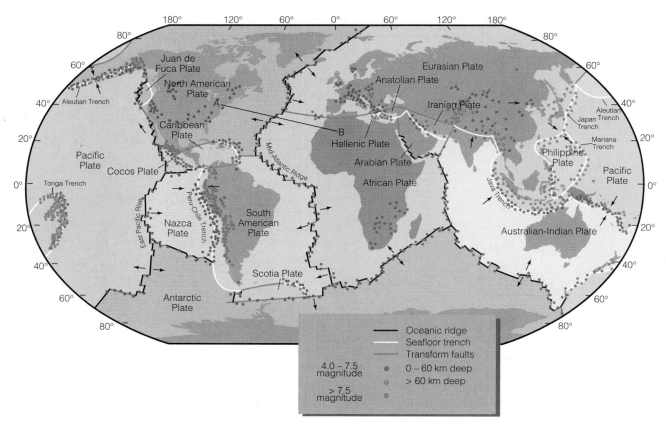

Figure 16.16 Earthquakes and Plate Margins The Earth's seismicity outlines plate margins. This map shows earthquakes of magnitude 4.0 or greater from 1960 to 1989.

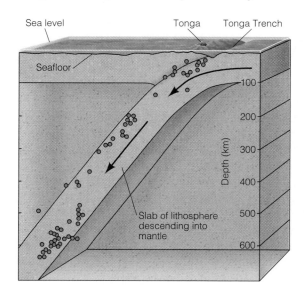

Figure 16.17 Benioff Zone Earthquake foci beneath the Tonga Trench, Pacific Ocean, during several months in 1965. Each circle represents a single earthquake. The earthquakes define the Benioff zone and are generated by downward movement of a comparatively cold slab of lithosphere.

This important observation strongly suggests that deep earthquakes originate within the relatively cold, downward-moving plate of lithosphere. Because some earthquake foci can be as deep as 700 km, it must be concluded that rapidly descending lithosphere can retain at least some brittle properties to that depth. The reason why earthquakes have not been detected at depths below 700 km remains unexplained. It may be that a rapidly sinking slab of lithosphere is sufficiently hot by the time it reaches a depth of 700 km that it has become ductile rather than brittle.

Locations of earthquakes reveal a great deal about the shapes and structures of tectonic plates. But they provide a static picture, a sort of snapshot of the way things are at the moment. In order to discover how the plates move and respond to forces, it is helpful to include other observations that can be obtained from seismic studies—most importantly, first-motion studies.

First-Motion Studies of Earthquakes

By careful study of earthquake waves recorded by seismographs, it is possible to tell the direction of motion of the fault that caused the earthquake. The information that is needed is contained in the arrival records of the seismic body waves.

A.

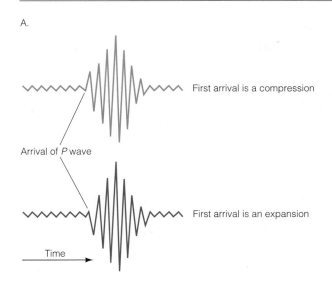

First arrival is a compression

Arrival of *P* wave

First arrival is an expansion

Time

B.

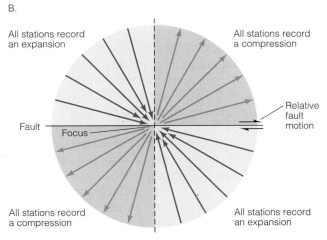

All stations record
an expansion

All stations record
a compression

Fault

Focus

Relative
fault
motion

All stations record
a compression

All stations record
an expansion

Figure 16.18 Motion on a Fault Initial motion of seismic body waves used to determine the direction of movement on a fault. A. Initial motion of a P wave detected by a seismograph is either a push away from the focus (that is, arrival of a compression) or a pull toward the focus (arrival of an expansion). B. Plotting the first motions detected at a number of seismograph stations allows the direction of movement on a fault to be uniquely determined. The example shown is for right-lateral motion on a strike-slip fault.

Consider the P-wave record. If the first arrival is a compressive pulse, the release of elastic energy and the fault motion must be *toward* the seismograph (Fig. 16.18A). If it is an expansion, the fault motion must be *away* from the seismograph. Figure 16.18B shows the effect of an earthquake caused by movement on a strike-slip fault. It is apparent that the first P-wave motion observed depends on the location of the seismograph. The fault movement can be determined by plotting first motions from several seismographs. The radiation pattern of body waves is in three dimensions. It is not possible to distinguish up-down movement from back-forth movement using P waves

alone. A 90° ambiguity is inherent in the P-wave radiation pattern. Because two possibilities exist—up-down versus back-and-forth—two independent measurements are needed. The measurement needed is the strike and dip of the fault, and that can be determined from the locations of several foci on the same fault surface.

S-wave oscillations also carry the signature of the direction of the first motion and can provide independent data that can be used to resolve any ambiguity.

EARTHQUAKES AND PLATE TECTONICS

The combined evidence of quake locations and the first motions of earthquakes provide some of the most detailed proof we have that the plate tectonics theory is correct. As shown in Figure 16.16, plate boundaries are outlined by narrow seismic belts. Plate boundary *motions* can be determined by first-motion studies.

We learned in Chapter 2 that there are three kinds of plate boundaries: (1) *divergent boundaries*, or *spreading centers*, which coincide with rift valleys and midocean ridges; (2) *transform fault boundaries*; and (3) *convergent boundaries*, which coincide with oceanic trenches and/or continental collision zones. Each kind of boundary is characterized by earthquakes that are distinctive as to fault motions and to the depth of foci. Refer back to Figure 2.7 which shows the kinds of boundaries and associated earthquakes.

Spreading Centers

Along a spreading center, two plates move apart from each other, and the lithosphere is thin and it is stretched by tensional stresses (Fig. 2.7A). The kinds of faults associated with tensional stresses are normal faults. First-motion studies confirm that earthquakes associated with spreading centers are due to normal faulting.

Earthquakes at spreading centers tend to have low Richter magnitudes and foci that are invariably less than 100 km deep and commonly less than 20 km. This indicates that the lithosphere is thin beneath a spreading center and that the ductile asthenosphere must come close to the surface.

Convergent Boundaries

As we discussed in Chapter 2, convergent plate boundaries are of two kinds: (1) subduction boundaries where lithosphere capped by oceanic crust is subducted into the asthenosphere and mesosphere; and (2) collision boundaries where two continents collide. Each kind of convergent boundary has a characteristic pattern of seismic activity, and each is complicated.

When oceanic lithosphere is subducted, it is subjected to complex stresses and different kinds of earthquakes can occur (see Fig. 2.7B). Lithosphere bends downward as it is subducted, and the bending causes normal faults in the upper part of the plate. Earthquakes associated with these faults all have very shallow foci, small Richter magnitudes, and normal fault first-motions. Overall, however, subduction involves one plate sliding beneath the other, and so the boundary between the two plates is a thrust fault. Earthquakes to a depth of at least 100 km (the region where the two plates of lithosphere are in contact) often have large Richter magnitudes and invariably have first motions that indicate thrust faulting. Below 100 km, where the subducted lithosphere is sinking through the asthenosphere, earthquakes occur *in* the subducted slab. Some earthquakes indicate tensional stresses (normal faults), whereas others indicate compressional stresses (reverse faults). The existence of deep earthquakes in subducted slabs due to tensional forces suggests that lithosphere sinks under its own weight.

Collision boundaries are the places where two continents collide. The Himalayan mountain chain, between India and Asia, is a present-day collision boundary. A zone of collision tends to be a region several hundred kilometers wide, and within it rocks are intensely compressed and thrust-faulted, as shown in Figure 2.7C. Within a collision zone the lithosphere is locally thickened. Earthquakes may have foci as deep as 300 km and may also have high Richter magnitudes and first motions that indicate thrust movements.

Transform Fault Boundaries

As we learned in Chapter 15, transform faults are huge, vertical, strike-slip faults that cut down through the lithosphere. They are boundaries where two plates slide past each other (Fig. 2.7D).

First-motion studies confirm that motion along transform fault boundaries is strike slip. Earthquakes always have shallow foci—that is, they are no deeper than 100 km—and they often have high Richter magnitudes. The location of earthquake foci suggests that when a transform fault cuts continental crust, a system of parallel faults rather than a single fracture can develop. This seems to be the case for the San Andreas Fault.

EARTHQUAKE PREDICTION

Some of the most dreadful natural disasters have been caused by earthquakes. It is hardly surprising, therefore, that a great deal of research around the world focuses on earthquakes. The hope is that through research we will be able to improve our forecasting ability.

Because China has suffered so many terrible earthquakes, Chinese scientists have made exhaustive efforts to predict quakes. They have even observed animal behavior, and on one occasion animals did successfully foretell a quake. On July 18, 1969, zookeepers at the People's Park in Tianjin observed highly unusual animal behavior. Normally quiet pandas screamed, swans refused to go near water, yaks did not eat, and snakes would not go into their holes. The keepers reported their observations to the earthquake prediction office, and at about noon on the same day a magnitude 7.4 earthquake struck.

There have been many informal reports of strange animal behavior before earthquakes, but the Tianjin quake is the only well-documented case. Unfortunately, most quakes do not seem to be preceded by anything odd. While scientists haven't given up on animals, they measure lots of other things besides animal behavior.

Most research on earthquake prediction now deals with changes that can be monitored in the properties of elastically strained rocks—properties such as rock magnetism, electrical conductivity, and porosity. Even simple observations, such as the level of well water, might indicate changes in porosity. Tilting of the ground or slow rises and falls in elevation may also indicate that strain is building up. Most significant are the small cracks and fractures that can develop in severely strained rock. These can cause swarms of tiny earthquakes—foreshocks—that may be a clue that a big quake is coming. One of the most successful cases of earthquake prediction, made by Chinese scientists in 1975, was based on slow tilting of the land surface, on fluctuations in the magnetic field, and on the swarms of small foreshocks that preceded the 7.3 Richter magnitude quake that struck the town of Haicheng. Half the city was destroyed, but authorities had evacuated more than a million people before the quake. As a result, only a few hundred were killed.

In places where earthquakes are known to occur repeatedly, such as along plate boundaries, patterns of recurrence can sometimes be discerned. If such a pattern suggests a recurrence interval of, say, a century, it may be possible to predict where and when a large quake may happen. Certainly it is possible to monitor such areas closely when a big quake is thought to be due. Studies of recurrence patterns have identified a number of *seismic gaps* around the Pacific Rim (Fig. 16.19). These are places where, for one reason or another, earthquakes have not occurred for a long time and where elastic strain is steadily increasing. Seismic gaps receive a lot of research attention from seismologists because they are considered to be the places most likely to experience large earthquakes.

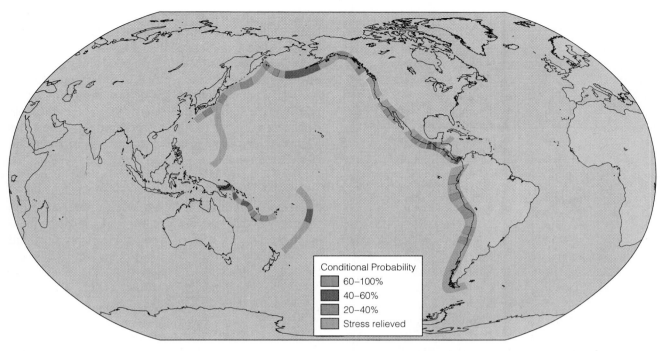

Figure 16.19 Where Big Earthquakes Are Expected Seismic gaps in the circum-Pacific belt. In the areas indicated, earthquakes of magnitude 7.0 or greater are known but have not occurred in recent times. Strain is now building up in each seismic gap, raising the probability that a large quake will occur soon.

USING SEISMIC WAVES AS EARTH PROBES

We have seen that P and S waves travel through rock at different speeds. These two types of waves also respond differently to changing rock properties. The arrival times of P and S waves at seismographs around the world provide records of waves that have traveled along many different paths. From such records, it is possible to calculate how the rock properties change and where distinct boundaries occur between layers having sharply different properties. Seismic waves are the most sensitive probes we have to measure the properties of the unseen parts of the crust, mantle, and core.

Layers of Different Composition

If the Earth's composition were uniform, and if no polymorphic changes occurred in the minerals present, the velocities of P and S waves would increase smoothly with depth. This is so because higher pressure leads to an increase in the density and rigidity of a solid, and it is these two properties that control the wave velocities. For an Earth of uniform composition, it is a straightforward matter to predict how long seismic waves would take to pass through. But observed travel times differ

greatly from such predictions. These differences can best be accounted for by supposing that velocities do not change smoothly with depth and that neither composition nor physical properties are constant throughout. Distinct boundaries (or discontinuities, as they are more commonly called) can be readily detected by refraction and reflection of body waves, as illustrated in Figure 16.7. From measurements of body waves, two major compositional boundaries have been detected. The first is the boundary between the crust and the mantle, and the second is that between the mantle and the core.

The Crust

Early in the twentieth century, the boundary between the Earth's crust and mantle was demonstrated by a scientist named Mohorovičić, who lived in what today is Yugoslavia. Mohorovičić noticed that, for earthquakes whose focus lay within 40 km of the surface, seismographs about 800 km from the epicenter recorded two distinct sets of P and S waves. He concluded that one pair of waves must have traveled from the focus to the station by a direct path through the crust, whereas the other pair represented waves that had arrived slightly earlier because they had been refracted by a boundary at some depth in the Earth. Evidently, the refracted waves had penetrated a zone of higher velocity below

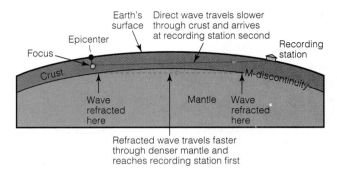

Figure 16.20 in diagram: Earth's surface, Epicenter, Focus, Direct wave travels slower through crust and arrives at recording station second, Recording station, Crust, M-discontinuity, Wave refracted here, Mantle, Wave refracted here, Refracted wave travels faster through denser mantle and reaches recording station first

Figure 16.20 Travel Paths Travel paths of direct and refracted seismic waves from shallow-focus earthquake to nearby seismograph station.

the crust, had traveled within that zone, and then had been again refracted upward to the surface (Fig. 16.20). Mohorovičić hypothesized that a distinct compositional boundary separates the crust from this underlying zone of different composition. Scientists now refer to this boundary as the **Mohorovičić discontinuity** and recognize it as the seismic discontinuity that marks the base of the crust. The feature is commonly called the **M-discontinuity**, and in conversation it is shortened still further to **moho**.

Thickness and Composition of the Crust
By seismic methods, it is possible to determine the thickness of the crust. Seismic wave speeds can be measured for different rock types in both the laboratory and

A.

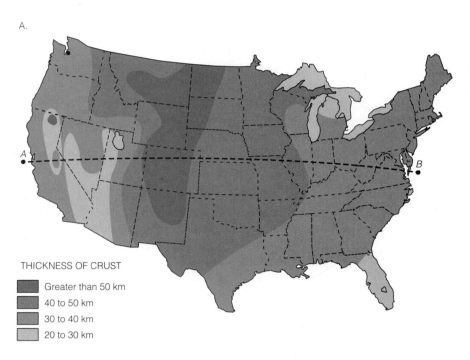

THICKNESS OF CRUST

- Greater than 50 km
- 40 to 50 km
- 30 to 40 km
- 20 to 30 km

Figure 16.21 Thickness of the Crust Crust beneath the United States. A. Thickness of crust is determined from measurements of seismic waves. B. Section through the crust along the line *A-B* (above). The crust tends to thicken beneath major mountain masses such as the Sierra Nevada, the Rocky Mountains, and the Appalachians. C. Profile of gravity traverse. The negative gravity anomalies over the Sierra, the Rockies, and the Appalachians are due to the roots of low-density rocks beneath these topographic highs.

B.

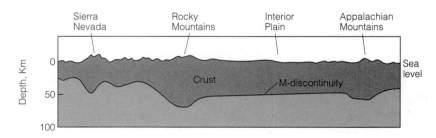

In diagram B: Sierra Nevada, Rocky Mountains, Interior Plain, Appalachian Mountains, Depth, Km, 0, 50, 100, Sea level, Crust, M-discontinuity

C.

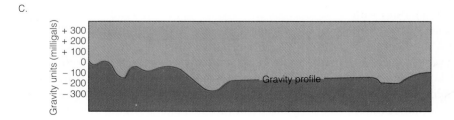

In diagram C: Gravity units (milligals), +300, +200, +100, 0, −100, −200, −300, Gravity profile

the field. When the speeds of waves received at a number of seismographs are calculated, laboratory measurements can be used to determine the depth of the moho and to estimate the probable composition of the crust. Beneath ocean basins the crust is less than 10 km thick. Elastic properties of the oceanic crust are those characteristic of basalt and gabbro. But in the continental crust both thickness and composition are very different. The continental crust ranges in thickness from 20 to nearly 60 km and tends to be thickest beneath major mountain masses (Fig. 16.21). Velocities in the continental crust are distinctly different from those in the oceanic crust. They indicate elastic properties like those of rock such as granite and diorite, although at some places just above the moho, velocities close to those of oceanic crust are often observed. These conclusions agree well with what is known about the composition of the crust from other lines of evidence, such as geological mapping and deep drilling. The agreement gives geologists confidence in drawing conclusions about the mantle where these other lines of evidence are scarce.

The Mantle

The mantle is something of an enigma. It is huge and controls much of what happens in the crust, but it cannot be seen. P-wave speeds in the crust range between 6 and 7 km/s. Beneath the moho, speeds are greater than 8 km/s. Laboratory tests show that rocks common in the crust, such as granite, gabbro, and basalt, all have P-wave speeds of 6 to 7 km/s. But rocks that are rich in dense minerals, such as olivine and pyroxene, have speeds greater than 8 km/s. We therefore infer that such rock, called peridotite, must be among the principal materials of the mantle. This inference is consistent with what little direct evidence is available concerning the composition of the upper part of the mantle. For example, some evidence can be obtained from rare samples of mantle rocks found in *kimberlite pipes—* narrow pipelike masses of intrusive igneous rock, sometimes containing diamonds, that intrude the crust but originate deep in the mantle (Fig. 16.22).

A.

B.

Figure 16.22 Samples from Deep in the Mantle The intrusion of kimberlite brings up fragments of rock from deep in the mantle. A. The shape and size of a kimberlite pipe can be judged from the hole left after the mining at Kimberley, South Africa. B. Kimberlite containing a crystal of diamond plus rounded fragments of mantle rock. The grayish-colored background material formed by crystallization of kimberlite magma.

The Core

Both P and S waves are strongly influenced by a pronounced boundary at a depth of 2900 km. When P waves reach that boundary, they are reflected and refracted so strongly that the boundary casts a P-wave shadow, which is an area of the Earth's surface, opposite the epicenter, where no P waves are observed (Fig. 16.23). Because this 2900-km boundary is so pronounced, geologists infer that it is the boundary between the mantle and the core. The same boundary casts an even more pronounced S-wave shadow. Here, however, the reason is not reflection or refraction, but the fact that shear waves cannot traverse liquids. Therefore, the huge S-wave shadow lets us conclude that the outer core is liquid.

Seismic waves cannot tell us the composition of the core, but they help us to deduce what it might be. Seismic-wave speeds calculated from travel times indicate that rock density increases slowly from about 3.3 g/cm^3 at the top of the mantle to about 5.5 g/cm^3 at the base of the mantle. The mean density of the whole Earth is 5.5 g/cm^3. Therefore, to balance the less dense crust and the mantle, the core must be composed of material with a density of at least 10 to 11 g/cm^3. The only common substance that comes close to fitting this

requirement is iron. Suggestive evidence also comes from meteorites. Iron meteorites are samples of material believed to have come from the core of an ancient, tiny planet, now disintegrated. All iron meteorites contain a little nickel, and the Earth's core presumably does too. Because S waves do not travel beyond the core-mantle boundary, it is inferred that the outer core is molten, that its composition is mostly iron, but that nickel and small amounts of other elements may be present too.

The Inner Core

P-wave reflections indicate the presence of a solid inner core enclosed within the molten outer core. The two cores appear to be essentially identical in composition. The reason for the change from liquid to solid probably relates to the effect of pressure on the melting temperature of iron. As the center of the Earth is approached, pressure rises to a value millions of times greater than atmospheric pressure. Temperature also rises but not steeply enough to offset the effect of pressure. From the base of the mantle (at a depth of 2900 km) to a depth of 5350 km, temperature and pressure are so balanced that iron is molten. But at a depth of 5350 km, another strong reflecting and refracting boundary occurs, one that has properties consistent

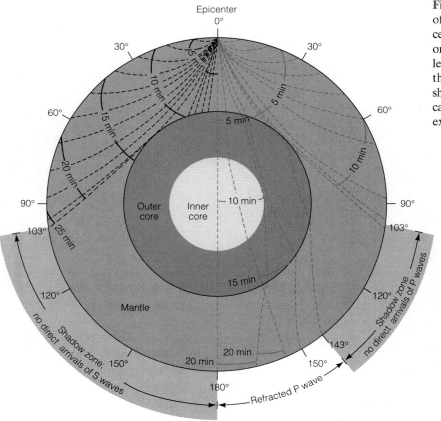

Figure 16.23 Seismic Shadow Zones Paths of P waves from an earthquake focus with epicenter at 0° shown in red in right half of section only. Paths of S waves shown in black in the left half. Reflection and refraction of P waves at the mantle-core boundary create a P-wave shadow zone from 103° to 143°. Because S waves cannot pass through a liquid, an S-wave shadow exists between 103° and 180°.

with a change from a liquid to a solid. Apparently, from 5350 km to the center of the Earth, rising pressure overcomes rising temperature, and iron is solid, creating the solid core.

Layers of Different Physical Properties in the Mantle

As far as we can determine, there are no pronounced compositional boundaries within the mantle. Nevertheless, as shown in Figure 16.24, seismic-wave velocities do not increase regularly from the base of the crust to the core-mantle boundary. There are sudden changes in speed that are apparently due to changes in the physical properties of the mantle.

The Low-Velocity Zone

The P-wave velocity at the top of the mantle is about 8 km/s, and it increases to 14 km/s at the core-mantle boundary. This increase is not smooth and constant, however. From the base of the crust to a depth of about 100 km, the P-wave velocity rises slowly to about 8.3 km/s. However, the velocity then starts to drop slowly to a value just below 8 km/s, and it remains low to a depth of about 350 km. This zone of reduced velocity between 100 and 350 km is not sharply defined and is better developed beneath the oceans than beneath the continents. This low-velocity zone can be seen as a small blip in both the P- and S-wave velocity curves in Figure 16.24. No evidence exists to suggest that the density decreases or the composition changes in this zone. To account for the velocity changes, therefore, we infer that the zone has a composition similar to that of the mantle immediately above and below, but that it is less rigid, less elastic, and more ductile than the adjacent regions. Evidence to support this idea comes from studies of the dispersion of surface waves. Long-wavelength surface waves that reach down more than 100 km are affected by a soft, ductile region in the mantle.

A possible explanation for the low-velocity zone is that between 100 and 350 km the geothermal gradient reaches temperatures close to the onset of partial melting of mantle rock. If this explanation is correct, either the rock strength drops sharply at temperatures close to melting, or melting starts and a small amount of liquid develops and forms very thin films around the mineral grains, thus serving as a lubricant. The amount of melting, if it occurs at all, must be very small, because the low-velocity zone does transmit S waves and we know that S waves cannot pass through liquids. Any liquid, like a thick film of oil, merely serves to lubricate the mineral grains in the mantle and at the same time to reduce wave velocities by reducing the elastic properties.

An integral part of the theory of plate tectonics is the idea that plates of lithosphere slide over a somewhat plastic zone in the mantle. The importance of the low-velocity zone for the theory is that such a zone proves the existence of the asthenosphere. The top of the low-velocity zone coincides with the base of the lithosphere. Thus, the low-velocity zone coincides with the asthenosphere.

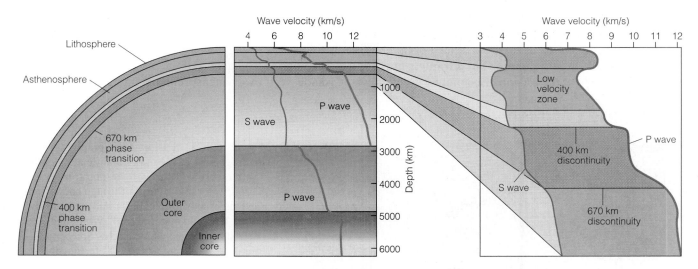

Figure 16.24 Velocity Changes with Depth Variation in seismic-wave velocity within the Earth. Changes occur at the boundaries between crust and mantle and between mantle and core, owing to change in composition. Another change occurs at a depth of 100 km and corresponds to the lithosphere-asthenosphere boundary. Changes also occur at 400 km and 670 km.

The 400-km Seismic Discontinuity

From the P- and S-wave curves in Figure 16.24 it is apparent that the velocities of both P and S waves increase sharply at about 400 km. Sharp though it seems, the increase is not sharp enough to be accounted for by a change in composition; the cause must be something else. A probable explanation is suggested by laboratory experiment. When olivine is squeezed at a pressure equal to that at a depth of 400 km, the atoms rearrange themselves into a denser polymorph. This process of atomic repacking caused by changes in pressure and temperatures is called a *polymorphic transition* (Chapter 3). In the case of olivine, the repacking involves a change to a structure resembling that found in a family of minerals called the spinels, of which magnetite is a well-known example. The structural repacking involves a 10 percent increase in density. The increase in seismic-wave velocities at 400 km is likely caused by the olivine-spinel polymorphic transition because the density increase, determined from seismic-wave velocities, is almost exactly 10 percent.

The 670-km Seismic Discontinuity

An increase in seismic-wave velocities—particularly P-wave velocity—also occurs at a depth of 670 km. This one is difficult to explain. The observed increase in mantle density is about 10 percent, but the boundary is diffuse. It is not yet possible to determine from either seismic evidence or laboratory experiment whether the boundary is due to a polymorphic transition, to a compositional change, or to both. Some scientists suggest that the increase at 670 km results solely from a polymorphic transition involving the repacking of atoms in the pyroxene minerals present in mantle rocks. Others suggest that the diffuse boundary indicates a polymorphic change affecting all silicate minerals present. One idea involves the rearrangement of silicon and oxygen to create denser anions in which each silicon atom is surrounded by six oxygen atoms rather than four. An intriguing hypothesis for the origin of the 670 km discontinuity involves evidence from earthquakes. The deepest earthquakes have foci of about 700 km, which is very close to the 670-km seismic discontinuity. Deep earthquakes are associated with sinking slabs of cool, oceanic lithosphere. Perhaps slabs of lithosphere are buoyant at 670 km and form a diffuse compositional boundary.

Opposing the point of view that slabs of lithosphere float buoyantly at 670 km is evidence from a research technique called *seismic tomography*. The method is similar to that used in medicine, where a three-dimensional picture of the interior of the human body is developed from slight differences in the intensities of X rays passing through in different directions. (CAT scan is the common name for X-ray tomography.) In a similar manner, heterogeneities in the mantle can be revealed by measuring slight differences in the velocities of seismic waves (Fig. 16.25). Seismic tomography does not reveal the presence of a compositional boundary at 670 km. For the present, we must conclude that the cause of the 670-km seismic discontinuity is unknown.

GRAVITY ANOMALIES AND ISOSTASY

The Earth is not a perfect sphere; careful measurement reveals that it is an ellipsoid that is slightly flattened at the poles and bulged at the equator.

The radius at the equator is 21 km longer than at the poles. Because the gravitational pull between two objects is inversely proportional to the square of the distance between their centers of mass, the pull exerted by the Earth's gravity on a body at the Earth's surface is slightly greater at the poles than it is at the equator. Thus, a man who weighs 90.5 kg at the north pole would observe his weight decreasing slowly and steadily to 90 kg simply by traveling to the equator. If the weight-conscious traveler made very exact measurements as he traveled, how-

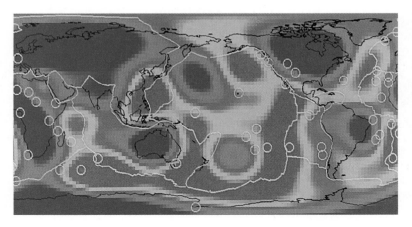

Figure 16.25 Seismic Tomography Lateral heterogeneity in the upper mantle at a depth of 150 km revealed through seismic tomography. Seismic waves travel faster through cooler, more rigid material (shown in blue), and more slowly in hotter, less rigid material (red). White lines show plate boundaries; white circles are centers of long-lived volcanic activity. Note that the red, low-velocity zones lie beneath spreading centers.

ever, he would observe that his weight changed irregularly rather than smoothly. From this he could conclude that the pull of gravity must change irregularly. If the traveler went one step further and carried a sensitive device called a **gravimeter** (or *gravity meter*) for measuring the pull of gravity at any locality, he would indeed find an irregular variation.

Gravity Anomalies

Gravimeters are similar to inertial seismographs. They consist of a heavy mass suspended by a sensitive spring (Fig. 16.26). When the ground is stable and free from vibrations due to earthquakes, the pull exerted on the spring by the heavy mass provides an accurate measure of the gravitational pull. Modern gravimeters are incredibly sensitive. The most accurate devices in operation can measure variations in the force of gravity as tiny as one part in a hundred million (10^{-8}).

In order to compare the pull of gravity from point to point on the Earth, corrections must be applied to gravimeter measurements for changes in latitude and topography. The idea behind the corrections is to know the pull of gravity at a constant distance from the center of the Earth. Then, if the rock mass between the gravimeter and the center of the Earth were everywhere the same, the adjusted figures for the force of gravity might be expected to be the same at every place on the Earth. In fact, the adjusted figures reveal large and significant variations called **gravity anomalies**. The anomalies are due to bodies of rock having differing densities. A simple example of an anomaly is shown in Figure 16.27. A great deal of important information can be derived from the anomalies.

The thickness of the crust beneath the United States, as determined from seismic measurements of the moho, is shown again in Figure 16.21B. Beneath the three major mountain systems (the Appalachians, the Rockies, and

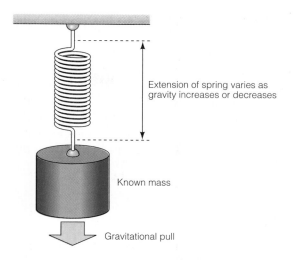

Figure 16.26 Measuring the Pull of Gravity A gravimeter is a heavy mass of metal suspended on a sensitive spring. The mass exerts a greater or lesser pull on the spring as gravity changes from place to place, extending the spring more or less. The mass of metal and the spring are contained in a vacuum together with exceedingly sensitive measuring devices.

the Sierra Nevada), the crust is thicker than in the non-mountainous regions of the country. In profile, the crust beneath the mountains resembles icebergs that have high peaks above the waterline but also massive roots below. The accuracy of this analogy is demonstrated by the gravity profile across the United States, shown in Figure 16.21C. Negative gravity anomalies are observed where the crust is thickest. The anomalies are caused by the masses of low-density rock in the mountains, just as the basin of low-density sediment produces the gravity anomaly shown in Figure 16.27.

The reason why a root of low-density rock forms in the first place provides some interesting insights into the Earth's physical properties. Mountains stand high and have iceberg-like roots beneath them because they are

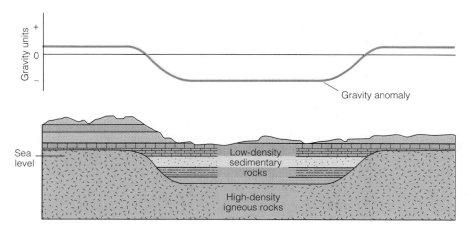

Figure 16.27 Gravity Anomaly Example of a gravity anomaly; a basin filled with low-density sedimentary rocks sitting on a basement of high-density igneous rocks. Gravity measurements reveal a pronounced gravity low throughout the basin. The magnitude of the anomaly can be used to calculate the thickness of the sedimentary rocks.

comprised of low-density rocks and are supported by the buoyancy of weak, easily deformed, but more dense rocks below. Mountains are, in a sense, floating. But it is not the crust that is floating on the mantle. Rather, it is the lithosphere, capped by a mass of thickened crust, that floats on the asthenosphere. Strange as it may seem, the topographic variations observed at the surface of the Earth arise not from the *strength* of the lithosphere but rather from its *weakness* and from the buoyancy of the asthenosphere.

Isostasy

The property of flotational balance among segments of the lithosphere is referred to as **isostasy**. The great ice sheets of the last glaciation provide an impressive demonstration of isostasy. The weight of a large continental ice sheet, which may be 3 to 4 km thick, will depress the lithosphere. When the ice melts, the land surface slowly rises again. The effect is very much like pushing a block of wood into a bucket of thick, viscous oil. When the wood is released, it slowly rises again to an equilibrium position determined by its density. The speed of its rising is controlled by the viscosity of the oil. Just like the block of wood, glacial depression and rebound mean that rock must flow laterally in the asthenosphere when the ice depresses the lithosphere, and then must flow back again when the deforming force is removed (Fig. 16.28). From the fact that the land surface in parts of northeastern Canada and Scandinavia is still rising, even though most of the thick ice sheets that covered these areas during the last glaciation had melted away by 7000 years ago (Fig. 16.29), we infer that the flow must be slow and therefore that the asthenosphere must be extremely viscous.

Continents and mountains are composed of low-density rock, and they stand high because they are thick and light; ocean basins are topographically low because the oceanic crust is composed of denser rock. Isostasy and the fact that the continental crust is less dense than the oceanic crust are the reason why the Earth has two pronounced topographic levels, as shown in Figure 2.4.

The important point to be drawn from this discussion of isostasy is that the lithosphere acts as if it were "floating" on the asthenosphere. (*Floating* is not exactly the correct word because the Earth is solid, but the lithosphere is buoyant and acts as though it were floating.) Sometimes gravity measurements suggest that a mountain has been pushed up so rapidly that it is top-heavy and has too little root of low-density rock to counterbalance its upper mass. Sometimes, as in the seafloor

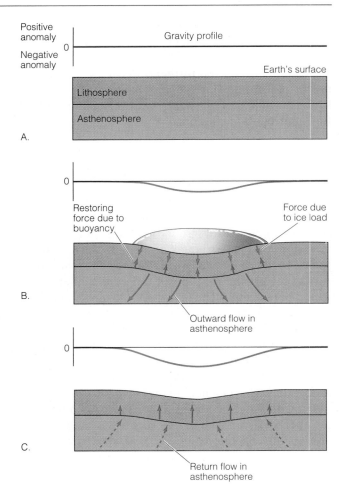

Figure 16.28 Ice Sheet Depresses the Lithosphere Depression of the lithosphere by a continental ice sheet. A. Prior to formation of the ice sheet, there is no gravity anomaly. B. When the ice sheet forms, it depresses the lithosphere. At some depth in the asthenosphere, material must slowly flow outward to accommodate the sagging lithosphere. C. When the ice melts, buoyancy slowly restores the lithosphere to its original level. A negative gravity anomaly continues until the depression is removed. The viscosity of the asthenosphere controls the rate of flow and therefore the rate of recovery.

trenches, it is observed that low-density crust has been dragged down so rapidly that it forms a root without a mountain mass above it. These and many other situations lead to local gravity anomalies. That the anomalies do not seem to become very large suggests that the Earth is always moving toward an isostatic balance. Indeed, isostasy is the principal explanation for vertical motions of the Earth's surface, just as plate tectonics is the principal explanation for lateral motions. In the next chapter we will discuss how the two combine to keep our dynamic Earth ever changing, ever active.

Figure 16.29 Ice Sheet Melts and Land Surface Rises Beach ridges raised by post-glacial uplift of the land in Nordaustlande, Svalbard, Norway. Such beaches provide clear evidence that the land has emerged from the sea after removal of the ice load.

SUMMARY

1. Abrupt movements of faults that release elastically stored energy cause earthquakes.

2. Earthquake vibrations are measured with seismographs.

3. Energy released at an earthquake's focus radiates outward as body waves, which are of two kinds—P waves (*Primary* waves, which are compressional) and S waves (*Secondary* waves, which are shear waves). Earthquake energy also causes the surface of the Earth to vibrate because of surface waves.

4. The focus and epicenter of an earthquake can be located by measuring the differences in travel times between P and S waves.

5. The amount of energy released during an earthquake is calculated on the Richter magnitude scale. The calculation is made from seismograph records of the seismic body waves.

6. Ninety-five percent of all earthquakes originate in the circum-Pacific belt (80%) and the Mediterranean-Himalayan belt (15%). The remaining 5 percent are widely distributed along the midocean ridges and elsewhere. Seismic belts outline tectonic plates; seismic-wave first-motion studies are used to determine the directions of movement across a plate boundary.

7. Seismic body waves can be refracted and reflected just as sound and light waves are. From the study of seimic-wave refraction and reflection, scientists infer the internal structure of the Earth by locating boundaries, or discontinuities, in its composition and physical properties. Pronounced compositional boundaries occur between the crust and mantle and between the mantle and outer core.

8. The base of the crust is a pronounced seismic discontinuity called the Mohorovičić discontinuity. Thick-

ness of the crust ranges from 20 to 60 km in continental regions but is less than 10 km beneath oceans.

9. Within the mantle there are two zones, at depths of 400 and 670 km, where sudden density changes produce seismic-wave discontinuities. The change at 400 km is probably produced by a polymorphic transition of olivine. The 670-km change might be due to either a polymorphic transition, a compositional change, or a combination of both.

10. The core has a high density and is inferred to consist of iron plus small amounts of nickel and other elements. The outer core must be molten because it does not transmit S waves. The inner core is solid.

11. From a depth of 100 km to 350 km there is a zone of low seismic-wave velocity that also causes pronounced dispersion of surface waves. This low-velocity zone coincides with the asthenosphere. The lithosphere, which is rigid and on average 100 km thick, overlies the asthenosphere.

12. The outer portions of the Earth are in approximate isostatic balance; in other words, like huge icebergs floating in water, the lithosphere "floats" on the asthenosphere.

THE LANGUAGE OF GEOLOGY

Benioff zone (p. 441)
body waves (p. 430)

earthquake focus (p. 430)
elastic rebound theory (p. 428)
epicenter (p. 430)

gravimeter (p. 451)
gravity anomalies (p. 451)

inertia (p. 429)
isostasy (p. 452)

M-discontinuity (p. 446)
Modified Mercalli Scale (p. 441)
moho (p. 446)
Mohorovičić discontinuity (p. 446)

P (for primary) waves (p. 431)

Richter magnitude scale (p. 434)

S (for secondary) waves (p. 431)
seismic belts (p. 441)
seismic sea waves (p. 440)
seismic waves (p. 430)
seismograph (p. 429)
seismology (p. 429)
surface waves (p. 430)

QUESTIONS FOR REVIEW

1. Explain how most earthquakes are thought to occur and why there seems to be a limit on earthquake magnitudes.

2. What is the relationship between an earthquake focus and the corresponding epicenter?

3. How are seismic waves recorded and measured? How would you locate an epicenter from seismic records? Explain how a focus is determined.

4. What are the differences between seismic body waves and surface waves? Identify two kinds of body waves and explain the differences.

5. Explain how seismologists use the Richter magnitude scale to estimate the energy released during an earthquake. In order for us to feel an earthquake, what minimum Richter magnitude must it have?

6. Earthquakes can cause damage in many ways; name four. Where on the Earth was the most disastrous earthquake on record and how did the people die? Where was the biggest known earthquake in the United States?

7. What are reflection and refraction, and how do they affect the passage of seismic waves? How can refraction and reflection be used to define the base of the crust? the core-mantle boundary?

8. Briefly describe how seismic waves can be used to infer that the outer core is molten while the inner core is solid. Why is the composition of the core thought to be largely metallic iron?

9. Under what circumstance is it possible to obtain samples of rocks from the mantle? How can such samples be used to check information about the mantle inferred from seismic-wave velocities?

10. What are seismic belts and how are they related to tectonic plates? Explain how seismic-wave records can be used to determine the motions of plate boundaries. How many kinds of plate boundaries are there, and what are their characteristic motions?

11. How do gravity anomalies arise, and how can they be measured?

12. Describe some evidence that proves that isostasy is operating in the Earth. How is the Earth's surface topography related to isostasy?

13. Draw an east-west profile of the shape of the crust under the United States and indicate how isostasy plays a role in what you have drawn.

Earthquake Hazards: The Wasatch Fault

The Wasatch Fault is a zone of major fault segments comprising one of the most active and longest normal fault zones in the world. Population growth in several adjacent communities has significantly increased over the past five years, resulting in the building of large subdivisions and related infrastructure at an alarming pace. Development is occurring in areas exposed to seismic hazards such as surface fault rupture and landslides.

OBJECTIVE: The primary objective of this case is to show how large surface-fault rupture earthquakes have impacted the landscape along Utah's Wasatch Front, specifically the formation of the WF, and how future earthquakes will affect the region's populace.

The Human Dimension: How do we coexist with earthquake hazards and mitigate the danger from a large earthquake?

Questions to Explore:

1. What are the physiographic, geologic, and tectonic characteristics of the Wasatch Fault?

2. What other hazards do surface fault rupture-producing earthquakes pose along Utah's Wasatch Range?

3. When will the next major earthquake along the Wasatch Fault occur?

17

Measuring
Our Earth

Understanding
Our Environment

GeoMedia

Climbing in the Southern Alps of New Zealand. The climber is scaling
Novara Peak. The snow-covered peak in the center is Mt. Cook, the one
to the right is Mt. Tasman. Hochstetter Glacier is between Mts. Cook
and Tasman; the glacier to the left of Mt. Cook is Ball Glacier. The photograph looks west.

Global Tectonics

Alpine Playgrounds

Everyone likes the mountains. As you hike along a trail through alpine meadows, you breathe the fresh, crisp mountain air. You look up to see rocky cliffs and majestic, snow-covered, cathedral-like peaks. Skiers thread their way down the powdery slopes. Across the valley, scoured by a river of glacial ice, you spot a rock-climber clinging to a sheer, near-vertical rock wall. Far below, canoeists test their skills against the turbulent waters of a glacier-fed stream. Wherever you look you see something different, something challenging, something majestic.

Mountains have awed, inspired, and challenged people throughout history. Some mountains are revered and held sacred; others are objects of superstition and fear. In the eighteenth century, a traveler described the Alps as producing in him "an agreeable kind of horror." His words evoke the sensation—common to many people—of being drawn to the mountains and yet, at the same time, feeling overwhelmed by their power. Today most people go to the mountains to feel renewed and revitalized. In a sense, people are drawn to the mountains because they confront us with the immense power and beauty of nature and the insignificance of humanity. Yet, at the same time, we feel a sense of closeness and connectedness with Earth processes when we are surrounded by mountainous peaks.

Each great mountain range in the world has its own unique characteristics, but they all have one important feature in common: they were created by the intense tectonic forces characteristic of collisions between crustal plates. Some—like the Andes—formed where an oceanic plate has been subducted under the edge of a continent, resulting in a chain of volcanic mountains. The Alps, the Appalachians, and the Himalayas, by contrast, formed as the result of grinding collisions between continents. If you go trekking in the Himalayas, don't be surprised to find fossils of shelled organisms high in the mountains, the remains of seafloor sediments uplifted by tectonic forces to form the highest peaks on the planet. Popular science writer John McPhee expressed the wonder and awesome power of the mountain-building process when he wrote: "If by some fiat I had to restrict all this writing to one sentence, this is the one I would choose: The summit of Mt. Everest is marine limestone."

Global tectonism has had fundamental consequences for the inhabitants of this planet, shaping the landscape, guiding the development of weather and climatic systems, and even influencing the course of the evolution of species. But nowhere is the power of tectonic forces more evident than in the great mountainous regions of the world.

CONTINENTAL DRIFT

People have wondered for a long time why continents have such irregular shapes and why ocean basins, mountain ranges, earthquake belts, and many other features occur where they do. When the first maps were made of the coastlines on either side of the Atlantic in the sixteenth century, it became apparent that the coasts were approximately parallel. People started to speculate why. They thought about a flood having cut an immense canyon—perhaps the great biblical flood. No realistic answers were forthcoming, but such speculations did get people thinking about why the Earth is the way it is. Scientists eventually began to think that there might be a single, underlying cause for the whole array of the Earth's major features. But what could that cause possibly be?

During the nineteenth century, people favored the idea that the Earth, originally a molten mass, had been cooling and contracting for centuries, with the crust being gradually compressed. These theorists pointed to mountain ranges full of folded strata as the places where past contraction had occurred and to seismic belts as places where contraction might be happening in the present. Contraction did explain some features, but it did not help with questions about the shapes and distribution of continents. Nor did it explain the great rift valleys and other features that are clearly caused by the crust having been stretched rather than compressed.

When scientists discovered at the beginning of the twentieth century that the Earth's interior is kept hot by radioactive decay, some of them suggested that the Earth might not be cooling but heating up (and therefore expanding). A much smaller Earth, they suggested, could once have been covered largely by continental crust. Heating would cause the Earth to expand, and the continental crust would then crack into fragments. As expansion continued, the cracks would grow into ocean basins, and through the cracks basaltic magma would rise up from the mantle to build new oceanic crust. The theory of an expanding Earth does offer a plausible explanation for the approximately parallel coastlines of adjacent continents, but it does not easily account for mountain ranges formed by compression.

To get around the flaws in both the expansion theory and the contraction theory, geologists began to examine the effects of other forces on the crust. By the middle of the twentieth century, however, all reasonable suggestions concerning the shapes and positions of continents seemed to have been exhausted. The time was ripe for a totally new approach. The new approach turned out be to plate tectonics. When great slabs of lithosphere—called plates—slide sideways across the asthenosphere, some parts of the slabs can be in compression, others in tension (that is, being pulled apart). When a plate splits in two, the broken edges of continental crust match perfectly. The energy needed to move plates turns out to be the Earth's internal heat energy, which causes great convective flows in the mantle.

Plate tectonics is the only hypothesis ever proposed that explains *all* of the Earth's major features.

A key proposal leading to the formulation of the plate tectonics theory was made early in the twentieth century, soon after the contraction theory collapsed. As we learned in Chapter 2, the German meteorologist Alfred Wegener proposed in 1912 that continents drift slowly across the surface of the Earth, sometimes breaking into pieces and sometimes colliding with each other. Collisions formed supercontinents, and breakages formed smaller continents. According to Wegener, today's continents are the broken fragments of the most recent supercontinent.

Pangaea

Wegener's theory of **continental drift** originated when he attempted, like many before him, to explain the striking match of the shorelines on the two sides of the

Atlantic, especially along Africa and South America. Wegener suggested that the most recent supercontinent existed during the Permian period when all the world's landmasses were joined together in a single continent, which he dubbed *Pangaea* (pronounced Pan-jeé-ah, meaning "all lands") (Fig. 17.1A). The northern half of Pangaea is called *Laurasia*, the southern half *Gondwanaland*. Laurasia is a name derived from *Lau-*

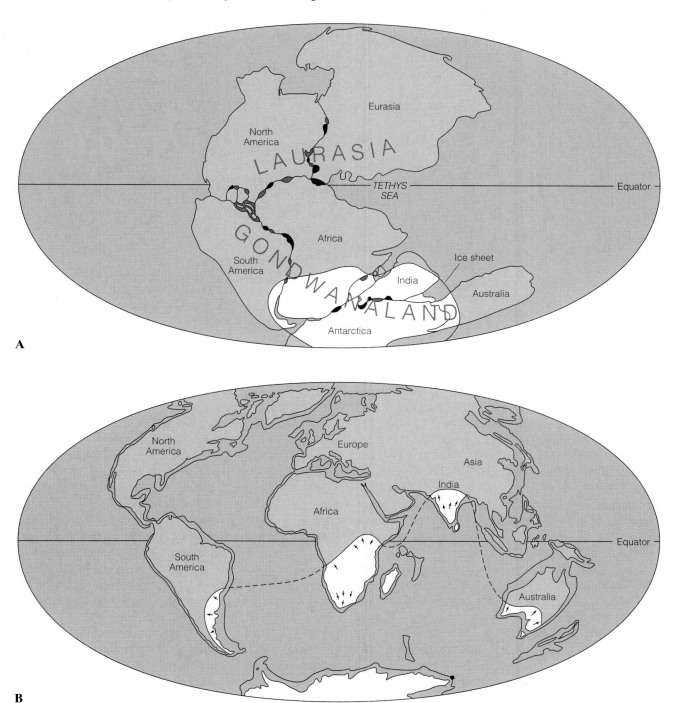

Figure 17.1 Pangaea The continents attained their present shapes when Pangaea broke apart 200 million years ago. A. The shape of Pangaea, determined by fitting together pieces of continental crust along a contour line 2000 m below sea level. (This is the line halfway down the continental slope, along which continental crust meets oceanic crust.) In a few places in this drawing, some overlap (black) occurs; elsewhere, small gaps (red) are found. These are places where post break-up events have modified the shapes of the continental margins. The white area is the region affected by continental glaciation 300 million years ago. B. Present continents and the 2000-m contour below sea level. The white areas are where evidence of the old ice sheets exists. Arrows show directions of movement of the former ice. The dashed line joining the glaciated regions indicates how large the ice sheet would have to have been if the continents were in their present positions at the time of glaciation.

rentia, an old name for the Precambrian core of Canada, and from *Eurasia*, a combined term for Europe and Asia. Gondwanaland is a name derived from a distinctive group of rocks found in central India. Similar rocks are found in Africa, Antarctica, Australia, and South America—this is one of the bits of evidence that suggest that India and today's southern hemisphere continents were once part of the same landmass. According to Wegener's hypothesis, Pangaea was somehow disrupted during the Mesozoic Era, and its fragments (the continents of today) slowly drifted to their present positions. Proponents of the theory likened the process to the breaking up of a sheet of ice that floats in a pond. The broken pieces, they argued, should all fit back together again, like pieces of a jigsaw puzzle. Figure 17.1A shows that a jigsaw reconstruction indeed works well.

One impressive line of evidence presented by Wegener that supports the former existence of Pangaea is that during the Late Carboniferous Period, about 300 million years ago, a continental ice sheet covered parts of South America, southern Africa, India, and southern Australia (Fig. 17.1B). However, if 300 million years ago continents were in the positions they occupy today, an ice sheet would have had to cover all the southern oceans and in places would even have had to cross the equator. Such a huge ice sheet could mean only that the world climate was exceedingly cold. Yet if the climate were cold, why has no evidence of glaciation at that time ever been found in the northern hemisphere? In the northern hemisphere thick coal measures are evidence of warm, tropical climates. This dilemma is explained neatly by continental drift: 300 million years ago, the regions covered by ice lay in high, cold latitudes surrounding the south pole and North America and Eurasia were close to the equator (Fig. 17.1A). No landmass covered the north pole, however, so there was no northern ice sheet. At that time, therefore, the Earth's climates need not have been greatly different from those of today.

Despite the impressive evidence supporting continental drift, many scientists remained unconvinced by Wegener's ideas, largely because no one could explain how the solid rock of a continent could possibly overcome friction and slide across the oceanic crust. The process is like trying to slide two sheets of coarse sandpaper past each other.

Apparent Polar Wandering

Wegener died in 1930, and although debate continued, its pace slowed down. A turning point came in the 1950s. From the mid-1950s to the mid-1960s, geophysicists made a number of remarkable discoveries. The first arose through studies of paleomagnetism. As we discussed in Chapter 8, certain igneous and sedimentary

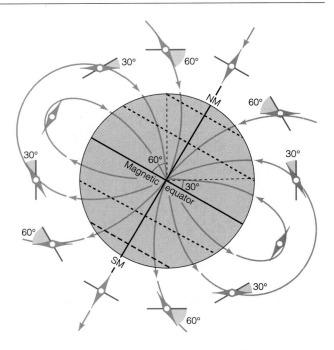

Figure 17.2 Magnetism and Latitude Change of magnetic inclination with latitude. The solid red diamonds show the magnetic inclinations of a free-swinging magnet. The solid blue line indicates a horizontal surface at each point.

rocks can become weakly magnetized and therefore preserve a fossil record of the Earth's magnetic field at the time and place the rocks formed. Three essential bits of information are contained in that fossil magnetic record. The first is the Earth's polarity—whether the magnetic field was normal or reversed at the time of rock formation. The second is the location of the magnetic poles at the time the rock formed. Just as a free-swinging magnet today will point toward today's magnetic poles, so too does paleomagnetism record the direction of the magnetic poles at the time of rock formation. The third piece of information, and the one that provides the data needed to say how far from the point of rock formation the magnetic poles lay, is the magnetic inclination, which is the angle with the horizontal assumed by a freely swinging bar magnet (Fig. 17.2). Note in Figure 17.2 that the magnetic inclination varies regularly with latitude, from zero at the magnetic equator to 90° at the magnetic pole. The paleomagnetic inclination is therefore a record of the place between the pole and the equator (that is, the **magnetic latitude**) where the rock was formed.

In the 1950s, geophysicists studying paleomagnetic pole positions found evidence suggesting that the poles wandered all over the globe. They referred to the strange plots of paleopole positions as *apparent polar wandering*. The geophysicists were puzzled by this evidence because the Earth's magnetic poles and the poles of rotation are close together. Determination of the mag-

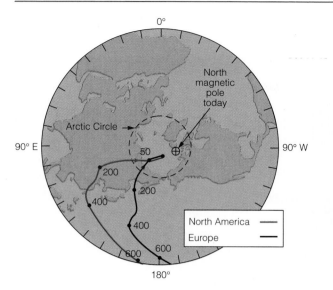

Figure 17.3 Apparent Polar Wandering Curves tracing the apparent path followed by the north magnetic pole through the past 600 million years. Numbers are millions of years before the present. The curve determined from paleomagnetic measurements in North America (red) differs from that determined from measurements made in Europe (black). Wide-ranging movement of the pole is unlikely; therefore, geologists conclude that it was the continents, not the pole, that moved.

netic latitude of any rock should therefore be a good indication of the geographic latitude at which the rock was formed. When it was discovered that the path of apparent polar wandering measured in North America differed from that in Europe (Fig. 17.3), geophysicists were even more puzzled. Somewhat reluctantly, they concluded that, because it is unlikely that the magnetic poles moved, the continents and the magnetized rocks moved. In this way, the hypothesis of continental drift was revived, but a mechanism to explain how the movement occurred was still lacking.

Seafloor Spreading

Help came from an unexpected quarter. All the early debate about continental drift, and even the data on apparent polar wandering, had centered on evidence drawn from the continental crust. But if continental crust moves, why shouldn't oceanic crust move too?

In 1962, Harry Hess of Princeton University hypothesized that the topography of the seafloor could be explained if the seafloor moves sideways, away from the oceanic ridges. His hypothesis came to be called the theory of **seafloor spreading** and was soon proved correct. Once again it was paleomagnetism that provided the proof.

Hess postulated that magma rose from the interior of the Earth and formed new oceanic crust along the mid-

ocean ridges. He could not explain what made the crust move away from the ridges, but he nevertheless proposed that it did and that as a consequence the oceanic crust far from any ridge was older than any crust nearer the ridge. A powerful test of the Hess theory was proposed by three geophysicists: Frederick Vine (who was a student at the time), Drummond Matthews (Vine's mentor), and Lawrence Morley (a Canadian scientist who made an independent discovery). The Vine–Matthews–Morley suggestion concerned the magnetism of the oceanic crust.

When lava is extruded at any midocean ridge, the rock it forms becomes magnetized and acquires the magnetic polarity that exists at the time the lava cools. If new lava is continuously making new oceanic crust, and if the crust is continuously moving away from the oceanic ridge, then this crust should contain a continuous record of the Earth's changing magnetic polarity. The oceanic crust is, in effect, a very slowly moving magnetic tape recorder. In fact, two oceanic tape recorders commence at each midocean ridge, one on each side of the ridge, in which successive strips of oceanic crust are magnetized with normal and reversed polarity (Fig. 17.4). It was a straightforward matter to match the sort of magnetic pattern observed in Figure 17.4 with a record of magnetic polarity reversals, such as that shown in Figure 8.19. The magnetic striping allowed the age of any place on the seafloor to be determined.

Because the ages of magnetic polarity reversals had been so carefully determined, magnetic striping also provided a means of estimating the speed with which the seafloor had moved. In some places, such movement was found to be remarkably fast: as high as 10 cm/yr.

PLATE TECTONICS: A NEW PARADIGM

Proof that the seafloor moves was the spur needed for the emergence of the theory of plate tectonics. The two essential points in formulating a theory of plate tectonics were, first, that the zone of low seismic wave velocities between 100 and 350 km deep (as discussed in Chapter 16) is exceedingly weak and has viscous fluid-like properties. It was quickly realized that the asthenosphere, which had been postulated many years earlier in order to explain isostasy but had never been proved, and the low-velocity zone must be one and the same. The second point was that the rigid lithosphere is strong enough to form coherent slabs (plates) that can slide sideways over the weak, underlying asthenosphere. These two points answered the main objection to Wegener's ideas— movement must occur with minimal resistance from friction. The lithosphere is much thicker than the crust,

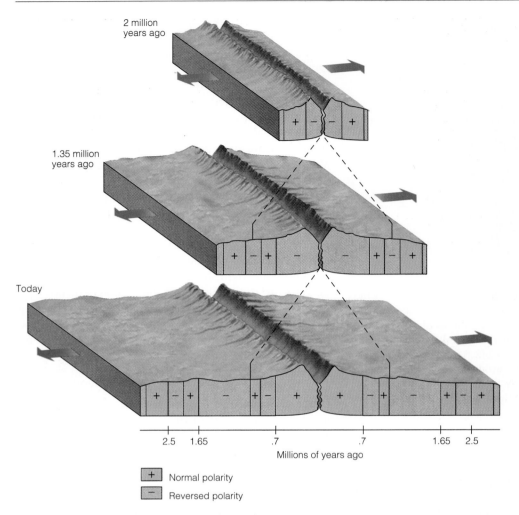

2 million years ago

1.35 million years ago

Today

2.5   1.65   .7   .7   1.65   2.5

Millions of years ago

| + | Normal polarity |
| - | Reversed polarity |

Figure 17.4 Oceanic Crust Records Magnetic Reversals Schematic diagram of oceanic crust. Lava extruded along an oceanic ridge forms new oceanic crust. As the lava cools, it becomes magnetized with the polarity of the Earth's magnetic field. Successive strips of oceanic crust have alternate normal (green) and reversed (brown) polarity. The ages of the reversals are the same as those in the magnetic polarity time scale shown in Figure 8.19.

however, and so one consequence of plate tectonics is that as the lithosphere moves, the crust is rafted along as a passenger. Continents move, to be sure, but they do so only as portions of larger plates, not as discrete entities.

Another consequence of the theory of plate tectonics provided a solution for one of the puzzles raised by seafloor spreading. If, as the theory of seafloor spreading required, new oceanic crust is being created along the midocean ridges, either the Earth must be expanding and the ocean basins getting larger, or else an equal amount of old crust must be being destroyed. The answer to the puzzle was provided by the previously unexplained Benioff zones (Chapter 16). These slanting zones of deep earthquake foci are the places where old, cold lithosphere is sinking back into the asthenosphere.

Destruction of old oceanic crust and the creation of new oceanic crust are in balance.

Structure of a Plate

The surface of the Earth is covered by six large and many small plates of lithosphere, each about 100 km thick (Fig. 1.15). The plates are rigid, or nearly so, and they move as single coherent units; that is, the plates do not crumple and fold like wet paper but act more like semirigid sheets of plywood floating on water. The plates may flex slightly, causing gentle up or down warping of the crust, but the only places where intense deformation occurs are at any edges along which plates

| TABLE 17.1 | Kinds of Plate Margins and Characteristic Features |

| Crust on Each Plate | Feature | Kind of Margin | | |
		Divergent	Convergent	Transform Fault
Oceanic–Oceanic	Topography	Oceanic ridge with central rift valley	Seafloor trench	Ridges and valleys created by oceanic crust
	Earthquake	All foci less than 100 km deep	Foci from 0 to 700 km	Foci as deep as 100 km
	Volcanism	Basaltic pillow lavas	Andesitic volcanoes in an arc of islands parallel to trench	Volcanism rare; basaltic along "leaky" faults
	Example	Mid-Atlantic Ridge	Tonga-Kermadec Trench: Aleutian Trench	Kane Fracture
Oceanic–Continental	Topography	—	Seafloor trench	—
	Earthquake	—	Foci from 0 to 700 km deep	—
	Volcanism	—	Andesitic volcanoes in mountain range parallel to trench	—
	Example	(No examples)	Western Coast of South America	(No examples)
Continental–Continental	Topography	Rift valley	Young mountain range	Fault zone that displaces surface features
	Earthquake	All foci less than 100 km deep	Foci as deep as 300 km over a broad region	Foci as deep as 100 km throughout a broad region
	Volcanism	Basaltic and rhyolitic volcanoes	No volcanism; intense metamorphism and intrusion of granitic plutons	No volcanism
	Example	African Rift Valley	Himalaya, Alps	San Andreas Fault

impinge on each other. Such plate margins are *active zones;* plate interiors are *stable regions.*

As we learned in Chapter 2, plates have three kinds of margins: divergent margins, or spreading centers along which two plates move apart from each other; convergent margins, along which two plates move toward each other; and transform fault margins, along which two plates simply slide past each other. Each type of margin creates distinctive topography in its vicinity and is associated with a distinctive kind of earthquake activity and volcanism. The features are summarized in Table 17.1.

Divergent Margins

As explained in Chapter 2, a divergent margin marks the growing edge of plates along which new lithosphere is created. It is here, at the spreading center, that newly formed oceanic crust becomes magnetized.

Magnetic Records and Plate Velocities

The most recent magnetic reversal recorded near the crest of a midocean ridge (in other words, the one closest to the ridge) occurred 730,000 years ago (Fig. 8.19). The oldest reversals so far found in oceanic crust date back to the middle Jurassic, about 175 million years ago. From the symmetrical spacing of magnetic time lines on the two sides of a midocean ridge (Fig. 17.5) it appears that both plates move away from a spreading center at equal rates. Appearances can be deceiving, however. The same pattern of magnetic time lines shown in Figure 17.5 would be observed if the African Plate (green) were stationary and both the Mid-Atlantic Ridge and the North American Plate (orange) were moving westward. (Later in this chapter, evidence will be presented to substantiate the suggestion that midocean ridges do indeed move.) In fact, all that can be deduced from magnetic time lines is the *relative velocity* of two plates. An answer to the question of *absolute velocities* requires

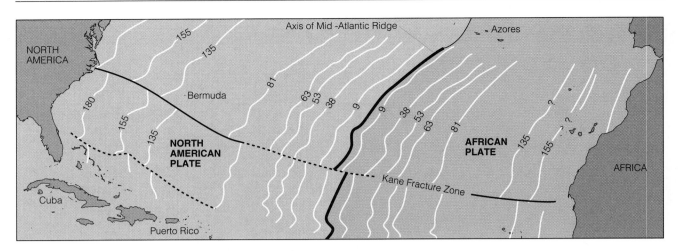

Figure   17.5   Age of North Atlantic   Age of the ocean floor in the central North Atlantic, deduced from magnetic striping. Numbers give ages in millions of years before the present. The Kane Fracture Zone, which is a transform fault, continues across the Atlantic and causes consistent displacement of the age contours.

more information, and we will return to this question too, later in the chapter.

Variations in Plate Velocities

The relative velocities of some plates are much greater than those of others (Fig. 17.6). The differences in speed appear to be related to the amount of continental lithosphere in a plate. Plates with only oceanic lithosphere tend to have high relative velocities. This is the case for the Pacific and Nazca plates. Plates with lots of thick continental lithosphere, such as the African, North American, and Eurasion plates, have low relative velocities.

A second reason plate velocities vary has to do with the geometry of motion on a sphere. One might think, intuitively, that all points on a plate move with the same velocity, but that is incorrect. Our intuition would be correct only if plates of lithosphere were flat and moved over a flat asthenosphere (like plywood floating on

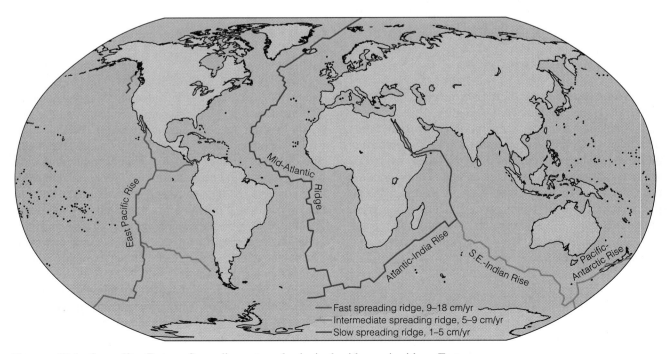

Figure   17.6   Spreading Rates   Spreading rates of principal midoceanic ridges. Fast spreading rates mean plates move away from each other between 9 and 18 cm/yr. Intermediate rates are 5 to 9 cm/yr.; slow rates are 1 to 5 cm/yr.

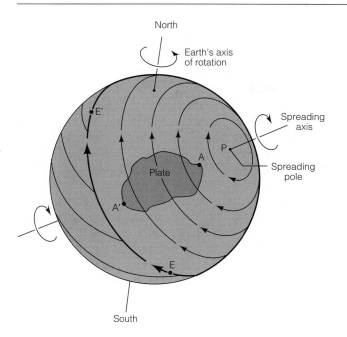

Figure 17.7 Movement on a Sphere Movement of a curved plate on a sphere. The movement of each plate of lithosphere on the Earth's surface can be described as a rotation about the plate's own spreading axis. Point *P* has zero velocity because it is the fixed point around which rotation occurs. Point *A´*, at the edge of the plate closest to the equator *EE´*, has a high velocity. Point *A*, closest to the pole, has a low velocity.

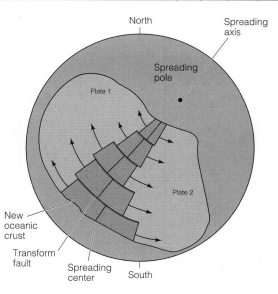

Figure 17.8 Geometry of Plate Motions Relationship between spreading axis, oceanic ridge, and transform faults in two adjacent plates. Plates 1 and 2 have a common spreading center (black) displaced by transform faults (red). Each segment of the oceanic ridge lies on a line of longitude that passes through the spreading pole. Each transform fault lies on a line of latitude with respect to the spreading pole. The width of new oceanic crust increases away from the spreading pole.

water). If this were true, then all points on the plate would move with the same velocity. However, plates of lithosphere are pieces of a shell on a spherical Earth; they are curved, not flat. In the geometry of a sphere, any movement on the surface is a rotation about an axis of the sphere. A consequence of such rotation is that different parts of a plate move with different velocities, as shown in Figure 17.7.

Plate A in Figure 17.7 moves independently of the Earth's rotation and instead rotates about an axis of its own, colloquially called a **spreading axis**. In the figure, point *P*, where the spreading axis reaches the surface, is a **spreading pole**.

The motion of each of the Earth's plates can be described in terms of rotation around a spreading axis, and the velocity of each point on the plate depends on its distance from the spreading pole. One consequence of different plate velocities is that the width of new oceanic crust bordering a spreading center increases with distance from the spreading pole (Fig. 17.8). A second consequence is that the projection of a spreading center passes through the spreading pole. Such a projection is analogous to a line of longitude. A third consequence is that each transform fault lies on a line analogous to a line of latitude around the spreading pole.

Topography of the Seafloor

The topography of the seafloor is controlled by the growth and movement of plates. Two prominent features in particular are related to spreading centers.

The first feature is the midocean ridges. The shape of any ridge is strongly influenced by the rate of spreading. Fast spreading rates, 9 to 18 cm/yr, mean that new oceanic crust is created very rapidly. This in turn means that magma must rise rapidly and continuously from below and that large magma chambers must lie at shallow depths below the center of the ridge. As a result, a fast-spreading center is thermally inflated and stands high above the seafloor. By contrast, a slow-spreading center is cooler and less inflated. The overall ridge at a slow center still stands high above the deep ocean floor, but the central rift is wider and more pronounced than the central rift in a fast-spreading center.

The second prominent feature is the ocean floor itself. A large fraction of the heat that escapes from the Earth's interior does so along spreading centers. As a result, not only the midocean ridges but also adjacent portions of the seafloor are high points because the lithosphere beneath them is thermally expanded. As lithosphere moves away from a ridge, it cools and contracts. As contraction occurs, the distance of the seafloor below sea level increases. A constant depth below sea level is reached after about 100 million years, by which time oceanic lithosphere has cooled and reached ther-

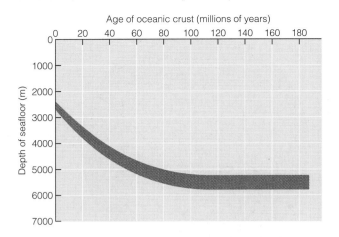

Age of oceanic crust (millions of years)

Figure 17.9 Seafloor Deepens with Age Depth of the seafloor in the world oceans as a function of the age of the oceanic crust. Near the spreading center, young lithosphere is thermally expanded. As it moves away from the midocean ridge, the lithosphere cools and contracts. The ocean becomes deeper as a result.

mal equilibrium (Fig. 17.9). To a first approximation, therefore, the depth of the ocean floor provides an estimate of the age of the oceanic crust.

Convergent Margins

Subduction zones were defined in Chapter 2 as the edges along which plates of lithosphere turn down into the mantle. Because subduction removes a plate from view, what happens as the plate sinks into the asthenosphere can only be inferred. Some of the evidence on which we can base inferences comes from seismic studies, some from the deformed rocks where the two converging plates meet, and a good deal from the kind of volcanism associated with subduction zones.

Magmatic Arcs

As a subducted plate descends, it is heated and eventually hydrous minerals such as amphiboles and chlorites breakdown and release water which, as discussed in Chapter 4, causes wet partial melting to start in the mantle above the subducted plate. This process forms andesitic magma. Rising to the surface, the magma forms a chain of stratovolcanoes like those shown in Figures 4.13 and 4.36. The arc-shaped region of magmatic activity is called a **magmatic arc**. It is parallel to the seafloor trench and separated from it by a distance of 100 to 400 km, the distance depending on the angle of dip of the descending plate. If the stratovolcanoes form on oceanic crust, the magmatic arc is also known as an **island arc**, but if the stratovolcanoes are built on continental crust, the magmatic arc is called a **continental vol-**

canic arc. The Japanese islands and the Aleutians are modern-day island arcs. The Andes are a continental volcanic arc. Some arcs are curved only slightly because they are parts of a circle with a large radius, and some are highly curved because they are parts of a circle having a smaller radius. The radius of curvature is an indication of the angle at which lithosphere is plunging back into the mantle. If the plunge is perpendicular to the Earth's surface, an island arc will be straight. If the angle of plunge is very shallow, the arc has a pronounced curvature.

Determination of the age of oceanic crust being subducted shows that old, cold, and therefore dense crust forms island arcs that have a large radius of curvature (implying a steep plunge). Young oceanic crust that has not reached thermal equilibrium and is less dense than older, colder crust forms arcs with short radii of curvature (implying a shallow plunge angle). This observation is informative because it indicates that oceanic lithosphere must be sinking under its own weight through the hot, weak asthenosphere. This means that old, cold lithosphere, when capped by oceanic crust, must be more dense than the hot, plastic asthenosphere. The older and colder the lithosphere, the faster the rate of sinking and the steeper the angle of the Benioff zone.

As lithosphere sinks, it must heat up. Earthquakes can occur in the downgoing slab as long as it is cool enough to be brittle. With a rapid sinking rate of 8 cm/yr, calculations show that lithosphere retains some brittle properties down to a depth of 700 km (Fig. 17.10). This is probably the reason that a few earthquake foci are as deep as 700 km.

Mélange

Many features on the Earth's surface occur as a result of deformation along convergent margins. A distinctive feature of some margins is the development of a **mélange**, a chaotic mixture of broken, jumbled, and thrust-faulted rock. Once a subduction zone forms and a seafloor trench is created, sediment accumulates in the trench. A sinking plate drags the sedimentary rock formed from this accumulated sediment downward beneath the overriding plate. Sedimentary rock has a low density. As a result, it is buoyant and cannot be dragged down very far. Caught between the overriding plate and the sinking plate, the sediment becomes shattered, crushed, sheared, and thrust-faulted to form a mélange (Fig. 17.11). As the mélange thickens, it becomes metamorphosed. The cold sedimentary rocks are dragged down so rapidly that they remain cooler than adjacent rock at the same depth. The kind of metamorphism that is common in many mélange zones, therefore, is that which occurs along curve C of Figure 7.14—a high-pressure, low-temperature, metamorphism distinguished by blue schists and eclogites. The blue color comes from a bluish amphibole called glaucophane.

Figure 17.10 Thermal Structure of Descending Lithosphere Computer-aided calculations of the temperature of a descending slab of cool lithosphere. Contours depict the temperature. A plate 100 km thick, descending at an angle of 45° and a rate of 8 cm/yr, will cool the surrounding mantle (as indicated by the bending of the temperature contours) but will slowly become heated as it sinks. Between 600 and 700 km, the temperature of the tip of the descending slab reaches the temperature of the adjacent mantle and earthquakes cease.

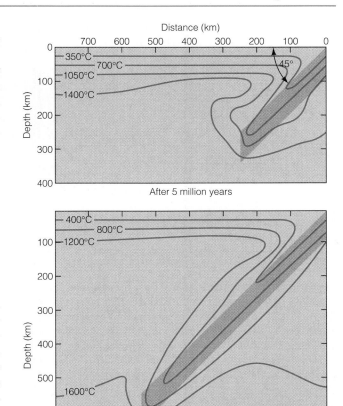

After 5 million years

After 10 million years

Fore-Arc Ridges and Fore-Arc Basins

Between the trench and the magmatic arc, both of which are prominent topographic features, two less prominent features are present along many convergent margins: the fore-arc ridge and the fore-arc basin (Fig. 17.12). A **fore-arc ridge** is commonly underlain by mélange and is caused by a local thickening of the crust due to thrust faulting at the edge of the overriding plate. A **fore-arc basin** is a low-lying region between the fore-arc ridge and the magmatic arc. The island of Sumatra in Indonesia is part of a magmatic arc that is flanked by a fore-arc ridge and basin (Fig. 17.13).

Back-Arc Basins

When the sinking rate of a subducting plate is faster than the forward motion of the overriding plate, part of the overriding plate can be subjected to tensional (pulling) stress. The leading edge of the overriding plate must remain in contact with the subduction edge or else a

huge void will open. What happens is that the overriding plate grows slowly larger at a rate equal to the difference in velocities between the two plates. Most

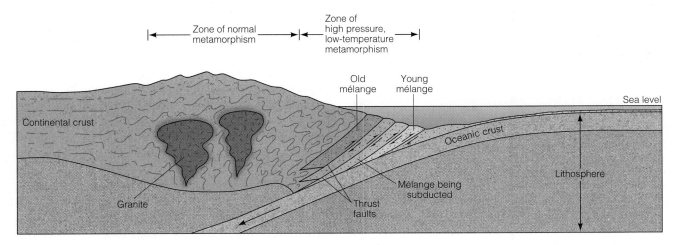

Figure 17.11 Mélange of Broken Rock A mélange is formed when young sediment in a trench is smashed by moving lithosphere and dragged downward in slices bounded by thrust faults. As successive slices are dragged down, older mélange, closer to the overriding plate, is pushed back up. The process is like lifting a deck of cards by adding new cards at the base of the deck.

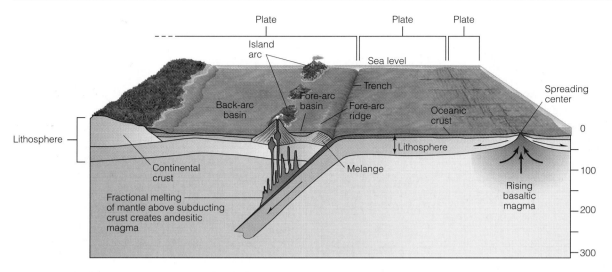

Figure 17.12   Features of a Convergent Plate Margin
Structure of a tectonic plate at a convergent margin. Lithosphere is capped by oceanic crust formed by basaltic magma rising from the asthenosphere. Moving laterally, the lithosphere accumulates a thin layer of marine sediment and eventually starts sinking into the asthenosphere. Along the line of subduction, an oceanic trench is formed, and sediment deposited in the trench, plus sediment from the moving plate, is compressed and deformed to create a mélange. The sinking oceanic crust eventually reaches the temperature where wet partial melting commences and forms andesitic magma, which then rises to form an island arc of stratovolcanoes on the adjacent plate. Behind the island arc, tensional forces lead to the development of a back-arc basin.

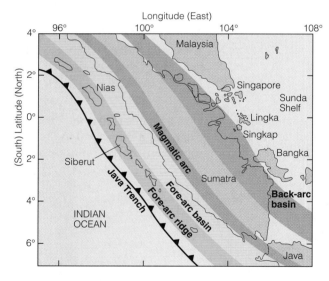

Figure 17.13   Modern Convergent Margin   Map of portion of Sumatra showing the positions of the major topographic features in a present-day convergent plate boundary. Compare with Figure 17.12. The fore-arc ridge is often underlain by mélange. The fore-arc and back-arc basins both tend to be filled with sediment derived from the magmatic arc.

commonly, this process is manifested by a thinning of the crust and an opening of an arc-shaped basin behind and parallel to the magmatic arc (Figs. 17.12 and 17.13). Basaltic magma may rise into such a **back-arc basin,** and a small region of new oceanic crust may even form.

Transform Fault Margins

The faults at plate margins are transform faults. As discussed in Chapter 16, these are huge, vertical, strike-slip faults cutting down into the lithosphere. They can form when either a spreading center or a convergent margin fractures the lithosphere (Fig. 17.14). The plates on either side of transform fault margins smash and abrade each other like two strips of sandpaper. As a result, the faults are marked by zones of intensely shattered rock. Where the faults cut oceanic crust, they make elongate zones of narrow ridges and valleys on the seafloor. They also influence topography by forming narrow valleys where they cut continental crust, but here the features are less pronounced than on the seafloor.

The sliding movement of transform faults causes a great many shallow-focus earthquakes, some of them of high magnitude. Most transform faults do not have any volcanic activity associated with them. Occasionally, however, a small amount of plate separation does occur and a "leaky" transform fault results in a small amount of volcanism.

The best known transform fault in North America is the San Andreas Fault in California. It is the largest of several transform faults that join segments of the mid-ocean ridge called the East Pacific Rise. Notice in Figure 2.14 how the transform faults and segments of the east Pacific Rise separate the North American Plate from the Pacific Plate. Figure 2.14 also shows that the San Andreas is only one of several faults that break the continental crust. The others are subsidiary to the

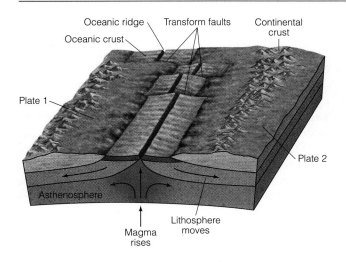

Oceanic ridge   Transform faults   Continental crust

Oceanic crust

Plate 1

Plate 2

Asthenosphere

Magma rises   Lithosphere moves

Figure 17.14 Transform Faults Transform faults form when an oceanic ridge forms. The shapes of the continental margins reflect faults now found along the oceanic ridge.

San Andreas, however, and are part of a fault zone that is roughly parallel to the main fault. Movement along the San Andreas Fault arises from movement between the North American and Pacific plates. Baja California and the portion of the state of California west of the fault are on the Pacific Plate. That plate is moving northwest, relative to the North American Plate, at a rate of several centimeters per year. In about 10 million years Los Angeles will have moved far enough north so as to be opposite San Francisco. In about 60 million years, at the present rate of movement, the segment of continental crust on which Los Angeles lies will have become

separated completely from the main mass of continental crust that makes up North America.

HOT SPOTS AND ABSOLUTE MOTIONS

It was pointed out earlier in this chapter that plate motions determined from magnetic time lines are only relative motions. In order to determine absolute motion, an external frame of reference is necessary. A familiar example of absolute versus relative motion occurs when one automobile overtakes another. If observers in the two automobiles could see only each other and could not see the ground or any fixed objects outside their cars, they could judge only the *difference* in velocity between the two cars. One car could be traveling 50 km/h and the overtaking car at 55 km/h, but all the observers could determine is that the *relative velocity* is 5 km/h. On the other hand, if the observers could measure velocity with respect to a stationary reference, such as the ground surface, they could determine that the *absolute velocities* were 50 and 55 km/h.

We would be constrained to determine only relative plate velocities if a fixed reference framework did not exist. Fortunately, a reasonable framework does exist. During the last century, the American geologist James Dwight Dana observed that the age of volcanoes in the Hawaiian chain increases from southeast to northwest (Fig. 17.15). When Harry Hess proposed that the seafloor moves, J. T. Wilson, the Canadian scientist who named

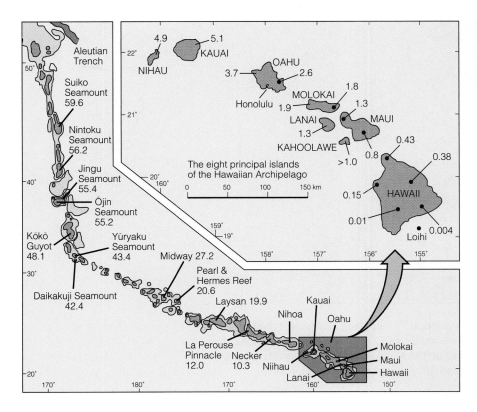

Figure 17.15 Chain of Shield Volcanoes Hawaiian-Emperor chain, showing the oldest reliable ages (in millions of years) for basaltic rocks of the shield volcanoes.

transform faults, pointed out that Dana's observation provided a possible test of the Hess hypothesis. Wilson postulated that a deep, long-lived magma source (a **hot spot**) lies somewhere far down in the mantle. If the seafloor moves, a volcano could remain in contact with the magma source only for about a million years. A chain of volcanoes should therefore be a consequence of lithosphere moving over a fixed hot spot. The Hawaiian islands and the chain of seamounts to the northwest provide compelling evidence in favor of Wilson's suggestion. (The evidence also indicates that about 40 million years ago the Pacific Plate apparently changed its direction of motion.)

Wilson made his suggestion as a way of testing seafloor spreading. However, it was not long before he realized that if long-lived hot spots do exist deep in the mantle, they might provide a series of fixed points against which absolute plate motions can be measured. More than a hundred hot spots have now been identified (Fig. 17.16). Using them for reference, geologists have found that the African Plate is very nearly stationary (evidenced by the fact that volcanoes there seem to be very long lived). Because the African Plate is almost completely surrounded by spreading edges, and because the relative velocities along the encircling ridges are closely matched, we must conclude that the Mid-Atlantic Ridge is moving westward and that the oceanic ridge that runs up the center of the Indian Ocean is moving to the east. If the absolute motion of the African Plate is zero or nearly so, the Mid-Atlantic Ridge in the southern Atlantic Ocean must be moving westward at the rate of about 2 cm/yr, and the absolute velocity of the South American Plate must be 4 cm/yr.

The Australian–Indian Plate is moving almost directly northward. All other plates, with the exception of the nearly stationary African Plate, are moving approximately eastward or westward. Several plates do not have convergent margins and must therefore be increasing in size. Most of the modern subduction zones are to be found around the Pacific Ocean along the edge of the Pacific Plate, and thus much of the oceanic lithosphere now being destroyed is in the Pacific. It follows then that the Indian Ocean, the Atlantic Ocean, and most other oceans must be growing larger, while the Pacific Ocean must be steadily getting smaller.

CAUSE OF PLATE TECTONICS

Just as Wegener could not explain what made continents drift, we are still unable to say exactly how convection currents make plates of lithosphere move. The situation is analogous to knowing the shape, color, size, and speed of an automobile, and knowing that gasoline supplies the energy needed for movement, but not knowing how the gasoline makes the engine work. Until the driving mechanism is explained, plate tectonics must

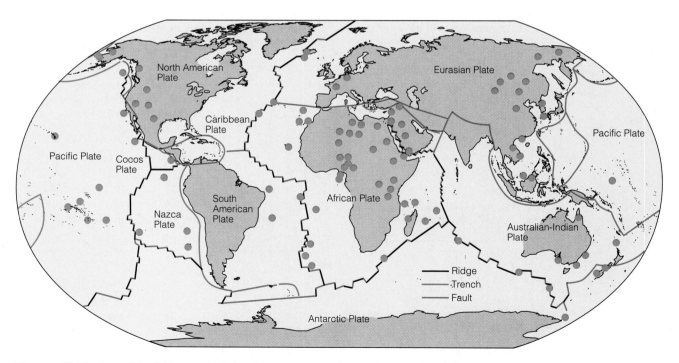

Figure 17.16 Long-Lived Centers of Volcanism Long-lived hot spots at the Earth's surface, each a center of volcanism, are believed to lie above deep-seated sources of magma in the mantle. Because the magma sources lie far below the lithosphere and so do not move laterally, hot spots can be used to determine the absolute motions of plates.

remain only an approximate description. Meanwhile, we can hypothesize about the causes of the motion and test the hypotheses by making detailed calculations based on the laws of nature.

The lithosphere and asthenosphere are closely bound together. If the asthenosphere moves, the lithosphere must move too, just as the movement of sticky molasses moves a piece of wood floating on the surface of the molasses. Also, movement of the lithosphere causes movement in the asthenosphere. Such is our state of uncertainty, however, that we cannot yet separate the relative importance of the two effects. However, on two points we can be quite certain: (1) the lithosphere must have kinetic energy in order to move, and (2) the source of this kinetic energy is the Earth's internal heat. We know, too, that the heat energy reaches the surface by convection in the mantle. What has not yet been figured out is the precise way convection and plate motions are linked.

Convection in the Mantle

The mantle is solid rock; however, below the lithosphere it is hot and weak enough so that, under low strain rates, even small stresses make it flow like a viscous liquid. Like a liquid, too, the mantle is subject to convection currents when a local source of heat causes a mass of rock to become heated to a higher temperature than surrounding rock. The heated mass expands, becomes less dense, and rises very slowly. To compensate for the rising mass, rock that is cooler and denser must sink downward. The rate at which heat reaches the Earth's surface can be accounted for only if convection in the mantle brings heat from the deep interior.

Any discussion about convection is speculative. Evidence from seismic tomography and heat flow indicates that convection of some sort does occur beneath the lithosphere. Even so, it is difficult to see how plate motions can be due entirely to convective motions. For this reason, most scientists believe that lithosphere motion is due to a combination of processes and that convection is only one of the processes. One important thing that convection must do is keep the asthenosphere hot and weak by bringing up heat from the deep mantle and core. In this sense at least, convection is essential for plate tectonics.

Movement of the Lithosphere

Three forces might play a role in moving the lithosphere. The first is a push away from a spreading center. Rising magma at a spreading center creates new lithosphere and in the process pushes the plates sideways (Fig. 17.17A). Once the process is started, it tends to keep

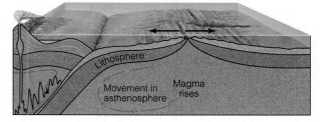

A. New lithosphere pushes plates apart

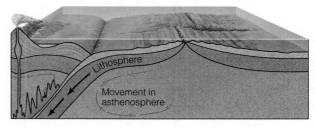

B. Cold dense lithosphere sinks and drags plate sideways

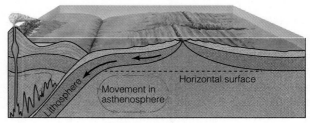

C. Plate of lithosphere slides down gentle slope

Figure 17.17 Why Lithosphere Moves Three suggested mechanisms by which lithosphere might move over the asthenosphere. A. Magma rising at a spreading edge exerts enough pressure to push the plates of lithosphere apart. B. A tongue of cold, dense lithosphere sinks into the mantle and pulls the rest of the plate behind it. C. A plate of lithosphere slides down a gently inclined surface of asthenosphere.

itself going. The problem is that pushing involves compression, but the existence of normal faults along a midocean ridge indicates a state of *tension*.

A second way by which lithosphere could be made to move is by dragging. Proponents of the dragging idea point out that a descending tongue of old, cold lithosphere is more dense than the surrounding hot asthenosphere. Because rock is a poor conductor of heat, they argue, the temperature at the center of a descending slab can be as much as 1000°C cooler than the mantle at depths of 400 to 500 km. The dense slab of lithosphere must therefore sink under its own weight and exert a pull on the entire plate. This is like a heavy weight that hangs over the side of a bed and is tied to the edge of a sheet. The weight falls and pulls the sheet across the bed. As lithosphere sinks and pulls the plate down, it places the sinking slab in tension. As described in Chapter 16, first-motion studies in Benioff zones do indicate some tensional forces. To compensate for the descending lithosphere, rock in the asthenosphere must flow slowly back toward the spreading edge (Fig. 17.17B).

Both the pushing and the dragging mechanism have problems, however. Plates of lithosphere are brittle, and they are much too weak to transmit large-scale pushing and pulling forces without major deformation occurring in their middle. Deformation is not present, however, and midplate seismicity, which would be expected for a plate undergoing deformation, is infrequent.

The third possible mechanism for lithosphere movement is for the plate to slide downhill away from the spreading center. The lithosphere grows cooler and thicker away from a spreading center. As a consequence, the boundary between lithosphere and asthenosphere slopes away from the spreading center. The slope is only 1 part in 100, but its own weight can cause the lithosphere to slide at a rate of several centimeters per year (Fig. 17.17C).

At present, there is no way to be sure of the relative importance of the three lithosphere mechanisms. Calculations suggest that each operates to some extent, so that the entire process is possibly more complicated than we now imagine. The prevailing idea at present is that subduction starts when old, cold lithosphere breaks and begins to sink, thereby pulling on the plate and starting the movement, and then the other processes combine to keep the movement going. Only future research will resolve the question.

PLATE TECTONICS, CONTINENTAL CRUST, AND MOUNTAIN BUILDING

In a sense, continental crust is simply a passenger being rafted on large plates of lithosphere. But it is a passenger that is buffeted, stretched, fractured, and altered by the ride.

Someone once characterized continental crust as the product of bump-and-grind tectonics. Each bump between two crust fragments forms a mountain belt, each grind a strike-slip fault, each stretch a rift valley. Scars left in the continental crust by bump-and-grind tectonics are evidence of former plate motions. That this evidence exists is fortunate because the most ancient crust known to exist in the ocean dates only from the mid-Jurassic Period. Indeed, the only direct evidence concerning geological events more ancient than the mid-Jurassic comes from the continental crust. In order to discuss how the continental crust has been affected by plate tectonics, it is helpful to look first at the large-scale structure of continents.

Regional Structures of Continents

Cratons

On the scale of a continent, two kinds of structural units can be distinguished within the continental crust. The first is a core of very ancient rock called a **craton** (Fig. 17.18). The term is applied to any portion of the Earth's crust that has attained tectonic and isostatic stability. Rocks within cratons may be deformed, but the deformation is invariably ancient.

Orogens

Draped around cratons are the second kind of crustal building unit, **orogens**, which are elongate regions of crust that have been intensely folded and faulted during continental collisions. Crust in an orogen is commonly thicker than crust in a craton, and many orogens—even some very old ones—have not yet attained isostatic equilibrium. Orogens are the eroded roots of ancient mountain ranges that formed as a result of collisions between the cratons. Orogens differ from each other in age, history, size, and details of origin; however, all were once mountainous terrains, and all are younger than the cratons they surround. Only the youngest orogens are mountainous today; ancient orogens, now deeply eroded, reveal their history through the kinds of rock they contain and the kind of deformation present.

Continental Shields

An assemblage of cratons and ancient orogens that has reached isostatic equilibrium is called a **continental shield**. That portion of a continental shield that is covered by a thin layer of little-deformed sediments is called a *stable platform*. North America has a huge continental shield at its core, and around the shield are four young orogens (Fig. 17.18). Because the North American shield crops out in Canada (especially Ontario and Quebec), but is mostly covered by flat-lying sedimentary rocks in the United States, geologists often refer to it as the Canadian Shield.

Through careful mapping and radiometric dating, geologists have identified several ancient cratons and orogens in the Canadian Shield (Fig. 17.18). Within the cratons, all rocks are older than 2.0 billion years. Such rocks can be observed in many places in eastern Canada, but within the United States cratonic rocks crop out only in a small region around Lake Superior. Nevertheless, by drilling through the cover of sedimentary rocks on the stable platform, geologists have discovered that cratons and orogens similar to those that surface in eastern Canada also lie below much of the central United States and part of western Canada.

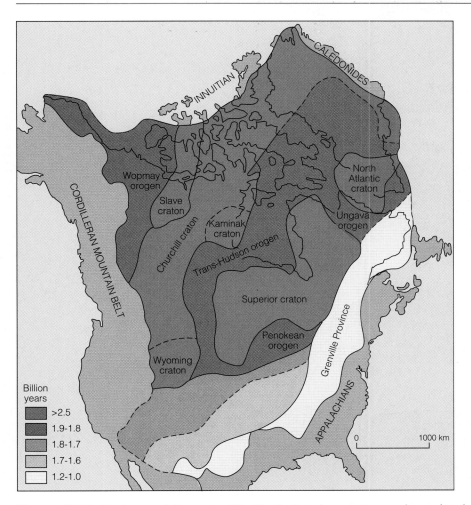

Figure 17.18 Cratons and Orogens   The North American cratons and associated orogens. The Caledonide, Appalachian, Cordilleran, and Innuitian orogens are each younger than 600 million years. The assemblage of ancient cratons and orogens that is surrounded by the four young orogens is the Canadian Shield.

The small cratons within the Canadian Shield shown in Figure 17.18 were probably minicontinents during the Archean Eon. By about 1.6 billion years ago, these minicontinents had become welded together to form the assemblage of cratons and ancient orogens we see in North America today. Each time two cratonic fragments collided, an orogen was formed between them. The existence of ancient collision belts—orogens—is the best evidence available to support the idea that plate tectonics operated at least as far back as 2 billion years ago.

Continental Margins

The fragmentation, drift, and welding together of pieces of continental crust are inevitable consequences of plate tectonics. Various combinations of these processes are responsible for the five types of continental margins we know of today: passive, convergent, collision, transform fault, and accreted terrane. Before we discuss additional evidence for plate tectonics, it will be helpful to review briefly the features associated with each of the continental margins.

Passive Continental Margins
A **passive continental margin** is one that occurs in the stable interior of a plate. The Atlantic Ocean margins of the Americas, Africa, and Europe are each passive continental margins. The eastern coast of North America, for example, is in the stable interior of the North American Plate, far from the plate margins. Passive continental margins develop when a new ocean basin forms by the rifting of continental crust, as illustrated in Figure 17.19. This process can be seen in the Red Sea, which is a young ocean with an active spreading center running down its axis (Fig. 17.20). New, passive continental margins have formed along both edges of the Red Sea.

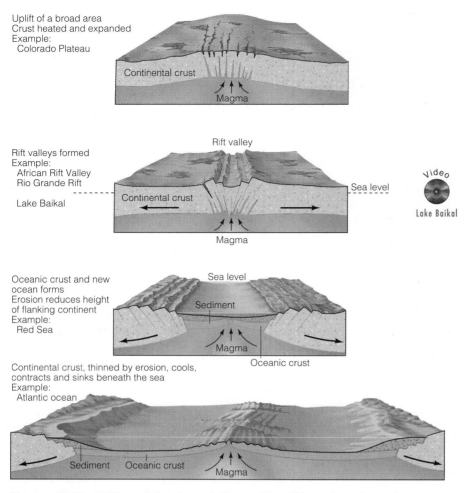

Uplift of a broad area
Crust heated and expanded
Example:
 Colorado Plateau

Rift valleys formed
Example:
 African Rift Valley
 Rio Grande Rift

Lake Baikal

Oceanic crust and new
ocean forms
Erosion reduces height
of flanking continent
Example:
 Red Sea

Continental crust, thinned by erosion, cools,
contracts and sinks beneath the sea
Example:
 Atlantic ocean

Figure 17.19 Rifting of Continental Crust The rifting of continental crust to form a new ocean basin bounded by passive continental margins. The rifting can cease at any stage. It is not necessarily correct to conclude that the African Rift Valley, for example, will open to form a new ocean.

Passive continental margins are places where great thicknesses of sediment accumulate. The kinds of sediment deposited are distinctive, and the Red Sea provides an example. Deposition commenced with clastic, nonmarine sediments followed by evaporites and then clastic marine shales. The sequence apparently arises in the following manner. Basaltic magma, associated with formation of the new spreading edge that splits the continent, heats and expands the lithosphere so that a plateau forms with an elevation of as much as 2.5 km above sea level. Tensional forces cause normal faults and form a rift so that there is a pronounced topographic relief between the plateau and the floor of the rift. The earliest rifting of the Red Sea must have looked very much the way the African Rift Valley looks today, although the sense of movement in the African Rift Valley is more complicated than simply expansion, as discussed in the essay *Watching a Continent Splinter*. Before the rift floor sank low enough for seawater to enter, clastic nonmarine sedi-

ments, such as conglomerates and sandstones, were shed from the steep valley walls and accumulated in the rift. Associated with these sediments are basaltic lavas, dikes, and sills, all formed by magma rising up the normal faults. As the rift widened, a point was reached where seawater entered. The early flow was apparently restricted, and the water was shallow, resembling a shallow lake more than an ocean. The rate of evaporation would have been high, and as a result strata of evaporite salts were laid down on top of the clastic nonmarine sediments. Finally, as rifting continued and the depth of the seawater increased, normal clastic marine sediments were deposited. This is the stage the Red Sea is in today. Eventually, as further rifting exposes new oceanic crust, the Red Sea will evolve into a younger version of the Atlantic Ocean.

Notice in Figure 17.20 that the Gulf of Aden, the Red Sea, and the northern end of the African Rift Valley meet at angles of 120°. Such a meeting point formed

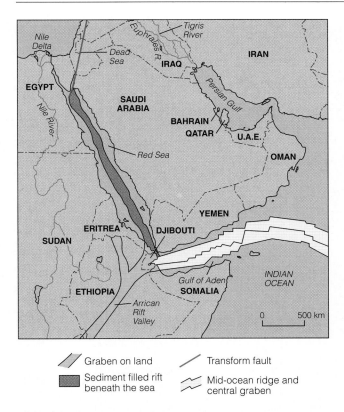

Graben on land

Sediment filled rift beneath the sea

Transform fault

Mid-ocean ridge and central graben

Figure 17.20 Triple Junctions Three spreading centers meet at a triple junction. Two, the Gulf of Aden and the Red Sea, are actively spreading, and there are passive continental margins along the adjacent coastlines. The African Rift, however, appears to be a failing rift that will not develop into an open ocean.

by three spreading edges is called **plate triple junction**. Two of the edges, the Gulf of Aden and the Red Sea, are active and still spreading. The third, the African Rift Valley, is apparently no longer spreading and possibly will not evolve into an ocean. What will remain on the African continent is a long, narrow sequence of grabens filled primarily with nonmarine sediment. The formation of three-armed rifts with one of the arms not developing into an ocean is apparently a characteristic feature of passive continental margins. This can be seen from Figure 17.21, which shows the reassembled positions of the continents flanking the Atlantic Ocean prior to breakup. Note that some of the world's largest rivers flow down valleys formed by undeveloped rifts associated with the opening of the Atlantic Ocean.

Continental Convergent Margins
A continental convergent margin is one where the edge of a continent coincides with a convergent plate margin in which oceanic lithosphere is being subducted beneath continental lithosphere. The Andean coast of South America is an example. On this coast, the Nazca Plate (capped by oceanic crust) is being subducted beneath the

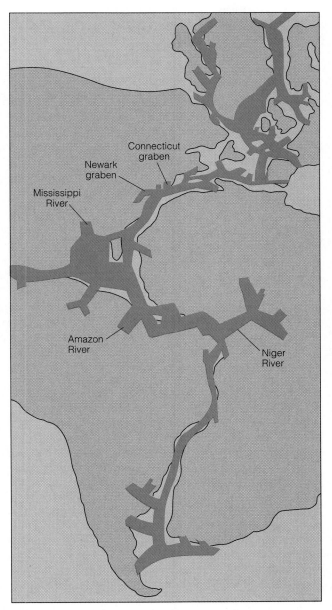

Figure 17.21 Rifts Around the Atlantic Map of a closed Atlantic Ocean showing the rifts formed when Pangaea was split by a spreading center. The rifts on today's continents are now filled with sediment. Some of them serve as the channelways for large rivers.

South American Plate (capped by continental crust). Wet partial melting in the mantle activated by water released by the subducted Nazca Plate produced the andesitic magma that formed the Andes (a continental volcanic arc).

Subduction produces intense deformation of a continental margin (together with characteristic magmatic activity and a distinctive style of metamorphism and deformation of sediments deposited in the trench). Sediment subjected to deformation in such a setting forms a mélange. The tectonic setting in which sediments are subjected to high-pressure, low-temperature metamor-

Measuring Our Earth

Watching a Continent Splinter

Travelers in East africa are awed by the Rift Valley. There, two sections of the Earth's crust are pulling apart, creating an enormous gash—more than 1800 meters deep in some places—that runs lengthwise from the Red Sea at the top of Africa to Mozambique in the south (Figure B17.1). For years geoscientists wondered why the rift disappears south of Mozambique. They now have an answer.

The Rift Valley marks the boundary between two tectonic plates: the Nubian Plate to the west and the Somalian Plate to the east. Such rifts are usually linked to other plate boundaries, but researchers were unable to trace the East African rift beyond the point where it disappeared in Mozambique. Unlike conditions along the rift to the north, there are no earthquakes or other signs of disturbance to the south. How, then, do these plates connect to the rest of the Earth's tectonic system?

Two researchers, Dezhi Chu of Exxon Production Research Company and Richard G. Gordon of Rice University, solved the mystery by studying rocks of the ocean floor south of Mozambique. In the seafloor between Africa and Antarctica there is a rift system that produces new igneous rock. The rift pushes the African plates apart but it also pushes them away from Antarctica. By using Antarctica as a reference point, the researchers were able to plot the motion of the Nubian and Somalian plates. They found that the two plates do not pull apart in a straight line; instead, they rotate around a pivot point east of South Africa. To the north, in the region of East Africa rift, they separate by 6 millimeters per year. To the south of the pivot point, they move toward each other at a rate of 2 millimeters per year, so slowly that they don't produce many large earthquakes.

Armed with a new understanding of the motion, or *kinematics*, of these two plates, geoscientists are in a better position to figure out how other tectonic plates interact. In so doing, they will create a fuller picture of the Earth's overall plate-tectonic scheme.

Figure B17.1 A Split in a Continent The African Rift Valley seen from the air. The view is looking north in northern Kenya. The cliff is a fault escarpment. In the distance is a small volcanic cone.

phism is in a mélange produced at a subduction zone. Adjacent and parallel to the belt of mélange, the crust beneath the continental volcanic arc is thickened and also metamorphosed, but here the metamorphism is regional metamorphism of the kind that occurs along curve B in Figure 7.14. One distinctive feature of a continental convergent margin, therefore, is a pair of parallel metamorphic belts.

The most distinctive feature of a continental convergent margin is the continental volcanic arc. Modern examples are in the chains of volcanoes in the Andes and the Cascade Range (Fig. 17.22). Where the volcanoes have been eroded and the deeper parts of the underlying magmatic arc exposed, granitic batholiths can be observed. They are remnants of the magma chambers that once fed stratovolcanoes far above. The strings of huge elongate batholiths that run from southern California to northern British Columbia (Fig. 4.34) provide a striking example.

Continental Collision Margins

A continental collision margin is one where the edges of two continents, each on a different plate, come into collision (Fig. 17.23). A modern example of a continental

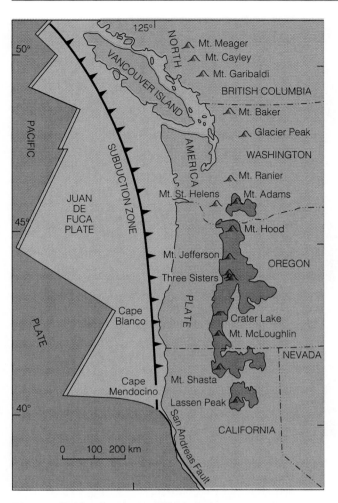

Figure 17.22 Continental Volcanic Arc Volcanoes of the Cascade Range, a continental volcanic arc. Each volcano has been active during the last 2 million years. Magma to form the volcanoes comes from wet partial melting above the oceanic crust on the Juan de Fuca Plate as it is subducted beneath the North American Plate.

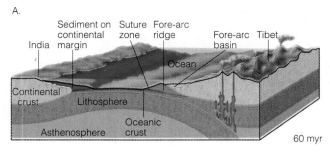

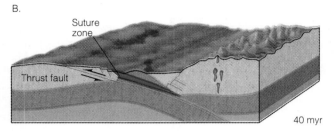

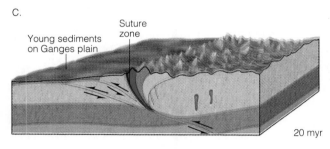

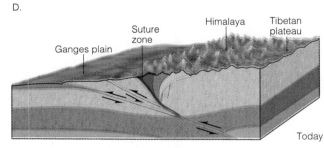

Figure 17.23 Collision of India with Asia Collision between two fragments of continental crust shown schematically for the collision between India and Tibet. Sediment is folded and faulted, and the lithosphere is thickened. The downward-moving plate of lithosphere capped by oceanic crust breaks off and continues to sink, but the edge of the remaining segment of plate, which is capped by buoyant continental crust, is partly thrust under the edge of the overriding plate, causing further elevation of the collision zone.

collision margin is the line of collision between the Australian–Indian Plate and the Eurasian Plate. India, on the Australian–Indian Plate, and Asia, on the Eurasian Plate, have collided and the Himalayan mountain chain is the result. As the essay *Continental Collisions and the Monsoons* demonstrates, highlands raised by collision can have dramatic effects on climate.

When continental crust is carried on a plate of lithosphere that is being subducted beneath a continental convergent margin, the two continental fragments must eventually collide. The collision sweeps up and deforms any sediment that accumulated along the margins of both continents and forms a mountain system characterized by intense folding and thrust faulting.

All modern continental collision margins are young orogens that are fold-and-thrust mountain systems. Occurring in great arc-shaped systems a few hundred

kilometers wide, fold-and-thrust mountain systems commonly reach several thousand kilometers in length. Within a fold-and-thrust mountain system, strata are compressed, faulted, folded, and crumpled, commonly in an exceedingly complex manner. Metamorphism and

Understanding Our Environment

Continental Collision and the Monsoons

For more than 2000 years, Arab traders have sailed to India during the hot summer months, a time when India and Southeast Asia are drenched by rain. They sail back home again during the pleasant, dry, winter months. They do so because summer winds blow from the west while winter winds blow from the east. The Arab name for this seasonal reversal of wind and weather is *mausim*, and from it we derive our word *monsoon*. The monsoon winds of India and Southeast Asia are a consequence of the uplift of the Himalaya and the Tibetan Plateau.

Computer modeling has shown that atmospheric circulation in the absense of the Himalaya and Tibetan Plateau would be very different from what it is today (Figure B17.2): the modern monsoon climate would not exist. The high mountains and plateau now divert the normal flow of westerly winds and are the primary factor in the present monsoon circulation. Summer heating of the high plateau causes warm air to rise, and this creates a low-pressure region that draws moist air inland from the adjacent ocean to produce the summer monsoon rainy season—the Arabs sail to India. In winter, the snow-covered surface of the plateau reflects solar radiation and cools down, creating a region of high pressure from which cold, dry air flows outward and downward—the Arabs sail home.

The history of uplift and erosion of the Himalaya and Tibetan Plateau shows how closely land, ocean, and atmospheric records are linked. Chinese geologists have discovered that these high lands are rather recent topographic features. They base their conclusion in part on evidence of Pliocene Epoch (5.3 to 1.6 million years ago) plant fossils collected at altitudes of 4000 to 6000 m that include many subtropical forms which today exist only at altitudes below

A. Before uplift

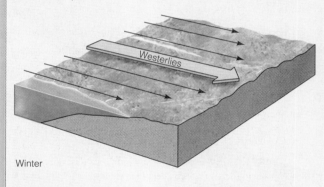

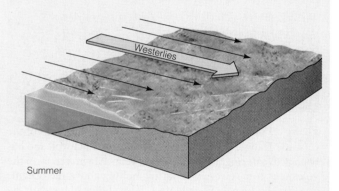

B. Post uplift

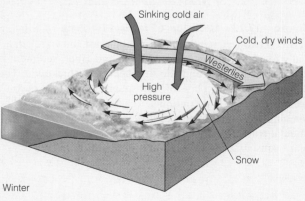

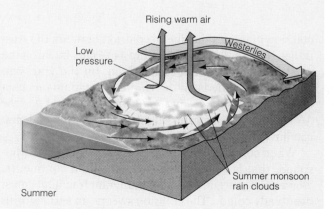

Figure   B17.2   Development of a monsoon climate system Before uplift, westerly winds pass directly across a landscape, the axis of flow shifting seasonally north and south. As a large plateau is uplifted, the westerly flow is diverted around the upland, a winter high-pressure system develops over the high snow-covered plateau and cold, dry winds blow clockwise off the upland region. In summer, as the plateau surface warms, the warm air rises, drawing moist marine air inland. The summer monsoon clouds move inland toward the plateau margin and lose their moisture as heavy rainfall.

2000 m. Additional evidence is provided by studies of changing rates of *denudation* (the continually destructive effects of weathering, mass-wasting, and erosion).

The calculation of long-term denudation rates require knowledge of how much rock debris has been removed from an area over a specified length of time. The total sediment removed from a mountain range includes sediment currently in transit, sediment temporarily stored on land en route to the sea, and sediment deposited in the ocean basin, mostly on or adjacent to the continental shelf. When the volume of all this sediment is measured, its solid rock equivalent and the average thickness of rock eroded from the source region can be calculated. Then, if the duration of the erosional interval can be dated, the average denudation rate can be determined.

For the Himalaya and Tibetan Plateau most of the sediment ultimately is deposited in deltas, as a blanket across the continental shelves, or in vast submarine sediment fans. The volume of sediment deposited during a specific time interval can be estimated using drill-core and seismic records of the seafloor. Calculating the equivalent rock volume of this sediment and averaging it over the area from which it was derived then provide both an average denuda-

tion rate and indications of any rate changes that may have occurred.

Sediment cores from the northern Indian Ocean show low rates of sedimentation, implying reduced erosion rates and low-altitude source areas, until about 10 million years ago when sediment accumulation rates rose sharply as a result of rapid uplift. Two peaks in sediment supply (9-6 and 4-2 million years ago), seen both in the deep-sea cores and in alluvium on land, are evidence of major intervals of Himalayan uplift.

There is suggestive evidence that monsoon winds started only during the second major phase of uplift. About 2.5 million years ago, fine dust, eroded by strong winds from desert basins north of the Himalaya, began accumulating widely over central and eastern China. The dust also was carried far out into the Pacific Ocean where it settled to form deep-sea clay. The onset of this vigorous wind erosion seems to have coincided with the beginning of the present monsoon climate of Asia, as well as with the first widespread glaciation of the Asian highlands. The posibility that the uplift of the Himalaya and the Tibetan Plateau may had a wider, even global, influence on the climate is a matter of current research.

igneous activity are always present. Examples are widespread: the Alps, the Himalayas, and the Carpathians are all young fold-and-thrust mountain systems formed during the Mesozoic and Cenozoic eras, while the Appalachians and the Urals are older, Paleozoic-aged systems.

All fold-and-thrust mountain systems develop from thick piles of sedimentary strata, commonly 15,000 m or more in thickness. The strata are predominantly marine. Some strata, as in the Appalachians, accumulated mainly in shallow water, while others, as in the Alps, accumulated mainly in deep water. Today we recognize that the sediments accumulated along passive continental margins, such as the modern Atlantic margin of North America. Before the theory of plate tectonics, however, the problem of huge catchment sites for sediment was a puzzle. The American geologist J. D. Dana coined the term *geosyncline* to describe what he perceived to be a great trough that received thick deposits of sediment during slow subsidence through long geologic periods. The site of sediment deposition is not a trough, however. Rather, it is the passive margin of a continent where oceanic and continental crust are joined. It is here that sediment derived by weathering of the adjacent continent accumulates. Shallow-water sediment forms on the continental shelf, and deep-water sediment accumulates on the continental slope and rise.

A passive continental margin eventually becomes a continental convergent margin when old oceanic lithosphere fractures close to the join between oceanic and continental crust and subduction commences. The

Atlantic margin of North America will probably become a continental collision margin at some time in the future.

The collision line between the two masses of deformed sediment caught up in a continental collision is commonly marked by the presence of *serpentinites*. These are rocks consisting mainly of the mineral serpentine that are formed by alteration of highly deformed fragments of oceanic crust, together with bits of the mantle rock from just below the Mohorovičić discontinuity caught up in the collision.

One distinctive feature of mountain systems formed by collision is that the new system lies in the interior of a major landmass. Modern examples are the previously mentioned Himalayan mountain chain, formed by the collision of India with Asia, and the Alps, which were formed by the collision of Africa and Europe starting in early Mesozoic time. The Ural Mountains in Russia and Kazakhstan and the Appalachians in eastern North America were both formed by Paleozoic collisions, and both mountain systems were originally in the interior of Pangaea. The Urals are still in the center of a great landmass—they separate Europe from Asia—but the Appalachians are again near a continental margin following the breakup of Pangaea. Orogens are such distinctive and important features of the Earth's crust that we will return to them later in the chapter and discuss the features of several mountain systems.

Transform Fault Margins
A *transform fault* continental margin occurs when the margin of a continent coincides with a transform fault

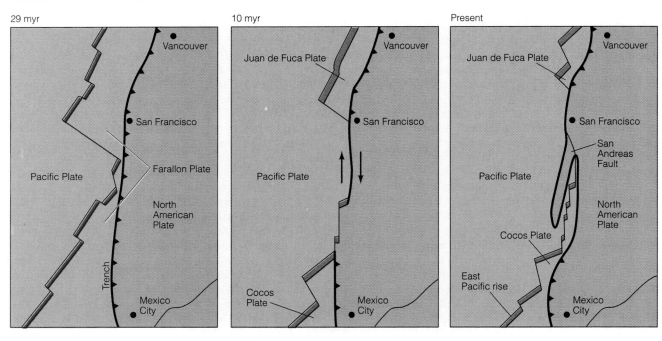

Figure 17.24 How the San Andreas Fault Began   Origin of the San Andreas Fault. Twenty-nine million years ago, the edge of North America overrode a portion of the Farallon Plate, creating two smaller plates in the process, the Cocos Plate and the Juan de Fuca Plate. The San Andreas Fault is the transform fault that connects the remaining pieces of the severed spreading ridge.

boundary of a plate. The most striking example of a modern transform fault boundary is the western margin of North America, from the Gulf of California to San Francisco, where it is bounded by the San Andreas Fault.

The San Andreas Fault apparently arose when the westward-moving North American continent overrode part of the East Pacific Rise, as shown in Figure 17.24. The San Andreas is the transform fault that connects the two remaining segments of the old spreading center.

Accreted Terrane Margins

An **accreted terrane** continental margin is a former convergent or transform fault margin that has been further modified by the addition of rafted-in, exotic fragments of crust such as island arcs. They are the most complex of the five kinds of continental margin. The northwestern margin of North America, from central California to Alaska, is an example of an accreted terrane margin.

Plate motion can raft fragments of crust tremendous distances. Eventually, any fragment not consumed by subduction is added (accreted) to a larger continental mass. Some of the fragments are island arcs formed by subduction of oceanic crust beneath oceanic crust. Other fragments form when they are sliced off the margin of a large continent by a transform fault, much as the San

Andreas Fault is slicing a fragment off North America today. Other combinations of volcanism, rifting, faulting, and subduction can also form fragments of crust that are too buoyant to be subducted. In the western Pacific Ocean, there are many such small fragments of continental crust; examples include the island of Taiwan, the Philippine islands, and the many islands of Indonesia. Each fragment, called a *terrane*, is a geological entity characterized by a distinctive stratigraphic sequence and structural history. The ultimate fate of all terranes is to be accreted to a larger continental mass. An accreted terrane is always fault-bounded and differs so markedly in its geological features from adjacent terranes that many geologists still use the term *suspect terranes*. The term dates from the time when the terranes had been recognized as being unusual but their origin had not yet been deciphered.

The western margin of North America, where a large number of suspect terranes have now been recognized, is the most carefully studied example of a young, accreted terrane margin (Fig. 17.25). Using a combination of lithologic and paleontologic studies, structural analysis, and paleomagnetism, geologists have recently identified the sources of many of the terranes, all of which have been accreted since the early Mesozoic, about 200 million years ago. One of the terranes is an ancient seamount; another is an older limestone platform

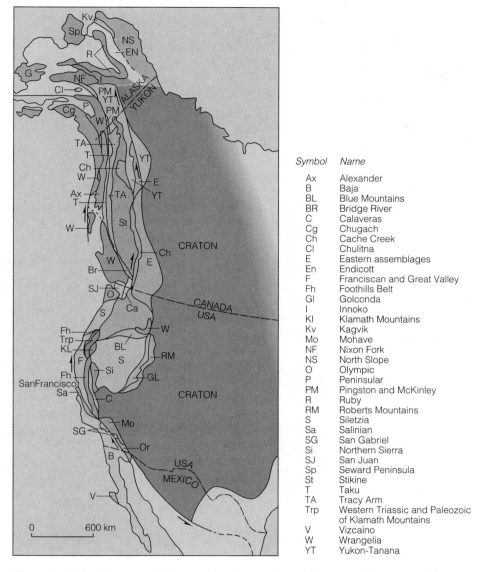

Symbol	Name
Ax	Alexander
B	Baja
BL	Blue Mountains
BR	Bridge River
C	Calaveras
Cg	Chugach
Ch	Cache Creek
Cl	Chulitna
E	Eastern assemblages
En	Endicott
F	Franciscan and Great Valley
Fh	Foothills Belt
Gl	Golconda
I	Innoko
Kl	Klamath Mountains
Kv	Kagvik
Mo	Mohave
NF	Nixon Fork
NS	North Slope
O	Olympic
P	Peninsular
PM	Pingston and McKinley
R	Ruby
RM	Roberts Mountains
S	Siletzia
Sa	Salinian
SG	San Gabriel
Si	Northern Sierra
SJ	San Juan
Sp	Seward Peninsula
St	Stikine
T	Taku
TA	Tracy Arm
Trp	Western Triassic and Paleozoic of Klamath Mountains
V	Vizcaino
W	Wrangelia
YT	Yukon-Tanana

Figure 17.25 Terranes of Western North America The western margin of North America is a complex jumble of terranes accreted during the Mesozoic and Cenozoic eras. Some terranes, such as Wrangellia (W), were fragmented during accretion and now occur in several different fragments.

formed on some other continental margin; still others are fragments of island arcs and even fragments of old metamorphic rocks. Paleomagnetic studies suggest that some terranes have moved 5000 km or more and that, once accreted, the process of movement did not necessarily cease. Later motion along transform faults caused still further reorganizations.

The recognition of accreted terrane margins of continents is a relatively new discovery. In a sense, it is a second stage of complexity in the plate-tectonic revolution. A great deal still remains to be discovered concerning terranes. How to recognize a terrane, how to work out where it came from, and how it was moved are all chal-

lenges facing geologists. Although it is a new and exciting concept, accreted terrane continental margins seem to be yet another distinctive piece of supporting evidence for ancient plate motions.

Mountain Building

Today's fold-and-thrust mountain ranges are the orogens that formed during the last few hundred million years. They are such distinctive and impressive features of the Earth's crust that we will briefly describe some examples.

The Appalachians

The Appalachians are a Paleozoic fold-and-thrust mountain system 2500 km long that border the eastern and southeastern coasts of North America (Fig. 17.26) and that continue offshore, as eroded remnants, beneath the sediment of the modern continental shelf. The sedimentary strata in the ranges contain mud cracks, ripple marks, fossils of shallow-water organisms, and, in places, fresh-water materials such as coal. Evidence is strong for deposition of sediment on the continental shelf of an old, passive continental margin. The sedimentary strata, which thicken from west to east, are underlain by a basement of metamorphic and igneous rocks.

Most but not all of the old strata of the Appalachians have now been deformed. Today, if we approach the central Appalachians from western New York or western Pennsylvania, we see first the former sediment occurring as essentially flat-lying, undisturbed strata (Fig. 17.27). Continuing eastward, we notice that the same strata thicken and become gently folded and thrust faulted. Many of Pennsylvania's oil pools are found in these gently folded strata. In eastern Pennsylvania, in the region known as the Valley and Ridge Province, the strata have been bent into broad anticlines and synclines. (Figure 1.16 is an elegant satellite image of the Valley and Ridge Province.)

Approaching the Appalachians of Tennessee and the Carolinas from the west, we see a different style of deformation. Here, thrust faults predominate (Fig. 17.28). Huge, thin slices of sedimentary strata were pushed westward, each successive slice riding upward and over earlier slices. In the west the strata are nearly flat-lying (Fig. 17.29), but as we move further east the strata dip more steeply to the east. The surface along which movement occurred is known as a **detachment surface**, and the slice that moved is commonly referred to by its French name, **décollement**. A distinctive feature of a décollement is that the style of deformation above the detachment surface is usually different from that below. That is, the weaker, more brittle, sedimentary strata above the detachment surface have been fractured, moved along thrust faults, and stacked like a series of thin cards, while the older rocks below tend to have resisted faulting and large-scale translation and to have been deformed by ductile deformation.

Proceeding east, toward the region from which the thrust slices came, we find the core of the Appalachians. Here the ancient basement rocks and the deep-water sediments that were deposited on the old continental rise have been pushed upward and can be examined. The strata are increasingly metamorphosed, and deformation becomes more intense as the line of collision is approached. Folds become isoclinal and then recumbent, and faulting is prevalent. In places, fragments of the old basement can be seen to have been thrust up over younger sedimentary strata. Finally, we reach a region

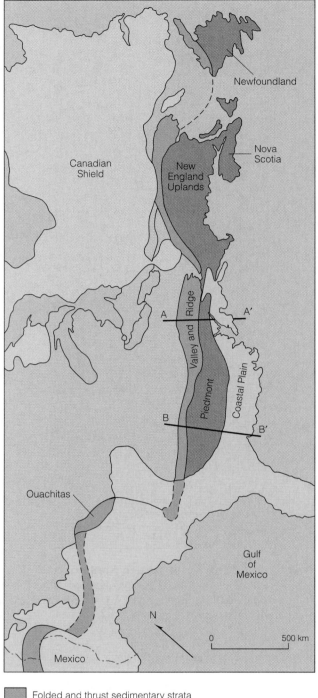

Folded and thrust sedimentary strata

Metamorphic and igneous rocks

Mesozoic and Cenozoic rocks of the Coastal Plain

Figure 17.26 The Appalachians The Appalachian Mountain System runs from Newfoundland to the Mexican border. Parts of the eastern and southern margins of the system are covered by younger sediments of the coastal plain. Figures 17.27 and 17.28 are cross sections drawn approximately along the lines A–A' and B–B', respectively.

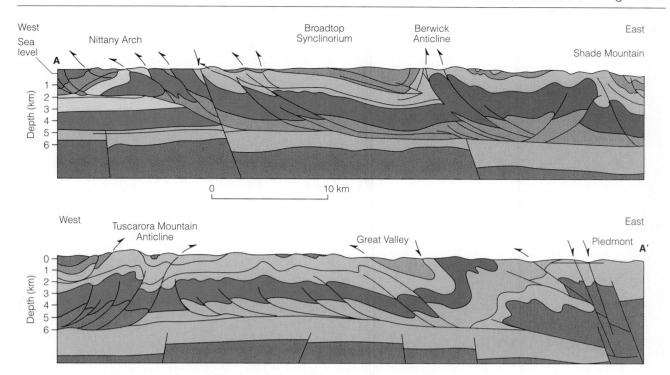

Figure 17.27 Valley and Ridge in Pennsylvania Section through the Valley and Ridge Province of the Appalachians in Pennsylvania, along the A–A' line in Figure 17.26. The structure includes both folding and thrusting. The prominent stratum is a limestone of middle Cambrian to early Ordovician age. (Note that A–A' has been cut in two to fit on the pages so that the left-hand edge of the bottom section joins the right-hand edge of the top section.)

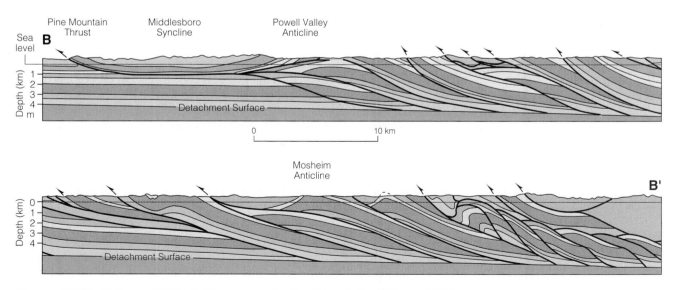

Figure 17.28 Valley and Ridge in Tennessee Section through the Valley and Ridge Province of the Appalachians in Tennessee and North Carolina, along the line B-B' in Figure 17.26 (again cut in two to fit on the page). Compare with Figure 17.27. Development of décollement by thrust faulting is predominant in the southern Appalachians.

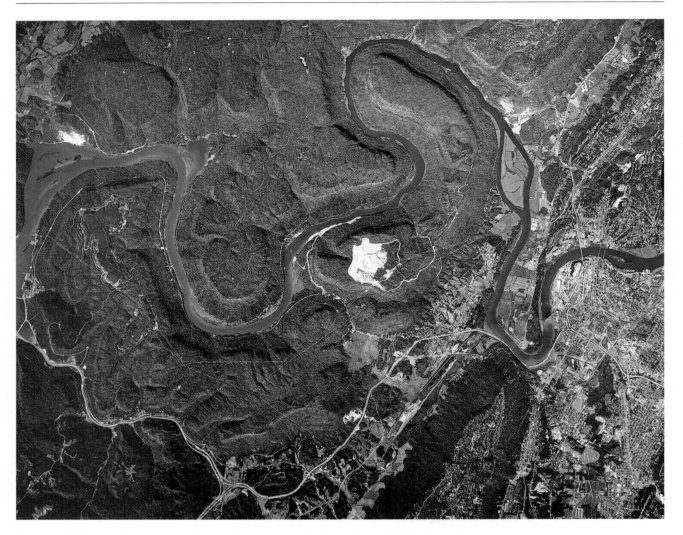

Figure 17.29 Southern Appalachians Southern Appalachians. An aerial photograph of Chattanooga and the Tennessee River corresponding to the left-hand side of Figure 17.28, near the letter B.

where intense metamorphism has occurred and where granite batholiths have been emplaced.

The Appalachians have a complex history that started more than 600 million years ago, as demonstrated in Figure 17.30. Notice that three collision events are postulated. The section drawn in Figure 17.28 is in the southeastern United States, where the rocks that form today's continental margin are thought by some geologists to be an accreted terrane of what was once a fragment of Africa.

The Alps

We naturally ask how well the Appalachian picture can be applied to other fold-and-thrust mountains. The answer is that similar features are found in all of them. The Alps and associated mountain ranges in southern Europe (Fig. 17.31) were formed later than the Appalachians, during the Mesozoic and Cenozoic eras,

as a consequence of a collision between the European and African plates. Nevertheless, the two systems have many features in common. For instance, the Jura Mountains, which mark the northwestern edge of the Alps, have the same folded form and origin as the Valley and Ridge Province (Fig. 17.32A). Also, the Jura Mountains were formed from shallow-water sediments deposited on an ancient continental shelf. In the high Alps, which correspond to the now deeply eroded Appalachians in Connecticut, Vermont, Virginia, and Maryland, thrusting appears to have developed on a much grander scale than in the Appalachians. The high Alps are composed of deeper-water marine strata that were deposited on an ancient continental slope and continental rise. Just as the collision that formed the Appalachians left a bit of Africa behind (Fig. 17.30), so the Alpine collision left some of Africa behind.

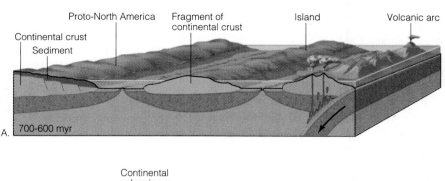

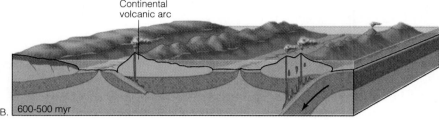

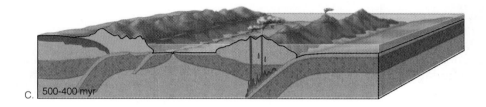

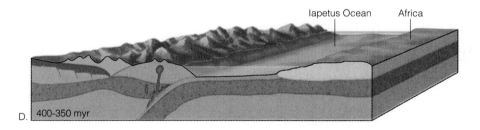

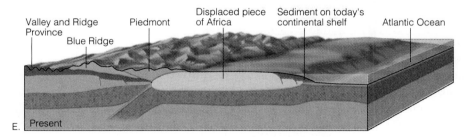

Figure 17.30 How the Southern Appalachians Were Formed
A suggested sequence of events explaining the evolution of the southern Appalachians in terms of plate tectonics. A sequence of subduction, collision, thrusting, and accretion produced the present-day structure. Iapetus is the posthumous name of the ocean that disappeared about 350 million years ago when Africa collided with North America. Iapetus was one of the minor Greek gods and father of Atlas and Prometheus.

The Canadian Rockies
The Canadian Rocky Mountains are a magnificent fold-and-thrust system that can also be compared with the Appalachians. A section at about the latitude of Calgary, Alberta, reveals all the features described for the Appalachians (Fig. 17.33). A central zone has been

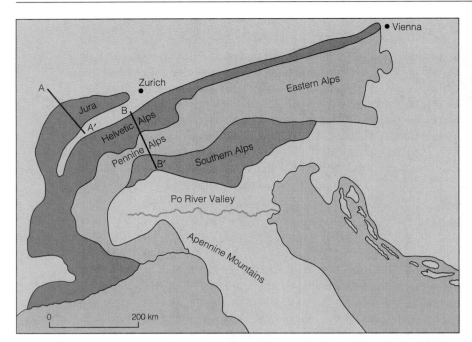

Figure 17.31 The Alps Major units of the Alps in Switzerland and Austria. The mountains were formed as a result of compressive forces operating in a southeastern to northwestern direction. Figures 17.32 A and B are southeast along the lines A–A' and B–B', respectively.

intensely metamorphosed. In it, parts of the older basement rocks have been thrust upward, and folding and thrust faulting are evident on the margins. The thrust sheets in the Canadian Rockies moved eastward away from the core zone (Fig. 17.34). Each sedimentary unit becomes thinner from west to east, indicating that the eastern portion formed from shallow-water sediments, while the core zone coincides with the thickest section of sediments, which were deposited in deep water.

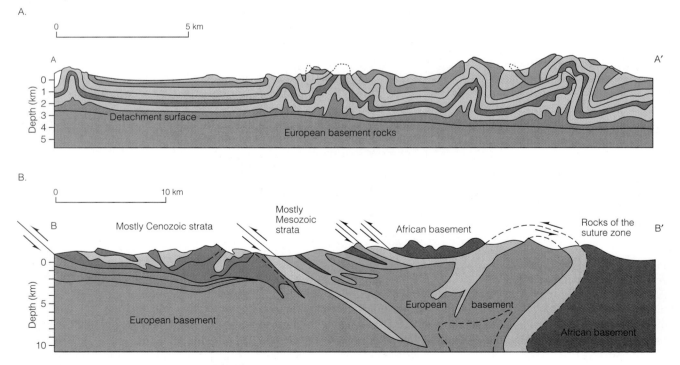

Figure 17.32 Anatomy of the Jura and the Alps Two sections through portions of the Alps. A. Section through the Jura Mountains along line A–A' in Figure 17.31. The weak sedimentary rocks were deformed by slippage along the detachment surface. B. Section in central Switzerland along the line B–B' in Figure 17.31. Strata have moved northward along great thrust faults that later were themselves folded. Part of the old African basement has been thrust over the European strata as a result of the collision.

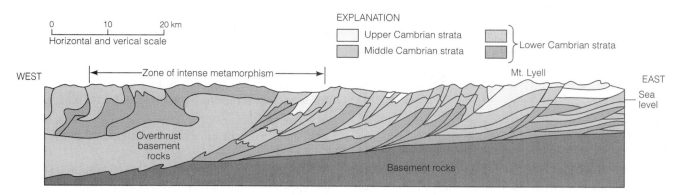

Figure 17.33 Anatomy of the Rockies in Canada Section through the Canadian Rocky Mountains at about the latitude of Calgary, Alberta. The zone of intense metamorphism coincides with the region of maximum uplift and maximum deformation. Farther east, where strata become progressively thinner, the pile has been greatly thickened by movement along thrust faults. The sense of movement is such that each fault block has moved toward the east, riding over the block beside it.

Figure 17.34 Thrust Sheets of the Rockies The Canadian Rockies. This early-morning photograph is looking west in the area of Banff National Park, Alberta. The ranges are the upthrust edges of thrust sheets. The sense of movement on the thrusts is toward the viewer.

SUMMARY

1. The lithosphere is broken into six large and many smaller plates, each about 100 km thick and each slowly moving over the top of the weak asthenosphere beneath it.

2. Three kinds of margins are possible between plates. Divergent margins (spreading centers) are those where new lithosphere forms; plates move away from them. Convergent margins (subduction zones) are lines along which plates compress each other and along which lithosphere capped by oceanic crust is subducted back into the mantle. Transform fault margins are lines where two plates slide past each other.

3. Plate movement can be described in terms of rotation across the surface of a sphere. Each plate rotates around a spreading axis. The spreading axis does not necessarily coincide with the Earth's axis of rotation.

4. Because plate movement is a rotation, the velocity varies from place to place on the plate.

5. Each segment of oceanic ridge that marks a divergent margin between two plates lies on a line of longitude passing through the spreading pole. Each transform fault margin between two plates lies on a line of latitude of the spreading pole.

6. The mechanism that drives a moving plate is not known, but apparently it results from a combination of convection in the mantle plus either pushing, dragging, or sliding forces that act on the plate.

7. Two major structural units can be discerned in the continental crust. Cratons are ancient portions of the crust that are tectonically and isostatically stable. Separating and surrounding the cratons are orogens of highly deformed rock, marking the site of mountain ranges.

8. An assemblage of cratons and deeply eroded orogens that forms the core of a continent is a continental shield.

9. There are five kinds of continental margins: passive, convergent, collision, transform fault, and accreted terrane.

10. Passive margins develop by rifting of the continental crust. The Red Sea is an example of a young rift; the Atlantic Ocean is a mature rift.

11. A characteristic sequence of sediments forms along a passive continental margin, starting with clastic nonmarine sediments, followed by marine evaporites, and then marine clastic sediments.

12. Continental convergent margins are the locale of paired metamorphic belts, chains of stratovolcanoes (magmatic arc), and linear belts of granitic batholiths.

13. Collision margins are the locations of fold-and-thrust mountain systems.

14. Transform fault margins occur where the edge of a continent coincides with the transform fault boundary of a plate.

15. Accreted terrane margins arise from the addition of blocks of crust brought in by subduction and transform fault motions.

THE LANGUAGE OF GEOLOGY

accreted terrane (p. 480)

back-arc basin (p. 468)

continental drift (p. 458)
continental shield (p. 472)
continental volcanic arc (p. 466)
craton (p. 472)

décollement (p. 482)
detachment surface (p. 482)

fore-arc basin (p. 467)
fore-arc ridge (p. 467)

hot spot (p. 470)

island arc (p. 466)

magmatic arc (p. 466)
magnetic latitude (p. 460)
mélange (p. 466)

orogen (p. 472)

passive continental margin (p. 473)
plate triple junction (p. 475)

seafloor spreading (p. 461)
spreading axis (p. 465)
spreading pole (p. 465)

QUESTIONS FOR REVIEW

1. Long ago it was suggested that the location and shape of mountain ranges and other major topographic features were the result of contraction as the Earth cooled. Why is that explanation incorrect?

2. Who was Alfred Wegener, and what revolutionary idea did he suggest? Why were scientists reluctant to accept Wegener's idea when it was first proposed?

3. Explain how the apparent wandering of magnetic poles throughout geologic history can be used to help prove continental drift.

4. What are the main features of seafloor spreading? What critical test proved that the seafloor does move?

5. What are the three kinds of margins that bound tectonic plates?

6. How are the velocities of tectonic plates determined? Do magnetic time lines on plates provide relative or absolute plate velocities?

7. What is a spreading pole and how does plate velocity depend on the position of a plate relative to the spreading pole?

8. Why does the radius of curvature differ from one island arc to another?

9. What is a mélange and what kind of metamorphism is associated with mélanges? Where might mélanges be forming today?

10. Draw a cross section through the lithosphere at a convergent plate margin and mark the positions of the fore-arc basin, the back-arc basin, and the magmatic arc.

11. What are cratons and how do they differ from orogens? Name three orogens in North America that are less than a billion years old.

12. Name three fold-and-thrust mountain ranges.

13. What is the tectonic environment of an island arc? Name two examples of modern oceanic island arcs.

14. What is the origin of the Cascade Range? the Sierra Nevada? the Appalachians?

15. What evidence indicates that plate tectonics has been operating for at least the last 2 billion years of Earth history?

16. Name the five kinds of continental margins and describe how they form.

17. Describe the sequence of events that leads to the opening of a new ocean basin flanked by two passive continental margins.

18. How does an accreted terrane margin form? Name a continental margin that was modified by terrane accretion.

19. What is a hot spot? Give an example of a hot spot.

20. How are the absolute velocities and directions of tectonic plates determined?

21. Name three forces that act on lithosphere that might cause it to slide over the asthenosphere.

For an interactive case abstract, virtual tours, activities, and additional learning resources, go to
GEOSCIENCES TODAY: www.wiley.com/college/skinner

Landscape and Life along the East African Rift: the Virunga Mountains, Rwanda

The Virunga Mountains are a region where D.R. Congo, Rwanda, and Uganda meet. The broad region is known as the East African rift system. The area boasts a beautiful landscape where one can observe the forces of nature at work. It is also the last habitat of the mountain gorilla—a most endangered species best known through the work of Diane Fossey and others. The country of Rwanda is familiar due to the horrific news of human tragedy and genocide, which occurred there in 1994, bringing personal insecurity, political instability, and environmental degradation to the surrounding area.

OBJECTIVE: The primary goals of the case are to show how tectonism and volcanism have molded the landscape around the Virunga Mountains; how those processes have influenced the distribution of plants, water, soil, animals, and other natural resources; and to understand the many facets of life which reflect the underlying physical realities.

The Human Dimension: How have humans interacted with this landscape, and how are human vulnerability and risk monitored and mitigated using "earth system science" tools and techniques?

Questions to Explore:

1. Where is the East African rift system, and how was it formed?

2. What types of landforms, landscapes, and drainage patterns can be observed in the East African rift region that are related to tectonism and volcanism?

3. How are vegetation, climate, and soils influenced by the underlying geomorphology, geology, and hydrology?

18

Understanding
Our Environment

GeoMedia

Intense erosion of the volcanic rocks of Kauai, most northerly of the main Hawaiian Islands. This scene is on the NaPali coast on the north-western side of the island.

The Changing Face of the Land

Eroding Volcanoes in the Mid-Pacific Ocean

We cannot directly observe the evolution of large-scale landscape features on the continents because of the immense time involved, but an ideal opportunity to visualize long-term landscape changes is provided by midocean volcanic island chains. Early geologists observed that the Hawaiian Islands appear more dissected northwestward from the island of Hawaii, the site of active volcanism, and inferred that this change reflected the increasing age of the islands in that direction. K/Ar dating has now shown that the volcanoes indeed increase in age from less than half a million years on Hawaii to about 5 million years on the oldest major islands farther up the chain.

The island chain is believed to have evolved as the Pacific Plate moved slowly northwestward across a midocean hot spot, above which frequent and voluminous eruptions built a succession of large volcanoes. Once formed, each volcanic island was carried slowly away from the hot spot in the direction of plate movement. Each of the emergent islands has approximately the same history. By examining the erosional landscapes on successively older volcanoes, one can obtain an understanding of how the oldest Hawaiian landscapes likely developed. During the initial constructional phase of a volcano, the rate of lava extrusion greatly exceeds the rate of erosion, and an island rises upward from the seafloor in the form of a broad shield volcano. As the volcanic pile accumulates, the localized added weight on the seafloor causes the underlying ocean crust to subside isostatically. This helps to limit the maximum altitude to which a volcanic island can rise (4–5 km above sea level) before the moving Pacific Plate carries it beyond the magma source. High Hawaiian volcanoes intercept the moist tradewinds, resulting in abundant precipitation on windward slopes. The resulting high rainfall has enabled streams to cut deep canyons that indent the island flanks. Progressive dissection of the land, as major streams cut downward and backward, gradually eliminates the original constructional slopes of the volcanic shield. As extrusive activity slackens and finally ceases, continuing erosion ultimately reduces the dissected volcano to sea level. In the case of the Hawaiian chain, this entire evolutionary process takes about 10 million years.

THE EARTH'S VARIED LANDSCAPES

The major components of the Earth's erosional system meet and interact at the land surface. As a result, the evolution of the Earth's landscapes involves a complex of processes related to the solid Earth, the atmosphere, the hydrosphere, and the biosphere. Because of this complexity, gaining an understanding of landscape evolution involves taking an *interdisciplinary* approach; that is, scientists in a number of related Earth Science disciplines must collaborate in order to gain meaningful answers about the evolution of landscapes.

One of the most striking properties of the Earth is its amazing variety of natural landscapes. Even a casual look at the land surface can raise some basic questions in our minds: How can the Earth's varied landscapes be explained? If landscapes have changed through time, what processes have controlled the changes? What clues do landscapes hold about the history of the Earth's mobile lithosphere and past climates?

The processes that produce changes in the land surface operate on many different time scales and spatial scales. Some processes act rapidly, even abruptly, and may cause only local changes. Sometimes these changes have a direct and adverse impact on people (for example, earthquakes, volcanic eruptions, storms, floods, landslides, and other natural hazards). Other processes

operate on geologic time scales and are difficult or impossible to observe directly. Nevertheless, their effects can be seen and measured both in surviving ancient landscapes and in the stratigraphic record (see The essay entitled *The Bridge of the Gods*); from measurements, computer models can be used to help understand how slowly acting processes influence landscape evolution.

Not only can surface processes affect people going about their daily lives, but everywhere they live people, too, are changing the face of the land. The construction of dams across streams, excavations for buildings or waste disposal sites, and the building of highways, cities, and airports all modify the landscape. Although such individual actions may seem insignificant, on the global scale they can quickly add up until, over the span of several human generations, their effect on the landscape is substantial.

Competing Geologic Forces

The evidence is overwhelming that the Earth's surface is constantly changing, although the rate and magnitude of change vary considerably from region to region. The changes reflect an ongoing contest between forces that raise the lithosphere (tectonic and isostatic uplift, volcanism) and the force of gravity which, aided by physical and chemical weathering that break down rock, causes various erosional agents to transfer rock debris from high places to low places. The net result is the progressive sculpture of the land into a surface of varied topography and **relief** (the difference in altitude between the highest and lowest points on a landscape) (Fig. 18.1). Relief varies regionally because surface geologic processes operate not only under different climatic conditions, but also on rocks of different type and structure that have differing resistances to erosion. As we have seen in Chapters 10 and 11, streams and glaciers efficiently erode and transport rock debris and move large quantities of sediment downslope and toward the sea. However, another group of surface processes also is responsible for moving vast amounts of rock debris downslope under the influence of gravity, as we saw in Chapter 9.

Factors Controlling Landscape Development

We have already seen that distinctive landscape elements result from the activity of various surface processes. A sand dune has a form that is different from that of a moraine. We can also distinguish between an alluvial fan and a delta on the basis of their form. In each case, the active process and depositional environment

A

B

Figure 18.1 Varied Landscapes A. The precipitous southern flank of Cerro Aconcagua (6959 m; 22,834 ft), the highest peak of the Andes and the southern hemisphere, rises 3600 m (11,800 ft) above the glaciated valley at its base, making it one of the areas of greatest local relief on the Earth. B. Subdued low-relief landscape of the Western Australia shield, with Tertiary laterite at surface.

lead to a unique end product, or landform. *Process*, then, is one factor that helps dictate the character of landforms.

Climate, in turn, helps determine which surface processes are active in any area. In humid climates, streams may be the primary agent that moves and deposits sediment, whereas in an arid region wind may locally assume the dominant role. Glaciers and periglacial phenomena are largely restricted to high latitudes and high altitudes where frigid climates prevail. Because climate also controls vegetation cover, it further controls the effectiveness of some important erosive processes; for example, a hillslope stabilized by plant roots that anchor the soil may become prone to mass-wasting if the plant cover is destroyed. Thus, hillslope erosion is linked to vegetation cover, which in turn is controlled by soil moisture, which is determined by climate—a complex linkage in which the atmosphere, the hydrosphere, and the biosphere all play a role in determining how surface processes affect the land surface.

In the same way that distinctive climatic regions of the Earth can be identified, it is also possible to identify regions containing distinctive landforms produced by one or more surface processes. However, because climates have changed through time, the active surface processes in some regions also have changed. As a result, some modern landscapes largely reflect former conditions rather than those of the present.

Within any climatic zone, a certain surface process may interact with various exposed rocks differently, depending on their *lithology* (Fig. 18.2). Some rock types are less erodible than others and will produce

Figure 18.2 Differential Erosion A false color satellite image of the region near Harrisburg, Pennsylvania, reveals a complicated series of northeast-trending ridges and valleys produced by differential erosion of sedimentary rocks. Ridges are underlain by resistant sandstones and conglomerates, whereas valleys are underlain by more erodible shales. The folded structure of the rocks is clearly visible due to the pronounced topographic relief between the less erodible and more erodible strata.

The Bridge of the Gods

As Meriwether Lewis and William Clark paddled down the Columbia River on the final leg of their momentous "voyage of discovery" that led them to the Pacific Ocean, they encountered a sudden constriction of the river, marked by a set of rapids. Here the river dropped 10–15 m along a 400 m reach of the channel. According to Native American oral history, people formerly had been able to cross the river at this point "without getting their feet wet." To the native population, the site was known as the "Bridge of the Gods." However, they reported that the rapids were not ancient, for their "fathers" in earlier days had voyaged upstream by canoe as far as The Dalles without encountering any obstruction. Their history stated that the damming of the river had caused the water to rise to great height upstream, and that the rapids appeared as the water overtopped the dam and began cutting down through it.

Located near the site of the ancient dam is a new dam, completed in 1938. Bonneville Dam now impounds the river, backing up the water for many kilometers upstream. In constructing the new dam, geologists determined that the ancient damming was caused by a massive landslide from the north side of the Columbia canyon. The slide occurred where a thick clayey zone is developed on altered rocks of the Ohanepecosh Formation that underlie sand and gravel (Eagle Creek Formation) and the Columbia River Basalt.

The clayey zone, acting as a lubricating layer when saturated, was the locus of landslide movement. As the slide moved downslope, it displaced the river to the south side of the valley, constricting its width to less than 400 m.

Lewis and Clark reported encountering dead tree stumps along the river channel nearly 100 km upstream from the Bridge of the Gods. Geologists later determined that the trees had been drowned as water behind the landslide dam rose as much as 15 m above the former river level. Radiocarbon dating of the outer wood of several trees implies a greater than 66% probability that they were killed sometime between about 1530 and 1770 AD, perhaps no more than a century before the Lewis and Clark voyage of discovery (1804-1806).

Geologists continue to speculate on what may have triggered the massive Bonneville landslide. Was it due to an interval of unusually rainy weather that saturated unstable ground? Could it have been a large earthquake? The latter possibility is tantalizing, for recent studies of past subduction-zone earthquakes in the Pacific Northwest (See Chapter 16, *Understanding Our Environment: A Great Earthquake in the Pacific Northwest?*) show that the last giant quake likely occurred in January, 1770 within the window of time suggested by the radiocarbon dates from the drowned forest.

greater relief and more prominent landforms than those more susceptible to erosion. Any single rock type, however, may behave differently under different climatic conditions. For example, limestone may underlie valleys in areas of moist climate where dissolution is very effective, but in dry desert regions the same kind of rock may form bold cliffs.

The ease with which a formation is eroded by streams depends chiefly on its composition and structure. Because of differential erosion, folded or faulted beds may stand in relief or control the drainage in such a way that they impart a distinctive pattern to the landscape, thereby disclosing the underlying structure (Fig. 18.2). Figure 10.32 shows some of the most common drainage patterns and the geologic structures that control them. An experienced geologist can often use drainage patterns to infer rock type, the orientation of a dipping rock unit, the manner in which the rocks are folded or offset, and the pattern and spacing of joints.

Relief of the land, another factor in landscape development, is related to tectonic environment. Tectonically active regions may have high rates of uplift, leading to high summit altitudes and steep slopes. Such landscapes tend to be extremely dynamic and generate high erosion rates (Fig. 18.3). In areas far away from active

tectonism, relief typically is low, erosion rates are much lower, and landscape changes take place more gradually. Even in nontectonic areas, however, a rapid rise or fall of sea level, or regional isostatic movements related to changing ice or water loads, may produce significant changes in streams and the landscapes they influence.

Finally, the concept of landscape evolution necessarily involves the element of *time*. Although some landscape features can develop rapidly, even catastrophically, others develop only over long geologic intervals. We know this, or at least we infer this, from measurements of the present rates of surface processes and by dating deposits that place limits on the ages of specific landforms or land surfaces.

LANDSCAPES PRODUCED BY RUNNING WATER

Most of the Earth's land areas, except those continuously covered by ice sheets, show the effects of running water. Even in extremely dry deserts, we can find evidence that ancient streams have had a hand in shaping the land.

Figure 18.3 Erosion in High-relief Terrain A stream cutting a deep gorge in the Hindu Kush of northern Pakistan transports a coarse gravel bed load, contributed largely by rockfalls and landslides from the steep adjacent cliffs.

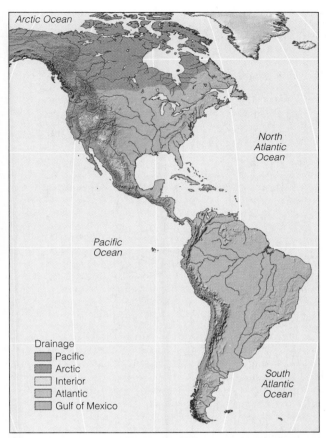

Figure 18.4 Major Drainage Divides Map of the western hemisphere showing the location of major drainage divides. The continental divide separating streams draining to the Pacific, Arctic, and Gulf of Mexico in North America and to the Pacific and Atlantic in South America follows the crest of the high cordillera in both hemispheres. In eastern North America, the divide separating Atlantic and Gulf of Mexico drainage follows the Appalachian Mountains and the approximate limit of ice-sheet glaciation south of the Great Lakes.

Tectonic and Climatic Control of Continental Divides

All the continents except ice-covered Antarctica can be divided into large regions from which major through-flowing rivers enter one of the world's major oceans. The line separating any two such regions is a **continental divide**, one of the major landscape elements of our planet. In North America, continental divides lie at the head of major streams that drain into the Pacific, Atlantic, and Arctic oceans (Fig. 18.4). In South America, a single continental divide extends along the crest of the Andes and divides the continent into two regions of unequal size. Streams draining the western (Pacific) slope of the Andes are steep and short, whereas to the east the streams take much longer routes along more gentle gradients to reach the Atlantic shore.

Because continental divides often coincide with the crests of mountain ranges and because mountain ranges are the result of uplift associated with the interaction of tectonic plates, a close relationship must exist between plate tectonics and the location of primary stream divides and drainage basins.

When a divide lies close to a continental margin, and there is a strong climatic gradient across it, an unequal distribution of precipitation can lead to a marked landscape asymmetry across the divide. Streams draining the windward (wet) side commonly have steeper gradients than those draining the lee (dry) side, and erosion rates are higher. Over time, the net effect will be the headward growth of channels on the wet side, leading to a shift in the divide toward the drier side.

A striking example is seen in the case of the Andes in southern South America. Through nearly 60 degrees of latitude, from the northern tip of the continent to well

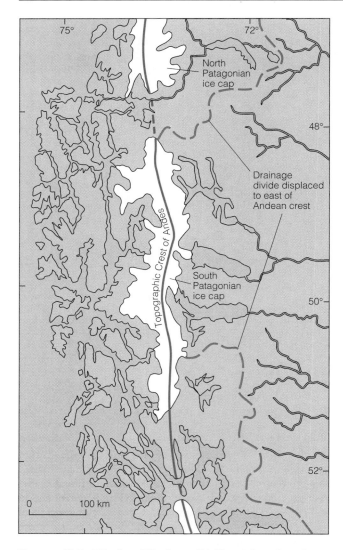

Figure 18.5 Displaced Drainage Divide Map of southernmost Andes showing coastal fjords of Chile, Andean ice caps, and topographic crest of range (bold solid line). In two sectors, the continental drainage divide (dashed line) lies well east of the range crest and partly coincides with a belt of Pleistocene end moraines.

south of Santiago, Chile, the drainage divide coincides with the topographic crest of the range. However, south of 45°, where easterly winds have given way to westerlies flowing off the Pacific, the continental divide has shifted eastward in several sectors. During glaciations, the mountain ice cap of the southern Andes apparently was thickest over the fjord region of southern Chile, which received abundant snowfall from the moist westerly winds flowing onshore. Outlet glaciers of a large ice cap flowed eastward across saddles in the main Andean crest, progressively deepening and ultimately eliminating these topographic barriers. In several sectors, the continental divide between the Pacific and Atlantic oceans now lies along the crest of end moraines marking the

limit of piedmont glaciers that terminated near or beyond the eastern front of the mountains (Fig. 18.5).

The Geographic Cycle of W. M. Davis

Mountain ranges and high plateaus are major landforms of the Earth's crust, and it is natural to ask: How long can such features persist? As soon as mountain building commences, the forces of erosion begin to attack the rising land. The average altitude of the land at any time, therefore, should reflect this contest between uplift and erosion.

One of the most influential theories of landscape evolution was proposed by the American geographer William Morris Davis in the late nineteenth century. Davis called his model the **geographic cycle**, implying that it had a beginning and an end. According to this concept, a cycle began with rapid uplift of a landmass, with little accompanying erosion, so that the initial relief was large. Erosion then progressively sculptured the land and reduced its altitude until it was worn down close to sea level (Fig. 18.6). Davis deduced that a landscape passed through a series of stages. During the initial stage, streams cut down vigorously into the uplifted landmass and produce sharp V-shaped valleys, thereby increasing the local relief. Gradually, the original gentle upland surface is consumed as the drainage system expands and valleys become deeper and wider. During the next stage, the land achieves its greatest local relief. Streams now begin to meander in their valleys, and valley slopes are gradually worn down by mass-wasting and erosion. In the final stage, the landscape consists of

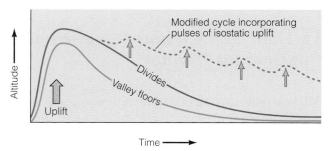

Figure 18.6 The Geographic Cycle The geographic cycle postulated by W. M. Davis begins with a pulse of uplift, during which the uplift rate increases, then diminishes. Initially, the landscape increases rapidly in altitude. As the uplift rate decreases, the land is lowered by erosion. Valleys deepen and local relief (the distance between divides and valley floors) reaches a maximum. During later stages of the cycle, the land is progressively lowered toward sea level and relief becomes lower and lower. If isostatic adjustments are also considered, a long time is required to erode the landscape to lower altitude (dashed line).

broad valleys containing wide floodplains, stream divides are low and rounded, and the landscape is slowly worn down ever closer to sea level. Davis also envisaged interruptions of erosion cycles related to climatic fluctuations or to renewed uplift. Others later pointed out that the crust's isostatic response to the progressive transfer of sediment from the land to the ocean should lead to pulses of uplift, requiring that a cycle be longer than Davis envisaged (Fig. 18.6).

Davis's theory attracted wide attention and formed the basis for most interpretations of landscape evolution during the decades following publication of his ideas. However, with renewed interest in surface process studies and the widespread acceptance of the theory of plate tectonics, it has become increasingly difficult to reconcile Davis's concept of an erosion cycle with what we know about global tectonics and Earth history. The major landscape features of the Earth have developed over long intervals of time as the lithosphere has evolved and continental arrangements have continually been reorganized. The lateral motions and resulting collisions of lithospheric plates, leading to the generation of orogenic belts, have provided much of the driving force for landscape change over hundreds of millions of years. Today's focus on landscape evolution has shifted from the concept of landscape cycles to questions about landscape equilibrium, studies of the relative importance of uplift rates and erosion rates, and investigations of the complex relationships and feedbacks between the atmosphere, hydrosphere, biosphere, and lithosphere, and the way the relationships and feedbacks dictate the character and evolution of the major landscape elements of the planet.

Landscape Equilibrium

Change is implicit in landscape evolution. Landscapes presumably will evolve if a change takes place in any of the controlling variables (process, climate, lithology, structure, relief). Change may be started by a tectonic event that causes a landmass to be uplifted or by a substantial fall of sea level that causes streams to assume new gradients. It can begin with a shift in climate that may modify the relative effectiveness of different surface processes. A change may also result as a stream, eroding downward through weak, erodible rock, suddenly encounters massive, hard rock beneath.

Over short intervals of time, rates of change may vary because of natural fluctuations in the magnitude and intensity of surface processes, but overall the landscape can be close to a steady state (Fig. 18.7A). Over longer intervals, the average rate of change may increase if the rate of tectonic uplift increases, or experience a gradual decrease as a land surface is progressively worn down toward sea level (Fig. 18.7B).

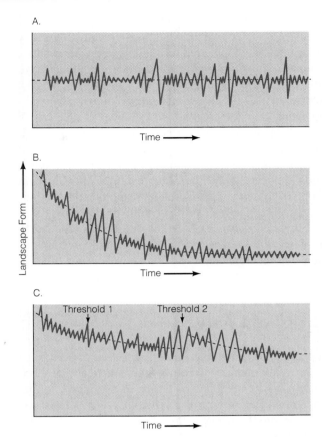

Figure 18.7 Three Conditions of Landscape Evolution Three possible equilibrium conditions in landscape evolution. A. Steady-state equilibrium, with the landscape fluctuating about some average condition. B. Dynamic equilibrium, with the landscape oscillating about an average value that is declining with time. C. Dynamic equilibrium, with sudden changes to the landscape occurring whenever a critical threshold is reached.

Does a landscape, then, ever achieve a state of complete equilibrium in which no change takes place? The answer apparently is no, for we have abundant evidence that the Earth's surface is now, and very likely always has been, a dynamic surface, constantly experiencing changes in response to the natural motions of the lithosphere, hydrosphere, and atmosphere. Nevertheless, it is apparent that conditions of near-equilibrium can be reached.

One reason why a state of perfect equilibrium in landscape systems is difficult to achieve is that the responses to change are often complex. Within a landscape unit such as a large drainage basin, a change in one of the controlling factors may at first impact only a small part of the basin. A sudden vertical movement along a fault that crosses the basin near its mouth will affect the long profile of the stream at that point and begin a series of compensating adjustments both in the upstream and downstream directions. Areas in the headward part of the basin may initially be unaffected, but as the stream system adjusts, changes will progressively affect other parts of the basin, including tributary streams, valley slopes, and ultimately valley heads, until a near-equilibrium condition is once more achieved. The change may move through the system in a complex manner, with many lags and minor readjustments taking place.

Several changes may affect a stream system simultaneously. For example, a sudden intense rainstorm may be concentrated in one tributary and a massive landslide occur in another. The response of the stream to one of these events may lag its response to the other, resulting in a complex adjustment of the entire system. It is not surprising, therefore, that geologists working with the depositional and erosional products of such changes can have difficulty in sorting out the events that caused them.

Threshold Effects

In many natural systems, sudden changes can take place without any outside stimulus when a critical threshold condition is reached. Sand grains in a stream channel may remain at rest until a critical current velocity is reached, at which point they begin to move. Certain glaciers apparently start to surge when the buildup of water pressure at their base attains a critical threshold value that forces the ice to float off its bed (Chapter 11).

The concept of thresholds is also applicable to landscapes. It implies that the development of landscapes, rather than being progressive and steady, can be punctuated by occasional abrupt changes. A landscape in near-equilibrium may experience a sudden change in form if a process operating on it reaches a threshold level. The condition of equilibrium is thereby affected, and a change in the landscape takes place, as it moves toward a new equilibrium condition (Fig. 18.7C). Studies in the western United States have shown, for example, that the development of gullies on the floors of some alluvial valleys is related to valley slope. Where valley floor slopes reach a certain critical value, which depends on the area of the drainage basin, they become unstable and gullying begins. Where the slope is gentler, valley floors remain ungullied. The critical slope value therefore constitutes a threshold; when it is exceeded, erosion of the valley floor begins.

UPLIFT AND DENUDATION

Obtaining reliable geologic information about long-term rates of landscape change is not simple. First, we must determine how mobile a landmass has been through time. If it is rising tectonically, we must ask how rapidly it is rising and whether rates of uplift have changed through time. Second, we must be able to calculate changing rates of **denudation**, the combined destructive effects of weathering, mass-wasting, and erosion. Knowing uplift and denudation rates for a region, we can then infer something about how the landscape may have evolved and anticipate how it may evolve in the future.

Uplift Rates

One way geologists attack the problem of calculating uplift rates is to measure how much local uplift occurs during historic large earthquakes, try to determine the recurrence interval of such earthquakes, and then extrapolate the recent rates of uplift back in time. This approach assumes that the brief historic record is representative of longer intervals of geologic time, an assumption that may not always be valid.

A second approach is to measure the warping, or vertical dislocation, of originally horizontal geologic surfaces of known age. Examples are flood basalts that have been deformed since extrusion, and uplifted coral reefs along a tectonically active coast (Fig. 12.33). In each case, we need to know both the difference between the present altitude of a basalt flow or coral reef and its altitude at the time of formation, as well as a radiometric age for the rock or coral.

Successive intervals of stream incision into a mountain range or plateau may produce a series of terraces that record uplift events. If the age of the terraces can be obtained, then uplift rates can be calculated. However, because terracing can also result from changes in stream activity caused by worldwide sea-level fluctuations, changes of climate, or variations in the structure of rocks across which a stream flows, relating river terraces to tectonic events is not always straightforward.

Finally, rocks that formed deep within the crust and subsequently were exposed at the surface by uplift and erosion can provide uplift rates. When the mineral zircon crystallizes in a plutonic rock, the subsequent decay of radioactive isotopes trapped in the mineral damages the internal arrangement of atoms, leaving tiny tracks (called *fission tracks*) that can be detected under a microscope. At high temperatures, fission tracks can form, but they quickly anneal and disappear, leaving no record of their former existence. Only when the cooling mineral falls below a certain critical temperature (its closure temperature) is the annealing rate so slow that fission tracks will be retained. The number of tracks forming in a mineral increases with time, and so track density can be used to date the time elapsed since a rock containing the mineral cooled below the closure temperature. In the example shown in Figure 18.8, we assume that the closure temperature for the mineral zir-

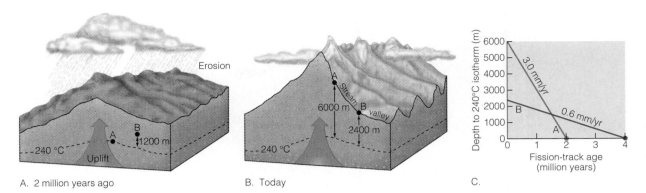

Figure 18.8 Uplift Rates Uplift rate across a mountain range calculated using fission-track ages. A. Two million years ago, a zircon crystal (A) in a cooling pluton passes the closure isotherm of 240°C and begins to acquire fission tracks. Another crystal (B) began acquiring tracks 2 million years earlier and since then has been uplifted 1200 m above the 240°C isotherm. B. Rock samples containing zircon crystals A and B collected from a stream valley eroded into the rising mountain range have fission-track ages of 2 and 4 million years, respectively, and lie 6000 and 2400 m, respectively, above the closure isotherm of 240°C. C. By using these data, the average uplift of samples A and B are calculated as 3.0 and 0.6 mm/year, respectively.

con is 240°C (464°F). By measuring the geothermal gradient (we'll assume in the example it is 40 C°/km), we then calculate that the 240°C isotherm lies at a depth of 6000 m (3.7 mi). The average rate of uplift can then be calculated because we know the fission track age of the rock and the total uplift since the rock first acquired fission tracks (i.e., the depth beneath the sample site at which the 240°C isotherm lies).

The current local uplift rate for one high region of the western Himalaya, based on fission-track measurements, is as high as 5 mm/year (0.2 in/yr). If sustained for only 2 million years, the total uplift would be 10 km (33,000 ft). Such a high rate of uplift is generally associated with steep slopes and high relief. This is certainly true in the Himalaya, where glacially eroded mountain slopes near the crest of the range typically exceed angles of 30° and local relief can exceed 5 km (16,500 ft). At lower altitudes, where streams are the dominant erosional force, slopes generally decline to between 15 and 20°. Mean erosion rates in these high mountains are estimated to be about 3 to 4 mm/year (0.1 to 0.2 in/yr), which is close to the average uplift rate.

Measured uplift rates in this and other tectonic belts are quite variable (ca. 1 to 10 mm/yr, averaged over intervals of several thousand to several million years). We can assume that for any region, the average values have changed through time as rates of seafloor spreading and plate convergence have varied. Each such change is likely to lead to compensating adjustment in landscapes as they begin to evolve toward a new condition of equilibrium.

Denudation Rates

The calculation of long-term denudation rates requires knowledge of how much rock debris has been removed from an area during a specified length of time. The total sediment removed from a mountain range, for example, will include sediment currently in transit, sediment temporarily stored on land on its way to the sea, and sediment deposited in the adjacent ocean basin, mostly on or near the continental shelf. If the volume of all this sediment can be measured, its solid rock equivalent and the average thickness of rock eroded from the source region can be calculated. Finally, if the duration of the erosional interval can be determined, the average denudation rate can be calculated.

For areas drained by major throughflowing streams, the volume of sediment reaching the ocean each year is a measure of the modern erosion rate. Most of this sediment ultimately is deposited in deltas, as a blanket across the continental shelves, and in vast submarine sediment fans. The volume of sediment deposited during a specific time interval can be estimated using drill-core and seismic records of the seafloor. Calculating the equivalent rock volume of this sediment and averaging it over the area of the drainage basin(s) from which it was derived then give an average denudation rate for the source region.

The highest measured sediment yields are from the humid regions of southern Asia and Oceania, and from basins that drain steep, high-relief mountains of young

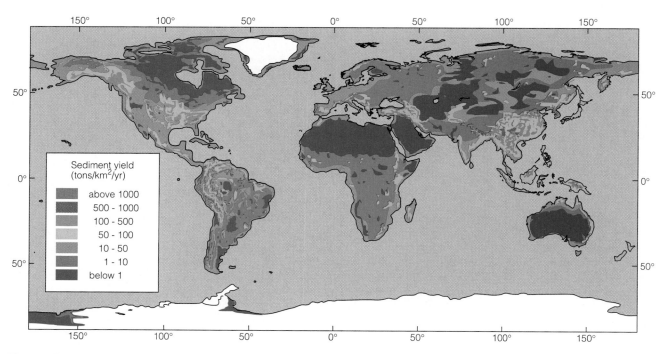

Figure 18.9 Sediment Yields Estimated rates of sediment yield on the continents. The highest yields are from southeastern Asia, which receives high precipitation related to the summer monsoon climate, and major high-relief mountain belts, including the Himalaya, the Alps, the Andes, and the coastal ranges of Alaska and British Columbia.

orogenic belts, such as the Himalaya, the Andes, and the Alps (Fig. 18.9). Low sediment yields characterize deserts and the polar and subpolar sectors of the northern continents. As one might expect, rates are high on steep slopes, and much higher in areas underlain by erodible clastic sediments or sedimentary rocks, or by low-grade metamorphic rocks, than in areas where crystalline or highly permeable carbonate rocks crop out. Structural factors also play a role, for rocks that are more highly jointed or fractured are more susceptible to erosion than massive ones. Denudation is surprisingly high in some dry climate regions, an important reason being that the surface often lacks a protective cover of vegetation. Soil protection by vegetation is likely a threshold phenomenon; below a certain vegetation density, erosion rates may increase.

The dominant erosional process can also strongly influence sediment yield. Measurements have shown that sediment yields generally increase with increasing glacier cover in a drainage basin. Values are unusually high in places like south-coastal Alaska, the most extensively glacierized temperate mountain region in the world, where denudation rates exceed those of comparable-sized nonglacierized basins in other regions by an order of magnitude. Under climates favorable for the expansion of temperate valley glaciers, both chemical and physical weathering rates tend to be substantially higher than the global average.

The continent discharging the most sediment to the ocean is Asia; it is also the continent on which the greatest average stream sediment loads have been measured. Asian rivers entering the sea between Korea and Pakistan are believed to contribute nearly half the total world sediment input to the oceans (Fig. 18.10). Second to Asia is the combined area of the large western Pacific islands of Indonesia, Japan, New Guinea, New Zealand, the Philippines, and Taiwan. Taiwan is especially remarkable because it produces only slightly less sediment than the entire contiguous United States!

One important factor that influences denudation rates is human activity, especially the clearing of forests, development of cultivated land, damming of streams, and construction of cities. Each of these activities has affected erosion rates and sediment yields in the drainage basins where they have occurred. Sometimes the results are dramatic: in parts of eastern United States, areas cleared for construction produce between 10 and 100 times more sediment than comparable rural areas or natural areas that are vegetated. On the other hand, in urbanized areas sediment yield tends to be low because the land is almost completely covered by buildings, sidewalks, and roads that protect the underlying rocks and sediments from erosion.

In many drainage basins, the measured and estimated sediment yields reflect conditions that are probably quite different from those of only a few decades ago

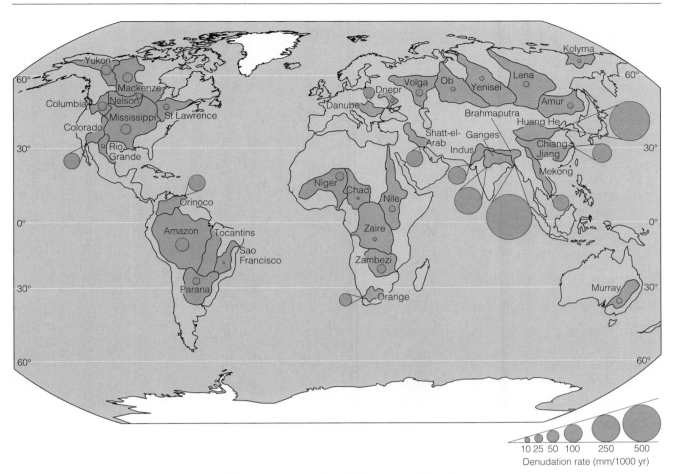

Figure 18.10 Rates of Denudation Present rates of denudation for the Earth's major drainage basins. Compare with Figure 18.9.

because much of the sediment that formerly reached the sea is now being trapped in reservoirs behind large dams. For example, the high Aswan Dam on the Nile river, in Egypt, now intercepts most of the sediment that formerly was carried by the Nile to the Mediterranean Sea.

Ancient Landscapes of Low Relief

The ultimate reduction of a landmass to low altitude, as envisioned by Davis, is likely to occur only if changes in plate motion lead to cessation of tectonic uplift. Denudation can then gradually lower the relief. Examples of such landscapes can be found in the world's shield areas, where the roots of ancient mountain systems have been exposed as the crust has thinned through the action of long-continued erosion and compensating isostatic adjustment.

Australia is widely believed to have some of the oldest landscapes of any of the continents. In places these ancient landscapes are dominated by inselbergs, (Fig. 13.31). Measurements of nuclides, produced by the bombardment of cosmic rays, which are concentrated just beneath the surface of the granitic inselbergs, have shown that the tops of these domes are eroding at a rate of less than a meter per million years. A combination of a relatively arid climate, long-term tectonic stability, and low erodibility of the exposed granite contribute to the exceptional stability of these landforms. Based on these measurements, it seems likely that this landscape predates the Quaternary Period (Fig. 8.10) and may have originated well back in the Tertiary Period, or even earlier.

Widespread erosional landscapes that have low relief and low altitude are not common. This must either mean that the Earth's crust has been very active in the recent geologic past or that such landscapes take an extremely long time to develop. Estimates have been made of the time it would take to erode a landmass to or near sea level by extrapolating current denudation rates into the past. Such estimates must take into account two important factors. Studies have shown that rates of denudation are strongly related to altitude, implying

Figure 18.11 Angular Unconformity Nearly flat-lying sedimentary strata in the upper walls of the Grand Canyon of the Colorado River rest on tilted older strata. The angular unconformity separating the two groups of rocks is a surface of low relief that can be traced for considerable distances, and represents a subdued ancient land surface.

that as the land is lowered the rate of denudation will decline. Furthermore, the eroding land will rise isostatically as the lithosphere adjusts to transfer of sediment from the land to the oceans. Both factors tend to increase the time it takes for the final reduction of a landmass to low altitude. Estimates of the time it would take to reduce a landmass about 1500 m high to near sea level, assuming no tectonic uplift, range from about 15 to 110 million years.

Although one might argue that much of the Earth's crust has been unusually active during the last 15 million years or more, could there have been earlier intervals of relative crustal quiet when low-relief surfaces did develop? Many ancient land surfaces are preserved in the geologic record as unconformities (Chapter 7). Some can be traced over thousands of square kilometers and can be shown to possess only slight relief (Fig. 18.11). Associated with such surfaces are weathering profiles that imply that the land surface was continuously exposed for long intervals at relatively low altitude. Such buried ancient landscapes are evidence of times when broad areas were eroded to low relief.

CONTINENTAL UPLIFT AND CLIMATIC CHANGE

Video Himalaya

The height and form of the continents can strongly influence both regional and world climate. Of singular importance to the evolution of the world's present climate system has been the uplift of the Himalaya and the Tibetan Plateau in the lower middle latitudes of central Asia over the last 50 million years (since the early Eocene), caused by the collision of the Indian Plate with the Asian Plate. The resulting high-altitude terrain may be the largest such feature produced on the Earth during Phanerozoic time (the last 540 million years) (Fig. 18.12). This vast uplifted region has had a major effect on the climate of Asia, as well as global circulation patterns. Since the collision began, more than 2000 km of crustal convergence has been accommodated by both lateral and vertical tectonic movements within Asia. During the first 20 million years of collision, the convergence was accommodated mainly by movement of crustal blocks to the east and southeast along

A.

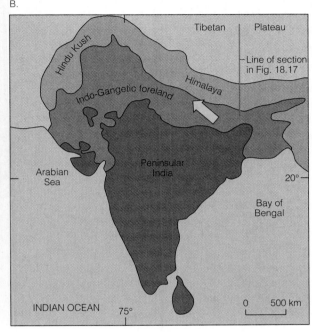

B.

Figure 18.12 Himalaya and Tibetan Plateau   A. An oblique view of the snow-covered Himalaya and the adjacent Tibetan Plateau rising above the Gangetic lowland (dark green) of India, as seen from an orbiting spacecraft.   B. Map of India and adjacent parts of Asia showing relationship of the Himalaya and Hindu Kush mountain systems to the Tibetan Plateau and the Indian subcontinent. Arrow shows direction of view of Figure 18.12A.

major strike-slip faults. Subsequently, the accommodation involved major north-dipping faults in the Himalaya and southern Tibet, together with many minor faults, which led to uplift of the Himalaya and the adjacent plateau. A variety of evidence indicates that by about 8 million years ago (late Miocene), the Tibetan Plateau had reached an altitude high enough to intensify the Asian monsoon system and affect global climate.

Central Asia, though clearly the most important, is not the only continental region to have experienced major regional uplift in the recent geologic past. Other areas include the high plateaus and mountains of the American West, the high Andes and Altiplano of South America, and the extensive plateaus of eastern and southern Africa that have mean altitudes of more than 1000 m.

To explore the effects of large-scale uplift on the Earth's climate system, global climate models have been used to compare conditions with the Tibetan Plateau (i.e., today) and without (prior to uplift). The climate simulations, supported by geologic evidence, point to some major global changes:

- Uplift has led to cooling of the midlatitude high plateaus. Raising the plateaus has led to local cooling of more than 16°C over Tibet (Fig. 18.13) and more than 8°C across the American West. The average temperature lowering between 30 and 40° N exceeds 10°C, and the mean hemispheric change over land areas is 8°C.

- Under no-plateau conditions, the westerlies would

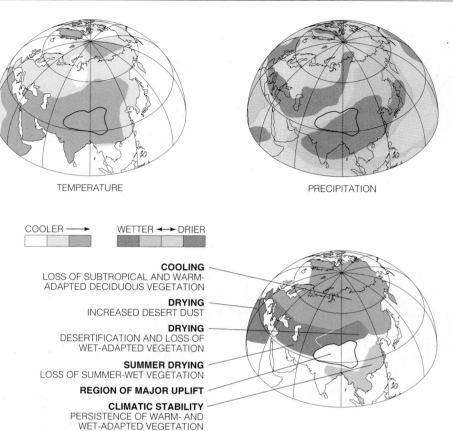

TEMPERATURE PRECIPITATION

COOLER ⟶   WETTER ⟷ DRIER

COOLING
LOSS OF SUBTROPICAL AND WARM-
ADAPTED DECIDUOUS VEGETATION

DRYING
INCREASED DESERT DUST

DRYING
DESERTIFICATION AND LOSS OF
WET-ADAPTED VEGETATION

SUMMER DRYING
LOSS OF SUMMER-WET VEGETATION

REGION OF MAJOR UPLIFT

CLIMATIC STABILITY
PERSISTENCE OF WARM- AND
WET-ADAPTED VEGETATION

Figure 18.13 Effects of Uplift on Climate Computer simulation of environmental changes in Asia resulting from uplift of the Tibetan plateau. A. Changes in temperature. B. Changes in precipitation. C. Overall changes in major geographic regions.

display strong zonal (latitudinal) flow (Fig. 18.14). Uplift in western North America and central Asia has created obstacles to low-level atmospheric flow, diverting the westerlies northward around the uplifted regions, with a return southward flow to the lee (downwind) of the high land. In North America, enhanced meandering of the jet stream brings cold air farther south, a condition that would favor ice-sheet glaciation in northern middle latitudes.

- An uplifted plateau and the bordering Himalaya generate intense precipitation on the windward flanks of the uplifted region as warm, moist summer air rises against the high topographic obstacle and cools. Increased rainfall in southeastern Asia, for example, contrasts with the formation of an arid rainshadow inland from the Himalaya, leading to desertification of the plateau surface (Fig. 18.14).

- Uplift has increased the seasonal contrast in precipitation related to an enhanced monsoon system. In southern Asia, summer heating of the high plateau causes warm air to rise, creating a low-pressure region that draws moist air inland from the adjacent ocean and produces the famous summer monsoon rains of the southern Asian countries (Fig. 18.14). In winter, the snow-covered surface of

the plateau reflects solar radiation and cools down, creating a region of high pressure from which cold, dry air flows outward and downward. Periods of maximum precipitation and runoff occur during times of maximum solar insolation that are dominated by orbital precession (Chapter 13).

- Increased precipitation related to strengthening of the monsoon means more frequent large floods during the summer monsoon season.

- Uplift has caused regional drying of continental interiors at middle latitudes. As a result of restructured atmospheric circulation, hot, dry air from interior Asia moves southwestward over Arabia and northern Africa, leading to summer drying (Fig. 18.13).

- Changes in ocean salinity and temperature related to changed atmospheric circulation would cause changes to deep and intermediate ocean circulation in the North Pacific and North Atlantic (Chapter 14).

The variety and magnitude of environmental and climatic changes linked to uplift of the mountains and plateau of central Asia point to the complexity of the interlinked components of the Earth System. We see that a major tectonic event resulting from the collision of two

A. Before uplift

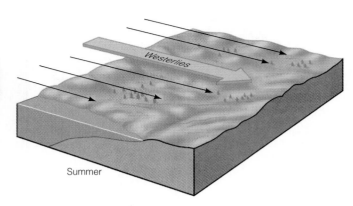

B. After uplift

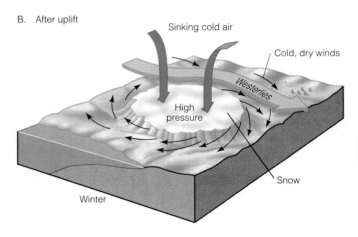

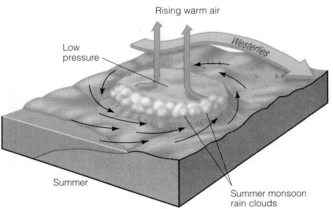

Figure 18.14 A Monsoon Climate Development of a monsoon climate system. Before uplift, westerly winds pass directly across a landscape, the axis of flow shifting seasonally north and south. As a large plateau is uplifted, the westerly flow is diverted around the upland, a winter high-pressure system develops over the high, snow-covered plateau, and cold, dry winds blow clockwise off the upland region. In summer, as the plateau surface warms, the warm air rises, drawing moist marine air inland. The summer monsoon clouds move inland toward the plateau margin and lose their moisture as heavy rainfall.

continents strongly affected not only the lithosphere in and beyond the belt of collision, but also the global atmosphere, hydrosphere, cryosphere, and biosphere. This continental impact has been a major influence on the shaping of the Earth's surface over the last 50 million years and no doubt will continue to be important far into the future.

UPLIFT, WEATHERING, AND THE CARBON CYCLE

The long-term trend of climate during the last 100 million years bears a general relationship to the reconstructed trend in atmospheric CO_2 concentration (Fig. 19.11). Explaining this long-term trend is important in trying to understand how the Earth's climate system has evolved. Of special interest is an explanation of the way the climate maintains a degree of equilibrium. Why doesn't atmospheric CO_2 rise to such high values that the Earth becomes a "hothouse" like Venus, or fall so low that the Earth cools down to become a frigid planet like Mars? According to one hypothesis, the average global rate of seafloor spreading controls the rate of CO_2 generation along midocean ridges and along zones of collision and subduction (Fig. 18.15). An additional and possibly larger source of atmospheric CO_2 than volcanic outgassing is metamorphism of carbon-rich ocean sediments carried downward in subduction zones. Because subduction and seafloor spreading are fundamental aspects of plate tectonics, variations in plate motion necessarily must play a large role in controlling

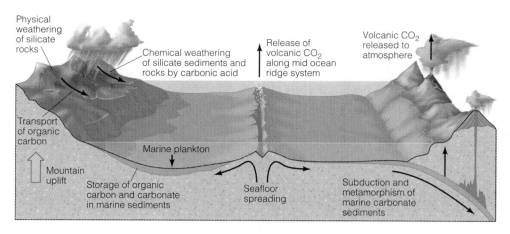

Figure 18.15 The Carbon Cycle Diagram of carbon cycle showing primary natural sources of atmospheric CO_2 (volcanism and metamorphism), removal of CO_2 due to surface weathering, and transport and long-term storage of organic carbon in marine sediments.

the rate of CO_2 buildup in the atmosphere. However, at the same time, CO_2 is removed from the atmosphere by the weathering of surface silicate rocks. As we have seen in Chapter 7, rainwater combines with CO_2 to form carbonic acid (H_2CO_3), and this acid weathers silicate rocks, as in the following generalized reaction:

$$CaSiO_3 + H_2CO_3 \rightarrow CaCO_3 + SiO_2 + H_2O$$
(silicate rock)(carbonic acid)(weathering products)(water)

The weathering products are carried by streams to the ocean. There the carbonate and silica are used by marine organisms, whose remains accumulate on the seafloor and are stored as sediment (Figs. 5.10 and 18.15). Weathering of silicate rocks, therefore, is a negative feedback in the climate system that can help move global climate toward an equilibrium condition. If seafloor spreading speeds up, more CO_2 enters the atmosphere, the Earth warms, the water vapor content of the atmosphere rises, and vegetation density increases. This speeds up the rate of chemical weathering of silicate rocks, which removes CO_2 from the atmosphere, thereby keeping the system in a more balanced state. If spreading slows, less CO_2 enters the atmosphere, the climate cools, weathering rates decrease, and the system adjusts to a near-balanced condition.

A further important negative feedback is burial of organic carbon. High erosion rates can lead to rapid sedimentation in sedimentary basins and in extensive submarine fans. Isolated from the weathering environment at the land surface, the carbon is quickly stored rather than being returned to the atmosphere by surface weathering. Recent studies suggest that carbon burial may be even more important than silicate weathering in long-term carbon storage.

An alternative hypothesis has been proposed in which tectonic uplift is the driving force behind changes in atmospheric CO_2 levels. Uplift could increase the rate of removal of atmospheric CO_2 because faulting exposes fresh, fractured rock; earthquakes promote landslides that dislodge rock from steep slopes; glacial and periglacial (physical) weathering is dominant on unvegetated high-altitude slopes; and runoff from high summer rainfall transfers physically weathered products from high and middle altitudes, where slopes are steepest, to lower altitudes, where they are chemically weathered in floodplains and deltas in warm, moist climates. Instead of being a relatively weak negative feedback, chemical weathering now becomes the major factor in controlling the long-term concentration of CO_2 in the atmosphere, and it focuses attention on areas of rapid uplift as primary sites influencing the Earth's carbon budget.

RECENT MOUNTAIN UPLIFT AND POSITIVE FEEDBACKS

The great altitude and relief of major mountain ranges like the Himalaya and Andes have led most geologists to assume that these impressive topographic features are the result of ongoing tectonic forces that have led to accelerated uplift during the late Cenozoic. But is tectonism the most likely or only cause? An alternative hypothesis questions this long-standing assumption and proposes that uplift may also be driven by important feedbacks in the Earth System. If a mountain system is uplifted by tectonic forces (e.g., plate subduction or collision), erosion and mass-wasting will tear away at the

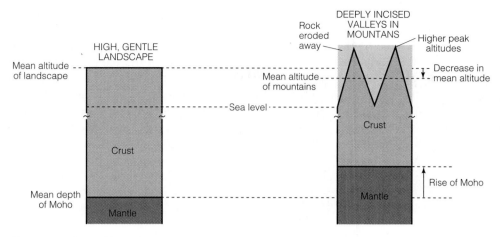

Figure 18.16 Effect of Erosion on Mean Altitude Simplified sections through the continental crust showing the effects of erosion on mean altitude, the depth of the Moho (crust/mantle boundary; see Chapter 16), and uplift of rock. If a change in climate causes a high, gentle landscape to be deeply incised by streams that erode valleys toward sea level, the remaining crustal rocks and the Moho will rise (due to isostasy), the mean altitude of the mountainous area will decrease slightly, and the highest peaks will be much higher than before.

rising landmass, transferring sediment to adjacent basins and to the ocean. Removal of this mass of rock creates an isostatic imbalance, the adjustment to which will cause the mountains to rise because of their reduced load on the underlying mantle (Fig. 18.16). Higher mountains provide an increased barrier to rain clouds, which will therefore lead to deep incision by rivers on the windward flank, further dissecting the mountains. The interesting consequence is that as the base of the crust rises due to a reduction in overlying load, and the

mountains increase in altitude, the mean altitude of the mountains actually decreases. This somewhat surprising result is confirmed by recent studies of the topography of the Tibetan plateau and the Himalaya. Although the mountain range rises to altitudes of more than 8000 m, its mean altitude of ca. 5000–5500 m is nearly identical to that of the adjacent relatively undissected plateau (Fig. 18.17).

An added factor in the equation is glaciation. If the isostatically rising mountains exceed the altitude of the

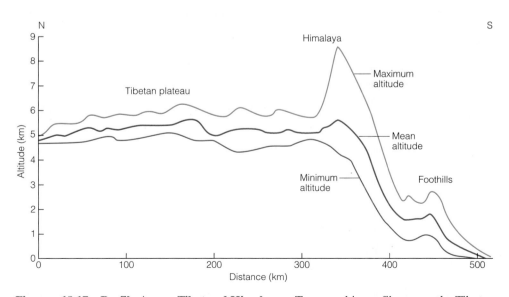

Figure 18.17 Profile Across Tibet and Himalaya Topographic profile across the Tibetan Plateau and the Himalaya (see Figure 18.12B for orientation) showing that in spite of much higher altitudes, the average altitude of the Himalaya is nearly equal to that of the plateau.

snowline, glaciers will form and the rate of denudation may increase. This positive feedback link may lead to additional positive feedbacks which, as we have seen, include increased physical and chemical weathering, a reduction in atmospheric CO_2, a cooling of global climate, and a further lowering of the snowline. Repeated lowering of the snowline during dozens of Quaternary glaciations (Fig. 15.16) likely accelerated erosion at high altitudes and may have helped maintain a relatively constant mean altitude.

These contrasting hypotheses that attempt to explain landscape and climate evolution illustrate the complexity of the problem. The entire Earth System is involved, and not only the basic mechanisms but also the complex positive and negative feedbacks leave us with difficult questions to answer and challenging research problems to pursue. The answers will not come from any single field, but rather from collaborative investigations by scientists in many different fields of Earth Science.

SUMMARY

1. Natural processes are constantly changing the character of the landscape, but so, too, are human activities. Although generally operating at slow rates, natural processes can produce dramatic changes on geologic time scales.

2. The Earth's surface topography reflects sculpture by different erosional processes and deposition of resulting sediment. Water, ice, and wind are the principal agents that erode the land and transfer sediment toward the ocean basins. The principal factors influencing landscape development are process, climate, lithology, relief, and time.

3. A close relationship exists between rock structure resulting from the movement of crustal plates and the location of major drainage basins and continental divides.

4. Landscapes are constantly adjusting to changes in the factors that control their development and likely are never in a complete state of equilibrium.

5. Landscapes evolve through time as tectonic forces raise crustal rocks and erosional agents wear them away. Rates of denudation in some mountain areas appear to be approximately equal to rates of uplift.

6. Ancient landscapes with low relief imply long intervals of erosion that slowly lowered the land surface to low altitude. Although not common in modern landscapes, extensive surfaces of low relief are preserved in some ancient rocks as widespread unconformities.

7. Late Cenozoic uplift of high mountains and plateaus has influenced world climate and led to local and hemispheric cooling, changed the path of the jet streams, created rainshadows that promoted desertification in continental interiors, and intensified monsoon circulation. Other effects include more frequent major summer monsoon floods and changes in the intermediate and deep-water circulation of the oceans.

8. Silicate weathering and burial of organic carbon remove CO_2 from the atmosphere and place it in long-term storage. Removal of CO_2 from the atmosphere reduces the greenhouse effect and influences world climate.

9. Mountain uplift may be the result both of tectonic activity and isostatic response to the erosion and transfer of rock debris away from mountain areas. In high, glaciated mountains rates of uplift and denudation may remain nearly balanced, resulting in relatively constant mean altitudes.

THE LANGUAGE OF GEOLOGY

continental divide (p. 495)   geographic cycle (p. 496)   relief (p. 492)

denudation (p. 498)

QUESTIONS FOR REVIEW

1. In what ways does mass-wasting differ from stream erosion? Name three geologic factors that make high mountain regions especially prone to landslides.

2. How might one calculate the volume of sediment removed from the land surface during the last 2 million years by a major stream that has built a large delta where it enters the ocean?

3. How might drainage patterns provide clues to rock structure in an area of dense tropical forest?

4. How would you determine the total relief of an area appearing on a topographic map?

5. How do bedrock geology and structural history control the position of the continental divide in western North America?

6. What geologic evidence points to periods of relative landscape equilibrium at times in the past?

7. How does the concept of *thresholds* apply to landscape evolution?

8. Describe two ways that you might be able to determine the uplift rate in a coastal mountain system.

9. Do glaciers increase or decrease denudation rates? Why?

10. How can human activity influence sediment yield from a drainage basin?

11. How might biologic factors help control the denudation rate of a drainage basin?

12. How could mountain uplift be related to geologic factors other than tectonism?

19

Measuring
Our Earth

Understanding
Our Environment

GeoMedia

Ozone in the upper atmosphere acts as a shield against harmful ultra-violet radiation. Record low concentrations of this gas in recent years, believed to be the result of human activity, places sunbathers at crowded beaches like this one at Malia, on the north coast of Crete, at increased risk of developing skin cancer.

Our Changing Planet

The Ozone Hole

Ozone, a pale-blue gas with a pungent odor, is present in the atmosphere in very small amounts (only 20–40 parts per billion by volume near the land surface). However, without this gas, life on the Earth would be very different, for ozone provides living organisms with a protective shield against harmful ultraviolet radiation from the Sun. For humans, direct exposure to ultraviolet light will damage the immune system, produce cataracts, substantially increase the frequency of skin cancer, and cause genetic mutations. Maximum concentrations of ozone are found in a layer between 25 and 35 km above the Earth's surface, a region where ultraviolet radiation breaks down molecules of oxygen (O_2) into oxygen atoms that are then able to combine with other O_2 molecules to form molecules of ozone (O_3). The ozone, in turn, is broken down by ultraviolet radiation, thereby creating a balance among O, O_2, and O_3.

In 1985, British scientists reported a startling discovery: a vast hole, about the size of Canada, had developed in the ozone layer above the Antarctic region. By 1987, measurements showed that concentrations of this life-protecting gas over Antarctica had dropped more than 50 percent since 1979 and that between altitudes of 15 and 20 km the depletion had reached 95 percent. Studies have shown that the productivity of oceanic phytoplankton, which lie at the base of the marine food chain, has decreased 6 to 12 percent in the region of the ozone hole, where record amounts of ultraviolet radiation have been measured. Record-low ozone values have also been measured over Australia and New Zealand, and continuing surveys showed that ozone values at all latitudes south of 60° decreased by 5 percent or more after 1979. By early 1993, the globally averaged total ozone concentration, measured by satellite, had decreased to unprecedented low values.

What had happened to upset the natural atmospheric balance among the three gaseous forms of oxygen? A decade earlier, it was recognized that a group of synthetic industrial gases, the *chlorofluorocarbons* (or *CFCs*), were entering the lower atmosphere and spreading rapidly around the world. As the CFCs ultimately rise into the upper atmosphere, ultraviolet radiation breaks them down, releasing chlorine. It is chlorine that does the damage: the chlorine atoms destroy the ozone in a catalytic reaction, with each chlorine being capable of destroying as many as 100,000 ozone molecules before other chemical reactions remove the chlorine from the atmosphere. The sunlight and very cold springtime temperatures (280°C or lower) in the upper atmosphere that are critical to the ozone destruction process are present in the south polar region, which is why the ozone hole is especially pronounced over Antarctica; in the Arctic,

Understanding Our Environment

The Clean-Up of Lake Washington

Lake Washington is a large, urban lake; Seattle extends along its western shoreline, and smaller cities and suburbs surround the rest. Although Seattle's sewage was diverted into Puget Sound in the 1930s, the growing suburban communities continued to route sewage into the lake. The sewage became a major concern in 1955, when University of Washington zoologist W. T. Edmondson recognized a large population of *Oscillatoria rubescens,* a planktonic cyanobacterium that can cause serious deterioration in water quality.

By the 1950s, the lake no longer supplied drinking water, but it was used extensively for swimming, boating, and fishing. Water quality is poor for such uses when clarity is reduced, foul odors are produced by large populations of algae, or bacterial concentrations pose a health risk. Moreover, large amounts of algae can lead to low levels of dissolved oxygen in the water, which can cause fish kills.

In northern temperate lakes, low levels of phosphorus limit the growth of algae. Sewage, however, is rich in phosphorus. As the communities around Lake Washington grew, the amount of sewage entering the lake increased dramatically, and the algae responded by growing rapidly. In 1957, studies of the chemistry of the streams flowing into the lake showed that sewage was the major source of phosphorus in the lake; diversion of the sewage effluent therefore would significantly reduce phosphorus levels in the lake.

The effluent could be diverted to nearby Puget Sound, as had been done in the case of Seattle. However, many jurisdictions were involved, and most had no access to Puget Sound. It was necessary to create a new govern-

mental agency, the Municipality of Metropolitan Seattle (Metro), to supervise the sewage diversion project. However, some people expressed alarm at the formation of a regional agency, fearing "creeping socialism" or a major land grab by Seattle. Efforts were made to discredit the scientific evidence. When the legislation authorizing the creation of Metro came to a vote in spring 1958, it failed in the smaller lakeside communities.

That summer, beaches along the lake were closed for the first time due to high bacterial counts. Both the public and elected officials got the message. That fall a modified bill was resubmitted and passed easily.

It took five years to build the immense network of sewer lines, and during that time residents watched the lake deteriorate as the amount of *Oscillatoria* in the water increased each summer. The algal blooms decreased water clarity and created foul odors. Diversion of the sewage was completed in 1968. The lake responded almost immediately. By 1971 the water was as clear as it had been in 1950. Change did not stop there, however. By 1976 *Oscillatoria* had been virtually eliminated from the lake, and water clarity far exceeded expectations.

The recent history of Lake Washington serves as a model of cooperation among scientists, government, and citizens in studying the causes of an environmental problem and forging a workable solution. It involved recognition of environmental warning signals, clear communication about their meaning, governmental willingness to change in order to implement a solution, and an active group of citizens who, in spite of the costs, were committed to restoring their lake and preserving it for future generations.

the period of the critical spring conditions is much shorter. With the documentation of ozone depletion in the upper atmosphere, scientists for the first time could show that human activity was having a detrimental global effect on one of the Earth's natural systems.

The CFCs responsible for the ozone hole are used in the manufacture of plastic foams, as propellants in aerosol cans, and as refrigerants. In 1987, scientific concerns about ozone destruction led 49 industrial nations to ratify the Montreal Protocol on Substances That Deplete the Ozone Layer, an agreement calling for a reduction in CFC production to 50 percent of 1986 amounts by the end of the century. However, because even such cutbacks in CFC use would permit continued degradation of the ozone shield, in 1990 the United States and many other nations pledged to eliminate CFC production entirely by the year 2000.

THE DYNAMIC EARTH

By now it should be clear to you why *The Dynamic Earth* was selected as the title for this book: the Earth is always changing. This fact, introduced in the opening pages, was underscored repeatedly as we examined the natural internal and external processes that shape our planet. However, we also have stressed that not all the changes taking place around us are natural. As we can see in the example of the ozone hole, many are the direct result of human activities.

Over many millennia prior to the industrial revolution, people slowly changed the Earth's natural landscapes as they built villages and cities, converted forests to agricultural land, and locally dammed and diverted streams. Then, as industrial technology developed, mineral and energy resources were needed to fuel an increasingly populous and demanding society. The exploitation of fossil fuels helped raise the standard of living for most people well beyond that of their forebears of only a century or two before.

In spite of the obvious benefits involved, the exploitation of our planet's rich natural resources has not been without cost (see the boxed essay: *The Clean-up of Lake Washington*). In many parts of the world, environmental deterioration is epidemic. In addition to scarring and poisoning the Earth's land surface, we have also unwittingly polluted the oceans and groundwater and changed the composition of the atmosphere. Even in places long considered to be the most remote on the planet—the frigid ice sheets of Antarctica, the vast Amazon rainforest, the trackless desert of Saudi Arabia, the lofty summits of the Himalaya—the impact of human activities is being felt. Today, human and natural geologic activities are inextricably intertwined, and it is increasingly apparent that *Homo sapiens* has become a major factor—a global factor—in geological change.

The Earth as an Interactive, Dynamic System

The geologic study of the Earth has made it abundantly clear that our planet is a complex, dynamic system. The solid, liquid, gaseous, and organic realms of the Earth are closely interlinked (see the boxed essay *The Younger Dryas Event and the End of the Last Ice Age*). A change in one part of the system is likely to affect other parts. For example, a massive earthquake might raise an extensive zone of coastal land, exposing and destroying nearshore marine habitats. A volcano might erupt lava that dams a river, thereby affecting other streams in the drainage system, while tephra and gases ejected into the atmosphere could lead to a hemisphere-wide drop in air temperatures.

Human action also can perturb a natural system, and not always in a predictable manner. In Chapter 10 we saw how construction of the Aswan Dam on the Nile River halted the natural supply of sediment to the Nile delta and how this disruption deprived the rich agricultural delta lands of their annual input of nutrients and left the delta front vulnerable to increased wave erosion. We learned in Chapter 11 how artificial irrigation of naturally arid land in California's San Joaquin Valley led to poisoning of wildlife habitats as selenium, flushed from the surface soils, reached toxic levels. In Chapter 14 we saw that tampering with the natural stream and estuarine system of San Francisco Bay has led to unanticipated consequences: disappearance of commercial fisheries, concentration of industrial poisons, and widespread destruction of wetland ecosystems. In each case, people upset the natural balance, causing the system to respond in unforeseen ways.

Time Scales of Change

Realization that the Earth is a dynamic planet and always changing has come slowly. While it is apparent to anyone that the weather can change from year to year, it is less obvious that climate can change over a person's lifetime. It is far more difficult to comprehend, from personal experience, that the solid Earth is also changing. A person might witness a volcanic eruption or experience a major earthquake, yet not guess that the ground underfoot is slowly and continuously moving, driven by forces deep beneath the surface. We now can measure the rates at which lithospheric plates move; we know that plate motion generally is in the range of 1 to 12 cm/yr. This means that over a lifetime of a 70-year-old person

Measuring Our Earth

The Younger Dryas Event and the End of the Last Ice Age

At the end of the last glaciation about 11,000 to 10,000 radiocarbon years ago, the climate in the North Atlantic and adjacent lands experienced a rapid and remarkable change. For 2000 years, the climate had been warming, causing ice sheets in North America and Europe to retreat and allowing plants and animals to reoccupy the deglaciated landscape. Many mountain glaciers in Britain and Scandinavia had disappeared, and the southern limit of sea ice in the North Atlantic had shifted far north, close to its present limit. By all indications, the glacial age was drawing to a rapid close. Then, very abruptly, the climate cooled. Water temperatures in the North Atlantic fell as the southern limit of polar water shifted southward, nearly to its full-glacial extent. The retreating ice sheets halted, then readvanced, and mountain glaciers were reborn in formerly ice-free cirques. Forests in northwestern Europe were rapidly replaced by low-growing herbaceous plants typical of full-glacial conditions. Among these plants was a distinctive flowering species, *Dryas octopetala*, now limited to polar latitudes and high altitudes. *Dryas* pollen is found abundantly in organic deposits dating to this interval and has provided the name used to identify this cold episode—the Younger Dryas.

Oxygen-isotope data obtained from Swiss lake sediments and a Greenland ice core show that the onset of the Younger Dryas episode was rapid. These records also indicate that the event terminated equally rapidly (Fig. B19.1). In fact, the ice core indicates that the climate over Greenland warmed about 7°C in only 40 years, a rate that exceeds even the unusually rapid average global rate of temperature rise that climate models project for the coming century.

The effects of Younger Dryas cooling are most pronounced around the North Atlantic, and so the search for a cause has focused on this region. As pieces of the puzzle have been assembled, it has become clear that the solution likely lies in interactions of the Earth's natural systems: the cryosphere, the oceans, the atmosphere, and the biosphere.

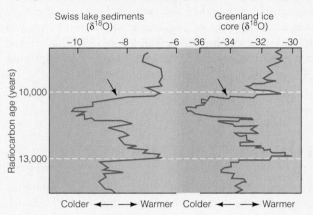

Figure B19.1 The Younger Dryas Measurements of oxygen isotopes in the sediments of a Swiss lake and an ice core from the Greenland Ice Sheet show an abrupt and rapid change of climate at the end of the last glaciation (arrows). The curves, which can be viewed as recording changes in temperature, show a sudden shift to colder climate followed by an abrupt return to warmer postglacial climate. Detailed studies of the ice core indicate that, at the end of the Younger Dryas event, the average temperature in Greenland abruptly climbed about 7°C in only 40 years.

the plate on which that person has lived traveled less than 10 m, a displacement far too small to be recognized without sophisticated scientific measurement.

Recognizable changes that affect the natural equilibrium of the Earth globally generally take place very slowly—over intervals far longer than an individual's lifetime. However, there are several exceptions to this generalization. One exception involves the impact of a large meteorite like the one that struck Arizona about 50,000 years ago (see Chapter 1). Fortunately, such events are rare, and no even-larger impact having global consequences has occurred in human history. A second exception is a major explosive volcanic eruption, which ejects so much gas and dust into the stratosphere that much of the incoming solar radiation is reflected back into space, thereby cooling the Earth's surface for sev-

eral years (Chapters 4 and 12). Changes we cause ourselves are a third exception. In the following pages, we will see how human activities have brought about unprecedented global changes that present us with major challenges for the coming decades.

THE CHANGING ATMOSPHERE

Probably since earliest times, people have asked important questions about the Earth: How did the major features of our planet originate? Has it always been the way we see it now? How can we explain the diversity of life

Glacial-geologic studies have shown that as the ice sheet over eastern North America retreated, vast meltwater lakes were ponded beyond the glacier margin. When the retreating ice uncovered a natural drainageway between these lakes and the North Atlantic, meltwater flowed rapidly into the ocean, where it formed a fresh-water lid over the denser salty marine water. The cold surface meltwater reduced evaporation from the ocean surface, thereby shutting down the ocean's thermohaline circulation system (Chapter 14). Eastward-flowing air masses traveling across the colder North Atlantic cooled and moved across Western Europe, bringing a return to frigid ice-age conditions (Fig. B19.2). As the huge meltwater lakes drained and meltwater flow from the ice sheet eventually slowed, ocean circulation resumed and warmer climate returned to the North Atlantic region, heralding the rapid termination of the ice age.

As we learn more about this remarkable natural climatic event, and see how the Earth's physical and biological systems responded to it, important insights will be gained that may help us anticipate future environmental changes linked to a rapidly warming world.

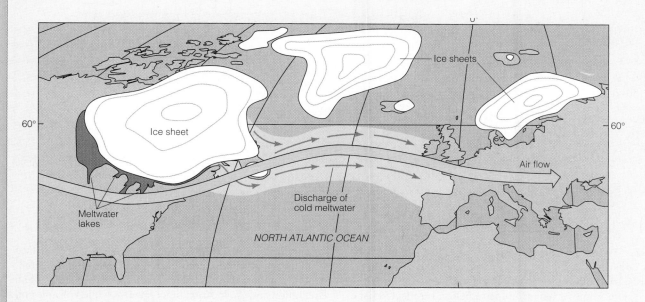

Figure B19.2   Causes of the Younger Dryas Event   Distribution of ice sheets in the North Atlantic region and North American ice-margin lakes during the Younger Dryas event. Rapid drainage of large volumes of meltwater into the western North Atlantic cooled the ocean surface and reduced its salinity, shutting down the thermohaline conveyor system. Air passing over the cold North Atlantic brought colder conditions to northwestern Europe that led to the growth of glaciers and a major change in vegetation communities.

on the Earth? What is the place of *Homo sapiens* in the multitude of living things that inhabit our planet? What causes changing patterns of weather? Has climate on the Earth always been like we find it today?

These questions have been a focus of study ever since people began to acquire and organize knowledge about the physical world around them—in other words, ever since science became a human endeavor. The questions pertain to three fundamental theories about the Earth. The first theory, about how the solid Earth works, is what we now call plate tectonics. The secondary theory, dealing with life since the earliest times, is the theory of evolution. The final theory deals with the atmosphere and can be called the theory of climatic change. Today's scientists feel rather confident that in plate tectonics and evolution they have found reasonable and consistent explanations for the first two sets of questions. A comprehensive theory of climatic change is not yet in hand, although major advances have been made during the last two decades. In many ways, this theory is more difficult to deal with than the other two, for it requires collaboration among scientists in many varied disciplines.

The Earth's climate system consists of a number of interacting subsystems that involve the atmosphere, the hydrosphere, the solid Earth, and the biosphere (Fig. 19.1). Climate operates on many time scales, and explanations of climatic change on a million-year time scale may have little or nothing to do with changes that take place on millennial or decadal time scales. So complicated are the interacting systems that only with the advent of supercomputers have we begun to answer

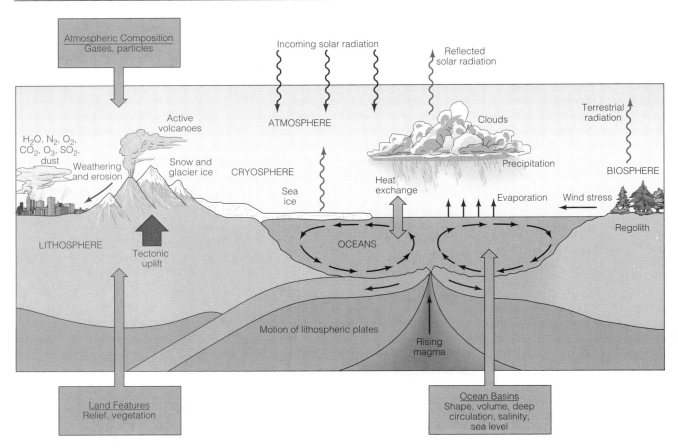

Figure 19.1 The Climate System The Earth's climate system consists of four interacting subsystems: atmosphere, hydrosphere, lithosphere, and biosphere. Solar energy drives the system. Some of the incoming radiation is reflected back into space from clouds, atmospheric pollutants, ice, snow, and other reflective surfaces. Tectonic movements affect surface relief (which influences atmosphere circulation) and the geometry of continents and ocean basins (which controls ocean circulation and the location of ice sheets), while volcanic and industrial gases affect atmospheric composition.

some of the basic questions about how the climate system works.

Geology plays an important role in this enterprise, for geologists have within their grasp a record of the Earth's changing climates that extends into the remote past. Climate can be read from the stratigraphic record in many ways. For example, paleontologists, by invoking the Principle of Uniformitarianism, infer past climates from the assemblages of fossil plants and animals they find. Sedimentologists and stratigraphers can infer many things about past climates from the nature of the sediments they study: the present distribution of these sediments, their mineralogy, the varied facies represented, features that indicate the agencies of transport and deposition, and the soils that represent former land surfaces. Isotope geologists can determine past surface temperatures from studies of terrestrial and marine sediments and polar ices cores. By reconstructing past climates, geologists can determine the range of climatic variability on different time scales and use their data to test the accuracy of computer models that try to simulate past climate conditions.

One important reason for these studies is to learn how the climate system behaves, what controls it, and how it is likely to change in the future. We know it *will* change, for the geologic record of climatic change is abundantly clear. What we still lack is a clear view of how it will change and at what rate. The answers are important not only scientifically, but socially and politically as well, for in the process of extracting and burning the Earth's immense supply of fossil fuels, people have unwittingly begun a great geochemical "experiment" that is likely to have a significant impact on our planet and its inhabitants.

The Carbon Cycle

A basic chemical substance involved in the experiment is carbon, an element that is essential to all forms of life. Before discussing the experiment, we need to understand how carbon moves through the biosphere, the lithosphere, the hydrosphere, and the atmosphere in a major biogeochemical cycle (a natural cycle describing the movements and interactions, through these several spheres, of the chemicals essential to life; see chapter 2).

Carbon occurs in four reservoirs: (1) in the atmosphere it occurs in carbon dioxide; (2) in the biosphere it occurs in organic compounds; (3) in the hydrosphere it occurs as dissolved carbon dioxide; and (4) in the crust it occurs both in the calcium carbonate of limestone and in decaying and buried organic matter such as peat, coal, and petroleum. Each reservoir is involved in the carbon cycle (Fig. 19.2).

The key to the carbon cycle is the biosphere, where plants continually extract CO_2 from the atmosphere and then break the CO_2 down by the process of photosynthesis to form organic compounds. When plants die, the organic compounds decay by combining with oxygen from the atmosphere to form CO_2 again. The passage of material through the biosphere is so rapid that the entire content of CO_2 in the atmosphere cycles every 4.5 years. However, the biospheric and atmospheric pathways interact with pathways in the hydrosphere and crust as well.

Not all dead plant matter in the biosphere decays immediately back to CO_2. A small fraction is transported and redeposited as sediment; some is then buried and incorporated in sedimentary rock. The buried organic matter joins the slower moving rock cycle and will reenter the atmosphere only when uplift and erosion have exposed the rock in which the organic matter is trapped.

Carbon dioxide from the atmosphere also is dissolved in the waters of the hydrosphere. There it is used by aquatic plants in the same way that land plants use CO_2 from the atmosphere. In addition, aquatic animals extract calcium and carbon dioxide from the water to make shells of $CaCO_3$. When the animals die, the shells accumulate on the seafloor, mixing with any $CaCO_3$ that may have been precipitated as chemical sediment. When compacted and cemented, the $CaCO_3$ forms limestone. In this way, too, some carbon joins the rock cycle. Eventually, the rock cycle will bring the limestone back to the surface where weathering and erosion will break it down; the calcium returns in solution to the ocean, and the carbon escapes as CO_2 to the atmosphere.

Next, we must consider what happens when human activities start to change these four carbon reservoirs and influence the fluxes between them. When we mine coal

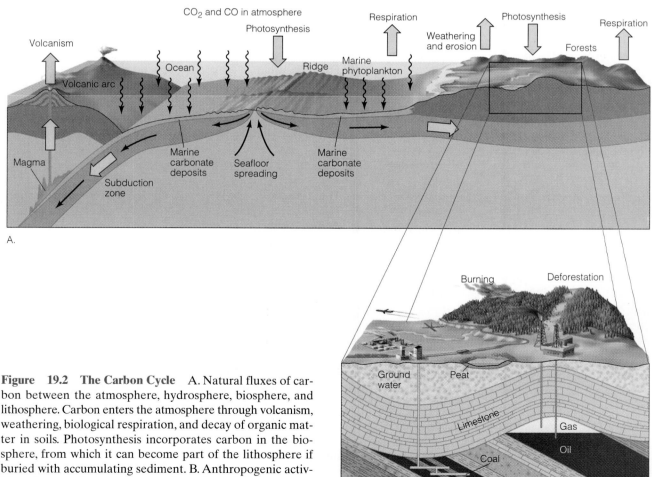

Figure 19.2 The Carbon Cycle A. Natural fluxes of carbon between the atmosphere, hydrosphere, biosphere, and lithosphere. Carbon enters the atmosphere through volcanism, weathering, biological respiration, and decay of organic matter in soils. Photosynthesis incorporates carbon in the biosphere, from which it can become part of the lithosphere if buried with accumulating sediment. B. Anthropogenic activities release carbon to the atmosphere through deforestation and burning of wood, and the combustion of fossil fuels.

and extract oil from the crust and then burn them, we convert organic matter to CO_2, thereby speeding up the rate of the carbon cycle. When we clear vast forests or allow deserts to expand when overgrazing kills off the vegetation cover, we are reducing the size of the biosphere reservoir. Although any individual action may appear insignificant, the cumulative effect of all human activities is now so great as to have a measurable effect. Both the burning of fossil fuels and the clearing of land cause CO_2 to be released to the atmosphere at rates that are faster than the natural rate. Unless this CO_2 is either dissolved in the hydrosphere or buried rapidly in sediments, the CO_2 content of the atmosphere inevitably must increase. The rate at which CO_2 dissolves in the hydrosphere is known to be slower than the rate at which human activities are adding it to the atmosphere. Therefore, the CO_2 content of the atmosphere should be increasing.

The Greenhouse Effect

The atmosphere is the engine that drives the Earth's climate system, and the Sun provides the energy that allows the engine to work (Fig. 19.3). Some of the solar radiation that reaches the atmosphere is reflected off clouds and dust and bounces back into space. Of the radiation that reaches the planet's surface, some is absorbed by the land and oceans, and some is reflected into space by water, snow, ice, and other highly reflective surfaces. This visible reflected solar radiation has a short wavelength. The Earth also emits long-wave, infrared radiation. However, a portion of the outgoing long-wave radiation encounters gases in the atmosphere which have chemical properties that prevent this energy from escaping. Instead, the energy is retained in the lower atmosphere, causing the temperature at the Earth's surface to rise. A comparable effect also explains why the air temperature in a glass greenhouse is warmer than the air outside: the glass, acting in much the same way as the atmospheric gases, prevents the escape of radiant energy. Hence, we refer to this phenomenon as the **greenhouse effect**.

It is the greenhouse effect that makes the Earth habitable. Without it, the surface of our planet would be as inhospitable as those of the other planets in the solar system. If the Earth lacked an atmosphere, its surface environment would be like that of the Moon, on the sunlit side of which temperatures are close to the boiling point of water, while on the dark side the temperature is far below freezing. The nearly airless surface of Mars is a frigid landscape whose closest earthly analogs are the

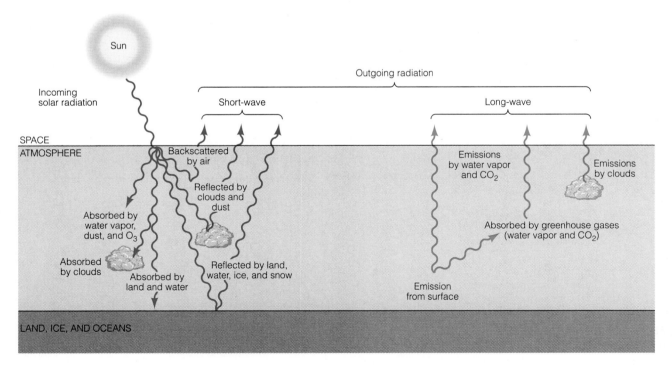

Figure 19.3 Greenhouse Gases Some of the short-wave (visible) solar radiation reaching the Earth is absorbed by the land and oceans, while some is bounced back into space off reflective surfaces that include snow, ice, clouds, and atmospheric dust. The Earth also radiates long-wave radiation back into space. Greenhouse gases trap some of the outgoing long-wave radiation, causing the atmosphere to heat up. With increasing concentration of these trace gases, the air temperature of the lower atmosphere rises.

frozen polar deserts. By contrast, Venus is much closer to the Sun, and its atmosphere is so dense that the greenhouse effect generates surface temperatures hot enough to melt lead.

Greenhouse Gases

Dry air consists mainly of three gases: nitrogen (79%), oxygen (20%), and argon (1%). However, water vapor is usually present in the Earth's atmosphere in concentrations of up to several percentage points and accounts for about 80 percent of the natural greenhouse effect. The remaining 20 percent is due to other gases present in very small amounts. Despite their very low concentrations, measurable in parts per billion by volume (ppbv) of air, these *trace gases* contribute significantly to the greenhouse effect (Table 19.1).

Chief among the trace gases is carbon dioxide, which has an average concentration in surface air of about 366,000 ppbv. Other significant gases, each a basic part of natural biogeochemical cycles and efficient in absorbing infrared radiation, are methane, nitrous oxide, and ozone. The commercially produced CFCs are an additional important group of greenhouse gases.

Trends in Greenhouse Gas Concentrations

During the past two decades, the greenhouse gases have received increasing scientific and public attention as it has become clear that the atmospheric concentration of each is rising.

Carbon Dioxide. In 1958, in conjunction with the International Geophysical Year, measurements were made of carbon dioxide concentration in the atmosphere near the top of Mauna Loa Volcano in Hawaii. This site was chosen because of its altitude and remote location, far from sources of atmospheric pollution. The measurements are still being made, and they show two remarkable things. First, the amount of CO_2 fluctuates

TABLE 19.1	**Atmospheric Trace Gases Involved in the Greenhouse Effect**					
	Carbon Dioxide (CO_2)	Methane (CH_4)	Nitrous Oxide (N_2O)	Chlorofluoro-carbons (CFCs)	Tropospheric Ozone (O_3)	Water Vapor (H_2O)
Greenhouse role	Heating	Heating	Heating	Heating	Heating	Heats in air; cools in clouds
Effect on stratospheric ozone	Can increase or decrease	Can increase or decrease	Can increase or decrease	Decrease	None	Decrease
Principal anthropogenic sources	Fossil fuels; deforestation	Rice culture; cattle; fossil biomass burning	Fertilizer land-use conversion	Refrigerants aerosols; industrial processes	Hydrocarbons (with NO_x) biomass burning	Land conversion; irrigation
Principal natural sources	Balanced in nature	Wetlands	Soils; tropical forests	None	Hydrocarbons	Evapotranspiration
Atmospheric lifetime	50–200 yr	10 yr	150 yr	60–100 yr	Weeks to months	Days
Atmospheric concentration at surface (ppbv)	366,000	1720	310	CFC-11: 0.28 CFC-12: 0.48	20–40	3000–6000 in stratosphere
Preindustrial (1750–1800) concentration at surface (ppbv)	280,000	790	288	0	10	Unknown
Annual rate of increase	0.5%	1.1%	0.3%	5%	0.5–2.0%	Unknown
Relative contribution to the anthropogenic greenhouse effect	60%	15%	5%	12%	8%	Unknown

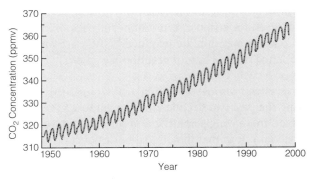

Figure 19.4 Increasing Amount of Carbon Dioxide in the Air Concentration of carbon dioxide in dry air, since 1958, measured at the Mauna Loa Observatory, Hawaii (given in ppmv [parts per million by volume] = ppbv/1000). Annual fluctuations reflect seasonal changes in biologic uptake of CO_2, while the long-term trend shows a persistent increase in the atmospheric concentration of this greenhouse gas.

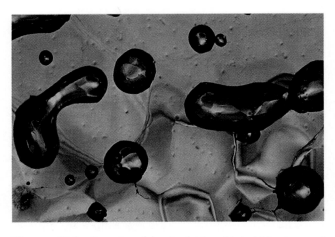

Figure 19.5 Samples of Ancient Air Air bubbles trapped in glacier ice. By melting the ice in a laboratory and collecting the gas, the content of CO_2 and other trace gases in these ancient samples of the atmosphere can be measured.

regularly with an annual rhythm (Fig. 19.4). In effect, the Earth, or more correctly, the biosphere, is breathing. During the growing season, CO_2 is absorbed by vegetation, and the atmospheric concentration falls. Then, during the winter dormant period, more CO_2 enters the atmosphere than is removed by vegetation, and the concentration rises. Second, the long-term trend is unmistakable and rising (Fig. 19.4). Since 1958, the CO_2 has risen from 315,000 to 366,000 ppbv; moreover, the rise is not linear but exponential (i.e., the rate is increasing with time).

The rising curve of atmospheric CO_2 immediately raises two questions: Is the observed rise in CO_2 unusual? And how can it be explained? To answer the first question, we must turn to the geologic record. Ice cores obtained from the Antarctic and Greenland ice sheets contain samples of ancient atmosphere. These samples exist in tiny air bubbles that were trapped when snow falling on the glaciers was slowly compressed and transformed into glacier ice (Fig. 19.5; Chapter 12). If the ice is melted in a laboratory and the air collected and analyzed, the amount of CO_2 and other trace gases in the ancient atmosphere can be measured. Ice from Vostok Station in Antarctica contains such a record extending back 160,000 years. The measurements show that during glacial ages, the atmosphere typically contained about 200,000 ppbv of CO_2, whereas during interglacial times the concentration rose to about 280,000 ppbv (Fig. 12.32). The glacier records also show that the preindustrial levels of CO_2 were close to 280,000 ppbv, the typical value for an interglacial age. The subsequent rapid increase to 366,000 ppbv by 1998 (Fig. 19.4) is unprecedented in the ice core record and implies that something very unusual is taking place.

A possible explanation for the extraordinary recent rise in atmospheric CO_2 is immediately suggested if

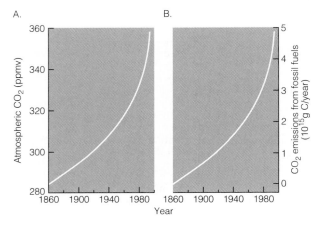

Figure 19.6 Carbon Dioxide and Fossil Fuels A. Since the beginning of the industrial revolution, the atmospheric concentration of CO_2 has risen at an increasing rate. B. The increase matches the growing rate at which CO_2 has been released through the burning of fossil fuels.

we examine the rate at which CO_2 has been added to the atmosphere since the beginning of the industrial revolution (Fig. 19.6). The curve tracking the increase in CO_2 closely resembles the curve showing the increase in carbon released into the atmosphere by the burning of fossil fuels. Because no known natural mechanism can explain such a rapid increase in CO_2, the inescapable conclusion is that anthropogenic (human-generated) burning of fossil fuels must be a primary reason for the observed increase in atmospheric CO_2. Additional contributing factors must be widespread deforestation, with its attendant burning and decay of cleared vegetation, and the use of wood as a primary fuel in many underdeveloped countries that have rapidly growing populations.

Surprisingly, over the past several years, the rate of increase of atmospheric carbon dioxide has slowed, but the reasons are unclear. The change must reflect either a decrease in CO_2 emissions, increased uptake of CO_2 by the oceans and terrestrial biota, or both. Despite recent worldwide economic recession, measurements show that fossil fuel consumption has not decreased enough to account for the change, nor are the oceans taking up the difference. Thus, increased storage of CO_2 in the biosphere is the most likely explanation. Ultimately, however, the explanation of such changes must await a more thorough understanding of the carbon cycle.

Methane.   Methane (CH_4) absorbs infrared radiation 25 times more effectively than CO_2, making it an important greenhouse gas despite its relatively low concentration (1720 ppbv). Since the late 1960s, when measurements of atmospheric methane began, the concentrations have increased at a rate of about 1 percent per year (although since 1984 the rate has decreased, possibly related to changes in human-generated emissions). Methane levels for earlier times have been obtained from ice core studies that show an increase that essentially parallels the rise in the human population. This relationship is not surprising, for much of the methane now entering the atmosphere is generated (1) by biological activity related to rice cultivation, (2) by leaks in domestic and industrial gaslines, and (3) as a byproduct of the digestive processes of domestic livestock, especially cattle. The global livestock population increased greatly in the twentieth century, and the total acreage under rice cultivation has increased more than 40 percent since 1950.

In prehistoric times, methane levels, like CO_2 levels, increased and decreased with the glacial/interglacial cycles (Fig. 12.32).

Other Trace Gases.   CFC-12, used mostly as a refrigerant, has 20,000 times the capacity of carbon dioxide to trap ultraviolet radiation, whereas CFC-11, which is widely used in making plastic foams and as an aerosol propellant, has 17,500 times the capacity. Both compounds are increasing in the atmosphere at an annual rate of about 5 percent. As we have already seen, the increasing atmospheric concentration of CFCs has produced worldwide concern because the scientific consensus is that these gases destroy ozone in the upper atmosphere, thereby leading to the formation of the Antarctic ozone hole.

Although ozone in the upper atmosphere is beneficial because it traps harmful infrared solar radiation, when this gas builds up in the troposphere (lower atmosphere) it contributes to the greenhouse effect. Both ozone and nitrous oxide, another effective greenhouse gas involved in biochemical cycles, are increasing annually at rates of 0.5 to 2 percent and 0.3 percent, respectively. Together, they account for about 13 percent of the anthropogenic greenhouse effect. Tropical forests are important in photosynthetically removing excess tropospheric ozone, which is largely produced by the combustion of fossil fuels. However, the wholesale destruction of these forests could lead to further concentration of this gas in the atmosphere. Nitrous oxide, released by microbial activity in soil, the burning of timber and fossil fuels, and the decay of agricultural residues, has a long lifetime in the atmosphere. Accordingly, atmospheric concentrations are likely to remain well above preindustrial levels even if emission rates stabilize.

Global Warming

If the atmospheric concentration of the greenhouse gases is rising, what does this portend for future climate? Does it mean that the Earth's surface temperature is warming, and, if so, by how much and at what rate? To try and answer these questions, we can first look at the historical record of climate and then see how forecasts of the future can be made using the newest powerful tool available: the supercomputer.

Historical Temperature Trends
Correctly assessing recent global changes in temperature is a very complicated task. The difficulty arises from the fact that few instrumental measurements were made before 1850, and the majority have been made since World War II. The earliest records are from western Europe and eastern North America. Data for oceanic areas, which encompass 70 percent of the globe, are sparse and decrease dramatically in number prior to 1945. Therefore, most "global" temperature curves are reconstructed primarily from land stations that are located mainly in the northern hemisphere. Numerous curves of average annual temperature variations since the mid- or late nineteenth century have been published, and although they differ in detail, they all show one characteristic feature: a long-term rise in temperature during the past century (Fig. 19.7) Although many departures from this trend are evident (wiggles in the curve), the total temperature increase since the late nineteenth century is about 0.6°C. Because the interval of rising temperatures coincides with the time of rapidly increasing greenhouse gas emissions, it is tempting to assume that the two phenomena are causally related. However, the temperature reconstructions prior to 1950 are based on relatively few data that are unequally distributed across the globe, so a convincing case is difficult to make. Even if the curves do approximate actual trends, it might be argued that the modest rise in global temperature during the past century falls within the natural variability of the Earth's climate system and would have occurred even if the greenhouse gas concentrations had not increased.

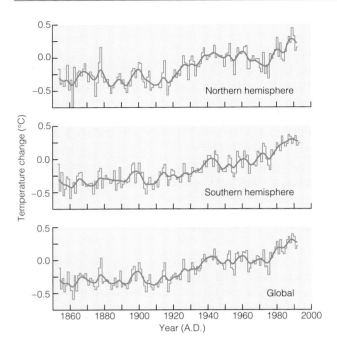

Figure 19.7 Historic Temperature Changes Hemispheric and global mean temperature changes since the mid-nineteenth century, incorporating both land and ocean data. Annual averages (thin line) and smoothed averages (bold line) are shown. In each of the curves, the long-term trend is a rise of temperature, with the total global rise for this interval being about 0.5°C.

If the historical temperature record is judged inconclusive, we can still explore the linkage between greenhouse gas emissions and present and future climate by turning to models of the climate system.

Climate Models
Global climate models are three-dimensional mathematical models of the Earth's climate system and are an outgrowth of efforts to forecast the weather. The most sophisticated are **general circulation models (GCMs)** that attempt to link processes in the atmosphere, the hydrosphere, and the biosphere. The sheer complexity of these natural systems means that such models, of necessity, are greatly simplified representations of the real world. Furthermore, many of the linkages and processes in the climate system are still poorly understood and therefore difficult to model. The models do not yet adequately portray the dynamics of ocean circulation or cloud formation, two of the most important elements of the climate system. Also absent from the models are many of the complex biogeochemical processes that link climate to the biosphere. Despite these limitations, GCMs have been very successful in simulating the general character of present-day climates and have greatly improved weathering forecasting. This success encourages us to use these models to gain a general global picture of future climates as the Earth's physical and chemical balance changes.

To run a modeling experiment that simulates the climate, a set of **boundary conditions** is specified. Boundary conditions are mathematical expressions of the physical state of the Earth's climate system at the period of interest for the experiment. Thus, an experiment designed to simulate the present climate would prescribe as boundary conditions the orbitally determined solar radiation reaching the Earth, the geographic distribution of land and ocean, the position and heights of mountains and plateaus, the concentrations of atmospheric trace gases, sea-surface temperatures, the limits of sea ice, the snow and ice cover on the land, the *albedo* (reflectivity) of land, ice, and water surfaces, and the effective soil moisture (the sum of water input and water loss).

Because the solution of the complex mathematical equations of a GCM requires considerable amounts of computer time, the three-dimensional grid spacing (the distance between points on and above the globe for which the solutions are calculated) in the model experiments is large in order to keep costs manageable. As a result, the resolution of the models is rather coarse: grid points commonly are separated by 4 or 5 degrees of latitude, or 450–550 km. Therefore, although these models can generate a reasonable picture of global and hemispheric climatic conditions, they are poor at resolving conditions at the scale of small countries, states, or counties. Until more powerful computers are built, or the cost of running a model experiment decreases, the spatial resolution of the GCMs is likely to remain relatively coarse.

Model Estimates of Greenhouse Warming
Predictions of climatic change related to greenhouse warming are based mainly on the results of GCMs developed in the United States and in the United Kingdom. The models differ in detail, as well as in the assumptions they employ. Nevertheless, they all predict that the anthropogenically generated greenhouse gases already in the atmosphere will lead to an average global temperature increase of at least 0.5 to 1.5°C. This prediction is consistent with the 0.6°C rise in temperature inferred from the instrumental record. They further predict that if the greenhouse gases continue to build up until their combined effect is equivalent to a doubling of the preindustrial CO_2 concentration, then average global temperatures will rise between 1 and 5°C. This does not mean, however, that the temperature will increase uniformly all over the Earth. Instead, the projected temperature change varies geographically, with the greatest change occurring in the polar regions (Fig. 19.8).

The rate at which the projected warming will occur depends on a number of basic uncertainties: How rapidly will concentrations of the greenhouse gases increase? How rapidly will the oceans, a major reservoir of heat and a fundamental element in the climate system (Fig. 19.1), respond to changing climate? How will changing climate affect ice sheets and cloud cover? What is the

A.

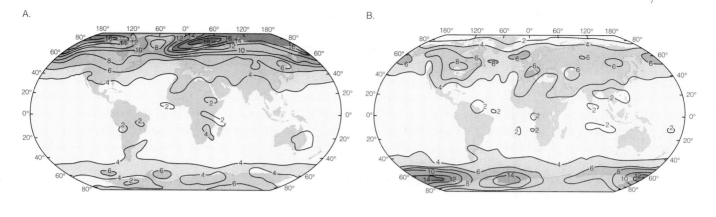

C.

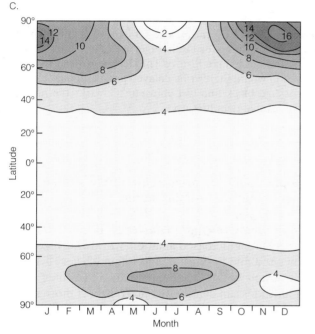

Figure 19.8 Temperature Rise If Carbon Dioxide Doubles
A forecast of future changes in surface air temperature (in °C) that would result from an effective doubling of atmospheric CO_2 concentration relative to that of the present. A. Temperature increases for winter (December, January, February). For example, along the lines labeled 4, the projected temperature increase is everywhere 4°C. B. Temperature increases for summer (June, July, August). C. A latitudinal cross section showing changes in zonal average air temperature through the year. This graph is a summary of the map patterns shown in A and B, but includes the spring and autumn months as well. Greenhouse warming is greatest at high latitudes, where temperature increases as great as 16°C are forecast by the model for the northern hemisphere winter.

range of natural variations in the climate system on the century time scale? The potential complexity is well illustrated by clouds. If the temperature of the lower atmosphere increases, more water will evaporate from the oceans. The increased atmospheric moisture will create more clouds, but clouds reflect solar energy back into space, which will have a cooling effect on the surface air, thereby having a result opposite that of the greenhouse effect.

Because of such uncertainties about the climate system, scientists tend to be cautious in their predictions. Nevertheless, there is a general consensus that (1) human activities have led to increasing atmospheric concentrations of carbon dioxide and other trace gases that have enhanced the greenhouse effect; (2) global mean surface air temperature has increased by 0.3 to 0.6°C during the last 100 years, and this increase may be the direct result of the enhanced greenhouse effect; and (3) during the next century global average temperature will likely increase at about 0.3°C per decade, assuming emission rates do not change. This projected

increase will lead to a global average temperature about 1°C warmer than present by the year 2025 and as much as 3°C warmer by the end of the next century. If governmental controls lead to lower emission rates, the decadal rise in temperature may be only 0.1–0.2°C. Nevertheless, the temperature increase related to the continued release of greenhouse gases will be larger and more rapid than any experienced in human history. In effect, we may be about to experience a "super interglaciation" warmer than any interglaciation of the past 2 million years (Fig. 19.9).

Environmental Effects of Global Warming

An increase in global surface air temperature by a few degrees does not sound like much. Surely, you say, we can put up with this rather insignificant change. However, if we stop and consider that the difference in average global temperature between the present and the coldest part of the last ice age was only about 5°C, we can begin to see how even a slight temperature change of a degree or two could well have global repercussions.

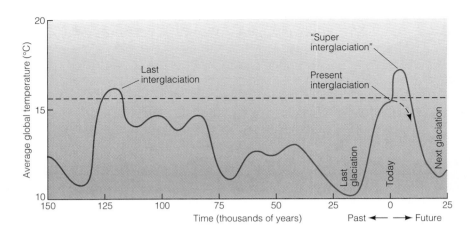

Figure 19.9 Temperature Changes Past and Future The course of average global temperature during the past 150,000 years and 25,000 years into the future. The natural course of climate (dashed arrow) would be declining temperatures leading to the next glacial maximum, about 23,000 years from now. With greenhouse warming, a continuing rise of temperature may lead to a "super interglaciation" within the next several centuries. The temperature may then be warmer than during the last interglaciation and warmer than at any time in human history. The decline toward the next glaciation would thereby be delayed a millennium or more. The dashed black line just above 15°C marks the present average global temperature.

Global warming is just one result of our great geochemical "experiment." There are many physical and biological side effects that are of considerable interest and concern.

Global Precipitation Changes. A warmer atmosphere will lead to increased evaporation from oceans, lakes, and streams, and to greater precipitation. However, the distribution of the increased precipitation will be uneven. Climate models suggest that the equatorial regions will receive more rainfall, in part because warmer temperature will increase rates of evaporation over the tropical oceans and promote the formation of rainclouds, whereas some interior portions of large continents, which are distant from precipitation sources, will become both warmer and drier.

Changes in Vegetation. Shifts in precipitation patterns are likely to upset ecosystems, causing vegetation communities, and animals dependent on them, to adjust to new conditions. Forest boundaries may shift during coming centuries in response to altered temperature and precipitation regimes. Some prime midcontinental agricultural regions are likely to face increased droughts and substantially reduced soil moisture that will have a negative impact on crops. Higher-latitude regions with short, cool growing seasons may see increased agricultural production as summer temperatures increase.

Increased Storminess. The shift to a warmer, wetter atmosphere favors an increase in tropical storm activity. Regions that now are impacted by hurricanes and typhoons could see an increase in the size and frequency of these devastating storms.

Melting (and Growing?) Glaciers. Because warmer summers favor increased ablation, worldwide recession of low- and middle-latitude mountain glaciers is likely in a warmer world. On the other hand, warmer air in high latitudes can evaporate and transport more moisture from the oceans to adjacent ice sheets, which may cause them to grow larger.

Reduction of Sea Ice. The greatly enhanced heating projected for high northern latitudes (Fig. 19.8) favors the shrinkage of sea ice. A reduction in polar sea ice, which has a high albedo, should complement the greenhouse effect by reducing the amount of solar radiation reflected back into space. Models show much less heating in the high-latitude southern hemisphere, suggesting little change in sea-ice cover there.

Thawing of Frozen Ground. Rising summer air temperatures will begin to thaw vast regions of perennially frozen ground at high latitudes. The thawing will likely affect natural ecosystems as well as cities and engineering works built on frozen ground.

Rise of Sea Level. As the temperature of ocean water rises, its volume will expand, causing world sea level to rise. This rise in sea level, supplemented by meltwater from shrinking mountain glaciers, is likely to increase calving along the margins of tidewater glaciers and ice sheets, thereby leading to additional rise in sea level. The rising sea will inundate coastal regions where millions of people live and will make the tropical regions even more vulnerable to larger and more frequent cyclonic storms.

Changes in the Hydrologic Cycle.   Shifting patterns of precipitation and warmer temperatures will likely lead to some significant local and regional changes in stream runoff and groundwater levels.

Decomposition of Soil Organic Matter.   As the temperature rises, the rate of decomposition of organic matter in soil will increase. Soil decomposition releases CO_2 to the atmosphere, thereby further enhancing the greenhouse effect. If world temperature rises by 0.3°C per decade, during the next 60 years soils will release an amount of CO_2 equal to nearly 20 percent of the projected CO_2 release due to combustion of fossil fuels, assuming the present rate of fuel consumption continues.

Breakdown of Gas Hydrates.   Gas hydrates are ice-like solids in which gas molecules, mainly methane, are locked in the structure of water. They are found in ocean sediments and beneath frozen ground. By one estimate, gas hydrates worldwide may hold 10,000 billion metric tons of carbon, twice the carbon in all the coal, gas, and oil reserves on land. They accumulate in ocean sediments beneath a water depth of 500 m, where the temperature is low enough and the pressure high enough to permit their formation. They also accumulate beneath frozen ground, which acts as a seal to prevent upward migration and escape of the gas. When gas hydrates break down, they release methane. Global warming at high latitudes will result in thawing of frozen ground that may well destabilize the hydrates there, release large volumes of methane, and thus amplify the greenhouse effect.

Although our present knowledge of how the Earth works, coupled with computer modeling, enables us to make educated projections of surface environmental changes that will result from greenhouse warming, most scientists are reluctant to make firm forecasts. Instead, they hedge their bets with qualifying adjectives like "possible," "probable," and "uncertain." Their understandable caution emphasizes the gap between what we know about the Earth and what we would like to know and points to the many challenges that still face scientists studying global change.

THE PAST AS A KEY TO THE FUTURE

To increase the likelihood that our predictions about the changing global environment are correct, we can invoke "Ayer's Law". This useful tenet of geology states: "Any-

thing that did happen, can happen." In other words, we can use the geologic record, which provides an invaluable archive of the past history of natural environmental changes on the Earth, as a key to understanding what the future may hold in store.

Lessons from the Past

Examining changes to physical and biological systems that occurred when human influence was absent or minimal allows us to see how these systems responded to sudden changes of climate. We can use this information to help us anticipate the character of environmental changes that may happen on a warming Earth. One example is a very rapid warming event that occurred in the North Atlantic region at the end of the last glaciation.

In addition to studying times of rapid climatic change, we can also explore the geologic record for information about times when the climate was warmer than now. Of particular interest are the early Holocene, from 10,000 to about 6000 years ago, when average temperatures were 0.5 to 1°C warmer; the warmest part of the last interglaciation, about 120,000 years ago, when global temperatures were about 1 to 2°C higher; and the middle Pliocene, about 4.5 to 3 million years ago, when average temperatures may have been 3 to 4°C warmer than present. These periods do not provide perfect analogs for present global warming because the distribution of solar radiation reaching the Earth's surface was different then (see Figs. 12.B2 and 12.B3). Nevertheless, these intervals enable us to see how plants and animals responded to climatic conditions that may have been broadly similar to those we may experience in the near future.

Still older intervals of unusually warm climate pose special problems for interpretation but have generated some imaginative solutions. They include the middle Cretaceous Period (about 100 million years ago) and the Eocene Epoch (about 40 to 50 million years ago).

Why Was the Middle Cretaceous Climate So Warm?

It's probably a good thing we did not live during the Middle Cretaceous Period. Not only was the world inhabited by huge carnivorous dinosaurs, but also the climate was one of the warmest in the Earth's history. Evidence that the world was much warmer than today is compelling (Fig. 19.10). Warm-water marine faunas were widespread, coral reefs grew 5 to 15° closer to the poles than they do now, and vegetation zones were displaced about 15° poleward of their present positions. Peat deposits that would give rise to widespread coal formations formed at high latitudes, and dinosaurs, which are generally thought to have preferred warm climates, ranged north of the Arctic circle. Sea level was 100 to 200 m higher, implying the absence of polar ice sheets, and isotopic measurements of deep-sea deposits indicate that intermediate and deep waters in the oceans were 15 to 20° warmer than now.

GCM simulations of the Middle Cretaceous world suggest that several factors likely were involved in producing such warm conditions: geography, ocean circulation, and atmospheric composition. The modeling simulations show that the Middle Cretaceous arrangement of continents and oceans (Fig. 19.10A), which influenced ocean circulation and planetary albedo, could account for nearly 5°C of the warming; of this amount, about a third is attributable to the absence of polar ice sheets. However, geography alone is inadequate to explain warmer year-round temperatures at high latitudes. Could the poleward transfer of heat be the answer? The oceans now account for about a third of the present poleward heat transfer, but modeling shows that even with the geography and ocean circulation rearranged as they were in the Middle Cretaceous, oceanic heat transfer cannot explain the greater high-latitude warmth. If the geologic data have been correctly interpreted, and the modeling results are reliable, some other factor must be involved. This factor appears to be CO_2, the major greenhouse trace gas.

Can an enhanced greenhouse effect be the key to explaining the exceptionally warm Middle Cretaceous climate? GCM experiments show that by rearranging the geography and also increasing carbon dioxide six to eight times above present concentrations, the warmer temperatures can be explained. Geochemical reconstructions of changing atmospheric CO_2 levels over the past 100 million years point to at least a tenfold increase in CO_2 during the Middle Cretaceous, leading to average temperatures as much as 8°C higher than now (Fig. 19.11). Compared to the average conditions forecast for the twenty-first century (2–4°C hotter), the Middle Cretaceous climate must have been a scorcher!

If CO_2 was an important factor in Middle Cretaceous warming, we still are faced with explaining how this gas increased so substantially. Unlike the modern world, combustion of fossil fuels cannot provide the answer. The most likely source is volcanic activity, which today constitutes a major source of CO_2 entering the atmosphere. Prior to the industrial revolution, volcanism probably replenished the atmosphere with as much as 65 percent of the carbon lost to sedimentation, mostly by slow, noneruptive degassing of CO_2 from magmas in the upper crust.

Geologic evidence points to an unusually high rate of volcanic activity in the Middle Cretaceous. Rates of continental drift were then about three times as great as now, implying increased extrusion rates at spreading ridges. In addition, vast outpourings of lava created a succession of great undersea plateaus across the southern Pacific Ocean between 135 and 115 million years ago, the time of maximum Cretaceous warmth. One of these—the Ontong-Java plateau in the southwestern Pacific—has more than twice the area of Alaska and reaches a

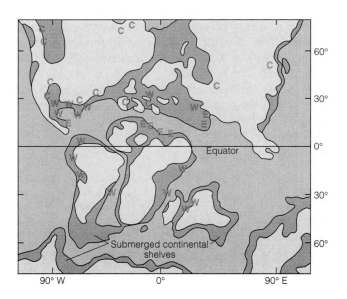

Figure 19.10 Warm Climate of the Middle Cretaceous During the Middle Cretaceous, sea level was 100 to 200 m higher than now and flooded large areas of the continents, producing shallow seas. Warm-water faunas (W) and evaporites (E) were present at low to middle latitudes, while coal deposits (C) developed in arctic regions, implying warmer year-round temperatures.

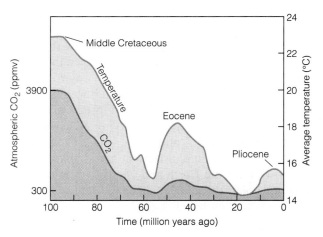

Figure 19.11 Changing Carbon Dioxide in the Atmosphere over Geologic Time Geochemical reconstruction of changing atmospheric CO_2 concentration and average global temperature over the past 100 million years. High CO_2 values and temperatures in the Middle Cretaceous contrast with much lower values of the present. Other intervals of higher temperature and CO_2 occurred during the Eocene and the Middle Pliocene.

thickness of 40 km. Such a massive outpouring of lava likely released massive amounts of CO_2. Could this gas emission have been sufficient to warm the climate to unprecedented levels? By one calculation, the eruptions could have released enough CO_2 to raise the atmospheric concentration to 20 times the preindustrial value, in the process raising average global temperature as much as 10°C. Other estimates range from 8 to 12 times the preindustrial value.

Recently, geologists have proposed that these vast lava outpourings are the result of super hot spots. Hot spots are thought to be due to rising plumes of hot rock (Chapter 17). A plume like that responsible for the Hawaiian hot spot, which is about 200 to 300 km across, may originate somewhere in the mesosphere, possibly at a depth of 670 km, where a distinct seismic discontinuity occurs (Chapter 16). A super hot spot is thought to arise from a *superplume*, which is conceived as being a plumelike mass of unusually hot rock that rises from the base of the mantle at a rate of 10 to 20 cm/yr and spreads out in a mushroom shape as it reaches shallower depths where confining pressures are lower. Modeling studies suggest that the size of a plume head depends on the depth at which the plume originates (Fig. 19.12) and that the Hawaiian plume must originate at a depth of less than 1000 km. The huge Middle Cretaceous lava plateaus, however, are thousands of kilometers across, implying that they could have formed only from superplumes rising from the core–mantle boundary, nearly 2900 km deep. Such a plume would be about 1000 km wide when it reached the upper mantle and would flatten to two or three times this width as it approached the base of the lithosphere. A superplume would be an efficient mechanism for allowing heat to escape from the Earth's core. If this hypothesis is correct, then the plate-tectonic cycle cools the mantle both by heat loss at spreading ridges and by the downward plunge of plates of cool lithosphere, while superplumes cool the core. By this reasoning, the core and atmosphere are linked dynamically, and the warm Middle Cretaceous climate was a direct consequence of the cooling of the Earth's deep interior.

Eocene Warmth and Lithosphere Degassing

By the end of the Cretaceous, world temperatures had fallen substantially below levels reached earlier in that period (Fig. 19.11). A reversal in this trend then brought temperatures to a new peak during the Early Eocene, about 50 million years ago. Varied evidence points to warmer conditions: alligator fossils on Ellesmere Island (at about 78°N latitude), tropical vegetation at up to 45°N latitude, tropical marine surface-water organisms at about 55°N in the Atlantic Ocean, and lateritic soils (indicating warm climate with seasonal rainfall) at up to 45° latitude in both hemispheres. Isotopic measurements of bottom-dwelling organisms show that the temperature of the deep ocean reached its highest Cenozoic value in the Early Eocene. The average atmospheric warming in the Eocene, relative to modern conditions, is estimated to be in the range of 1–4 C° and likely close to 2 C°.

Scientists speculate that two processes may have contributed to the higher Eocene temperatures: greater poleward transfer of heat by the oceans and an increased concentration of CO_2 in the atmosphere. Enhanced poleward heat transfer could have prevented the formation of polar sea ice which, because it is highly reflective, bounces solar radiation back into space, thereby keeping high-latitude surface temperatures cold. An open ocean, which is much less reflective than sea ice, would absorb solar radiation, thus warming the high latitudes.

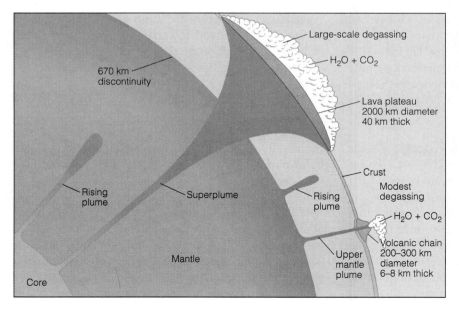

Figure 19.12 Superplumes Rise Through the Mantle New evidence has led to the hypothesis that slowly rising superplumes from the core–mantle boundary build huge lava plateaus when they reach the lithosphere and may give rise to large-scale degassing of CO_2 that greatly enhances the greenhouse effect. By contrast, plumes rising from the base of the upper mantle at 670 km produce much smaller hot spots that generate series of volcanoes like those of the Hawaiian-Emperor chain.

Estimates of Cenozoic atmospheric CO_2 concentrations place the highest value—about two to six times the modern value—during the Eocene (Fig. 19.11). The big question is where the excess CO_2 originated. Two likely sources have been proposed: volcanic outgassing, both from continental volcanoes and midocean ridges, and degassing of the upper lithosphere during metamorphism.

When carbonates and shales undergo regional metamorphism in orogenic belts, CO_2 is released as a byproduct of the metamorphic reactions. Most of the CO_2 is thought to be released under low-grade (greenschist facies) metamorphic conditions because such conditions occur in the upper part of an orogen, where brittle fracturing and faulting can provide an avenue for gases to escape to the surface. Studies show that regional metamorphism of the Himalaya (Fig. 19.13) may have been contemporaneous with the Eocene warming. In addition, regional metamorphism at this time has been documented in the Mediterranean region (Greece, Turkey) and in the circum-Pacific region (New Caledonia, Japan, western North America). Although the interval of metamorphism in these regions is not yet tightly dated, calculations suggest that the quantity of CO_2 generated from these regions is likely sufficient in itself to explain the unusual buildup of this gas in the Eocene atmosphere.

An additional factor to consider is the removal of atmospheric CO_2 during weathering of orogenically uplifted silicate rocks. CO_2 is removed from the atmosphere when carbonic acid is produced (see Table 8.1). As the weak acid decomposes the rocks, bicarbonate released by the weathering reactions is carried by streams, in solution, to the sea and is there converted to calcium carbonate by marine organisms. When the organisms die, their skeletal remains become stored in carbonate sediments at the seafloor. The net effect is to isolate in marine sediments much of the atmospheric CO_2 involved in weathering. It is difficult to estimate how much CO_2 was removed from the atmosphere in this way as a result of widespread mountain uplift, but it seems possible that the general late Cenozoic cooling trend which ultimately led to the glacial ages may be at least partly explained by high weathering rates in orogenic mountain belts and the consequent storage of the carbon dioxide in ocean sediments.

Modeling Past Global Changes

Paleoclimatic reconstructions offer a means of testing the accuracy of climate models. Our confidence in using GCMs for predicting the future will be strengthened if the models not only can accurately reproduce the present climate, but also can reproduce past climates. We can test, or "validate," these climate simulations by comparing the model results against independent geologic evidence of the conditions then prevailing.

Figure 19.13 Metamorphism and the Release of Carbon Dioxide Folded metasedimentary rocks from the Swat region of the northwestern Himalaya. Regional metamorphism of marine sedimentary rocks during collision of India with Asia is regarded as a possible source of carbon dioxide that contributed to the high concentration of this gas in the Eocene atmosphere.

The same GCMs that are used to model the present global weather are also used to model ancient climates. The main difference is in the boundary conditions. For example, one set of experiments has attempted to simulate the Earth's climates at 3000-year intervals since the last glacial maximum about 18,000 years ago. The specified boundary conditions changed substantially over this time interval (Fig. 19.14): solar radiation varied by as much as 8 percent as the Earth went through one precessional cycle; the area of glacier ice shrank, and airborne dust decreased as the ice age came to an end; both sea-surface temperature and atmospheric CO_2 increased as the climate moved toward its present interglacial state. During each of the seven time periods modeled, the boundary conditions were different from those of the other periods. As a result, the successive simulations change as the model Earth passes from a glacial age to an interglacial age.

As an example of such a simulation, we can examine the results of an experiment focusing on the eastern hemisphere 9000 years ago (Fig. 19.15A). You can see what the specified boundary conditions are by noting the values at 9000 years ago in Figure 19.14. The model "predicts" that 9000 years ago the climate in a broad belt across northern Africa and southern Asia experienced warmer and wetter summers and cooler winters; the Mediterranean region at that time had warmer, drier summers, southern Africa had colder, drier winters, and Australia had colder winters and warmer summers. The

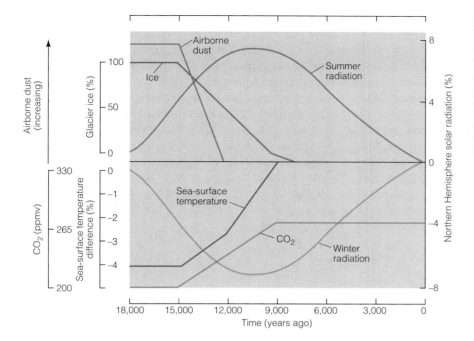

Figure 19.14 Global Climate Model
Boundary conditions used in one set of GCM simulations of world climate between the last glacial maximum (18,000 years ago) and the present. Solar radiation values are based on known astronomical motions of the Earth (Fig. B11.3), while ice, dust, CO_2, and sea-surface temperature values are based on geological data.

model also indicates that the increase in summer moisture was largely due to an increase in the strength of the summer monsoon in western Africa and southern Asia. Figure 19.14 shows us that the boundary conditions for 9000 years ago were different from those of today mainly with respect to northern hemisphere solar radiation, which was about 8 percent greater in summer and 8 percent less in winter. This difference was a major factor in strengthening the summer monsoon.

To see how reasonable these model predictions are, we can examine geologic evidence of conditions about 9000 years ago (Fig. 19.15B). The evidence points to a broad belt of land across northern Africa, the Middle East, and China that had greater effective moisture (i.e., precipitation minus evaporation) than now. In Africa and western China, the increased moisture is evidenced by high water levels in closed-basin lakes and the widespread occurrence in China of an early Holocene soil that indicates an increase in monsoon precipitation. In this example, the results of the model simulation are generally consistent with the paleoclimatic data assembled by geologists.

A.

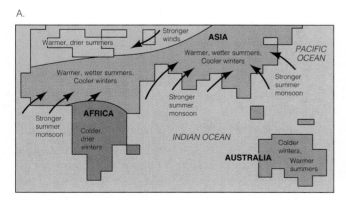

B.

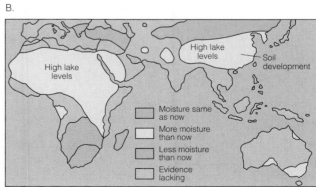

Figure 19.15 What the Climate Was Like 9 Millennia Ago Climates of Africa and southern Asia during the early Holocene (9000 years ago). A. GCM simulation indicates that a belt crossing northern Africa and southeastern Asia had warmer, wetter summers and cooler winters 9000 years ago. These conditions resulted from a strengthening of summer monsoon winds, which brought precipitation to the continents from the warm tropical oceans. At that time, the Mediterranean region and Australia had warmer, drier summers, while southern Africa and Australia had colder winters. B. Geologic evidence generally bears out the GCM results: a belt running from northern Africa through the Middle East and China received more moisture 9000 years ago, as shown by high lake levels and paleosols that indicate increased soil moisture. Evidence also shows that southern Africa and Madagascar, as well as the northeastern Mediterranean region, were drier than now.

PERSPECTIVES OF GLOBAL CHANGE

Most of us would find it fascinating, and no doubt convenient, if we could look into a crystal ball and see the future. It certainly would be helpful if we had a clear vision of our climatic future. At present, the best we can conclude is that the force of scientific evidence and theory makes it very probable that the climate is warming up and will continue to warm as we add greenhouse gases to the atmosphere. There also is a high probability that average global temperatures ultimately will increase by 2 to 4°C, leading to widespread environmental changes.

It is less probable that the temperature will increase steadily, for there are natural, and as yet largely unpredictable, modulations of the climate system on the time scale of years to decades. As an example, the huge eruption of Krakatau in 1883 led to a 0.5°C drop in average global temperature (Chapter 4), and, earlier, the even larger Tambora eruption of 1815 was followed by the "year without a summer," during which midsummer snow and frost caused severe hardships in Europe and New England. The huge eruption of the Philippine volcano Pinatubo in June 1991 introduced a vast quantity of fine ash and sulfurous gas into the stratosphere, where it quickly began to spread into the northern and southern hemispheres. As the veil of dust and gas spread throughout the atmosphere and reflected incoming solar radiation, average surface temperatures fell about half a degree C. The volcanic effect may well reverse temporarily any upward trend in average global temperature attributable to continued emission of greenhouse gases.

While the short-term prospect (on the scale of human generations) is for a warmer world, if we stand back and look at our great geochemical "experiment" from a geological perspective, we will perceive that it is only a brief, very rapid, yet nonrepeatable perturbation in the Earth's climatic history. It is nonrepeatable because once the Earth's store of easily extractable fossil fuels is used up, most likely within the next several hundred years, the human impact on the atmosphere will inevitably decline, and the climate system should revert to a more natural state. The greenhouse perturbation may well last a thousand years, and perhaps more, but ultimately the changing geometry of the Earth's orbit will propel the climate system inexorably into the next glacial age (Fig. 19.9).

SUMMARY

1. Synthetic chlorofluorocarbon (CFC) gases entering the upper atmosphere break down and release chlorine, which destroys the protective ozone layer. Discovery of a vast and recurring ozone hole over Antarctica has led to international efforts to eliminate CFC production by the end of the century.

2. The Earth's climate system involves the atmosphere, hydrosphere, lithosphere, and biosphere. Changes affecting it operate on time scales ranging from decades to millions of years.

3. Using information from the geologic record, geologists can measure the magnitude and geographic extent of past climatic changes, determine the range of climate variability on different time scales, and test the accuracy of computer models that simulate past climatic conditions.

4. The carbon cycle is among the most important of the Earth's biogeochemical cycles. Carbon resides in the atmosphere, the biosphere, the hydrosphere, and in the crust and cycles through these reservoirs at different rates.

5. The anthropogenic extraction and burning of fossil fuels perturb the natural carbon cycle and have led to an increase in atmospheric CO_2 since the start of the industrial revolution.

6. The greenhouse effect, caused by the trapping of long-wave infrared radiation by water vapor and trace gases in the atmosphere, makes the Earth a habitable planet.

7. The increase in atmospheric trace gases (CO_2, CH_4, O_3, N_2O, and the CFCs) due to human activities is projected to warm the lower atmosphere by 2° to 4°C by the end of the next century.

8. A probable 0.5°C increase in average global temperature since the mid-nineteenth century may reflect the initial part of this warming. The rate of warming is likely to reach 0.3°C per decade and may lead to a "super interglaciation," making the Earth warmer than at any time in human history.

9. Potential physical and biological consequences of global warming include global changes in precipitation and vegetation patterns; increased storminess; melting of glaciers, sea ice, and frozen ground; worldwide rise of sea level; local and regional changes in the hydrologic cycle; and increased rates of organic decomposition in soils.

10. Evidence of past intervals of rapid environmental change in the geologic record and reconstructions of past warmer intervals provide insights into physical and biological responses to global warming. Such reconstructions also permit evaluation of general circulation models.

11. Although the Earth's surface environments may change substantially during the next several centuries in response to greenhouse warming, viewed from the geologic perspective, this interval will appear as only a brief perturbation in the Earth's climate history.

THE LANGUAGE OF GEOLOGY

boundary conditions (p. 522)   *general circulation model (p. 522)*   *greenhouse effect (p. 518)*

QUESTIONS FOR REVIEW

1. Why does chlorine have such an adverse effect on the ozone layer, despite the fact that it is released into the atmosphere in very small amounts?

2. Describe the carbon cycle and indicate why we regard it as one of the most important biogeochemical cycles.

3. In what ways can carbon be trapped in the Earth and become part of the rock cycle? How can such stored carbon once again find its way into the atmosphere?

4. If atmospheric CO_2 can be dissolved in streams, lakes, groundwater, and the oceans, and also efficiently absorbed by vegetation, why is the burning of fossil fuels causing the CO_2 content of the atmosphere to increase?

5. How is the Earth's atmosphere similar to a garden greenhouse, and why?

6. What are the anthropogenic sources of the principal greenhouse gases?

7. What geologic evidence indicates that the present concentrations of carbon dioxide and methane in the atmosphere are exceptional compared to those of the last several hundred thousand years?

8. What are the major boundary conditions that must be specified in GCM climate simulation experiments? How have these conditions changed since the maximum of the last glaciation?

9. Why do we remain uncertain about the extent to which average global temperature will rise in the next century as a result of greenhouse warming?

10. Give an example of an environmental effect arising from global warming that could enhance the greenhouse effect and lead to additional warming.

11. What factors are likely to cause world sea level to rise in a warming climate? In what ways is rising sea level likely to impact the human population?

12. Why is the geologic record important in helping predict the environmental effects of greenhouse warming? Give two examples.

 For an interactive case abstract, virtual tours, activities, and additional learning resources, go to
GEOSCIENCES TODAY: www.wiley.com/college/skinner

Using GIS and Remote Sensing to Monitor Urban Habitat Change

The Salt Lake Valley of Utah is the major population center of the state. Bounded in four directions by mountains as well as by Great Salt Lake on the northwest, the area has the potential of limiting human population growth and urban sprawl by placing restrictions on the amount of developable land. A greater population density resulting from these limits has impacted a number of interesting, but fragile ecosystems around the Great Salt Lake.

OBJECTIVE: The primary objective is to use land use and land cover change as a framework to explore urban habitat change while focusing on the fundamentals of geographical information systems and remote sensing techniques for modeling and monitoring such change. Emphasis is placed on monitoring the biological, ecological, and human implications of urban growth in a suburban environment.

The Human Dimension: What implications does urbanization have on global environmental change? How does this affect your city or region?

Questions to Explore:

1. What is the unique situation of the Salt Lake Valley in regard to land use and land cover change in an urban-to-rural transition zone?

2. What are the biophysical characteristics of a city in a transect from core to periphery?

3. What are the "new tools" of Earth Systems Science and how are these tools used to monitor rural-to-urban land use and land cover conversion?

20

Understanding Our Environment

Using Resources

GeoMedia

Mining salt from Karum Salt Lake, in the Danakil Depression, Eritrea. The miners shape salt into smooth, manageable slabs to be loaded on camels, mules, and donkeys, for transport into the interior of Africa.

Resources of Minerals and Energy

Natural Resources and Human History

Civilization and natural resources—the former would not have been possible without the latter. Millions of years ago our ancestors started using resources when they picked up suitably shaped stones and used them as hunting aids. After a while they discovered that flint, chert, obsidian, and other tough stones were best for knives and spear points. Because the most desired stones could be found in only a few restricted places, trading started. Next our ancestors started gathering and trading salt. Originally, dietary needs for salt were satisfied by eating the meat brought home by hunters. When farming started, however, diets became cereal-based, and extra salt was needed. We don't know when or where the mining of salt started, but long before recorded history salt routes criss-crossed the globe.

Metals were first used more than 20,000 years ago. Copper and gold are both found as native metals, and these were the earliest metals to be used. But native copper is rare, and so eventually other sources of copper were needed. By 6000 years ago our ancestors had learned how to extract copper from certain minerals by the process called smelting. Before another thousand years had passed, they had discovered how to smelt minerals of lead, tin, zinc, silver, and other metals. The technique of mixing metals to make alloys came next; bronze (copper and tin) and pewter (tin, lead, and copper) came into use. The smelting of iron is more difficult than the smelting of copper, and so development of an iron industry came much later—about 3300 years ago.

The first people to use oil instead of wood for fuel were the Babylonians, about 4500 years ago. The Babylonians lived in what is now Iraq, and they used oil from natural seeps in the valleys of the Tigris and Euphrates rivers. The first people to mine and use coal were the Chinese, about 3100 years ago. At about the same time, the Chinese drilled the first wells for natural gas—some nearly 100 m deep.

By the time that first the Greek and then the Roman empires came into existence about 2500 years ago, our ancestors had come to depend on a very wide range of resources—not just metals and fuels, but also processed materials such as cements, plasters, glasses, and porcelains. The list of materials we mine, process, and use has grown steadily larger ever since. Today we have industrial uses for almost all of the naturally occurring chemical elements, and more than 200 kinds of minerals are mined and used.

MINERAL RESOURCES

Can you imagine a world without machines? Our modern world with its 6.0 billion inhabitants couldn't operate without them. Machines are used to produce our food, to make our clothes, to transport us around, and to help us communicate. The metals needed to build machines and

the fuels needed to run them are dug from the Earth. Deposits of metallic minerals and fuels are formed by geological processes. Geological processes, therefore, influence the daily lives of each and every one of us.

In many of the previous chapters, we pointed out how geological processes such as weathering, sedimentation, and volcanism can, under suitable conditions, form valuable mineral and energy deposits. The "suitable conditions" are not common, however, and for this reason mineral and energy resources are limited in quantity and hard to find. No geological challenge is more difficult or more rewarding than the search for, and discovery of, new resources of minerals and energy. In this chapter we focus on how and where the different kinds of mineral and energy deposits form.

We turn first to the mineral substances that provide materials from which machines and a myriad other necessary things can be made. The number and diversity of such substances are so great that to make a simple classification covering all of them is almost impossible. Nearly every kind of rock and mineral can be used for something. A society such as ours, which depends so much on energy-intensive industry, not only requires a diverse group of metals for machines, but also demands a host of nonmetallic mineral products, such as shale and limestone for making cement, gypsum for making plaster, salt for making chemical compounds, and calcium phosphate (apatite) for making fertilizer.

Supplies of Minerals

Many industrialized nations are rich in many kinds of **mineral deposits** (any volume of rock containing an enrichment of one or more minerals), which they are exploiting vigorously. Yet no nation is entirely self-sufficient in mineral supplies, and so each must trade with other nations to fulfill its needs (Fig. 20.1).

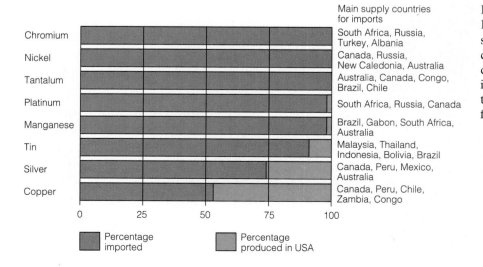

Main supply countries for imports

Mineral	Main supply countries for imports
Chromium	South Africa, Russia, Turkey, Albania
Nickel	Canada, Russia, New Caledonia, Australia
Tantalum	Australia, Canada, Congo, Brazil, Chile
Platinum	South Africa, Russia, Canada
Manganese	Brazil, Gabon, South Africa, Australia
Tin	Malaysia, Thailand, Indonesia, Bolivia, Brazil
Silver	Canada, Peru, Mexico, Australia
Copper	Canada, Peru, Chile, Zambia, Congo

■ Percentage imported ■ Percentage produced in USA

Figure 20.1 Commodities That Must Be Imported Selected mineral substances for which the United States' consumption exceeds production. The difference must be supplied by imports. Data are plotted for 1989, but the percentages have changed little from year to year.

Mineral resources have three distinctive aspects. First, occurrences of usable minerals are limited in abundance and distinctly localized at places within the Earth's crust. This is the main reason no nation is self-sufficient in mineral supplies. Because usable minerals are localized, they must be searched out, and the search ranges over the entire globe. The special branch of geology concerned with discovering new supplies of usable minerals is called, appropriately, **exploration geology**.

Second, the quantity of a given material available in any one country is rarely known with accuracy, and the likelihood that new deposits will be discovered is difficult to assess. As a result, production over a period of years can be difficult to predict. Thus, a country that today can supply its need for a given mineral substance may face a future in which it will become an importing nation. Britain, for example, once supplied most of its own mineral needs but can no longer do so today. A little more than a century ago, Britain was a great mining nation, producing and exporting such materials as tin, copper, tungsten, lead, and iron. Today, the known deposits have been worked out.

Third, unlike plants and animals, which are cropped yearly or seasonally and then replenished, deposits of minerals are depleted by mining and eventually exhausted. This disadvantage can be offset only by finding new occurrences or by using the same material repeatedly—that is, by making use of scrap.

The peculiarities of the mineral industry place a premium on the skills of exploration geologists and engineers, who play the essential roles in finding and mining mineral deposits. The finding is accomplished through application of the basic geological principles set forth in this book. Much ingenuity has been expended in bringing the production of minerals to its present state. Because known deposits are being rapidly exploited while demands for minerals continue to grow, we can be sure that even more ingenuity will be needed in the future.

Ore

Minerals for industry are sought in deposits from which the desired substances can be recovered least expensively. The more concentrated the desired minerals, the more valuable the deposit. In some deposits the desired minerals are so highly concentrated that even very rare substances such as gold and platinum can be seen with the naked eye. For every desired mineral substance, a *grade* (level of concentration) exists, below which the deposit cannot be worked economically (Fig. 20.2). To distinguish between profitable and unprofitable mineral deposits, the word **ore** is used, meaning an aggregate of minerals from which one or more minerals can be extracted profitably. It is not always possible to say exactly what the grade must be, or how much of a given mineral must be present, in order to constitute an ore. Two deposits may have the same grade and be the same

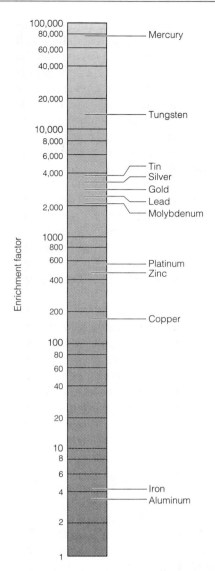

Figure 20.2 Enrichment Needed to Form an Ore Before a mineral deposit can be worked profitably, the percentage of valuable metal in the deposit must be greatly enriched above its average percentage in the Earth's crust. The enrichment is greatest for metals that are least abundant in the crust, such as gold and mercury. As mining and mineral processing become more efficient and less expensive, it is possible to work leaner ore, and enrichment factors decline. Note that the scale is a magnitude (logarithmic) scale, in which the major divisions increase by multiples of ten.

size, but only one of them is ore. There could be many reasons for the difference; for example, the uneconomic deposit could be too deeply buried or located in so remote an area that the costs of mining and transport would be high enough to make the final product noncompetitive with the same product from other deposits. Furthermore, as costs and market prices fluctuate, a particular aggregate of minerals may be an ore at one time but not another.

The ore challenge is twofold: (1) to find the ores (which all together underlie an infinitesimally small proportion of the Earth's land area), and (2) to mine the ores and get rid of the contaminants as cheaply as possible. Both steps are technical problems; engineers have been so successful in solving them that some deposits now considered ore are only one-sixth as rich as were the lowest-grade ores 100 years ago. Copper provides an interesting example. The lowest-grade ores ever mined— about 0.5 percent copper—were worked at a time of high metal prices, in the 1970s. In the 1990s, the lowest grade of mineable copper ore is closer to 1 percent copper. The reason for this is that overproduction of copper around the world, combined with an economic recession, produced a glut of newly mined copper. This, in turn, drove the price of copper down and led to the closing of many mines, in particular those mines with the lowest grades.

Gangue

Ore minerals such as sphalerite, galena, and chalcopyrite (Chapter 3) from which desired metals can be extracted are usually mixed with other, nonvaluable minerals, collectively termed **gangue** (pronounced gang). Familiar minerals that commonly occur as gangue are quartz, feldspar, mica, calcite, and dolomite.

ORIGIN OF MINERAL DEPOSITS

All ores are mineral deposits because each of them is a local enrichment of one or more minerals or mineraloids. The reverse is not true, however. Not all mineral deposits are ores. "Ore" is an economic term, whereas "mineral deposit" is a geological term. How, where, and why a mineral deposit forms is the result of one or more geological processes. Whether or not a given mineral deposit is an ore is determined by how much we human beings are prepared to pay for its content. Fascinating though the economics of ores and mining is, this topic cannot be explored in this book. Instead, discussion is limited to the origin of mineral deposits without regard to questions of economics.

In order for a deposit to form, some process or combination of processes must bring about a localized enrichment of one or more minerals. A convenient way to classify mineral deposits is through the principal concentrating process. Minerals become concentrated in five ways:

1. Concentration by hot, aqueous solutions flowing through fractures and pore spaces in crustal rock to form **hydrothermal mineral deposits** (Chapter 7).

2. Concentration by magmatic processes within a body of igneous rock to form **magmatic mineral deposits** (Chapter 4).

3. Concentration by precipitation from lake water or seawater to form **sedimentary mineral deposits** (Chapter 6).

4. Concentration by flowing surface water in streams or along the shore, to form **placers** (Chapter 10).

5. Concentration by weathering processes to form **residual mineral deposits** (Chapter 5).

Hydrothermal Mineral Deposits

Many of the most famous mines in the world contain ores that were formed when their ore minerals were deposited from hydrothermal solutions. As we discussed in Chapter 7, more mineral deposits have probably been formed by deposition from hydrothermal solutions than by any other mechanism. However, the origins of hydrothermal solutions are often difficult to decipher. Some solutions originate when water dissolved in a magma is released as the magma rises and cools. Other solutions are formed from rainwater or seawater that circulates deep in the crust.

An example of the way a hydrothermal solution can form from deeply circulating seawater is shown in Figure B7.5. Because the heat source for seawater hydrothermal solutions of the kind illustrated in this figure is midocean ridge volcanism, and because the ore minerals deposited are always sulfides, mineral deposits formed from such solutions are called **volcanogenic massive sulfide deposits**.

The ore–mineral constituents in volcanogenic massive sulfide deposits derive from the igneous rocks of the oceanic crust. Heated seawater reacts with the rocks it is in contact with, causing changes in both mineral composition and solution composition. For example, feldspars are changed to clays and epidote, and pyroxenes are changed to chlorites. As the minerals are transformed, trace metals such as copper and zinc, present by atomic substitution, are released and become concentrated in the slowly evolving hydrothermal solution.

Hydrothermal solutions having similar compositions can form in many different ways. Which ore constituents are carried in solution depends on the kinds of rocks involved in the formation of the solution. For example, copper and zinc are present in pyroxenes by atomic substitution, so that the pyroxene-rich rocks of the oceanic crust yield solutions charged with copper and zinc. As a result, volcanogenic massive sulfide deposits are rich in copper and zinc.

The most important question concerning hydrothermal solutions is not *where* the water and dissolved mineral constituents came from, but rather what made the solutions precipitate their soluble mineral load and form a mineral deposit?

Causes of Precipitation

When a hydrothermal solution moves slowly upward, as with groundwater percolating through an aquifer, the solution cools very slowly. If dissolved minerals were precipitated from such a slow-moving solution, they would be spread over great distances and would not be sufficiently concentrated to form an ore. But when a solution flows rapidly, as in an open fracture, or through a mass of shattered rock, or through a layer of porous tephra where flow is less restricted, cooling can be sudden and can occur over short distances. Rapid precipitation and a concentrated mineral deposit are the result. Other effects—such as boiling, a rapid decrease in pressure, composition changes of the solution caused by reactions with adjacent rock, and cooling as a result of mixing with seawater—can also cause rapid precipitation and form concentrated deposits. When valuable minerals are present, an ore can be the result.

Examples of Precipitation

Veins form when hydrothermal solutions deposit minerals in open fractures, and many such veins are found in regions of volcanic activity (Fig. 20.3). The famous gold deposits at Cripple Creek, Colorado, were formed in fractures associated with a small caldera, and the huge tin and silver deposits in Bolivia are in fractures that are localized in and around stratovolcanoes. In each case, the fractures formed as a result of volcanic activity, and the magma chambers that fed the volcanoes served as the sources of the hydrothermal solutions that rose up and formed the mineralized veins.

A cooling granitic stock or batholith is a source of heat just as the magma chamber beneath a volcano is—and it can also be a source of hydrothermal solutions. Such solutions move outward from a cooling stock and will flow through any fracture or channel, metasomatically altering the surrounding rock in the process and commonly depositing valuable minerals. Many famous ore bodies are associated with intrusive igneous rocks. The tin deposits of Cornwall, England, and the copper deposits at Butte, Montana; Bingham, Utah; and Bisbee, Arizona, are examples. For examples of hydrothermal deposits forming today, see the accompanying essay, *Hydrothermal Mineral Deposits Forming Today.*

Magmatic Mineral Deposits

The processes of partial melting and fractional crystallization discussed in Chapter 4 are two ways of separating some minerals from others. Fractional crystallization, in particular, can lead to the creation of valuable mineral deposits. The processes involved are entirely magmatic, and so such deposits are referred to as magmatic mineral deposits.

Pegmatites formed by fractional crystallization of granitic magma (Chapter 4) commonly contain rich concentrations of such elements as lithium, beryllium, cesium, and niobium. Much of the world's lithium is mined from pegmatites such as those at King's Mountain, North Carolina, and Bikita in Zimbabwe. The great Tanco pegmatite in Manitoba, Canada, produces much of the world's cesium, and pegmatites in many countries yield beryl, one of the main ore minerals of beryllium.

Crystal setting, another process of fractional crystallization, can also form valuable mineral deposits. The process is especially important in low-viscosity basaltic magma. When a large chamber of basaltic magma crys-

Figure 20.3 Mining a Rich Vein A rich vein in Potosi, Bolivia, containing chalcopyrite, sphalerite, and galena cutting andesite. The andesite has been altered metasomatically by the hydrothermal solution that deposited the ore minerals.

Hydrothermal Mineral Deposits Forming Today

Three extraordinary discoveries over a 15-year period have changed our thinking about mineral deposits. The first discovery, in 1962, was accidental. Until that year no one was sure where to look for modern hydrothermal solutions or even how to recognize one when it was found. Drillers seeking oil and gas in the Imperial Valley of southern California were astonished when they struck a 320°C brine at a depth of 1.5 km. As the brine flowed upward, it cooled and precipitated minerals it had been carrying in solution. Over three months, the well deposited 8 tons of siliceous scale containing 20 percent copper and 8 percent silver by weight. The drillers had found a hydrothermal solution that could, under suitable flow conditions, form a rich mineral deposit.

The Imperial Valley is a sediment-filled graben covering the join between the Pacific and North American plates, where the East Pacific Rise passes under North America (Fig. B20.1). Volcanism is the source of heat for the brine solution discovered in 1962. These brines provided the first unambiguous evidence that hydrothermal solutions can leach metals from ordinary sediments.

Before geologists had a chance to fully absorb the significance of the Imperial Valley discovery, a second remarkable find was announced. In 1964, oceanographers discovered a series of hot, dense, brine pools at the bottom of the Red Sea. The brines are trapped in the graben formed by the spreading center between the Arabian and African plates (Fig. B20.2), and they are so much saltier, and therefore more dense, than seawater that they remain ponded in the graben even though they are as hot as 60°C. Many such brine pools have now been discovered.

The Red Sea brines rise up the normal faults associated with the central rift of a spreading center and, like the Imperial Valley brines, have evolved to their present compositions through reactions with the enclosing rocks. The Red Sea brine discovery was surprising, but even more surprising was the discovery that sediments at the bottom of the pools contained ore minerals such as chalcopyrite,

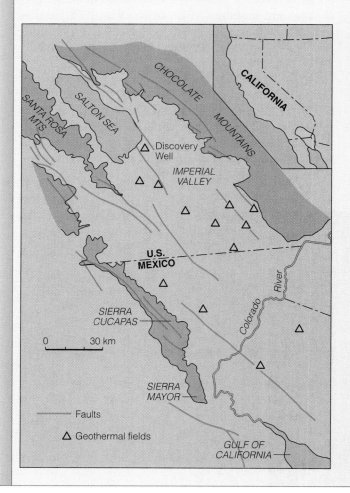

Figure B20.1   Present-day Hydrothermal Solution   The Imperial Valley graben (also known as the Salton Trough). The graben is bounded by the Chocolate Mountains on the east and the Santa Rosa Mountains on the west. Hydrothermal solutions were discovered in a well drilled on the southern end of the Salton Sea. Places where geothermal activity is known, and where other hydrothermal solutions may be present at depth, are marked with triangles.

galena, and sphalerite. In other words, the oceanographers had discovered modern stratabound mineral deposits in the process of formation.

The third remarkable discovery was really a series of discoveries that commenced in 1978. Scientists using deep-diving submarines made a series of dives on the East Pacific Rise at 21°N latitude. To their amazement, they found 300°C hot springs emerging from the seafloor 2500 m below sea level. Around the hot springs lay a blanket of sulfide minerals. The submariners watched a modern volcanogenic sulfide deposit forming before their eyes.

Each of the discovery sites—Imperial Valley, Red Sea, and 21°N—is on a spreading center, and so there is no doubt that the deposits are forming as a result of plate tectonics. Soon the hunt was on to see if seafloor deposits could be found above subduction zones. In 1989, a joint German-Japanese oceanographic expedition to the western Pacific discovered the first modern subduction-related deposits. No longer are geologists limited to speculating about how certain mineral deposits *might* have formed. Today we can study them as they grow.

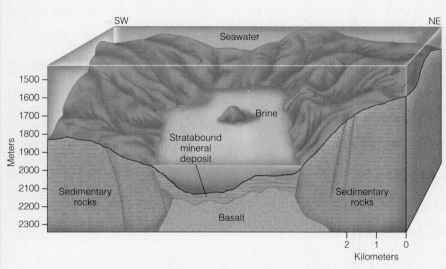

Figure B20.2   Ore Deposit Forming on the Seafloor   Topography of the Red Sea graben near the Atlantis II brine pool. Hot, dense brines rise up normal faults, pond on the floor of the graben, and form stratabound deposits rich in copper and zinc.

tallizes, one of the first minerals to form is chromite, the main ore mineral of chromium. As shown in Figure 4.41, settling of the dense chromite crystals to the bottom of the magma chamber can produce almost pure layers of chromite. The world's principal ores of chromite, in the Bushveld Igneous Complex in South Africa and the Great Dike of Zimbabwe, were both formed as a result of crystal settling.

Sedimentary Mineral Deposits

The term **sedimentary mineral deposit** is applied to any local concentration of minerals formed through processes of sedimentation. Any process of sedimentation can form localized concentrations of minerals, but it has become common practice to restrict use of the term *sedimentary* to those mineral deposits formed through precipitation of substances carried in solution.

Evaporite Deposits

The most direct way in which sedimentary mineral deposits form is by evaporation of lake water or seawater. The layers of salts that precipitate as a consequence of evaporation are called evaporite deposits (Chapter 6).

Examples of salts that precipitate from lake waters of suitable composition are sodium carbonate (Na_2CO_3), sodium sulfate (Na_2SO_4), and borax ($Na_2B_4O_7 \cdot 1OH_2O$). Huge evaporite deposits of sodium carbonate were laid down in the Green River basin of Wyoming during the Eocene Epoch. This is the same lake in which the rich Green River oil shales were deposited (Fig. 6.17). Borax and other boron-containing minerals are mined from evaporite lake deposits in Death Valley and Searles and Borax lakes, all in California, and in Argentina, Bolivia, Turkey, and China.

Much more common and important than lake water evaporites are the marine evaporites formed by evap-

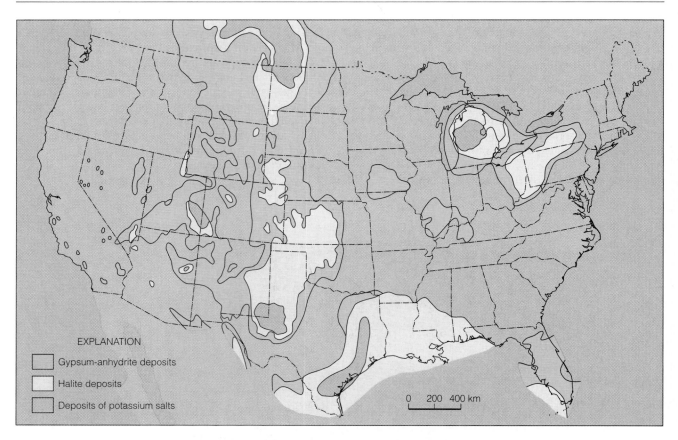

Figure 20.4 Evaporite Deposits Portions of the United States known to be underlain by marine evaporite deposits. The areas underlain by gypsum and anhydrite do not contain halite. The areas underlain by halite are also underlain by gypsum and anhydrite. The areas underlain by potassium salts are also underlain by halite and by gypsum and anhydrite.

oration of seawater. The most important salts that precipitate from seawater are gypsum ($CaSO_4 \cdot 2H_2O$), halite (NaCl), and carnallite ($KCl.MgCl_2 \cdot 6H_2O$). Low-grade metamorphism of marine evaporite deposits causes another important mineral, sylvite (KCl), to form from carnallite. Marine evaporite deposits are widespread; in North America, for example, strata of marine evaporites underlie as much as 30 percent of the land area (Fig. 20.4). Most of the salt that we use, plus the gypsum used for plaster and the potassium used in plant fertilizers, is recovered from marine evaporites.

Iron Deposits

Sedimentary deposits of iron minerals are widespread, but the amount of iron in average seawater is so small that such deposits cannot have formed from seawater that is the same as today's seawater.

All sedimentary iron deposits are tiny by comparison with the class of deposits characterized by the *Lake Superior-type iron deposits*. These remarkable deposits, mined principally in Michigan and Minnesota, were long the mainstay of the United States' steel industry but are declining in importance today as imported ores

replace them. The deposits are of early Proterozoic age (about 2 billion years or older) and are found in sedimentary basins on every craton, particularly in Labrador, Venezuela, Brazil, the former USSR, India, South Africa, and Australia. Every aspect of the Lake Superior-type deposits indicates chemical precipitation, as we discussed in Chapter 6. The deposits are interbedded layers of chert and several different kinds of iron minerals. Because the deposits are so large, it is inferred that the iron and silica must have been transported in surface water, but the cause of precipitation remains unknown. Many experts suspect that Lake Superior-type deposits may be ancient evaporites that formed from seawater of a different composition than today's seawater.

Lake Superior-type iron deposits are not ores. The grades of the deposits range from 15 to 30 percent Fe by weight, and the deposits are so fine grained that the iron minerals cannot be easily separated from the gangue. Two additional processes can form ore. First, as discussed in the essay on mineral deposits formed by weathering in Chapter 5, leaching of silica during weathering can lead to secondary enrichment and can produce ores containing as much as 66 percent Fe. Compare

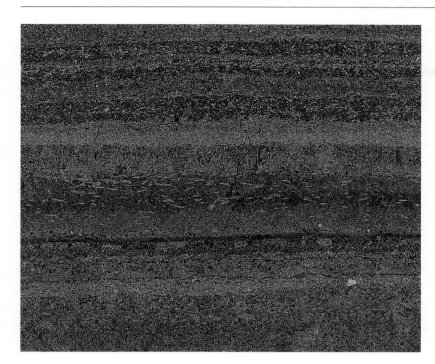

Figure 20.5 A Stratabound Ore Stratabound ore of lead and zinc from Kimberley, British Columbia. The layers of pyrite (yellow), sphalerite (brown), and galena (grey) are parallel to the layering of the sedimentary rock in which they occur. The specimen is 4 cm across.

Figure 6.13 which is a Lake Superior-type iron deposit in the Hamersley Range, Western Australia, with Figure B5.3, a sample of ore developed by secondary enrichment in the Hamersley Range. The rocks in Figure 6.13 contain about 25 percent Fe, whereas those in Figure B5.3 have had most of the silica leached out and contain about 60 percent Fe.

The second way a Lake Superior-type iron deposit can become an ore is through metamorphism. Two changes occur as a result of metamorphism. First, grain sizes increase so that separating ore minerals from the gangue becomes easier and therefore cheaper. Second, new mineral assemblages form, and iron silicate and iron carbonate minerals originally present can be replaced by magnetite or hematite, both of which are desirable ore minerals. The grade is not increased by metamorphism. It is the changes in grain size and mineralogy that transforms the sedimentary rock into an ore. Iron ores formed as a result of metamorphism are called *taconites*, and they are now the main kind of ore mined in the Lake Superior region.

Stratabound Deposits

Some of the world's most important ores of lead, zinc, and copper occur in sedimentary rocks. The ore minerals—galena, sphalerite, chalcopyrite and pyrite—occur in such regular, fine layers that they look like sediments (Fig. 20.5). The sulfide mineral layers are enclosed by and parallel to the sedimentary strata in which they occur, and for this reason such deposits are called *stratabound mineral deposits*. They look like sediments but are not, strictly speaking, sediments. It is more correct to consider most stratabound deposits as being diagenetic in origin.

Stratabound deposits form when a hydrothermal solution invades and reacts with a muddy sediment. Reactions between sediment grains and the solution cause deposition of the ore minerals. Deposition commonly occurs before the sediment has become a sedimentary rock.

The famous copper deposits of Zambia, in central Africa, are stratabound ores, as are the great Kupferschiefer deposits of Germany and Poland. The world's largest and richest lead and zinc deposits, at Broken Hill and Mount Isa in Australia and at Kimberley in British Columbia, are also stratabound ores.

Placers

The way in which a mineral with a high specific gravity can become concentrated by flowing water was discussed in Chapters 10 and 14. Deposits of minerals having high specific gravities are *placers*. The most important minerals concentrated in placers are gold, platinum, cassiterite (SnO_2), and diamond. Typical locations of placers are illustrated in Figure 20.6.

Gold is the most valuable mineral recovered from placers; more than half of the gold recovered throughout all of human history has come from placers. This is the result of the huge gold production from South Africa, which has come from placers.

The South African gold deposits are really fossil placers, and they have many unusual features. Most placers are found in stream gravels that are geologically young. The South African fossil placers are a series of gold-bearing conglomerates (Fig. 20.7) that were laid

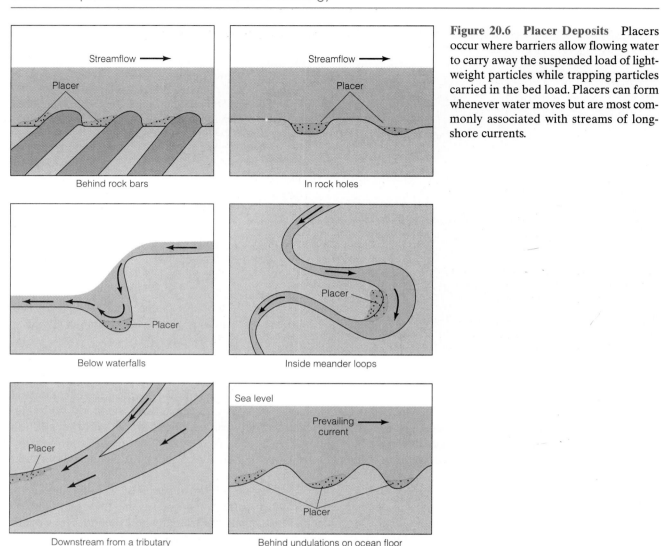

Figure 20.6 Placer Deposits Placers occur where barriers allow flowing water to carry away the suspended load of lightweight particles while trapping particles carried in the bed load. Placers can form whenever water moves but are most commonly associated with streams of longshore currents.

Streamflow →

Placer

Behind rock bars

Streamflow →

Placer

In rock holes

Placer

Below waterfalls

Placer

Inside meander loops

Placer

Downstream from a tributary

Sea level

Prevailing → current

Placer

Behind undulations on ocean floor

Figure 20.7 Paleoplacer Gold is recovered from ancient fossil deposits of the Witwatersrand, South Africa. The gold is found at the base of conglomerate layers interbedded with finer-grained sandstone, here seen in weathered outcrop at the site where gold was first discovered in 1886.

down 2.7 billion years ago as gravels in the shallow marginal waters of a marine basin. Associated with the gold are grains of pyrite and uranium minerals. As far as size and richness are concerned, nothing like the deposits in the Witwatersrand basin has been discovered anywhere else. Nor has the original source of all the placer gold been discovered, and so it is not possible to say why so much of the world's mineable gold should be concentrated in this one sedimentary basin.

Mining in the Witwatersrand basin has reached a depth of 3600 m (11,800 ft). This is the deepest mining in the world, and there are plans to continue mining to depths as great as 4500 m. Despite such ambitious plans, the heyday of gold mining in South Africa has probably passed because the deposits are running out of ore.

Through the middle years of the 1980s, the price of gold fluctuated between about $14 and $16 a gram. This led to a boom in gold prospecting and to the discovery of a large number of new ore deposits in the United States, Canada, Australia, the Pacific islands, and elsewhere. Most of the new discoveries are hydrothermal deposits. Despite all the new discoveries, South Africa with its huge fossil placers continues to dominate the world's gold production. In 1998, South Africa still supplied about 25 percent of all the gold produced in the world, but the production rate is dropping slowly.

Residual Mineral Deposits

Weathering occurs because newly exposed rock is not chemically stable when it is in contact with rainwater and the atmosphere. Chemical weathering, in particular, leads to mineral concentration through the removal of soluble materials in solution and the concentration of a less soluble residue. A common example of a deposit formed through residual concentration is laterite, and the most important kind of laterite is bauxite (Chapter 5).

Bauxites are the source of the world's aluminum. They are widespread in the world, but they are concentrated in the tropics because that is where lateritic weathering occurs. Where bauxites are found in present-day temperate conditions, such as France, China, Hungary, and Arkansas, it is clear that the climate was tropical when the bauxites formed.

All bauxites are vulnerable to erosion. They are not found in glaciated regions, for example, because overriding glaciers scrape off the soft surface materials. The vulnerability of bauxites means that most deposits are geologically young. More than 90 percent of all known deposits formed during the last 60 million years, and all of the very large deposits formed less than 25 million years ago.

Metallogenic Provinces

Many kinds of mineral deposits tend to occur in groups and to form what exploration geologists call **metallogenic provinces**. These are defined as limited regions of the crust within which mineral deposits occur in unusually large numbers. A striking example is the metallogenic province shown in Figure 7.17, which runs along the western side of the Americas. Within the province is the world's greatest concentration of large hydrothermal copper deposits. These deposits are associated with intrusive igneous rocks that are invariably porphyritic (Chapter 4), and they are therefore called **porphyry copper deposits**. The intrusive igneous rocks, and therefore the deposits themselves, were formed as a consequence of subduction because they are in, or adjacent to, old stratovolcanoes.

Metallogenic provinces form as a result of either climate control (as in the formation of bauxite deposits in the tropics) or plate tectonics. Magmatic, hydrothermal, and stratabound deposits all form near present or past plate boundaries (Fig. 20.8). This is hardly surprising as the deposits are related directly or indirectly to igneous activity, and most igneous activity we now know is related to plate tectonics.

USEFUL MINERAL SUBSTANCES

It is convenient for purposes of discussion to group mineral products on the basis of the way they are used rather than the way they occur. Excluding substances

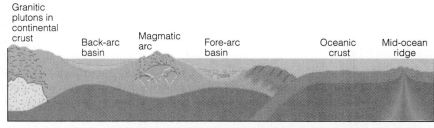

	Granitic plutons in continental crust	Back-arc basin	Magmatic arc	Fore-arc basin		Oceanic crust	Mid-ocean ridge
Metals	Tin Tungsten Bismuth Copper	Copper Zinc Gold Chromium	Copper Gold Silver Tin Lead Mercury Molybdenum	Lead Zinc Copper	Chromium	Manganese Cobalt Nickel	Copper Zinc
Deposits	Vein: contact metamorphic	Volcanogenic massive sulfide, stratabound, evaporites.	Porphyry copper; veins	Stratabound in sediments	Magmatic chromite	Manganese nodules	Volcanogenic massive sulfide

Figure 20.8 Plate Tectonics and Mineral Deposits Locations of certain kinds of mineral deposits in terms of plate structures.

TABLE 20.1	Principal Mineral Substances, Grouped According to Use

Metals
    Geochemically abundant metals
        Iron, aluminum, magnesium, manganese, titanium
    Geochemically scarce metals
        Copper, lead, zinc, nickel, chromium, gold, silver, tin, tungsten, mercury, molybdenum, uranium, platinum, palladium, and many others
Nonmetallic substances
    Used for chemicals
        Sodium chloride (halite), sodium carbonate, sulfur, borax, fluorite
    Used for fertilizers
        Calcium phosphate (apatite), potassium chloride (sylvite), sulfur, calcium carbonate (limestone), sodium nitrate
    Used for building
        Gypsum (for plaster), limestone, clay (for brick and tile), asbestos, sand gravel, crushed rock of various kinds, shale (for cement)
    Used for ceramics and abrasives
        Ceramics: Clay, feldspar, quartz
        Abrasives: Diamond, garnet, corundum, pumice, quartz

used for energy, there are two broad groups: (1) minerals from which metals such as iron, copper, and gold can be recovered, and (2) nonmetallic minerals, such as salt, gypsum, and clay, used not for the metals they contain but for their properties as chemical compounds. The nonmetallic substances can be further subdivided on the basis of more specialized uses (Table 20.1).

Without exception, the useful metals are present in the crust in such small amounts that we can mine and recover them only when rich mineral deposits can be located.

Geochemically Abundant Metals

Metals can be usefully subdivided on the basis of their average percentage in the crust. Those present in such abundance that they make up 0.1 percent by weight or more of the crust are considered to be **geochemically abundant**. These metals are iron, aluminum, manganese, magnesium, and titanium.

Geochemically abundant metals require comparatively small enrichment factors to form extremely large deposits. The minerals that are concentrated in deposits of the geochemically abundant metals tend to be oxides and hydroxides. The most important kinds of deposits are residual, sedimentary, and magmatic deposits.

Geochemically Scarce Metals

Metals that make up less than 0.1 percent by weight of the crust are said to be **geochemically scarce**.

With the exception of copper, zinc, and chromium, minerals of the geochemically scarce metals are not found in common rocks. However, chemical analysis of any rock will reveal that even though the minerals are absent, geochemically scarce metals are certainly present. Further research will reveal that the scarce metals are present exclusively as a result of atomic substitution (Chapter 3). Atoms of the scarce metals (such as nickel, cobalt, and copper) can readily substitute for more common atoms (such as magnesium and calcium). In order for a mineral deposit to form, therefore, some gathering and concentrating agent, such as a hydrothermal solution, must react with the rock-forming minerals and leach the scarce metals from them. The solution must then transport the metals in solution and deposit them as separate minerals in a localized place. With such a complicated chain of events, it is hardly surprising that deposits of geochemically scarce metals are rarer and much smaller than deposits of the abundant metals.

Most ore minerals of the scarce metals are sulfides; a few, such as the ore minerals of tin and tungsten, are oxides. In the case of gold, platinum, palladium, and a few less common elements, the metal itself is the most important mineral. Most scarce metal deposits form as hydrothermal or magmatic mineral deposits. In the case of gold and platinum, placer concentration is also important.

ENERGY RESOURCES

A healthy, hard-working person can produce just enough muscle energy to keep a single 75-watt light bulb burning for 8 hours a day. It costs about 10 cents to purchase the same amount of energy from the local electrical utility. Viewed strictly as machines, humans aren't worth much. By comparison, the amount of mechanical and electrical energy used each 8-hour working day in North America could keep four hundred 75-watt bulbs burning for every person living there.

To see where all this energy is used, it is necessary to sum up all the energy employed to grow and transport food, make clothes, cut lumber for new homes, light streets, heat and cool office buildings, and do myriad other things. The uses can be grouped into three categories: transportation, home and commerce, and industry (meaning all manufacturing and raw material processing plus the growing of foodstuffs). The present-day uses and sources of energy in the United States are summarized in Figure 20.9.

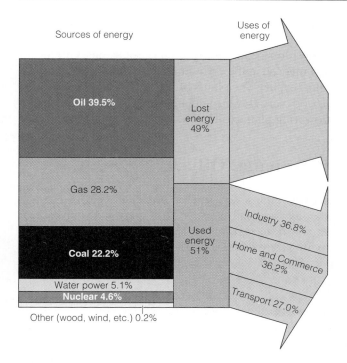

Sources of energy

Oil 39.5%

Gas 28.2%

Coal 22.2%

Water power 5.1%

Nuclear 4.6%

Other (wood, wind, etc.) 0.2%

Uses of energy

Lost energy 49%

Used energy 51%

Industry 36.8%

Home and Commerce 36.2%

Transport 27.0%

Figure 20.9 How Energy Is Used Uses and sources of energy in the United States. Lost energy arises both from inefficiencies of use and from the fact that the laws of thermodynamics impose a limit to the efficiency of any engine and therefore a limit on the fraction of available energy that can be usefully employed.

How much energy do all the people of the world use? The total is enormous. The energy drawn annually from major fuels—coal, oil, and natural gas—plus that from nuclear power plants, is 2.6×10^{20} J. Nobody keeps accurate accounts of all the wood and animal dung burned in the cooking fires of Africa and Asia, but the amount has been estimated to be so large that when it is added to the 2.6×10^{20} J figure, the world's total energy consumption rises to about 3.0×10^{20} J annually. This is equivalent to the burning of 2 metric tons of coal or 10 barrels of oil for every living man, woman, and child each year! Energy consumption around the world is very uneven, however. In less developed countries such as India and Tanzania, energy use is equivalent to burning only 3 or 4 barrels of oil per person per year, whereas in a developed country such as the United States, energy use is equivalent to burning more than 50 barrels of oil per person per year.

Supplies of Energy

The chief sources of the energy consumed in highly industrialized nations are few: the fossil fuels (coal, oil, natural gas), hydroelectric and nuclear power, wood, wind, and a very small amount of muscle energy. As recently as a century ago, wood was an important fuel in industrial societies, but now it is used mainly for space heating in some dwellings.

FOSSIL FUELS

The term *fossil fuel* refers, as we discussed in Chapter 6, to the remains of plants and animals trapped in sediment that can be used for fuel. The kind of sediment, the kind of organic matter trapped, and the changes in the organic matter that take place as a result of burial and diagenesis determine the kind of fossil fuel that forms.

In the ocean, microscopic photosynthetic phytoplankton and bacteria are the principal sources of trapped organic matter. Shales do most of the trapping. Once bacteria and phytoplankton are trapped in the shale, the organic compounds they contain—*proteins*, *lipids*, and *carbohydrates*—become part of the shale, and it is these compounds that are transformed (mainly by heat) to oil and gas.

On land, it is trees, bushes, and grasses that contribute most of the trapped organic matter to shales; these large land plants are rich in resins, waxes, and lignins, which tend to remain solid and form coal rather than oil or natural gas.

In many marine and lake shales, burial temperatures never reach the levels at which the original organic molecules are converted to the organic molecules found in oil and natural gas. Instead, an alteration process occurs in which waxlike substances containing large molecules are formed. This material, which remains solid, is called *kerogen*, and it is the substance in so-called *oil shales*. Kerogen can be converted to oil and gas by mining the shale and heating it in a retort.

Coal

The combustible sedimentary rock we call coal is the most abundant fossil fuel. Most of the coal mined is eventually burned either under boilers to make steam for electric generators, or else converted to coke, an essential ingredient in the smelting of iron ore and the making of steel. In addition to its use as a fuel, coal is a raw material for nylon and many other plastics, plus a multitude of other organic chemicals. The conditions under which organic matter accumulates in swamps as peat, then during burial and diagenesis is converted to coal, are discussed in Chapter 5. Coalification involves the loss of volatile materials such as H_2O, CO_2, and CH_4 (methane). As the volatiles escape, the remaining coal is increasingly enriched with carbon. Through coalification, peat is converted successively to *lignite* (one type of coal), *subbituminous coal*, and *bituminous coal* (Fig. 6.16). These coals are sedimentary rocks. However, *anthracite*, a still later phase in the coalification process, is a metamorphic rock.

Because of its low volatile content, anthracite is hard to ignite, but once alight it burns with almost no smoke. In contrast, lignite is rich in volatiles, burns smokily, and ignites so easily that it is dangerously subject to spontaneous ignition.

In regions where metamorphism has been intense, coal has been changed so thoroughly that it has been converted to graphite, in which all volatiles have been lost. Graphite will not burn in an ordinary fire.

Occurrence of Coal

A coal seam is a flat, lens-shaped body having the same surface area as the swamp in which it originally accumulated. Most coal seams tend to occur in groups. In western Pennsylvania, for example, 60 seams of bituminous coal are found. This clustering indicates that the coal must have formed in a slowly subsiding site of sedimentation.

Coal swamps seem to have formed in many sedimentary environments, of which two types predominate. One consists of slowly subsiding basins in continental interiors and the swampy margins of shallow inland seas formed at times of high sea level. This is the home environment of the bituminous and subbituminous coal seams in Utah, Montana, Wyoming, and the Dakotas. The second sedimentary environment consists of continental margins with wide continental shelves (that is, continental margins in plate interiors) that were flooded at times of high sea level. This is the environment of the bituminous coals of the Appalachian region.

Coal-Forming Periods

Although peat can form under even subarctic conditions, it is clear that the luxuriant plant growth needed to form thick and extensive coal seams develop most readily in a tropical or semitropical climate. The Great Dismal Swamp in Virginia and North Carolina is one of the largest modern peat swamps. It contains an average thickness of 2 m of peat. However, unless this swamp lasts millions of years, even that dense growth is insufficient to produce a coal seam as thick as some of the seams in Pennsylvania.

Peat formation has been widespread and more or less continuous from the time land plants first appeared about 450 million years ago, during the Silurian Period. The size of peat swamps has varied greatly, however, and so, as a consequence, has the amount of coal formed. By far the greatest period of coal swamp formation occurred during the Carboniferous and Permian periods, when Pangaea existed. The great coal beds of Europe and the eastern United States formed at this time, when

the plants of coal swamps were giant ferns and scale trees (gymnosperms). The second great period of coal deposition peaked during the Cretaceous Period but commenced in the early Jurassic and continued until the mid-Tertiary. The plants of the coal swamps during this period were flowering plants (angiosperms) much like flowering plants today.

Petroleum: Oil and Natural Gas

As was mentioned in the introductory essay to this chapter, rock oil is one of the earliest resources our ancestors learned to use. However, the major use of oil really started about 1847 when a merchant in Pittsburgh, Pennsylvania, started bottling and selling rock oil from natural seeps to be used as a lubricant. Five years later, in 1852, a Canadian chemist discovered that heating and distillation of rock oil yielded kerosene, a liquid that could be used in lamps. This discovery spelled doom for candles and whale-oil lamps. Wells were soon being dug by hand near Oil Springs, Ontario, in order to produce oil. In Romania in 1856, using the same hand-digging process, workers were producing 2000 barrels a year.* In 1859, the first oil well was drilled in Titusville, Pennsylvania. On August 27, 1859, at a depth of 21.2 m, oil-bearing strata were encountered and up to 35 barrels of oil a day were pumped out. Oil was soon discovered in West Virginia (1860), Colorado (1862), Texas (1866), California (1875), and many other places.

The earliest known use of natural gas was about 3000 years ago in China, where gas seeping out of the ground was collected and transmitted through bamboo pipes to be ignited and used to evaporate saltwater in order to recover salt. It wasn't long before the Chinese were drilling wells to increase the flow of gas. Modern use of gas started in the early seventeenth century in Europe, where gas made from wood and coal was used for illumination. Commercial gas companies were founded as early as 1812 in London and 1816 in Baltimore. The stage was set for the exploitation of an accidental discovery at Fredonia, New York, in 1821. A water well drilled in that year produced not only water but also bubbles of a mysterious gas. The gas was accidentally ignited and produced such a spectacular flame that a new well was drilled on the same site and wooden pipes were installed to carry the gas to a nearby hotel, where 66 gas lights were installed. By 1872, natural gas was being piped as far as 40 km from its source.

Origin of Petroleum

Petroleum is a term used for both oil and natural gas. That petroleum is a product of the decomposition of

---

* A barrel is equal to 42 U.S. Gal and is the volume generally used when commercial production of oil is discussed.

organic matter trapped in sediment was discussed in Chapter 6. That petroleum migrates through aquifers and becomes trapped in reservoirs was discussed in the essay at the end of Chapter 15.

The migration of petroleum deserves further discussion. The sediment in which organic matter is accumulating today is rich in clay minerals, whereas most of the strata that constitute oil or gas pools are sandstones (consisting of quartz grains), limestones and dolostones (consisting of carbonate minerals), and much-fractured rock of other kinds. Long ago, geologists realized that oil and gas form in one kind of material (shale) and at some later time migrate to another (sandstone or limestone).

Petroleum migration is analogous to groundwater migration. When oil and gas are squeezed out of the shale in which they originated and enter a body of sandstone or limestone, they can migrate more easily than before because most sandstones and limestones are more permeable than any shale. The force of molecular attraction between oil and quartz or carbonate minerals is weaker than that between water and quartz or carbonate minerals. Hence, because oil and water do not mix, water remains fastened to the quartz or carbonate grains while oil occupies the central parts of the larger openings in the porous sandstone or limestone. Because it is lighter than water, the oil tends to glide upward past the carbonate- and quartz-held water. In this way it becomes segregated from the water; when it encounters a trap, it can form a pool.

Most of the petroleum that forms in sediments does not find a suitable trap, and eventually makes its way, along with groundwater, to the surface. It is estimated that no more than 0.1 percent of all the organic matter originally buried in a sediment is eventually trapped in an oil or a gas pool. It is not surprising, therefore, that the highest ratio of oil and gas pools to volume of sediment is found in rock no older than 2.5 million years, and that nearly 60 percent of all the oil and gas discovered so far has been found in strata of Cenozoic age (Fig. 20.10). This does not mean that older rocks produced less petroleum. It simply means that oil in older rocks has had a longer time in which to escape.

Distribution of Oil

Petroleum deposits, like coal, are frequent but are distributed unevenly. The reasons for uneven petroleum distribution are not as obvious as they are with coal. Suitable source sediments for petroleum are very widespread and seem as likely to form in subarctic waters as in tropical regions. The critical controls seem to be a supply of heat to effect the conversion of solid organic matter to oil and gas and the formation of a suitable trap before the petroleum has leaked away.

Solid organic matter is converted to oil and gas within a specific range of depth and temperature defined by the geothermal gradients shown in Figure 20.11. If a thermal

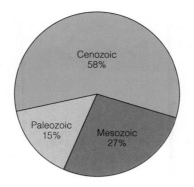

Figure 20.10 **Age of Strata and Yield of Oil** Percentage of world's total oil production from strata of different ages.

gradient is too low (less than 1.8°C/100 m), conversion does not occur to either oil or gas. If the gradient is above 5.5°C/100 m, conversion to gas starts at such shallow depths that little trapping occurs. Once oil and gas have been formed, they will accumulate in pools only if suitable traps are present. Most oil and gas pools are found beneath anticlines; the timing of the folding event is therefore a critical part of the trapping process. If folding occurs after petroleum has formed and migrated, pools cannot form. The great oil pools in the Middle East arose through the fortunate coincidence of the

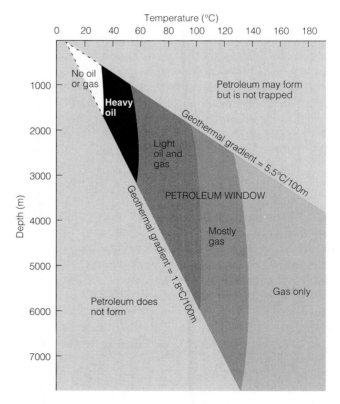

Figure 20.11 **Where Oil and Gas Form** Regions of depth and temperature within which oil and gas are generated and trapped.

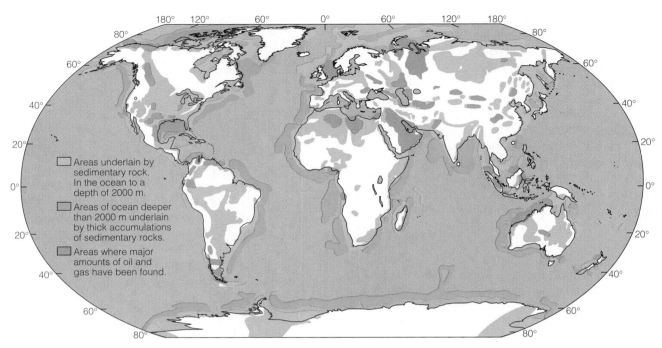

Figure 20.12   Where Oil and Gas Occur   Areas underlain by sedimentary rock and regions where large accumulations of oil and gas have been located. Where the ocean is deeper than 2000 m, sedimentary rock has yet to be tested for its oil and gas potential.

right thermal gradient and the development of anticlinal traps during the collision of Europe and Asia with Africa.

How much oil is there in the world? This is an extremely controversial question. Approximately 700 billion barrels have already been pumped out of the ground. A lot of additional oil has been located by drilling but is still waiting to be pumped. Probably a great deal more oil remains to be found by drilling. Unlike coal, for which the volume of strata in a basin of sediment can be accurately estimated even before drilling, the volume of undiscovered oil can only be surmised. The way guesses are made is to use the accumulated experience of a century of drilling. Knowing how much oil has been found in an intensively drilled area, such as eastern Texas, experts make estimates of probable oil volumes in other regions where rock types and structures are similar to what we find in eastern Texas. Using this approach, and considering all the sedimentary basins of the world (Fig. 20.12), experts estimate that 1500 to 3000 billion barrels of oil will eventually be discovered.

Tars
Oil that is exceedingly viscous will not flow easily and cannot be pumped. Colloquially called **tar**, heavy, viscous oil acts as a cementing agent between mineral grains in an oil pool. The tar can be recovered only if the sandstone is mined and heated enough to make the tar flow. The resulting tar must then be processed to recover the valuable gasoline fraction. The cost of mining and treating "tar sands," as heavy, viscous oil deposits are

called, is high, but it is technologically possible and someday tar sands may be an important source of fuel. The largest known occurrence of tar sands is in Alberta, Canada, where the *Athabasca Tar Sand* covers an area of 5000 km^2 and reaches a thickness of 60 m (Fig. 20.13). Similar deposits almost as large are known in Venezuela and in the former USSR.

Oil Shale
Another potential source of petroleum is kerogen in shale. If the kerogen is heated, it breaks down and forms liquid and gaseous hydrocarbons similar to those in oil and gas. All shales contain some kerogen, but to be considered an energy resource the kerogen must yield more energy than is required to mine and heat it. Only those shales that yield 40 or more liters of distillate per ton can be considered because the energy needed to mine and process a ton of shale is equivalent to that created by burning 40 liters of oil.

The world's largest deposit of rich oil shale is in the United States. During the Eocene Epoch, many large, shallow lakes existed in basins in Colorado, Wyoming, and Utah; in three of them a series of rich organic sediments were deposited that are now the Green River Oil Shales (Fig. 6.17). The richest shales were deposited in the lake in Colorado now called the Piceance basin. These shales are capable of producing as much as 240 liters of oil per ton. Scientists of the U.S. Geological Survey estimate that, in the Green River Oil Shales alone, oil-shale resources capable of producing 50 liters

Figure 20.13 Tar Sands Athabasca tar sand being mined in Alberta, Canada. After mining, the tar-cemented sands are heated in order to soften and remove the tar prior to processing.

or more of oil per ton of shale can ultimately yield about 2000 billion barrels of oil.

Rich deposits of oil shale in other parts of the world have not been adequately explored, but there is a huge deposit in Brazil called the Irati Shale. Another very large deposit is known in Queensland, Australia, and others have been reported in such widely dispersed places as South Africa and China. Although oil shales have been mined and processed in an experimental fashion in the United States, the only countries where extensive commercial production has been tried are the former USSR and China. Production expenses today make exploitation of oil shales in all countries unattractive by comparison with oil and gas. Most experts believe, however, that large-scale mining and processing of oil shale will eventually occur.

How Much Fossil Fuel?

Are supplies of fossil fuels adequate to meet future demands? If we use a barrel of oil as our unit of measurement, we can compare quantities of all fossil fuels directly. Approximately 0.22 ton of coal produces the same amount of heat energy as one barrel of oil. Thus, the world's recoverable coal reserves of $13,800 \times 10^9$ tons are equivalent to about 63,000 billion barrels of oil.

Considering the approximate world-use rate of oil (30 billion barrels a year), and comparing the estimated recoverable amounts of fossil fuels (Table 20.2), we see that only coal seems to have the capacity to meet long-term demands.

OTHER SOURCES OF ENERGY

Three sources of energy other than fossil fuels have already been developed to some extent: the Earth's plant life (so-called biomass energy), hydroelectric

TABLE 20.2	Amounts of Fossil Fuels Possibly Recoverable Worldwide (Unit of Comparison is a Barrel of Oil)	
Fossil Fuel	Total Amount in Ground (billions of barrels)	Amount Possibly Recoverable (billions of barrels)
Coal	About 100,000	62,730[a]
Oil and gas (flowing)	1500–3000	1500–3000
Trapped oil in pumped-out pools	1500–3000	0–?
Viscous oil (tar sands)	3000–6000	500–?
Oil shale	Total unknown: much greater than coal	1000–?

[a] 0.22 ton of coal = 1 barrel of oil

energy, and nuclear energy. Five others—the Sun's heat, winds, waves, tides, and the Earth's internal heat (geothermal power)—have been tested and developed on a limited basis, but none has yet been shown to be developable on a large scale. Many experts believe geothermal heat is the energy source most likely to be further exploited in the near future.

Biomass Energy

Scientists working for the United Nations estimate that wood and animal dung used for cooking and heating fires now amount to energy production of 4×10^{19} J annually. This is approximately 14 percent of the world's total energy use. The greatest use of wood as a fuel occurs in developing countries, where the cost of fossil fuel is very high in relation to income.

Measurements made on living plant matter indicate that new plant growth on land equals 1.5×10^{11} metric tons of dry plant matter each year. If all of this were burned, or used in some other way as a biomass energy source, it would produce almost nine times more energy than the world uses each year. Obviously, this is a ridiculous suggestion because in order to do so all the forests would have to be destroyed, plants could not be eaten, and agricultural soils would be devastated. Nevertheless, controlled harvesting of fuel plants could probably increase the fraction of the biomass now used for fuel without serious disruption to forests or to food supplies. In several parts of the world, such as Brazil, China, and the United States, experiments are already under way to develop this energy source.

Hydroelectric Power

Hydroelectric power is recovered from the potential energy of stream water as it flows to the sea. As discussed in Chapter 10, in order to convert the power of flowing water to electricity efficiently, it is necessary to dam streams. Unfortunately, reservoirs behind dams fill with silt, and so even though water power is continuous, dams and reservoirs have limited lifetimes.

Water power has been used in small ways for thousands of years, but only in the twentieth century has it been used to any significant extent for generating electricity. All the water flowing in the streams of the world has a total recoverable energy estimated at 9.2×10^{19} J/yr, an amount equivalent to burning 15 billion barrels of oil per year. Thus, even if all the possible hydropower in the world were developed, we could satisfy only about one-third of the present world energy needs. We have to conclude that, for those fortunate countries with large rivers and suitable dam sites,

hydropower is very important, but for most countries hydropower holds limited potential for development.

Nuclear Energy

Nuclear energy is the heat energy produced during controlled transformation of suitable radioactive isotopes (a process called **fission**). Three of the radioactive atoms that keep the Earth hot by spontaneous decay—^{238}U, ^{235}U, and ^{232}Th—can be mined and used to obtain nuclear energy. Fission is accomplished by bombarding the radioactive atoms with neutrons, thus accelerating the rate of decay and the release of heat energy. The device in which this operation is carried out is called a **pile**.

When ^{235}U fissions, it not only releases heat and forms new elements but also ejects some neutrons from its nucleus. These neutrons can then be used to induce more ^{235}U atoms to fission, and a continuous chain reaction occurs. The function of a pile is to control the flux of neutrons so that the rate of fission can be controlled. When a chain reaction proceeds without control, an atomic explosion occurs. Controlled fission, therefore, is the method used by nuclear power plants, and a tremendous amount of energy can be obtained in the process. The fissioning of one gram of ^{235}U produces as much heat as the burning of 13.7 barrels of oil. Unfortunately, however, ^{235}U is the only natural radioactive isotope that will maintain a chain reaction, and it is the least abundant of the three radioactive isotopes that are mined for nuclear energy. Only one atom of each 138.8 atoms of uranium in nature is ^{235}U. The remaining atoms are ^{238}U, which will not sustain a chain reaction. However, if ^{238}U is placed in a pile with ^{235}U that is undergoing a chain reaction, some of the neutrons will bombard the ^{238}U and convert it to plutonium-239 (^{239}Pu). This new isotope can, under suitable conditions, sustain a chain reaction of its own. The pile in which the conversion of ^{238}U takes place is called a **breeder reactor**. The same kind of device can be used to convert ^{232}Th into ^{233}U, which also will sustain a chain reaction. Unfortunately, breeder reactors and nuclear power plants based on them are more complex and less safe than ^{235}U plants, so all the present nuclear power plants use ^{235}U.

Already there are more than 300 piles in nuclear power plants operating around the world. They utilize the heat energy from fission to produce steam that drives turbines and generates electricity. Approximately 7.6 percent of the world's electrical power is derived from nuclear power plants. In France, more than half of all the electrical power comes from nuclear plants; the fraction is rising sharply in some other European countries and in Japan, too. The reason for the increase is obvious. Japan and most European countries do not have adequate supplies of fossil fuels in order to be self-sufficient.

Many problems are associated with nuclear energy. The isotopes used in power plants are the same isotopes used in atomic weapons, and so a security problem exists. The possibility of a power plant failing in some unexpected way creates a safety problem. The dreadful Chernobyl disaster in 1986 in the former USSR is an example of such an event. Finally, the problem of safe burial of dangerous radioactive waste matter must be faced. Some of the waste matter will retain dangerous levels of radioactivity for thousands of years, an issue that is discussed in the essay *The Waste Disposal Problem: Geology and Politics*.

Understanding Our Environment

The Waste Disposal Problem: Geology and Politics

Waste materials containing substances that are hazardous to human health are produced in ever-increasing amounts in industrial societies. The major kinds of waste are poisonous metals like mercury and lead, organic chemicals from industrial processes, and radioactive elements generated in nuclear reactors. When toxic wastes are disposed of, it is necessary to ensure that if these substances are accidentally released into the environment they will not become concentrated in unsafe levels in food, drinking water, or other materials used in ordinary life.

The problem of controlling release is especially difficult in the case of radioactive waste. Radioactive waste contains one or more chemical elements whose atoms are continually emitting radiation in the form of small atomic particles and X rays. In disposing of this material, the goal is not to eliminate the radiation completely but to make certain that amounts of radioactive elements escaping into the environment are so small that the general background level of radiation is not raised appreciably.

If radioactive waste, or any other kind of toxic waste, is to be disposed of by burial, a major worry is the effect of underground water. Rocks below the land surface are saturated with water, and water is the principal means by which toxic substances might escape into surface environments. Dissolved poisonous substances might be carried to the surface in springs or seepages, or through wells drilled by farmers.

Some low-level waste (which contains small amounts of dangerous materials) can be safely disposed of by shallow landfill, a method that is commonly used to remove nontoxic industrial waste. The problems posed by high-level wastes are much more difficult. Concentrations of toxic substances in these materials are so great that even small quantities escaping into the environment can be a menace to living creatures. The most abundant type of high-level waste consists of spent fuel rods from nuclear reactors. For the most part these are stored underwater at reactor sites, where the radiation they produce is absorbed by the water and the metal-containing basins. Fuel rods can also be kept in dry storage, surrounded by metal or some other solid material that will prevent the escape of radiation.

Why, then, should we worry about the disposal of high-level waste? It is not harming the environment, so why not just leave it where it is? The difficulty is that constant vigilance is needed to make sure nothing goes wrong. Water can evaporate, pumps malfunction, metals crack or corrode.

If any of these should happen, someone has to be present to prevent waste from leaking into the environment. This job would not be so difficult were it not for the times involved. Radioactive elements in waste will eventually decay to nonradioactive isotopes, but some remain dangerously active for a very long time. Is there any place where these wastes can be stored with a guarantee that none will leak out in 24,000 years, or in 100,000 years?

It seems likely that sites can be found deep in solid rock where waste could be placed with a very small chance of any appreciable amount escaping, even after a thousand centuries. Such a site must be a few hundred meters underground and consist of rock that is solid, durable, and free of large cracks. The site must also be geologically stable— a place where erosion is slow, earthquakes are infrequent, and there is no sign of recent volcanic activity. But the chief worry will still be groundwater. Is the water in the rock noncorrosive and slow moving? Does it travel a long way before it emerges at the surface? These are standard geological questions, and with a little investigation a geologist can find answers to them.

Most geologists agree that the best way to accomplish disposal is to sink a shaft a few hundred meters deep at a carefully selected site, excavate tunnels from the base of the shaft, and drill holes in the tunnel floors into which the waste containers would be placed. The holes and the tunnels would then be filled with clay so that the waste would be completely isolated even if the containers eventfully corroded. In the United States a site of this kind is being prepared in an area of southern Nevada known as Yucca Mountain. However, no high-level waste has yet been disposed of there.

What has gone wrong? The difficulty lies in the very long times for which geologists must give assurance that no significant amount of radioactive material will escape the Earth's surface. How sure is it that waste buried at the Yucca Mountain site would remain in place for a hundred thousand years? Very sure, say geologists, but they cannot provide an absolute guarantee. For people living in the area, the lack of such a guarantee makes them suspicious and inclined to oppose the disposal project with all the political means at their command.

Thus, an environmental problem, seemingly solved from a geological standpoint, has become a political problem that remains unsolved.

Geothermal Power

Geothermal power, as the Earth's internal heat flux is called, has been used for more than 50 years in New Zealand, Italy, and Iceland (Fig. 20.14) and more recently in other parts of the world, including the United States. How this is done was discussed in the essay on geothermal energy in Chapter 4.

Most of the world's geothermal steam reservoirs are close to plate margins because plate margins are where most recent volcanic activity has occurred. A depth of 3 km seems to be a rough lower limit for big geothermal steam and hot-water pools. It is estimated that the world's geothermal reservoirs could yield about 8×10^{19} J—equivalent to burning 13 billion barrels of oil. This estimate incorporates the observation that, in New Zealand and Italy, only about 1 percent of the energy in a geothermal reservoir is recoverable. If the recovery efficiency were to rise, the estimate of recoverable geothermal resources would also rise. But even if the efficiency rose to 50 percent, geothermal power, like hydropower, could satisfy only a small part of human energy needs. For this reason, as we discussed in the essay on geothermal power in Chapter 4, a good deal of attention is being given to creating artificial geothermal steam fields. So far, experiments have been only partially successful.

Energy from Winds, Waves, Tides, and Sunlight

The most obvious source of energy is the Sun. The amount of energy reaching the Earth each year from the Sun is approximately 4×10^{24} J—that is, ten thousand times more than we humans use. We already put some of the Sun's energy to work in greenhouses and in solar

Figure 20.14 Geothermal Power in Iceland Iceland does not have coal or oil, but it does have lots of geothermal power. At Nesjavellir in southwestern Iceland, steam from geothermal wells drives an electrical power plant. After passing through the turbine that generates electricity, the steam is used to heat cold water from Lake Thingvallavatna. At a temperature of 83°C the heated water is sent, via an insulated 27-km-long pipeline, to Reykjavik, Iceland's main city. The hot water is used for space heating and washing.

homes, but the amount so used is tiny. The major challenge is to convert solar energy directly to electricity (Fig. 20.15). Devices that effect such a conversion, called photovoltaic devices, have been invented. So far their

Figure 20.15 Capturing Solar Energy Photovoltaic panels convert the energy of sunlight to electricity at a power-generating site in southern California.

Figure 20.16 Wind Energy A field of windmills near Palm Springs, California. The windmills generate electricity using the kinetic energy of wind.

costs are too high and efficiencies too low for most uses, although they are already widely used in small calculators, radios, and other devices that use very little power.

Winds and waves are both secondary expressions of solar energy. Winds, in particular, have been used as an energy source for thousands of years through sails on ships and windmills. Today, huge farms of windmills are being erected in suitably windy places (Fig. 20.16). Although there are problems and high costs with windmills, it seems very likely that by the year 2000 or sooner, windmills will be cost-competitive with coal-burning electrical power plants. Unfortunately, much of the wind energy is in very-high-altitude winds. Steady surface winds have only about 10 percent of the energy the human race now uses. As with hydro- and geothermal power, therefore, wind power may become locally significant but will probably not be globally important.

Waves, which arise from winds blowing over the ocean, contain an enormous amount of energy. We can see how powerful waves are along any coastline during a storm. Wave power has been used to ring bells and blow whistles as navigational aids for centuries, but so far no one has discovered how to tap wave energy on a large scale. Devices that have been designed to do this tend to fail because of corrosion or storm damage.

Tides arise from the gravitational forces exerted on the Earth by the Moon and the Sun. If a dam is put across the mouth of a bay so that water can be trapped at high tide, the outward flowing water at low tide can drive a turbine (Fig. 20.17). Unfortunately, the process is inefficient, and few places around the world have tides high enough to make tidal energy feasible.

It is clear that numerous sources of energy exist and that far more energy is available than we can use. What is not yet clear is when, or even whether, we will be clever enough to learn how to tap the different energy sources in nonhazardous ways that do not disrupt the environment.

Figure 20.17 Tidal Power La Rance tidal power plant in southern France. With the tide rising, water is seen flowing through the barrage across the mouth of the River Rance. At high tide the barrage is closed. At low tide, the barrage is opened again, and the water flowing back to the sea drives electricity-generating turbines.

SUMMARY

1. When a mineral deposit can be worked profitably, it is called an ore. The waste material mixed with ore minerals is gangue.

2. Mineral deposits form when minerals become concentrated in one of five ways: (1) precipitation from hydrothermal solutions to form hydrothermal mineral deposits; (2) concentration through crystallization to form magmatic mineral deposits; (3) concentration from lake water or seawater to form sedimentary mineral deposits; (4) concentration in flowing water to form placers; and (5) concentration through weathering to form residual deposits.

3. Hydrothermal solutions are brines, and they can either be given off by cooling magma or else form when either groundwater or seawater penetrates the crust, becomes heated, and reacts with the enclosing rocks.

4. Hydrothermal mineral deposits form when hydrothermal solutions deposit dissolved minerals because of cooling, boiling, pressure drop, mixing with colder water, or through chemical reactions with enclosing rocks.

5. Chromite, the main ore mineral of chromium, is the most important mineral concentrated by fractional crystallization.

6. Sedimentary mineral deposits are varied. The largest and most important are evaporites. Marine evaporite deposits supply most of the world's gypsum, halite, and potassium minerals.

7. Gold, platinum, cassiterite, diamonds, and other minerals are commonly found mechanically concentrated in placers.

8. Bauxite, the main ore of aluminum, is the most important kind of residual mineral deposit. Bauxite forms as a result of tropical weathering.

9. The distribution of many kinds of mineral deposits is controlled by plate tectonics because most magmas and most sedimentary basins are where they are because of plate tectonics.

10. Geochemically abundant metals, which make up 0.1 percent or more of the crust, tend to form residual, sedimentary, or magmatic mineral deposits. The amounts available for exploitation in such deposits are enormous.

11. Deposits of geochemically scarce metals, present in the crust in amounts less than 0.1 percent, form mainly as hydrothermal, magmatic, and placer deposits. Amounts of scarce metals available for exploitation are limited and geographically restricted.

12. Nonmetallic substances are used mainly as chemicals, fertilizers, building materials, and ceramics and abrasives.

13. Coal originated as plant matter in ancient swamps and is both abundant and widely distributed.

14. Oil and gas originated as organic matter trapped in shales and decomposed chemically due to heat and pressure following burial. Later, these fluids moved through reservoir rocks and were caught in geologic traps to form pools.

15. When heated, part of the solid organic matter found in shale—called kerogen—will convert to oil and gas. Oil from shales is the world's largest resource of fossil fuel. Unfortunately, most shale contains so little kerogen that more oil must be burned to heat the shale than is produced by the conversion process.

16. Biomass energy, mainly the burning of wood and animal dung, accounts for about 14 percent of the world's energy.

17. Hydropower, the energy of flowing streams, is exploited at many places around the world, but only locally does it have the potential to be a major energy source.

18. Nuclear energy, derived from atomic nuclei of radioactive isotopes, chiefly uranium, accounts for 7.6 percent of the world's electrical power but presents problems of plant safety and waste disposal.

19. Minor amounts of energy are derived from the Sun's heat, tides, waves, winds, and geothermal power.

THE LANGUAGE OF GEOLOGY

breeder reactor (p. 550)

exploration geology (p. 535)

fission (p. 550)

gangue (p. 536)
geochemically abundant metals (p. 544)
geochemically scarce metals (p. 544)

hydrothermal mineral deposit (p. 536)

magmatic mineral deposit (p. 536)
metallogenic province (p. 543)
mineral deposit (p. 534)

ore (p. 535)

pile (p. 550)
porphyry copper deposit (p. 543)

residual mineral deposit (p. 536)

sedimentary mineral deposit (p. 539)

tar (p. 548)

volcanogenic massive sulfide deposit (p. 536)

QUESTIONS FOR REVIEW

1. What are mineral deposits? Describe five ways by which a mineral deposit can form.

2. If there are any mineral deposits in the area where you live or study, what kind of deposits are they and how did they form?

3. What factors determine whether or not a mineral deposit is ore?

4. How do hydrothermal solutions form and how do they form mineral deposits?

5. Briefly describe the formation of three kinds of sedimentary mineral deposits.

6. What factors control the concentration of minerals in placers? Name four minerals mined from placers.

7. Compare and contrast mineral deposits of the geochemically abundant and geochemically scarce metals.

8. What is a fossil fuel? Name four kinds of fossil fuel.

9. During what two periods in the Earth's history was most coal formed? Explain why coal formed, when and where it did, and why the coal is now found where it is.

10. What kind of rocks serve as source rocks for petroleum? In what kinds of rocks does petroleum tend to be trapped? Why?

11. Oil drillers find more petroleum per unit volume of rock in Cenozoic rocks than in Paleozoic rocks of the same kind. Explain.

12. Oil shales are rich in organic matter. Explain why such shales have not served as source rock for petroleum.

13. Discuss the relative amounts of energy available from the different fossil fuels. What is your opinion about how fossil fuels will be used in the future?

14. What is nuclear energy? How is it used to make electricity, and what possible dangers are there in developing it?

15. What limitations does the development of hydroelectric power present? of wave and wind power?

For an interactive case abstract, virtual tours, activities, and additional learning resources, go to
GEOSCIENCES TODAY: www.wiley.com/college/skinner

Petroleum Geology: Persian Gulf vs. Overthrust Belt

The Persian Gulf, currently the world's major producer of oil, contains more than two-thirds of the world's known oil reserves and over one-quarter of known natural gas reserves. Located on the opposite side of the Earth are the Overthrust Belt and the adjacent Greater Green River Basin in the Rocky Mountains of the western United States. This area, covering portions of Utah, Wyoming, and Idaho, is a regionally important producer of oil and natural gas, but it does not contribute significantly to world production.

OBJECTIVE: The primary focus of this case is to foster an understanding of the origin and distribution of oil and gas in two very different geologic environments and to explore the social, economic, and environmental issues associated with hydrocarbon exploration and production in both regions.

The Human Dimension: How has demand for Persian Gulf oil affected economic and cultural relationships with the region's petroleum-producing countries?

Questions to Explore:

1. How do the geologic histories of the two regions relate to their reserves of oil and natural gas?

2. What are the principal geologic similarities and differences between these two important oil-producing regions?

3. What are the major environmental impacts of hydrocarbon utilization?

Understanding
Our Environment

Using
Resources

GeoMedia

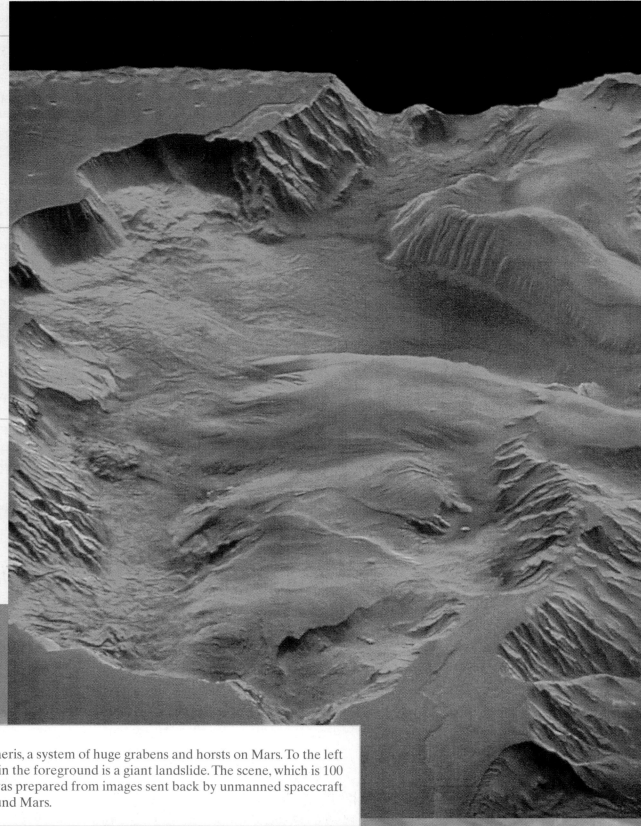

Valles Marineris, a system of huge grabens and horsts on Mars. To the left of the horst in the foreground is a giant landslide. The scene, which is 100 km across, was prepared from images sent back by unmanned spacecraft in orbit around Mars.

Beyond Planet Earth

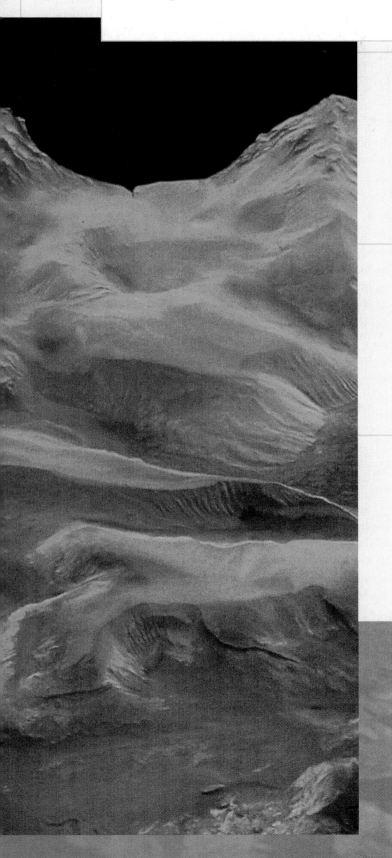

The Goldilocks Problem

In the early 1960s, before space vehicles started to probe the limits of the solar system, the Earth's was the only atmosphere for which we had reliable data. The atmospheres of Mars and Venus, the two most Earthlike planets, were thought to be composed principally of nitrogen, as the Earth's atmosphere is. In retrospect, the idea seems foolish, and it demonstrates how far off the mark a well-reasoned hypothesis can be. In fact, the atmospheres of both Mars and Venus are composed mostly of carbon dioxide, and nitrogen concentrations account for only a small percentage.

The Earth is the odd planet because nearly all of its carbon dioxide has been removed from the atmosphere and deposited as carbonate minerals or organic carbon in sedimentary rocks. If all of this carbon were in the atmosphere, the ratios of carbon to nitrogen would be similar in the atmospheres of all three planets. Why has the Earth's carbon dioxide been almost completely extracted from the atmosphere? Probably because the Earth has abundant water, which has made possible the weathering reactions that extract carbon dioxide from the atmosphere and the development of a biosphere that leads to the burial of organic carbon in sediments. The atmospheres of Mars and Venus are both very dry. As far as we know, the Earth, Mars, and Venus were assembled out of more or less the same material with more or less the same complements of water and other volatile compounds. Why, then, are these atmospheres so dry? Probably because Mars is too cold, Venus is too hot, and the Earth is just right. That's the Goldilocks problem.

Mars is farther from the Sun than is the Earth, and it has a thin atmosphere with a surface pressure $\frac{1}{160}$ that of the Earth. Because temperatures on the surface never rise as high as 0°C, most of Mars's water is presumed to be preserved in permanent deposits of subsurface ice; much of Mars's carbon may also be locked in solid form as dry ice. The thin atmosphere is apparently a consequence of the low temperature.

Venus is closer to the Sun than the Earth is and has a massive atmosphere with a surface pressure 70 times larger than ours. Because of the greenhouse effect of this massive atmosphere, the surface temperature is a searing 480°C, well above the boiling point of water. Perhaps Venus has always been too hot for water to condense. Instead, water in the upper atmosphere seems to have been broken apart into hydrogen and oxygen by ultraviolet radiation from the Sun. The light atoms of hydrogen could have escaped into space, while the heavier oxygen could have reacted with rocks to become incorporated into the rocks of the solid planet.

From the study of planetary atmospheres, we have learned that a habitable planet like ours is improbable. Small differences in planetary origins led to widely divergent evolutionary paths. A planet a little too close to the Sun becomes hot and dry like Venus; a planet a little too far away grows cold and dry like Mars. Too large a planet captures a massive atmosphere like those of Jupiter and Saturn; too small a planet ends up with no atmosphere at all, like Mercury and the Moon. The requirements for habitability are stringent indeed.

THE SOLAR SYSTEM

Interest in visible stars dates back to prehistory. Our ancestors noticed that a few stars seem to wander in completely different paths from the annual progression of most of their fellows across the sky. The Greeks mapped the paths of these strange objects and called them *planetai*, or wanderers. The Romans named the objects after their gods: Saturn, Jupiter, Mars, Venus, and Mercury. As described in Chapter 1, we now know that the curious wandering paths of the planets result from orbital motions around the Sun and that the other objects in the heavens are suns, so far distant that they seem to occupy fixed places in the sky. Three planets—Uranus, Neptune, and Pluto—were unknown to our ancestors because they are visible only through telescopes. We now know, too, that other planets besides the Earth have moons of their own. A few vital statistics of the planets and their moons are given in Figure 1.6.

In 1957, when the first artificial satellite was placed in orbit around the Earth, a brand-new scientific specialty emerged. **Planetology**, devoted to a comparative study of the Earth with the Moon and other planets, has taught us much about the Earth's earliest days—that time, before 4.0 billion years ago, for which most of the record seems to have been erased by the operation of the rock cycle. In the present chapter we present a brief account of the enormous successes of planetology. The discoveries of the past 35 years have motivated scientists to seek a unifying theory of origin for all suns and their planets. If the scientists are successful, it may one day be possible to say which of the billions of other suns visible in the heavens have planets like the Earth, and perhaps even to say which, if any, of those planets may have biospheres of their own.

Planets and Moons

As we learned in Chapter 1, the solar system consists of the Sun, 9 planets, 61 moons, plus unknown numbers of asteroids, comets, and small rocky fragments called meteoroids. The planets are divided into two groups: the terrestrial planets, which have densities of 3 g/cm^3 or more, and the jovian planets, which all have densities below 2 g/cm^3.

The four planets closest to the Sun are the terrestrial planets. They are small, rocky bodies, and although they share many similarities there are important differences, too. In order of increasing distance from the Sun, the terrestrial planets are Mercury, Venus, Earth, and Mars. Unmanned spacecraft have visited each of the terrestrial planets, and Mars and Venus have had several spacecraft land on their surfaces. As a consequence, we know a good deal about the terrestrial planets.

The jovian planets, Jupiter, Saturn, Uranus, Neptune, and Pluto, all lie farther from the Sun than Mars. With the exception of Pluto, the jovian planets have such thick atmospheres that it is impossible to see what lies below the atmosphere.

The planet farthest from the Sun, Pluto, is the smallest of all the planets, being little larger than the Moon. Pluto lacks the massive atmosphere of the large jovian planets and has a density of 2.06 g/cm^3, intermediate between the high densities of the terrestrial planets and the low densities of the other jovian planets. The most probable explanation of Pluto's density is that the planet has a rocky center but a thick outer layer of ice.

All the jovian planets have been visited by spacecraft, but no craft has ever landed. Nevertheless, we can infer a lot about the structure of the jovian planets from measurements made by passing spacecraft. We have learned, for example, that inside each jovian planet there seems to be solid core, and we can infer that it is probably rocky, like the terrestrial planets.

The orbits of all the planets around the Sun are elliptical in shape. The planets all revolve around the Sun in the same direction, and the orbits lie in nearly the same plane as the orbit of the Earth around the Sun. The orbit of Pluto is an exception because it is tilted at 17° to the **ecliptic**, which is the name given to the plane of the Earth's orbit. Most of the moons revolve around the planet in the same direction as the planets revolve around the Sun. The reason for the similarity of orbital motions is thought to be the way planets and moons form—by condensation of a rotating gas cloud (Fig. 1.8). The rotations of the planets and moons today are inherited from the rotation of the ancient gas cloud.

Meteoroids and Asteroids

Meteoroids are small stony or metallic objects from interplanetary space. They can be observed in motion only when they flash into view as they plunge through Earth's atmosphere. They, too, follow orbits around the Sun. The solar system occupies a region in space that is at least 12 billion km in diameter. Measuring outward from the Sun, the Titius-Bode rule states that the distance to each planet is approximately twice as far as the next inner one. Thus, the distance from the Sun to Mercury is 58 million km, while the distance to Venus, the next planet, is 107 million km.

The Titius-Bode rule seems to break down between Mars and Jupiter. Mars is 226 million km from the Sun, and Jupiter is 775 million km. But this rule suggests that a planet should be found about 450 million km from the Sun. No planet is present here, however. What are found instead are at least 100,000 **asteroids**, small, irregularly shaped rocky bodies that have orbits lying between the orbits of Mars and Jupiter. The asteroids are either fragments of a planet that once existed and was somehow broken up, or they are rocky fragments that failed to gather into a planetary mass.

The asteroids and meteoroids are dense, rocky objects. They are believed to have formed in the inner regions of the solar system, the same way and at the same time as the terrestrial planets.

Cratered Surfaces

The most important processes that shape the surface of the Earth can be divided into three groups: tectonic processes, magmatic processes, and the *surficial* processes of weathering, mass-wasting, and erosion. To varying degrees, each group of processes plays, or has played, a role in shaping the surfaces of all the rocky planets and moons in the solar system. However, in a planetary context, a fourth process must be added—the process of impact cratering.

The process by which a planetary surface is deformed as a result of a transfer of energy from a bolide to the planetary surface is known as **impact cratering**. A **bolide** is defined as an impacting body; it can be a meteoroid, an asteroid, or a comet. Impact cratering on the Earth was discussed in Chapter 1 and illustrated in Figure 1.4. Many ancient impact craters have been discovered on the Earth. However, no large, natural impact crater has been observed as it formed. What is known about the process of impact cratering comes largely from laboratory experiments. The sequence of events is illustrated in Figure 21.1. As a high-speed bolide impacts and penetrates the surface of a planet, it causes a jet of debris to be ejected at high velocity away from the point of impact. At the same time, the impact compresses the underlying rocks and sends intense shock waves outward. The pressures produced by the shock waves from a large bolide are so great that the strength of the rock is exceeded and a large volume of crushed and brecciated material results. In very large impact events, local melting and even vaporization may occur. Once the compressive shock waves have passed, rapid decompression occurs. Decompression causes more material to be ejected from the impact crater and produces a blanket of ejecta that surrounds the crater and thins away from the rim.

Note in Figure 21.1 that the stratigraphy of layers in the ejecta blanket adjacent to the crater rim has been overturned. In the largest impact structures, the central crater is circled by one or more raised rings of deformed rock. The outer rings are presumed to form as a result of the initial compression (Fig. 21.2). Following the immediate impact event, a number of postimpact events tend to modify the crater.

Crater walls may slump (Fig. 21.3), isostatic rebound may produce changes in the floor and rim of the crater, and erosion may fill the crater with debris. In some instances, magma may rise along fractures produced by the impact, and lava may fill the crater.

Approximately 200 impact craters have been identified on the Earth. However, impact events must have been much more common than this small number implies. The reason so few craters have been found is that weathering and the rock cycle continually erase the evi-

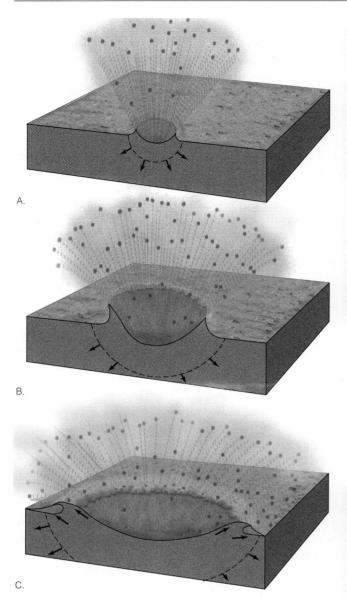

Figure 21.1   Impact Crater   The shapes of impact craters are the same regardless of the angle of entry of the impacting bolide. The shape develops in three stages.   A. The initial contact ejects a high-velocity jet of near-surface material.   B. The passage of shock waves through the bedrock produces high pressures and the compression of strata. In places the rock strength is exceeded, and fracturing and brecciation result. Decompression throws broken rock out of the crater.   C. Strata along the rim of the crater are folded back and overturned by decompression. The ejected debris forms a circular blanket around the crater.

dence. On most planetary surfaces a very different situation prevails. On most of the terrestrial planets, and on many of the rocky moons, atmospheres are absent, and tectonic activity has either ceased or is so slow that a rock cycle probably does not operate. As a result, the most striking features of the solar system are impact craters, some of which are more than 4 billion years old.

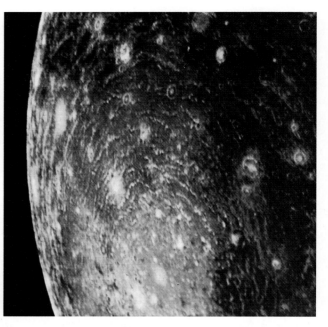

Figure 21.2   Rings Around a Crater   Multiple concentric rings around Valhalla, an impact crater on Callisto, a moon of Jupiter. The crater (central bright spot) is 600 km in diameter. The rings were formed as a result of thrust faults caused by the compression of the giant impact. The surface of Callisto is ice. Rings are also observed around giant craters on rocky planetary bodies, but they are neither so pronounced nor so numerous as the rings around Valhalla.

Figure 21.3   Impact Crater on the Moon   A large impact crater on the Moon. The crater is more than 200 km in diameter; the oblique photo was taken from a manned spacecraft. The highlands in the far distance are part of the ejecta blanket. The stepped terraces in the middle distance were formed as a result of postcratering collapse of the crater rim. The hills in the foreground lie at the center of the crater and were formed by rebound of the crater floor during decompression.

The largest craters are 2000 km or more in diameter, but they range down to the size of a pin's head. Such tiny craters are produced by dust-sized meteoroids.

THE TERRESTRIAL PLANETS

Besides impact cratering, volcanism has been widespread on each of the terrestrial planets and the Moon. On Mercury, the Moon, Mars, and Venus, it is clear that volcanism is predominantly basaltic. As discussed later in the chapter, only on Venus is it likely that other kinds of volcanism, similar to those found on the Earth, may have occurred. However, Venus is cloud-covered and difficult to study, so the question remains open. What space exploration has made clear, and what analysis of lunar and martian rocks has confirmed, is that basaltic volcanism is a common process in the solar system. This means, presumably, that each planet has gone through (or is still going through) a stage in its developmental history when internal heating caused partial melting, resulting in the formation of basaltic magma. The products of partial melting are apparently similar on all of the rocky bodies in the solar system, suggesting that the parent bodies all have similar compositions. This means, in turn, that the difference observed today between the rocky planets must reflect factors such as size and distance from the Sun rather than composition.

Mercury

Mercury, the innermost planet, is so close to the Sun that it can be seen only just before sunrise or just after sunset. Telescopic viewing is difficult under such circumstances, and as a result almost everything that is known about Mercury comes from observations made during fly-by missions of unmanned spacecraft. Mercury has a diameter of 4880 km and is just a little larger than the Moon. It rotates slowly about its axis and has a density of 5.4 g/cm^3.

To account for the high density of such a small planet, it is necessary to conclude that Mercury has a metallic core about 3600 km in diameter. Mercury's core alone is the size of the Moon (Fig. 21.4). Images of the surface sent back by spacecraft show that Mercury is heavily pockmarked by ancient craters, that it lacks an atmosphere, and that there is no evidence of moving plates of lithosphere. The largest impact basins are filled with basaltic lava flows; this means that Mercury had a period of magmatic activity. However, the lava plains are not crumpled and deformed; thus, the magmatic activity was not followed by tectonic activity. There is no evidence that anything is happening on Mercury today. It seems to be a dead planet.

The largest impact structure on Mercury is the Caloris basin, 1300 km in diameter. On the far side of Mercury, exactly opposite Caloris, the surface is jumbled into a weird, hilly terrain. Scientists who have studied Mercury believe that the weird terrain was produced by com-

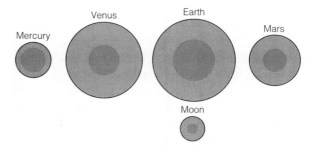

Figure 21.4 Cores of the Terrestrial Planets Comparative sizes of the cores of the Moon and the terrestrial planets. Mercury, nearest the Sun, where only the highest temperature materials could condense, has a huge core.

pressive shock waves from the bolide that formed Caloris. The waves passed completely through the planet and disrupted the far side (Fig. 21.5).

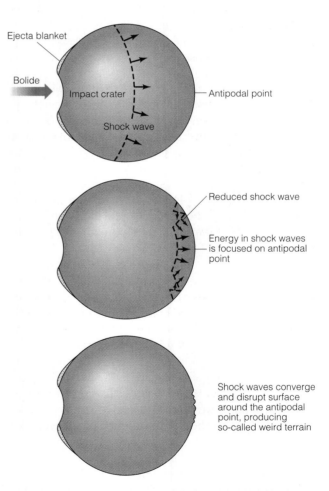

Figure 21.5 "Weird" Terrain Origin of the "weird" terrain on Mercury. The terrain is exactly opposite the giant impact crater, Caloris. Compressive waves caused by the impact traveled through the body of the planet, and surface waves traveled around the edge. Both sets of waves were focused at the antipodal point, and they severely disrupted the surface.

Mercury's most extraordinary feature is the presence of a magnetic field about one one-hundredth as strong as that of the Earth. The field is dipolar, and the magnetic axis coincides with the axis of rotation. The origin of the magnetic field is a puzzle. If Mercury's core is molten, the magnetic field could be caused by the fluid motions of the planet's rotation. There are two problems with this idea, however. The first is that Mercury rotates so slowly—once every 59 days—that it is difficult to see how fluid motions strong enough to produce such a magnetic field could exist in the core. The second problem concerns tectonism: If Mercury has a huge core of molten iron, the interior must be very hot. Why then is Mercury not tectonically active? The puzzle of the magnetic field remains unanswered.

The Moon

The Earth's Moon is very like Mercury—a stark, impact-cratered body with abundant evidence of basaltic volcanism. Because its diameter is 3476 km, only a little less than that of Mercury, the Moon is often described as a small terrestrial planet, and the Earth–Moon pair is described as a double planet. The largest moons of Jupiter and Saturn, though bigger than our Moon, are tiny in comparison with the sizes of their giant neighbors.

Because the Moon is a small, dense, rocky object, it probably formed in the inner regions of the solar system, just as the other terrestrial planets did. If ideas about the way planets form are even partly correct, the Moon's structure and composition should be similar to those of the other terrestrial planets. Information about the Moon, therefore, has been very helpful in developing our ideas about the origins and development of the terrestrial planets.

Structure
Each time astronauts visited the Moon, the measurements they made yielded clues about its structure. The most informative measurements came from seismic waves. Compared to earthquakes, moonquakes are weak and infrequent. Moonquakes large enough to be detected by instruments placed on the Moon by astronauts number fewer than 400 per year; on the Earth, the same instruments would record nearly a million quakes per year. Despite the low frequency of moonquakes and the weakness of the seismic waves they generate, the quakes reveal a good deal about the lunar structure (Fig. 21.6). Some of the most important points are (1) on the side of the Moon that faces the Earth (at least on those parts that have been visited by astronauts) there is a crust about 65 km thick; (2) the Moon is layered; (3) covering the surface is a layer of regolith that ranges from a few meters to a few tens of meters thick; (4) below the regolith is a layer (about 2 km thick) of shat-

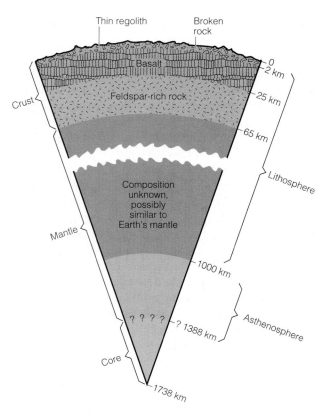

Figure 21.6 Structure of the Moon A section through the Moon showing the layered structure. The crust is known with certainty only in the vicinity of the astronauts' landing sites, and the existence of the small core is still uncertain.

tered and broken rock produced by the continual rain of large and small bolides; (5) below the broken-rock zone is about 23 km of basalt, then 40 km of a feldspar-rich rock; and (6) at a depth of 65 km the velocities of seismic waves increase rapidly, indicating that the lunar crust overlies mantle.

The scarcity of moonquakes (and their weakness) immediately suggests that processes such as volcanism and plate tectonics, the causes of most earthquakes, are not happening on the Moon. Some moonquakes are caused by meteorites hitting the Moon. However, most quakes occur in groups and at the times when the Moon's elliptical orbit brings it closest to the Earth—the very moment when gravitational forces between the Earth and the Moon are strongest. This suggests that most moonquakes result from the gravitational pull that the Earth exerts on the Moon. The pull causes slight movement along cracks, each one causing a tiny moonquake. Foci of moonquakes have been recorded as deep as 1000 km; this observation, too, is informative. It means that, unlike the Earth, the Moon is rigid enough at 1000 km for elastic deformation to store energy and for brittle failure to occur. This, in turn, means that the lunar lithosphere must be at least 1000 km thick and that

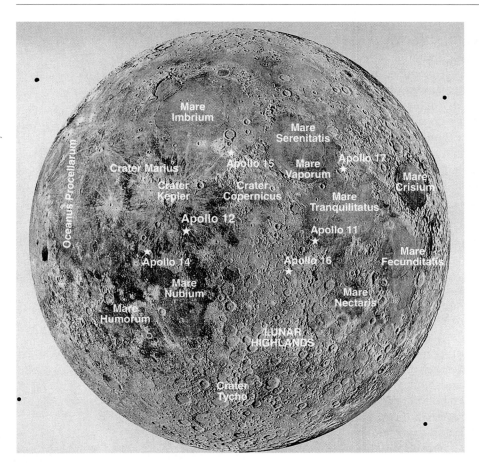

Figure 21.7 Cratered Surface of a Luran "Sea" Face of the Moon as seen from Earth. Light-colored areas pitted with craters are lunar highlands. Dark-colored lowland areas are maria formed when basaltic lava flowed out to fill the craters made by exceptionally large impacts. Copernicus, Kepler, and Tycho are three prominent young meteorite craters that formed after the mare basins were filled. The impacts that made them splashed bright-colored rays of rocky debris over ancient basalt. The sites of the six *Apollo* lunar landing missions are indicated.

the Moon's asthenosphere, if it exists at all, lies very deep within the Moon's body. The thick, rigid lithosphere makes it impossible for there to be any present-day tectonic activity on the Moon. The Moon, like Mercury, seems to be dead.

The Moon's density is 3.3 g/cm³, whereas that of the Earth is 5.5 g/cm³. Assuming that the Moon's core, like the Earth's, is mainly iron, it can be calculated that the radius of the Moon's core can be no larger than about 700 km. If the core were larger, the Moon's overall density would have to be higher.

The Lunar Surface

The Moon's surface can be divided into two general categories: **highlands**, mountainous areas that appear to us as light-colored patches; and **maria** (Latin for seas; singular is **mare**, pronounced mah-ray), smooth lowland areas that appear to us as dark-colored regions (Fig. 21.7). The highlands are regions of intense cratering. The maria are covered by basaltic lava flows (Fig. 21.8). In places, the lavas fill and cover the impact craters. The highlands must therefore be older than the maria. Note, in Figure 21.7, that each mare is circular or nearly so. Detailed study shows that they are actually giant impact craters that predate the basaltic flows that now fill them.

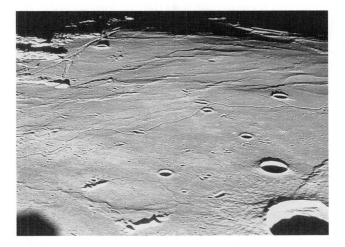

Figure 21.8 What the Moon Looks Like A photograph of the surface of a mare. Taken by an astronaut during the *Apollo 11* mission, the photograph shows portions of Mare Tranquillitatis. The irregular ridges, looking like long sand dunes, are ancient basaltic lava flows. The impact crater on the lower right-hand side is called Maskelyne.

In the highlands, mountains soar tens of thousands of meters above the maria and in places stand even higher than the Earth's mountains. Because lunar mountains are big, one way to obtain clues about the Moon's interior regions is to learn whether isostasy operates as it does on the Earth and whether the great lunar mountains have roots. The test is best made by measuring variations in the Moon's gravitational pull. Such measurements were made while spacecraft orbited the Moon; they indicate that the mountains do have roots but also that the highlands are isostatically balanced. We can infer, then, that when the lunar mountains formed, at least one of the Moon's outer layers must have been sufficiently fluidlike and plastic to enable the highlands to float.

Rock and Regolith

Astronauts returned to the Earth with a treasure trove of rock and regolith. From these have come many clues about the Moon's composition and history. The most interesting samples are the igneous rocks; in terms of age and composition, three different kinds are found. The first and oldest consists of feldspar-rich rocks such as anorthosite, a variety of igneous rock formed by extreme magmatic differentiation and consisting largely of calcium-rich plagioclase. Radiometric dates of these oldest rocks, which come from the highlands, indicate they could have been formed as long as 4.5 billion years ago, only 100 million years after the formation of the Moon. The second kind of igneous rock is basalt that contains high concentrations of potassium and phosphorus; this, too, is 4 billion or more years old. The potassium-rich basalts likewise come from highland areas and are lava flows that seem to represent the last igneous activity in those areas. The third kind of igneous rock is also basalt, but it is rich in iron and titanium rather than potassium. Titanium- and iron-rich lavas have been found only in the maria (Fig. 21.9A) and are dated radiometrically at 3.2 to 3.8 billion years. We infer that such material underlies each mare to a depth of about 25 km. Mare basalt, then, seems to have formed several hundred million years after the highlands formed. Even though magmas do not seem to form on the Moon today, the mare basalts prove that magmas similar to those on the Earth existed on the Moon for a long time.

The lunar regolith is a mixture of gray pulverized rock fragments and small particles of dust, many of which are glassy. Its composition is essentially that of the lunar igneous rocks. Regolith covers all parts of the lunar surface like a gray-brown shroud (Fig. 21.9B), as though giant hammers had crushed the surface rock. Indeed, the surface has been hammered but by the bolides that continually strike the Moon. The impacts are unhindered by the moderating influence of an atmosphere.

History of the Moon

The history of the Moon begins about 4.6 billion years ago, by which time the Moon had formed as a solid body. The available evidence does not prove exactly

A.

B.

Figure 21.9 Rock Samples from the Moon Two kinds of lunar samples brought back by astronauts. A. Basalt, containing numerous vesicles formed when gases escaped during cooling and crystallization of lava. Collected during the Apollo 12 mission, this sample is typical of the kinds of basalt found in the maria. B. Regolith, a mixture of many rock and mineral types, together with glassy fragments (spheres) produced by the bombardment of the lunar surface by meteorites. The largest particles in the pictures are about 1 mm in diameter.

how the Moon was formed, but the presence of ancient impact craters does indicate that the final stages of formation involved the accretion of innumerable bolides, large and small.

As the Moon grew larger by accretion, the strength of its gravitational attraction increased so that the speeds at which accreting bolides reached the lunar surface grew ever greater. Eventually, speeds became so great that each impact generated a large amount of heat. Near the end of the accretion process, so much heat was generated that an outer layer, 150 to 200 km thick, of the Moon's body was apparently melted. Thus, the Moon had a solid interior with a molten outer shell, in effect a **magma ocean**. The period of rapid accretion soon ended, and the magma ocean began to cool and crystallize. The first mineral to crystallize was plagioclase feldspar, and because it was lighter than the parent magma, it floated. Soon a thick crust, rich in feldspar crystals, floated in the remaining liquid. We see that old crust today in the highlands. The liquid in which the feldspar crust floated eventually became the upper part of the lunar mantle. Bolides fall on the lunar surface today, so it can be inferred that bolides must also have been falling while the crust was forming. The largest impacts must have broken the crust, letting liquid from below ooze out to form the ancient potassium-rich lava flows. The astronauts found samples of those ancient products of magmatic differentiation in the lunar highlands.

The magma ocean would have crystallized within about 400 million years. By about 4 billion years ago, therefore, the highland crust had formed, and the major activity on the Moon had become the incessant rain of bolides that pitted and pocked the surface. Some bolides, larger than others, made exceptionally large impact scars. These circular cavities eventually became the **mare basins** (Fig. 21.7).

While the Moon's surface was being bombarded by bolides, its interior regions were slowly heating up. If we use the composition of meteorites that fall on the Earth as a way to judge the amount of radioactivity present in the materials from which the Moon and the terrestrial planets formed, the amount of radioactivity in the Moon was small. Nevertheless, it was sufficient to cause partial melting in the upper mantle to start about 3.8 billion years ago. So formed, the magma worked its way up to the surface along fractures caused by impacts of the largest meteorites. When it reached the surface, the magma filled the impact basins and formed the basalt flows now seen in the mare basins. By 3 billion years ago, the extrusion of mare basalt flows had ceased.

Except for the continuing rain of bolides, the Moon has remained a tectonically and magmatically dead planet.

Lessons for the Earth

Did the Earth ever look like the Moon? It probably did. The Moon supports the idea that the terrestrial planets formed by accretion of solid bodies. The Earth is larger than the Moon, so its gravitational pull is stronger. Presumably, therefore, an even thicker magma ocean covered the Earth at the end of the accretion process. Perhaps the Earth's earliest crust began to form through cooling of that magma. However, all traces of the Earth's primitive crust have been lost. The reason is not hard to find. Because radioactive heating on the Earth has continued to create magma, probably the earliest crust has long since been remelted and reabsorbed.

Mars

Mars, with a diameter of 6787 km, has only one-tenth of the Earth's mass. Despite its size, Mars is Earthlike in many ways. It rotates once every 24.6 h, so the length of the martian day is nearly the same as the length of an Earth day. Also, Mars has an atmosphere, although it is only one one-hundred and sixtieth as dense as the Earth's and consists largely of carbon dioxide. Mars has polar "ice" caps consisting mostly of frozen carbon dioxide ("dry ice"), as well as a small amount of water ice. The Martian ice caps are more like frost coatings than the thick ice sheets of the Earth. Like the Earth, Mars has seasons, and the diameters of the ice caps alternately grow and shrink with the coming of winter and summer.

Mars lacks a dipolar magnetic field. Like the Moon it has a density somewhat less than that of the Earth. However, when allowance is made for the fact that Mars is smaller than the Earth and that internal pressures are less, the difference in density is small. This means that the composition of the Earth and Mars must be similar and that Mars has a core. Presumably the core is solid. If it were molten, Mars would probably have a magnetic field.

Unfortunately, little information is available about the compositional layering of Mars. One of the spacecraft that landed on Mars in 1976 carried a seismometer with it, but no marsquakes occurred during the months the seismometer was active, so seismic evidence is unavailable.

Surface Features

The topography of Mars is extraordinary, and many features have yet to be adequately explained. The most obvious surface features can be seen in Figure 21.10.

Approximately half the martian surface, the southern hemisphere, is densely cratered and resembles the surfaces of the Moon and Mercury. The largest impact crater discovered in the solar system, Hellas, with a diameter of nearly 2000 km, is on Mars. Scientists who have studied the images of the Martian surface sent back by unmanned spacecraft conclude that the southern hemisphere is covered by ancient crust similar to the crust of the lunar highlands, and presumably of the same age. The dense population of craters records a rain of bolides prior to 4 billion years ago.

The northern hemisphere presents a very different picture. Craters are sparse, and large areas are relatively smooth. The obvious conclusion to draw is that some process (or processes) has produced a much younger surface. The probable answer to the puzzle is provided by the widespread evidence of volcanism in the northern hemisphere.

At least 20 huge shield volcanoes and many smaller cones have been discovered. The giants among them are

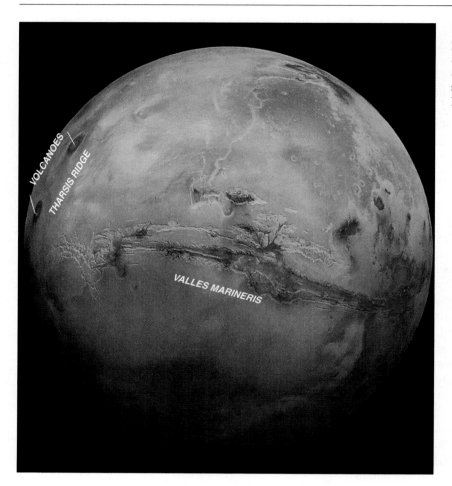

Figure 21.10 Image of Mars Mars, prepared from a mosaic of images made by *Mariner* spacecraft. The canyon running across the middle is Valles Marineris. Three shield volcanoes are visible on the lower left; the lowermost is Olympus Mons.

in the Tharsis Ridge (Fig. 21.10), and the largest of all is Olympus Mons (Fig. 21.11) whose basal diameter is 600 km, approximately the distance from Boston to Washington, D.C. Olympus Mons stands 27 km above the surrounding plains and is capped by a complex caldera that is 80 km across (Fig. 21.12). Mauna Loa, the largest volcano on the Earth, is also a shield volcano, but it is only 225 km in diameter and 9 km high above the adjacent seafloor. It is estimated that the amount of volcanic rock in Olympus Mons exceeds all of the volcanic rock in the entire Hawaiian chain of volcanoes.

The presence of a huge volcanic edifice such as Olympus Mons implies several things. First, in order for such a huge volcano to form, long-lived sources of magma must be present in the martian interior. Second, the magma source must remain connected to the volcanic vent for a very long time. This, in turn, means that the martian lithosphere must be stationary. In short, plate tectonics is not operating on Mars. A third implication is that the martian lithosphere must be thick and strong. If it were not, it would be bowed down by the weight of Olympus Mons. Isostasy is apparently not operating, or if it is operating, the isostatic response is very slow.

Are volcanoes active on Mars today? This intriguing question cannot be answered exactly, but it is possible that some volcanism persists. The evidence comes from the volcanic cones. If they were old features, they would be pitted by impact craters. Craters are rare on the volcanic slopes, leading some experts to conclude that the youngest flows on Olympus Mons are probably less than 100 million years old.

Adjacent to Olympus Mons is Valles Marineris, a system of canyons that dwarf the Grand Canyon in Arizona (see opening photo, Chapter 21). If the same feature were present on the Earth, it would stretch from San Francisco to New York. The great canyons are a series of giant grabens, but what formed them is not known. Possibly they formed when great crustal up-warping occurred, or when the crust subsided into an opening left empty when magma was extruded to build the huge volcanoes.

Erosion

As remarkable as the giant volcanoes and canyons are, even more remarkable features can be seen in the images returned by orbiting spacecraft and by the two Viking landers. *Viking 1* landed in a region of the northern hemisphere called Chryse Planitia, and *Viking 2* landed in a somewhat similar region called Utopia Planitia. In both cases, the images sent back showed a reddish-brown surface covered by loose stones and

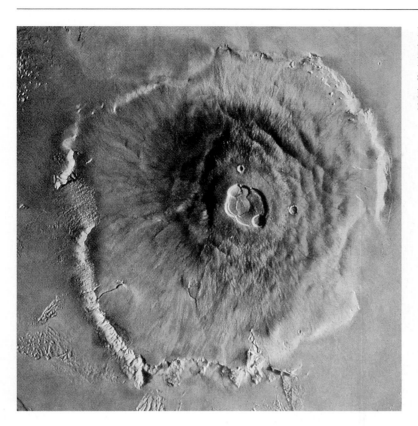

Figure 21.11   Largest Volcano in the Solar System   Olympus Mons, a giant shield volcano on Mars, is the largest volcanic edifice discovered so far in the solar system. The image is a computer-enhanced view made from a number of images taken by *Mariner* spacecraft.

Figure 21.12   Caldera on a Martian Volcano   Detail of the summit caldera of Olympus Mons. Several phases of caldera collapse have occurred. The irregular ridges in the center foreground are basaltic flows.

Figure 21.13 What Mars Looks Like The bleak surface of Mars; could it ever be made habitable? This is the first color picture taken by the spacecraft *Viking 2* after it landed on the Martian surface in 1976. The photo confirms what astronomers had suspected for a long time—that Mars is reddish colored because the Martian regolith is red. Experiments carried out by *Viking 2* seeking evidence of past or present forms of microscopic life in the Martian regolith were negative.

windblown sand (Fig. 21.13). Both landers made chemical analyses of the regolith on which they landed, and in both cases the results indicated clays and a sulfate mineral (probably gypsum). Weathering has obviously influenced the martian surface, and wind continually blows the weathering products around. Indeed, when the spacecraft *Mariner 9* was nearing Mars in 1971, a sandstorm of spectacular proportions was raging that continued unabated for several months. It seems likely that on Mars wind-driven sand is now the main agent of erosion.

Still more spectacular than the martian sand is evidence that water has influenced the surface of Mars (Fig. 21.14).

Valleys look much like those cut by intermittent desert streams on the Earth. They meander, they branch, and they have braided patterns and other features characteristic of valleys made by running water. Some of the features look as though they were caused by gigantic floods. Yet Mars now lacks rainfall, streams, lakes, and seas. The sparse H_2O that does occur on Mars is in the atmosphere or exists near the poles condensed as ice. The martian surface is too cold for liquid water to exist.

Some have suggested that ice might be present beneath the surface dust as frozen ground and that in former times the martian climate warmed up, causing the ice to melt and creating torrential floods.

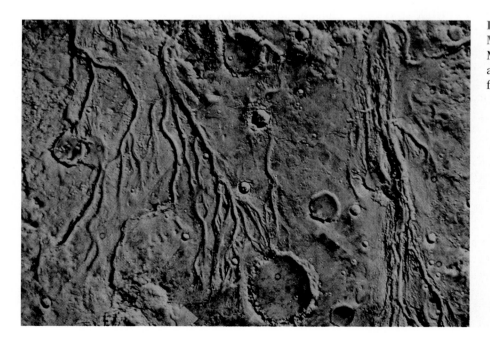

Figure 21.14 Evidence of Water on Mars Braided stream channels on Mars. The channels may have formed as a result of a sudden release of frozen groundwater.

Understanding Our Environment

Is There Life Out There?

As far as we know, only one planet in the solar system—the Earth—is, or ever has been, home to carbon-based life. Mercury, the Moon, and the satellites of the Jovian planets, with the exception of Titan, lack atmospheres. Venus is so hot nothing could live on its surface, and the fact that so much CO_2 is in Venus's atmosphere makes it unlikely life has ever existed there. The one possible exception is Mars. Mars has H_2O ice, and we know that at times in the past water has been present because we can see water-cut channels.

Mars today is very dry, and its atmosphere is about 0.007th of an Earth atmosphere. That is far too low for life. But Mars may once have had a denser atmosphere and may once have been warmer. If this is the case, perhaps life existed on Mars in the past. The two *Viking* spacecraft that landed on Mars carried a very sensitive instrument called a mass spectrometer. The instrument was designed to test samples of Martian soil to see if any organic compounds might be present.

On the Earth, rocks that are more than 2 billion years old still contain traces of organic compounds that indi-cate life forms. If similar organic molecules could be found in Martian rocks and soil, they would provide strong evidence that life once existed on Mars. Unfortunately, the *Viking* tests revealed nothing suggestive of past life.

Does life exist on planets that circle other suns out there in space? There are billions upon billions of suns, so it is almost inevitable that there are billions of planets too. Surely a few of the planets must be Earthlike and therefore might support life. The real problem is how we might ever find out about life on other planets. If we sent out radio messages and hoped to get a reply, we might have to wait hundreds of years because the distances to even the closer suns are so great. However, consider the possibility that a somewhat more advanced civilization than ours actually does exist somewhere out in space. Presumably they would already have sent messages. So far, we haven't heard or seen any messages because they would ask the same kind of questions we do. Such evidence is negative and not very good, but so far it does suggest that the Earth might well be unique and we might very well be alone.

Geological History

The early history of Mars must have been much like that of the Earth and the Moon. The terrain of the southern hemisphere is equivalent to the lunar highlands. Then, as in both the Earth and the Moon, radioactive heating inside Mars created magma by partial melting. Because Olympus Mons and its mates are shield volcanoes, we infer that the magma they extruded was of low viscosity and, therefore, was basalt. This means that partial melting has occurred. But we cannot yet say when the volcanism started. To be able to do that we must have radiometric dates.

Planetary differentiation and volcanism would have released volatiles such as H_2O and CO_2. Because Mars is compositionally like the Earth, it can be calculated that the amount of H_2O released would have been sufficient to cover a smooth, uniform Mars with water to a depth of 50 to 100 m. It is possible, therefore, that very early in martian history—a time corresponding to the Archean on the Earth—rains may have fallen, lakes and streams may have existed, and water erosion may have occurred. It is possible, too, that primitive life may have existed on Mars at this time, as suggested in the essay: *Is There Life Out There?* For some reason the martian atmosphere became thinner. Probably this happened because weathering and erosion locked up H_2O and CO_2 in clay and carbonate minerals. It was a one-way process: H_2O and CO_2 were not released to the atmosphere again because plate tectonics did not occur; thus, a rock cycle could not operate and so could not cause recycling. The temperature would have dropped as the atmosphere thinned. Liquid water eventually became unstable on the surface of Mars. However, H_2O could still exist as ice, and this is presumably the form it is in today, possibly existing as a cement between grains in the regolith. In short, Mars became a planet covered by frozen regolith. The torrential floods that caused the striking erosion features may have occurred when regions of frozen ground were subject to sudden melting, perhaps by near intrusion of magma or sudden changes in climate.

Venus

Venus is the planet most like the Earth in size and mass, but the differences outweigh the similarities. Venus is enveloped in a cloudy atmosphere of carbon dioxide that is 70 times more dense than the Earth's atmosphere. The clouds prevent direct visual observation of the venusian surface. Information comes from several spacecraft that have landed on the surface and radioed back information, and from radar measurements of the surface topography.

Landing spacecraft have reported that the surface temperature of Venus is astonishingly hot, about 480°C. At this temperature, metals such as lead, zinc, and tin are in a molten state. The explanation of the high temper-

Figure 21.15 Image of Venus A full-planet view of Venus produced from the radar imaging of the *Magellan* spacecraft. The actual color of the planetary surface is not known. Yellow areas are highlands.

ature is evident. First, Venus is closer to the Sun than the Earth. However, more important is the fact that the carbon dioxide in Venus's atmosphere acts like the glass of a greenhouse: It lets the Sun's rays through to heat the surface but serves as a barrier that prevents heat from leaving. Russian spacecraft have operated in this intense heat long enough to send back a few local images of the surface of Venus. The images show masses of broken rock fragments, each about 20 cm across, covering the surface. Several of the Russian spacecraft carried out quick chemical analyses before they were overcome by the high temperatures. In each case, the analyses suggest that the rocks analyzed were basaltic in composition. A few analyses indicated that some rocks on Venus have relatively high potassium content. This observation suggests that Earthlike magmatic differentiation, probably following Bowen's reaction series, may have been operating.

In 1991, a spacecraft called *Magellan* reached Venus. *Magellan* went into orbit around Venus and started examining the surface topography using radar (Fig. 21.15). The results have provided scientists with an unprecedented view of Venus. It will take years before all of the results have been studied, but already some important issues have been resolved.

Volcanism

The dominant process on the surface of Venus is volcanism. Vast volcanic plains cover 80 percent of the surface, and scattered across the plains are thousands of shield volcanoes ranging up to 500 km in diameter (Fig. 21.16). The volcanic plains and shield volcanoes were clearly formed from low-viscosity, basaltic magma.

Not all of the volcanism on Venus has been basaltic, however. A cluster of seven steep-sided volcanic domes has been discovered (Fig. 21.17). Each dome is about 25 km across, and except for the fact of their large size, the domes look much like the rhyolitic and dacitic lava domes that occur on the Earth. The lava domes indicate that high-viscosity magma occurs on Venus. This observation confirms a suspicion suggested by the chemical analysis performed by the Russian spacecraft: magmatic differentiation has occurred on Venus.

Tectonics

Sensitive radar measurements of Venus's shape show that the total relief from the bottom of the deepest chasm to the top of the highest peak is 13 km. By comparison, the total relief on Earth is 20 km. The radar measurements also show that Venus does not have a bimodal distribution of its surface topography the way the Earth

Figure 21.16   Shield Volcano on Venus   Sapas Mons, a basaltic large shield volcano on Venus. The volcano is about 400 km in diameter and rises about 2 km above the surrounding terrain. In the distance is Maat Mons, largest of the volcanoes on Venus, which stands 8 km above the surrounding plain. This computer image was generated from radar data gathered by the *Magellan* spacecraft.

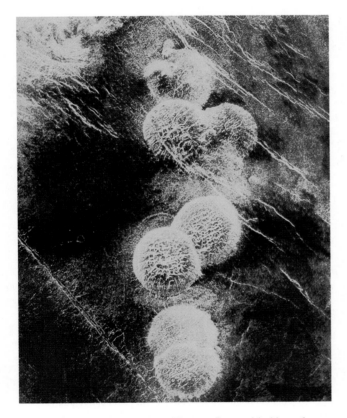

Figure 21.17   Lava Domes on Venus   Steep-sided lava domes on Venus seen by the *Magellan* spacecraft early in 1991. Each dome is about 25 km in diameter. The shapes and steep sides resemble rhyolitic lava domes on the Earth, and this suggests that high viscosity magma might be produced on Venus by some form of magmatic differentiation.

does (Fig. 2.4). The surface of Venus is unimodal. Because the Earth's bimodal surface topography is a result of plate tectonics, it seems unlikely that plate tectonics operate on Venus.

Elongate mountain ranges (Fig. 21.15) have been discovered on Venus. The ranges seem to have formed by a combination of both extension and compression. One popular hypothesis for their formation is that mantle plumes thrust the surface upward and produced extensional features. Then the highest areas slid downward along thrust faults as a result of gravity, thereby producing compressional features. The origin of Venus's mountains, the reasons for the lava plains being crisscrossed by giant faults, and the origin of huge grabens are all topics for further research. Venus may not have plate tectonics, but it certainly seems to have mantle convection and to be tectonically active. The more we learn about Venus, the more apparent it becomes that the Earth is more similar to Venus than to any other planet.

Comparison of the Terrestrial Planets

Despite the differences between the terrestrial planets, they also share many features in common. All the terrestrial planets seem to have metallic cores, all have undergone partial melting to form basaltic magma, and all have been modified by impact cratering. The differences between the terrestrial planets apparently arise

Using Resources

Terraforming

When astronauts Armstrong, Collins, and Aldrin returned from the Moon with the first lunar samples, a new scientific horizon was reached. The year was 1969, and for the first time the possibility that humans might live and work beyond planet Earth was a viable prospect. Many of the early hopes have been found impractical, but two continue to be considered: one is to make one of the other planets habitable, and the other is to mine other bodies in the solar system.

Some scientists consider the concept of changing a planet to make it habitable an environmental monstrosity, but others look on the notion with favor and have coined the term *terraforming* for changing a hostile planet or moon into something like the Earth. The most obvious candidate planet is Mars—Venus is too hot, the Moon and Mercury are too small, and the large moons of Jupiter and Saturn too far away. The time needed to effect a full transformation of Mars is estimated to be up to 100,000 years. The essential two steps are the creation of a suitable atmosphere and the raising of the temperature.

The first phase of the transformation would be the construction of factories on the surface of Mars. The factories would produce greenhouse gases and release them into the thin Martian atmosphere. As the gases accumulated, they would raise the atmospheric pressure and trap more of the Sun's heat, thereby heating the surface. This would cause the carbon dioxide snow on the Martian poles to melt and increase the CO_2 content of the atmosphere. Eventually, the surface temperature on Mars, presently a chilly -260°C, would warm to above 0° and water would be able to flow on the surface. Plants would be able to grow in a warm atmosphere rich in CO_2, but, of course, humans could not breathe the air until photosynthesis had built up the oxygen content of the atmosphere. Initially, humans would have to live in inflatable, domed cities in which a breathable atmosphere could be maintained. Whenever they went outside to tend to growing plants they would have to carry an oxygen supply with them. Eventually, after 100,000 years, Mars would have an atmosphere like that of the Earth.

Terraforming verges on the edge of science fiction, but getting resources from outside the Earth is not so far-fetched. Among the ideas that are being considered is the notion of mining nickel and iron from one of the asteroids. Because the asteroids are small, their gravitational pull is weak, and so it would be easy to lift material from the surface; thus, on large bodies it is practical only to use resources there, not to return then to the Earth. There is an even more difficult problem—it is not even clear that rich ores of the kind found on the Earth occur on the Moon or Mars.

On the Earth most of the rich ores formed when the atmosphere, hydrosphere, or biosphere, or all three, interacted with the lithosphere. Lacking such interactions—because the Earth is the only body in the solar system with a biosphere and a hydrosphere—there is no reason to think that other bodies house mineral treasures like those on Earth.

Fascinating though the ideas of extending our civilization into space may be, the likelihood of anything coming to fruition seems very remote. If anything does happen, it will surely be far in the future. In the meantime, we have to learn to live with the resources available on the Earth and to use those resources in ways that keep planet Earth habitable.

through the interaction of several factors. First, the planet size controls not only the atmosphere, but also the thermal properties. Small planets cool rapidly, and magmatic activity soon ceases. Large planets cool more slowly and remain magmatically active for much longer. Second, a planet's distance from the Sun determines whether or not H_2O can exist as a liquid.

The third factor is the presence or absence of life. The Earth's atmosphere is the way it is because living plants and animals play essential roles in the geochemical cycles that control the atmosphere's composition. If life had developed on Venus, that planet would probably have an atmosphere like the Earth's. However, life apparently did not develop, so all of the CO_2 is still in the atmosphere. On the Earth, plants and animals have been the means whereby carbon dioxide has been removed from the atmosphere, and the carbon is locked up in rocks as fossil organic matter and as calcium carbonate. For a discussion on the probability of making one of these terrestrial planets habitable, see the essay *Terraforming*.

THE JOVIAN PLANETS AND THEIR MOONS

Jupiter

The giant, gassy planets tell us little about the evolution of the Earth, but they provide the best preserved samples of the gases from which all of the planets are believed to have formed. Thus, they reveal much about how the solar system may have formed. Jupiter is the largest and best studied of the giant planets. It has an atmosphere composed of hydrogen, helium, ammonia, and methane, as well as other trace constituents. It is inferred that a rocky core exists inside the dense atmosphere.

Jupiter has about twice the mass of all the other planets combined. Had it been slightly larger, it would have reached an internal temperature high enough for nuclear burning to start, and as a result it would have been a sun. Jupiter is unusual in many ways. For example, it gives off twice as much energy as it receives from the Sun. The reason for this seems to be that Jupiter is still undergoing gravitational contraction, which gives off heat energy.

The Jovian Moons

One of the most interesting things about Jupiter is its moons, four of which are as large as, or larger, than the Earth's Moon. The moons closest to Jupiter, Io and Europa, have densities of 3.5 and 3.2 g/cm^3, respectively, indicating that they are rocky bodies.

The moon closest to Jupiter, Io, is extraordinary. It is a highly colored body with shades of yellow and orange predominating, suggesting that it is covered by sulfur and sulfurous compounds (Fig. 21.18). Impact craters are absent, and the reason is not hard to find—Io is volcanically active. Indeed, it is by far the most volcanically active body in the solar system. Impacts by bolides certainly must occur, but the craters are quickly covered up by volcanic debris.

Io's volcanism seems to be of two kinds. The first is the familiar basaltic volcano found so widely throughout the solar system. Lava plains and shield volcanoes are the result. One of the shield volcanoes, Ra Patera, is almost as large as Olympus Mons on Mars. Fresh lava flows can be seen on its slopes. The second kind of volcanism seems to involve sulfur and sulfur dioxide (SO_2). Huge orange-yellow flows of what is presumed to be molten sulfur have been seen, some as much as 700 km long. Most striking, however, are active volcanic plumes that throw sprays of sulfurous gases and entrained solid particles as high as 300 km above the surface of Io. Nine active plumes were observed by the two *Voyager* spacecraft as they flew by Io (Fig. 1.10A). The volcanic plumes seem to be geyserlike in origin, but the fluid that boils and erupts is SO_2, not H_2O. It has been estimated that the plumes eject 10^{16} g of fine, solid particles each year. This quantity is sufficient to bury the surface of Io with a layer of pyroclastic debris 100 m thick in a million years; it is no wonder no impact craters are to be seen. The process of surface renewal is much faster than it is on the Earth.

The amount of heat energy needed to drive Io's volcanoes is much greater than the heat that could be produced through radioactive decay in a stony planet. Io's volcanic heat comes from a different source—the gravitational pull exerted by the huge mass of Jupiter. As Io moves around Jupiter in an elliptical orbit, it is periodically stretched more or less by the gravitational pull—more during a close approach, less when far away. The

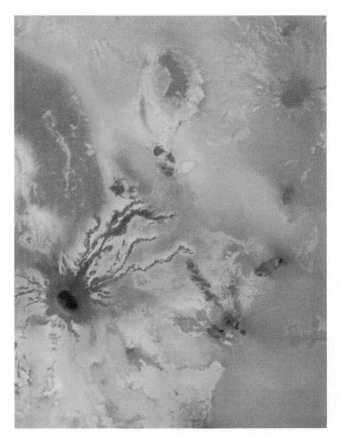

Figure 21.18 Evidence of Volcanism on Io The bright colors on the surface of Io are caused by sulfur and sulfurous compounds given off during volcanism. The feature in the lower left of the image is a volcanic cone; numerous lava flows, basaltic in composition, radiate out from the volcano. The width of the field of view is 1000 km. The image was taken by *Voyager I* in 1979 at a distance of 128,500 km.

bending and stretching due to the fluctuating gravitational pull generates heat, just as a copper wire becomes hot if it is bent back and forth. No other object in the solar system demonstrates the effect of tidal stresses so dramatically as Io.

Farther away from Jupiter than Io are the three large moons Europa, Ganymede, and Callisto. Their densities decrease the farther they are from Jupiter. Europa has a density of 3.2 g/cm^3, Ganymede, 2.0 g/cm^3, and Callisto, 1.8 g/cm^3. The reason for the lowered densities was discovered during the *Voyager* missions: each of the moons is sheathed with a layer of H_2O ice. The outer moons, and especially Europa, may have small metallic cores, but their densities suggest that their main masses lie in thick mantles consisting of ice and silicate minerals. Above the mantle are crusts of nearly pure ice 100 km or more thick.

Craters are rare on Europa; however, the surface is split and criss-crossed by an intricate network of fractures (Fig. 1.10B). Presumably, some tectonic process renews Europa's ice surface through fracture and upwelling.

The upwelling may involve melting of the ice. The source of energy that results in melting is probably tidal, caused, as with Io, by the gravitational pull of Jupiter.

Ganymede has a much thicker sheath of ice than does Europa. Ganymede is too far from Jupiter to be influenced by tides, and it is pitted by craters. However, on Ganymede some slow-acting process is appar-

Figure 21.19 The Icy Surface of Ganymede   The surface of Ganymede, largest of Jupiter's moons, viewed from a distance of 312,000 km by *Voyager 2* in 1979. Ganymede is covered by a thick crust of ice. The dark surface is ancient ice, presumably covered by dust and impact debris. It is split into continent-sized fragments that are separated by light-colored, grooved terrains of younger ice. Apparently, Ganymede is tectonically active, and the grooved terrains seem to be the places where new ice rises from below, but how this happens, and what causes the grooves, is not known. The field of view is approximately 1300 km across.

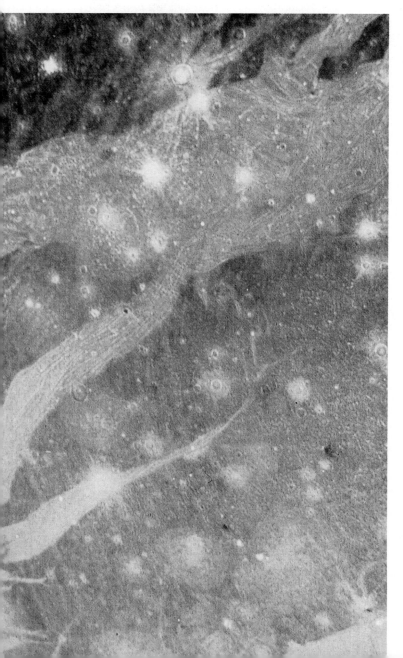

ently renewing and reworking the ice surface because some regions have few, if any, impact craters. The surface is divided into dark and light areas (Fig. 21.19). The dark areas are more heavily cratered than the light areas and are presumably older. Within the lighter (younger) areas there are striking grooves, fractures, and grabenlike structures. It is possible that a unique kind of plate tectonics may be operating on Ganymede. The dark regions are ancient ice continents, and the lighter areas are the places where ice rises convectively from the depths to create new ice crust.

The most distant moon, Callisto, must contain the greatest proportion of ice because its density is the least of the jovian moons. The surface is again icy, but it must be very ancient ice as the number of impact craters is great. No evidence has yet been found to suggest that the surface of Callisto is being renewed and reworked.

Saturn and Its Moons

Saturn has a composition like that of Jupiter, but it is not as large. Like Jupiter, Saturn radiates more energy than it receives from the Sun. Saturn's most striking feature is its immense ring system. Before the *Voyager* mission to Saturn in 1980, little was known about Saturn's moons. Less is known about them still than is known about the moons of Jupiter. Most of the moons are small and have low densities. In composition and structure they seem to resemble Ganymede and Callisto. They are ice-covered, and cratering is extensive. Two of the moons, like the surface on Ganymede, display evidence that ice flow may be renewing the surfaces.

The most distinctive of the Saturnian moons is Titan, a body larger than Mercury. It is the only moon in the solar system large enough to retain a substantial atmosphere. Unfortunately, the atmosphere is an opaque, orange-colored smog that shrouds the surface from view. The composition of Titan's atmosphere is mostly nitrogen; however, ethane, acetylene, ethylene, hydrogen cyanide, and other unpleasant substances are also present. The density of Titan is 1.9 g/cm^3, which suggests that it contains about 45 percent H_2O ice and 55 percent rocky matter. Sunlight working on the atmosphere has caused ethylene and acetylene to form through photochemical reactions, and it is these compounds that produce the smog that covers Titan. Because sunlight cannot penetrate the smog, the surface of Titan must be a cold and unfriendly place. Scientists who have studied the data sent back by *Voyager II* suggest that the surface temperature is -180°C, that Titan is covered by an ocean of liquid ethane and methane, and that continents of H_2O ice rise up from the ocean floor. Titan is even stranger than science fiction. Who would have imagined a planetary body with oceans of liquid hydrocarbons and continents of ice?

SUMMARY

1. Planets close to the Sun, such as Mercury, Mars, the Earth, and Venus, are small, dense, rocky bodies. Planets farther away, such as Saturn and Jupiter, are large, low-density bodies.

2. The Moon has a layered structure probably formed by differentiation.

3. The Moon probably has a small core surrounded by a thick mantle and is capped by a crust 65 km thick.

4. On the Moon, magma was formed early but is no longer generated. The Moon is a magmatically dead planet.

5. The highlands of the Moon are remains of ancient crust built by magmatic differentiation more than 4 billion years ago.

6. The maria (lunar lowlands) are vast basins created by the impacts of giant meteorites and later filled in by lava flows.

7. Each of the terrestrial planets went through a period of internal radioactive heating that led to generation of basaltic magma.

8. The Earth and possibly Mars and Venus are still producing magma from radioactive heating. Mars and Venus both appear to be one-plate planets, not multi-plate planets like the Earth.

9. Mars seems to be magmatically active. Olympus Mons, a shield volcano on Mars, is the largest volcano yet found in the solar system.

10. The principal eroding agent on Mars is wind-driven sand and dust. Water or some other flowing liquid cut stream channels at some time in the past.

11. Venus has about the same size and density as the Earth. It has a dense atmosphere of carbon dioxide and a surface temperature of about 500°C.

12. None of the other terrestrial planets has a climate hospitable to human life.

13. The large outer planets, Jupiter, Saturn, Neptune, and Uranus, have thick atmospheres that obscure their rocky cores. Pluto, the outermost planet, is probably covered by H ice.

14. The moons of Jupiter have progressively thicker outer layers of ice the farther away they are from Jupiter.

THE LANGUAGE OF GEOLOGY

asteroids (p. 559)

bolide (p. 559)

ecliptic (p. 559)

highlands (p. 563)

impact cratering (p. 559)

magma ocean (p. 565)
mare (p. 563)

mare basins (p. 565)
maria (p. 563)

planetology (p. 558)

QUESTIONS FOR REVIEW

1. What is planetology? How is planetology related to geology?

2. Name the four terrestrial planets in order of their positions out from the Sun.

3. Besides the terrestrial planets, what other solid objects are to be found in the inner reaches of the solar system?

4. What is the most important process for shaping planetary surfaces in the solar system?

5. Briefly describe the kinds of magmatic activity that have been observed on planets and moons other than the Earth.

6. What evidence leads scientists to conclude that in the earliest days of the Moon's history it had a molten outer layer and a solid interior?

7. What is the origin of the lunar maria?

8. How could it be possible for evidence of water erosion to exist on Mars without lakes, streams, or seas being present?

9. What is the source of the energy that made Io, one of Jupiter's moons, the most tectonically active object in the solar system?

10. Compare the structure of Ganymede, a moon of Jupiter, with the Earth's Moon. What explanation can you offer for the difference?

APPENDIX A
UNITS AND THEIR CONVERSIONS

ABOUT SI UNITS

Regardless of the field of specialization, all scientists use the same units and scales of measurement. They do so to avoid confusion and the possibility that mistakes can creep in when data are converted from one system of units, or one scale, to another. By international agreement the SI units are used by all, and they are the units used in this text. SI is the abbreviation of Système International d'Unités (in English, the International System of Units).

Some of the SI units are likely to be familiar, some unfamiliar. The SI unit of length is the meter (m), of area the square meter (m^2), and of volume the cubic meter (m^3). The SI unit of mass is the kilogram (kg), and of time the second (s). The other SI units used in this book can be defined in terms of these basic units. Three important ones are:

1. The newton (N), a unit of force defined as that force needed to accelerate a mass of 1 kg by 1 m/s^2; hence 1 N = 1 kg·m/s^2. (The period between kg and m indicates multiplication.)

2. The joule (J), a unit of energy or work, defined as the work done when a force of 1 newton is displaced a distance of 1 meter; hence 1 J = 1 N·m. One important form of energy as far as the Earth is concerned is heat. The outward flow of the Earth's internal heat is measured in terms of the number of joules flowing outward from each square centimeter each second; thus, the unit of heat flow is J/cm^2/s.

3. The pascal (Pa), a unit of pressure defined as a force of 1 newton applied across an area of 1 square meter; hence 1 Pa = 1 N/m^2. The pascal is a numerically small unit. Atmospheric pressure, for example (15 lb/in^2), is 101,300 Pa. Pressure within the Earth reaches millions or billions of pascals. For convenience, earth scientists sometimes use 1 million pascals (megapascal, or MPa) as a unit.

Temperature is a measure of the internal kinetic energy (expressed as movement) of the atoms and molecules in a body. In the SI system, temperature is measured on the Kelvin scale (K). The temperature intervals on the Kelvin scale are arbitrary, and they are the same as the intervals on the more familiar Celsius scale (°C). The difference between the two scales is that the Celsius scale selects 100°C as the temperature at which water boils at sea level, and 0°C as the freezing temperature of water at sea level. Zero degrees Kelvin, on the other hand, is absolute zero, the temperature at which all atomic and molecular motions cease. Thus, 0°C is equal to 273.15 K, and 100°C is 373.15 K. The temperatures of processes on and within the Earth tend to be at or above 273.15 K. Despite the inconsistency, earth scientists still use the Celsius scale when geological processes are discussed.

Appendix A provides a table of conversion from older units to Standard International (SI) units.

PREFIXES FOR MULTIPLES AND SUBMULTIPLES

When very large or very small numbers have to be expressed, a standard set of prefixes is used in conjunction with the SI units. Some prefixes are probably already familiar; an example is the centimeter (which is one hundredth of a meter, or 10^{-2} m). The standard prefixes are

tera	1,000,000,000,000 =	10^{12}
giga	1,000,000,000 =	10^9
mega	1,000,000 =	10^6
kilo	1,000 =	10^3
hecto	100 =	10^2
deka	10 =	10
deci	0.1 =	10^{-1}
centi	0.01 =	10^{-2}
milli	0.001 =	10^{-3}
micro	0.000001 =	10^{-6}
nano	0.000000001 =	10^{-9}
pico	0.000000000001 =	10^{-12}

One measure used commonly in geology is the nanometer (nm), a unit by which the sizes of atoms are measured; 1 nanometer is equal to 10^{-9} meter.

COMMONLY USED UNITS OF MEASURE

Length

Metric Measure

1 kilometer (km)	=	1000 meters (m)
1 meter (m)	=	100 centimeters (cm)
1 centimeter (cm)	=	10 millimeters (mm)
1 millimeter (mm)	=	1000 micrometers (μm) (formerly called microns)
1 micrometer (mm)	=	0.001 millimeter (mm)
1 angstrom (Å)	=	10^{-8} centimeters (cm)

Nonmetric Measure

1 mile (mi)	=	5280 feet (ft) = 1760 yards (yd)
1 yard (yd)	=	3 feet (ft)
1 fathom (fath)	=	6 feet (ft)

Conversions

1 kilometer (km)	=	0.6214 mile (mi)
1 meter (m)	=	1.094 yards (yd) = 3.281 feet (ft)
1 centimeter (cm)	=	0.3937 inch (in)
1 millimeter (mm)	=	0.0394 inch (in)
1 mile (mi)	=	1.609 kilometers (km)
1 yard (yd)	=	0.9144 meter (m)
1 foot (ft)	=	0.3048 meter (m)
1 inch (in)	=	2.54 centimeters (cm)
1 inch (in)	=	25.4 millimeters (mm)
1 fathom (fath)	=	1.8288 meters (m)

Area

Metric Measure

1 square kilometer (km^2)	=	1,000,000 square meters (m^2)
	=	100 hectares (ha)
1 square meter (m^2)	=	10,000 square centimeters (cm^2)
1 hectare (ha)	=	10,000 square meters (m^2)

Nonmetric Measure

1 square mile (mi^2)	=	640 acres (ac)
1 acre (ac)	=	4840 square yards (yd^2)
1 square foot (ft^2)	=	144 square inches (in^2)

Conversions

1 square kilometer (km^2)	=	0.386 square mile (mi^2)
1 hectare (ha)	=	2.471 acres (ac)
1 square meter (m^2)	=	1.196 square yards (yd^2)
	=	10.764 square feet (ft^2)
1 square centimeter (cm^2)	=	0.155 square inch (in^2)
1 square mile (mi^2)	=	2.59 square kilometers (km^2)
1 acre (ac)	=	0.4047 hectare (ha)
1 square yard (yd^2)	=	0.836 square meter (m^2)
1 square foot (ft^2)	=	0.0929 square meter (m^2)
1 square inch (in^2)	=	6.4516 square centimeter (cm^2)

Volume

Metric Measure

1 cubic meter (m^3)	=	1,000,000 cubic centimeters (cm^3)
1 liter (l)	=	1000 milliliters (ml)
	=	0.001 cubic meter (m^3)
1 centiliter (cl)	=	10 milliliters (ml)
1 milliliter (ml)	=	1 cubic centimeter (cm^2)

Nonmetric Measure

1 cubic yard (yd^3)	=	27 cubic feet (ft^3)
1 cubic foot (ft^3)	=	1728 cubic inches (in^3)
1 barrel (oil) (bbl)	=	42 gallons (U.S.) (gal)

Conversions

1 cubic kilometer (km^3)	=	0.24 cubic miles (mi^3)
1 cubic meter (m^3)	=	264.2 gallons (U.S.) (gal)
	=	35.314 cubic feet (ft^3)
1 liter (l)	=	1.057 quarts (U.S.) (qt)
	=	33.815 ounces (U.S. fluid) (fl. oz.)
1 cubic centimeter (cm3)	=	0.0610 cubic inch (in^3)
1 cubic mile (mi^3)	=	4.168 cubic kilometers (km^3)
1 acre-foot (ac-ft)	=	1233.46 cubic meters (m^3)
1 cubic yard (yd^3)	=	0.7646 cubic meter (m^3)
1 cubic foot (ft^3)	=	0.0283 cubic meter (m^3)
1 cubic inch (in^3)	=	16.39 cubic centimeters (cm^3)
1 gallon (gal)	=	3.784 liters (l)

Mass

Metric Measure

1000 kilograms (kg)	= 1 metric ton (also called a tonne) (m.t)
1 kilogram (kg)	= 1000 grams (g)

Nonmetric Measure

1 short ton (sh.t)	= 2000 pounds (lb)
1 long ton (l.t)	= 2240 pounds (lb)
1 pound (avoirdupois) (lb.)	= 16 ounces (avoirdupois) (oz) = 7000 grains (gr)
1 ounce (avoirdupois) (oz)	= 437.5 grains (gr)
1 pound (Troy) (Tr. lb)	= 12 ounces (Troy) (Tr. oz)
1 ounce (Troy) (Tr. oz)	= 20 pennyweight (dwt)

Conversions

1 metric ton (m.t)	= 2205 pounds (avoirdupois) (lb)
1 kilogram (kg)	= 2.205 pounds (avoirdupois) (lb)
1 gram (g)	= 0.03527 ounce (avoirdupois) (oz) 0.03215 ounce (Troy) (Tr. oz) = 15,432 grains (gr)
1 pound (lb)	= 0.4536 kilogram (kg)
1 ounce (avoirdupois) (oz)	= 28.35 grams (g)
1 ounce (avoirdupois) (oz)	= 1.097 ounces (Troy) (Tr. oz)

Pressure

1 pascal (Pa)	= 1 newton/square meter (N/m^2)
1 kilogram/square centimeter (kg/cm^2)	= 0.96784 atmosphere (atm) = 14.2233 pounds/square inch (lb/in^2) = 0.098067 bar
1 bar	= 0.98692 atmosphere (atm) = 105 pascals (Pa) = 1.02 kilograms/square centimeter (kg/cm^2)

Energy and Power

Energy

1 joule (J)	= 1 newton meter (N.m)
	= 2.390×10^{-1} calorie (cal)
	= 9.47×10^{-4} British thermal unit (Btu)
	= 2.78×10^{-7} kilowatt-hour (kWh)
1 calorie (cal)	= 4.184 joule (J)
	= 3.968×10^{-3} British thermal unit (Btu)
	= 1.16×10^{-6} kilowatt-hour (kWh)
1 British thermal unit (Btu)	= 1055.87 joules (J)
	= 252.19 calories (cal)
	= 2.928×10^{-4} kilowatt-hour (kWh)
1 kilowatt hour	= 3.6×10^6 joules (J)
	= 8.60×10^5 calories (cal)
	= 3.41×10^2 British thermal units (Btu)

Power (energy per unit time)

1 watt (W)	= 1 joule per second (J/s)
	= 3.4129 Btu/h
	= 1.341×10^{-3} horsepower (hp)
	= 14.34 calories per minute (cal/min)
1 horsepower (hp)	= 7.46×10^3 watts (W)

Temperature

To change from Fahrenheit (F) to Celsius (C)

$$°C = \frac{(°F - 32°)}{1.8}$$

To change from Celsius (C) to Fahrenheit (F)

$$°F = (°C \times 1.8) + 32°$$

To change from Celsius (C) to Kelvin (K)

$$K = °C + 273.15$$

To change from Fahrenheit (F) to Kelvin (K)

$$K = \frac{(°F - 32)}{1.8} + 273.15$$

APPENDIX B

TABLES OF THE CHEMICAL ELEMENTS AND NATURALLY OCCURRING ISOTOPES

| TABLE B.1 | Alphabetical List of the Elements |

Element	Symbol	Atomic Number	Crustal Abundance, Weight Percent	Element	Symbol	Atomic Number	Crustal Abundance, Weight Percent
Actinium	Ac	89	Human-made	Gold	Au	79	0.0000002
Aluminum	Al	13	8.00	Hafnium	Hf	72	0.004
Americium	Am	95	Human-made	Hahnium	Ha	105	Human-made
Antimony	Sb	51	0.00002	Hassium	Hs	108	Human-made
Argon	Ar	18	Not known	Helium	He	2	Not known
Arsenic	As	33	0.00020	Holmium	Ho	67	0.00016
Astatine	At	85	Human-made	Hydrogen[b]	H	1	0.14
Barium	Ba	56	0.0380	Indium	In	49	0.00002
Berkelium	Bk	97	Human-made	Iodine	I	53	0.00005
Beryllium	Be	4	0.00020	Iridium	Ir	77	0.00000002
Bismuth	Bi	83	0.0000004	Iron	Fe	26	5.80
Boron	B	5	0.0007	Krypton	Kr	36	Not known
Bromine	Br	35	0.00040	Lanthanum	La	57	0.0050
Cadmium	Cd	48	0.000018	Lawrencium	Lw	103	Human-made
Calcium	Ca	20	5.06	Lead	Pb	82	0.0010
Californium	Cf	98	Human-made	Lithium	Li	3	0.0020
Carbon[a]	C	6	0.02	Lutetium	Lu	71	0.000080
Cerium	Ce	58	0.0083	Magnesium	Mg	12	2.77
Cesium	Cs	55	0.00016	Manganese	Mn	25	0.100
Chlorine	Cl	17	0.0190	Meitnerium	Mt	109	Human-made
Chromium	Cr	24	0.0096	Mendelevium	Md	101	Human-made
Cobalt	Co	27	0.0028	Mercury	Hg	80	0.000002
Copper	Cu	29	0.0058	Molybdenum	Mo	42	0.00012
Curium	Cm	96	Human-made	Neodymium	Nd	60	0.0044
Dysprosium	Dy	66	0.00085	Neon	Ne	10	Not known
Einsteinium	Es	99	Human-made	Neptunium	Np	93	Human-made
Erbium	Er	68	0.00036	Nickel	Ni	28	0.0072
Europium	Eu	63	0.00022	Nielsbohrium	Ns	107	Human-made
Fermium	Fm	100	Human-made	Niobium	Nb	41	0.0020
Fluorine	F	9	0.0460	Nitrogen	N	7	0.0020
Francium	Fr	87	Human-made	Nobelium	No	102	Human-made
Gadolinium	Gd	64	0.00063	Osmium	Os	76	0.00000002
Gallium	Ga	31	0.0017	Oxygen[b]	O	8	45.2
Germanium	Ge	32	0.00013	Palladium	Pd	46	0.0000003

TABLE B.1	Alphabetical List of the Elements, *continued*

Element	Symbol	Atomic Number	Crustal Abundance, Weight Percent	Element	Symbol	Atomic Number	Crustal Abundance, Weight Percent
Phosphorus	P	15	0.1010	Sodium	Na	11	2.32
Platinum	Pt	78	0.0000005	Strontium	Sr	38	0.0450
Plutonium	Pu	94	Human-made	Sulfur	S	16	0.030
Polonium	Po	84	Footnote[d]	Tantalum	Ta	73	0.00024
Potassium	K	19	1.68	Technetium	Tc	43	Human-made
Praseodymium	Pr	59	0.0013	Tellurium[c]	Te	52	0.000001
Promethium	Pm	61	Human-made	Terbium	Tb	65	0.00010
Protactinium	Pa	91	Footnote[d]	Thallium	Tl	81	0.000047
Radium	Ra	88	Footnote[d]	Thorium	Th	90	0.00058
Radon	Rn	86	Footnote[d]	Thulium	Tm	69	0.000052
Rhenium	Re	75	0.00000004	Tin	Sn	50	0.00015
Rhodium[c]	Rh	45	0.00000001	Titanium	Ti	22	0.86
Rubidium	Rb	37	0.0070	Tungsten	W	74	0.00010
Ruthenium[c]	Ru	44	0.00000001	Uranium	U	92	0.00016
Samarium	Sm	62	0.00077	Vanadium	V	23	0.0170
Scandium	Sc	21	0.0022	Xenon	Xe	54	Not known
Seaborgium	Sg	106	Human-made	Ytterbium	Yb	70	0.00034
Selenium	Se	34	0.000005	Yttrium	Y	39	0.0035
Silicon	Si	14	27.20	Zinc	Zn	30	0.0082
Silver	Ag	47	0.000008	Zirconium	Zr	40	0.0140

Source: After K. K. Turekian, 1969.

[a] Estimate from S. R. Taylor (1964).

[b] Analyses of crustal rocks do not usually include separate determinations for hydrogen and oxygen. Both combine in essentially constant proportions with other elements, so abundances can be calculated.

[c] Estimates are uncertain and have a very low reliability.

[d] Elements formed by decay of uranium and thorium. The daughter products are radioactive with such short half-lives that crustal accumulations are too low to be measured accurately.

TABLE B.2	Naturally Occurring Elements Listed in Order of Atomic Numbers, Together with the Naturally Occurring Isotopes of Each Element, Listed in Order of Mass Numbers

Atomic Number[a]	Name	Symbol	Mass Numbers[b] of Natural Isotopes	Atomic Number[a]	Name	Symbol	Mass Numbers[b] of Natural Isotopes
1	Hydrogen	H	1, 2, 3[c]	20	Calcium	Ca	40, 42, 43, 44, 46, 48
2	Helium	He	3, 4	21	Scandium	Sc	45
3	Lithium	Li	6, 7	22	Titanium	Ti	46, 47, 48, 49 50
4	Beryllium	Be	9, 10	23	Vanadium	V	50, 51
5	Boron	B	10, 11	24	Chromium	Cr	50, 52, 53, 54
6	Carbon	C	12, 13, 14	25	Manganese	Mn	55
7	Nitrogen	N	14, 15	26	Iron	Fe	54, 56, 57, 58
8	Oxygen	O	16, 17, 18	27	Cobalt	Co	59
9	Fluorine	F	19	28	Nickel	Ni	58, 60, 61, 62, 64
10	Neon	Ne	20, 21, 22	29	Copper	Cu	63, 65
11	Sodium	Na	23	30	Zinc	Zn	64, 66, 67, 68, 70
12	Magnesium	Mg	24, 25, 26	31	Gallium	Ga	69, 71
13	Aluminum	Al	27	32	Germanium	Ge	70, 72, 73, 74, 76
14	Silicon	Si	28, 29 30	33	Arsenic	As	75
15	Phosphorus	P	31	34	Selenium	Se	74, 76, 77, 80, 82
16	Sulfur	S	32, 33, 34, 36	35	Bromine	Br	79, 81
17	Chlorine	Cl	35, 37	36	Krypton	Kr	78, 80, 82, 83, 84, 86
18	Argon	Ar	36, 38, 40	37	Rubidium	Rb	85, 87
19	Potassium	K	39, 40, 41	38	Strontium	Sr	84, 86, 87, 88

| TABLE B.2 | Naturally Occurring Elements Listed in Order of Atomic Numbers, Together with the Naturally Occurring Isotopes of Each Element, Listed in Order of Mass Numbers, *cont.* |

Atomic Number[a]	Name	Symbol	Mass Numbers[b] of Natural Isotopes	Atomic Number[a]	Name	Symbol	Mass Numbers[b] of Natural Isotopes
48	Cadmium	Cd	106, 108, 110, 111, 112, 113, 114, 116	68	Erbium	Er	162, 166, 167, 168, 170
49	Indium	In	113, 115	69	Thulium	Tm	169
50	Tin	Sn	112, 114, 115, 116, 117, 118, 119, 120, 122, 124	70	Ytterbium	Yb	168, 170, 171, 172, 173, 174, 176
				71	Lutetium	Lu	175, 176
51	Antimony	Sb	121, 123	72	Hafnium	Hf	174, 176, 177, 178, 179, 180
52	Tellurium	Te	120, 122, 123, 124, 125, 126, 128, 130	73	Tantalum	Ta	180, 181
53	Iodine	I	127	74	Tungsten	W	180, 182, 183, 184, 186
54	Xenon	Xe	124, 126, 128, 129, 130, 131, 132, 134, 136	75	Rhenium	Re	185, 187
				76	Osmium	Os	184, 186, 187, 188, 189, 190, 192
55	Cesium	Cs	133				
56	Barium	Ba	130, 132, 134, 135, 137, 138	77	Iridium	Ir	191, 193
57	Lanthanum	La	138, 139	78	Platinum	Pt	190, 192, 195, 196, 198
58	Cerium	Ce	136, 138, 140, 142	79	Gold	Au	197
59	Praseodymium	Pr	141	80	Mercury	Hg	196, 198, 199, 200, 201, 202, 204
60	Neodymium	Nd	142, 143, 144, 145, 146, 148, 150	81	Thallium	Tl	203, 205
62	Samarium	Sm	144, 147, 148, 149, 150, 152, 154	82	Lead	Pb	204, 206, 207, 208
				83	Bismuth	Bi	209
63	Europium	Eu	151, 153	84	Polonium	Po	210
64	Gadolinium	Gd	152, 154, 155, 156, 157, 158, 160	86	Radon	Rn	222
				88	Radium	Ra	226
65	Terbium	Tb	159	90	Thorium	Th	232
66	Dysprosium	Dy	156, 158, 160, 161, 162, 163, 164	91	Protactinium	Pa	231
67	Holmium	Ho	165	92	Uranium	U	234, 235, 238

[a] Atomic number = number of protons.

[b] Mass number = protons + neutrons.

[c] ☐ indicates isotope is radioactive.

APPENDIX C
TABLES OF THE PROPERTIES OF COMMON MINERALS

TABLE C.1	Properties of the Common Minerals with Metallic Luster					
Mineral	**Chemical Composition**	**Form and Habit**	**Cleavage**	**Hardness / Specific Gravity**	**Other Properties**	**Most Distinctive Properties**
Bornite	Cu_5FeS_4	Massive. Crystals very rare.	None. Uneven fracture.	3 / 5	Brownish bronze on fresh surface. Tarnishes purple, blue, and black. Grayish-black streak.	Color, streak.
Chalcocite	Cu_2S	Massive. Crystals very rare.	None. Conchoidal fracture.	2.5 / 5.7	Steel-gray to black. Dark gray streak.	Streak.
Chalcopyrite	$CuFeS_2$	Massive or granular.	None. Uneven fracture.	3.5–4 / 4.2	Golden yellow to brassy yellow. Dark green to black streak.	Streak. Hardness distinguishes from pyrite.
Chromite	$FeCr_2O_4$	Massive or granular.	None. Uneven fracture.	5.5 / 4.6	Iron black to brownish black. Dark brown streak.	Streak and lack of magnetism distinguishes from ilmenite and magnetite.
Copper	Cu	Massive, twisted leaves and wires.	None. Can be cut with a knife.	2.5–3 / 9	Copper color but commonly stained green.	Color, specific gravity, malleable.
Galena	PbS	Cubic crystals, coarse or fine-grained granular masses.	Perfect in three directions at right angles.	2.5 / 7.6	Lead-gray color. Gray to gray-black streak.	Cleavage and streak.
Gold	Au	Small irregular grains.	None. Malleable.	2.5 / 19.3	Gold color. Can be flattened without breakage.	Color, specific gravity, malleability.
Hematite	Fe_2O_3	Massive, granular, micaceous.	Uneven fracture.	5–6 / 5	Reddish-brown, gray to black. Reddish-brown streak.	Streak, hardness.
Ilmenite	$FeTiO_3$	Massive or irregular grains.	Uneven fracture.	5.5–6 / 4.7	Iron-black. Brown-reddish streak differing from hematite.	Streak distinguishes hematite. Lack of magnetism distinguishes magnetite.
Limonite (*Goethite* is most common.)	A complex mixture of minerals, mainly hydrous oxides.	Massive, coatings, botryoidal crusts, earthy masses.	None.	1–5.5 / 3.5–4	Yellow, brown, black, yellowish-brown streak.	Streak.

| TABLE C.1 | | Properties of the Common Minerals with Metallic Luster, *continued* | | | | |

Mineral	Chemical Composition	Form and Habit	Cleavage	Hardness / Specific Gravity	Other Properties	Most Distinctive Properties
Magnetite	Fe_3O_4	Massive, granular. Crystals have octahedral shape.	None. Uneven fracture.	5.5–6.5 / 5	Black. Black streak. Strongly attracted to a magnet.	Streak, magnetism.
Pyrite ("Fool's gold")	FeS_2	Cubic crystals with striated faces. Massive.	None. Uneven fracture.	6–6.5 / 5.2	Pale brass-yellow, darker if tarnished. Greenish-black streak.	Streak. Hardness distinguishes from chalcopyrite. Not malleable, which distinguishes from gold.
Pyrolusite	MnO_2	Crystals rare. Massive, coatings on fracture surfaces.	Crystals have a perfect cleavage. Massive, breaks unevenly.	2–6.5 / 5	Dark gray, black or bluish black. Black streak.	Color, streak.
Pyrrhotite	FeS	Crystals rare. Massive or granular.	None. Conchoidal fracture.	4 / 4.6	Brownish-bronze. Black streak. Magnetic.	Color and hardness distinguish from pyrite, magnetism from chalcopyrite.
Rutile	TiO_2	Slender, prismatic crystals or granular masses.	Good in one direction. Conchoidal fracture in others.	6–6.5 / 4.2	Reddish-brown (common), black (rare). Brownish streak. Adamantine luster.	Luster, habit, hardness.
Sphalerite (zinc blende)	ZnS	Fine to coarse granular masses. Tetrahedron shaped crystals.	Perfect in six directions.	3.5–4 / 4	Yellowish-brown to black. White to yellowish-brown streak. Resinous luster.	Cleavage, hardness, luster.
Uraninite	UO_2 to U_3O_8	Massive, with botryoidal forms. Rare crystals with cubic shapes.	None. Uneven fracture.	5–6 / 6.5–10	Black to dark brown. Streak black to dark brown. Dull luster.	Luster and specific gravity distinguish from magnetite. Streak distinguishes from ilmenite and hematite.

| TABLE C.2 | | Properties of Rock-Forming Minerals with Nonmetallic Luster | | | | |

Mineral	Chemical Composition	Form and Habit	Cleavage	Hardness / Specific Gravity	Other Properties	Most Distinctive Properties
Amphiboles (A complex family of minerals, *Hornblende* is most common.)	$X_2Y_5Si_8O_{22}(OH)_2$ where X = Ca, Na; Y = Mg, Fe, Al.	Long, six-sided crystals; also fibers and irregular grains.	Two; intersecting at 56° and 124°.	5–6 / 2.9–3.8	Common in metamorphic and igneous rocks. *Hornblende* is dark green to black; *actinolite*, green; *tremolite*, white.	Cleavage, habit.
Andalusite	Al_2SiO_5	Long crystals, often square in cross-section.	Weak, parallel to length of crystal.	7.5 / 3.2	Found in metamorphic rocks. Often flesh-colored.	Hardness, form.

| TABLE C.2 | Properties of Rock-Forming Minerals with Nonmetallic Luster, *continued* |

Mineral	Chemical Composition	Form and Habit	Cleavage	Hardness / Specific Gravity	Other Properties	Most Distinctive Properties
Anhydrite	$CaSO_4$	Crystals are rare. Irregular grains or fibers.	Three, at right angles.	3 / 2.9	Alters to gypsum. Pearly luster, white or colorless.	Cleavage, hardness.
Apatite	$Ca_5(PO_4)_3$ (F_3, OH, Cl)	Granular masses. Perfect six-sided crystals.	Poor. One direction.	5 / 3.2	Green, brown, blue, or white. Common in many kinds of rocks in small amounts.	Hardness, form.
Aragonite	$CaCO_3$	Massive, or slender, needle-like crystals.	Poor. Two directions.	3.5 / 2.9	Colorless or white. Effervesces with dilute HCl.	Effervescence with acid. Poor cleavage distinguishes from calcite.
Asbestos			See Serpentine			
Augite			See Pyroxene			
Biotite			See Mica			
Calcite	$CaCO_3$	Tapering crystals and granular masses.	Three perfect oblique angles to give a rhomb-shaped fragment.	3 / 2.7	Colorless or white. Effervesces with dilute HCl.	Cleavage, effervescence with acid.
Chlorite	$(Mg, Fe)_5$ $(Al, Fe)_2$ $Si_3O_{10}(OH)_8$	Flaky masses of minute scales.	One perfect; parallel to flakes.	2–2.5 / 2.6–2.9	Common in metamorphic rocks. Light to dark green. Greasy luster.	Cleavage—flakes not elastic, distinguishes from mica. Color.
Dolomite	$CaMg(CO_3)_2$	Crystals with rhomb-shaped faces. Granular masses.	Perfect in three directions as in calcite.	3.5 / 2.8	White or gray. Does not effervesce in cold, dilute HCl unless powdered. Pearly luster.	Cleavage. Lack of effervescence with acid.
Epidote	Complex Ca, Fe and Al.	Small elongate crystals. Fibrous.	One perfect, one poor.	6–7 / 3.4	Yellowish-green to dark green. Common in metamorphic rocks.	Habit, color. Hardness distinguishes from chlorite.
Feldspars: *Potassium* feldspar *(orthoclase)* is a common variety	$KAlSi_3O_8$	Prism-shaped crystals, granular	Two perfect, at right angles.	6 / 2.6	Common mineral. Flesh-colored, pink, white, or gray.	Color, cleavage.
Plagioclase	$NaAlSi_3O_8$ (albite) and $CaAl_2Si_2O_8$ (anorthite) and all compositions between.	Irregular grains, cleavable masses. Tabular crystals.	Two perfect, not quite at right angles.	6–6.5 / 2.6–2.7	White to dark gray. Cleavage planes may show fine parallel striations.	Cleavage. Striations on cleavage planes will distinguish from potassium feldspar.
Fluorite	CaF_2	Cubic crystals, granular masses.	Perfect in four directions.	4 / 3.2	Colorless, bluish green. Always an accessory mineral.	Hardness, cleavage does not effervesce with acid.

TABLE C.2 | Properties of Rock-Forming Minerals with Nonmetallic Luster, *continued*

Mineral	Chemical Composition	Form and Habit	Cleavage	Hardness / Specific Gravity	Other Properties	Most Distinctive Properties
Garnets	$X_3Y_2 (SiO_4)_3$; X = Ca, Mg, Fe, Mn; Y = Al, Fe, Ti, Cr.	Perfect crystals with 12 or 24 sides. Granular masses.	None. Uneven fracture.	6.5–7.5 / 3.5–4.3	Common in metamorphic rocks. Red, brown, yellowish-green, black.	Crystals, hardness, no cleavage.
Graphite	C	Scaly masses.	One, perfect. Forms slippery flakes.	1–2 / 2.2	Metamorphic rocks. Black with metallic to dull luster.	Cleavage, color. Marks paper.
Gypsum	$CaSO_4 \cdot 2H_2O$	Elongate or tabular crystals. Fibrous and earthy masses.	One, perfect. Flakes bend but are not elastic.	2 / 2.3	Vitreous to pearly luster. Colorless.	Hardness, cleavage.
Halite	NaCl	Cubic crystals.	Perfect to give cubes.	2.5 / 2.2	Tastes salty. Colorless, blue.	Taste, cleavage.
Hornblende			See Amphibole			
Kaolinite	$Al_2Si_2O_5 (OH)_4$	Soft, earthy masses. Submicroscopic crystals.	One, perfect.	2–2.5 / 2.6	White, yellowish. Plastic when wet; emits clayey odor. Dull luster.	Feel, plasticity, odor.
Kyanite	Al_2SiO_5	Bladed crystals.	One perfect. One imperfect.	4.5 parallel to blade, 7 across blade. / 3.6	Blue, white, gray. Common in metamorphic rocks.	Variable hardness, distinguishes from sillimanite. Color.
Mica: *Biotite*	$K(Mg, Fe)_3$-$AlSi_3O_{10}$-$(OH)_2$	Irregular masses of flakes.	One, perfect.	2.5–3 / 2.8–3.2	Common in igneous and metamorphic rocks. Black, brown, dark green.	Cleavage, color. Flakes are elastic.
Muscovite	$KAl_3Si_3O_{10} (OH)_2$	Thin flakes.	One, perfect.	2–2.5 / 2.7	Common in igneous and metamorphic rocks. Colorless, pale green or brown.	Cleavage, color. Flakes are elastic.
Olivine	$(Mg, Fe)_2SiO_4$	Small grains, granular masses.	None. Conchoidal fracture.	6.5–7 / 3.2–4.3	Igneous rocks. Olive green to yellowish-green.	Color, fracture, habit.
Orthoclase			See Feldspar			
Plagioclase			See Feldspar			
Pyroxene (A complex family of minerals. *Augite* is is most common.)	$XY(SiO_3)_2$ X = Y = Ca, Mg, Fe	8-sided stubby crystals. Granular masses.	Two, perfect, nearly at right angles.	5–6 / 3.2–3.9	Igneous and metamorphic rocks. *Augite*, dark green to black; other varieties white to green.	Cleavage.
Quartz	SiO_2	6-sided crystals, granular masses.	None. Conchoidal fractures.	7 / 2.6	Colorless, white, gray, but may have any color, depending on impurities. Vitreous to greasy luster.	Form, fracture, striations across crystal faces at right angles to long dimension.
Serpentine (Fibrous variety is *asbestos*.)	$Mg_3Si_2O_5 (OH)_4$	Platy or fibrous.	One, perfect.	2.5–5 / 2.2–2.6	Light to dark green. Smooth, greasy feel.	Habit, hardness.

| **TABLE C.2** | **Properties of Rock-Forming Minerals with Nonmetallic Luster,** *continued* | | | | | |

Mineral	Chemical Composition	Form and Habit	Cleavage	Hardness / Specific Gravity	Other Properties	Most Distinctive Properties
Sillimanite	Al_2SiO_5	Long, needle-like crystals, fibers.	Breaks irregularly, except in fibrous variety.	6–7 / 3.2	White, gray. Metamorphic rocks.	Hardness distinguishes from kyanite. Habit.
Talc	$Mg_3Si_4O_{10}(OH)_2$	Small scales, compact masses.	One, perfect.	1 / 2.6–2.8	Feels slippery. Pearly luster. White to greenish.	Hardness, luster, feel cleavage.
Tourmaline	Complex silicate of B, Al, Na, Ca, Fe, Li and Mg.	Elongate crystals, commonly with triangular cross section.	None.	7–7.5 / 3–3.3	Black, brown, red, pink, green, blue, and yellow. An accessory mineral in many rocks.	Habit.
Wollastonite	$CaSiO_3$	Fibrous or bladed aggregates of crystals.	Two, perfect.	4.5–5 / 2.8–2.9	Colorless, white, yellowish. Metamorphic rocks. Soluble in HCl.	Habit. Solubility in HCl and hardness distinguish amphiboles, kyanite, sillimanite.

| **TABLE C.3** | **Properties of Some Common Gemstones** | | | | | |

Mineral and Variety	Composition	Form and Habit	Cleavage	Hardness / Specific Gravity	Other Properties	Most Distinctive Properties
Beryl: *Aquamarine* (blue) *Emerald* (green) (green) *Golden beryl* (golden-yellow)	$Be_3Al_2Si_6O_{18}$	Six-sided, elongate crystals common.	Weak.	7.5–8 / 2.75	Bluish green, green, yellow, white, colorless. Common in pegmatites.	Form. Distinguished from apatite by its hardness.
Corundum: *Ruby* (red) *Sapphire* (blue)	Al_2O_3	Six-sided, barrel-shaped crystals.	None, but breaks easily across its crystal.	9 / 4	Brown, pink, red, blue, colorless. Common in metamorphic rocks. Star sapphire is opalescent with a six-sided light spot showing.	Hardness.
Diamond	C	Octahedron-shaped crystals.	Perfect, parallel to faces of octahedron.	10 / 3.5	Colorless, yellow; rarely red, orange, green, blue, or black.	Hardness, cleavage.
Garnet: *Almandite* (red) *Grossularite* (green, cinnamon-brown) *Andradite* (variety *demantoid* is green)	A rock-forming mineral — see Table C.2.					
Opal (A mineraloid)	$SiO_2 \cdot nH_2O$	Massive, thin coating. Amorphous.	None. Conchoidal fracture.	5–6 / 2–2.2	Colorless, white, yellow, red, brown, green, gray, opalescent.	Hardness, color, form.

TABLE C.3	Properties of Some Common Gemstones					

Mineral and Variety	Composition	Form and Habit	Cleavage	Hardness / Specific Gravity	Other Properties	Most Distinctive Properties
Quartz: (1) Coarse crystals *Amethyst* (violet) *Cairngorm* (brown) *Citrine* (yellow) *Rock crystal* (colorless) *Rose quartz* (pink) (2) Fine-grained *Agate* (banded, many colors) *Chalcedony* (brown, gray) *Heliotrope* (green) *Jasper* (red)	A rock-forming mineral — see Table C.2.					
Topaz	Al_2SiO_4 $(OH, F)_2$	Prism-shaped crystals, granular masses.	One, perfect.	8 / 3.5	Colorless, yellow, blue, brown.	Hardness, form, color.
Tourmaline	A rock-forming mineral — see Table C.2					
Zircon	$ZrSiO_4$	Four-sided elongate crystals, square in cross section.	None.	7.5 / 4.7	Brown, red, green, blue, black.	Habit, hardness.

GLOSSARY

Some definitions are not included in the glossary; *units of measurement* can be found in Appendix A, *chemical elements* are listed in Appendix B, and *mineral names* are given in Appendix C. Chapter numbers in parentheses indicate the chapter in which a chapter-ending keyterm is first defined.

A horizon. A soil horizon that either underlies the O horizon or is the uppermost horizon. Generally dark colored and characterized by an accumulation of organic matter. (Ch. 5)

Aa. A rubbly, rough-looking form of lava, usually basaltic in composition.

Ablation. The loss of mass from a glacier. (Ch. 12)

Ablation area. A region of net loss on a glacier characterized by a surface of bare ice and old snow from which the last winter's snowcover has melted away.

Absolute velocity (of a plate). The velocity of a plate of lithosphere measured against a fixed, external frame of reference. Compare with *relative velocity*.

Abyssal plain. A large flat area of the deep seafloor having slopes less than about 1 m/km, and ranging in depth below sea level from 3 to 6 km. (Ch. 1)

Accreted terrane. Block of crust moved laterally by strike-slip faulting or by a combination of strike-slip faulting and subduction, then accreted to a larger mass of continental crust. Also called a *suspect terrane*. (Ch. 16)

Accreted terrane continental margin. Continental margin modified by the addition of island arcs and other rafted-in, exotic fragments of crust.

Accretion. The process by which solid bodies gather together to form a planet or a continent.

Accumulation. The addition of mass to a glacier. (Ch. 12)

Accumulation area. An upper zone on a glacier, covered by remnants of the previous winter's snowfall and representing an area of net gain in mass.

Active zone. The margin of a tectonic plate where deformation is occurring.

Agglomerate. The pyroclastic rock consisting of bomb-sized tephra, that is, tephra in which the average particle diameter is greater than 64 mm.

Aggradation. Depositional upbuilding, as by a stream.

Albedo. The reflectivity of the surface of a planet.

Alluvial fan. A fan-shaped body of alluvium typically built where a stream leaves a steep mountain valley. (Ch. 10)

Alluvium. Sediment deposited by streams in non-marine environments. (Ch. 10)

Alpha particle (α-particle). An atomic particle expelled from an atomic nucleus during certain radioactive transformations, equivalent to an $^4_2$He nucleus stripped of its electrons.

Amorphous. A term applied to solids that lack internal atomic order.

Amphibolite. A metamorphic rock of intermediate grade, generally coarse-grained, containing abundant amphibole. (Ch. 7)

Amygdule. A vesicle filled by secondary minerals such as calcite and quartz deposited by groundwater.

Andesite. A fine-grained igneous rock with the composition of a diorite.

Andesite line. A line on a map that roughly surrounds the Pacific Ocean basin and inside of which andesite is not found.

Andesite magma. One of the three common magma types; a magma with an SiO_2 content of about 60 percent by weight.

Angle of repose. The steepest angle, measured from the horizontal, at which rock debris remains stable. (Ch. 8)

Angular unconformity. An unconformity marked by angular discordance between older and younger rocks. (Ch. 8)

Anhydrous. A term applied to a substance that is H_2O free. Opposite of hydrous.

Anion. An ion with a negative electrical charge. (Ch. 13)

Anorthosite. A coarse-grained igneous rock consisting largely of plagioclase.

Antecedent stream. A stream that has maintained its course across an area of the crust that was raised across its path by folding or faulting.

Anthracite. A metamorphic rock derived from coal by heat and pressure.

Anticline. An upfold in the form of an arch. (Ch. 15)

Aphanite. An igneous rock in which the constituent mineral grains are so small they can only be seen clearly by using some kind of magnification. (Ch. 4)

Apparent polar wandering. The apparent motions of the magnetic poles derived from measurements of pole positions using paleomagnetism.

Aquiclude. A body of impermeable or distinctly less permeable rock adjacent to an *aquifer*. (Ch. 11)

Aquifer. A body of permeable rock or regolith saturated with water and through which groundwater moves. (Ch. 11)

Archean Eon. The eon that follows the Hadean Eon.

Arête. A jagged, knife-edged ridge crest created where glaciers have eroded back into a ridge.

Arkose (-arkosic sandstone). A sandstone in which feldspar is a major mineral component.

Artesian aquifer. An aquifer in which water is under hydraulic pressure. (Ch. 11)

Artesian spring. A natural spring that draws its supply of water from an artesian aquifer.

Artesian system. An inclined aquifer that permits water confined in it to rise to the surface in a well or along a fissure.

Artesian well. A well in which water rises above the aquifer.

Ash. Tephra in which particles have an average diameter less than 2 mm. Also called *volcanic ash*.

Ash tuff. Pyroclastic rock in which the tephra particles are less than 2 mm in diameter.

Asphalt. See *tar*.

Asteroids. Irregularly shaped rocky bodies that have orbits lying between the orbits of Mars and Jupiter. (Ch. 21)

Asthenosphere. The region of the mantle where rocks become ductile, have little strength, and are easily deformed. It lies at a depth of 100 to 350 km below the surface. (Ch. 1)

Asymmetric fold. A fold in which one limb dips more steeply than the other.

Atmosphere. The mixture of gases, predominantly nitrogen, oxygen, carbon dioxide, and water vapor, that surrounds the Earth. (Ch. 1)

Atoll. A coral reef, often roughly circular in plan, that encloses a shallow lagoon. (Ch. 14)

Atom. The smallest individual particle that retains the distinctive properties of a given chemical element. (Ch. 13)

Atomic number. The number of protons in the nucleus of an atom. (Ch. 3)

Atomic substitution. See *ionic substitution*.

Axial plane. An imaginary plane that divides a fold as symmetrically as possible, and that passes through the axis.

Axis (of a fold). The median line between the limbs, along the crest of an anticline or the trough of a syncline. (Ch. 15)

B horizon. A soil horizon generally lying below an A horizon, usually brownish or reddish in color, and commonly enriched in clay and iron oxides. (Ch. 15)

Back-arc basin. An arc-shaped basin formed by crustal thinning behind a magmatic arc. (Ch. 16)

Backshore. A zone extending inland from a berm to the farthest point reached by waves.

Backwash. The seaward return of water down a beach following the swash of a wave.

Bajada. A broad alluvial apron composed of coalescing adjacent fans. (Ch. 13)

Bar. An accumulation of alluvium formed in a channel where a decrease in stream velocity causes deposition.

Barchan dune. A crescent-shaped sand dune with horns pointing downwind.

Barrier island. A long island built of sand, lying offshore and parallel to the coast. (Ch. 4)

Barrier reef. A reef separated from the land by a lagoon. (Ch. 14)

Basalt. A fine-grained igneous rock with the composition of a gabbro.

Basaltic magma. One of the three common types of magma; contains about 50 percent SiO_2 by weight.

Base level. The limiting level below which a stream cannot erode the land. (Ch. 10)

Batholith. The largest kind of pluton. A very large, igneous body of irregular shape that cuts across the layering of the rock it intrudes. (Ch. 4)

Bauxite. An aluminous laterite formed by tropical weathering. The preferred ore of aluminum. (Ch. 5)

Bay. A wide, open, curving indentation or inlet of a sea or lake into an adjacent land mass.

Bay barrier. A ridge of sand or gravel that completely blocks the mouth of a bay.

Beach. Wave-washed sediment along a coast, extending throughout the surf zone. (Ch. 14)

Beach drift. The irregular movement of particles along a beach as they travel obliquely up the slope of a beach with the swash and directly down this slope by the backwash. (Ch. 14)

Beach ridge. A low ridge of sand parallel to and on the landward side of a beach.

Bed. The smallest formal unit of a body of sediment or sedimentary rock. (Ch. 6)

Bedding. The layered arrangement of strata in a body of sediment or sedimentary rock. (Ch. 6)

Bedding plane. The top or bottom of a bed. (Ch. 6)

Bed load. Coarse particles that move along the bottom of a stream channel. (Ch. 10)

Bedrock. The continuous mass of solid rock that makes up the crust.

Benioff strain seismograph. See *strain seismograph.*

Benioff zone. A narrow, well-defined zone of deep earthquake foci beneath a seafloor trench. (Ch. 16)

Berm. A nearly horizontal or landward-sloping bench formed of sediment deposited by waves.

Beta particle. (β-particle). An electron expelled from an atomic nucleus during certain radioactive transformations.

Biogenic rock. Rock formed by lithification of biogenic sediment.

Biogenic sediment. Sediment composed mainly of fossil remains. (Ch. 6)

Biogeochemical cycles. Natural cycles describing the movements and interactions through the Earth's spheres of the chemicals essential to life. (Ch. 2, 19)

Biomass energy. The energy obtained through burning plant matter.

Biosphere. The totality of the Earth's organisms and, in addition, organic matter that has not yet been completely decomposed. (Ch. 1)

Bituminous coal. The highest grade of coal.

Blueschist. A metamorphic rock formed under conditions of high pressure and low temperature containing blue-colored amphiboles.

Body waves. Seismic waves that travel outward from an earthquake focus and pass through the Earth. (Ch. 16)

Bolide. An impacting body, either a meteorite, an asteroid, or a comet. (Ch. 21)

Bombs. Tephra particles having average diameters greater than 64 mm.

Bonding. The electrostatic forces that hold atoms together to form compounds by sharing and transfer of electrons. See also *covalent bonding, ionic bonding, metallic bonding* and *van der Waals bond.*

Bottomset layer. A gently sloping, fine, thin part of each layer in a delta.

Boundary conditions. A mathematical expression of the physical state of the Earth's climate system at the period of interest for a climate-simulation experiment. (Ch. 19)

Bowen's reaction series. A schematic description of the order in which different minerals crystallize during the cooling and progressive crystallization of a magma. See also *continuous* and *discontinuous reaction series.*

Braid delta. A delta composed of coarse-grained sediment built by a braided stream into a standing body of water.

Braided stream. A channel system consisting of a tangled network of two or more smaller branching and reuniting channels that are separated by islands or bars. (Ch. 10)

Breaker. An oversteepened wave that collapses in a mass of turbulent water against a shore or reef.

Breakwater. An offshore barrier built to protect a beach or anchorage from incoming waves. (Ch. 14)

Breccia. A coarse-grained rock composed of cemented angular fragments.

Breeder reactor. A nuclear reactor in which nonfissionable isotopes such as ^{238}U are converted to fissionable isotopes. (Ch. 20)

Brittle fracture. Rupture of a solid body that is stressed beyond its elastic limit.

Burial metamorphism. Metamorphism caused solely by the burial of sedimentary or pyroclastic rocks. (Ch. 7)

Butte. Isolated, often flat-topped, steepsided, desert hill. Smaller than a *mesa.*

C horizon. The deepest soil horizon, lying beneath the A horizon and/or B horizon of a soil profile; often yellowish-brown in color and consisting of weathered parent rock or sediment (Ch. 5).

Calcareous ooze. A deep-sea pelagic sediment composed largely of calcareous skeletal remains. See also *deep-sea ooze.*

Caldera. A roughly circular, steep-walled volcanic basin several kilometers or more in diameter. (Ch. 4)

Caliche. A solid, almost impervious layer of whitish calcium carbonate in a soil profile. (Ch. 5)

Calving. The progressive breaking off of icebergs from a glacier that terminates in deep water. (Ch. 12)

Capillary attraction. The adhesive force between a liquid, such as water, and a solid.

Carbonate shelf. A shallow marine shelf where sedimentation is dominated by carbonate-secreting organisms.

Carbonic acid. A weak acid resulting from the solution of small quantities of carbon dioxide in rain or groundwater. (Ch. 5)

Cataclastic metamorphism. Metamorphism that involves change of texture caused by mechanical effects such as crushing and shearing, but no change in mineral assemblage. (Ch. 7)

Catastrophism. The concept that all of the Earth's major features, such as mountains, valleys, and oceans, have been produced by a few great catastrophic events. (Ch. 1)

Cation. A positive ion. (Ch. 3)

Cave. A natural underground opening, generally connected to the surface and large enough for a person to enter.

Cavern. A large cave or system of interconnected cave chambers.

Celsius scale. A temperature scale in which the boiling point of water is 100° and the freezing point is 0°.

Cementation. The diagenetic process by which clastic sediments are converted to rock through deposition or precipitation of minerals in the spaces between the grains. (Ch. 6)

Cenozoic era. The youngest era of the Phanerozoic Eon. (Ch. 8)

Central rift valley. A long, narrow valley at the crest of a midocean ridge.

Chalk. Compacted carbonate shells of minute floating organisms.

Channel. The passageway in which a stream flows. (Ch. 10)

Chemical differentiation by partial melting.

Chemical elements. The most fundamental substances into which matter can be separated by chemical means.

Chemical sediment. Sediment formed by precipitation of minerals from solutions in water. (Ch. 6)

Chemical weathering. The decomposition of rocks through chemical reactions such as hydration and oxidation. (Ch. 5)

Chloroflurocarbons. Synthetic industrial gases that

destroy ozone in the upper atmosphere and contribute to the greenhouse effect. Also called CFCs.

Chrons. See *magnetic chrons.*

Cirque. A bowl-shaped hollow on a mountainside, open downstream, bounded upstream by a steep slope (headwall), and excavated mainly by frost wedging and by glacial abrasion and plucking. (Ch. 12)

Cirque glacier. A glacier that occupies a bowl-shaped hollow on the side of a mountain.

Clast. Any individual particle of clastic sediment.

Clastic sediment. See *detritus.*

Cleavage. The tendency of a mineral to break in preferred directions along bright, reflective plane surfaces. (Ch. 3)

Climate. The average weather conditions of a place or area over a period of years. (Ch. 13)

Closed system. Any system with a boundary that allows the passage in or out of energy but not of matter. (Ch. 1)

Coal. A black, combustible, sedimentary or metamorphic rock consisting chiefly of decomposed plant matter and containing more than 50 percent organic matter. (Ch. 4)

Coalification. The stages by which plant matter is converted first to peat, then lignite, subbituminous coal, and bituminous coal.

Col. A gap or pass in a mountain crest where the headwalls of two cirques intersect.

Collision zone. A convergent plate margin where two plates collide . (Ch. 4)

Colluvium. Loose, incoherent deposits on or at the base of slopes and moving mainly by creep. (Ch. 8)

Column. A stalactite joined with a stalagmite, forming a connection between the floor and roof of a cave.

Columnar joints. Joints that split igneous rocks into long prisms or columns.

Comet. A small celestial body that circles the Sun with a highly elliptical orbit.

Compaction. A decrease in porosity and bulk of a body of sediment as additional sediment is deposited above it, or due to pressures resulting from deformation. (Ch. 6)

Complex ion. A strongly bonded pair of ions that act in the same way as a single ion, forming compounds by bonding with other elements. (Ch. 3)

Composition (of a mineral). The proportions of the various chemical elements in a mineral. (Ch. 3)

Compound. A combination of atoms of different elements bonded together.

Compressional stress. Differential stress that squeezes and compresses a body. (Ch. 15)

Compressional waves. See *P waves.*

Conchoidal fracture. Breakage resulting in smooth, curved surfaces.

Concretion. A hard, localized body, having distinct boundaries, enclosed in sedimentary rock, and consisting of a substance precipitated from solution, commonly around a nucleus.

Conduction. The means by which heat is transmitted through solids without deforming the solid. (Ch. 1)

Cone of depression. A conical depression in the water table immediately surrounding a well. (Ch. 11)

Confined aquifer. An aquifer bounded by aquicludes. (Ch. 11)

Confining stress. See *uniform stress.*

Conformity. Strata that have been deposited layer upon layer without interruption. (Ch. 8)

Conglomerate. A sedimentary rock composed of clasts of rounded gravel set in a finer-grained matrix. (Ch. 4)

Consequent stream. A stream whose pattern is determined solely by the direction of slope of the land.

Contact metamorphism. (also called *thermal metamorphism*). Metamorphism adjacent to an intrusive igneous rock. (Ch 7)

Continental collision margin. A plate margin along which two continental masses collide.

Continental convergent margin. The margin of a continent that is adjacent to a subduction zone.

Continental crust. The part of the Earth's crust that comprises the continents, which has an average thickness of 45 km. (Ch. 1)

Continental divide. A line that separates streams flowing towards opposite sides of a continent, usually into different oceans. (Ch. 18)

Continental drift. The slow, lateral movements of continents across the surface of the Earth. (Ch. 16)

Continental rise. A region of gently changing slope where the floor of the ocean basin meets the margin of a continent. (Ch. 2)

Continental shelf. A submerged platform of variable width that forms a fringe around a continent. (Ch. 2)

Continental shield. An assemblage of cratons and orogens that has reached isostatic equilibrium. (Ch. 16)

Continental slope. A pronounced slope beyond the seaward margin of the continental shelf. (Ch. 2)

Continental volcanic arc. An arcuate chain of andesitic volcanoes on the continental crust formed as a result of subduction. (Ch. 16)

Continuous reaction series. The continuous change of mineral composition, through ionic substitution, as a magma crystallizes See also *discontinuous reaction series.*

Convection. The process by which hot, less dense materials rise upward, being replaced by cold, more dense, downward-flowing material to create a convection current. (Ch. 1)

Convection current. The flow of material as a result of convection. (Ch. 1)

Convergent margin. The zone where plates meet as they move toward each other. See also *subduction zone.* (Ch. 2)

Coquina. A limestone composed solely or chiefly of loosely aggregated shells and shell fragments.

Core. The spherical mass, largely metallic iron, at the center of the Earth. (Ch. 1)

Coriolis effect. An effect that causes any body that moves freely with respect to the rotating solid Earth to veer toward the right in the northern hemisphere and toward the left in the southern hemisphere, regardless of the initial direction of the moving body. (Ch. 13)

Correlation of strata. Determination in time stratigraphic age of the succession of strata found in two or more different areas. (Ch. 8)

Covalent bonding. The force between two atoms that have filled their energy-level shells by sharing one or more electrons. (Ch. 3)

Crater. A funnel-shaped depression, opening upward, at the top of a volcano from which gases, fragments of rock, and lava are ejected. (Ch. 4)

Craton. A core of ancient rock in the continental crust that has attained tectonic and isostatic stability. (Ch. 16)

Creep. The imperceptibly slow down slope movement of regolith. (Ch. 8)

Creep of glacier ice. Slow deformation of glacier ice, with movement occurring along the internal planes of ice crystals.

Crevasse. A deep, gaping fissure in the upper surface of a glacier. (Ch. 12)

Cross bedding. Beds that are inclined with respect to a thicker stratum within which they occur. (Ch. 6)

Cross section. See *geologic cross section.*

Crust. The outermost and thinnest of the Earth's compositional layers, which consists of rocky matter that is less dense than the rocks of the mantle below. (Ch. 1)

Cryosphere. The portion of the hydrosphere that is ice, snow, and frozen ground.

Crystal. A solid compound composed of ordered, three-dimensional arrays of atoms or ions chemically bonded together and displaying crystal form. (Ch. 3)

Crystal faces. The planar surfaces that bound a crystal. (Ch. 3)

Crystal form. The geometric arrangement of crystal faces. (Ch. 3)

Crystalline. See *crystal structure.*

Crystal settling. The process of fractional crystallization by which dense minerals sink and form segregated layers of one or more minerals in a magma chamber.

Crystal structure. The geometric pattern that atoms assume in a solid. Any solid that has a crystal structure is said to be *crystalline.* (Ch. 3)

Curie point. A temperature above which permanent magnetism is not possible. (Ch. 8)

Dacite. A fine-grained igneous rock with the composition of a granodiorite.

Darcy's law. The relationship between discharge, coefficient of permeability, and hydraulic gradient in percolating groundwater. (Ch. 11)

Daughter product. (-daughter). The product arising from radioactive decay Compare *parent*. (Ch. 8)

Debris avalanche. A granular flow of regolith moving at high velocity (≥10 m/s). (Ch. 8)

Debris fall. The relatively free fall or collapse of regolith from a steep cliff or slope. (Ch. 8)

Debris flow. The downslope movement of a mass of unconsolidated regolith more than half of which is coarser than sand. (Ch. 8)

Debris slide. The slow to rapid downslope movement of regolith across an inclined surface. Compare *rockslide*. (Ch. 8)

Décollement. A body of rock above the detachment surface of a thrust fault. (Ch. 16)

Decomposition. (of rocks). Chemical weathering.

Deep-sea fan. Huge fan-shaped body of sediment at the base of the continental slope that spreads downward and outward to the deep seafloor. (Ch. 6)

Deep-sea ooze. A muddy marine sediment composed mainly of the remains of microscopic marine organisms. See also *calcareous ooze and siliceous ooze*. (Ch. 6)

Deflation. The picking up and removal of loose particles by wind. (Ch. 13)

Dehydrate. The loss of water. (Ch. 5)

Delta. A body of sediment deposited by a stream where it flows into standing water. (Ch. 10)

Density. The average mass per unit volume.

Denudation. The sum of the weathering, mass-wasting, and erosional processes that result in the progressive lowering of the Earth's surface. (Ch. 18)

Desert. Arid land, whether "deserted" or not, in which annual rainfall is less than 250 mm (10 in) or in which the evaporation rate exceeds the precipitation rate. (Ch. 13)

Desertification. The invasion of desert into non-desert areas. (Ch. 13)

Desert pavement. A surface layer of coarse particles concentrated chiefly by deflation. (Ch. 13)

Desert varnish. A thin, dark, shiny coating consisting mainly of manganese and iron oxides, formed on the surfaces of stones and rock outcrops in desert regions after long exposure. (Ch. 13)

Detachment surface. The surface along which a large-scale thrust fault moves. (Ch. 16)

Detrital sediment. The accumulated particles of broken rock and skeletal remains of dead organisms (also called *clastic sediment*). (Ch. 6)

Detritus. (also called *clastic sediment* and *detrital sediment*). The loose fragmented debris produced by the mechanical breakdown of older rocks. (Ch. 16)

Diagenesis. Chemical, physical, and biological changes that affect sediment after its initial deposition and during and after its slow transformation into sedimentary rock. (Ch. 6)

Diatomite. A sedimentary rock formed by lithification of *siliceous ooze*. (Ch. 14)

Differential stress. Stress in a solid that is not equal in all directions. (Ch. 7)

Differential weathering. Weathering that occurs at different rates or intensity as a result of variations in the composition and structure of rocks.

Dike. Tabular, parallel-sided sheets of intrusive igneous rock that cut across the layering of the intruded rock. (Ch. 4)

Diorite. A coarse-grained igneous rock consisting mainly of plagioclase and ferromagnesian minerals. Quartz is sparse or absent.

Dip. The angle in degrees between a horizontal plane and an inclined plane, measured down from horizontal in a plane perpendicular to the strike. (Ch. 15)

Dip-slip fault. A normal or reverse fault on which the only component of movement lies in a plane normal to the strike of the fault surface.

Discharge. The quantity of water that passes a given point in a stream channel per unit time. (Ch. 10)

Discharge area. Area where subsurface water is discharged to streams or to bodies of surface water. (Ch. 11)

Disconformity. An irregular surface of erosion between parallel strata. (Ch. 8)

Discontinuous reaction series. The discontinuous sequence of reactions by which early formed minerals in a crystallizing magma react with residual liquid to form new minerals. See also *continuous reaction series*.

Dispersion. Waves of different wavelengths traveling at different velocities.

Dissolution. The chemical weathering process whereby minerals and rock material pass directly into solution. (Ch. 5)

Dissolved load. Matter dissolved in stream water. (Ch. 10)

Divergent margin (of a plate). A fracture in the lithosphere where two plates move apart. Also called a *spreading center*. (Ch. 2)

Divide. The line that separates adjacent drainage basins. (Ch. 10)

Dolostone. A sedimentary rock composed chiefly of the mineral dolomite.

Drainage basin. The total area that contributes water to a stream. (Ch. 10)

Drift. See *glacial drift*.

Dripstone. A deposit chemically precipitated by dripping water in an air-filled cavity. (Ch. 11)

Dropstone. A stone released from a melting iceberg that plunges into unconsolidated sediment on the seafloor or a lake bottom.

Drumlin. A streamlined hill consisting of glacially deposited sediment and elongated parallel with the direction of ice flow. (Ch. 12)

Ductile deformation. The irreversible deformation induced in a solid that is stressed beyond its elastic limit but before rupture occurs. (Ch. 15)

Dune. A mound or ridge of sand deposited by wind. (Ch. 13)

E horizon. Soil horizon, sometimes present below the A horizon, that is grayish or whitish in color.

Earthflow. A granular flow of regolith with velocities ranging from 10^{-5} to 10^{-1} m/s. (Ch. 8)

Earthquake focus. The point of the first release of energy that causes an earthquake. (Ch. 16)

Earthquake magnitude. See *Richter magnitude scale*.

Earth's gravity. An inward-acting force with which the Earth tends to pull all objects toward its center.

Eccentricity (of Earth's orbit). The degree to which the shape of the Earth's orbit departs from perfect circularity.

Ecliptic. Plane of the Earth's orbit around the Sun. (Ch. 21)

Eclogite. A metamorphic rock containing garnet and jadeitic pyroxene.

Ejecta blanket. Layer of broken rock surrounding an impact crater.

Elastic deformation. The reversible or nonpermanent deformation that occurs when an elastic solid is stretched and squeezed and the force is then removed. (Ch. 15)

Elastic limit. The limiting stress beyond which a body suffers irreversible deformation. (Ch. 15)

Elastic rebound theory. The theory that earthquakes result from the release of stored elastic energy by slippage on faults. (Ch. 16)

Electrons. Negatively charged atomic particles.

Emergence of a coast. An increase in the area of land exposed above sea level resulting from uplift of the land and/or fall of sea level. (Ch. 14)

End moraine. A ridgelike accumulation of drift deposited along the margin of a glacier.

Energy-level shell. The specific energy level of electrons as they orbit the nucleus of an atom. (Ch. 3)

Eolian. Pertaining to the wind, especially erosional and depositional processes, as well as landforms and sediments resulting from wind action. (Ch. 6)

Eon. The largest interval of geologic time. We are now in the fourth eon. (Ch. 8)

Epicenter. That point on the Earth's surface that lies vertically above the focus of an earthquake. (Ch. 16)

Epidote amphibolite. A metamorphic rock containing both amphibole and epidote as major constituents.

Epoch. The time during which a geologic series accumulates. (Ch. 8)

Equilibrium line. A line that marks the level on a glacier where net mass loss equals net gain. (Ch. 12)

Era. The primary time division of eons. (Ch. 8)

Erosion. The complex group of related processes by which rock is broken down physically and chemically and the products are moved. (Ch. 1)

Erratic. A glacially deposited rock fragment whose composition differs from that of the bedrock beneath it. (Ch. 12)

Eruption column. A mixture of ash and hot gases that rises upward as a column above an erupting volcano.

Esker. A long narrow ridge, often sinuous, composed of stratified drift.

Estuary. A semienclosed body of coastal water within which seawater is diluted with fresh water. (Ch. 6)

Evaporite. Sedimentary rock composed chiefly of minerals precipitated from a saline solution through evaporation. (Ch. 6)

Evaporite deposits. Layers of salts that precipitate as a consequence of evaporation.

Exfoliation. The spalling off of successive shells, like the "skins" of an onion, around a solid rock core. (Ch. 5)

Exploration geology. The special branch of geology concerned with discovering new supplies of usable minerals. (Ch. 20)

Exposure (also called an *outcrop*). A place where rock or sediment is exposed at the Earth's surface.

External processes. All the activities involved in erosion and in the transport and deposition of the eroded materials.

Extraterrestrial material. Material originating outside the Earth.

Extrusive igneous rock. Rock formed by the solidification of magma poured out onto the Earth's surface. (Ch. 4)

Facies. A distinctive group of characteristics, within a rock unit, that differs as a group from those elsewhere in the same unit. See also *sedimentary facies* and *metamorphic facies*. (Ch. 6)

Fan. See *alluvial fan*.

Fan delta. A gravel-rich delta formed where an alluvial fan builds outward into a standing body of water.

Fault. A fracture in a rock along which movement occurs. (Ch. 15)

Fault breccia. Crushed and broken rock adjacent to a fault.

Ferromagnesian minerals. The common rock-forming minerals that contain iron and/or magnesium as essential constituents.

Fiord. See *fjord*.

Fission. Controlled radioactive transformation. (Ch. 20)

Fissure eruption. Extrusion of lava or pyroclasts and associated gases along an extended fracture.

Fjord. A deep, glacially carved valley submerged by the sea. Also spelled *fiord*. (Ch. 12)

Fjord glacier. A glacier that occupies a fjord.

Flash flood. A local and sudden flood of water through a stream channel, generally of relatively great volume and short duration. (Ch. 13)

Flood. A discharge great enough to cause a stream to overflow its banks.

Floodplain. The part of any stream valley that is inundated during floods. (Ch. 10)

Flowstone. A deposit chemically precipitated from flowing water in the open air or in an air-filled cavity. (Ch. 11)

Fluvial. Of, or pertaining to, streams or rivers, especially erosional and depositional processes of streams and the sediments and landforms resulting from them.

Foliation. The planar texture of mineral grains, principally micas, produced by metamorphism. (Ch. 7)

Fold. An individual bend or warp in layered rock. (Ch. 15)

Folding. The bending of rocks or sediments.

Footwall block of a fault. The block of rock below an inclined fault. (Ch. 15)

Fore-arc basin. A basin parallel to a deep-sea trench and separated from the trench by a *fore-arc ridge*. (Ch. 16)

Fore-arc ridge. See *fore-arc basin*.

Foreset layer. The coarse, thick, steeply sloping part of each layer in a delta.

Foreshore. A zone extending from the level of lowest tide to the average high-tide level.

Formation. A body of rock distinctive enough on the basis of physical properties to constitute a basic unit for geologic mapping. The basic unit of rock stratigraphy. (Ch. 8)

Fossil. The naturally preserved remains or traces of an animal or a plant. (Ch. 6)

Fossil fuel. Remains of plants and animals trapped in sediment that may be used for fuel. (Ch. 6)

Fracture. Irreversible deformation of a rock in which the limits of both elastic and ductile deformation have been exceeded; breakage. (Ch. 15)

Fringing reef. A coral reef attached to or bordering the adjacent land. (Ch. 14)

Frost heaving. The lifting of regolith by the freezing of water contained within it. (Ch. 8)

Frost wedging. The formation of ice in a confined opening within rock, thereby causing the rock to be forced apart. (Ch. 5)

Fumarole. A volcanic vent that emits only gases.

Gabbro. A coarse-grained igneous rock in which olivine and pyroxene are the predominant minerals and plagioclase is the feldspar present. Quartz is absent.

Gamma rays (γ-rays). Very short wavelength electromagnetic radiation given off by an atomic nucleus during certain radioactive transformations.

Gangue. The nonvaluable minerals of an ore. (Ch. 20)

Garnet peridotite. A coarse-grained igneous rock consisting largely of olivine, garnet, and pyroxene.

Gelifluction. Downslope movement of the thawed surface layer of regolith in a region of perennially frozen ground. (Ch. 8)

General Circulation Model (GCM). A mathematical model used to simulate present and past climate conditions on the Earth. (Ch. 19)

Geochemically abundant elements. Those chemical elements that individually comprise 0.1 percent or more by weight of the crust. (Ch. 17)

Geochemically scarce elements. Those chemical elements that individually comprise less than 0.1 percent by weight of the crust. (Ch. 20)

Geographic cycle. A hypothesis of landscape evolution proposed by W.M. Davis in which an uplifted land area passes through sequential stages of development as it is eroded down to a surface of low relief. (Ch. 18)

Geologic column. A composite diagram combining in chronological order the succession of known strata, fitted together on the basis of their fossils or other evidence of relative or actual age. (Ch. 8)

Geologic cross section. A diagram showing the arrangement of rocks in a vertical plane.

Geologic map. A map that shows the distribution, at the surface, of rocks of various kinds or of various ages.

Geologic time scale. A sequential arrangement of geologic time units, as currently understood.

Geologists. Scientists who study the Earth. (Ch. 1)

Geology. The science of the Earth. (Ch. 1)

Geosphere. All of the solid Earth. (Ch. 1)

Geosyncline. A great trough that has received thick deposits of sediment during its slow subsidence through long geologic periods.

Geothermal gradient. The rate of increase of temperature downward in the Earth. (Ch. 1)

Geothermal power. Heat energy drawn from the Earth's internal heat.

Geyser. A thermal spring equipped with a system of plumbing and heating that causes intermittent eruptions of water and steam.

Glacial drift. Sediment deposited directly by glaciers or indirectly by meltwater in streams, in lakes, and in the sea. Also called *drift*. (Ch. 12)

Glacial grooves. See *glacial striations*.

Glacialmarine drift (also referred to as glacial-marine sediment). Terrigenous sediment dropped onto the seafloor from floating ice shelves or from icebergs. (Ch. 12)

Glacial striations. Subparallel scratches inscribed on a clast or a bedrock surface by rock debris embedded in the base of a glacier. Wider and deeper markings on bedrock are *glacial grooves*.

Glaciation. The modification of the land surface by the action of glacier ice. (Ch. 12)

Glacier. A permanent body of ice, consisting largely of recrystallized snow, that shows evidence of downslope or outward movement, due to the stress of its own weight. (Ch. 12)

Gneiss. A high-grade metamorphic rock, always coarse-grained and foliated, with marked compositional layering but with imperfect cleavage. (Ch. 7)

Gondwanaland. The southern half of Pangaea, consisting of present-day Australia, India, Madagascar, Africa, Antarctica, and South America.

Graben (also called a *rift*). A trenchlike structure bounded by parallel normal faults. See also *half-graben*. (Ch. 15)

Grade. A term for the level of concentration of a metal in an ore. Usually expressed as a percentage.

Graded bed. A layer in which the size of the sedimentary particles grades upward from coarse to finer. (Ch. 6)

Graded stream. A stream in which the slope has become so adjusted, under conditions of available discharge and prevailing channel characteristics, that the steam is just able to transport the sediment load available to it.

Gradient. A measure of the vertical drop over a given horizontal distance. (Ch. 10)

Grain flow. Mass-wasting of dry or nearly dry granular sediment with air filling the pore space.

Granite. A coarse-grained igneous rock containing quartz and feldspar, with potassium feldspar being more abundant than plagioclase.

Granitic. Any coarse-grained igneous or metamorphic rock having a texture and composition resembling that of a granite.

Granodiorite. A coarse-grained igneous rock resembling a granite, in which plagioclase is more abundant than potassium feldspar.

Granular flow. A type of flow in which the weight of the flowing mass is supported by grain-to-grain contact or repeated collision between grains. (Ch. 8)

Granular texture. The interlocking arrangements of mineral grains in granitic rocks.

Granulite. A high-grade metamorphic rock, usually coarse-grained and indistinctly foliated, containing pyroxenes as a major mineral.

Gravimeter (also called a *gravity meter*). A sensitive device for measuring the pull of gravity at any locality. (Ch. 16)

Gravity anomalies. Variations in the pull of gravity after correction for latitude and altitude. (Ch. 16)

Gravity meter. See *gravimeter*.

Greenhouse effect. The property of the Earth's atmosphere by which long-wavelength heat rays from the Earth's surface are trapped or reflected back by the atmosphere. (Ch. 19)

Greenhouse gases. The gases in the atmosphere, mainly H_2O, CO_2, CFCs, and CH_4, that cause the greenhouse effect.

Greenschist. A low-grade metamorphic rock rich in chlorite.

Greywacke. See *lithic sandstone*.

Groin. A low wall, built on a beach, that crosses the shoreline at a right angle. (Ch. 14).

Groundmass. The fine-grained matrix of a porphyry.

Ground moraine. Widespread drift with a relatively smooth surface topography consisting of gently undulating knolls and shallow closed depressions.

Groundwater. All the water contained in the spaces within bedrock and regolith. (Ch. 11)

Growth habit. A characteristic growth form of a mineral.

Gyre. A large subcircular current system. (Ch. 14)

Hadean eon. The oldest eon. (Ch. 8)

Half-graben. A trenchlike structure formed when the hanging-wall block moves downward on a curved fault surface. See *graben*. (Ch. 15)

Half-life. The time required to reduce the number of parent atoms by one-half as a result of radioactive decay. (Ch. 8)

Hand specimen. A rock sample of convenient size to hold in the hand for study.

Hanging valley. A glacial valley whose mouth is at a relatively high level on the steep side of a larger glaciated valley.

Hanging-wall block (of a fault). The block of rock above an inclined fault. (Ch. 15)

Hardness. Relative resistance of a mineral to scratching. (Ch. 3)

Hard water. Water containing an unusually high amount of calcium carbonate.

Headwall. The steep cliff that bounds the upslope side of a cirque.

Heat. The energy a body has due to the motions of its atoms.

Heat energy. The energy of a hot body.

Heat flow. The outward flow of heat from the Earth's interior.

High grade metamorphism. Metamorphism under conditions of high temperature and high pressure. (Ch. 7)

Highlands. See *lunar highlands*.

Hinge fault. A fault on which displacement dies out perceptibly along strike and ends at a definite point.

Historical geology. Study of the chronology of the Earth's past events, both physical and biological. (Ch. 1)

Homogeneous stress. See *uniform stress*.

Horn. A sharp-pointed peak bounded by the intersecting walls of three or more cirques.

Hornfels. A hard, fine-grained rock developed during contact metamorphism of a shale. (Ch. 7)

Horst. An elevated elongate block of crust bounded by parallel normal faults. See also *graben*. (Ch. 15)

Hot spot. A fixed point on the Earth's surface defined by long-lived volcanism. (Ch. 16)

Humus. The decomposed residue of plant and animal tissues.

Hurricane. A tropical cyclonic storm having winds that exceed 120 km/h. See also *typhoon*.

Hydration. The incorporation of water into a crystal structure. (Ch. 5)

Hydraulic gradient. The slope of the water table. (Ch. 11)

Hydrocarbons. Any organic compound (gaseous, liquid, or solid) consisting wholly of carbon and hydrogen. (Ch. 6)

Hydroelectric power. Energy recovered from the potential energy of rivers as they flow downward to the sea.

Hydrologic cycle. The day-to-day and longterm cyclic changes in the hydrosphere. (Ch. 2)

Hydrolysis. A chemical reaction in which the H^+ or OH^- ions of water replace ions of a mineral. (Ch. 5)

Hydrosphere. The totality of the Earth's water, including the oceans, lakes, streams, water underground, and all the snow and ice, including glaciers. (Ch. 1)

Hydrothermal mineral deposit. Any local concentration of minerals formed by deposition from a hydrothermal solution. (Ch. 20)

Hydrothermal solutions. Hot brines either given off by cooling magmas, or produced by reactions between hot rock and circulating water, that concentrate minerals in solutions. (Ch. 7)

Hydrous. A term applied to substances that contain H_2O or (OH).

Hypothesis. An unproved explanation for the way things happen. (Ch. 1)

Iapetus. Name given to the ocean that disappeared when North America and Europe collided during the Paleozoic Era.

Ice cap. A dome-shaped body of ice and snow that covers a mountain highland, or lower-lying land at high latitude, and that displays generally radial outward flow.

Ice-contact stratified drift. Stratified sediment deposited in contact with supporting glacier ice. (Ch. 12)

Ice field. A broad, nearly level area of glacier ice in a mountainous region consisting of many interconnected mountain glaciers.

Ice sheet. A continent-sized mass of ice that overwhelms nearly all the land surface within its margin.

Ice shelf. Thick glacier ice, connected to glaciers on land, that floats on the sea and commonly is located in large coastal embayments at high latitudes.

Igneous rock. Rock formed by the cooling and consolidation of magma. (Ch. 2)

Ignimbrite. The poorly sorted mass of tephra deposited by a pyroclastic flow. See also *welded tuff*.

Impact cratering. The process by which a planetary surface is deformed as a result of a transfer of energy from a bolide to the planetary surface. (Ch. 21)

Index fossil. A fossil that can be used to identify and date the strata in which it is found and is useful for local correlation of rock units.

Index mineral. A mineral whose first appearance marks the outer limits of a specific zone of metamorphism.

Inertia. The resistance a large mass has to sudden movement. (Ch. 16)

Inertial seismograph. A device for measuring earthquake waves based on inertia of a mass suspended on a sensitive spring.

Inner core. The central, solid portion of the Earth's core. (Ch. 1)

Inorganic compound. Chemical compounds that do not consist largely of the elements carbon and hydrogen.

Inselberg. Steep-sided mountain, ridge, or isolated hill rising abruptly from adjoining monotonously flat plains. (Chs. 13, 18)

Intergranular fluids. The fluids, both liquid and gas, that fill the tiny pore spaces in a rock. (Ch. 7)

Intermediate grade of metamorphism. Metamorphism under conditions of intermediate pressures and temperatures.

Internal processes. All activities involved in movement or chemical and physical change of rocks in the Earth's interior.

Intrusive igneous rocks. Any igneous rock formed by solidification of magma below the Earth's surface. (Ch. 4)

Ion. An atom that has excess positive or negative charges caused by electron transfer. (Ch. 3)

Ionic bonding. The electrostatic attraction between negatively and positively charged ions. (Ch. 3)

Ionic radius. The distance from the center of the nucleus to the outermost shell of the orbiting electrons. (Ch. 3)

Ionic substitution. (also called *atomic substitution*). The substitution of one ion for another in a random fashion throughout a crystal structure. (Ch. 3)

Island arc. An arcuate chain of andesitic stratovolcanoes sitting on oceanic crust, parallel to a seafloor trench, and separated from it by a distance of 150-300 km. (Ch. 16)

Isoclinal fold. A fold in which both limbs are parallel.

Isograd. A line on a map connecting points of first occurrence of a given mineral in metamorphic rocks. (Ch. 7)

Isolated system. Any system that has a boundary that prevents the passage in or out of energy and matter. (Ch. 1)

Isostasy, Principle of. The ideal property of flotational balance among segments of the lithosphere. (Chs. 2, 16)

Isotope. Atoms of an element having the same atomic number but differing mass numbers. (Chs. 3, 8)

Joints. Fractures in a rock on which no observable movement has occurred. (Ch. 5)

Jovian planets. Giant planets in the outer regions of the solar system that are characterized by great masses, low densities, and thick atmospheres consisting primarily of hydrogen and helium.

K horizon. A horizon, present in some arid-zone soils beneath the B horizon, that is impregnated with calcium carbonate.

Kame. A short, steep-sided knoll of stratified drift.

Kame terrace. A terrace of ice-contact stratified drift along a valley side.

Karst topography. An assemblage of topographic forms resulting from dissolution of carbonate bedrock and consisting primarily of closely spaced sinkholes. (Ch. 11)

Kerogen. Insoluble, waxlike organic matter found in sedimentary rocks, especially shales.

Kettle. A basin within a body of drift created by melting out of a mass of underlying ice.

Key bed. A thin and generally widespread bed with sedimentary characteristics so distinctive that it can be easily recognized but not confused with any other bed.

Kimberlite pipes. Narrow, pipelike masses of igneous rocks, sometimes containing diamonds that intrude the crust but originate deep in the mantle.

Kinetic energy. The energy possessed by a moving body.

Laccolith. A lenticular pluton intruded parallel to the layering of the intruded rock, above which the layers of the invaded country rock have been bent upward to form a dome.

Lacustrine. Pertaining to, produced by, or formed in a lake.

Lagoon. A bay inshore from an enclosing reef or island paralleling a coast.

Lake Superior-type iron deposit. Iron-rich chemical-sedimentary rocks in which chert and iron-rich layers are interbedded on a fine scale. All known deposits are early Proterozoic in age, or older.

Laminar flow. A pattern of flow in which fluid particles move in parallel layers. (Ch. 10)

Landslide. Any perceptible downslope movement of a mass of bedrock or regolith, or a mixture of the two. (Ch. 8)

Lapilli. Tephra with particles having an average diameter between 2 and 64 mm.

Lapilli tuff. Pyroclastic rock in which the average diameter of tephra particles ranges between 2 and 64 mm.

Lateral moraine. An end moraine built along the side of a valley glacier.

Laterite. A hardened soil horizon characterized by extreme weathering that has led to concentration of secondary oxides of iron and aluminum. (Ch. 5)

Latitude. Part of a grid used for describing positions on the Earth's surface, consisting of parallel circles called *parallels of latitude.*

Laurasia. The northern half of Pangaea, consisting of present-day Asia, Europe, and North America.

Lava. Magma that reaches the Earth's surface through a volcanic vent. (Ch. 4)

Lava dome. A dome-shaped mass of sticky, gas-poor lava erupted from a volcanic vent following a major eruption.

Law (scientific). A statement that some aspect of nature is always observed to happen in the same way and that no deviations have ever been seen. (Ch. 1)

Law of faunal succession. Fossil faunas and floras succeed one another in a definite, recognizable order.

Law of original horizontality. See *original horizontality.*

Leaching. The continued removal, by water solutions, of soluble matter from bedrock or regolith. (Ch. 5)

Left-lateral fault. A strike-slip fault in which relative motion is such that to an observer looking directly at the fault, the motion of the block on the opposite side of the fault is to the left. A *right-lateral fault* has right-handed movement.

Levee. See *natural levee.*

Lignite. A low-grade coal with a calorific value between that of peat and bituminous coal.

Limbs of a fold. The sides of a fold. (Ch. 15)

Limestone. A sedimentary rock consisting chiefly of calcium carbonate, mainly in the form of the mineral calcite. (Ch. 4)

Linear dune. A long, straight, ridge-shaped dune paralleling the wind direction.

Liquefaction. The rapid fluidization of sediment as a result of an abrupt shock such as earthquakes.

Lithic sandstone. A dark-colored sandstone containing quartz, feldspar, and a large amount of tiny rock fragments. Also called *greywacke.*

Lithification. The process that converts a sediment into a sedimentary rock. (Ch. 6)

Lithology. The systematic description of rocks in terms of mineral assemblage and texture.

Lithosphere. The outer 100 km of the solid Earth, where rocks are harder and more rigid than those in the plastic asthenosphere. (Ch. 1)

Little Ice Age. The interval of generally cool climate between the middle thirteenth and middle nineteenth centuries, during which mountain glaciers expanded worldwide.

Load. The material that is moved or carried by a natural transporting agent, such as a stream, the wind, a glacier, or waves, tides, and currents. (Ch. 10)

Local base level. Any base level, other than sea level, below which a stream controlled by that base level cannot erode the land.

Loess. Wind-deposited silt, sometimes accompanied by some clay and fine sand. (Ch. 13)

Long profile. A line drawn along the surface of a stream from its source to its mouth. (Ch. 10)

Longitude. Part of a grid used for describing positions on the Earth's surface, consisting of half circles joining the poles. The half circles are called *meridians.*

Longshore current. A current, within the surf zone, that flows parallel to the coast. (Ch. 14)

Low grade metamorphism. Metamorphism under conditions of low temperature and low pressure. (Ch. 7)

Low-velocity zone. A region in the mantle, approximately between a depth of 100 and 350 km, where seismic-wave velocities decrease.

Lunar highlands. Mountainous regions on the Moon, believed to consist of anorthosite and gabbro. (Ch. 21)

Luster. The quality and intensity of light reflected from a mineral. (Ch. 2)

Magma. Molten rock, together with any suspended mineral grains and dissolved gases, that forms when temperatures rise and melting occurs in the mantle or crust. (Chs. 2, 4)

Magma ocean. Molten outer layer of the Moon early formed as a result of intense rain of bolides. (Ch. 21)

Magmatic arc. An arcuate chain of magmatic activity lying above a subduction zone, parallel to and separated from the seafloor trench by 100 to 400 km. (Ch. 16)

Magmatic differentiation by fractional crystallization. Compositional changes that occur in magmas by the separation of early formed minerals from residual liquids. (Ch. 4)

Magmatic differentiation by partial melting. The process of forming magmas with differing compositions by the incomplete melting of rocks. (Ch. 3)

Magmatic mineral deposit. Any local concentration of minerals formed by magmatic processes in an igneous rock. (Ch. 10)

Magnetic chrons. Periods of predominantly normal polarity (as at present), or predominantly reversed polarity. (Ch. 8)

Magnetic declination. The clockwise angle from true north assumed by a magnetic needle.

Magnetic field. Magnetic lines of force surrounding the Earth.

Magnetic inclination. The angle with the horizontal assumed by a freely swinging bar magnet.

Magnetic latitude. The latitude of a place on the Earth with respect to the magnetic poles. (Ch. 16)

Magnitude (of an earthquake). See *Richter magnitude scale.*

Mantle. The thick shell of dense, rocky matter that surrounds the core. (Ch. 1)

Marble. A metamorphic rock derived from limestone and consisting largely of calcite. (Ch. 7)

Mare. Dark-colored lowland region of the Moon underlain by basalt. (Ch. 21)

Mare basins. Huge circular impact structures on the Moon that were later filled by basaltic flows. (Ch. 21)

Maria. Plural of *mare.*

Mass balance (of a glacier). The sum of the accumulation and ablation on a glacier during a year. (Ch. 12)

Mass number. The sum of the protons and neurons in the nucleus of an atom. (Chs. 3, 8)

Mass-wasting. The movement of regolith downslope by gravity without the aid of a transporting medium. (Ch. 8)

M-discontinuity. See *Mohorovičić discontinuity.*

Meander. A looplike bend of a stream channel. (Ch. 10)

Mechanical deformation. The changes in texture of a rock due to grinding, crushing and development of foliation during metamorphism.

Megascopic. Features that can be seen by the unaided eye, or by the eye assisted by a simple lens that magnifies up to 10 times.

Mélange. A chaotic mixture of broken, jumbled, and thrust-faulted rock above a subduction zone. (Ch. 16)

Mercalli Scale. See *Modified Mercalli Scale.*

Mesa. An isolated, flat-topped, steep-sided desert landform larger than a *butte.*

Mesosphere. The region between the base of the asthenosphere and the core-mantle boundary. (Ch. 1)

Mesozoic Era. The middle era of the planerozoic Eon. (Ch. 8)

Metallic bonding. A form of covalent bond between atoms in which electron sharing occurs with inner energy-level shells rather than the outermost shells. (Ch. 3)

Metallogenic provinces. Limited regions of the crust within which mineral deposits occur in unusually large numbers. (Ch. 20)

Metamorphic aureole. A shell of metamorphic rock, produced by contact metamorphism, surrounding an igneous intrusion. (Ch. 7)

Metamorphic facies. Contrasting assemblages of minerals that reach equilibrium during metamorphism within a specific range of physical conditions belonging to the same metamorphic facies. (Ch. 7)

Metamorphic rock. Rock whose original compounds or textures, or both, have been transformed to new compounds and new textures by reactions in the solid state as a result of high temperature, high pressure, or both. (Ch. 2)

Metamorphic zone. The region on a map between isograds. (Ch. 7)

Metamorphism. All changes in mineral assemblage and rock texture, or both, that take place in sedimentary and igneous rocks in the solid state within the Earth's crust as a result of changes in temperature and pressure. (Ch. 7)

Metasomatism. The process by which rocks have their composition distinctly altered by the addition or removal of materials in solution. (Ch. 7)

Meteorites. Small stony or metallic objects from interplanetary space that impact a planetary surface

Microscopic. Those features of rocks that require high magnification in order to be viewed.

Midocean ridge. Continuous rocky ridges on the ocean floor, many hundreds to a few thousand kilometers wide with a relief of more than 0.6 km. Also called *oceanic ridge* and *oceanic rise.* (Ch. 2)

Migmatite. A composite rock containing both igneous and metamorphic portions. (Ch. 7)

Mineral. Any naturally formed, crystalline solid with a definite chemical composition and a characteristic crystal structure. (Ch. 3)

Mineral assemblage. The variety and abundance of minerals present in a rock. (Ch. 3)

Mineral deposit. Any volume of rock containing an enrichment of one or more minerals. (Ch. 20)

Mineral group. A mineral that displays extensive ionic substitution without changing the cation-anion ratio.

Mineralogy. The special branch of geology that deals with the classification and properties of minerals.

Mineraloid. A naturally occurring mineral-like solid that lacks either a crystal structure or a definite composition, or both. (Ch. 3)

Modified Mercalli Scale. A scale used to compare earthquakes based on the intensity of damage caused by the quake. (Ch. 16)

Moho. See *Mohorovičić discontinuity.*

Mohorovičić discontinuity (also called *M-discontinuity* and *Moho*). The seismic discontinuity that marks the base of the crust. (Ch. 16)

Moh's relative hardness scale. A scale of relative mineral hardness determined by scratching, divided into 10 steps, each marked by a common mineral.

Molecule. The smallest unit that retains the distinctive properties of a compound. (Ch. 3)

Monocline. A local steepening in an otherwise uniformly dipping pile of strata. (Ch. 15)

Moraine. An accumulation of drift deposited beneath or at the margin of a glacier and having a surface form that is unrelated to the underlying bedrock. (Ch. 12)

Mountain chain. A large scale, elongate geologic feature consisting of numerous ranges or systems, regardless of similarity in form or equivalence in ages.

Mountain range. An elongate series of mountains forming a single geologic feature.

Mountain system. A group of ranges similar in general form, structure and alignment, and presumably owing their origin to the same general causes.

Mudcracks. Cracks caused by shrinkage of wet mud as its surface dries.

Mudflow. A flowing mass of predominantly fine-grained rock debris that generally has a high enough water content to make it highly fluid; a rapidly moving type of *debris flow.* (Ch. 8)

Mudstone. A clastic sedimentary rock composed of mineral fragments finer than those in a siltstone.

Natural gas. A gaseous component of petroleum; chiefly methane.

Natural levee. A broad, low ridge of fine alluvium built along the side of a stream channel by water that spreads out of the channel during floods. (Ch. 10)

Neutron. An electrically neutral particle with a mass 1833 times greater than that of the electron.

Nonconformity. Stratified rocks that unconformably overlie igneous or metamorphic rocks. (Ch. 8)

Normal fault. A fault, generally steeply inclined, along which the hanging-wall block has moved relatively downward. (Ch. 15)

Nuclear energy. The heat energy produced during controlled fission or fusion of atoms.

Nucleus (of an atom). The assemblage of protons and neutrons in the core of an atom.

O horizon. An accumulated layer of humus that is the uppermost horizon in many soil profiles.

Oblique-slip fault. A fault on which movement includes both horizontal and vertical components.

Obsidian. An extrusive igneous rock that is wholly or largely glass.

Oceanic crust. The crust beneath the oceans. (Ch. 1)

Oceanic ridge. See *midocean ridge.*

Oceanic rise. See *midocean ridge.*

Oil. The liquid component of petroleum.

Oilfield. A group of oil pools, usually of similar type, or a single pool in an isolated position.

Oil pool. An underground accumulation of oil and gas in a reservoir limited by geologic barriers.

Oil shale. A shale containing waxlike substances that will break down to liquid and gaseous hydrocarbons when heated.

Oolitic limestone. A sedimentary rock composed of accumulations of tiny, round, calcareous bodies called *oolites.*

Open fold. A fold in which the two limbs dip gently and equally, and away from the axis.

Open system. Any system that has a boundary that allows the passage in or out of both energy and matter (Ch. 1)

Ore. An aggregate of minerals from which one or more minerals can be extracted profitably. (Ch. 20)

Organic compound. Chemical compounds made from carbon and hydrogen, with or without other elements such as nitrogen and oxygen.

Original horizontality (law of). Waterlaid sediments are deposited in strata that are horizontal, or nearly horizontal, and parallel, or nearly parallel, to the Earth's surface. (Ch. 8)

Orogenic belts. See *orogens.*

Orogens. Elongate regions of the crust that have been intensively folded, faulted, and thickened as a result of continental collisions. (Ch. 16)

Orogeny. The process by which large regions of the crust are deformed and uplifted to form mountains.

Outcrop. See *exposure.*

Outcrop area. The area on a geological map shown as occupied by a particular rock unit.

Outer core. The outer portion of the Earth's core, which is molten. (Ch. 1)

Outwash. Stratified drift deposited by meltwater streams. (Ch. 12)

Outwash plain. A body of outwash that forms a broad plain.

Outwash terrace. A terrace formed by dissection of an outwash plain or valley train.

Overland flow. The movement of runoff in broad sheets or groups of small, interconnecting rills. (Ch. 10)

Overturned fold. A fold in which the strata in one limb have been tilted beyond vertical.

Oxbow lake. A crescent-shaped, shallow lake occupying the abandoned channel of a meandering stream.

Oxidation. A process in which a chemical element loses electrons. (Ch. 5)

Oxidizing environment. A sedimentary environment in which oxygen is present and organic remains are readily converted by oxidation into carbon dioxide and water.

Pahoehoe. A smooth, ropy-surfaced lava flow, usually basaltic in composition.

Paleomagnetism. Remanent magnetism in ancient rock recording the direction of the magnetic poles at some time in the past.

Paleosol. A soil that formed at the ground surface and subsequently was buried and preserved. (Ch. 5)

Paleozoic Era. The oldest era of the Phanerozoic Eon. (Ch. 8)

Pangaea. The name given to a supercontinent that formed by collision of all the continental crust during the late Paleozoic.

Parabolic dune. A sand dune of U-shape with open end of the U facing upwind.

Parallel of latitude. See *latitude.*

Parallel strata. Strata whose individual layers are parallel.

Parent (radioactive). An atomic nucleus undergoing radioactive decay. Compare *daughter product.* (Ch. 8)

Parent material (of a soil). The regolith from which a soil develops. (Ch. 5)

Passive continental margin. A continental margin in a plate interior. (Ch. 16)

Peat. An unconsolidated deposit of plant remains that is the first stage in the conversion of plant matter to coal. (Ch. 6)

Pediment. A sloping surface cut across bedrock and thinly or discontinuously veneered with alluvium that slopes away from the base of a highland in an arid or semiarid environment. (Ch. 13)

Pegmatite. An exceptionally coarse-grained intrusive igneous rock, commonly granitic in composition and texture.

Pelagic sediment. Sediment consisting of the remains of marine organisms living in the open ocean.

Perched water body. A water body perched atop an aquiclude that lies above the main water table.

Percolation. The movement of groundwater in the saturated zone. (Ch. 11)

Periodotite. A coarse-grained igneous rock consisting largely of olivine, with or without pyroxene.

Periglacial. A land area beyond the limit of glaciers where low temperature and frost action are important factors in determining landscape characteristics. (Ch. 12)

Period. The time during which a geologic system accumulated. (Ch. 8)

Permafrost. Sediment, soil, or bedrock that remains continuously at a temperature below 0°C for an extended time. (Ch. 12)

Permeability. A measure of how easily a solid allows a fluid to pass through it. (Ch. 11)

Petroleum. Gaseous, liquid, and semi-solid substances occurring naturally and consisting chiefly of chemical compounds of carbon and hydrogen. (Ch. 6)

Petrology. The special branch of geology that deals with the occurrence, origin, and history of rocks.

Phanerite. An igneous rock in which the constituent mineral grains are readily visible to the unaided eye. (Ch. 4)

Phanerozoic. The eon that follows the Proterozoic Eon. (Ch. 8)

Phase transition. Atomic repacking caused by changes in pressure and temperature.

Phenocrysts. The isolated large mineral grains in a porphyry.

Photosynthesis. The process by which plants combine water and carbon dioxide to make carbohydrates and oxygen.

Phyllite. A well-foliated metamorphic rock in which the component platy minerals are just visible. (Ch. 7)

Physical geology. The study of the processes that operate at or beneath the surface of the Earth, and the materials on which those processes operate. (Ch. 1)

Physical weathering. Disintegration of rocks by mechanical processes, such as frost-wedging. (Ch. 5)

Piedmont glacier. A broad glacier that terminates on a piedmont slope beyond confining mountain valleys and is fed by one or more large valley glaciers.

Pile. A device in which nuclear fission can be controlled. (Ch. 20)

Pillow basalt. Discontinuous, pillowshaped masses of basalt, ranging in size from a few centimeters to a meter or more in greatest dimension. (Ch. 4)

Placer. A deposit of heavy minerals concentrated mechanically. (Ch. 10)

Place of the ecliptic. See *ecliptic.*

Planet. A large celestial body that revolves around the Sun in an elliptical orbit.

Planetary accretion. The process by which bits of condensed solid matter were gathered to form the planets.

Planetary nebula. A flattened, rotating disk of gas surrounding a proto-sun.

Planetology. A comparative study of the Earth with the Moon and with the other planets. (Ch 21)

Plateau basalt. Flat plains of lava formed as a result of a fissure eruption of basalt. (Ch. 4)

Plate tectonics. The special branch of tectonics that deals with the processes by which the lithosphere is moved laterally over the asthenosphere. (Ch. 1)

Plate triple junction. Junction between three plate spreading edges. The angle between any two edges is 120°. (Ch. 16)

Playa. A dry lake bed in a desert basin. (Ch. 13)

Plunge (of a fold). The angle between a fold axis and the horizontal. (Ch. 15)

Plunging fold. A fold with an inclined axis.

Pluton. Any body of intrusive igneous rock, regardless of shape or size. (Ch. 4)

Point bar. An arcuate deposit of sand or gravel along the inside of the bend of a meander loop. (Ch. 15)

Polar (cold) glacier. A glacier in which the ice is below the pressure melting point throughout, and the ice is frozen to its bed. (Ch. 12)

Polar easterlies. Globe-encircling belts of easterly winds in the high latitudes of both hemispheres.

Polar front. The region where equatorward-moving polar easterlies meet poleward-moving westerlies.

Polarity reversals. Changes of the Earth's magnetic field to the opposite polarity.

Polymerization. The process of linking silicate tetrahedra into large anion groups. (Ch. 3)

Polymorph. A compound that occurs in more than one crystal structure. (Ch. 3)

Pores (*-pore space*). The innumerable tiny openings in rock and regolith that can be filled by water or other fluids.

Porosity. The proportion (in percent) of the total volume of a given body of bedrock or regolith that consists of pore spaces. (Ch. 10)

Porphyry. Any igneous rock consisting of coarse mineral grains scattered through a mixture of fine material grains. (Ch. 4)

Porphyry copper deposit. A class of hydrothermal mineral deposit associated with intrusions of porphyritic igneous rocks. (Ch. 20)

Potential energy. Stored energy.

Precession of the equinoxes. A progressive change in the Earth–Sun distance for a given date.

Pressure melting point. The temperature at which ice can melt at a given pressure. (Ch. 12)

Primary waves. See *P waves*.

Principle. A statement that some aspect of nature is always observed to happen in the same way and that no deviations have ever been observed. (Ch. 1)

Principle of stratigraphic superposition. See *stratigraphic superposition*.

Principle of Uniformitarianism. The same external and internal processes we recognize in action today have been operating unchanged, though at different rates, throughout most of the Earth's history. (Ch. 1)

Progradation. The outward extension of a shoreline into the sea or lake due to sedimentation.

Prograde metamorphic effects The metamorphic changes that occur while temperatures and pressures are rising. (Ch. 7)

Proterozoic Eon. The eon that follows the Archean. (Ch. 8)

Proton. A positively charged particle with a mass 1832 times greater than the mass of an electron.

Pumice. A natural glassy froth made by gases escaping through a viscous magma.

P waves. Seismic body waves transmitted by alternating pulses of compression and expansion. *P* waves pass through solids, liquids, and gases. (Ch. 16)

Pyroclast. A fragment of rock ejected during a volcanic eruption. (Ch. 7)

Pyroclastic flow. A hot, highly mobile flow of tephra that rushes down the flank of a volcano during an eruption. (Ch. 4)

Pyroclastic rocks. Rocks formed from pyroclasts. (Ch. 4)

Pyrometer. An optical device for measuring temperature. (Ch. 4)

Quartzite. A metamorphic rock consisting largely of quartz, and derived from a sandstone. (Ch. 7)

Radiation. Transmission of heat energy through the passage of electromagnetic waves.

Radioactivity. The process by which isotopes of one element transform spontaneously to other isotopes of the same or different elements.

Radiometric age. The length of time a mineral has contained its built-in radioactivity clock.

Rainshadow. A dry region on the downwind side of a mountain range where precipitation is noticeably less than on the windward side.

Recharge. The addition of water to the saturated zone of a groundwater system. (Ch. 11)

Recharge area. Area where water is added to the saturated zone. (Ch. 11)

Recrystallization. The formation of new crystalline minerals within a rock. (Ch. 6)

Recumbent fold. A fold in which the axial plane is horizontal.

Reducing environment. An environment in which oxygen is lacking and organic matter does not decay, but instead is slowly transformed into solid carbon.

Red giant. A large, cool star with a high luminosity and a low surface temperature (about 2500 K), which is largely convective and has fusion reactions going on in shells. (Ch. 3)

Reef. A generally ridgelike structure composed chiefly of the calcareous remains of sedentary marine organisms (e.g., corals, algae).

Reef limestone. A carbonate sedimentary rock formed of fossil reef organisms.

Reflection. The bouncing of a wave off the surface between two media.

Refraction. The change in velocity when a wave passes from one medium to another.

Regional metamorphism. Metamorphism affecting large volumes of crust and involving both mechanical and chemical changes. (Ch. 7)

Regolith. The irregular blanket of loose, noncemented rock particles that covers the Earth. (Ch. 1)

Relative velocity (of a plate). The apparent velocity of one plate relative to another.

Relief. The range in altitude of a land surface. (Ch. 18)

Replacement. The process by which a fluid dissolves matter already present and at the same time deposits from solution an equal volume of a different substance. (Ch. 11)

Reservoir rock. A permeable body of rock in which petroleum accumulates.

Residual mineral deposit. Any local concentration of minerals formed as a result of weathering. (Ch. 20)

Resurgent dome. The uplifting of the collapsed floor of a caldera to form a structural dome.

Retrograde metamorphic effects. Metamorphic changes that occurs as temperature and pressure are declining.

Reverse fault. A fault, generally steeply inclined, along which the hanging-wall block has moved relatively upward. (Ch. 15)

Rhyolite. A fine-grained igneous rock with the composition of a granite. (Ch. 3)

Rhyolite magma. One of the three common magma types. A magma with an SiO_2 content of about 70 percent by weight.

Richter magnitude scale. A scale, based on the recorded amplitudes of seismic body waves, for comparing the amounts of energy released by earthquakes. (Ch. 16)

Rift. See *graben* and *half-graben*.)

Right-lateral fault. See *left-lateral fault*.

Rind. See *weathering rind*.

Rip current. A high-velocity current flowing seaward from the shore as part of the backwash from a wave.

Ripple mark. One of a series of small, fairly regular, subparallel ridges preserved in rock and representing a former rippled sedimentary surface.

Rock. Any naturally formed, nonliving, firm, and coherent aggregate mass of mineral matter that constitutes part of a planet. (Ch. 1)

Rock cleavage (also called *slaty cleavage*). The property by which a rock breaks into platelike fragments along flat planes. (Ch. 5)

Rock cycle. The cyclic movement of rock material, in the course of which rock is created, destroyed, and altered through the operation of internal and external Earth processes. (Ch. 2)

Rockfall. The free falling of detached bodies of bedrock from a cliff or steep slope. (Ch. 8)

Rock flour. Fine rock particles produced by glacial crushing and grinding.

Rock glacier. A lobe of ice-cemented rock debris that

moves slowly downslope in a manner similar to glaciers. (Ch. 8)

Rockslide. The sudden and rapid downslope movement of detached masses of bedrock across an inclined surface. (Ch. 8)

Rock-stratigraphic unit. Any distinctive rock unit that can be distinguished from other strata on the basis of composition and physical properties.

Roof rock. A rock, such as shale, that is impermeable and caps a petroleum reservoir.

Runoff. The fraction of precipitation that flows over the land surface. (Ch. 10)

Salinity. The measure of the sea's saltiness; expressed in parts per thousand (per mil.). (Ch. 14)

Saltation. The progressive forward movement of a sediment particle in a series of short intermittent jumps along arcing paths. (Ch. 10)

Sand ripples. A series of small and rather regular ridges on the surface of a body of sand, such as a dune. (Ch. 13)

Sand sea. Vast tract of shifting sand. (Ch. 13)

Sandstone. A medium-grained clastic sedimentary rock composed chiefly of sand-sized grains.

Saturated zone. The groundwater zone in which all openings are filled with water. (Ch. 11)

Scale (of a map). The proportion between a unit of distance on a map and the unit it represents on the Earth's surface.

Schist. A well-foliated metamorphic rock in which the component platy minerals are clearly visible. (Ch. 7)

Schistosity. The parallel arrangement of coarse grains of the sheet-structure minerals, like mica and chlorite formed during metamorphism under conditions of differential stress. (Ch. 7)

Scientific method. The use of evidence that can be seen and tested by anyone who has the means to do so, consisting often of observation, formation of a hypothesis, testing of that hypothesis and formation of a theory, formation of a law, and continued reexamination. (Ch. 1)

Sea arch. An opening through a headland, generally produced by wave erosion, that forms a bridge of rock over water.

Sea cave. A cave at the base of a seacliff produced by wave erosion.

Seafloor spreading (theory of). A theory proposed during the early 1960s in which lateral movement of the oceanic crust away from midocean ridges was postulated. (Ch. 16)

Seamount. An isolated submerged volcanic mountain standing more than 1000 m above the seafloor.

Secondary enrichment. The process by which a sulfide mineral deposit is chemically weathered and enriched in its metal content as a result. (Ch. 5)

Secondary mineral. A mineral formed later than the rock enclosing it, usually at the expense of an earlier formed primary mineral.

Secondary waves. See *S waves*.

Sediment. Regolith that has been transported by any of the external processes. (Ch. 2)

Sediment drifts. Huge bodies of sediment, up to hundreds of kilometers long, deposited and shaped by deep ocean currents along a continental margin.

Sediment flows. Mass wasting of mixtures of sediment, water, and air. (Ch. 8)

Sedimentary facies. A distinctive group of characteristics within a sedimentary unit that differs, as a group, from those elsewhere in the same unit.

Sedimentary mineral deposit. Any local concentration of minerals formed through processes of sedimentation. (Ch. 20)

Sedimentary rock. Any rock formed by chemical precipitation or by sedimentation and cementation of mineral grains transported to a site of deposition by water, wind, ice, or gravity. (Ch. 2)

Seismic belts. Large tracts of the Earth's surface that are subject to frequent earthquake shocks. (Ch. 16)

Seismic sea waves (also called *tsunami*). Long-wavelength ocean waves produced by sudden movement of the seafloor following an earthquake. Incorrectly called tidal waves. (Ch. 16)

Seismic tomography. A way of revealing inhomogeneities in the mantle by measuring slight differences in the frequencies and velocities of seismic waves.

Seismic waves. Elastic disturbances spreading outward from an earthquake focus. (Ch. 16)

Seismograph. The device used to study the shocks and vibrations caused by earthquakes. (Ch. 16)

Seismology. The study of earthquakes. (Ch. 16)

Serpentinite. A rock composed largely of the mineral serpentine.

Setting time. The moment a mineral starts accumulating a daughter product produced by radioactive decay.

Shale. A fine-grained, clastic sedimentary rock.

Shard. See *volcanic shard*.

Shear strength. The internal resistance of a body to movement. (Ch. 8)

Shear stress (on a free-standing body). The force acting on a body that causes slippage or translation. (Chs. 9, 15)

Shear waves. See *S waves*.

Sheet erosion. The erosion performed by overland flow. (Ch. 10)

Shield volcano. A volcano that emits fluid lava and builds up a broad dome-shaped edifice with a surface slope of only a few degrees. (Ch. 4)

Shore profile. A vertical section along a line perpendicular to a shore.

Silicate (*-silicate mineral*). A mineral that contains the silicate anion. (Ch. 3)

Silicate anion. A complex ion $(SiO_4)^{-4}$, that is present in all silicate minerals. (Ch. 3)

Silicate mineral. See *silicate*.

Siliceous ooze. Any pelagic deep-sea sediment of which at least 30 percent consists of siliceous skeletal remains. See also *deep-sea ooze*.

Sill. Tabular, parallel-sided sheets of intrusive igneous rock that are parallel to the layering of the intruded rock. (Ch. 4)

Siltstone. A sedimentary rock composed mainly of silt-sized mineral fragments.

Sinkhole. A large solution cavity open to the sky. (Ch. 11)

Slate. A low-grade metamorphic rock with a pronounced slaty cleavage. (Ch. 7)

Slaty cleavage. See *rock-cleavage*. (Ch. 7)

Slickensides. Striated or highly polished surfaces on hard rocks abraded by movement along a fault.

Slip face. The straight, lee slope of a dune.

Slump. A type of slope failure in which a downward and outward rotational movement of rock or regolith occurs along a concave-up slip surface. (Ch. 8)

Slurry flow. A moving mass of sediment that is saturated with water trapped among the grains and transported with the flowing mass. (Ch. 8)

Snowline. The lower limit of perennial snow. (Ch. 12)

Soft water. Groundwater that contains little dissolved matter and no appreciable calcium.

Soil. The part of the regolith that can support rooted plants. (Ch. 5)

Soil horizons. The subhorizontal weathered zones formed as a soil develops. (Ch. 5)

Soil profile. A vertical section through a soil that displays its component horizons. (Ch. 5)

Solar system. The Sun and the group of objects in orbit around it. (Ch. 1)

Sole marks. Irregularities formed by currents together with tracks and other markings preserved on the bedding plane of sandstone or siltstone.

Solifluction. The very slow downslope movement of waterlogged soil and surficial debris. (Ch. 8)

Sorting. A measure of the range of particle size of sediment. (Ch. 6)

Source-rock. A sedimentary rock containing organic matter that is a source of petroleum.

Spall. Flakey material broken away from the surface of a rock exposed to intense heat. (Ch. 5)

Spatter cone. A cone-shaped pile of bits of lava surrounding a volcanic vent.

Spatter rampart. A linear pile of bits of lava erupted along a fissure.

Specific gravity. A number stating the ratio of the weight of a substance to the weight of an equal volume of pure water. A dimensionless number numerically equal to the density. (Ch. 3)

Spheroidal weathering. The successive loosening of concentric shells of decayed rock from a solid rock mass as a result of chemical weathering. (Ch. 5)

Spit. An elongate ridge of sand or gravel that projects from land and ends in open water. (Ch. 14)

Spreading axis. The axis of rotation of a plate of lithosphere. (Ch. 16)

Spreading center (also called a *divergent margin*). The new, growing edge of a plate. Coincident with a midocean ridge. (Ch. 2).

Spreading pole. The point where a spreading axis reaches the Earth's surface. (Ch 16)

Spring. A flow of groundwater emerging naturally at the ground surface. (Ch. 11)

Stable platform. That portion of a craton that is covered by a thin layer of little-deformed sediments.

Stable zone. The interior part of a tectonic plate.

Stack. An isolated rocky island or steep rock mass near a cliffy shore, detached from a headland by wave erosion.

Stalactite. An iciclelike form of dripstone and flowstone, hanging from cave floors.

Stalagmite. An "icicle" of dripstone and flowstone projecting upward from cave floors.

Star dune. An isolated hill of sand having a base that resembles a star in plan.

Steady state. A condition in which the rate of arrival of material or energy equals the rate of escape.

Stock. A small, irregular body of intrusive igneous rock, smaller than a batholith, that cuts across the layering of the intruded rock. (Ch. 4)

Stoping. The process by which a rising body of magma wedges off fragments of overlying rock that then sink through the magma chamber.

Strain. The measure of the changes in length, volume, and shape in a stressed material. (Ch. 15)

Strain rate. The rate at which a rock is forced to change its shape or volume. (Ch. 15)

Strain seismograph. A device for recording earthquake waves based on the flexure of a long, rigid rod.

Strata. See *stratum*.

Stratabound mineral deposits. Ores of lead, zinc, copper, and other metals enclosed in sedimentary rocks in such a way that they closely resemble primary sediments.

Stratification. The layered arrangement of sediments, sedimentary rocks, or extrusive igneous rocks. (Ch. 6)

Stratified drift. Glacial drift that is both sorted and stratified. (Ch. 12).

Stratigraphic superposition (principle of). In a sequence of strata, not later overturned, the order in which they were deposited is from bottom to top. (Ch. 8)

Stratigraphy. The study of strata. (Ch. 8)

Stratovolcano. Volcano that emits both tephra and viscous lava, and that builds up steep conical mounds. (Ch. 4)

Stratum (plural = *strata*). A distinct layer of sediment that accumulated at the Earth's surface. (Ch. 4)

Streak. A thin layer of powdered mineral made by rubbing a specimen on a nonglazed porcelain plate.

Stream. A body of water that carries detrital particles and dissolved substances and flows down a slope in a definite channel. (Ch. 10)

Streamflow. The flow of surface water in a well-defined channel. (Ch. 10).

Stress. The magnitude and direction of a deforming force.

Striations. See *glacial striations*.

Strike. The compass direction of a horizontal line in an inclined plane. (Ch. 15)

Strike-slip fault. A fault on which displacement has been horizontal and parallel to the strike of the fault. (Ch. 15)

Structural geology. The branch of geology devoted to the study of rock deformation.

Structure (of minerals). See *crystal structure*.

Subatomic particles. The small particles that combine to form an atom—electrons, protons, and neutrons.

Subduction. The process by which old, cold lithosphere sinks into the asthenosphere. (Ch. 2)

Subduction zone (also called a *convergent margin*). The linear zone along which a plate of lithosphere sinks down into the asthenosphere. (Ch. 2)

Submarine canyon. A steep-sided valley on the continental shelf or slope resembling a river-cut canyon on land.

Submergence (of a coast). A rise of water level relative to the land so that areas formerly dry are inundated. (Ch. 14)

Subsequent stream. A stream whose course has become adjusted so that it occupies belts of weak rock or other geologic structures.

Superposed stream. A stream that was let down, or superposed, from overlying strata onto buried bedrock having composition or structure unlike that of the covering strata.

Surf. Wave activity between the line of breakers and the shore. (Ch. 14)

Surface waves. Seismic waves that are guided by the Earth's surface and do not pass through the body of the Earth. (Ch. 16)

Surge. An unusually rapid movement of a glacier marked by dramatic changes in glacier flow and form. (Ch. 12)

Suspect terrane. See *accreted terrane*.

Suspended load. Fine particles suspended in a stream. (Ch. 10)

Swash. The surge of water up a beach caused by waves moving against a coast.

S waves. Seismic body waves transmitted by an alternating series of sideways (shear) movements in a solid. *S* waves cause a change of shape and cannot be transmitted through liquids and gases. (Ch. 16)

Symmetrical fold. A fold in which both limbs dip equally away from the axial plane.

Syncline. A downfold with a troughlike form. (Ch. 15)

System. The primary unit in a time-stratigraphic sequence of rocks. (Ch. 8)

Taconite. Iron ore found as a result of metamorphism of Lake Superior-type iron deposits.

Talus. The apron of rock waste sloping outward from the cliff that supplies it. (Ch. 8)

Tar (also called *asphalt*). An oil that is viscous and so thick it will not flow. (Ch. 20)

Tarn. A small, generally deep mountain lake occupying a cirque.

Tectonic cycle. A model that describes the movement and interactions of lithospheric plates, the internal processes that drive plate motion, and the types of rock and rock formations that develop as a result of tectonic movement and interactions. (Ch. 2)

Tectonics. The study of movement and deformation of the lithosphere. (Ch. 1)

Temperate (warm) glacier. A glacier in which the ice is at the pressure-melting point and water and ice coexist in equilibrium. (Ch. 12)

Tensional stress. The differential stress on a body that causes stretching and elongation. (Ch. 15)

Tephra. A loose assemblage of pyroclasts. (Ch. 4)

Tephra cone. A cone-shaped pile of tephra deposited around a volcanic vent. (Ch. 4)

Terminal moraine. An end moraine deposited at the front of a glacier.

Terminus. The outer, lower margin of a glacier.

Terrace. An abandoned floodplain formed when a stream flowed at a level above the level of its present channel and floodplain. (Ch. 10)

Terrane. A large piece of crust with a distinctive geological character.

Terrestrial planets. The innermost planets of the solar system (Mercury, Venus, Earth, and Mars), which have high densities and rocky compositions.

Tethys. The name of a narrow sea separating Gondwanaland from Laurasia.

Texture. The overall appearance that a rock has because of the size, shape, and arrangement of its constituent mineral grains. (Ch. 3)

Theory. A hypothesis that has been examined and found to withstand numerous tests. (Ch. 1)

Thermal metamorphism. See *contact metamorphism*.

Thermal plume. A vertically rising mass of heated rock in the mantle.

Thermal spring. A natural spring that emits hot water.

Thermohaline circulation. Global patterns of water circulation propelled by the sinking of dense cold and salty water. (Ch. 14)

Thin section. A thin slice of rock glued to a glass slide and used for microscopic examination.

Thrust faults (also called *thrusts*). Low-angle reverse faults with dips less than 15°. (Ch. 15)

Tidal bore. A large, turbulent, wall-like wave of

water caused by the meeting of two tides or by the rush of tide up a narrowing inlet, river, estuary, or bay.

Tidal bulge. A bulge in bodies of marine and fresh water, produced by the gravitational attraction of the Moon and Sun, that moves around the Earth as it rotates.

Tide. The twice-daily rise and fall of the ocean surface resulting from the gravitational attraction of the Moon and Sun.

Till. A nonsorted sediment deposited directly from glacier ice. (Ch. 12).

Tillite. A nonsorted sedimentary rock of glacial origin (i.e., a lithified till).

Tilt (of axis). The angle of the Earth's rotational axis with respect to the plane of the Earth's orbit.

Time-stratigraphic unit. All the rocks or sediments that formed during a specific interval of geologic time. Compare *rock-stratigraphic unit*.

Titius–Bode rule. The distance to each planet is approximately twice as far as the next inner one, measured from the Sun.

Tombolo. A ridge of sand or gravel that connects an island to the mainland or to another island.

Topography. The relief and form of the land.

Topset layer. A layer of stream sediment that overlies the forest layers in a delta.

Tradewind. A globe-encircling belt of winds in the low latitudes. They blow from the northeast in the northern hemisphere and the southeast in the southern hemisphere.

Transform. The junction point where one of the major deformation features—a midocean ridge, a seafloor trench, or a strike-slip fault—meets another.

Transform faults. The special class of strike-slip fault that links major structural features. (Ch. 15)

Transform fault continental margin. The margin of a continent that coincides with a transform fault..

Transform fault margin. A fracture in the lithosphere along which two plates slide past each other. (Ch. 2)

Transverse dune. A sand dune forming a wavelike ridge transverse to wind direction.

Trap. A reservoir rock plus a roof rock that serve to accumulate petroleum.

Trenches. Long, narrow, very deep, and arcuate basins in the seafloor. (Ch. 2)

Tributary. A stream that joins a larger stream. (Ch. 10)

Triple junction. See *plate triple junction*.

Tsunami. See *seismic sea waves*.

Tuff. A pyroclastic rock consisting of ash- or lapilli-sized tephra, hence *ash tuff* and *lapilli tuff.*

Turbidite. A graded layer of sediment deposited by a turbidity current. (Ch. 6)

Turbidity current. A gravity-driven current consisting of a dilute mixture of sediment and water having a density greater than the surrounding water. (Ch. 6)

Turbulent flow. A pattern of flow in which particles of fluid move in swirls and eddies. (Ch. 10)

Typhoon. A term used in the western Pacific Ocean for a tropical cyclonic storm. See *hurricane*.

Unconfined aquifer. An aquifer with an upper surface that coincides with the water table. (Ch. 11)

Unconformity. A substantial break or gap in a stratigraphic sequence that marks the absence of part of the rock record. (Ch. 8)

Unconformity-bounded sequence. A grouping of strata that is bounded at its base and top by unconformities of regional or interregional extent.

Uniform layer. A layer of sediment or sedimentary rock that consists of particles of about the same diameter.

Uniform stress. Stress that is equal in all directions. Also called *confining stress* or *homogeneous stress.* (Ch. 8)

Uniformitarianism. See *Principle of Uniformitarianism.* (Ch. 1)

Unsaturated zone (zone of aeration). The groundwater zone in which open spaces in regolith or bedrock are filled mainly with air.

Valley glacier. A glacier that flows from a cirque or cirques onto and along the floor of a valley.

Valley train. A body of outwash that partly fills a valley.

van der Waals bond. A weak electrostatic attraction that arises because certain ions and atoms are distorted from a spherical shape. (Ch. 3)

Varve. A pair of sedimentary layers deposited during the seasonal cycle of a single year. (Ch. 6)

Ventifact. Any bedrock surface or stone that has been abraded and shaped by wind-blown sediment. (Ch. 13)

Vesicle. A small opening, in extrusive igneous rock, made by escaping gas originally held in solution under high pressure while the parent magma was underground.

Viscosity. The internal property of a substance that offers resistance to flow. (Ch. 14)

Volcanic ash. See *ash.*

Volcanic breccia. A formed as a result of explosive volcanic activity.

Volcanic neck. The approximately cylindrical conduit of igneous rock forming the feeder pipe of a volcanic vent that has been stripped of its surrounding rock by erosion.

Volcanic sediment (in the ocean). Sediment from submarine volcanoes, together with ash from oceanic and nonoceanic volcanic eruptions.

Volcanic shard. A particle of ash-sized, glassy tephra.

Volcano. The vent from which igneous matter, solid rock, debris, and gases are erupted. (Ch. 4)

Volcanogenic massive sulfide deposit. A mineral deposit formed by deposition of sulfide materials from a submarine hot spring. (Ch. 20)

Wall-rock alteration. Changes produced in the mineral assemblages of rocks lining the flow channel of a hydrothermal solution.

Water quality. The fitness of water for human use, as affected by physical, chemical, and biological factors.

Water table. The upper surface of the saturated zone of groundwater. (Ch. 11)

Wave. An oscillatory movement of water characterized by alternate rise and fall of the surface.

Wave base. The effective lower limit of wave motion, which is half of the wavelength. (Ch. 14)

Wave-cut bench. A bench or platform cut across bedrock by surf. (Ch. 14)

Wave-cut cliff. A coastal cliff cut by surf.

Wavelength. The distance between the crests or troughs of adjacent waves. (Ch. 14)

Wave refraction. The process by which the direction of a series of waves, moving into shallow water at an angle to the shoreline, is changed. (Ch. 14)

Weathering. The chemical alteration and mechanical breakdown of rock materials during exposure to air, moisture, and organic matter. (Chs. 2, 5)

Weathering rind. A discolored rim of weathered rock surrounding an unweathered core. (Ch. 5)

Welded tuff (also called *ignimbrite*). Pyroclastic rocks, the glassy fragments of which were plastic and so hot when deposited that they fused to form a glassy rock. (Ch. 3)

Well. An excavation in the ground designed to tap a supply of underground liquid, especially water or petroleum.

Westerlies. Globe encircling belts of winds centered at about 45° latitude in both hemispheres.

Windchill factor. The heat loss from exposed skin as a result of the combined effects of low temperature and wind speed. (Ch. 13)

Xenoliths. Fragments of country rock still enclosed in a magmatic body when it solidifies.

Yardang. An elongate, streamlined, wind-eroded ridge. (Ch. 13)

Zone of aeration. See *unsaturated zone.*

SELECTED REFERENCES

Chapter 1

Alvarez, W., and Asaro, F., 1990, What caused the mass extinction? An extraterrestrial impact: Scientific American, October, p. 78–84.

Erwin, D.H., 1996, The mother of mass extinctions: Scientific American, July, p. 72–78.

National Academy of Sciences, 1989, On being a scientist: Washington, DC, National Academy Press.

Skinner, B.J., Porter, S.C., and Botkin, D.B., 1999, The Blue Planet: An introduction to earth system science: New York, John Wiley and Sons.

Chapter 2

Hallam, A., 1973, A revolution in the earth sciences: From continental drift to plate tectonics: New York, Oxford University Press.

Kearey, P., and Vine, F.J., 1990, Global tectonics: Oxford, Blackwell Scientific.

Macdonald, K.C., and Fox, P.J., 1990, The mid-ocean ridge: Scientific American, v. 262, p. 72–79.

Chapter 3

Brady, J., and Holum, J., 1996, Chemistry: The study of matter and its changes: New York, John Wiley and Sons.

Dietrich, R.V., and Skinner, B.J., 1990, Gems, granites, and gravels: Cambridge, Cambridge University Press.

Dietrich, R.V., and Skinner, B.J., 1979, Rocks and rock minerals: New York, John Wiley and Sons.

Klein, C., and Hurlburt, C.S. Jr., 1999, Manual of mineralogy, 21st ed.: New York, John Wiley and Sons.

Skinner, B.J., 1976, A second Iron Age ahead? American Scientist, v. 64, p. 258–269.

Chapter 4

Blatt, H., and Tracy, R.J., 1996, Petrology: Igneous, sedimentary, and metamorphic, 2nd ed.: New York, W.H. Freeman.

Decker, R., and Decker, B., 1989, Volcanoes: San Francisco, W.H. Freeman.

Foxworthy, B.L., and Hill, M., 1982, Volcanic eruption of 1980 at Mount St. Helens. The first 100 days: Professional Paper no. 1249, U.S. Geological Survey.

Haymon, R.M., and Macdonald, K.C., 1985, The geology of deep-sea hot springs: American Scientist, v. 73, p. 441–449.

Philpotts, A.R., 1990, Principles of igneous and metamorphic petrology: Englewood Cliffs, NJ, Prentice-Hall.

Rona, P.A., 1986, Mineral deposits from sea-floor hot springs: Scientific American, v. 254, January, p. 84–92.

Simkin, T., and Fiske, R.S., 1983, Krakatau, 1883: The volcanic eruption and its effects: Washington, DC, Smithsonian Institute Press.

Chapter 5

Birkeland, P.W., 1999, Soils and geomorphology: New York, Oxford University Press.

Brown, L.R., and Wolf, E.C., 1984, Soil erosion: Quiet crisis in the world economy: Worldwatch Institute, Worldwatch Paper 60.

Carroll, D., 1970, Rock weathering: New York, Plenum.

Nahon, D.B., 1991, Introduction to the petrology of soils and chemical weathering: New York, John Wiley and Sons.

Retallack, G., 1989, Soils of the past, an introduction to paleopedology: New York, HarperCollins Academic.

Robinson, D.A., and Williams, R.B.G., eds., 1994, Rock weathering and landform evolution: New York,

Chapter 6

Blatt, H., 1992, Sedimentary petrology, 2nd ed.: New York, W.H. Freeman.

Blatt, H., and Tracy, R.J., 1996, Petrology: Igneous, sedimentary, and metamorphic, 2nd ed.: New York, W.H. Freeman.

McLane, M., 1995, Sedimentology: New York, Oxford University Press.

Siever, R., 1988, Sand: New York, Scientific American Library.

Reineck, H.H., and Singh, I.B., 1980, Depositional sedimentary environments, 2nd ed.: New York, Springer-Verlag.

Chapter 7

Blatt, H., and Tracy, R.J., 1996, Petrology: Igneous, sedimentary, and metamorphic, 2nd ed.: New York, W.H. Freeman.

Dietrich, R.V., and Skinner, B.J., 1979, Rocks and rock minerals: New York, John Wiley and Sons.

Philpotts, A.R., 1990, Principles of igneous and metamorphic petrology: Englewood Cliffs, NJ, Prentice-Hall.

Yardley, B.W.D., 1989, An introduction to metamorphic petrology: London, Longman Scientific.

Yardley, B.W.D., MacKenzie, W.S., and Guilford, C., 1990, Atlas of metamorphic rocks and their textures: New York, John Wiley and Sons.

Chapter 8

Faure, G., 1986, Principles of isotope geology, 2nd ed.: New York, John Wiley and Sons.

Palmer, A.R., 1984, Decade of North American geology geologic time scale: Boulder, CO, Geologi-

cal Society of America, Map and Chart Series MC50.

Stanley, S.M., 1993, Exploring earth and life through time: New York, W.H. Freeman.

Tarling, D.H., 1983, Paleomagnetism: London, Chapman and Hall.

Chapter 9

Costa, J.E., and Wieczorek, G.F., 1987, Debris flows/avalanches: process, recognition, and mitigation: Boulder, CO, Geological Society of America, Reviews in Engineering Geology VII.

National Research Council, 1987, Confronting natural disasters. An International Decade for Natural Hazard Reduction: Washington, DC, National Academy Press.

Parsons, A.J., 1988, Hillslope form: London, Routledge.

Porter, S.C., and Orombelli, G., 1981, Alpine rockfall hazards: American Scientist, v. 69, p. 67–75.

Chapter 10

Dunne, Thomas, and Leopold, L.B., 1978, Water in environmental planning: San Francisco, W.H. Freeman.

McCullough, D.G., 1968, The Johnstown Flood: New York, Simon and Schuster.

Morisawa, M., 1986, Rivers, form and process: New York, John Wiley and Sons.

Chapter 11

Back, W., Rosenshein, J.S., and Seaber, P.R., eds, 1988, Hydrology. The geology of North America, vol. O–2: Boulder, CO, Geological Society of America.

Dolan, R., and Goodell, H.G., 1986, Sinking cities: American Scientist, v. 74, p. 38–47.

Jennings, J.N., 1983, Karst landforms: American Scientist, v. 71, p. 578–586.

Palmer, A.N., 1991, Origin and morphology of limestone caves: Geological Society of America Bulletin, v. 103, p. 1–21.

Trudgill, S., 1985, Limestone geomorphology: White Plains, NY, Longman.

Chapter 12

Benn, D.I., and Evans, D.J.A., 1998, Glaciers and glaciation: New York, Arnold.

Broecker, W.S., and Denton, G.H., 1990, What drives glacial cycles, Scientific American, v. 262, p. 48–56.

Dawson, A.G., 1992, Ice-Age Earth: New York, Routledge.

Imbrie, J., and Imbrie, K.P., 1979, Ice ages: Solving the mystery: Short Hills, NJ, Enslow.

Sharp, R.P., 1988, Living ice. Understanding glaciers and glaciation: Cambridge, Cambridge University Press.

Swiss National Tourist Office, 1981, Switzerland and her glaciers: Berne, Kummerly, and Frey.

Chapter 13

Greeley, R., and Iverson, J., 1985, Wind as a geological process: Cambridge, Cambridge University Press.

Pye, K., 1987, Aeolian dust and dust deposits: New York, Academic Press.

Pye, K., and Tsoar, H., 1990, Aeolian sand and sand dunes: New York, HarperCollins Academic.

Walker, A.S., 1982, Deserts of China: American Scientist, v. 70, p. 366–376.

Chapter 14

Carter, R.W.G., and Woodroffe, C.D., eds., 1994, Coastal evolution: Cambridge, Cambridge University Press.

Duxbury, A.C., and Duxbury, A.B., 1994, An introduction to the world's oceans, 4th ed.: Dubuque, Iowa, Wm. C. Brown.

Hardisty, J., 1990, Beaches, form and process: New York, HarperCollins Academic, 352 pp.

Pilkey, O.H., and Neal, W.J., 1988, coastal geologic hazards, in R.E. Sheridan and J.A. Grow, eds., The geology of North America, v. I-2, The Atlantic Continental Margin: U.S.: Boulder, CO, Geological Society of America, p. 549–556.

Chapter 15

Davis, G.H., and Reynolds, S.J., 1996, Structural geology of rocks and regions: New York, John Wiley and Sons.

Suppe, J., 1985, Principles of structural geology: Englewood Cliffs, NJ, Prentice-Hall.

Twiss, R.J., and Moores, E.M., 1992, Structural geology: New York, W.H. Freeman.

Chapter 16

Atwater, B.F., 1987, Evidence for great Holocene earthquakes along the outer coast of Washington State: Science, v. 236, p. 942–944.

Bolt, B.A., 1993, Earthquakes, 3rd ed.: New York, W.H. Freeman.

Bolt, B.A., 1993, Earthquakes and geological discovery: New York, Scientific American Library.

Johnston, A.C., and Kanter, L.R., 1990, Earthquakes in stable continental crust: Scientific American, v. 262, no. 3, p. 68–75.

McKenzie, D.P., 1983, The Earth's mantle: Scientific American, v. 249, March, p. 114–129.

Wuethrich, B., 1994, It's official: Quake danger in northwest rivals California's: Science, v. 265, p. 1802–1803.

Wysession, M., 1995, The inner workings of the Earth: American Scientist, v. 83, p. 134–146.

Chapter 17

Cox, A., and Hart, R.B., 1986, Plate tectonics: How it works: Palo Alto, CA, Blackwell.

Dalziel, I.W.D., 1995, Earth before Pangaea: Scientific American, January, p. 58–63.

Hoffman, P.F., 1988, United plates of America, the birth of a craton: Early Proterozoic assembly and growth of Laurentia: Annual Reviews of Earth and Planetary Sciences, v. 16, p. 543–603.

Kearey, P., and Vine, F.J., 1990, Global tectonics: Oxford, Blackwell Scientific.

Smith, D.K., and Cann, J.R., 1993, Building the crust at the Mid-Atlantic Ridge: Nature, v. 365, p. 707–714.

Solomon, S.C., and Toomey, D.R., 1992, The structure of mid-ocean ridges: Annual Review of Earth and Planetery Sciences, v. 20, p. 329–364.

Chapter 18

Chorley, R.J., Schumm, S.A., and Sugden, D.E., 1995, Geomorphology: New York, Methuen.

Ruddiman, W.F., ed., 1997, Tectonic uplift and climate change: New York, Plenum.

Ruddiman, W.F., and Kutzbach, J.E., 1991, Plateau uplift and climatic change: Scientific American, v. 264, no. 4, p. 66–75.

Chapter 19

Brown, B., and Morgan, L., 1990, The miracle planet: New York, Gallery Books.

Jones, P.D., and Wigley, T.M.L., 1990, Global warming trends: Scientific American, v. 263, p. 84–91.

Mungall, C., and McLaren, D.J., eds., 1990, Planet under stress: The challenge of global change: New York, Oxford University Press.

Robert, N., ed., 1994, The changing global environment: Oxford, Blackwell.

Schneider, S.H., 1987, Climate modeling: Scientific American, v. 256, p. 72–80.

Silver, C.S., and DeFries, R.S., 1990, One Earth, one future: our changing global environment: Washington, DC, National Academy Press.

Toon, O.B., and Turco, R.P., 1991, Polar stratospheric clouds and ozone depletion: Scientific American, v. 264, p. 68–74.

White, R.M., 1990, The great climate debate: Scientific American, v. 263, p. 36–44.

Chapter 20

Craig, J.R., Vaughan, D.J., and Skinner, B.J., 1988, Resources of the Earth: Englewood Cliffs, NJ, Prentice-Hall.

Davis, G.R., 1990, Energy for the planet Earth: Scientific American, v. 263, no. 3, p. 54–62.

Fulkerson, W., Judkins, R.R., and Sanghvi, M.K., 1990, Energy from fossil fuels: Scientific American, v. 263, no. 3, p. 128–135.

Häfele, W., 1990, Energy from nuclear power: Scientific American, v. 263, no. 3, p. 136–144.

Hodges, C.A., 1995, Mineral resources, environmental issues, and land uses: Science, v. 268, p. 1305–1312.

Sawkins, F.J., 1990, Metal deposits in relation to plate tectonics, 2nd ed.: Berlin, Springer-Verlag.

Skinner, B.J., 1986, Earth resources, 3rd ed.: Englewood Cliffs, NJ, Prentice-Hall.

Weinberg, Carl J., and Williams, R.H., 1990, Energy from the Sun: Scientific American, v. 263, no. 3, p. 146–155.

Chapter 21

Black, David C., 1991, Worlds around other stars: Scientific American, v. 264, no. 1, p. 76–82.

Greeley, R., 1985, Planetary landscapes: London, Allen and Unwin.

Greeley, R., and Iversen, J.D., 1985, Wind as a geological process: On Earth, Mars, Venus, and Titan: Cambridge, Cambridge University Press.

Grieve, R.A.F., 1990, Impact cratering on the Earth: Scientific American, v. 262, no. 4, p. 66–73.

PHOTO CREDITS

Chapter 1
Opener: Jerry Alexander/Tony Stone Images. Figure 1.1: ©Tom Bean. Figure 1.2a: ©John S. Shelton. Figure 1.2b: Francois Gohier/Ardea London. Figure 1.3: Courtesy K. Roy Gill. Figure 1.4: ©John S. Shelton. Figure 1.5: W. Alvarez/Photo Researchers. Figure 1.7a: Comstock, Inc. Figure 1.7b: Courtesy NASA. Figure 1.10: Courtesy Space Photography Lab of Arizona. Figure 1.11: ©Breck P. Kent. Figure 1.16: ©Earth Satellite Corporation. Figure B1.1: David Turnley/Black Star.

Chapter 2
Opener: Kevin Kelley/Tony Stone Images. Figure 2.9: Yann Arthus-Bertrand/Altitude/Peter Arnold, Inc. Figure 2.12: Ric Ergenbright. Figure 2.16: George Corster/Comstock, Inc. Figure 2.17a: ©Tom Bean/DRK Photo. Figure B2.3: Phil Degginger/Bruce Coleman, Inc.

Chapter 3
Opener: ©Index Photo. Figure 3.7a: Michael Hochella. Figures 3.9, 3.11, 3.12, 3.13, 3.15, 3.16, 3.18, 3.23, 3.26, 3.27, 3.28, & 3.29: William Sacco. Figure 3.10: Brian J. Skinner. Figure 3.14: ©Breck Kent/Earth Scenes. Figure 3.17: Courtesy The Natural History Museum, London. Figure 3.25: Courtesy Christie A. Callender, Texaco Exploration and Production Technology. Figure 3.22a: Brian J. Skinner. Figure 3.22b: ©Boltin Picture Library. Figures 3.22c-g: ©Breck P. Kent. Figure 3.30a: ©Breck P. Kent/Earth Scenes. Figures 3.30b &c: Courtesy The Natural History Museum, London. Figure 3.32: Brian J. Skinner. Figure B3.1: ©John Cancalosi/DRK Photo.

Chapter 4
Opener: Gary Braasch/CORBIS. Figure 4.2: G. Brad Lewis/Gamma Liaison. Figures 4.3 & 4.5: Courtesy J.D. Griggs, USGS. Figure 4.4: Krafft Explorer/Photo Researchers. Figure 4.6: ©Fred Hirschmann. Figure 4.7: Brian J. Skinner. Figure 4.8a: William E. Ferguson. Figure 4.8b: Courtesy J.P. Lockwood, USGS. Figure 4.8c: Steven L. Nelson/Alaska Stock Images. Figure 4.9: Roger Werth/Woodfin Camp & Associates. Figure 4.11: S.C. Porter. Figure 4.12a: Krafft Explorer/Photo Researchers. Figure 4.12b: ©Tom Bean/DRK Photo. Figure 4.13: Steve Vidler/Leo de Wys, Inc. Figures 4.14a & 4.15 : S.C. Porter. Figure 4.14b & 4.17: Brian J. Skinner. Figure 4.16: Rich Buzzelli/Tom Stack & Associates. Figure 4.19: Courtesy Lyn Topinka, USGS. Figure 4.20: Courtesy J.D. Griggs, USGS. Figure 4.21: Courtesy Dr. Kevin McDonald, Dept of Geology, University of California, Santa Barbara. Figure 4.22: David Hiser/Photographers/Aspen/PNI. Figure 4.23a: ©Fred Hirschmann. Figure 4.23b: ©William Felger/Grant Heilman Photography. Figures 4.24a, 4.29, 4.30 & 4.41b: Brian J. Skinner. Figure 4.24b, 4.26, & 4.27: William Sacco. Figure 4.25: ©Tony Waltham. Figure 4.31a: Ron Sanford/fSTOP Pictures. Figure 4.33a: ©Gary Ladd. Figure 4.33b: ©Tom Bean/DRK Photo. Figure 4.35: ©Fred Hirschmann. Figure 4.39: Craig Johnson. Figure B4.1a: ©Jeffrey Hutcherson/DRK Photo. Figure B4.1b: ©Fred Hirschmann.

Chapter 5
Opener: ©Georges Ollen/Sygma. Figure 5.1: ©Randall Schaetzl. Figure 5.2: ©Jeff Gnass. Figure 5.3: ©Tony Waltham. Figure 5.4: Bill Hatcher. Figure 5.5: Fiorenzo Ugolini. Figure 5.6: S.C. Porter. Figure 5.7: Stan Osolinski/Oxford Scientific Films, Ltd. Figure 5.8: Kenneth W. Fink/Ardea London. Figure 5.9a: S.C. Porter. Figure 5.9b: Telegraph Colour Library/FPG International. Figure 5.10: S.C. Porter. Figure 5.11: ©Natural History Photographic Agency. Figure 5.12: Douglas H. Clark. Figure 5.14: G.R. Roberts. Figures 5.19, 5.20, 5.22, & 5.26: S.C. Porter. Figure 5.23: Frans Lanting/Minden Pictures. Figure 5.24: Claus Meyer/Black Star. Figure 5.25: Larry Lefever/Grant Heilman Photography. Figure 5.27: Art Wolfe/Allstock/Tony Stone Images. Figure B5.1: Fletcher & Baylis/Photo Researchers. Figure B5.2: Josep Soler. Figures B5.3 & B5.4: Brian J. Skinner.

Chapter 6
Opener: ©Nanci Kahn, Institute of Human Origins. Figure 6.1: ©Josef Muench. Figures 6.3, 6.5, 6.7, 6.12, 6.18a, 6.19b, 6.24, & 6.27: S.C. Porter. Figure 6.6: David Muench Photography. Figure 6.8, 6.13, & 6.17: Brian J. Skinner. Figure 6.9: Francois Gohier/Photo Researchers. Figure 6.10: Courtesy Scripps Institute of Oceanography. Figure 6.11: Courtesy Richard J. Stewart. Figure 6.14: Martin G. Miller. Figure 6.15: Kristin Finnegan/Allstock/Tony Stone Images. Figure 6.18b: Stephen Trimble/DRK Photo. Figure 6.19a: Gordon Wiltsie/Bruce Coleman, Inc. Figure 6.20a: Carr Clifton/Minden Pictures. Figure 6.20b: ©Tom Bean. Figure 6.23: David Krinsley. Figure B6.1: G. Whitely/Photo Researchers. Figure B6.3: Robert Azzi/Woodfin Camp & Associates.

Chapter 7
Opener: Raphael Caillarde/Gamma Liaison. Figure 7.2: Joe McDonald/Visuals Unlimited. Figures 7.3 & 7.7: William Sacco. Figures 7.4, 7.5, 7.9, 7.15, & B7.1: Brian J. Skinner. Figure 7.6: ©Brenda Sirois. Figure 7.10: Craig Johnson. Figure 7.11: Courtesy Schalk W. vand der Merwe. Figure B7.3: H.P. Merten/The Stock Market. Figure B7.4: Courtesy John Allcock. Figure B7.6: Courtesy Dudley Foster, Woods Hole Oceanographic Institution.

Chapter 8
Opener: ©Brownie Harris/The Stock Market. Figure 8.1a: ©Jeff Gnass. Figure 8.5: ©Joseph Muench. Figure 8.8: T.A. Wiewandt/DRK Photo.

Chapter 9
Opener: J. Langevin/Corbis-Sygma. Figures 9.4, 9.5, 9.6, 9.7, 9.10, 9.13, 9.27 & 9.28: S.C. Porter. Figure 9.11: JimSugar/Black Star. Figure 9.12: Courtesy USGS, Vancouver. Figure 9.16: Courtesy A. Lincoln Washburn. Figure 9.18: Steve McCutcheon/Allstock/Tony Stone Images. Figure 9.20: Courtesy Steven Brantley, USGS. Figure 9.22: Steven McCutcheon. Figure B9.1: Galen Rowell/Mountain Light Photography, Inc.

Chapter 10
Opener: Sue Ogrocki/Reuters/CORBIS. Figure 10.1: ©G. Haling/Photo Researchers. Figure 10.6: Andrew Holbrooke/Matrix. Figures 10.9a & 10.15: Courtesy NASA. Figures 10.9b & 10.12: S.C. Porter. Figure 10.19: William Sacco. Figure 10.20: Hiroji Kubota/Magnum Photos, Inc. Figure 10.25: ©Landform Slides. Figure 10.26: ©Martin Miller. Figure 10.27: Earth Satellite Corporation. Figure B10.1: Lloyd Cluff.

Chapter 11
Opener: ©Tom Bean/DRK Photo. Figure 11.13: Troy L. Pewe. Figures 11.14, 11.19 & 11.28: S.C. Porter. Figure 11.18: Heidi Bradner/Panos Pictures. Figure 11.20: Gary Ladd Photography. Figure 11.21: Jeff Gnass/The Stock Market. Figure 11.22: Davis Hiser/Photographers Aspen. Figure 11.24: Sam C. Pierson Jr./Photo Researchers. Figure 11.25: Jim Tuten/Black Star. Figure 11.27: Hiroji Kubota/Magnum Photos, Inc.

Chapter 12
Opener: ©Frans Lanting/Minden Pictures. Figure 12.2: Sam Abell/National Geographic Society. Figure 12.3: Courtesy USGS. Figure 12.4: Courtesy John Price, USDA. Figure 12.6: Courtesy NASA. Figures 12.7, 12.18, 12.19: S.C. Porter. Figure 12.13: John Lythgoe/Planet Earth Pictures. Figure 12.17: Courtesy USGS. Figure 12.20: William Thompson/National Geographic Society. Figure 12.21: James Wilson/Woodfin Camp & Associates. Figure 12.23: ©Tom Bean/Allstock/Tony Stone Images. Figures 12.22, 12.24, 12.25, 12.28, & 12.33b: S.C. Porter. Figure 12.26: Courtesy Gerald Osborn. Figure B12.1: Jack Sayers/Australian Antarctic Division. Figure B12.2: Steve McCutcheon. Figure B12.3: ©William R. Sallaz/Liaison Agency. Figure 12.33a: From Deonton & Porter "Neoglaciation", *Scientific American*, (1970), Vol 222. p. 100.

Chapter 13
Opener: Lauren Goodsmith/The Image Works. Figure 13.4: David Muench Photography. Figures 13.8, 13.10, & 13.12: S.C. Porter. Figure 13.13: Wesley A. Ward. Figure 13.15: Galen Rowell/Mountain Light Photography, Inc. Figure 13.16: George Ger-

ster/Comstock, Inc. Figure 13.17: Courtesy Ari Zhisheng. Figures 13.18, 13.19, & 13.26: S.C. Porter. Figure 13.24: Courtesy Eric S. Cheney. Figure 13.25: John Buitenkant/Photo Researchers. Figure 13.28: Stephen J. Krasemann/Peter Arnold, Inc. Figure 13.29: ©John S. Shelton. Figure 13.31: Mannfred Gottschalk/Tom Stack & Associates. Figure 13.32: Victor Engleberg/Photo Researchers. Figure B13.1: Courtesy NASA.

Chapter 14

Opener: ©Sandra Dawes/The Image Works. Figure 14.1: Courtesy Walter H.F. Smith and David T. Sandwell. Figure 14.9: Greg Scott/Masterfile. Figures 14.12 & 14.19: ©John S. Shelton. Figure 14.20: Ray Mahley/SUPERSTOCK. Figure 14.23: ©Spaceshots, Inc. Figure 14.24a: C.C. Lockwood/Earth Scenes. Figure 14.25a & b: Nicholas Devore III/Bruce Coleman, Inc. Figure 14.25c: Michael Friede/Woodfin Camp & Associates. Figure 14.27: Courtesy Georg Westermann, USGS. Figure 14.29a: Arthur Bloom. Figure 14.32: James Stanfield/National Geographic Society. Figure 14.33: Dick Davis/Photo Researchers. Figure B14.1: Courtesy Advanced Satellite Productions, Inc., and Earth Observation Satellite Corporation.

Chapter 15

Opener: H. Veiller/©Explorer. Figure 15.5a: ©Landform Slides. Figure 15.5b: Helmut Gritscher/Peter Arnold, Inc. Figure 15.8 & 15.17: ©John S. Shelton. Figure 15.15: David Muench Photography. Figure 15.19a: ©Betty Crowell. Figures 15.19b & 15.20a: ©Tom Bean. Figure B15.2: Jim Lukoski/Black Star.

Chapter 16

Opener: ©Bill Nation/Corbis-Sygma. Figure 16.6: ©The Exploratorium. Figure 16.12: Francois Gohier/Photo Researchers. Figure 16.13: Boris Yuchenko/AP/Wide World Photos. Figure 16.14: Lysaght/Gamma Liaison. Figure 16.15: Steve McCutcheon. Figure 16.22a: Bob Gossington/Bruce Coleman,

Inc. Figure 16.22b: William E. Ferguson. Figure 16.25: Brian J. Skinner. Figure 16.29: Wes Blake, Jr. Figure B16.3: Courtesy USGS.

Chapter 17

Opener: ©Colin Monteath/Oxford Scientific Films, Ltd. Figure 17.29: Grant Heilman Photography. Figure 17.34: James Balog/Black Star. Figure B17.1: ©Clive Jones/Environmental Images.

Chapter 18

Opener: ©Anne Heimann/The Stock Market. Figures 18.1a, 18.3 & 18.11: S.C. Porter. Figure 18.1b: Brian J. Skinner. Figure 18.2: ©Earth Satellite Corporation. Figure 18.12a: Courtesy NASA.

Chapter 19

Opener: ©David Hanson/Tony Stone Images. Figures 19.5 & 19.13: S.C. Porter.

Chapter 20

Opener: ©Victor Englebert. Figure 20.3 & 20.7: Brian J. Skinner. Figure 20.5: William Sacco. Figure 20.13: Peter Christopher/Masterfile. Figure 20.14: Simon Fraser/SPL/Photo Researchers. Figure 20.15: Charles Krebs/Allstock/Tony Stone Images. Figure 20.16: Ron Sanford/Black Star. Figure 20.17: Ronald Toms/Oxford Scientific Films, Ltd.

Chapter 21

Opener: ©NASA/TSADO/Tom Stack & Associates. Figures 21.2, 21.3, 21.7, 21.8, 21.12, & 21.13: Courtesy NASA. Figure 21.9a: Brian J. Skinner. Figure 21.9b: Roger Ressmeyer/CORBIS. Figure 21.10: Courtesy USGS. Figure 21.11: USGS/Tom Stack & Associates. Figure 21.14: Courtesy Astronomical Society of the Pacific. Figure 21.15: NASA/JPL/Starlight/CORBIS. Figure 21.16: David P. Anderson/Photo Researchers. Figure 21.17: NASA/Mark Martin/Science Source/Photo Researchers. Figures 21.18 & 21.19: Courtesy Space Photography Laboratory, Arizona State University.

Line Drawings

Chapter 9

Figure 9.25: Adapted by permission form J.G. Moore, W.R. Normark, and R.T. Holcomb (1994), Giant Hawaiian underwater landslides, *Science*, col. 264, pp. 46-47.

Chapter 10

Figure 10.31: Reprinted by permission from Stanley A. Schumm, M. Paul Mosley, and William E. Weaver (1987), *Experimental fluvial geomorphology* (New York, John Wiley & Sons), Fig. 2.18.

Chapter 12

Figure 12.34: Based in part on C. Lorius et al. (1900), The ice-core record: Climate sensitivity and future greenhouse warming. *Nature*, Vol 347, pp. 139-145; J.T. Houghton, G.J. Jenkins, and J.J. Ephraums (1990), *Climate Change: The IPCC scientific assessment* (Cambridge University Press), Fig. 2.

Chapter 14

Figure 14.3a: Adapted by permission of Houghton Mifflin from *The Times Atlas of the World*, Comprehensive edition (1967), plate 3 (mean surface salinity). Figure 14.3b: Adapted by permission of Andrew McIntyre from CLIMAP Project Members (1981). Map and Chart Series MC-36, Map 6A, published y the Geological Society of America. Figure 14.5: Adapted by permission from A.L. Gordon (1990/91). The role of thermocline circulation in global climate change. *Lamont-Doherty Geological Observatory (Columbia University) Report 1990/91*, Fig. 3, p. 48. Figure 14.6A: Adapted by permission from S.J. Lehman and L.D. Keigwin (1992). Deep circulation revisited, *Nature*, vol. 358, pp. 197-198, Copyright © 1992 Macmillan Magazines Limited. Figure 14.6B: Based on information from Lamont-Doherty Geological Observatory (Columbia University) Report 1990/91, Fig. 4, p. 50; J. Imbrie et al. (1992). On the structure and origin of major glaciation cycles, 1: Linear responses to Milankovitch forcing. *Paleoceanography* 7, Fig 1b, p. 704, published by the American Geophysical Union. Figure 14.30: adapted from N.J. Shackleton (1987). Oxygen isotopes, ice volume, and sea level, *Quaternary Science Reviews*, vol. 6, Fig 5, p. 187, by permission of Elsevier Science Ltd., Pergamon Imprint Oxford, England.

INDEX